Advanced EPR

Applications in Biology and Biochemistry

Advanced EPR

Applications in Biology and Biochemistry

Edited by

A. J. Hoff

Huygens Laboratory, State University of Leiden,
Niels Bohrweg 2, P.O. Box 9504, 2300 RA Leiden,
The Netherlands

ELSEVIER

Amsterdam — Oxford — New York — Tokyo   1989

ELSEVIER SCIENCE PUBLISHERS B.V.
Sara Burgerhartstraat 25
P.O. Box 211, 1000 AE Amsterdam, The Netherlands

Distributors for the United States and Canada:

ELSEVIER SCIENCE PUBLISHING COMPANY INC.
655, Avenue of the Americas
New York, NY 10010, U.S.A.

ISBN 0-444-88050-X

Printed and bound by Antony Rowe Ltd, Eastbourne
Transferred to digital printing 2005

Dedicated to George Feher on the occasion of his 65th birthday, in recognition of his seminal work on the development of EPR and ENDOR and their application to biological systems.

PREFACE

In 1945 E. Zavoisky published the first electron paramagnetic resonance experiment, the first successful magnetic resonance experiment ever [Fiz. Zh.,9(1945)211–245]. Three decades later, EPR had reached full maturity. So much so that a decline appeared to set in and many chemistry departments, regarding EPR primarily as an analytic tool, shifted their research effort in spectroscopy to other techniques. Only in a few niches EPR still flourished, and it is truly amazing that we are now witnessing such a strong revival. This, in my opinion, is due to two prime causes: Firstly, applications of EPR and, parallel with it, electron-nuclear double resonance in biology and biochemistry, although started early enough, had a much longer incubation time than those in chemistry and only now appear to have entered an exponential growth phase. Secondly, about ten years ago pulsed EPR techniques, albeit devised two decades earlier, began to take off with the general availability of low-cost microwave switches and solid-state amplifiers and with the development of special cavities. The penetration of EPR into biological research and the development of new techniques have lead to, and are the result of, the founding of several EPR centers devoted to the biological applications of EPR and ENDOR. This in turn has led to a rather spectacular growth in the number of applications in which advanced EPR techniques are used to tackle problems of biological interest.

From the above it will be clear, I think, that now is a particularly propitious time to publish a book reporting on the new developments in EPR and ENDOR, with emphasis on applications in biology and biochemistry. The present volume is broadly organized into four parts. In Chapters 1–6 pulsed EPR is discussed in detail. Peisach and Mims give a general introduction to Electron Spin Echo (ESE) and Electron Spin Echo Envelop Modulation (ESEEM) spectroscopy, pointing out their capabilities and pitfalls and reviewing applications primarily in metalloenzyme studies. Dikanov and Astashkin discuss ESEEM of disordered systems in depth, emphasizing simple, approximate relations for extracting information on hyperfine and quadrupole couplings and on the number and geometry of coupled nuclei, while Singel describes the advantages of multi-frequency ESEEM. De Beer and van Ormondt give a thorough and much needed treatment of recent sophisticated mathematical techniques for resolution enhancement in the time-domain analysis of ESEEM. Freed and coworkers give a detailed account of Fourier-transform ESE and of 2D-ESE and -ELDOR (electron-electron double resonance), with applications to nitroxide spin labels, from which it is clear that EPR now approaches the sophistication of NMR techniques. Schweiger reviews a number of his novel schemes for pulsed EPR and ENDOR, which allow higher selectivity and time resolution than the standard techniques.

In the second part of the book Chapters 7–12 provide detailed discussions of a number of novel experimental methods. Hyde and Froncisz treat the design of a variety of loop-gap resonators and their applications, including cw and pulsed EPR, ENDOR and *in situ* electrochemical generation of radicals. The construction and applications of a very high frequency (250 GHz) cw-EPR spectrometer is described by Freed and coworkers. A paper by Feher and coworkers (reprinted with permission of the American Institute of Physics) calls attention to the method of temperature modulation, a technique that, although presently seldom utilized, is nevertheless eminently useful for the registration of very broad lines of, for example, metalloproteins. In three chapters, McLauchlan and Stehlik and coworkers review the theory and applications of time-resolved and transient-nutation EPR while Hore discusses the detailed analysis of spin-polarized spectra recorded with these two techniques.

The third part of the book sees seven chapters on double-resonance techniques, five on ENDOR and two on optically- and reaction yield-detected resonance. Möbius and coworkers give a detailed account of ENDOR and TRIPLE resonance of radicals in liquid solution with emphasis on photosynthetic pigments and reaction centers. Hoffman and coworkers introduce the theory of solid state ENDOR and describe its application to the study of metalloproteins, except hemoglobin, which is treated by Kappl and Hüttermann. Brustolon and Segre describe ENDOR of disoriented systems, with applications to nitroxide spin labels, while Dinse discusses the exciting possibilities of pulsed ENDOR. The editor of this book reviews ODMR with emphasis on the applications to biochemistry and especially photosynthesis, whereas the theory and applications of RYDMR are outlined by Lersch and Michel-Beyerle.

The fourth and final part of the book is devoted to a thorough discussion of a number of new developments in the application of EPR to various biological and biochemical problems. Marsh and Horvath treat the modern, refined methods of the simulation of the spectra of nitroxide spin labels, incorporated in lipid membranes under different regimes of motion. Swartz and Glockner introduce the use of spin labels for the measurement of the concentration of oxygen in cells and tissue and give a few examples of the fascinating possibilities of EPR-imaging. Hagen discusses his breakthrough work on g-strain broadening of metalloproteins. Moura et al. review their and others' extensive work on a variety of iron-sulfur proteins, native and with substituted Fe-clusters. Brudvig treats EPR spectroscopy of Mn-proteins, culminating in a description of the work done on the oxygen-evolving complex of plant photosynthesis. Finally, in the last chapter, Solomon and coworkers give a comprehensive account of the spectroscopy of copper proteins, interrelating EPR and ENDOR with a number of other spectroscopic techniques.

The present volume is, I believe, not only an up-to-date survey of existing EPR techniques and their applications in biology and biochemistry, but also gives many ideas for future developments in instrumentation and theory. It is clear that the future of pulsed EPR is very bright indeed, and extension to other frequencies, both lower and (much) higher than 9 GHz, has been achieved or is underway. New microwave-bridge and cavity designs are revolutionizing the EPR spectrometer. Very high frequency cw-EPR is an extremely promising tool for biochemical and biophysical studies. The development of the theory of spin-label EPR in all motional regimes and of g-strain broadening goes hand-in-hand with tremendous advances in data reduction and simulation algorithms. The field is alive and lively indeed and we owe the contributors much for communicating so well a sense of wonder and expectation.

The figures in this volume are original or are reproduced with the permission of the author and copyright holder. I am indebted to the American Institute of Physics for permission to reproduce an article by Feher et al. (Chapter 9). I am grateful to all authors for consenting to submit camera-ready manuscripts, even if this meant spending hours trying to induce refractive word-processing equipment to cooperate. This has reduced the production costs of the book considerably. All have made a real effort to make the presentation of the various chapters as uniform as possible and, although overlap in a multi-author volume is to some extent inevitable and even useful, to avoid unnecessary duplication by cross-referencing to related chapters.

Last but not least, I acknowledge a special debt to Zina for allowing me to sacrifice so many hours to the preparation of this book that would have been better spent bicycling with her through the tulip fields around Leiden. Through Zina I salute all authors' partners for their support.

A.J. Hoff   Leiden, May 1989

CONTENTS

Chapter 8   Electron Paramagnetic Resonance at 1 Millimeter Wavelengths
            by D.E. Budil, K.A. Earle, W.B. Lynch and J.H. Freed

Chapter 9   Observation of EPR Lines Using Temperature Modulation
            by G. Feher, R.A. Isaacson and J.D. McElroy

Chapter 10 Time-Resolved EPR
            by K.A. McLauchlan

XIV

Chapter 20 Spin Label Studies of the Structure and Dynamics of Lipids and Proteins in Membranes
by D. Marsh and L.I. Horváth

Chapter 21 Measurements of the Concentration of Oxygen in Biological Systems Using EPR Techniques
by H.M. Swartz and J.F. Glockner

Chapter 22 g-Strain: Inhomogeneous Broadening in Metalloprotein EPR
by W.R. Hagen

Chapter 23 EPR of Iron-Sulfur and Mixed-Metal Clusters in Proteins
by I. Moura, A. Macedo and J.J.G. Moura

Chapter 24 EPR Spectroscopy of Manganese Enzymes
by G.W. Brudvig

Chapter 25 EPR Spectra of Active Sites in Copper Proteins
       by E.I. Solomon, A.A. Gewirth and T.D. Westmoreland

LIST OF CONTRIBUTORS

A.V. Astashkin — Institute of Chemical Kinetics and Combustion, Academy of Sciences of the USSR, Novosibirsk 630090, U.S.S.R.

C.H. Bock — Fachbereich Physik, Freie Universität Berlin, Arnimallee 14, D-1000 Berlin 33, B.R.D.

G.W. Brudvig — Department of Chemistry, Yale University, New Haven, CT 06511, U.S.A.

M. Brustolon — Dipartimento di Chimica Fisica, Università di Padova, Via Loredan 2, I-35131 Padova, Italy

D.E. Budil — Department of Chemistry, Cornell University, Ithaca, NY 14853, U.S.A.

R. de Beer — Department of Applied Physics, University of Technology, P.O. Box 5046, 2600 GA Delft, The Netherlands

S.A. Dikanov — Institute of Chemical Kinetics and Combustion, Academy of Sciences of the USSR, Novosibirsk 630090, U.S.S.R.

K.-P. Dinse — Institut für Physik, Universität Dortmund, Otto Hahnstrasse 30, D-4600 Dortmund 50, B.R.D.

K.A. Earle — Department of Chemistry, Cornell University, Ithaca, NY 14853, U.S.A.

G. Feher — Department of Chemistry, University of California - San Diego, La Jolla, CA 92093, U.S.A.

J.H. Freed — Department of Chemistry, Cornell University, Ithaca, NY 14853, U.S.A.

W. Froncisz   Department of Biophysics, National Biomedical ESR
                     Center, Medical College of Wisconsin, Milwaukee, WI
                     53226, U.S.A.

A.A. Gewirth   Department of Chemistry, Stanford University,
                     Stanford, CA 94305-5080, U.S.A.

J.F. Glockner   College of Medicine, University of Illinois, 190
                     Medical Sciences Building, 506 S. Mathews Avenue,
                     Urbana, IL 61801, U.S.A.

J. Gorcester   Rijksuniversiteit Leiden, Huygens Laboratorium, Niel:
                     Bohrweg 2, Postbus 9504, 2300 RA Leiden, The
                     Netherlands

R.J. Gurbiel   Institute of Molecular Biology, Jagiellonian
                     University, al. Mickiewicza 3, 31-120 Krakow, Poland

W.R. Hagen   Department of Biochemistry, Agricultural University,
                     Dreijenlaan 3, 6703 HA Wageningen, The Netherlands

A.J. Hoff   Department of Biophysics, Huygens Laboratory, State
                     University of Leiden, P.O. Box 9504, 2300 RA Leiden,
                     The Netherlands

B.M. Hoffman   Department of Chemistry, Northwestern University,
                     Evanston, IL 60201, U.S.A.

P.J. Hore   Physical Chemistry Laboratory, Oxford University,
                     Oxford OX1 3QZ, England

L.I. Horváth   Institute of Biophysics, Biological Research Centre,
                     H-6701 Szeged, Hungary

J. Hüttermann   Biophysik und Physikalische Grundlagen der Medizin,
                     Universität des Saarlandes, D-6650 Homburg/Saar,
                     B.R.D.

J.S. Hyde   Department of Biophysics, National Biomedical ESR Center, Medical College of Wisconsin, Milwaukee, WI 53226, U.S.A.

R.A. Isaacson   Department of Chemistry, University of California - San Diego, La Jolla, CA 92093, U.S.A.

R. Kappl   Biophysik und Physikalische Grundlagen der Medizin, Universität des Saarlandes, D-6650 Homburg/Saar, B.R.D.

W. Lersch   Institut für Physikalische und Theoretische Chemie der Technischen Universität München, Lichtenbergstrasse 4, D-8046 Garching, B.R.D.

W. Lubitz   Physikalisches Institut, Universität Stuttgart, Teilinstitut 2, Pfaffenwaldring 57, D-7000 Stuttgart 80, B.R.D.

W.B. Lynch   Advanced Materials Corp., c/0 Mellon Institute, 4400 Fifth Avenue, Pittsburgh, PA 15213, U.S.A.

A. Macedo   Centro de Química Estrutural, Complexo Interdisciplinar, Av. Rovisco Pais - 1096 Lisboa Codex, Portugal

D. Marsh   Max-Planck-Institut für Biophysikalische Chemie, Abteiling Spektroskopie, D-3400 Göttingen, B.R.D.

J.D. McElroy   AT&T Bell Laboratories, 260 Cherry Hill Road, Parsippany, NJ 07054, U.S.A.

K.A. McLauchlan   Physical Chemistry Laboratory, University of Oxford, South Parks Road, Oxford OX1 3QZ, England

M.E. Michel-Beyerle   Institut für Physikalische und Theoretische Chemie der Technischen Universität München, Lichtenbergstrasse 4, D-8046 Garching, B.R.D.

G.L. Millhauser   Department of Chemistry, University of California, Santa Cruz, CA 95064, U.S.A.

W.B. Mims   Exxon Research and Engineering Company, Annandale, NJ 08801, U.S.A.

K. Möbius   Institut für Molekülphysik, Fachbereich Physik, Freie Universität Berlin, Arnimallee 14, D-1000 Berlin 33, B.R.D.

I. Moura   Centro de Química Estrutural, Complexo Interdisciplinar, Av. Rovisco Pais - 1096 Lisboa Codex, Portugal

J.J.G. Moura   Centro de Química Estrutural, Complexo Interdisciplinar, Av. Rovisco Pais - 1096 Lisboa Codex, Portugal

J. Peisach   Departments of Molecular Pharmacology and Physiology and Biophysics, Albert Einstein College of Medicine, 1300 Morris Park Avenue, Bronx, NY 10461, U.S.A.

M. Plato   Institut für Molekülphysik, Fachbereich Physik, Freie Universität Berlin, Arnimallee 14, D-1000 Berlin 33, B.R.D.

A. Schweiger   Laboratory of Physical Chemistry, Swiss Federal Institute of Technology, ETH-Zentrum, CH-8092 Zurich, Switzerland

U. Segre   Dipartimento di Chimica Fisica, Università di Padova, Via Loredan 2, I-35131 Padova, Italy

D.J. Singel   Department of Chemistry, Harvard University, 12 Oxford Street, Cambridge, MA 02138, U.S.A.

M. Sivaraja   Department of Chemistry, Northwestern University, Evanston, IL 60201, U.S.A.

E.I. Solomon   Department of Chemistry, Stanford University,
                           Stanford, CA 94305-5080, U.S.A.

D. Stehlik   Fachbereich Physik, Freie Universität Berlin,
                           Arnimallee 14, D-1000 Berlin 33, B.R.D.

H.M. Swartz   College of Medicine, University of Illinois, 190
                           Medical Sciences Building, 506 S. Mathews Avenue,
                           Urbana, IL 61801, U.S.A.

M.C. Thurnauer   Chemistry Division, Argonne National Laboratory, 9700
                           South Cass Avenue, Argonne, IL 60439, U.S.A.

D. van Ormondt   Department of Applied Physics, University of
                           Technology, P.O. Box 5046, 2600 GA Delft, The
                           Netherlands

M.M. Werst   Department of Chemistry, Northwestern University,
                           Evanston, IL 60201, U.S.A.

T.D. Westmoreland   Department of Chemistry, Stanford University,
                           Stanford, CA 94305-5080, U.S.A.

CHAPTER 1

ESEEM AND LEFE OF METALLOPROTEINS AND MODEL COMPOUNDS[*]

WILLIAM B. MIMS and JACK PEISACH

1. INTRODUCTION

It is surprising, perhaps, that experiments which involve coherent excitation should have played such a minor role in EPR, all the more so since the
sources of excitation used in EPR have always been coherent, obviating any
need for developing a new kind of signal source as in coherent optics. The
reason, on closer examination, lies in the short intrinsic coherence times of
the physical systems examined by the microwave resonance method, and in the
excess broadening of many resonance lines caused by crystal strains. At low
temperatures, where lattice relaxation times cease to affect linewidth, this
excess broadening or inhomogeneous broadening, still often produces linewidths of the order of 10 G (28 MHz), even in the ideal case of lightly
doped, single crystal samples. In the time domain this corresponds to a
lifetime of 11 nanoseconds.

Fortunately, the spin echo technique provides a means of overcoming inhomogeneous line broadening, and by using this method one can observe spin
coherence for times of several microseconds, or more. In the frequency
domain this corresponds to a linewidth of less than 0.03 MHz or 10 milli-
gauss. The electron spin coherence times, or phase memory times, remain
several microseconds, even in crystalline samples with very broad inhomogeneous lines, or in frozen solution samples where the EPR spectrum consists
of a continuum, many hundreds of gauss wide. The technical price for making
use of this convenient property is the acquisition of a short-pulse, high-
power microwave transmitter system, a low-noise, non-overloading receiver,
and a timing and pulse programming system. All but the first of these are
becoming cheaper and more readily available, and thirty or more laboratories
world-wide are now equipped to perform coherent EPR experiments of this kind
(1).

Coherent EPR methods have been found especially useful for investigating
materials that are not available in single crystal form. The very broad
spectral distributions found in powders and frozen solutions pose no new
problems, other than those of signal intensity, since the coherence times are

[*]This work was supported in part by Grants RR-02583 and GM-40168 from the
United States Public Health Service.

no shorter, and the corresponding spectral widths, or "spin packet" widths, are not greater than in single crystals. Such methods would indeed seem to be intrinsically destined to find a major application in metalloprotein studies. Several different types of coherent EPR experiments have been proposed, but, out of these only two have found a biological application, and they will form the subject of this article. A third type of experiment, pulsed ENDOR in its several forms (see the Chapter by K.P. Dinse), has been performed on non-biological materials and may be expected to find a biological application in the future.

The first method to be discussed here, the electron spin echo envelope modulation (ESEEM) method, forms a natural complement to ENDOR, either in its non-coherent continuous wave mode or in its pulsed mode. The ESEEM method is particularly well suited for detecting nuclei that are weakly coupled to nearby electrons -- for example the nuclei of a ligand atom or of a substrate atom in a metalloprotein complex. Since this appears to be a promising field of application, we shall spend much time discussing ways of performing this kind of experiment. For more general discussions of ESEEM as a spectroscopic tool, reference can be made to review articles in this (the Chapters by S.A. Dikanov and D. Singel) and in other volumes (2-5).

The second method to be discussed here is based on a measurement of the linear electric field effect (LEFE) in EPR, which is useful for investigating the symmetry and bonding of paramagnetic complexes. Although there have been numerous LEFE experiments performed on inorganic single crystal samples, this method has been used on metalloprotein samples only by the authors of the present article, who offer apologies here for constant references to their own work. The paucity of LEFE work in the biological area is due to the fact that a short-pulse, high-power pulsed EPR system is indispensable for LEFE measurements on metalloprotein samples in frozen solution, whereas older, non-coherent continuous wave methods are, to a certain degree, successful with inorganic single crystals. We should like to stress the fact that any laboratory already possessing a short-pulse electric spin echo (ESE) system -- for example, an instrument acquired in order to perform ESEEM studies -- can, for several thousand dollars and some extra shop work, gain access to the LEFE method. Although LEFE measurements may be needed only occasionally to answer some particular questions, the investment would seem to be worthwhile, and we hope that the present review may encourage laboratories possessing ESE equipment to take the necessary upgrading step. Some practical details regarding LEFE techniques are found in reference (6). Earlier review articles dealing generally with the application of ESE techniques related to biological materials are in refs. 2-5.

2. APPLICATION INVOLVING THE NUCLEAR MODULATION EFFECT (ESEEM)

The electron spin echo envelope modulation (ESEEM) effect is described in detail elsewhere in this volume. We can, therefore, confine ourselves here to stating the essential formulae and to outlining those features which bear directly on the application of ESEEM measurements to the study of metal-ion-ligand complexes, such as those encountered in metalloprotein samples and in model compounds.

The modulation effect arises as a result of the heterodyning of the allowed and semiforbidden microwave transitions between the levels of a coupled electron-nuclear system (Fig. 1). For this to occur, both kinds of transitions must be excited simultaneously by the microwave transmitter pulses. If only one transition is induced, as is the case when the electron-nuclear coupling energy is much larger than the Zeeman energy, there is no modulation effect. There is also no modulation effect if the microwave field H_1 in the pulses is too weak, since then only one or the other of the heterodyning transitions can be driven by the pulses. As a rough criterion, H_1 should exceed the distance (in Gauss) between the relevant allowed and semiforbidden transitions. Even then, unless H_1 substantially exceeds the minimum criterion, the depth of modulation, which is an important parameter in some of the experiments described here, may be smaller than that predicted by simple formulae.

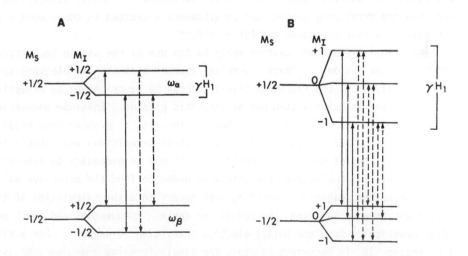

Fig. 1.   A. Magnetic energy level scheme for an I = 1/2 nucleus coupled to an S = 1/2 electron spin.   B.   A level scheme for an I = 1 nucleus interacting with an S = 1/2 electron spin. Solid lines indicate allowed transitions; broken lines indicate semiforbidden transitions.

Two minor experimental points are worth noting. The description of the echo process in terms of 90° and 180° pulses, which is usually given in NMR texts, is largely irrelevant in the case of ESE experiments on frozen metallo-protein solution samples. The range of microwave excitation (~ ±H_1 on either side of resonance) is usually much less than the width of the overall spectral distribution, so that during the microwave pulses, most spins are undergoing motions of a much more complex kind. Fortunately, this makes no major difference to the modulation pattern, provided that H_1 is large enough to excite both allowed and semiforbidden transitions in the same way for the systems falling within the 2H_1 bandwidth. It has no effect on the observed modulation frequencies, which depend on nuclear precession occurring in the intervals between the pulses, although it can result in a small shift in the zero of the time scale, by an amount not exceeding half the trans-mitter pulse width.

A second practical point is that it does not usually matter if the echo repetition rate is too fast to allow complete relaxation between echo cycles. The echo signal amplitude will be reduced by the partial saturation of the spin system, but this disadvantage may be more than offset by the greater speed of data accumulation. Changes in the modulation depth could, in principle, be caused by differences in the relaxation rates of allowed and semiforbidden transitions, but any such effect is generally obscured by cross relaxation in samples having the usual paramagnetic concentrations. Only when non-cross-relaxing species are simultaneously excited is there some risk of observing a rate-dependent modulation effect.

Some final words of caution apply to the use of the simple theoretical formulae given below. These formulae assume that the two electron spin states involved in the resonance transition can be described by an effective spin S=1/2 and that the system has an isotropic g value. Envelope modulation is observed in many other cases, but the theoretical problem then becomes more complex. For instance, if the electron transitions are between the magnetic substates of the S = 5/2 Mn(II) ion, it may be necessary to diagonal-ize the whole 6 x 6 electron spin matrix in order to find the effective value of the electron magnetic moment M_z that enters into the calculation of the modulation frequencies and the modulation depth. Frequencies and depth may differ drastically from one Mn(II) electron transition to another. For a sim-ilar reason it is incorrect to apply the simple formulae given here to sys-tems with a large g anisotropy, since in this case, M_z is not necessarily aligned parallel to the magnetic field H_0 and may be larger than the value one would calculate from the standard g formula (7). The modulation frequen-cies remain the same as the "ENDOR" frequencies for the system, but the calcu-lation of these frequencies and of the modulation depth is less straightforward. This particular problem is often overlooked in biological experiments, and, indeed, the errors may not be large in comparison with those which result from spherical averaging together with any other convenient approximations. It should receive attention, though, if g is highly anisotropic and if modulation depth measurements play an important part in the interpretation of data.

2.1 Two-pulse echoes

When a nucleus with spin $I = 1/2$ is coupled with an electron with spin $S = 1/2$, and when $g = 2$, the envelope of two-pulse electron spin echoes (Fig. 2A) is modulated by a function:

$$V_{mod}(\tau) = 1 - k/2 + (k/2)[\cos(\omega_\alpha\tau) + \cos(\omega_\beta\tau)$$
$$- (1/2)\cos(\omega_\alpha + \omega_\beta)\tau - (1/2)\cos(\omega_\alpha - \omega_\beta)\tau] \qquad [2.1]$$

where ω_α ω_β are the superhyperfine (ENDOR) frequencies associated with the upper and lower electron spin states. If the electron nuclear coupling is purely dipolar, the modulation depth factor k is given by:

$$k = (\omega_I B/\omega_\alpha\omega_\beta)^2 \qquad [2.2]$$

where $\omega_I = (1/\hbar)(g_n\beta_n H_0)$ $\qquad [2.3]$

and $B = (1/\hbar)(gg_n\beta\beta_n)(3\cos\theta\sin\theta)/r^3$. $\qquad [2.4]$

The quantities g, g_n, β, β_n, are the g factors and the values of the magneton for the electron and nucleus, H_0 is the Zeeman field, r is the distance between the electron and nucleus, and θ the angle between the line joining the electron and nucleus and H_0. In Eqn. 2.3 ω_I is the free precession frequency (in radian units) of the nucleus (i.e. the frequency that might be observed in the absence of any electron-nuclear coupling). The frequencies ω_α, ω_β are given (in the case of pure dipolar coupling) by:

$$\omega_\alpha^2 = (B/2)^2 + (A_d/2 + \omega_I)^2 \qquad [2.5]$$
$$\omega_\beta^2 = (B/2)^2 + (A_d/2 - \omega_I)^2 \qquad [2.6]$$

where

$$A_d = (1/\hbar)(gg_n\beta\beta_n)(3\cos^2\theta - 1)/r^3. \qquad [2.7]$$

For proton modulation as commonly observed in an X-band ESEEM experiment, the dipolar coupling coefficients A_d and B are several times smaller than ω_I. Then, to the first order of approximation, $\omega_\alpha \simeq \omega_I + A_d/2$, $\omega_\beta \simeq \omega_I - A_d/2$, $k \simeq (B/\omega_I)^2$, and the modulation depth varies as $\sin^2(2\theta)/r^6$. The maximum modulation depth occurs at $\theta = 45°$, i.e. at an angle for which the A_d term is relatively small. The sum frequency term in Eqn. 2.1 is independent of A_d and reduces, in first order of approximation, to

$$-1/4k\cos(2\omega_I\tau). \qquad [2.8]$$

Because of the variation of A_d with the orientation of the centers in a frozen solution sample the fundamental frequency term is spread out over a frequency band, and the corresponding modulation pattern is rapidly damped out.

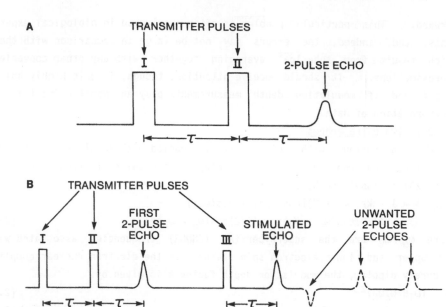

Fig. 2.   (A) 2-pulse and (B) 3-pulse electron spin echo sequences In 2-pulse experiments, the echo envelope is obtained by gradually increasing τ in successive pulse sequences, the amplitudes being measured by sample hold cir-cuitry.   In 3-pulse studies, τ is set to a fixed value and τ' is varied.   The "unwanted" two pulse echoes in B result from the failure of ideal 90-90 pulse conditions in broad line samples. They correspond to different 2-pulse com-binations of I, II, III, and the first 2-pulse echo.

The modulation pattern due to the harmonic term in [2.8] persists for a longer time since it is not, to the first order, dependent on A_d.

A more elaborate calculation taking into account the nuclear quadrupolar interaction energy is generally needed for nuclei with I>1/2. However, if the nuclear quadrupolar energy is small compared with the Zeeman energy, it remains possible to use the I = 1/2 formulae with some modifications. This approximation can often be used in X-band experiments that involve coupling with molecules containing deuterium (I = 1), or that involve coupling with other nuclei such as ^{23}Na (I = 3/2) in environments that minimize the effects of nuclear quadrupolar coupling. The case of deuterium is of special importance in experiments aimed at investigating the accessibility of enzyme active sites to solvents and in substrate labelling experiments.

To the first order of approximation the four ENDOR transition frequencies for weakly coupled deuterium nuclei are given in radian units by

$$\omega_1 = \omega_D + 1/2\ A_d + \omega_Q$$
$$\omega_2 = \omega_D + 1/2\ A_d - \omega_Q$$
$$\omega_3 = \omega_D - 1/2\ A_d + \omega_Q$$
$$\omega_4 = \omega_D - 1/2\ A_d - \omega_Q \tag{2.9}$$

where $\omega_D = g_n \beta_n H_o / \hbar$ is the free nuclear precession frequency for deuterium nuclei (without dipolar or quadrupolar coupling),

$$A_d = (1/\hbar) g g_n \beta \beta_n (3\cos^2\theta - 1)/r^3 \tag{2.10}$$

and where

$$\omega_Q = (3/8) e^2 qQ (3\cos^2\theta_Q - 1)/\hbar. \tag{2.11}$$

For D_2O, (which will serve as a model for the purposes of this discussion), the quadrupolar field is very nearly axial; θ_Q is the angle between the quadrupolar axis and H_o, and the zero field nuclear quadrupolar frequencies v^+ and v^- (in Hz units) are approximately given by

$$v^+ = v^- = (3/4) e^2 Qq/h. \tag{2.12}$$

The modulation depth parameter k_D for deuterium is 8/3 times that for protons, as prescribed by the formula: $k \propto I(I + 1)$ (8-10). This formula holds good when the quadrupolar interaction is small compared with the Zeeman interaction, so that, for the initial part of the modulation pattern at least, the frequencies ω_1, ω_2, ω_3, ω_4 in Eqn. 2.9 can all be considered together, i.e. in this case approximated by ω_D. A detailed calculation for the $I = 1$ case also yields a number of terms in k_D^2 (11), which correspond to state mixing of the $M_I = \pm 1$ superhyperfine levels by the dipolar interaction. These terms result from the partial relaxation of the selection rule excluding the doubly forbidden $\delta M_I = 2$ nuclear quantum jump. However, they can be ignored in first order, leaving the simplified form of the two-pulse modulating function

$$V_{mod} = (1 - (1/2)k_D) + (1/4)k_D(\cos\omega_1\tau + \cos\omega_2\tau + \cos\omega_3\tau + \cos\omega_4\tau)$$
$$- (1/8)k_D[\cos((\omega_1 + \omega_3)\tau) + \cos((\omega_2 + \omega_4)\tau)$$
$$+ \cos((\omega_1 - \omega_3)\tau) + \cos((\omega_2 - \omega_4)\tau)]. \tag{2.13}$$

As may be seen by substituting frequencies from Eqn. 2.9, the A_d term disappears in first order from the sum frequency terms, and ω_Q disappears from the difference frequency terms, leaving

$$V_{mod} = (1 - (1/2)k_D) + k_D \cos(\omega_D\tau)\cos((1/2)A_d\tau)\cos(\omega_Q\tau)$$
$$- (1/4)k_D\cos(2\omega_D\tau) \cos(2\omega_Q\tau) - (1/4)k_D \cos(A_d\tau). \tag{2.14}$$

These formulae are easily extended to the case where the electron-nuclear spin Hamiltonian includes a small Fermi contact term of the form $A_c \underline{I} \cdot \underline{S}$. In Eqn. 2.9, ω_1 and ω_2 are increased by $(1/2)A_c$ whereas ω_3 and ω_4 are reduced by $(1/2)A_c$. The factors $\cos((1/2)A_d\tau)$, $\cos(A_d\tau)$ in Eqn. 2.14 become, respectively, $\cos((1/2)((A_c + A_d)\tau))$, $\cos((A_c + A_d)\tau)$.

Metal ion centers in proteins are surrounded by numerous weakly coupled hydrogen atoms, and often also by other magnetic nuclei such as ^{14}N. All of these nuclei contribute simultaneously to the modulation pattern. If nuclei I_1, I_2, ...I_n are coupled to the same electron spin, then the overall modulation function is given by the product of the individual modulating functions, as indicated by the formula

$$V_N(I_1,I_2,...I_n) = V_1(I_1) \times V_2(I_2) \times ...V_n(I_n). \qquad [2.15]$$

When the overall modulation depth in a particular experiment is not too great, it may be simpler to replace the product by the sum obtained by expanding the $V_r(I_r)$ terms Eqn. 2.15 to the first order in the individual depth parameters k_r. This linear approximation is sometimes a convenient way of representing the modulation function due to coupled nuclei that lie outside of the first coordination sphere. But caution is required when treating large numbers of nuclei, since the accumulation of higher order products $k_q k_r$ will eventually offset the smallness of the products considered individually. Higher order product terms may also include useful structural information as shown in section 2.9.

There are two disadvantages of the two-pulse ESEEM method relative to ENDOR. As can be seen from Eqns. 2.1 and 2.13 the function V_{mod} contains sum and difference frequency terms as well as terms in the individual superhyperfine frequencies. These sum and difference terms occur to the first order in the depth parameter k, and, although they are distinguished by their negative phase, and appear as distinct inverted lines in the cosine Fourier transform (see e.g., Fig. 9B) they can become a source of confusion in complex spectra. A second major disadvantage of the two-pulse ESEEM method is the shortness of the phase memory time, and the associated broadening of the superhyperfine lines that make the spectrum much harder to resolve than an ENDOR spectrum. In organic samples the phase memory time cannot be increased very much beyond $\simeq$ 2 microsec, either by lowering the temperature or by reducing the concentration of paramagnetic centers, since it is due to internal magnetic field fluctuations caused by spin-spin flips of hydrogen nuclei. These flips occur at a rate that, in the liquid helium range, is virtually independent of temperature. On the other hand, there are certain advantages of the ESEEM method. It allows one to detect electron-nuclear coupling for weakly coupled nuclei or for nuclei with very small moments, and it provides, via the modulation depth, an additional source of information. The insensitivity to the magnitude of the nuclear moment can be seen by setting $\omega_\alpha \simeq \omega_\beta \simeq \omega_I$ in Eqn. 2.2 and substituting values from Eqns. 2.3 and 2.4. The way in which modulation depth depends on the electron-nuclear coupling can be seen from Eqns. 2.2 - 2.4. The modulation depth is independent of instrumental settings, provided that the pulsed microwave transmitter can deliver adequate power, and generate the H_1 required in order to excite all the relevant allowed and semiforbidden transitions.

Fortunately, it is possible to retain the advantages of the ESEEM method and avoid the disadvantages associated with two-pulse ESEEM experiments by performing experiments in the three-pulse "stimulated echo" mode. Although much of the early work was done in the two-pulse mode, it has subsequently been found that, in many instances, it is preferable to use the three-pulse method for metalloprotein studies. The two-pulse method is mainly useful for making a rapid check on the quality of a sample, and for obtaining a rough preview of its modulation properties before resorting to a more detailed and time consuming three-pulse study.

2.2 Three-pulse echoes

Under the conditions stated in conjunction with Eqn. 2.1, the modulating function in the three pulse case (see Fig. 2B) is given by

$$V_{mod} = 1 - (1/2)k\{\sin^2(\omega_\alpha\tau/2)[1-\cos(\omega_\beta\tau')]$$
$$+ \sin^2(\omega_\beta\tau/2)[1-\cos(\omega_\alpha\tau')]\} \qquad [2.16]$$

where τ is the time between pulses I and II, τ' is the time between pulses I and III and the other quantities are as defined in the two-pulse case. The echo, often described as a "stimulated echo", appears at a time τ after pulse III.

The stimulated echo signal is, in fact, a form of free induction signal. The first two pulses, which in the idealized narrow-line case are 90° pulses, produce a toothed pattern in the resonance line (i.e. a toothed pattern in M_z, the z component of electron spin magnetization; see e.g., Fig. 12 in ref. 4). How this pattern is generated can be seen as follows. Immediately after pulse II, the spins exactly on resonance with the microwave transmitter frequency are inverted. Spins whose frequencies lie $\pm \pi/\tau$ on either side of exact resonance precess by $\pm 180°$ between pulses I and II and are returned to the normal state by pulse II. Spins with frequencies $\pm 2\pi/\tau$ away from exact resonance precess by 360° between the pulses and are inverted, just like the spins on exact resonance. Spins offset by $\pm 3\pi/\tau$ are returned to the normal state, and so on. In the more realistic wide-line case, where ideal 90° spin nutations do not occur, and where spin reorientation follows less simple rules, a toothed pattern of periodicity $2\pi/\tau$ in the frequency domain is still generated, but the pattern becomes shallower to either side of resonance, the principal effect being limited to a range $\pm 2H_1$. Because of the deviations from precise 90° spin nutation, there is some residual M_x and M_y magnetization, which results in precessional coherence and causes the appearance of a two-pulse echo at time τ after pulse II (see Fig. 2B) but this extra echo can be ignored for the moment. The important property of the toothed pattern in M_z is that it can persist for times that are very much longer than the phase memory time. Eventually it is erased by cross relaxation or by lattice relaxation, but it can last as long as 100 μseconds in magnetically dilute samples at helium temperatures, even for those materials that contain large numbers of hydrogen nuclei. The actual lifetime of the pattern depends in part, on the value chosen for τ. Thus, the longer the time τ, the closer the teeth, and the more easily the M_z pattern is erased by cross relaxation.

The toothed pattern in M_z could, in principle, be measured by continuous wave EPR, using a rapid field sweep after pulse II, but it is easier to use the free induction technique. A third 90° pulse generates the Fourier transform of the M_z pattern, which consists of a pulse offset from pulse III by time τ. This is the stimulated echo. One can see as follows that this echo, like the two-pulse echo, varies in amplitude according to the nuclear modulation effect. Between pulses II and III the precession of nuclei, which are polarized by the sudden changes in local field caused by pulses I and II, produces cyclic variations in the local magnetic field acting on the electron spins. So, depending on the point reached in the nuclear precession cycle when pulse III is applied, the toothed pattern is either fully resolved or partially blurred in comparison with what it would be without any nuclear coupling. The stimulated echo amplitude thus varies according to the nuclear precession period and the pulse II to pulse III timing. The important difference between this and the modulation effect in two-pulse experiments is that the modulation patterns can now extend over many nuclear cycles, being limited only by the widths of the associated superhyperfine lines and not by the electron spin echo phase memory.

A further advantage of working in the three-pulse rather than the two-pulse mode will be apparent from Eqn. 2.16. If the experiment is conducted by setting τ to a constant value and by varying τ', then only the frequencies ω_α, ω_β, and not their sums or differences, appear in the result. However, the two frequencies ω_α and ω_β are not fully independent of one another, but are connected in a way that enables one to correlate them. For example, if one sets τ so that it comprises a whole number of cycles of the frequency ω_α, then ω_β is excluded from the envelope and vice versa. This "suppression effect", in which the frequencies belonging to the upper set of levels ($M_S = +1/2$) are controlled by the frequencies belonging to the lower set ($M_S = -1/2$) and vice versa, also occurs with $I > 1/2$. Although in this case frequency suppression may only be partial, it is sometimes useful for deciding when two superhyperfine transitions occupy corresponding places in the upper and lower manifolds of superhyperfine levels (see e.g., Fig. 14 in ref. 4). The suppression effect is also useful for eliminating proton modulation when, as often happens in biological studies, this is of no particular experimental interest.

The product formula corresponding to Eqn. 2.15 for the three-pulse case

(12) is

$$V_{mod}(I_1, I_2, \ldots I_n) = (1/2) \prod_{i=1}^{n} V_i(\tau, \tau') + (1/2) \prod_{i=1}^{n} V_i(\tau', \tau) \qquad [2.17]$$

where $\prod$ is the product operator and the functions V_i are obtained by writing $V_{mod}(\tau, \tau')$ as the sum of two terms

$$V_{mod}(\tau, \tau') = (1/2)[V(\tau, \tau') + V(\tau', \tau)]. \qquad [2.18]$$

For instance, in Eqn. 2.16 V_{mod} can be written in the form [2.18] with $V(\tau, \tau')$ given by

$$V(\tau, \tau') = 1 - (1/2)k[1-\cos(\omega_\alpha \tau)][1-\cos(\omega_\beta \tau')]. \qquad [2.19]$$

If the electron is coupled with a nucleus with $I>1/2$ it is generally necessary to compute the three-pulse ESEEM function numerically (though if only the frequencies and not the modulation depths are required, an ENDOR type calculation is sufficient). However, some simplification can be made in the case of deuterium because of the small nuclear quadrupolar energies encountered in most deuterium-containing molecules. Assuming that the dipolar interaction, the quadrupolar interaction, and any contact interaction are all small in relation to the Zeeman energy, we can, as in the two-pulse case, write down first order expressions for the four ENDOR frequencies, and treat the modulation depth as being governed by a parameter k_D which is 8/3 larger than the parameter k in Eqn. 2.2. Terms in k_D^2 (i.e. terms involving $\delta M_z = 2$ transitions) can be ignored, and we have

$$
\begin{aligned}
V_{mod}(\tau, \tau') = 1 - (1/4)k_D(&[\sin^2(\omega_1 \tau/2)][1-\cos(\omega_3 \tau')] \\
&+ [\sin^2(\omega_2 \tau/2)][1-\cos(\omega_4 \tau')] \\
&+ [\sin^2(\omega_3 \tau/2)][1-\cos(\omega_1 \tau')] \\
&+ [\sin^2(\omega_4 \tau/2)][1-\cos(\omega_2 \tau')]).
\end{aligned}
\qquad [2.20]
$$

If the time τ is set to one deuterium cycle or less, it is often possible to approximate ω_1, ω_2, ω_3, $\omega_4 \simeq \omega_D$ in the $\sin^2(\omega\tau/2)$ factors, thus simplifying [2.20] still further. (See section 2.5.)

  2.3 Application of formulae to frozen solution samples:Spherical averaging

It should be remembered that the formulae given in the previous sections of this article all refer to single crystal samples, in which the dipolar parameters A_d, B, the quadrupolar parameters ν_Q, θ_Q, etc., have specific values depending on the orientation of the molecules in the complex with respect to H_0. Before these formulae can be used to interpret data obtained with frozen solution samples, it is necessary to form a suitable average over all the orientations of the complex that contribute to the echo signal at a particular selected value of H_0 in the EPR spectrum. If all the relevant information were to be available, this average would be readily obtained by computation. More commonly, however, the problem presents itself in the inverse form. Very little is known at the outset, and the parameters have, for the most part, to be inferred by fitting computations based on sets of trial values to the experimental results.

Fortunately, it is possible to simplify the inverse problem by making some fairly coarse approximations. One of the most useful of these is to assume that the directions of the electron-nuclear line, and of the principal quadrupolar axes, are spherically distributed and uncorrelated with one another. Thus, for instance, in the case of a frozen solution of a paramagnetic ion in H_2O it may be assumed that the angle θ in Eqns. 2.4 and 2.7 takes on all values from 0° to 180°, and that a summation of computed echo envelopes, weighted by the spherical factor $\sin\theta$, will provide a curve that is good enough for comparison with experimental data. For the same complex in D_2O a double summation, incorporating a summation over θ and a second summation over θ_Q, the latter weighted by $\sin\theta_Q$, must be made. Where the modulation pattern is due to ^{14}N, an even coarser approximation has been found useful in some situations. The factor governing the depth of modulation is simply ignored, and a spherically averaged ENDOR frequency spectrum is computed without it. This spectrum has been found helpful in assigning ^{14}N transitions, although, since the amplitude values are incorrect, it cannot be transformed into a time wave to be compared with the echo envelope itself. A spherical averaging approximation is also useful when applying the product formula [2.15]. It is assumed here that the angular coordinates of each nucleus are uncorrelated with each other, so that the echo envelope in the multinuclear case can be found by multiplying the spherically averaged envelopes for each nucleus considered separately.

Of course, approximations such as these should be tested whenever possible by comparing the results of the calculation with experimental data on well characterized complexes, or by comparing the computations with others made at higher levels of sophistication. It should also be remembered that in using the spherical approximation one may sacrifice useful experimental information. By making measurements at a number of different H_0 settings one can work in an intermediate realm, somewhere between single crystal spectroscopy and spherically averaged frozen solution spectroscopy. For example, in the limiting case, one can approximate to single crystal conditions, by setting H_0 at the g_{max} or g_{min} ends of the frozen solution spectrum of a center with 3 different principal g values. Analyses along these lines have proved useful in ENDOR experiments, and are being tried out in ESEEM experiments.

When computing the frozen solution modulation pattern for deuterium, one can avoid lengthy simulations and take a short cut that involves the assumption that the angles θ and θ_Q are uncorrelated. The rationale of this pro-

cedure is as follows. We assume that the magnetic dipolar interaction and
the quadrupolar interaction are both small in relation to the Zeeman energy
so that the ENDOR frequencies can be approximated to the first order as in
Eqn. 2.9. If there is, in addition, a small contact interaction of the form
$A_c\underline{I}\cdot\underline{S}$ this can be included by adding a further first order correction of
$+(1/2)A_c$ to ω_1,ω_2 and $-(1/2)A_c$ to ω_3,ω_4. Let us suppose that
envelope simulations that take everything except the quadrupolar interaction
into account, have already been made. Then the effect of introducing the
terms $\pm\omega_Q$ in Eqn. 2.9 is to spread each monochromatic element in the frequency spectrum that corresponds to these simulations (other than the d.c.
term) into a spectrum described by the function (13)

$$\phi_Q(f) = \phi_{Q1}(f) + \phi_{Q2}(f)$$

where

$\phi_{Q1}(f) = 1/(0.5v^\pm - f)^{1/2}$; ($f = -v^\pm$ to $0.5v^\pm$, $\phi_{Q1} = 0$ elsewhere),
$\phi_{Q2}(f) = 1/(0.5v^\pm + f)^{1/2}$; ($f = -0.5v^\pm$ to $v^\pm$, $\phi_{Q2} = 0$ elsewhere),   [2.21]
and $v^\pm \simeq v^+ \simeq v^-$ as in Eqn. 2.12.

This convolution of functions in the frequency domain becomes a multiplication in the time domain. The existing simulations can therefore be converted
into simulations that include the deuterium quadrupolar interaction by multiplying them by the Fourier transform of $\phi_Q(f)$, which is denoted in the following discussion by $D_Q(t)$. Moreover, since ω_Q is not dependent on the
electron nuclear distance r, the multiplication need only be performed once,
on the products of the time waves for all similarly bonded (e.g. D_2O) deuterons in the problem. The required multiplying function $D_Q(t)$ is shown in
Fig. 3. For deuterated ice this function crosses zero at $\simeq$ 4.2 μseconds and
again at $\simeq$ 10.8 μseconds; between these two times the quadrupolar interaction
causes a phase reversal in the deuteron modulation pattern.

The short cut method outlined above can be applied without further modification to three-pulse envelope computations. (The effect of introducing ω_Q
into the $\sin^2(\omega_i\tau/2)$ factors in Eqn. 2.20 is small enough to be ignored
if, as in many experiments, τ is set to a constant value of approximately
half a deuterium cycle.) Two-pulse experiments involve an additional complication since the sum and difference terms give rise to distributions centered on zero frequency and on the second harmonic of the NMR frequency. The
quadrupolar contribution cancels out for the former and is doubled for the
latter (see Eqn. 2.14). The component in the time waveform corresponding to
the very low frequency distribution is therefore unaffected by the introduction of the deuterium quadrupolar interactions, whereas the time waveform
corresponding to the second harmonic must be multiplied by the function in
Fig. 3 with time in units of $1/v\pm$ instead of $2/v\pm$ on the horizontal scale.
This explains a frequently noted difference between two-pulse ^{1}H and [2]H

14

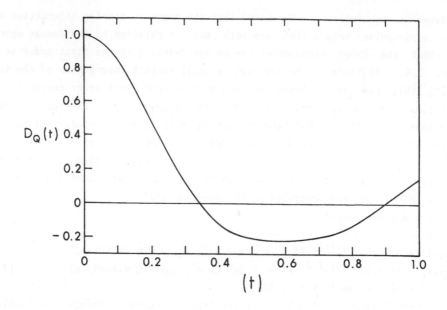

Fig. 3. A 3-pulse deuterium ESEEM pattern, computed without taking the nuclear quadrupolar interaction into account, can be converted into one which includes the nuclear quadrupolar interaction by multiplying it by $D_Q(t)$. (The d.c. term and other modulating functions that do not involve deuterium must first be removed.) Time is in units of $2/v\pm$ where $v\pm \simeq v+ \simeq v-$ (the quadrupolar field is assumed to be roughly axial). For deuterium in deuterated ice the graph covers times from 0 to approximately 12 microseconds.

modulation patterns. After a certain number of cycles ^{1}H patterns tend to be dominated by the second harmonic frequency, since the broadening effect of the dipolar interaction is absent for this harmonic component [see Eqn. 2.8]. But, in the case of ^{2}H patterns, both harmonics are broadened by the quadrupolar frequency offsets, the decay for the second harmonic being twice as fast as for the fundamental, so that there is no longer a tendency for the second harmonic to prevail at later times in the two-pulse envelope. Only if A_d or A_c are large do they become the major determinants and cause the ^{1}H type of modulation pattern to be seen in a deuterated sample.

2.4 <u>Isotopic labelling and the method of waveform division</u>

As pointed out earlier, the ESEEM method is especially useful for detecting nuclei that are weakly coupled to an electron spin. This suggests applications in which a substrate molecule is labelled with a nucleus such as ^{2}H or ^{13}C, so that the substrate to enzyme bonding can be studied by observing the magnetic interactions between the labelling nucleus and the metal ion at the enzyme active site. The problem is to disentangle the ESEEM component due to the nucleus in question from components due to all the other nuclei present.

As an introductory illustration we cite an early experiment in which the binding between the metalloprotein complex Cu(II)-conalbumin and ^{13}C labelled oxalate anion was studied in this way (14) (Fig. 4A). The measurements were done in the two-pulse mode, though later experience suggests that a three-pulse study would probably have yielded better data and been more informative. Two two-pulse envelopes were recorded, one for Cu(II) conalbumin bound by normal C^{12} oxalate and one for Cu(II) conalbumin bound by ^{13}C-labelled oxalate. Certain minor differences between these two envelopes could, with some difficulty, be detected, but both patterns were dominated by modulation due to protons and due to ^{14}N nuclei belonging to an imidazole ligand (see section 2.7).

In order to enhance these small differences the echo envelope obtained with the ^{13}C-labelled oxalate was divided by the envelope obtained with the normal C^{12} oxalate (Fig. 4B). If the samples had been single crystals,

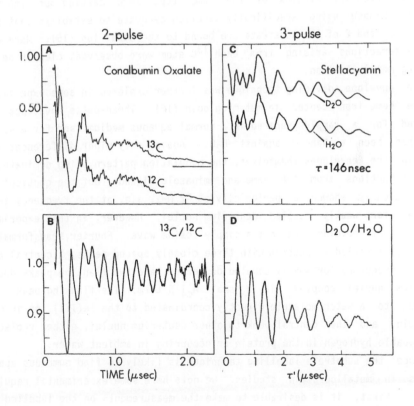

Fig. 4. A. ESEEM patterns for Cu(II) conalbumin with ^{13}C- and ^{12}C-oxalate. B. Ratio of envelopes in A. C. ESEEM patterns for stellacyanin in D_2O and H_2O. D. Ratio of envelopes in C.

then, according to Eqn. 2.15, one would in this way have factored out all the modulation components due to the ^{1}H and ^{14}N nuclei, since these nuclei were present in both samples alike, and the quotient waveform would have contained only those $V_i(I_i)$ components associated with the ^{13}C nuclei in oxalate. For the frozen solution samples that were actually used, this simple conclusion does not necessarily follow, however. Each of the two envelope measurements corresponded to an average over complexes with many different orientations, and the effect of the division was to give the quotient of the two averages. This is not the same thing as the average of the quotient waveforms for each orientation of the complex considered separately. In practice, however, the envelope-dividing procedure turned out to be reasonably successful in removing modulation due to ^{1}H and ^{14}N nuclei, and in revealing the ESEEM component due to ^{13}C. In spite of the smallness of the ^{13}C contribution in the two pulse envelope, it was easily established that oxalate was closely bound to the metal ion in Cu(II)-conalbumin.

More recently, studies of the same type were carried out with VO^{2+} pyruvate kinase using specifically labelled pyruvate to establish that both carbons 1 and 2 of the substrate are bound to the metal ion (15). Here contact interactions arising from each ^{13}C atom were observed, clearly demonstrating metal ligation.

The envelope dividing technique was further explored in some experiments on the heme-iron center in met-myoglobin (16). Three-pulse envelopes were measured for a sample prepared in a normal aqueous medium, and for a sample that had been exchanged against D_2O. Again, only minor differences were seen in the envelopes themselves, the modulation pattern being dominated by ^{14}N interactions from the heme and imidazole ligands, but the quotient envelope clearly showed a periodicity at $\cong$2 MHz, i.e. at the frequency to be expected for weakly coupled deuterium nuclei. However, in this experiment the quotient envelope was not a simple cosine wave. Fourier transformation (Fig. 5) revealed a spectrum with three closely spaced peaks, a central peak at the frequency for weakly coupled deuterium, and two flanking peaks due to deuterium nuclei coupled via a small $A_c\underline{I}\cdot\underline{S}$ term. The flanking peaks were assigned to a water molecule directly coordinated to the Fe(III) ion in met-myoglobin, and the central peak to other deuterium nuclei, either replacing exchangeable hydrogen in the protein or occurring in ambient water.

Since the substrate labelling technique is likely to find numerous applications in metalloprotein studies, we note here some experimental requirements. First, it is desirable to make the measurements on the labelled and unlabelled materials under conditions that are as nearly identical as possible. Various minor errors, due to insufficient H_1 fields, decay of the three-pulse echo amplitude because of cross relaxation effects, etc., will

MYOGLOBIN

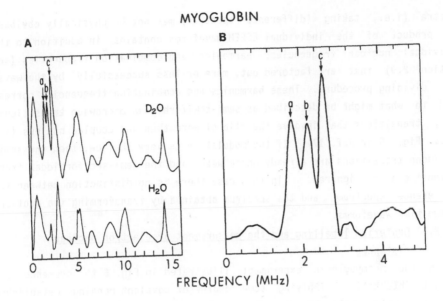

Fig. 5. (A). Cosine Fourier transform of the echo envelopes for metmyo-globin in D_2O (upper trace) and in H_2O (lower trace). Differences can be seen in the D_2O spectrum in the vicinity of 2 MHz. (B). Spectrum obtained by Fourier transforming the quotient of time wave forms for metmyoglobin in D_2O and in H_2O.   Lines due to ^{14}N and ^{1}H are greatly reduced in in-tensity in the quotient spectrum. Peaks a and c are due to ^{2}H on a water molecule directly coordinated to heme Fe(III). Peak b is due to other weakly coordinated deuterium atoms.

then tend to cancel in the quotient. It is also important to have reliable baseline measurements.   (This would matter less if one were simply making a subtraction instead of calculating a quotient.) The baseline can be record-ed, together with each modulation tracing, by programming the ESE spectro-meter so that the boxcar gate is moved off the echo signal for the last few percent of the envelope recording (see e.g., Figs. 11,12). Proton modulation can be eliminated by filtering e.g. by applying a Gaussian notch filter to the transformed envelope. But, it is better to eliminate as much of the pro-ton modulation as possible by setting τ (in a three-pulse experiment) to a multiple of the proton period, thus using the suppression effect [see Eqn. 2.16].   This is not merely an alternative way of filtering the waveform. By suppressing proton modulation one (ideally) leaves the echo amplitude un-affected, whereas filtering leaves it reduced by half the proton depth para-meter.

Although it can be shown to follow from the fundamental ESEEM formulae, the necessity for dividing time waveforms rather than subtracting frequency spectra (i.e., taking 'difference spectra') may not be physically obvious. The product of the individual ESEEM waveforms contains, in addition to the individual nuclear frequencies, harmonics and combination frequencies (see Section 2.9) that are factored out, more or less successfully, by the waveform dividing procedure. These harmonics and combination frequencies correspond to what might be described as semi-semiforbidden microwave transitions, i.e., transitions that involve the flip of more than one coupled nucleus (see e.g., Fig. 5 in Ref. 17). If the modulation is very shallow, then the semiforbidden transitions are already quite weak, and the semi-semiforbidden transitions can be ignored. In this case there is no distinction between the 'difference spectrum' and the spectrum obtained by transforming the quotient of two time waveforms.

2.5 Deuterium labelling and the measurement of electron nuclear distances

In the metmyoglobin experiment illustrated in Fig. 5 the presence of a contact interaction, implying some degree of covalent bonding, established that the water molecule was directly coordinated with Fe(III). Often, however, there is no measurable contact interaction for hydrogen or for deuterium in a ligating or substrate molecule. If the magnetic dipolar interaction were large enough, the frequency offsets due to A_d might be used to establish proximity and to estimate the nuclear distance (see Eqn. 2.7), but these offsets are often too small, and in frozen solution samples are distributed too smoothly over the spectrum to be reliably interpreted. The problem is worse for ESEEM than for ENDOR since the modulation depth, which depends on the factor $\sin^2 2\theta$, approaches zero for the angles $\theta=0$ and $\theta=90$ where the coefficient A_d reaches its extremal values. At the angle $\theta=50.8°$, where the $\sin^2 2\theta$ factor weighted by the spherical factor $\sin\theta$ reaches its maximum, the $3\cos^2\theta-1$ factor in A_d has the value 0.2. Quadrupolar broadening tends to obscure the contribution due to A_d still further in the case of deuterium.

Fortunately, there is a good alternative method. Measurements may be made on the modulation depth, which is a more sensitive indicator of the electron-nuclear distance than A_d, since, for weakly coupled nuclei, it varies as $1/r^6$ (Eqns. 2.2, 2.4) rather than as $1/r^3$. For the best results and for simplicity of interpretation, experiments should be performed in the three-pulse mode with τ set equal to one half of the free deuterium nuclear precession period (i.e. $\tau \simeq 250$ nsec for X-band experiments). The deuterium modulation depth is then at or near its maximum value, the factors $\sin^2(\omega_1\tau/2)$ are effectively unity, and the small discrepancies between ω_1, ω_2, ω_3, ω_4 and ω_d only affect the result to the second order for the first few modulation cycles. Eqn. 2.20 becomes

$$V_{mod}(\tau,\tau') \simeq 1-(1/4)k_D[4-\cos(\omega_1\tau')-\cos(\omega_2\tau')-\cos(\omega_3\tau')-\cos(\omega_4\tau')]. \qquad [2.22]$$

(A correction factor, representing the spherical average of the deviations from unity of the factors $\sin^2(\omega_1\tau/2)$ etc., can easily be computed, but the errors introduced by omitting this factor are probably too small to matter in comparison with errors resulting from other approximations.)

The function V_{mod} might be simplified still further by setting $\omega_1,\omega_2,\omega_3,\omega_4$ to a common value ω_D in [2.22]. We then should have

$$V_{mod}(\tau,\tau') \simeq 1-k_D(1-\cos\omega_D\tau') \qquad\qquad [2.23]$$

where substitution of the spherically averaged value

$$k_D = (16/5)(g\beta/H_0)^2(1/r^6) \qquad\qquad [2.24]$$

in place of k_D would lead at once to the required value of r. (To obtain this mean value, approximate all frequencies to ω_D, apply the $k\alpha I(I+1)$ rule, and average over the sphere.) However, somewhat more care must be exercised in approximating the $\cos(\omega_i\tau')$ terms than in approximating the $\sin^2(\omega_i\tau/2)$ terms in [2.20], since τ' is usually several times longer than τ and the error is substantially larger. Errors which arise as a result of ignoring the quadrupolar interaction in [2.23] can be allowed for by applying a correction factor based on the spherical averaging assumption as in Fig. 3. It will be clear from this figure that there is a great advantage in making the measurement of modulation depth on the earliest available deuterium cycle. The size of the correction factor, and errors in the assumptions used to derive it, increase rapidly with time τ'. Measurements of modulation depth made on a single early cycle, or made on a least squares fit to this cycle, are likely to yield a more reliable result than a least squares fit to the whole of the deuterium modulation pattern.

The metmyoglobin result (Fig. 4) illustrates another problem often encountered in a deuterium labelling experiment. One peak in the spectrum obtained in this experiment was assigned to deuterons that did not belong to the D_2O molecule bound to the Fe(III) ion, but it was not possible to obtain any detailed structural information from the amplitude of this deuteron peak (i.e. from the depth of the associated modulation pattern), since it was not clear how many nuclei were involved. In the weakly coupled case, one nucleus at a small radial distance r modulates the envelope as deeply as several nuclei situated farther away. The modulation depth yields, in effect, the sum $\Sigma n_i r_i^6$ where n_i is the number of nuclei at distance r_i. Small displacements of the spectral line and observations of the accelerated decay of the modulation pattern, caused by the dipolar term A_d, offer an alternative means for estimating nuclear distances. But it is better to make measurements of this kind on protonated samples, free from the complications arising from the deuterium quadrupolar interaction.

In spite of these ambiguities, the sum $\Sigma n_i/r_i^6$ is often useful as an indicator of the degree to which the metal ion active site is accessible to water or other components in the ambient medium, although the interpretation then depends on assumptions made about the form of the protein surface and should be supported by comparisons with model systems (18).

2.6 Problems arising from contact and pseudodipolar coupling

As pointed out earlier, the electron nuclear dipolar coupling must often be considered in conjunction with a contact coupling of the form $A_c \underline{I} \cdot \underline{S}$ when interpreting ESEEM data. In ^{13}C labelling experiments, such as the Cu(II)-conalbumin oxalate experiment used as an introductory example in section 2.4, and in experiments that involve the interpretation of ESEEM patterns due to ^{14}N and ^{15}N nuclei (see section 2.7), the observed contact term is often as large as the Zeeman term. In such cases, it is necessary to review the simple assumptions made hitherto.

The first and most obvious effect of a large contact term is to shift the nuclear frequencies. If the contact term is appreciable but still not as large as the Zeeman term, the ESEEM spectrum for an I=1/2 nucleus will contain two lines at approximately $g_n \beta_n H_0 \pm (1/2)A_c$, as in the D_2O coordinated met-myoglobin experiment cited in section 2.5. If the contact term is of roughly the same order of magnitude as the Zeeman term, as in the Cu(II)-conalbumin oxalate experiment, one of the frequencies will be very low and may be hard to detect. If the contact term is comparable with, but larger than the Zeeman term, then there will be lines at approximately $(1/2)A_c \pm g_n \beta_n H_0$. When the contact term is very much larger than the Zeeman term, the modulation effect becomes shallow or vanishes altogether since, in this limit, the "allowed" transitions are almost 100% allowed, and the "semiforbidden" transitions are exceedingly weak (see the introduction to section 2).

A secondary effect of the shift in nuclear frequencies is to change the modulation depth. This can be seen from Eqn. 2.2. If, because of the contact interaction, one of the two frequencies $\omega_\alpha, \omega_\beta$ becomes very small, the depth factor k is much larger than it would be otherwise. But there is a different reason why the modulation depth may be anomalously large in cases such as this, and to understand the reason it is necessary to take into account the full expression for the electron-nuclear interaction.

The interaction associated with the partial covalency of a coordinating bond is described by the product $\underline{I} \cdot \underline{\underline{A}} \cdot \underline{S}$, where $\underline{I}$ is a vector made up of the nuclear spin operators I_x, I_y, I_z, $\underline{S}$ is a vector made up of the electron spin operators S_x, S_y, S_z, and $\underline{\underline{A}}$ is a 3 x 3 matrix containing the elements $A_{xx}, A_{xy}, A_{xz}, A_{yx}, A_{yy}$, etc. If the z axis is parallel to H_0, and if the electron spin is aligned along this axis (i.e., if H_0 is set at the g_z end of the EPR spectrum), the component $A_{zz} I_z S_z$ in the expansion of $\underline{I} \cdot \underline{\underline{A}} \cdot \underline{S}$, together with the dipolar and Zeeman inter-actions determine (to the first order) the superhyperfine frequencies. The terms $A_{xz}I_xS_z$ and $A_{yz}I_yS_z$, together with terms of similar form in the dipolar interaction determine the modulation depth. If, however, H_0 is moved from the g_z end of the EPR spectrum to the g_x end, then the component $A_{xx}I_xS_x$ replaces $A_{zz}I_zS_z$ in the determination of the super-hyperfine frequencies, and $A_{zx}I_zS_x$ together with $A_{yx}I_yS_x$ replace $A_{xz}I_xS_z$ and $A_{yz}I_yS_z$ in the determination of modulation depth. Changes of this kind are usually described by rotating the x,y,z axes to new axes x′,y′,z′. The interaction then becomes $\underline{I}' \cdot \underline{\underline{A}}' \cdot \underline{S}'$, where $\underline{I}', \underline{S}'$ are new spin operators, and where $\underline{\underline{A}}'$ is a matrix derived from $\underline{\underline{A}}$ by making a similarity transformation. Once the transformation of axes is made, it is convenient to drop the primes, remembering that the numbers in the $\underline{\underline{A}}$ matrix will be different after the transformation (i.e., after the rotation of H_0 relative to the principal g axis).

The contact term $A_c\underline{I} \cdot \underline{S}$ represents only a part of $\underline{I} \cdot \underline{\underline{A}} \cdot S$, the number A_c being 1/3 x the trace of the matrix A, i.e., $A_c = (1/3)(A_{xx} +A_{yy}+A_{zz})$. The $A_c\underline{I} \cdot \underline{S}$ term has the useful property that it remains unaltered when the axis system is rotated (i.e., when the H_0 setting is changed). In the case of the electron-nuclear coupling this contact term often accounts for most of the coupling energy. But, the residual term $\underline{I} \cdot \underline{\underline{A}} \cdot \underline{S} - A_c\underline{I} \cdot \underline{S}$ also contributes to the superhyperfine frequencies, and can cause line broadening, because of the inhomogeneous distribution of orientations, as well as shifts in the ENDOR line positions as the field H_0 is scanned from one end of the spectrum to another. Also important to note is that this term contains components $A_{xz}I_xS_z$ and $A_{yz}I_yS_z$ (i.e., modulation depth terms) that vary with the rotation of the axes (or with the H_0 setting), and which can outweigh the effect of the magnetic dipolar interaction, thus invalidating the formula for the coefficient B in equation 2.4. Together with the changes in the frequencies ω_α, ω_β caused by A_c, etc., these components of $\underline{\underline{A}}$ can radically modify the depth parameter k (see Eqns. 2.2 and 2.4), rendering it no longer a usable measure of the electron-nuclear distance r.

The residue $\underline{I} \cdot \underline{\underline{A}} \cdot \underline{S} - A_c\underline{I} \cdot \underline{S}$ is sometimes referred to as the "pseu-dodipolar coupling". This is an easily remembered name, but it should be noted that the pseudodipolar coupling involves covalency effects and is not merely an enhancement of the "classical" dipolar coupling used in deriving Eqns. 2.4, 2.7, etc. "Classical" dipolar coupling signifies the electron magnetic interaction between small bar magnets with centers a distance r apart. For low symmetry electron nuclear coupling geometries such as those that occur in metalloproteins, (for example, in the oxalate coordinated

Cu(II)-transferrin complex discussed in section 2.4, where coupling is transmitted from Cu(II) through oxygen to ^{13}C in the coordinated ligand) it is
quite unlikely that the interaction has the simple cylindrical symmetry of
the classical dipolar interaction. There is also no reason to suppose that
the principal axes of $\underline{\underline{A}}$ and of the classical dipolar interaction are the
same. In general, caution is needed when interpreting modulation depths for
materials that show a sizeable $A_c\underline{I}\cdot\underline{S}$ term. This term may be symptomatic
of the presence of a significant $\underline{I}\cdot\underline{\underline{A}}\cdot\underline{S}$ coupling, containing pseudodipolar elements that enhance (or possibly reduce) the modulation depth in
unforeseeable ways. In order to fit the full $\underline{I}\cdot\underline{\underline{A}}\cdot\underline{S}$ to experimental
data, 6 new variables are needed (5 in addition to A_c), and these still do
not give the one structurally important quantity, namely the electron-nuclear
distance.

Similar reservations apply to the interpretation of modulation depths due
to other labelling nuclei, including deuterium. Fortunately, deuterium label-
ling often occurs at positions where the $\underline{I}\cdot\underline{\underline{A}}\cdot\underline{S}$ term is small compared
with the Zeeman term, as evidenced by the lack of any large displacement of
the NMR frequency. But ^{15}N, which often coordinates a metal ion directly,
or which can occur in the structural backbone of a labelled molecule, tends
to resemble ^{13}C in being subject to large $\underline{I}\cdot\underline{\underline{A}}\cdot\underline{S}$ couplings, equalling
or exceeding the Zeeman energy (see section 2.8).

2.7 Coupling with nitrogen nuclei

The ^{14}N (I=1) nucleus has a quadrupole moment that, in many organic
molecules, leads to zero field splittings which are comparable with the ^{14}N
Zeeman energy encountered in X-band ESE experiments (0.92 MHz at 3000 G).
There can also be a contact interaction of the same order of magnitude, e.g.
for nitrogen directly coordinating Fe(III), or for nitrogen nuclei indirectly
coordinating Cu(II) such as from bound imidazole. These different terms in
the electron nuclear interaction may be separated and analyzed by substituting ^{15}N (I=1/2) for ^{14}N. The situation is then similar to that described
for ^{13}C in the previous section. But more information can often be obtained from the nitrogen modulation pattern in unsubstituted material, in which
it is possible, through the nuclear quadrupole parameters, to identify the
ligand molecule chemically.

This can be illustrated by the much studied case of the Cu(II)-imidazole
complex (19), which occurs in many copper proteins. Imidazole consists of a
five membered ring containing two nitrogens and three carbons, the two nitrogens being separated by a carbon. The Cu(II) ion coordinates one nitrogen,
the other, the "remote" nitrogen usually being protonated. The Cu(II):^{14}N
interaction for the directly coordinating nitrogen is dominated by a large A
term and cannot be observed by the ESEEM method (although it can be studied

by ENDOR).   For the remote nitrogen the term A_c is (in X-band experiments) comparable in magnitude with the Zeeman interaction.   The full $\underline{I} \cdot \underline{\underline{A}} \cdot \underline{S}$ term consists of a pseudodipolar component (see section 2.6), which enhances the modulation depth, so that the modulation is easy to observe in spite of the considerable electron nuclear distance (4.16 Å), and a contact term which, as will shortly appear, greatly facilitates the measurement of the ^{14}N quadrupolar parameters.

As was pointed out in the previous section, it is the component $A_{zz}S_z I_z$ in the spin Hamiltonian operator $\underline{I} \cdot \underline{\underline{A}} \cdot \underline{S}$ and not the con- tact term $A_c \underline{I} \cdot \underline{S}$ that concerns us here.   However, if the value of $A_{zz}I_z S_z$ does not vary greatly as the z axis is rotated (i.e as the electron spin reorients itself in response to changes in the direction of H_0) then the effect on the superhyperfine frequencies is largely determined by the $A_c \underline{I} \cdot \underline{S}$ component in the electron nuclear interaction, and we can for convenience speak of the contact term, bearing in mind that, in careful measurements, it might become necessary to take the anisotropic part of $\underline{I} \cdot \underline{\underline{A}} \cdot \underline{S}$ into account.

In the case of the Cu(II)-imidazole complex the description in terms of a simple contact interaction will suffice for the purpose of the present dis- cussion (Fig. 6).   For the remote nitrogen of the imidazole ligand, in the protonated state, A_c has a value of approximately 2 MHz. The $M_s = +1/2$ superhyperfine states are therefore shifted up by 1 MHz, and the $M_s = -1/2$ states are shifted down by the same amount.   For one set of states the $0.5A_c$ shift adds to the ^{14}N Zeeman component (0.92 MHz at 3000 G), and for the other set, it almost cancels the Zeeman component. The situation can be described in physical terms by saying that the A_c term generates a local magnetic field at the ^{14}N nucleus that roughly doubles the Zeeman field for one set of states and reduces the Zeeman field almost to zero for the other set.   This results in an interesting situation. The energy levels for the latter set of superhyperfine states are almost the same as they would be in a zero-field nuclear quadrupole spectroscopy experiment, (if indeed, an experi- ment could be performed in the presence of the Cu(II) paramagnet).

In an experiment on the copper protein stellacyanin (Fig. 7), the fre- quencies obtained by Fourier transforming an experimental three-pulse echo envelope recorded at 3170 Gauss, were 0.7, 1.47 and 4.0 MHz (20). Of these the first two are close to the zero field quadrupolar frequencies v^- and v^+ for protonated imidazole, $(v^+, v^-, v^0) = (1.42, 0.72, 0.70)$. The high frequency corresponds to a transition between the $M_I = +1$ and $M_I = -1$ levels of ^{14}N in protonated imidazole in a field $H_0 \simeq 6000$ G, i.e. twice the Zeeman field.   This accounts for all the ^{14}N ENDOR transi- tions except the two corresponding to $M_I = +1 \rightleftharpoons 0$ and $M_I = -1 \rightleftharpoons 0$ at the

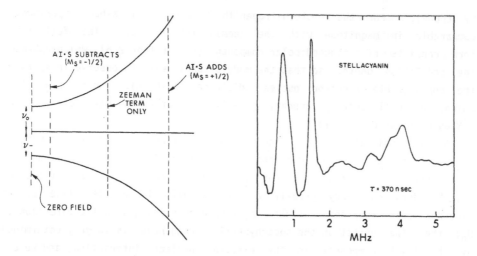

Fig. 6. (left) Nuclear level separations as a function of Zeeman field amplitude for the remote ^{14}N coupled to the electron spin in a Cu(II)-imidazole complex. For one of the $M_S = \pm 1/2$ submanifolds, the contact term $A_c\underline{I}\cdot\underline{S}$ almost cancels the nuclear Zeeman term, thus causing the zero field quadrupolar frequencies to appear in the echo envelope.

Fig. 7. (right) Fourier cosine transform of the 3-pulse ESEEM pattern for stellacyanin. The observed frequencies correspond to the protonated form of the imidazole ligand. Lines at 0.7 and 1.4 MHz belong to the $M_S = -1/2$ submanifold where the contact and nuclear Zeeman terms cancel. For the $M_S = +1/2$ submanifold in which the contact and nuclear Zeeman energies add, the interval between the inner two levels varies with the orientation of the Zeeman field, and no clearly resolved lines corresponding to these transitions are observable in the frozen solution spectrum; the interval between the outer two levels is relatively insensitive to the Zeeman field orientation and a broad $\delta M_I = 2$ line is seen at $\simeq 4$ MHz.

high ($\simeq 6000$ G) effective field value, which could not be recognized in the Fourier transform of the envelope, presumably because they are too broad in a frozen solution sample to yield any easily identifiable spectral feature. It may be noted that the spectral lines observed, at 1.4 MHz and at 4.0 MHz correspond to $\delta M_I = 2$ transitions, which are normally forbidden. They yield strong lines here because the ^{14}N magnetic substates are mixed by the quadrupolar interaction.

Results for Cu(II)-imidazole, taken in conjunction with a number of other studies in which Fe(III):^{14}N interactions have been observed (21), constitute a strong argument for constructing an ESE spectrometer with a wide range of microwave operating frequencies. With such a spectrometer, it should be possible to adjust H_0 so as to bring about the cancellation of contact and Zeeman energies for a wide variety of ^{14}N-containing molecules, and thus to identify ligands by means of the zero-field nuclear quadrupolar frequencies. A precise adjustment of the Zeeman energy is not required, since the ^{14}N nuclear frequencies change only quadratically with H_0 (in this case quadratically with the net discrepancy between Zeeman and contact contributions) near the cancellation point. An approach to the cancellation point can sometimes be detected by the narrowing of one or more lines in the spectrum (i.e. a lengthening of the modulation pattern). Further confirmation can be obtained by measuring the high frequency line, or by repeating the experiment at a new field and a new microwave frequency.

An interesting consequence of the near-cancellation of Zeeman and contact energies is that the corresponding ^{14}N lines are unobservable by ENDOR methods (22). Physically, this is explained by considering the way in which electron spin orientation follows the resultant of the Zeeman field and the ENDOR r.f. field (electron spin orientation follows the resultant field adiabatically, since the ENDOR field is oscillating at a much lower frequency than the electron spin resonance frequency). These movements of the electron spin, occurring at the ENDOR frequency, generate, via the $A_c \underline{I} \cdot \underline{S}$ interaction, a varying local field at the ^{14}N nucleus that cancels the field applied directly by the ENDOR coils so that it cannot induce transitions. This screening mechanism is related to, but opposite in effect, to the enhancement mechanism that often facilitates ENDOR experiments for samples with a large electron-nuclear contact coupling.

The ^{14}N modulation patterns observed in the early ESEEM experiments on Cu(II) proteins were easy to interpret because the relevant ^{14}N nuclear quadrupole parameters were known from NQR experiments on pure imidazole and on the imidazole group in histidine. For many other metalloproteins the problem is harder to solve and comparisons must be made with systems that have only a limited resemblance to the metal ion center in the protein. Such comparisons are facilitated if, as is frequently done in the NQR literature, we express the results in terms of the overall strength of the quadrupolar interaction $e^2 Qq/h$ and the asymmetry parameter η. The two parameters can then be plotted against one another on a diagram as in Fig. 8. They are related to the zero field quadrupolar frequencies v^+, v^-, v^0 (where $v^0 = v^+ - v^-$) by the equations

$$e^2 Qq/h = 2/3(v^+ + v^-); \quad \eta = 3v^0/(v^+ + v^-),\qquad [2.25]$$

the quantity of $e^2 Qq/h$ being in frequency (MHz) units. To save unnecessary duplication of data, the frequencies observed in NQR, or in ESEEM experiments have been converted into $e^2 Qq/h$ and η units in the discussion that follows.

ESEEM spectra for stellacyanin (20), $(e^2 Qq/h, \eta) = (1.49, 0.94)$ yielded results very close to those measured for the protonated nitrogen site in 1-histidine (1.44, 0.91), and in pure imidazole (1.43, 0.98) (23,24), thus

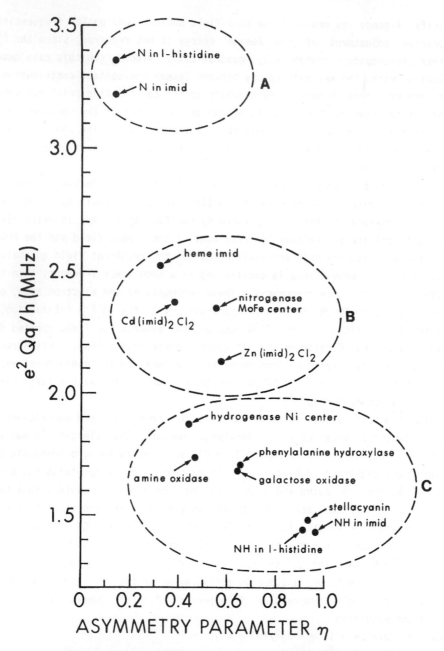

Fig. 8. ^{14}N quadrupolar parameters for imidazole, metal imidazole complexes and for metalloproteins. Three regions in the figure are set off. Region C contains the quadrupolar parameters for protonated ^{14}N of imidazole (imid) as in l-histidine and in copper proteins. Region A is for the deprotonated ^{14}N in histidine and imidazole. Region B contains the parameters for metal imidazole models in which the observed ^{14}N is directly coordinated by a metal ion, and for similar metalloprotein centers.

providing good evidence for the assignment discussed earlier in this section. However, the parameters corresponding to the frequencies observed for the Cu(II) binding site in galactose oxidase (1.69, 0.64) (25) and for the Cu(II) binding site in phenylalanine hydroxylase (1.70, 0.65) (26) lie some way off. These parameters are still quite far from the parameters for the deprotonated nitrogen site in 1-histidine (3.36, 0.13) and in imidazole (3.23, 0.13) (23,24), so a ligand assignment to protonated imidazole still remains reasonable, but some kind of perturbation appears to be involved -- possibly hydrogen bonding between the -NH group of imidazole and the protein backbone or a structurally forced change in the orientation of the imidazole of a type not found in simple Cu(II)-imidazole complexes.

In the case of the Cu(II)-imidazole complex, and for the analogous centers in copper proteins, the ^{14}N nucleus that gives rise to the ESEEM pattern is not the ^{14}N nucleus directly coordinated to Cu(II), but the "remote" nitrogen of the imidazole ligand, situated at a distance of ≈ 4.16 A. The directly coordinated ^{14}N is coupled by a large $A_c\underline{I}\cdot\underline{S}$ term and is not observed at all in ESEEM, though it can be observed by ENDOR. However, in some other cases, it is possible to see ESEEM patterns due to the ^{14}N atom that is directly coordinated to the metal ion.

In heme pyridine complexes, where the pyridine nitrogen is directly coordinated to low spin Fe(III), both the porphyrin ^{14}N nuclei and the pyridine ^{14}N give ESEEM patterns (21). The patterns for the pyridine ^{14}N (though not for the porphyrin ^{14}N nuclei) are once again characterized by near cancellation of contact and Zeeman energies and yield NQR parameters (3.13, 0.32). Near cancellation also occurs for imidazole nitrogen in the heme-imidazole complexes, yielding parameters (2.53, 0.32). Since these are a long way from the parameters observed for the remote nitrogen of imidazole, and nearer to those for deprotonated imidazole and for heme-pyridine, it seems reasonable to assign the heme-imid pattern to directly coordinated imidazole ^{14}N in this particular case. This interpretation is further supported by NQR measurements on Zn(II)-imid and Cd(II)-imid complexes (27). The heme-imid parameters are in the same region of the diagram as the parameters for Zn(II)(imid)$_2$Cl$_2$ (2.13, 0.57), and Cd(II)(imid)$_2$Cl$_2$ (2.37, 0.38).

A diagram such as that in Fig. 8 is especially useful for interpreting ^{14}N ESEEM spectra that cannot immediately be identified by comparing the quadrupolar parameters with those of a compound for which NQR data are available. Thus, for the Mo-Fe polynuclear center in nitrogenase (28), the ESEEM spectrum yields the quadrupolar parameters (1.87, 0.43). Since the nearest surrounding points are those for heme-imid, Cd(II)(imid)$_2$Cl$_2$, and Zn(imid)$_2$Cl$_2$, it is probable that this center is coordinated by an imidazole ligand and that the ^{14}N responsible for the pattern is directly coordinated. The ESEEM spectra for the Ni center in the hydrogenases from Methanobacterium thermoautrophicum (29) and Desulfvibrio gigas (30) yields the parameters (1.87, 0.43) and (1.87 and 0.41), respectively. The point on the diagram for the nickel center lies some way off from the cluster of points for stellacyanin and for the -NH group in imidazole and 1-histidine. But it is relatively close to the points for galactose oxidase (25), phenylalanine hydroxylase (26), and for the copper protein bovine serum amine oxidase (31,32), thus making an imidazole ^{14}N assignment likely.

ESEEM data involving ^{14}N other than from imidazole are still relatively scarce. Porphyrin ^{14}N has been observed to yield ESEEM patterns in low-spin heme compounds (16), but the contact energy appears to be too large, by about a factor of two, to cancel the Zeeman energy in X-band experiments, and the results therefore do not readily yield the NQR parameters. (These might, perhaps, be obtained by performing experiments in the 18 GHz range.)

For hexacoordinate heme-NO or O_2-Co(II)-tetraphenylporphyrin complexes with pyridine as the axial ligand trans to NO or O_2, nuclear modulation from pyrrole ^{14}N is observed (33). Computer simulations using the values e^2qQ/h = 1.8 MHz, η = 0.55, a contact interaction A = 1.3 MHz, and assuming a common set of principal axes for the nqi and the dipolar tensors, give good agreement with experiment.

ESEEM due to coupling between peptide nitrogen and a polynuclear iron center has been observed in fumarate reductase (34). The ESEEM spectrum yields NQR parameters (3.3, 0.5) similar to those for the backbone nitrogen in small di- and tri-peptides (24).

2.8 Coupling with other nuclei

The ^{31}P (I=1/2) nucleus is easy to observe in ESEEM since it has no quadrupolar interaction, and the nuclear free precession frequency (5.17 MHz at 3000 G) is in a convenient range (35). For ^{31}P, occurring in bound phosphate or polyphosphate groups (Fig. 9), and coordinating a paramagnetic metal ion, a contact coupling term $A_c\underline{I}\cdot\underline{S}$ several times smaller than the X-band Zeeman energy has been observed (36). Since it is unlikely that this coupling energy is associated with a large pseudo-dipolar term, estimates of the electron-nuclear distance based on modulation depth probably remain valid in this case. (The point could be checked by making measurements on models.) Isotopic substitution, as in the labelling experiments described earlier, is not possible for ^{31}P, but the envelope division method is still useful for removing modulation components due to ^{14}N and ^1H in those cases where a comparison is made between spectra obtained with and without phosphate binding. Caution is needed when interpreting results, since phosphate binding might exclude or modify other contributions to the modulation pattern.

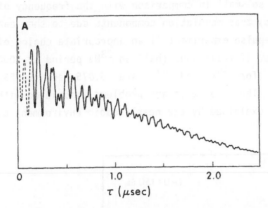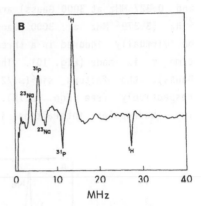

Fig. 9. Electron spin echo decay envelope (A) and Fourier cosine transform (B) for Nd^{3+}-ATP complex prepared with Na_3ATP. The continuous line in A shows the experimental data and the broken line an extension of the envelope to time $\tau = 0$. In (B), the Fourier transform demonstrates that Na^+ interacts with Nd^{3+} and is likely a component of a bimetallic ATP complex.

Modulations due to ^{95}Mo have been observed in complexes of molybdate and the pink acid phosphatase, uteroferrin, a protein that contains a redox-active binuclear iron center (37). The molybdate anion is an analogue of phosphate and inhibits enzymatic activity. The nuclear spin of the two isotopes ^{95}Mo and ^{97}Mo is 5/2 and the NMR frequencies, being only 2% apart, yield only one ESEEM pattern. The natural abundances of the odd isotopes ^{95}Mo and ^{97}Mo are 15.9% and 9.6% respectively, the even isotopes ^{92}Mo, ^{94}Mo, ^{96}Mo, ^{98}Mo, ^{100}Mo (all NMR silent) making up the remaining 74.5%. The two nuclear quadrupole moments differ by a large factor (^{97}Mo having a larger moment), but nuclear quadrupolar effects are probably not significant here, since the field at the Mo nucleus is approximately cubic for the tetrahedral MoO_4 ion.

In the study of molybdate and uteroferrin an isotopic labelling technique was employed to resolve the Mo contribution in the presence of larger contributions due to ^1H and ^{14}N nuclei in the protein. Data were collected for two identical samples, one being prepared with ^{95}Mo and the other with natural abundance molybdate. The quotient waveform obtained by dividing one echo envelope into the other yielded the modulation component associated with ^{95}Mo (Section 2.4).

ESEEM for ^{23}Na (I=3/2) has been observed for Nd(III)- and Mn(II)-ATP complexes in the presence of Na^+, and observations have been made (35,38) on samples where Na^+ was replaced by K^+ (I=3/2 for ^{39}K and ^{41}K), the purpose being to isolate the Na component by the envelope division method

(Section 2.4). The free precession frequencies of ^{39}K and ^{41}K (0.596 MHz and 0.327 MHz at 3000 Gauss) are so small in comparison with the frequency of ^{23}Na (3.379 MHz at 3000 Gauss) that modulation components due to them can be virtually ignored in a three-pulse experiment if an appropriate choice of time τ is made (Fig. 10). Thus, if τ=148 nsec (half an ^{23}Na period at 3000 Gauss), the factors $\sin^2(\omega\tau/2)$ for ^{39}K and ^{41}K are 0.075 and 0.023, respectively (see Eqn. 2.16). The absence of any problems associated with the quadrupolar interaction is explained by the nearly cubic environment of

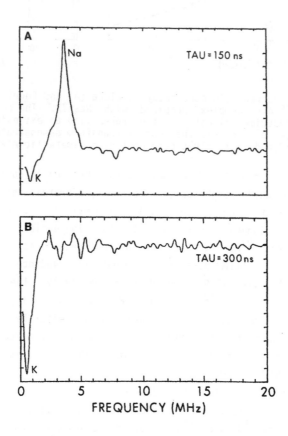

Fig. 10. Ratio spectra for Mn(II)-ATP complex demonstrating magnetic inter-action with ^{23}Na and ^{39,41}K. Three-pulse ESEEM data were obtained with τ value of 150 and 300 ns for identical samples that only differed in that one was prepared with Na$_3$ATP and the other with K$_3$ATP. ESEEM ratios were Fourier transformed and the spectra are shown. As these are ratio spectra, the ^{23}Na spectral contribution appears as a positive peak and that from ^{39,41}K, a negative peak. Note that with the appropriate use of τ, individ-ual spectral components are emphasized. Thus in (A), where the τ value of 150 nsec corresponds to roughly half the ^{23}Na periodicity, one clearly ob-serves the ^{23}Na line. In (B), where the τ value of 300 nsec corresponds to a multiple of the ^{23}Na period, the ^{23}Na line is suppressed.

the ^{23}Na nucleus in the closed electron shell of the Na$^+$ ion. In addition to ^{23}Na, ^{133}Cs (I=7/2, abundance 100%) has been observed (15) in experiments on pyruvate kinase with vanadyl ion (VO^{2+}) as the paramagnetic probe (Fig. 11). Here again the absence of any observed quadrupolar interaction is due to the cubic environment of the ^{133}Cs nucleus.

Labelling with ^{17}O nuclei would clearly be a useful technique if the practical difficulties of making observations on frozen solution samples could be resolved. ^{17}O has a nuclear spin I=5/2 and a quadrupolar moment that, in ice, results in the three zero field splittings $\simeq$ 3.4 MHz, 1.8 MHz, and 1.7 MHz (24). Zero field splittings somewhat larger than these have been observed in a number of organic molecules. The oxygen nucleus is not generally found in an environment with cubic symmetry, and the quadrupolar interaction is therefore likely to lead to serious complications in an ESEEM experiment.

2.9 <u>Finding the number of coupled nuclei</u>

The product formulae [2.15, 2.17] offer a means for finding out how many equivalent nuclei are coupled to an electron spin. This application of the product formula was pointed out in section 2.5 where it was noted that a measurement of the deuterium modulation depth could provide an estimate for the quantity $\Sigma n_i/r^6$, where n_i is the number of nuclei in the ith coordinating shell. The assumption in section 2.5 was that the modulation depth

VANADYL PYRUVATE KINASE + PHOSPHOENOLPYRUVATE + ^{133}Cs$^+$

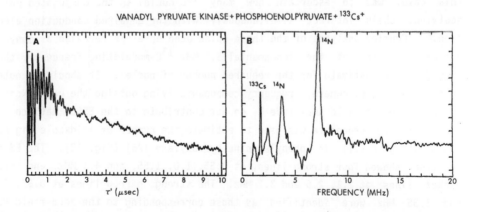

Fig. 11.   ESEEM data (left) and corresponding Fourier transform (right) for pyruvate kinase with vanadyl as paramagnetic probe, Cs$^+$ the univalent cation, and phosphoenol pyruvate. For time τ' less than 150 nsec, the echo decay envelope has been reconstructed according to the method of Mims (36). The line in the spectrum at 1.75 MHz is at the Larmor frequency of ^{133}Cs. Lines at 3.78 and 7.25 MHz are ascribed to amino ^{14}N from a protein lysine amino residue bound to VO^{+2} (15). The drop at the end of the time waveform in A occurs when the boxcar is suddenly shifted off the echo signal in order to sample the baseline level.

depended primarily on the magnetic dipolar interaction and could be calculated theoretically for any values of n_i and r_i.

Unfortunately, the measurement of modulation depth cannot be used to estimate the number of nuclei in many other cases of interest -- for instance
when ^{13}C or ^{14}N are the nuclei concerned -- because of the likelihood of
there being a substantial pseudodipolar interaction (see section 2.6), which
cannot be calculated a priori. Model studies are not to be relied on either,
since small differences in the ligating mechanism between models and metalloproteins can lead to significant errors in estimating the value of the pseudodipolar interaction.

If the modulation depth is not too small, the product formula can be used
in an alternative way. The product in Eqn. 2.15 contains factors of the form
$a + b\cos\omega\tau$. If there are n such factors, the resulting modulation function
contains harmonics $\cos2\omega\tau$, $\cos3\omega\tau$, ... $\cos n\omega\tau$. Factors $a_1 + b_1\cos\omega_1\tau$,
$a_2 + b_2\cos\omega_2\tau$, etc., corresponding to different nuclear frequencies,
combine to give terms in $\cos(\omega_1+\omega_2)\tau$, $\cos(\omega_1-\omega_2)\tau$, and so on. Identification of these harmonics and combination terms in the cosine transform of
a three-pulse echo envelope is therefore evidence for the presence of two or
more similarly coupled nuclei.

Although it falls outside the realm of metalloprotein studies, the work
of Thomann and others (40) on ^{13}C labelled polyacetylene provides a particularly striking example of this use of the product formula. The problem, in
this case, was to ascertain how many ^{13}C nuclei in the conjugated polyacetylene chain interact simultaneously with the unpaired conduction electron. Fourier analysis of the three-pulse echo envelope revealed as many as
eleven harmonics of the fundamental 3.1 MHz ^{13}C-modulating frequency, thus
providing an estimate for the required number of nuclei. It should be noted
that here, as elsewhere, frequency components lying outside the band excited
by the microwave field amplitude H_1 do not contribute to the ESEEM pattern.

A similar procedure was used to estimate the number of imidazole ligands
coordinating Cu(II) in phenylalanine hydroxylase (26) (Fig. 12). The ESEEM
spectrum showed four strong lines, at 0.55, 1.0, 1.55, and 4.1 MHz, and three
weaker lines at 2.1, 2.6 and 3.1 MHz. The strong, narrow lines at 0.55, 1.0
and 1.55 MHz were identified as those corresponding to the zero field NQR
transition for protonated ^{14}N in an imidazole ligand, and the broader line
at 4.1 MHz to the $\delta M_I=2$ transition for the same nucleus (see discussion in
section 2.7). The weak line at 3.1 MHz was interpreted as the second harmonic of the 1.55 MHz line, the line at 2.1 MHz as the combination frequency due
to lines at 0.55 and 1.55, and the line at 2.6 MHz as the combination frequency due to lines at 1.0 and 1.55. These harmonics and combination lines were
weaker than the other since they contained the depth parameter k to the sec-

Phenylalanine Hydroxylase

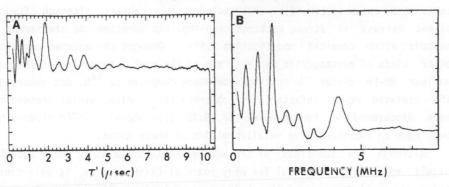

Fig. 12. ESEEM data (left) and corresponding Fourier transform (right) for the Cu(II) center of phenylalanine hydroxylase from <u>Chromobacterium</u> <u>viola-ceum</u>. The three weak lines in the spectrum are combination lines arising from the product theorem (Eqns. 2.15,2.17) and demonstrate that two equivalent ^{14}N are coupled to the electron spin. The drop at the end of the time waveform in A denotes the baseline level (see Fig. 11).

ond order.

The conclusion drawn from these results was that Cu(II) in the enzyme was coordinated by two imidazole ligands (but with the NQR frequencies displaced some distance from those found in free imidazole). In order to demonstrate experimentally the validity of the argument underlying the use of the product formula, the authors repeated the experiment using a Cu(II)-(N-methylimidazole) compound which, although not a true chemical model for the enzymatic active site, was found to mimic the phenylalanine hydroxylase spectrum reasonably well. ESEEM spectra corresponding to one or more N-methylimidazole ligands were obtained by forming Cu(II) complexes as described in the paper, the spectra obtained with multiple ligand complexes showing the appropriate harmonic and combination lines.

It would be possible to adduce further examples of such harmonic and combination frequency spectra by reexamining earlier ESEEM results. Reference may be made to the ESEEM spectrum obtained for galactose oxidase (Fig. 3 in Ref. 25) and to spectra obtained for superoxide dismutase (Figs. 1 and 2 in Ref. 41).

2.10 <u>ESEEM with substituent ions as paramagnetic probes</u>

Relatively few EPR active metal-ion centers occur naturally in metalloproteins, most of them involving either iron or copper. High spin (S=5/2) (42) and low spin (S=1/2) (43-45) iron centers occur in a number of frequently studied enzymes and can be observed in methemoglobin and metmyoglobin (42). Polynuclear Fe centers occur in ferredoxins and related electron transfer proteins (46,47). EPR active copper occurs in some electron transfer proteins and enzymes (48,49). Polynuclear copper centers, although often EPR silent because of strong exchange coupling, can sometimes be studied by EPR methods after chemical modification (50). Amongst the enzymes containing other kinds of paramagnetic centers, one of the most interesting is the polynuclear Mo-Fe center in nitrogenase where coupling to ^{14}N, not observed in the isolated Mo-Fe cofactor, is observed (51). Also, nickel centers have been discovered in hydrogenases (52,29,30) (see above). ESEEM studies have been made on representative metalloproteins of these kinds.

Although the sub-class of EPR active metalloproteins is large enough in itself to provide material for many years of ESEEM studies, it only constitutes a small fraction of the whole class of biologically important metalloproteins, which, in a majority of cases, contain non-paramagnetic ions such as Mg, Zn and Ca. This limitation can, as has been demonstrated in EPR studies, be overcome in part by making paramagnetic substitutions. For example, Mn^{2+} (S=5/2) and vanadyl (S=1/2) can be substituted for Mg^{2+}, Zn^{2+} or Ca^{2+}, and, as is well known from studies of inorganic single crystals, the rare earth ions can be substituted for Ca^{2+}.

The vanadyl ion is a particularly good choice for ESEEM experiments (15) since it is characterized by only two electron spin levels (53) and thus gives rise to fairly simple ESEEM patterns (Fig. 11). (There are no substantial differences between the modulation patterns associated with the eight ^{51}V hyperfine transitions.) Mn^{2+} is a more satisfactory probe from the biological standpoint (54), since it has enzymatic activity in many instances. Mn^{2+} has six electron spin-levels and five different electron spin transitions (ignoring the ^{55}Mn hyperfine structure). Overlapping of these transitions in frozen solution samples tends to blur the Mn^{2+} ESEEM patterns.

Nevertheless, ESEEM studies have demonstrated coupling to ^{14}N in a Mn(II) concanavalin A complex, due most probably to a histidine imidazole (55). As in the case of low spin Fe(III) complexes (21), and in contrast to the Cu(II) case (20), the modulation pattern is due here to the directly coordinated, rather than the remote, ^{14}N. This interpretation is confirmed by studies of the imidazole and N-methyl imidazole complexes of Cu(II) and of Mn(II) (55). For Cu(II), N-alkylation perturbs the quadrupolar field of the remote ^{14}N, thus changing the zero field quadrupolar parameters and greatly altering the ^{14}N frequencies observed in the ESEEM spectrum. With Mn(II), however, the remote ^{14}N of the imidazole ligand is clearly not the nitrogen nucleus responsible for the ESEEM pattern, since N-alkylation of imidazole has little effect on the observed ^{14}N frequencies.

Co^{2+} has also been used as a substitute for iron in studies of myoglobin since, unlike iron, it is EPR active in the divalent state. Studies were carried out with 6-coordinate oxy cobaltous tetraphenyl porphyrin (TPP) complexes, models for oxy cobalt myoglobin (33). It was found that the Fermi contact couplings for a series of axially coordinated, substituted pyridines and imidazoles showed a trend towards stronger hyperfine interaction with weaker bases, while the coupling to pyrrole ^{14}N remain virtually the same across the series. In addition, ^{14}N pyrrole couplings were the same as in isoelectronic hexacoordinate NO-Fe(II)-TPP complexes with axial nitrogenous bases.

Of the rare earth ions that may be of use as a replacement for calcium, Ce^{3+}, Nd^{3+}, Er^{3+}, and Yb^{3+} all have effective spin $S'=1/2$ and therefore only two electron spin levels. Their advantages and disadvantages are, in brief, as follows. Ce^{3+} has no nuclear hyperfine structure, Nd^{3+} usually has the narrowest range of g-values, Er^{3+} relaxes rapidly and may have to be studied at temperatures below the helium lambda point ($<2.2^{o}$K), and Yb^{3+} has the smallest ionic radius. It will be noted that all of these ions are trivalent whereas calcium is divalent. This would not prevent them from replacing calcium easily, but it may result in other problems, for instance in the coordination of additional anions. Eu^{2+}, which would otherwise be a good choice, unfortunately has a spin S=7/2, and therefore has a large number of different electron transitions that overlap in experiments on frozen solution samples. The same is true for Gd^{3+} (S=7/2).

2.11 ESEEM and radicals

The ESEEM method is eminently well suited to the study of radical species that, because of their relatively small g broadening give strong echo signals. Spin lattice relaxation times are longer than for metal ions, and studies can often be made at temperatures higher than those in the liquid helium range. Also, since g is isotropic and the g values are close to 2, data can readily be analyzed by the simple formulae given earlier in section 2.

ESEEM patterns due to ^{14}N and to ^{15}N have been extensively studied in radical cations of chlorophyll a, in bacterial chlorophyll a, and in other radical centers occurring in photosynthetic systems, as described elsewhere in this volume. ^{14}N modulation patterns have also been observed in the pyrroloquinoline quinone (PQQ) radical (56), which is a cofactor of amine oxidase (57) and other copper oxidases as well. Whether these patterns arise from PQQ or from the protein moiety is not as yet clear.

More complex experiments have been performed on the flavin radical that is a cofactor in flavoproteins. Here, ESEEM studies were carried out on the semiquinone form of _Azotobacter_ flavodoxin, and on protein samples reconstituted with various ^{15}N- and ^{13}C-labelled flavin mononucleotides (FMN)

(58). It was shown that ^{14}N modulations arise from both protein nitrogen and from either N(1) or N(3) of the flavin pyrimidine ring. Substitution of ^{13}C at the 2-position of the FMN leads to the appearance of a line at the [13]C Larmor frequency, with no observable line splitting from a contact interaction.

2.12 Conclusion

The examples and references cited in this section illustrate the kinds of experimental studies, analogous to ENDOR studies, which can be performed with an ESE spectrometer having the requisite pulse power and time resolution. Interpretations based on the simplified explanations and formulae given here, often provide useful biological information with a minimum of complex analysis. It should also be noted that because of weak electron nuclear coupling, many of these studies, which are easily performed by the ESEEM method, would be difficult by conventional c.w. ENDOR. Amongst the further developments of the method which can be anticipated, perhaps the most important would be its extension to single crystals of proteins. The experiments of Liao and Hartmann (59) on single crystals of Al_2O_3:Cr(III) (ruby) demonstrate the resolution and the wealth of analytical detail obtainable in ESEEM studies of this kind.

3. THE LINEAR ELECTRIC FIELD EFFECT (LEFE) IN EPR

The first experiment of this type (60) reported in the literature (61) was performed on a single crystal of silicon doped with Fe. This material normally yields a single EPR line. With an electric field of 10 kV/cm applied to the sample, the line splits into two lines of equal intensity separated by 14.7 Gauss, the separation being proportional to the applied field (Fig. 13).

Similar experiments were subsequently performed on a number of different metal ion centers in inorganic single crystals (see data tabulations in Ref. 60), but in most cases the line splittings were harder to observe. Larger electric fields were needed, and the splittings were sometimes smaller than the linewidth itself. The reason why can easily be understood if one considers the order of magnitude of the local electrostatic fields already acting on a metal ion in a crystal. The field due to a single electronic charge at a distance of 2.5 Å in vacuo is 2.3×10^8 V/cm, a field that might be reduced by dielectric screening effects to 10^8 V/cm in an actual crystal. It is therefore apparent that for laboratory fields in the generally accessible range of 10^5 V/cm the change in crystal field parameters, and hence the change in g-factor or in a D term, is not likely to exceed 0.1%. For example, one might expect an EPR line, observed at X-band in a magnetic field of 3000 G, to shift by less than 3 G. Experiments have shown that the situa-

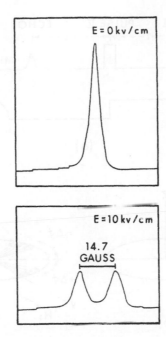

Fig. 13. The LEFE as observed in a single-crystal sample of Fe in silicon (61). The line is split because there are two sites that are indistinguishable in EPR but that undergo equal and opposite shifts in an applied electric field.

tion is no more favorable in cases where the bonding is partially covalent, and where the electrostatic "crystal field" becomes a "ligand field". Indeed, even where the effect itself is large, it may still be difficult to observe, since large electric field effects tend to be found in materials with wide resonance lines. This is understandable when one considers that EPR parameters that are sensitive to applied electric fields tend also to be sensitive to internal strains and to local irregularities in the crystalline ionic charge configuration.

3.1. <u>LEFE measurements by the spin echo method</u>

The difficulties posed by the inhomogeneous broadening of EPR lines, and by the smallness of the LEFE, are overcome by using an electron spin echo (ESE) method (62). It is then not necessary to shift the resonance by an amount comparable to the linewidth, (for example, in single crystals by about 10 G), but only by an amount comparable to the spin packet width, which can be defined in terms of the Fourier transform of the two-pulse echo decay envelope and is typically 0.02 G or less. The experimental procedure is as follows (Fig. 14). At the midpoint of a two-pulse electron spin echo sequence, and in coincidence with pulse II, an electric field step is suddenly applied to the sample. This causes a shift $\Delta\omega$ (in radian units) in the pre-

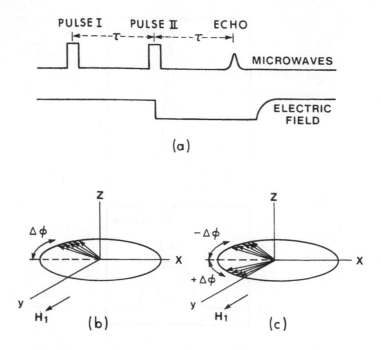

Fig. 14. Procedure for making an LEFE measurement by the spin echo method. In (a) the electric field is applied at the same time as transmitter pulse II and is held on until after the appearance of the echo signal. (b) Phase convergence of spins showing the phase change that occurs when the precession frequencies are shifted as a result of applying the field. (c) If there are two sites that undergo equal and opposite frequency shifts, the net magnetization responsible for generating the echo signal is the resultant of two vectors shifted with phases $\pm\Delta\phi$. Partial or complete cancellation of the echo can occur.

cession frequency of the spin packets so that, at a time τ later, when the spin packets refocus to generate the echo signal, they have accumulated an extra precessional phase of $\tau\Delta\omega$. This phase shift might be observed directly in a phase coherent ESE spectrometer, but in most cases the measurement is yet simpler. Many crystals contain two kinds of sites, the one being the "inversion image" of the other. (To derive one site geometry from the other it is necessary to move each of the atoms surrounding the paramagnetic center through it and out to an equal distance on the opposite side.) These inversion image sites give identical EPR spectra but undergo equal and opposite shifts in an applied electric field. Hence the splitting of the resonance line in early LEFE experiments (Fig. 13). In the case of an ESE experiment on a single crystal sample, the two groups of spins can be made to produce precession magnetization vectors shifted by $\pm\tau\Delta\omega = 90°$, $270°$, etc., which interfere destructively. The measurement can then be made by varying either τ or the electric field E, and by locating null points where phase cancellation occurs. Overall, the ratio $R(E\tau)$ of the echo amplitudes with and without the applied electric field E is given by

$$R(E\tau) = \cos(\Delta\omega\tau). \qquad [3.1]$$

In the case of a powder or frozen solution sample it can be shown that for every center undergoing an upward frequency shift there is present, with equal probability, a center undergoing a downward frequency shift. For each such pair the resultant echo signal is given by (3.1). Generalizing to include pairs with a distribution of shifts $S(\Delta\omega)$ we have

$$R(E\tau) = \int S(\Delta\omega)\cos(\Delta\omega\tau)d\Delta\omega \qquad [3.2]$$

Actually, since it is often easier to do the experiment by varying E than by varying τ, and since $\Delta\omega$ is proportional to E, it is more convenient to write Eqn. 3.2 in the form

$$R(E\tau) = \int S(\Delta\bar{\omega})\cos(\Delta\bar{\omega}E\tau)d(\Delta\bar{\omega}) \qquad [3.3]$$

where $\Delta\bar{\omega} = \Delta\omega/E$ is the radian/sec shift per unit electric field.

It should be noted that not all paramagnetic centers undergo electric field induced shifts. Free radicals are insensitive to the ligand field environment and are therefore negligibly affected by electric field induced changes in it. Metal ions in sites having inversion symmetry (centrosymmetric sites) are also unaffected to the first order in E. This absence of an LEFE for centrosymmetric sites is one example of the way in which the LEFE is restricted by the local site symmetry and can be understood as follows. Let us tentatively assume that when an electric field is applied, the resonance frequency is shifted to $\omega + \Delta\omega$, where $\Delta\omega$ is given by the power series $\Delta\omega = \alpha E + \beta E^2 + \gamma E^3 + \dots$. Now, let us consider the direction of E in relation to the arrangement of atoms about a centrosymmetric active site. (The direction of H_0 is unimportant here since it makes no difference to the EPR frequency.) If we reverse E, its geometric relation to the atomic configuration remains the same, and the shift cannot therefore alter. But the expression for $\Delta\omega$ becomes $\Delta\omega = -\alpha E + \beta E^2 - \gamma E^3 \dots$. So, equating the two algebraic expressions for $\Delta\omega$ before and after the reversal of E, we see that the coefficients α, γ, of the terms in odd powers of E must be zero. The lowest order electric effect for a centrosymmetric site is the quadratic shift $\Delta\omega = \beta E^2$, and this, in most practical situations, is too small to detect (63). By similar arguments one can show that the LEFE is associated with the odd-harmonic components of the ligand field. The property of detecting and characterizing the odd ligand field is one important feature of LEFE measurements, which sets them apart from measurements made by conventional EPR methods.

In LEFE measurements on single-crystal samples that contain low symmetry sites the magnetic field is rotated about two orthogonal axes, its orientation being described by the polar and azimuthal angles θ and ϕ, and the experiments are then repeated for three orthogonal electric field orientations along x, y, an z axes. A full description of the resulting variations in the shift may involve as many as 18 independent coefficients. Measurements on frozen solution samples are much simpler, but in order to understand the relationship between the data and the symmetry properties of the site, and in order to make computer simulations where these are needed, one must first of all be able to describe the shift in the single crystal case.

For a low symmetry site the g variation as a function of magnetic field orientation (in the absence of electric fields) is expressed algebraically by the formula

$$g^2 = G_1 l^2 + G_2 m^2 + G_3 n^2 + 2G_4 mn + 2G_5 ln + 2G_6 lm \qquad [3.4]$$

where $l = \cos\phi\sin\theta$, $m = \sin\phi\sin\theta$, $n = \cos\theta$, are the direction cosines of H_0. (θ is the angle between the z axis and H_0; ϕ is the angle between the x-axis and the projection of H_0 in the xy plane.) The G coefficients, or "g^2 coefficients", are related to the elements in the 3 x 3 symmetric matrix for g by the equations

$$G_1 = g_{11}^2, \quad G_2 = g_{22}^2, \quad G_3 = g_{33}^2, \quad G_4 = g_{21}g_{31} + g_{22}g_{32} + g_{23}g_{33},$$

$$G_5 = g_{11}g_{31} + g_{12}g_{32} + g_{13}g_{33}, \quad G_6 = g_{11}g_{21} + g_{12}g_{22} + g_{13}g_{23}.$$

The subscripts 1,2,3 denote the x,y,z axes with respect to which the g matrix is defined. This formulation, though the most convenient in numerical computations, is often replaced by a more intuitive geometric formulation when describing the results of EPR experiments. The g matrix is referred to a set of principal axes in which the terms g_4, g_5, g_6 vanish, and the results are described in terms of three principal g values ($g_1 = \sqrt{G_1}'$ where G_1' is the new value for G_1 in the principal axis system, etc.). The full description includes three angles that relate this principal axis system to the crystalline axes, so that the total number of parameters is again altogether six, just as it is in Eqn. 3.4.

The effect of applying electric fields E_x, E_y, E_z along x, y, z axes can be visualized in terms of the geometric picture as follows. The electric field components modify the three principal g values and tilt the set of principal axes (i.e. they tilt the "g-ellipsoid" and twist it in a way which can be described by three angular parameters). However, although this picture can be intuitively helpful at times, it does not lend itself easily to computation or to data analysis, which can be done more straightforwardly by using the algebraic description in Eqn. 3.4. Differentiating Eqn. 3.4 we have

$$\delta g^2 = l^2 \delta G_1 + m^2 \delta G_2 + n^2 \delta G_3 + 2mn\delta G_4 + 2ln\delta G_5 + 2lm\delta G_6. \qquad [3.5]$$

Substituting

$$\delta G_1 = \left(\frac{\delta G_1}{\delta E_x}\right) E_x + \left(\frac{\delta G_1}{\delta E_y}\right) E_y + \left(\frac{\delta G_1}{\delta E_z}\right) E_z$$

for the LEFE shift in G_1 and writing similar expressions for δG_2 etc. we have

$$\delta(g^2) = Ex\{B_{11}l^2 + B_{12}m^2 + B_{13}n^2 + 2B_{14}mn + 2B_{15}ln + 2B_{16}lm\}$$
$$\qquad + Ey\{B_{21}l^2 + B_{22}m^2 + B_{23}n^2 + 2B_{24}mn + 2B_{25}ln + 2B_{26}lm\} \qquad [3.6]$$
$$\qquad + Ez\{B_{31}l^2 + B_{32}m^2 + B_{33}n^2 + 2B_{34}mn + 2B_{35}ln + 2B_{36}lm\}$$

where $B_{11} = dG_1/dE_x$, $B_{21} = dG_1/dE_y$, $B_{31} = dG_1/dE_z$, $B_{12} = dG_2/dE_x$, etc. Altogether 18 coefficients are needed to specify the shifts $\delta(g^2) = 2g\delta g$ for a site with the lowest possible point symmetry C_1. LEFE shifts affecting the crystal field coefficients D and E that appear in the spin Hamiltonian of ions with $S>1/2$, and the tilting of the principal axis system associated with D and E, can be described in a somewhat similar manner, there being 15 coefficients for the lowest, C_1 point symmetry in this case.

In going from the single crystal case to powders and frozen solutions, we note first of all that the axis system can no longer be referred to X-ray crystallographic data, and that, for want of anything more specific, one has to adopt an x,y,z axis system based on the three principal g values $g_1=g_x$, $g_2=g_y$, $g_3=g_z$. This simplifies Eqn. 3.4 since G_4, G_5, G_6 now vanish, but it does not result in the elimination of δG_4, δG_5, δG_6 in Eqn. 3.5, since the applied electric field can still produce a "tilt" in the principal axes of g in addition to changes in the magnitude of the principal g values. For each individual center in a frozen solution sample Eqn. 3.6 therefore has the same number of terms as in the single crystal case.

Frozen solution simulations with such a large number of independent parameters would seem to pose a major problem, but fortunately, as will be shown in subsequent illustrations, we can often reduce the number of coefficients by fitting the data approximately to a higher symmetry. Also, it is sometimes useful to focus attention on those coefficients that can be extracted directly from the experimental observations as we now show (64). By setting H_o at either end of the EPR spectrum for a center with "rhombic" symmetry one can select complexes which have either the g_x axis or the g_z axis parallel to H_o. To be more specific, let us assume that g_x and g_z are the largest and smallest of the principal g values, respectively. Then, at the low H_o (g_z) end, l=m=o and by Eqn. 3.6

$$\delta(g^2) = B_{13}E_x + B_{23}E_y + B_{33}E_z. \qquad [3.7]$$

If H_o is parallel to the applied electric field, $E_x = E_y = 0$, and Eqn. 3.7 reduces to

$$\delta(g^2) = B_{33}E_z = B_{33}E_{||} \qquad [3.8]$$

where $E_{||}$ denotes an electric field parallel to H_o. The situation here is analogous to the one encountered in single-crystal experiments, and the ratio function $R(E\tau)$ describing the echo amplitude before and after the application of the electric field step is ideally a cosine as in Eqn. 3.1. Substituting

$\Delta\omega/\omega = \delta g/g$, and $\delta(g^2) = 2g\delta g$, in Eqn. 3.8 we obtain $\Delta\omega = B_{33}E_{||}$ $\omega/2g^2$, and, making the appropriate substitutions in Eqn. 3.1, we have

$$R(E\tau)=\cos[B_{33}(\omega/2g_z^2)E_{||}\tau]. \qquad [3.9]$$

If H_0 is perpendicular to E, then E lies in the x-y plane and

$$\delta(g^2) = B_{13}E_x+B_{23}E_y. \qquad [3.10]$$

In this latter case the x and y axes of the complexes selected by the low-field H_0 setting are not specified and the LEFE shifts have a distribution of values. The ratio function $R(E\tau)$ is derived from Eqn. 3.3 and is given by

$$R(E\tau) = J_0(\Delta\omega_{max}E_\perp\tau) \qquad [3.11]$$

where $\Delta\omega_{max} = (\omega/2g_z^2) (B_{13}^2+B_{23}^2)^{1/2}$.

$E\perp$ indicates an electric field perpendicular to H_0, and J_0 is the zero order Bessel function. (J_0 is similar in general appearance to a damped cosine). Corresponding results can be derived for an H_0 setting at the high-field g end of the EPR spectrum, and are obtained from Eqns. 3.9 and 3.11 by changing subscripts z→x, 1→3, 3→1.

In practice the ideal forms of the function $R(E\tau)$ predicted by Eqns. 3.9, 3.11 are not always observed. The main reason for this (other than possible electric field inhomogeneity in the sample) is that the selection of complexes made by setting H_0 at the top end or at the bottom end of the EPR spectrum is less than perfect. Although both functions in Eqn. 3.9, 3.11 cross the zero axis, making it possible in principle to determine the LEFE shift by searching for a value of E which results in complete destructive interference of the echo signal, the experimental ratio functions $R(E\tau)$ may deviate from the ideal form, or may fall asymptotically to zero in a somewhat inconclusive fashion. A better alternative is to determine the value $(E\tau)_{1/2}$ of the product $E\tau$ which halves the echo amplitude. This value is easy to locate and is less sensitive to inhomogeneity of the selected group of complexes. We can define shift parameters σ in terms of $(E\tau)_{1/2}$ by the equation

$$\sigma = 2\pi/[6\omega(E\tau)_{1/2}] \qquad [3.12]$$

which, in the case of the ideal ratio functions in Eqns. 3.9, 3.11, are related to the B_{ij} coefficients by

$$\sigma_{||,z} = B_{33}/2g_z^2 \qquad [3.13]$$
$$\sigma_{||,x} = B_{11}/2g_x^2 \qquad [3.14]$$
$$\sigma_{\perp,z} = (B_{13}^2+B_{23}^2)^{1/2}/2.9g_z^2 \qquad [3.15]$$
$$\sigma_{\perp,x} = (B_{31}^2+B_{21}^2)^{1/2}/2.9g_x^2 \qquad [3.16]$$

the σ subscripts $||$, $\perp$ denote, as before, the orientation of E in relation to H_0; the σ subscripts x,z denote whether σ was measured at the g_x or g_z end of the spectrum. (For the $E_{||}H_0$ settings in [3.13, 3.14] the parameter $\sigma = (|\delta g/g|)/E$, i.e., it defines the fractional g shift per unit electric field amplitude.)

The above discussion covers only six coefficients and only four independent numbers. Though these numbers may be quite useful in themselves -- for instance if the LEFE is associated with a major ligand that determines the orientation of the g_z axis -- there is a considerable amount of additional information to be obtained by making LEFE measurements throughout the EPR range of fields. A more complete picture can be obtained by plotting the shift parameters $\sigma_{||}$ and $\sigma_\perp$ defined as in [3.12], as a function of H_0 throughout the EPR spectrum. The LEFE curves obtained in this way have various characteristic forms and can sometimes be used to assign metalloproteins to their correct class, as described later. The LEFE curves can also be compared with computer simulations based on trial values of the B_{ij} coefficients.

3.2. Mechanisms responsible for the LEFE shifts

Theoretical analyses of the LEFE shifts have been made in a number of cases, most concerning centers in ionic single-crystal samples. The problem of calculating the shifts falls into two parts: a) a calculation of the internal electrostatic field that is actually seen by the paramagnetic center when a voltage is applied across the sample, including any higher order harmonic components, and b) a calculation of the effect of this increment in the crystal field on the terms in the spin Hamiltonian. Physical displacements of the atoms in the complex, and the polarization of the electronic wave functions belonging to the paramagnetic ion have both been taken into account in theoretical papers. It should be noted here that shifts of electronic charge do not in themselves give rise to an LEFE. Displacement of charge must be associated with a change in g or in the crystal field splitting.

The theoretical results show that it is possible to account for the LEFE shifts more or less as adequately as one can account for the values of other EPR parameters. But much of the work is not directly applicable to situations encountered in metalloproteins, where electron sharing with ligands tends to be more important than simple crystal field potentials, and where ionic charge configurations in the environment play a less significant role. A model in which a simple Lorentz internal electric field, (given by E(internal)=1/3(ϵ+2)E(applied)), polarizes the molecular orbital of the unpaired electron in the metal-ion-ligand complex, would seem to be appropriate in this case. In discussing ionic crystal data, some authors have also considered a model in which the internal electric field acts in two ways: by displacing the ion relative to its ligands, thus altering the degree of covalent bonding ("ionic effect"), and by directly polarizing the metal-ion-ligand mixed wave functions ("electronic-effect"). Since this area of research is still largely unexplored as far as organically bonded metal atoms are concerned, it is not possible to say just which approach will prove most useful.

The calculation of the electronic effect involves a detailed consideration of the wave functions, spin-orbit coupling mechanisms, etc., which enter into the calculation of the g factor or the crystal field splitting. It also requires consideration of wave functions of opposite parity to those primarily responsible for the paramagnetic properties. The non-centro-symmetric environment mixes the two kinds of wave functions, e.g., for the iron group the 3d and 4p wave functions. The applied electric field shifts the g values by altering this mixture. It is not possible to go into more detail here, since each paramagnetic ion, and each complex, poses a somewhat different problem. Reference may be made to a recent theoretical paper by Gerstman and Brill (65), dealing with the LEFE in "blue" copper proteins, and also to LEFE studies of half met hemocyanin derivatives by Solomon et al., in this volume.

3.3. "Fingerprint" experiments

In spite of the dearth of theoretical studies it has, nevertheless, been possible to extract some useful biological information from LEFE measurements. As a first example we note the "LEFE fingerprint" method of identifying and classifying metal ion centers in proteins (66). LEFE measurements were made on a series of iron-sulfur proteins, the $\sigma_{||}$ and $\sigma_{\perp}$ shift parameters being plotted as a function of the H_0 setting as described earlier. The proteins studied in these early experiments were known to contain one of three types of clusters: a) two-iron clusters giving EPR signals when in the reduced state, e.g. spinach ferredoxin and bovine adrenodoxin; b) four-iron clusters giving EPR signals in the reduced state, e.g. clostridial ferredoxin, and c) four-iron clusters giving EPR signals in the oxidized state, i.e. high potential iron-sulfur proteins (HIPIP). Those with two-iron clusters formed a class apart. Their LEFE curves are discussed subsequently. Of the remaining iron-sulfur proteins, those with four-iron centers that were EPR active when reduced gave LEFE curves of a more or less similar shape, starting from low values of $\sigma_{||}$ and $\sigma_{\perp}$ at low H_0, and rising to values of $\sigma_{||}$ and $\sigma_{\perp}$ several times larger at high H_0 (as for example, in Fig. 15A). The HIPIP curves showed common features distinguishing them markedly from curves for ferredoxins with a four-iron cluster; at low H_0, $\sigma_{||}$ was about twice $\sigma_{\perp}$, but towards higher H_0, $\sigma_{||}$ fell below $\sigma_{\perp}$ (e.g., Fig. 15B).

Merely in order to classify iron-sulfur proteins as ferredoxins or HIPIP's, these LEFE measurements would scarcely justify the effort, since the two groups are already distinguished by their range of g values, and by the

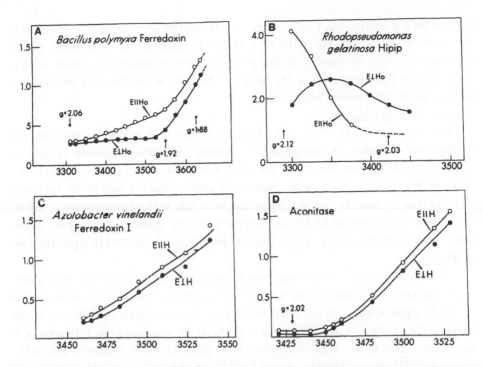

Fig. 15. LEFE shift parameters σ plotted as a function of magnetic field for A. _Bacillus_ _polymyxa_ ferredoxin; B. _Rhodopseudomonas_ _gelatinosa_ HIPIP; C. _Azotobacter_ _vinelandii_ ferredoxin 1; and D. bovine aconitase. Experiments were carried out with the electric field E parallel to the magnetic field (E∥H) or the electric field perpendicular to the magnetic field (E⊥H).

oxidation states required to give a resonance signal. However, a group of iron-sulfur proteins was later discovered that resembled the HIPIP's in being EPR active only when oxidized, but which differed in having a range of g values like that of the four-iron ferredoxins. LEFE measurements on these proteins (Fig. 15 C,D) (67,68) gave curves which were closer to those for the four-iron ferredoxins than to those for the HIPIP's, but there were certain distinctive features such as an almost exact superposition of the $\sigma_{\parallel}$ and $\sigma_{\perp}$ curves and an increase by 4:1 or more in going from low H_0 to high H_0. The evidence pointed to a new class of iron-sulfur proteins with a different site symmetry, and, subsequently, one member of the group, Azotobacter vinelandii ferredoxin I (Fig. 15C), was identified by X-ray crystallographic analysis as containing a three-iron cluster (69). Subsequently, a three-iron cluster was discovered in aconitase (Fig. 15D) (70), while more recently (34) one of the iron-sulfur centers in E. coli fumarate reductase was shown by the LEFE method to consist of a three-iron cluster also. Since chemical methods, either by iron analysis or by cluster extrusion are often not accurate enough to distinguish between three- and four-iron atoms per molecule, the LEFE method provides a useful and simple means for assigning an iron-sulfur protein to the correct group. The value of a finger print check of this kind, either for the purpose of identification or as a means of detecting sample impurities, obviously increases as the body of information on the LEFE behavior of various metal-ion centers accumulates.

3.4. Symmetry tests

The simplest test that can be made by the LEFE method is one aimed at finding out whether a paramagnetic complex is centrosymmetric or not; if there is a shift then the center does not have inversion symmetry. Most centers in frozen solution samples have a low point symmetry and do in fact give LEFE shifts. However, we shall mention here two instances in which a center might have been expected to be centrosymmetric, and in which its actual deviation from centrosymmetry could provide useful clues as to its structure and bonding.

The redox center in the two-iron ferredoxins consists of two iron atoms, each tetrahedrally coordinated by four sulfur atoms, two of them being common to both iron atoms. It has been shown that the coordinating geometry is approximately cubane, i.e., it can be arrived at by constructing a framework of two cubes having a face in common, iron atoms at their centers, and sulfur atoms on alternate corners (see Fig. 1 in Ref. 66). In the reduced, EPR active state, an electron is shared between the two iron atoms. If the electron were equally shared between the two metal ions, the complex would be centrosymmetric about a point midway between them (with the point symmetry D_{2h}), and there would be no LEFE. Experiments show that this is very nearly the case for one such two-iron protein, adrenodoxin (66). The shifts, which are very small and almost impossible to detect, might be ascribed to deviations from the ideal cubane structure or to a non-centrosymmetric configuration of atoms outside the first coordination shell. However, for spinach ferredoxin, which has the same nominal two-iron cubane active site geometry, the LEFE shifts are several times larger, suggesting that the electron is not equally shared between the two iron atoms in this case. The reason for this distinction and its possible relation to the redox properties of the two ferredoxins raise interesting questions, which could not be explored further at the time the experiments were made. It was also noted that, for spinach ferredoxin, the largest shifts were obtained with E parallel to H_0 at high H_0 (g_x) end, and also with E perpendicular to H_0 at the low H_0 (g_z) end of the EPR spectrum. This strongly suggests that the paramagnetic (and redox) electron, (i.e. the electron most likely to be displaced by the electric field), is easily polarized along the g_x axis, and that this axis lies along the line joining the two iron atoms.

The case of Cu(II) complexes in aqueous solution constitutes another

instance in which an LEFE observation yielded an unexpected piece of structural information. Though these Cu complexes are only of peripheral concern
in biological studies, where they are encountered either as impurities or as
models for more complex systems, they may be of some interest in connection
with copper-oxygen coordination, a topic which is relevant to high-temperature superconductors. It is frequently stated in textbooks of inorganic
chemistry that the normal stereochemical environment for hexa-coordinate,
divalent copper consists of a centrosymmetric square planar array for four
close lying ligands with copper at their center, and a pair of further removed equidistant ligands lying along an axis perpendicular to the plane. In
this case the point symmetry would be D_{4h} and all LEFE parameters would be
zero. However, this assumption was immediately contradicted for the Cu(II)-
aquo complex (Fig. 16A) and for a series of other simple Cu(II) complexes in
frozen solution (71). Only for Cu(II) in the compound copper-uroporphyrin,
where square planar coordination was enforced by the geometry of the porphyrin ring, was it possible to approximate to the expected results.
Otherwise for Cu(II) with monodentate ligands, LEFE shifts were easily ob-
served, the $\sigma_{||}$ and $\sigma_{\perp}$ curves being similar in scale and in form for all
the complexes studied.

Two possible explanations for this "anomaly" were considered. There
could be a disparity in the distances of the two axial ligands, resulting in
a C_{4v} point symmetry, or there could be a small tetrahedral distortion in
the "square planar" coordination, resulting in a D_{2d} point symmetry. Both
C_{4v} and D_{2d} are axial point groups, and the EPR spectrum would therefore
remain axial and not reveal either of these non-centrosymmetric distortions.

Further interpretation of the LEFE results for the Cu(II) complexes involves consideration of the B_{ij} (i.e. g^2 shift) coefficients defined earlier (Eqn. 3.6), and a computer simulation of their effects on the shapes of
the $\sigma_{||}$ and $\sigma_{\perp}$ vs H_0 curves. Fortunately, both of the point groups
concerned involve only a small number of B_{ij} coefficients, two for D_{2d}
and three for C_{4v}. The simulations (which also took the large copper
hyperfine splitting into account) showed conclusively that the LEFE shifts
arose from a D_{2d} distortion of the nominally square planar Cu(II) coordination. This would appear to be the natural, lowest energy, coordination for
the cupric ion, as evidenced by these experiments and also by subsequent LEFE
investigations of the Cu(II) ion in a series of oxide glasses (72).

This last illustration suggests that LEFE experiments in conjunction with
computer simulations might be used more generally to determine the point symmetry of metalloprotein centers. Computation is now very much cheaper than
it was when some of the earlier LEFE experiments were performed, and it need
not be excessively time consuming if, as in the case of copper complexes, an

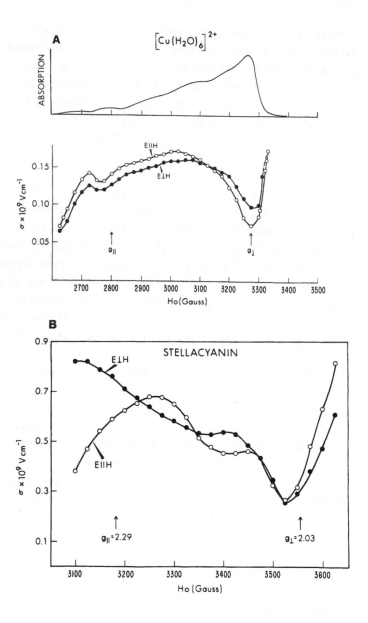

Fig. 16. A. (upper) Electron spin echo amplitude as a function of magnetic field setting; (lower) LEFE shift parameters for the Cu(II) aquo complex. The curve in the upper trace is similar to that which one would obtain by integrating a typical EPR spectrum. Shift parameters in the lower trace are measured at a series of magnetic field settings with the electric field parallel (E||H) or perpendicular (E⊥H). The position in the EPR spectrum that corresponds to each of these measurements can be seen from the upper curve. B. LEFE for stellacyanin.

attempt is made to find relatively high trial symmetry that models the major bonding mechanisms. Large measured values for one or more B_{ij} coefficients would point to the existence of a highly polarizable electronic wave function associated with the unpaired spin, and would help to establish the orientation of a related ligand. This is illustrated by the results obtained in an LEFE study of the myoglobin azide complex (64) (Fig. 17). At the low H_0 end of the EPR spectrum, i.e., with H_0 along the g_z axis, a large shift was observed for $E||H_0$ and a small shift for $E{\perp}H_0$. At the high H_0 end, i.e., with H_0 along the g_x axis, a small shift was observed for $E||H_0$ and a large shift for $E{\perp}H_0$. These shifts correspond to large values for

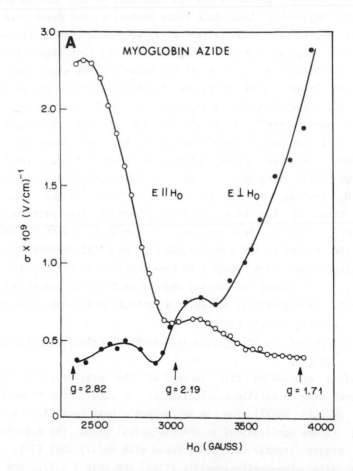

Fig. 17. LEFE of sperm whale myoglobin azide. Shifts were determined with the electric field aligned parallel ($E||H_0$) and perpendicular ($E{\perp}H_0$) to the Zeeman field. The principal g values are indicated by the vertical arrows.

the coefficients B_{31}, B_{33}. Physically, they suggest the presence of a strong polarizable bond oriented in the z-direction. For myoglobin azide, this is the direction perpendicular to the plane of the porphyrin (73).

LEFE experiments on proteins containing a Type I or "blue" copper center provide another illustration (74). The "blue" copper proteins are characterized by an extraordinarily intense optical absorption near 600 nm, and by a value of the hyperfine coupling constant $A_{||}$ that is smaller than that found for copper in simple metal-ion-ligand complexes. LEFE measurements on three such proteins have yielded shifts that are several times larger than those in other copper complexes, and have shown that there is a direct correlation between the shift and the magnitude of the $\sim$600 nm absorption coefficient, suggesting that both these properties are associated with the same bonding mechanism. Chemical and X-ray data show that Cu is coordinated by a mercaptide ligand, partial charge transfer to this ligand being responsible for the intense blue color of the protein. One might therefore expect to find the largest LEFE shifts when the electric field was oriented along the Cu-S bond, since there the polarizing effect would be greatest.

The experimental curves of $\sigma_{||}$ and $\sigma_{\perp}$ for the blue copper proteins (Fig. 16B) not only showed larger shifts than were found for the Cu(II)-aquo complexes mentioned earlier, but had one major distinguishing feature, namely a large shift parameter $\sigma_{\perp}$ at the low $H_0(g_{||})$ end of the spectrum. At this H_0 setting, $\sigma_{\perp}$ was two or more times larger than $\sigma_{||}$. On the other hand, at the high field ($g_{\perp}$) end of the spectrum, $\sigma_{||}$ was larger than $\sigma_{\perp}$. Taking these results in conjunction with the optical data and with the evidence for a mercaptide bond we might reasonably assume that the mercaptide bond lies roughly in the direction of the $g_{||}$ axis and that this bond is polarizable to varying degrees in the blue copper proteins studied. It would be interesting to derive a connection between these results and the redox properties of the center.

3.5. Applied electric fields as a means of resolving EPR and ESEEM spectra

Before concluding this survey of the relatively few LEFE studies that have been made on metalloprotein samples, we should like to mention an observation on the Mn(II) ion in an inorganic host crystal, which could lead to useful future applications in the biological area. The experiments were made on a single crystal of $CaWO_4$ doped with Mn(II) ions (75). The Mn ion occupies axial non-centrosymmetric sites, the spin S = 5/2, and the axial crystal field splitting coefficient D = 413 MHz is small compared with the X-band resonance frequency. Depending on the orientation of the electric field, the LEFE manifests itself in this material as either a shift in D and the generation of a small rhombic E term, or as a small tilt of the principal axis system with respect to which D and E are defined. The g-value ($g \simeq 2.0$) and the nuclear hyperfine coupling coefficients are negligibly affected by the electric field.

The single crystal EPR spectrum in $(Ca,Mn)WO_4$ consists of 30 lines (some of which may overlap) arranged in five groups of six hyperfine lines associated with the $I=5/2$ nuclear spin of ^{55}Mn. Since D is so much smaller than the Zeeman energy at X band the electronic levels can, to a fair approximation, be described by the M_S quantum numbers $\pm 1/2$, $\pm 3/2$, $\pm 5/2$, and the transitions obey the $\Delta M = 1$ selection rule, i.e. the allowed transitions are $+1/2 \rightleftharpoons -1/2$, $\pm 1/2 \rightleftharpoons \pm 3/2$ and $\pm 3/2 \rightleftharpoons \pm 5/2$. To this order of approximation the LEFE shifts in the $\pm 3/2 \rightleftharpoons \pm 5/2$ set are twice as large as large as in the $\pm 1/2 \rightleftharpoons \pm 3/2$ set of lines, while the shifts in the $\pm 1/2 \rightleftharpoons -1/2$ are zero. Actually, in spite of the small value of D, some state mixing does occur, and LEFE shifts can also be observed for the $\pm 1/2 \rightleftharpoons -1/2$ transitions, but the shifts here are several times smaller than the shifts for the other sets of lines. A series of "boxcar" or "field swept ESE spectra", with and without an applied electric field step, were recorded for this material. (A "boxcar spectrum" is the EPR spectrum obtained by setting the "boxcar" or sample/hold unit on the echo signal and varying H_o.) It was found, as expected, that by choosing suitable values of τ and of the electric field one could eliminate either the $\pm 5/2 \rightleftharpoons \pm 3/2$ or the $\pm 3/2 \rightleftharpoons \pm 1/2$ set of lines from the spectrum.

This suggests an application in which the Mn(II) spectrum for a metalloprotein sample might be considerably simplified, or, alternatively, an application aimed at resolving a compound spectrum arising from centers with a large shift [e.g. low-spin Fe(III)] and centers with a small or zero shift [such as Cu(II) or radicals] into its component parts. The method could be used not only to resolve the EPR spectra, but also to separate out the ESEEM patterns associated with different components. In the latter case, the experiment would be done in the three-pulse mode with τ fixed and τ' variable, and with the electric field step applied either in coincidence with pulse II, or with the field step applied at some convenient time between pulses II and III. (The applied field is without effect on the echo amplitude during the pulse II to pulse III interval.) It will be recognized, of course, that the technique suggested here is analogous to the familiar laboratory trick of partially saturating one of two EPR species in order to simplify a spectrum.

4. SOME EXPERIMENTAL NOTES

Although it is not feasible here to include a comprehensive guide to ESE measurements on biological materials, we append some short notes in the hope that they will be useful to those entering the field.

A.   Experiments involving metal ions generally have to be performed in the liquid helium range ($\leq 4.2°K$) and some may even require temperature below the lambda point ($\leq 2.2°K$). Low temperatures are needed in order to obtain adequate phase memory times. The requirement is much more stringent than for conventional EPR.   It should be also noted that a fast relaxing impurity is likely to shorten the phase memory for all other paramagnetic centers in the sample, even when the resonance frequencies are entirely different, and when there is no possibility of energy transfer by cross relaxation (76).

B.   The electron spin concentration in metalloprotein frozen solution samples is usually too low to shorten the phase memory, which is limited primarily by proton-proton spin flip-flops. Magnetic dilution may be needed, however, for small proteins and must be considered when experimenting on chemical model compounds.   Protein samples may be diluted with glycerol. For a test sample consisting of Cu(II) aquo in water:glycerol mixtures, a concentration of 5 mM has been found to be the upper limit.

C.   It is important to use a microwave resonator with a high filling factor.   This inexpensive component can increase the power of an ESE system by an order of magnitude, improving signal to noise and greatly reducing the time needed to acquire good quality ESE data. A variety of designs are available (the Chapter by Hyde and Froncisz, for example). The choice among them will depend on trade offs between convenience and filling factor, and on any special features of the experiment, such as optical excitation, ENDOR, rapid magnetic field steps, and electric field steps.

D.   Resonator Q values have to be low in order to admit short microwave pulses and in order to avoid long ringing times. Values of $Q \simeq 120$ are typical.   Low Q values can be obtained by overcoupling. An adjustable coupling device is useful for making temporary reductions in Q so as to limit the ringing time when recording the very early part of an echo envelope (77).

E.   In a three-pulse experiment on a wide-line sample, a series of two-pulse echoes are generated in addition to the required stimulated echo (see Fig. 2).   These echoes overlap the stimulated echo at time $\tau' = 2\tau$, $\tau' = 3\tau$ and will cause glitches to appear in the recorded echo envelope unless precautions are taken. One way of eliminating these glitches is to reverse the phase of pulses I and II in alternate echo cycles. This has no effect on the stimulated echo signal but inverts the unwanted echoes so that they alternately add and subtract at the overlap points, leaving a zero net contribution to the signal level.

F.   It can be seen from Eqns. 2.1 and 2.16 that the function modulating the echo envelope consists of a sum of cosines. The associated frequencies can therefore, in principle, be extracted by taking the Fourier cosine transform of the envelope from time τ (or τ') = 0 to τ (or τ') = ∞, or what amounts to the same thing, by taking the Fourier transform from τ (or τ') = $-\infty$ to τ (or τ') = $+\infty$ of the even function obtained by mirroring the envelope about τ (or τ') = 0. Unfortunately, it is not possible in practice to measure the envelope over the mathematically required range. There is no problem at long times, since the envelope is normally recorded out to a point where the modulation becomes undetectable, and, insofar as is necessary for computational purposes, the envelope can be extended beyond this point with a straight line or exponential curve. However, at short times there will be a missing portion between τ (or τ') = 0 and τ (or τ') $\simeq$ 100 to 300 nanosec., this being the time required to allow microwave ringing in the resonator to decay, plus (in the three pulse case) the time required to complete the pulse sequence.* This missing portion of the echo envelope can introduce major artifacts into ESEEM spectra obtained for metalloproteins, since, in these materials, the total duration of the modulation pattern often does not exceed a few microseconds.

It may be felt that, in such a situation, the scientifically correct strategy is somehow to transform the piece of the envelope that one can record, and to ignore the rest. This is to misunderstand the nature of the Fourier transformation. One inevitably makes an assumption about the form of the missing portion of the time wave. "Ignoring" it is equivalent to constructing an arbitrary function to fill in the gap in time. The procedure adopted might, for example, imply joining the first recorded point back to τ (or τ') = 0 by taking a straight line, or might imply mirroring the recorded envelope about the initial recording time instead of about time zero.

The construction of a "best guess" continuation of the waveform to time zero lies in the realm of "spectral estimation", and is one of the topics discussed in the chapter by de Beer and van Ormondt. Failing any such sophisticated treatment of the data, it is often more realistic to extend the echo envelope back to time zero by means of a freehand sketch (e.g. on a digitizing pad), than it is to adopt other rudimentary expedients, which could involve less reasonable assumptions about the unrecorded portion of the envelope. For illustrations of the spectral artifacts arising from dead time errors see Figs. 2-4 in Ref. 39.

G. An ESEEM experiment is performed by sampling the echo amplitude at a succession of times τ or τ'. If we assume that the echo envelope function remains the same throughout the experiment (although only one value of the

*The stimulated echo modulation pattern can in fact be extended into the range of times τ' shorter than τ by moving pulse III backward in time, so that it crosses and then precedes pulse II. What one has done is to switch to a new stimulated echo sequence, but the modulation pattern is continuous. A large unwanted bump occurs when pulses II and III cross.

function is actually measured in any given echo cycle), then the procedure is
analogous to that of digitizing a time waveform, and is subject to the usual
sampling rate constraints.   Thus, the sampling rate must exceed twice the
highest frequency in the spectrum of the time wave. For example, if samples
are taken at 10 nanosecond increments, the effective sampling rate SR is 100
MHz, and the highest permitted frequency is 50 MHz. If a higher frequency F
is present, then the sampling procedure will generate an "alias" or spurious
frequency SR-F.

Since the highest NMR frequency encountered in a three-pulse X-band ESEEM
experiment is usually the proton frequency at $\sim$ 13 MHz, or in a two-pulse
X-band experiment, the proton sum frequency at $\sim$ 26 MHz, there would seem to
be ample headroom with a 100 MHz effective sampling rate. However, in experiments aimed at extracting weak modulation components from a complex spectrum,
it may become necessary to take into account the higher harmonics and combination frequencies (see section 2.9) generated as a result of the product theorem (Eqns. 2.15, 2.17). For instance, in a sample with four protons coupled
to the same electron, the theoretical two-pulse spectrum in X-band experiments extends to $\sim$ 104 MHz and would, if sampled at 100 MHz, lead to alias
frequencies at 48 MHz, 35 MHz, 22 MHz, 9 MHz and 4 MHz.

In this particular case the alias frequencies could almost certainly be
ignored, since modulation due to one coupled proton is usually quite shallow,
and the higher harmonics introduced by the product theorem would have negligible amplitudes.   But problems might arise in experiments at higher H_0
values or with materials showing deep modulation. In order to avoid aliasing, all frequencies above SR/2 should be removed from the time wave before
it is sampled. Since there is no physically real waveform in an ESEEM experiment, until after sampling (by measurement of the echo amplitude), this rules
out most filtering techniques, including such procedures as lengthening the
receiver time constant, speeding up the scan rate, etc. These procedures filter the frequency spectrum after the introduction of the alias components and
cannot eliminate them selectively. A form of low pass filtering before sampling can be introduced by limiting the microwave field amplitude H_1 so
that frequencies above SR/2 are not excited to any significant extent, but
the frequency cut off associated with this method is not very sharp. A method of obtaining the benefit of a faster sampling rate, and hence of reducing
aliasing, without a major upgrade of the pulse timing system, is to repeat
the envelope recording with the sample times offset (using e.g. a fixed delay
line) by half the standard sample to sample interval (e.g. 5 nsec). For the
envelope obtained by combining both recordings, the effective sampling rate
is doubled. (This method is sometimes called 'interleaving'.)

H.   Frozen solutions, including aqueous solutions, become good insulators at low temperatures and, in our experiments, have withstood fields of 10^5 V/cm. Liquid helium, above or below the lambda point, is likewise a good insulator, but helium gas is a very poor insulator. An immersion dewar system, in which gas or gas bubbles are kept away from regions near exposed electrodes has proven satisfactory in LEFE studies.

REFERENCES

1   A recently compiled listing of these facilities can be obtained by writing Dr. J. Peisach, Albert Einstein College of Medicine, Bronx, NY 10461.
2   L. Kevan and R.N. Schwartz (Eds), Time Domain Electron Spin Resonance, John Wiley and Sons, New York, 1970 Chapters 5-8.
3   W.B. Mims and J. Peisach, in: R.G. Shulman (Ed.) Biological Applications of Magnetic Resonance, Academic Press, New York, 1979, pp. 221-269.
4   W.B. Mims and J. Peisach, in: L.J. Berliner and J. Reuben (Eds), Biological Magnetic Resonance, Plenum Press. New York, 1981, pp. 213-263.
5   Yu. D. Tsvetkov and S.A. Dikanov, in: Helmut Sigel (Ed.) Metal Ions in Biological Systems, Vol. 22, Marcel Dekker, Inc., New York, 1987, pp. 207-263.
6   W.B. Mims, Rev. Sci. Instrum., 45 (1974) 1583-1591.
7   A. Abragam and B. Bleaney, "Electron Paramagnetic Resonance of Transition Ions", Clarendon Press, Oxford, p. 174, Fig. 3.10 (1970). See also ref. 2, p. 295.
8   V.F. Yudanov, K.M. Salikhov, G.M. Zhidomirov and Yu.D. Tsvetkov, Zh. Strukt. Khem., 10 (1969) 732 [J. Struct. Chem., 10, (1969) 625].
9   W.B. Mims, J. Peisach and J.L. Davis, J. Chem. Phys., 66 (1977) 5536-5550.
10 S.A. Dikanov, A.A. Shubin and V.N. Parmon, J. Mag. Res., 42 (1981) 474-487.
11 W.B. Mims, Phys. Rev. B5, (1972) 2409-2419.
12 S.A. Dikanov, V.F. Yudanov and Yu. D. Tsvetkov, J. Mag. Res., 34 (1979) 631-645.
13 A. Abragam, Principles of Nuclear Magnetism, Clarendon Press, Oxford, 1961, see p. 237.
14 J.J. Zweier, J. Peisach and W.B. Mims, J. Biol. Chem., 257 (1982) 10314-10316.
15 P. Tipton, J. Cornelius, J. McCracken and J. Peisach, Biochemistry, (1988) submitted.
16 J. Peisach, W.B. Mims and J.L. Davis, J. Biol. Chem., 259 (1984) 2704-2706.
17 W.B. Mims and J.L. Davis, J. Chem. Phys., 64 (1976) 4836-4846.
18 W.B. Mims, J.L. Davis and J. Peisach, Biophys. J. 45 (1984) 755-766.
19 W.B. Mims and J. Peisach, J. Chem. Phys., 69 (1978) 4921-4930.
20 W.B. Mims and J. Peisach, J. Biol. Chem., 254 (1979) 4321-4323.
21 J. Peisach, W.B. Mims and J.L. Davis, J. Biol. Chem., 254 (1979) 12379-12389.
22 H.L. van Camp, R.H. Sands and J.A. Fee, J. Chem. Phys., 75 (1981) 2098-2107.
23 M.J. Hunt, A.L. Mackay and D.T. Edmonds, Chem. Phys. Lett., 34 (1975) 473-475.
24 D.T. Edmonds, Physics Reports, 29 (1977) 233-290.
25 D.J. Kosman, J. Peisach and W.B. Mims, Biochemistry, 19 (1980) 1304-1308.
26 J. McCracken, S. Pember, S.J. Benkovic, J.J. Villafranca, R.J. Miller and J. Peisach, J. Am. Chem. Soc., 110 (1988) 1069-1074.

27   C.I.H. Ashby, C.P. Cheng and T.L. Brown, J. Am. Chem. Soc., 100 (1978) 6057-6067.

28   H. Thomann, T.V. Morgan, H. Jin, S.J.N. Burgmayer, R.E. Bare and E.I. Stiefel, J. Am. Chem. Soc., 109 (1987) 7913-7914.

29   S.L. Tan, J.A. Fox, N. Kojima, C.T. Walsh and W.H. Orme-Johnson, J. Am. Chem. Soc., 106 (1984) 3064-3066.

30   A. Chapman, R. Cammack, C.E. Hatchikian, J. McCracken and J. Peisach, FEBS Lett., (1988), in press.

31   J. McCracken, J. Peisach and D.M. Dooley, J. Am. Chem. Soc., 109 (1987) 4064-4072.

32   B. Mondovi, L. Morpurgo, E. Agostinelli, O, Befani, J. McCracken and J. Peisach, Eur. J.Biochem., 168 (1987) 503-507.

33   R.S. Magliozzo, J. McCracken and J. Peisach, Biochemistry, 26 (1987) 7923-7931.

34   R. Cammack, A. Chapman, J. McCracken, J.B. Cornelius, J. Peisach and J.H. Weiner, Biochim. Biophys. Acta, (1988) in press.

35   T. Shimizu, W.B. Mims, J. Peisach and J.L. Davis, J. Chem. Phys., 70 (1979) 2249-2254.

36   R. LoBrutto, G.W. Smithers, G.H. Reed, W.H. Orme-Johnson, S.L. Tan and J.S. Leigh, Biochemistry, 25 (1986) 5654-5660.

37   K. Doi, J. McCracken, J. Peisach and P. Aisen, J. Biol. Chem., 263 (1988) 5757-5763.

38   J. McCracken and J. Peisach, unpublished observations.

39   W.B. Mims, J. Magn. Res., 59 (1984) 291-306.

40   H. Thomann, H. Jin and G.L. Baker, Phys.Rev. Lett., 59 (1987) 509.

41   J.A. Fee, J. Peisach and W.B. Mims, J. Biol. Chem., 256 (1981) 1910-1914.

42   J. Peisach, W.E. Blumberg, S. Ogawa, E.A. Rachmilewitz and R. Oltzik, J. Biol. Chem., 256 (1971) 3342-3355.

43   W.E. Blumberg and J. Peisach, in: R. Dessy, J. Willard and L. Taylor (Eds), Bioinorganic Chemistry, Advances in Chemistry Series, 100, Amer. Chem. Soc., Washington, 1971, pp. 271-291.

44   T.D. Smith and J.R. Pilbrow, in: L.J. Berliner and J. Reuben (Eds), Biological Magnetic Resonance, Vol. 2, Plenum Press, New York, 1980, pp. 85-168.

45   G. Palmer, Biochem. Soc. Trans., 13 (1985) 548-560.

46   W.E. Blumberg and J. Peisach, Arch. Biochem. Biophys., 162 (1974) 502-512.

47   R. Cammack, D.S. Patil and V.M. Fernandez, Biochem. Soc. Trans., 13 (1985) 572-578.

48   J. Peisach and W.E. Blumberg, Arch. Biochem. Biophys., 165 (1974) 691-708.

49   J.F. Boas, in: R. Lontie (Ed.) Copper Proteins and Copper Enzymes, Vol. 1, CRC Press, Boca Raton, 1984, pp. 63-92.

50   E.I. Solomon, in: L. Que, Jr. (Ed.) Metal Clusters in Proteins, ACS Symposium Series 372, Amer. Chem.Soc., Washington, 1988, pp. 116-151.

51   H. Thomann, T.V. Morgan, H. Jin, S.J.N. Burgmayer, R.E. Bare and E.I. Stiefel, J. Am. Chem. Soc., 109 (1987) 7913-7914.

52   E.G. Graf and R.K. Thauer, FEBS Lett., 136 (1981) 165-169.

53   N.D. Chasteen, in: L.J. Berliner and J. Reuben (Eds), Biological Magnetic Resonance, Vol. 3, Plenum Press, New York, 1980, pp. 53-120.

54   G.H. Reed and G.D. Markham, in: L.J. Berliner and J. Reuben (Eds), Biological Magnetic Resonance, Vol. 6, Plenum Press, New York, 1984 pp. 73-142.

55   C.F. Brewer, J. McCracken and J. Peisach, unpublished observations.

56   J.McCracken, D.M. Dooley and J. Peisach, unpublished observations.

57   C.L. Lobenstein-Verbeek, J.A. Jongejan, J. Frank and J.A. Duine, FEBS Lett., 170 (1984) 305-309.

58   D.E. Edmonson, R. DeFrancesco, J. McCracken and J. Peisach, XIII International Conference on Magnetic Resonance in Biological Systems, Madison, WI, August (1988).

59   P.F. Liao and S.R. Hartmann, Phys. Lett., 38A (1972) 295; Phys. Rev. B8 (1972) 69.

60   For a detailed discussion of the theoretical and experimental aspects of the LEFE see W.B. Mims, The Linear Electron Field Effect in Paramagnetic Resonance, Clarendon Press, Oxford, (1976).

61   G.W. Ludwig and H.H. Woodbury, Phys. Rev. Lett., 7 (1961) 240-241.

62   W.B. Mims, Phys. Rev., 133 (1964) A835-A840.

63   For an experiment on the quadratic electric effect see A.B. Roitsin, Sov. Phys. Sol. St. 10 (1968) 751.

64   W.B. Mims and J. Peisach, J. Chem. Phys., 64 (1976) 1074-1091.

65   B.S. Gerstman and A.S. Brill, Phys. Rev., A37 (1988) 2151-2164.

66   J. Peisach, N.R. Orme-Johnson, W.B. Mims and W.H. Orme-Johnson, J. Biol. Chem., 252 (1977) 5643-5650.

67   J. Peisach, H. Beinert, M.H. Emptage, W.B. Mims, J.A. Fee, W.H. Orme-Johnson, A.R Rendina and N.R. Orme-Johnson, J. Biol. Chem., 258 (1983) 13014-13016.

68   B.A.C. Ackroll, C.B. Kearney, W.B. Mims, J. Peisach and H. Beinert, J. Biol. Chem., 259 (1984) 4015-4018.

69   D. Ghosh, W. Furey, Jr., S. O'Donnell and C.D. Stout, J. Biol. Chem., 256 (1981) 4185-4192.

70   T.A. Kent, J.-L. Dreyer, M.C. Kennedy, B.A. Huynh, M. Emptage, H. Beinert and E. Munck, Proc. Natl. Acad. Sci. U.S.A., 79 (1982) 1096-1100.

71   J. Peisach and W.B. Mims, Chem. Phys. Lett., 37 (1976) 307-310.

72   W.B. Mims, G.E. Peterson and C.R. Kurkjian, Phys. and Chem. of Glasses, 19 (1978) 14.

73   J.S. Griffith, Nature, 180 (1957) 30-31.

74   J. Peisach and W.B. Mims, Eur. J. Biochem., 84 (1978) 207-214.

75   A. Kiel and W.B. Mims, Phys. Rev., 153 (1967) 378-385.

76   W.B. Mims, Phys. Rev., 168 (1968) 370-389.

77   R.D. Britt and M.P. Klein, J. Mag. Res., 74 (1987) 535-540.

CHAPTER 2

ESEEM OF DISORDERED SYSTEMS: THEORY AND APPLICATIONS

SERGEI A. DIKANOV and ANDREI V. ASTASHKIN

1.   INTRODUCTION

Paramagnetic centers are known to be present in many biological systems. They exist there continuously or may be formed under external influences. This allows to solve some biological problems using several techniques of paramagnetic resonance that were greatly developed in the past few decades.

The samples of biological interest subjected to magnetic resonance investigations are usually amorphous, with the paramagnetic centers randomly oriented. As a result, structural studies of these centers by cw EPR spectroscopy are often hampered by inhomogeneous broadening of EPR absorption lines masking the spectral hyperfine structure. A definite success of magnetic resonance in the study of the paramagnetic centers' nearest environment in disordered solids may be attributed to the development of a pulsed EPR technique: electron spin echo (ESE) (1-3).

In the ESE method the inhomogeneous broadening is excluded by the effect of reversible spin dephasing and the unresolved hyperfine structure manifests itself as a modulation of the ESE signal decay envelope. One obtains in this way data on the types of the magnetic nuclei surrounding the centers, their number and the parameters of their magnetic interactions - hyperfine (hfi) and quadrupole (nqi). This provides information on the structure of the nearest environment, coordination with certain molecules, the bond types, etc.

The present review is concerned with the general properties inherent in ESE modulation in disordered systems in the time and frequency domains, and the methods of ESE modulation analysis. This 'dry matter' is illustrated by examples of the experimental investigation of some concrete systems.

The ESE signals result from the action of certain microwave pulse sequences on a paramagnetic sample placed in a constant magnetic field B_0. The microwave frequency is typically in the range 9-10 GHz (X-band) and B_0 equals some 300 mT to satisfy the magnetic resonance condition $\nu_0 = (\gamma_e/2\pi)B_0$. In the simplest type of the ESE experiment two short (~ 10-100 ns) powerful microwave pulses with magnetic component $\vec{B}_1 \perp \vec{B}_0$ are applied, which are separated by the time interval τ. At the time 2τ a spontaneous emission - a spin echo pulse - emerges. This variant of spin echo has been termed a two-pulse or primary echo.

Besides the two-pulse, a three-pulse sequence is often employed in ESE experiments. In this case the spin system is excited by three microwave pulses

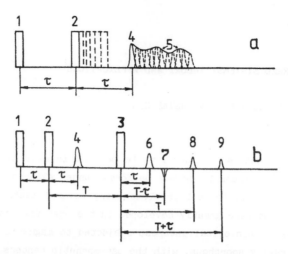

Fig. 1. Schemes of the pulse sequence and the formation of the echo signal in the two-(a) and three-pulse (b) spin echo experiments: 1,2,3 - microwave pulses, 4 - primary echo signal, 5 - electron spin echo envelope modulation, 6 - stimulated echo signal, 7,8,9 - unwanted primary echo signals appearing as a result of the combination of the third pulse with primary echo signal 4 and the microwave pulses 2 and 1.

separated by time intervals τ and T, giving rise to a number of primary echo signals formed by different pairs of pulses (Fig. 1). Another type of signal is formed at the time $2\tau + T$ by the combined action of all three microwave pulses, and just this one is called a three-pulse or stimulated echo signal.

Generally, the dependence of echo signal amplitude on the time interval - τ or T - is recorded in the experiment. This dependence is influenced by two different factors. One is the existence of magnetic relaxation processes that return the spin system perturbed by microwave pulses to the equilibrium state. This leads to a monotonous decay of the ESE signal as the time interval is incremented. The second is the interaction of unpaired electrons with magnetic nuclei causing oscillations in the echo signal decay, which are frequently observed and are termed electron spin echo envelope modulation (ESEEM). These two factors are mostly statistically independent and hence result in the product of monotonously decaying and oscillating functions.

Both simple qualitative and rigorous quantum mechanical models for the origin of the echo signal and its modulation are described in the literature (1-9). Therefore, the present review deals only with the final result of the formalized quantum mechanical analysis, which allows us to formulate conclusions on the frequencies of ESE modulations and their amplitudes.

In the system of one electron with $S = \frac{1}{2}$ and one nucleus with spin I, the ESE modulation in a general case is governed by the nuclear Zeeman, isotropic and anisotropic hyperfine and nuclear quadrupole interactions (the latter equals zero for nuclei with $I = \frac{1}{2}$). The interactions enumerated define the direction of the nuclear spin quantization axis and its change with a sudden change in electron spin orientation under the influence of microwave pulses. This nuclear spin reorientation disturbs the effect of phase compensation of the electron spin precession and leads finally to the appearance of the ESE modulation. A calculation within the density matrix formalism developed in (10,11) gives the following general expressions for the modulation in such a spin system:

$$V(\tau) = k_o + \sum_{i<j} k_{ij}^{(\alpha)} \cos 2\pi\nu_{ij}^{(\alpha)}\tau + \sum_{l<n} k_{ln}^{(\beta)} \cos 2\pi\nu_{ln}^{(\beta)}\tau$$

$$+ \sum_{i<j} \sum_{l<n} k_{ij,ln}^{(\alpha,\beta)} [\cos 2\pi(\nu_{ij}^{(\alpha)} + \nu_{ln}^{(\beta)})\tau + \cos 2\pi\nu(\nu_{ij}^{(\alpha)} - \nu_{ln}^{(\beta)}\tau)] \qquad [1]$$

in the primary ESE and:

$$V(\tau,T) = \frac{1}{2}\{V_\alpha(\tau,T) + V_\beta(\tau,T)\} = k_o/2 + \frac{1}{2} \sum_{i<j} k_{ij}^{(\alpha)} \cos 2\pi\nu_{ij}^{(\alpha)}\tau$$

$$+ \sum_{i<j} \cos 2\pi\nu_{ij}^{(\alpha)}(\tau + T)\{\frac{1}{2}k_{ij}^{(\alpha)} + \sum_{l<n} k_{ij,ln}^{(\alpha,\beta)} \cos 2\pi\nu_{ln}^{(\beta)}\tau\}$$

$$+ k_o/2 + \frac{1}{2} \sum_{l<n} k_{ln}^{(\beta)} \cos 2\pi\nu_{ln}^{(\beta)}\tau$$

$$+ \sum_{l<n} \cos 2\pi\nu_{ln}^{(\beta)}(\tau + T)\{\frac{1}{2}k_{ln}^{(\beta)} + \sum_{i<j} k_{ij,ln}^{(\alpha,\beta)} \cos 2\pi\nu_{ij}^{(\alpha)}\tau\} \qquad [2]$$

in the stimulated one. In these expressions $\nu_{ij}^{(\alpha)}$ and $\nu_{ln}^{(\beta)}$ are the nuclear transition frequencies for positive and negative electron spin projections onto the direction of the external magnetic field $\vec{B}_o$. The coefficients k with different indexes, defining the amplitude of modulation harmonics, depend only on the matrix elements of the operator $\hat{S}_x$.

It follows from the above relations that the primary ESE modulation frequencies are those of nuclear transition frequencies and their linear combinations, referred to the different electron spin projections. The stimulated ESE signal decay recorded at a constant value of τ as a function of T reveals only the nuclear transition frequencies. The amplitude of different harmonics in stimulated ESE depends on the choice of τ which gives the means to intensify some harmonics and to suppress the others in this type of experiment. This effect is termed in the literature the suppression effect.

The problem to express the modulation amplitudes and frequencies in an analytical form is solved only for nuclei with zero quadrupole moment. The common examples of such nuclei are the nuclei with spin $I = \frac{1}{2}$. In this case the modulation is described by the expressions:

$$V(\tau) = 1 - k/2 + \frac{1}{2}k[\cos 2\pi\nu_\alpha\tau + \cos 2\pi\nu_\beta\tau - \frac{1}{2}\cos 2\pi(\nu_\alpha + \nu_\beta)\tau$$

$$- \frac{1}{2}\cos 2\pi(\nu_\alpha - \nu_\beta)\tau] \tag{3}$$

in the primary ESE and:

$$V(\tau,T) = \frac{1}{2}[V_\alpha(\tau,T) + V_\beta(\tau,T)]$$

$$V(\tau,T) = \frac{1}{2}\{1 - \frac{1}{2}k[(1 - \cos 2\pi\nu_\alpha\tau)(1 - \cos 2\pi\nu_\beta(\tau + T))]\}$$

$$+ \frac{1}{2}\{1 - \frac{1}{2}k[(1 - \cos 2\pi\nu_\alpha(\tau + T))(1 - \cos 2\pi\nu_\beta\tau)]\} \tag{4}$$

in the stimulated ESE. Here ν_α and ν_β are the nuclear transition frequencies for $S = \frac{1}{2}$:

$$\nu_{\alpha(\beta)} = \sqrt{(\nu_I \mp A/2)^2 + B^2/4}$$

and the factor k, which determines the modulation amplitude, is:

$$k = \left((\nu_I B)/(\nu_\alpha \nu_\beta)\right)^2 .$$

The parameters A and B are equal to $A = T_{zz}$, $B = (T_{zx} + T_{zy})^{\frac{1}{2}}$, where T_{zi} are hfi tensor components in the laboratory coordinate system ($z // \vec{B}_o$).

Simple "exact" relations expresssing the primary and stimulated ESE modulation in the case of hfi with a nucleus of arbitrary spin in terms of the formulae for $I = \frac{1}{2}$ were derived in (14). This result is really approximate, however, because it has been obtained neglecting the nuclear quadrupole interaction for nuclei with $I > \frac{1}{2}$.

In real systems the unpaired electron usually interacts with more than one nucleus. In this case the modulation is described by the products (10):

$$V_N(\tau) = \prod_i V_i(\tau) \tag{5}$$

in the primary ESE and:

$$V_N(\tau,T) = \frac{1}{2}[\prod_i V_{i\alpha}(\tau,T) + \prod_i V_{i\beta}(\tau,T)] \tag{6}$$

in the stimulated one (15). As a result, the modulation may exhibit harmonics
with frequencies that are linear combinations of the transition frequencies of
different nuclei. However, the sets of these combination frequencies in the
primary and stimulated ESE are different due to the different multiplication
laws.

In the experiment the electron spin echo signal decay is recorded not
within a whole range of successively increasing time intervals between microwave pulses, but beginning from some nonzero interval, termed the dead time.
The main reason for the presence of the dead time is related to the finite time
of microwave power damping in the cavity after the application of pulses - the
so-called cavity ringing. After switching off the microwave pulses, the amplitude of this ringing is so high that for some time t_r the receiver is overload-
ed and fails to respond to a useful echo signal. The ringing time is governed
by the Q-value of the cavity and usually amounts to several hundreds of nanoseconds. In the two-pulse technique the echo signal arises in a time 2τ after
the second pulse, which is equal to the interval between the pulses themselves.
Therefore, at $\tau < t_r$ the primary ESE signal is not registered by the spectrometer and the dead time τ_d in the primary echo equals t_r.

In the three-pulse technique one studies the dependence of the stimulated
ESE signal amplitude on the sum of intervals $\tau' = \tau + T$ at a constant value of
τ. Since the stimulated ESE signal emerges in a time τ after the third pulse,
the τ-value is limited from below by t_r. In addition, due to the overlap of the
unwanted primary echo signals with the stimulated echo signal, the minimum
value of the variable time interval T at which the undistorted stimulated echo
signal can be recorded is equal to $T_{min} = 2\tau$. Thus, the dead time in the three-
pulse technique amounts to the value $\tau'_d = \tau + T_{min} = 3\tau$ with $\tau \geq t_r$.

To decrease the dead time, various procedures have been proposed (see
chapter 4 of this volume). But even these fail to overcome completely the dead
time caused by the cavity ringing. Therefore, the existence of the dead time is
a real peculiarity of the ESE spectroscopy to be faced in both the experiments
and in the analysis of the experimental results.

2. ESEEM IN THE TIME AND FREQUENCY DOMAINS IN DISORDERED SYSTEMS

2.1 ESEEM in the Weak hfi Limit

It is convenient to start a discussion of the time domain analysis of ESE
modulation in disordered systems with the simplest case of extremely weak axially symmetric anisotropic hfi of one electron spin with a nuclear spin $I = \frac{1}{2}$.
The above described parameters A and B now have the following orientational
dependences:

$$A = T_\perp (1 - 3 \cos^2\theta) \; , \; B = 3T_\perp \sin\theta \cos\theta \; . \tag{7}$$

Assuming A and B to be small compared to a nuclear Zeeman frequency ν_I, the frequencies of the nuclear transitions may be rewritten as $\nu_{\alpha(\beta)} = \nu_I \mp A/2$. Hence, after simple transformation, [3] yields for the primary ESE modulation:

$$V(\tau) = 1 - \frac{1}{2} \frac{B^2}{\nu_I^2} [1 - 2 \cos 2\pi\nu_I\tau \cos 2\pi(\frac{A\tau}{2}) + \frac{1}{2} \cos 2\pi(2\nu_I\tau) + \frac{1}{2} \cos 2\pi(A\tau)]. \quad [8]$$

In a disordered sample, $V(\tau)$ should be averaged over all orientations of the paramagnetic center, giving:

$$\langle V(\tau) \rangle = 1 - \frac{3}{10} \frac{T_\perp^2}{\nu_I^2} [2 - 4 \cos 2\pi\nu_I\tau \; F(T_\perp\tau) + \cos 2\pi(2\nu_I\tau) + F(2T_\perp\tau)] \quad [9]$$

where:

$$F(T_\perp t) = \frac{15}{4} \int_0^\pi \sin^2\theta \cos^2\theta \cos[\pi T_\perp t(1 - 3\cos^2\theta)] \sin\theta \, d\theta . \quad [10]$$
$$t = \tau, \; 2\tau$$

The result obtained means that the initial time interval of the primary ESE signal decay is modulated with frequencies ν_I and $2\nu_I$. The amplitude of the ν_I harmonic is four times as large as that of the $2\nu_I$ harmonic. As the parameter $T_\perp t$ increases with increasing time interval t, the value $F(T_\perp t)$ decreases monotonically. Numerical calculations show that it becomes zero at $\pi T_\perp t \approx 3.6$ and then becomes negative with absolute value not exceeding 0.05. Hence for large times, a decay of the ν_I harmonic should be observed and the $2\nu_I$ harmonic with an amplitude independent of τ will dominate in the modulation.

In the range of small $T_\perp t$ values, the cosine function under the integral in [10] can be expanded in the series expansion up to the second term, thus reducing the expression for $F(T_\perp t)$ to:

$$F(T_\perp t) = 1 - \frac{4\pi^2}{14} T_\perp^2 t^2 . \quad [11]$$

To estimate the range of validity of this truncation, we will use the point-dipole approximation for the electron-nuclear anisotropic hyperfine interaction. Then $T_\perp = g g_I \beta \beta_I / r^3 h$ and the relative value for the hyperfine and nuclear Zeeman interactions may be characterized by the ratio:

$$T_\perp / \nu_I = g\beta / B_o r^3 , \quad [12]$$

which does not depend on the nuclear magnetic moment and, for known B_o, is defined only by the distance r. In X-band experiments, for $B_o = 330$ mT and $g \approx 2$, the ratio $T_\perp / \nu_I$ has the value 0.1 - 0.2 for r = 3 - 4 A. At Q-band (35 GHz) the ratio has the same value for r = 1.9 - 2.5 A.

For the point dipoles, as follows from [6,8], the modulation amplitude coefficient depends on the electron-nucleus distance as $1/r^6$. The time t satisfying the equation $F(T_\perp t) = 0$, is given by $t = (3.6 \text{ hr}^3)/(\pi g g_I \beta \beta_I)$, i.e. t (F = 0) increases as r^3.

As a rule, the environment of a paramagnetic center involves more than one magnetic nucleus. In the weak hfi approximation the contributions to the ESE modulation from different nuclei of one type are summed resulting in:

$$\langle V_N(\tau) \rangle = 1 - \frac{3}{10} \frac{g^2 \beta^2}{B_o^2} \Sigma \frac{1}{i \; r_i^6} [2-4 \cos 2\pi \nu_I \tau \; F(T_{\perp i}\tau) + \cos 2\pi(2\nu_I \tau) + F(2T_{\perp i}\tau)]. \qquad [13]$$

Relation [13] indicates that the coefficient defining the initial amplitude of the modulation harmonics and the damping time of the ν_I harmonics becomes dependent on the spatial distribution of the nuclei. The time of complete damping of the ν_I harmonic is governed by the distance to the most remote nuclei measurably contributing to the modulation, and the time dependence of this harmonic's amplitude reduction is defined by the electron-nuclear distances distribution. These features of the ESEEM point to the possibility to extract information about the nearest environment of the paramagnetic center by determining separately the number of magnetic nuclei contributing to the modulation and their distance to the center.

For the stimulated ESE one obtains in the same way:

$$\langle V(\tau, T) = 1 - \frac{6}{5} \frac{g^2 \beta^2}{B_o^2 r^6} \sin^2 \pi \nu_I \tau \{1 - \cos 2\pi \nu_I (\tau + T) \; F[T_\perp(\tau + T)]\} . \qquad [14]$$

Hence, in disordered samples a weak hfi induces in the stimulated ESE only one modulation with the Zeeman frequency ν_I.

The existence of the above-mentioned suppression effect clearly follows from the form of [14]. The amplitude of the ν_I harmonic will be maximum at τ values satisfying the condition $\sin^2 \pi \nu_I \tau_{max} = 1$, i.e. at $\tau_{max} = (n + \frac{1}{2})/\nu_I$. Therefore, to observe the maximum modulation amplitude in the stimulated ESE, one should choose the time τ corresponding to the signal minimum in the primary ESE; alternatively, at τ corresponding to the maximum of the primary ESE signal, the stimulated ESE modulation is lacking.

The dependence of the ν_I harmonic's amplitude on time $\tau + T$ in the stimulated ESE is defined by the function $F[T_\perp(\tau + T)]$ and is analogous to the dependence of the ν_I harmonic's amplitude on τ in the primary ESE. For example, Fig. 2 shows the primary and stimulated ESEEM for a CH_2COO^- radical trapped in a powder of glycine containing ^{15}N. This matrix has two types of magnetic nuclei, i.e. ^1H and ^{15}N. The cosine Fourier spectrum of the primary ESE has two peaks, corresponding to the proton and nitrogen Zeeman frequencies, 14.4 and 1.45 MHz in the external magnetic field B_o = 339 mT, respectively, and two

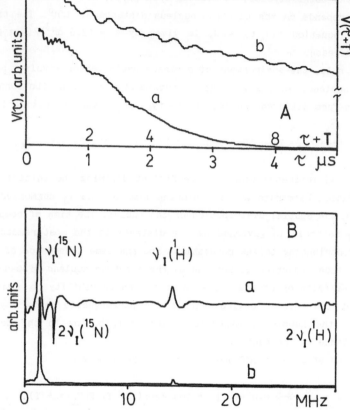

Fig. 2. The primary (a) and stimulated (b) ESEEM envelopes for the CH_2OO^- radical in a powder of glycine with ^{15}N (A). The cosine Fourier spectrum of primary ESEEM (a) and modulus Fourier spectrum (b) of stimulated ESEEM (B).

peaks twice the Zeeman frequencies of these nuclei with negative amplitudes. The stimulated ESE spectrum presents only the proton and nitrogen Zeeman frequencies.

2.2 Numerical Analysis of the ESEEM in Disordered Systems

The above analytical consideration of the modulation effects in disordered systems in the weak hfi limit ($\nu_I \gg A,B$) allows one to interpret some experimental results. For many realistic situations, however, the limiting condition is not fulfilled. For example, in X-band ESE experiments a less rigid condition, $\nu_I \gtrsim A,B$ is often encountered. Since for arbitrary hfi parameters values the averaging over different orientations of the paramagnetic center cannot be performed analytically, the problem is solved with the aid of numerical calculations.

The effect of different parameters on the form of ESE modulation may be demonstrated using model calculations for a nucleus with $I = \frac{1}{2}$. The modulation envelopes of Fig. 3 have been calculated by averaging [3] over the angle θ between the axis of the axially symmetric hfi tensor and the direction of the external magnetic field:

$$\langle V(\tau)\rangle_\theta = \frac{1}{2} \int_0^\pi V(\theta,\tau) \sin\theta \, d\theta \, . \qquad\qquad [15]$$

The calculations were carried out for $\nu_I = 1$ to exclude the dependence on a particular nucleus. All other parameters of magnetic interactions (for example, hfi) are measured in ν_I units and time intervals are measured in $1/\nu_I$ units.

Figs. 3a-c present the calculations for the primary ESE modulation at a = 0, $T_\perp = 0.05$, 0.1, 0.15. These sets of parameters correspond to the weak anisotropic hfi approximation, and the curves derived confirm the results of the analytical analysis. The calculated curves explicitly display the dominant ν_I harmonic in the low-time region of the primary ESE envelopes. As the time interval τ increases, the amplitude of this harmonic decreases, and oscillations with the $2\nu_I$ frequency become dominant. The decay time of the ν_I harmonic increases with decreasing τ, and the modulation amplitude decreases. In the stimulated echo, there is in this case only the ν_I harmonic and the dependence of its amplitude on τ and T obeys the regularities mentioned in the previous section.

With increasing $T_\perp$ and in the presence of isotropic hfi the form of the primary ESE modulation envelopes becomes more complicated. In the low-time region the modulation is now represented by the superposition of the averaged frequencies $\nu_{\alpha(\beta)}$ and $\nu_\alpha \pm \nu_\beta$. With increasing τ, the $\nu_\alpha \pm \nu_\beta$ harmonic becomes dominant (for example, see Fig. 3d-f). This can be seen from a comparison between the modulation decay times of the primary and stimulated ESE envelopes.

For complex envelopes, knowledge about the frequencies most distinctly manifesting themselves in the ESEEM may be gained through the analysis of the modulation spectra resulting from some kind of frequency transformation of the envelopes. This approach to the ESEEM analysis will be discussed below. On the other hand, a marked difference in envelopes obtained for different isotropic and anisotropic hfi parameters allows one to develop alternative methods based on the comparison of calculated envelopes with experimental ones in a time domain optimized to obtain the best agreement between them. This is performed usually in the approximation of uncorrelated nuclei, neglecting the real geometric arrangement of the nuclei around a paramagnetic center and assuming their relative position to be uncorrelated. In this case, for example, the primary ESE modulation is given by $V_N(\tau) = \prod_{i=1}^{N} \langle V_{I_i}(\tau,\theta)\rangle_\theta$. For equivalent nuclei this formula assumes the form $V_N(\tau) = \langle V_{I_i}(\tau,\theta)\rangle_\theta^N$.

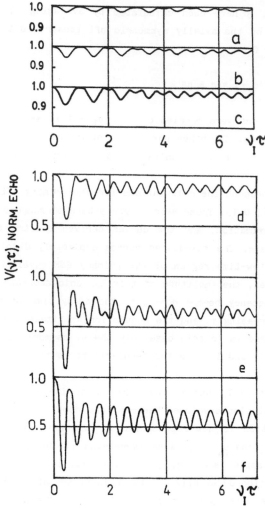

Fig. 3. The calculated primary ESEEM in a disordered system for a = 0, $T_\perp$ = 0.05 (a), 0.1 (b), 0.15 (c), 0.5 (d), 1 (e), 2 (f) (ν_I = 1).

2.3 ESEEM Spectra and Their Properties

In a disordered sample the ESE modulation spectrum is the superposition of spectra at different orientations. The shapes and the maxima positions of the lines in this spectrum are defined by the angular dependence of the modulation frequencies and their amplitudes, and the statistical weight of each orientation.

The line shapes in the ESE modulation spectra of a disordered system consisting of one electron with S = ½ and one nucleus with I = ½ have been analyzed in (17-19) with the aid of numerical calculations and have been compared with the line shapes in the ENDOR spectra. It has been shown that the compo-

nents of modulation spectra are smooth bell-shaped lines with one maximum, the
position of which is determined by the relative value of isotropic and anisotropic components of the hyperfine interaction.

At first glance, the line shape in modulation spectra does not permit the
determination of the isotropic and anisotropic hfi parameters directly from the
positions of maxima, without numerical calculations of the spectra. It is noteworthy, however, that the spectra considered in (17-19) were ideal in the sense
that they did not take into account the spectrometer dead time. Its presence
changes radically the conclusions made, which has first been noted in (20) and
later on analyzed in detail in (21).

Let us consider the influence of the dead time on the ESE modulation spectra in the case of axially symmetric hfi. Fig. 4a shows the transformation of
the calculated stimulated ESE modulation spectrum with non-overlapping lines ν_α
and ν_β, with increasing dead time t_d. At zero dead time each line is approximately bell-shaped and has a single maximum. As t_d increases, the amplitudes of
different parts of the spectral lines decrease non-uniformly. The amplitudes of
the gentle sloping inner parts of the lines corresponding to relatively fast-
damping modulation, decrease faster than the amplitudes of the steep edges related to rather slow-damping modulation. As a result, at a sufficiently long
dead time, when $1/t_d$ becomes comparable with the initial linewidth, two lines
appear in the spectrum instead of the single bell-shaped one, with maxima close
to the frequencies of nuclear transitions for perpendicular and parallel relative orientations of external magnetic field and the axis of the axial hfi tensor:

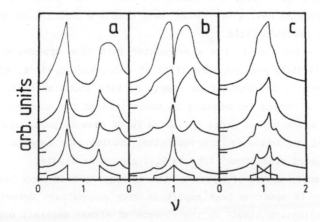

Fig. 4. Modulus Fourier spectra of stimulated ESE in the case of an axially
symmetric hfi ($\tau = 0.5$): $T_\perp = -0.3$, a = 1 (a), 0.3 (b), 0 (c). The dead time
increases from top to bottom: 0, 0.5, 1, 1.5. In Figs. 4 - 7 a and $T_\perp$ are
measured in ν_I units and the dead time in $1/\nu_I$ units. ENDOR spectra are shown
schematically at the bottom of Figs. 4 - 6. From (21).

$$\nu_{\alpha(\beta)}^{\perp} = |\nu_I \, (\overset{-}{+}) \, \frac{a}{2} \, (\overset{-}{+}) \, \frac{T_{\perp}}{2}|$$

$$\nu_{\alpha(\beta)}^{//} \, |\nu_I \, (\overset{-}{+}) \, \frac{a}{2} \, (\overset{+}{-}) \, T_{\perp}| \; . \tag{16}$$

The amplitudes of the $\nu^{//}$ components are substantially less than those of the $\nu^{\perp}$ components. The form of the ENDOR spectral lines is schematically depicted below the modulation spectra at the bottom of the figure. The anisotropic ENDOR absorption lines ν_{α} and ν_{β} lie within the range limited by the frequencies following from [16].

The above shows that the line shapes in the ESE modulation spectra at a sufficiently long dead time become "ENDOR-like" with characteristic maxima, allowing us to determine the hfi parameters directly from the spectra, without numerical calculations. The result obtained above is confirmed by calculating modulation spectra with other hfi parameters. The spectra in Fig. 4b are obtained using an isotropic hfi constant $a = -T_{\perp}$. In this case, instead of two lines, at sufficiently large t_d three lines are observed in the spectrum, one of them being an intensive peak at the nuclear Zeeman frequency. Fig. 4c demonstrates how a line with almost unresolved structure, formed by two overlapping anisotropic components ν_{α} and ν_{β} with zero isotropic hfi constant, transforms into a resolved quadruplet as the dead time increases, with maxima determined by expression [16].

Fig. 5 shows the change in the modulation spectra with dead time in the case of non-axial hfi tensors. A comparison demonstrates the coincidence of maxima in the modulation spectra at large t_d with singularities in the ENDOR spectra. Thus, generally, the maxima in modulation spectra with a long dead time correspond to having hfi tensor axes that are parallel to the direction of the external magnetic field $\vec{B}_o$.

In the above aspect, the spectrometer dead time appears to be a positive factor facilitating the interpretation of modulation spectra, allowing data on hfi parameters to be obtained as simply as with ENDOR spectra. It is achieved at the cost of losing information on the origin of the separate peaks observed in modulation spectra at large t_d, which is of essence particularly for systems with several magnetically non-equivalent nuclei. To ascertain this origin, spectra at short dead times $(1/t_d \gg T_{zi})$ should be studied.

The transformation of spectra with increasing dead time leads to an overall decrease in spectrum peak amplitude with concomitant deterioration of its quality, manifesting itself by the absence of either explicit maxima or even by the masking of separate peaks (most commonly of $\nu_{\alpha,\beta}^{//}$) by the noise level. Under these conditions, the spectral information extracted from the position of the $\nu_{\alpha(\beta)}^{\perp}$ components is often insufficient for a separate determination of the

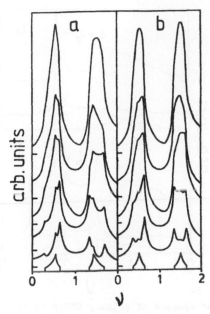

Fig. 5. Modulus Fourier spectra of stimulated ESEEM in the case of non-axial hfi (τ = 0.5): a = 1, T_{11} = -0.3, T_{22}/T_{11} = 0.4 (a), 0 (b). The dead time increases from top to bottom: 0, 0.5, 1.0, 1.5, 2.5. From (21).

isotropic and anisotropic hfi components, and the necessity arises to obtain additional independent data to allow one to fill the information gap. Such data may be obtained by the analysis of the position of the maximum of the combination harmonic $\nu_\alpha + \nu_\beta$ in the primary ESE.

The primary ESE modulation includes the combination harmonics $\nu_\alpha + \nu_\beta$ and $|\nu_\alpha - \nu_\beta|$ as well as the frequencies ν_α and ν_β of the nuclear transitions. The harmonic $|\nu_\alpha - \nu_\beta|$ is highly anisotropic and, as a rule, is not detectable in modulation spectra. Therefore we shall ignore this harmonic and consider only the harmonic $\nu_\alpha + \nu_\beta$. As follows from an expansion of the square root in the expression for $\nu_{\alpha(\beta)}$, under the condition:

$$|\nu_I \mp T_{zz}/2| \geq \sqrt{T_{zx}^2 + T_{zy}^2}/2 \ , \tag{17}$$

this harmonic has a low anisotropy:

$$\nu_\alpha + \nu_\beta \approx 2\nu_I \left(1 + \frac{0.125(T_{zx}^2 + T_{zy}^2)}{\nu_I^2 - (T_{zz}/2)^2}\right) \tag{18}$$

and at small values of T_{zi} ($T_{zi} \ll \nu_I$) its frequency is close to the double Zeeman frequency.

At hfi anisotropy values comparable with $\nu_I (T_{zi} \sim \nu_I)$ the frequency

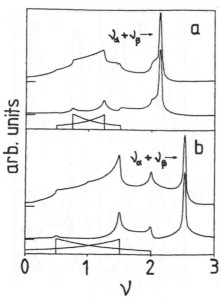

Fig. 6. Modulus Fourier spectra of primary ESEEM in the case of axially symmetric hfi: a = 0, $T_\perp$ = -0.5 (a), -1 (b). The dead time increases from top to bottom: 0, 1. From (21).

$\nu_\alpha + \nu_\beta$ can be considerably shifted from the $2\nu_I$ position. Calculations of primary ESE modulation spectra analogous to those of stimulated ESE modulation spectra indicate that the position of the maximum of this combination line is practically independent of the dead time (Fig. 6) and that it corresponds well with the maximum of the coefficient k (21). The analytical expression for the maximum of the $\nu_\alpha + \nu_\beta$ harmonic for axial symmetric hfi has the form (21):

$$\nu_\alpha + \nu_\beta \approx 2\nu_I \left(1 + \frac{0.25\ T_\perp^2}{\nu_I^2 - ((a - T_\perp/2)/2)^2}\right) . \qquad [19]$$

Model calculations of primary ESE modulation spectra show also that the condition [17] is a prerequisite for the pronounced $\nu_\alpha + \nu_\beta$ harmonic maximum appearing in the spectra. This is not surprising, because under any other condition this harmonic becomes comparatively strongly anisotropic. Hence, we can simplify [19] even more:

$$\nu_\alpha + \nu_\beta = 2\nu_I + \frac{0.5}{\nu_I}\ T_\perp^2 , \qquad [20]$$

neglecting the contribution to the denominator in [19] of the hyperfine interaction terms, which approximation is valid under the conditon [17]. Expression [20] shows that the position of the $\nu_\alpha + \nu_\beta$ harmonic maximum should be practically independent of the value of the isotropic hfi constant.

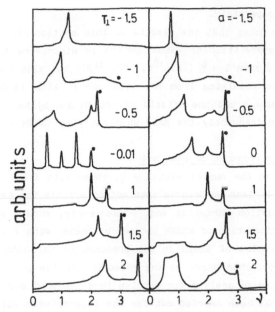

Fig. 7. Modulus Fourier spectra of primary ESEEM. In the left column a = 1 and $T_\perp$ changes, in the right column $T_\perp$ = 1 and a changes. Black dots show the position of the ν_α + ν_β component.

Examples of the primary ESE modulation spectra calculated for different values of isotropic and anisotropic hfi are given in Fig. 7. This figure demonstrates the dependence of the maximum position of the ν_α + ν_β harmonic on the anisotropic hfi value and its near-independence of the isotropic hfi constant, observable in the spectra.

Expression [20] gives a simple way to estimate the value of the anisotropic hfi tensor component $T_\perp$ through the shift $\Delta\nu$ of the ν_α + ν_β harmonic maximum from the $2\nu_I$ position:

$$|T_\perp| = \sqrt{2\nu_I \Delta\nu} .$$ [21]

For example, for Fig. 7a, bottom spectrum, Eq. [21] yields $T_\perp$ = 1.8 which is close to the exact value $T_\perp$ = 2 used in the calculation, despite the fact that in this case condition [17] is violated for certain orientations. Hence, the maximum shift of the recombination harmonic ν_α + ν_β from $2\nu_I$ in the primary ESE modulation spectra in disordered samples can be used as an independent experimental parameter for the estimation of the anisotropic hfi value $T_\perp$. Moreover, in the presence of at least one peak corresponding to the nuclear transition frequency $\nu_{\alpha(\beta)}$ at a certain orientation, this provides a possibility for the independent determination, directly from the spectrum, of isotropic and anisotropic hfi parts separately.

It should be noted that the results of this section are of importance not only for the interpretation of ESEEM spectra in disordered systems reflecting hfi with nuclei of spin $I = \frac{1}{2}$ (^{1}H, ^{15}N, ^{19}F, ^{31}P), but also for the analysis of modulation spectra resulting from hfi with nuclei with larger spin when the quadrupole interaction and the effects of fourth and higher orders in hfi are negligible, as is frequently the case when studying ESEEM involving deuterium nuclei ($I = 1$).

2.4 ESEEM in the Presence of Weak nqi

In contrast to the nuclei with $I = \frac{1}{2}$, those with $I \geq 1$ possess a quadrupole moment. The nuclear quadrupole interaction affects the amplitudes and frequencies of modulation harmonics and, consequently, should be taken into account in the consideration of ESEEM induced by nuclei with $I \geq 1$.

Since the problem of ESEEM in the presence of nqi cannot be solved in a general form, approaches based on the selection of the prevailing interaction have been developed to analyze the problem in disordered systems. The most detailed analysis has been carried out for the case of weak nqi. It was revealed that even a weak nqi leads to new effects in the modulation of disordered paramagnetic centres. Neglecting it in the analysis of experimental data may in some cases result in an incorrect estimation of the hfi parameters.

The problem of taking nqi into account in the analysis of ESEEM was first raised, and solved using first order perturbation theory, in (22). According to (22), the primary ESE modulation induced by weak quadrupole and hyperfine (as compared with nuclear Zeeman) interactions of a nucleus with arbitrary spin I is described by the expression:

$$V_I^Q(\tau) = 1 - \frac{2}{3} I(I+1) \frac{B^2}{\nu_I^2} [1 - 2 \cos 2\pi\nu_I\tau \cos 2\pi(\tfrac{A}{2}\tau) G(\gamma t)$$

$$+ \tfrac{1}{2} \cos 2\pi(2\nu_I\tau) G(2\gamma\tau) + \tfrac{1}{2} \cos 2\pi(A\tau)] , \qquad [22]$$

where

$$\gamma = \frac{3}{4} \frac{eQV_{zz}}{hI(2I-1)} ,$$

$$G(x) = \frac{3}{2I(I+1)(2I+1)} \sum_{m=-I}^{I-1} [I(I+1)-m(m+1)]\cos 2\pi(2m+1)x ,$$

eQ is the nuclear quadrupole moment and V_{zz} is a component of the electric field gradient tensor. For the value of the nqi approaching zero, $\gamma \to 0$, $G(\gamma\tau) \to 1$ and [22] turns into a formula similar to [8] with the prefactor $\frac{1}{2}$ in the second term of the righthand side replaced by $\frac{2}{3} I(I+1)$.

From [22] it follows that the weak nqi changes the time dependence of the amplitudes of the modulation harmonics ν_I and $2\nu_I$. The amplitude of the ν_I harmonic is affected by the additional factor $G(\gamma\tau)$, while the $2\nu_I$ harmonic, which in the presence of hfi only was time independent, also acquires a time dependence in the form of the factor $G(2\gamma\tau)$. In disordered samples the damping of ν_I will now be determined by the result of the averaging of the product $B^2 G(\gamma\tau)$ $\cos 2\pi(A\tau/2)$ over different orientations of the paramagnetic center, and the damping of $2\nu_I$ by the averaging of the function $B^2 G(2\gamma\tau)$.

The criterion for the applicability of the weak nqi approximation is the fulfillment of the condition $\nu_I \gg 3/8 \, (e^2qQ)/hI$.

For $I = 1$, the function $G(\gamma t)$ equals $\cos 2\pi\gamma t$. Using this one may estimate the decay time of the $2\nu_I$ harmonic due to the nqi effect for deuterium nuclei ($I = 1$), which are frequently studied in ESE experiments. Averaging of $<\cos 2\pi\gamma t>$ over all orientations testifies that depending on the mutual orientation of the hfi and nqi tensors, the modulation decay time has a value of the order of:

$$t_Q \approx \frac{4y}{3\pi}\left(\frac{h}{e^2qQ}\right), \qquad t = \tau, \, 2\tau \qquad [23]$$

where $y = 2 - 3$ depending on the relative orientation of the hfi and nqi tensors. Assuming e^2Qq/h to be equal to about 0.2 MHz for deuterium nuclei (23, 24), one can estimate the damping of the $2\nu_I$ harmonic in the primary ESE: $\tau_Q \approx 2 - 3$ μs.

The ν_I harmonic is damped by the joint action of hfi and nqi; the damping depends on the relative values of $T_\perp$ and e^2Qq/h. For anisotropic hfi the decay time increases in proportion to $\sim 1/T_\perp$. When e^2Qq/h is small compared with $T_\perp$, the ν_I decay time depends mainly on the hfi. A decrease in $T_\perp$ relative to e^2qQ leads to a faster decay of ν_I than expected from the dependence on $1/T_\perp$. And finally, if e^2Qq/h exceeds $T_\perp$, the decay time is determined primarily by the nqi. This conclusion is valid for both the primary and the stimulated echo. Hence, the weak nqi in disordered systems affects the decay time of the ν_I harmonic modulation, which irrespective of the configuration of deuterium nuclei around the paramagnetic center will be no more than $4 - 6$ μs.

The influence of the weak deuterium nqi on the ESEEM is illustrated in Figs. 8a,b. Fig. 8a exhibits the decay of the primary ESE signal of hydrogen atoms trapped at 77 K in a γ-irradiated 8M H_2SO_4/H_2O glass. In this system, the matrix contains only one type of magnetic nuclei, namely protons (ν_I = 15.43 MHz in B_0 = 362.5 mT, corresponding to the high-field EPR spectrum component). The proton Zeeman frequency is observed in the initial part of the modulation curve and further on there occurs a gradual doubling of the modulation frequency. The decay of the primary ESE signal of deuterium atoms trapped in the 8M

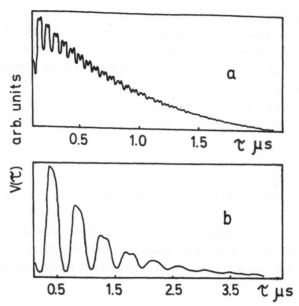

Fig. 8. The decay of the primary ESE signal for H atoms in H_2SO_4/H_2O (a), and D atoms in D_2SO_4/D_2O (b).

D_2SO_4/D_2O glass is shown in Fig. 8b. In this sample, the environment of a trapped atom involves deuterons rather than protons, which in the external field B_o = 346.7 mT have the Zeeman frequency ν_I = 2.27 MHz. Along with the change in the modulation frequency, an increase in the modulation amplitude is observed in the initial part of the envelope, in accordance with the spin factor I(I + I). In addition, in the sample with deuterium nuclei the $2\nu_I$ modulation manifests itself only in the initial part of the ESE signal decay and is of minor amplitude.

The effect of nqi displays itself more vividly in the stimulated ESE modulation, in which the combination harmonics are lacking and only the ν_I frequency is present. This modulation decays in a time τ + T = 4 μs (Fig. 9b). If this decay would be determined by the hfi, then in the ESE of hydrogen atoms the decay of proton modulation would occur in a time τ + T = 0.6 μs, due to its inverse proportionality to the nuclear g_I-factor. In the stimulated ESE of hydrogen atoms (Fig. 9a), however, we observe a weakly decaying modulation of ν_I frequency to substantially longer times τ + T. This result indicates that in disordered samples, the decay of the ESE modulation with the Zeeman frequency of the deuterons will occur in a time t_Q, irrespective of their distance to the paramagnetic center. Thus, the structural information on nuclei with an anisotropic hfi not exceeding the nqi will be lost.

Following the theoretical analysis of the weak nqi effect in (23,26), with

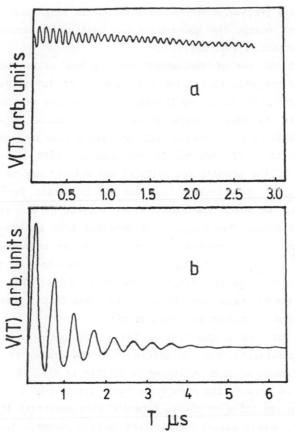

Fig. 9. The decay of the stimulated ESE signal for H atoms in H_2SO_4/H_2O (a) and for D atoms in D_2SO_4/D_2O (b).

the above-presented estimates and their qualitative experimental verificiation, a number of investigations has been carried out that treat the problem of the weak nqi effect. Calculations of the ESE modulation for an arbitrary spin, which take into account the effect of hfi and nqi on the frequencies with second and third order perturbation theory, are discussed in (27,28) and the effect of only nqi on the frequencies and amplitudes to first order, are described in (29). A special analysis (30) is devoted to the manifestation of nqi non-axiality in the ESE modulation. An expression deduced by Mims (10) that accounts for the effect of the nqi on the frequencies of nuclear transitions, was extended by taking the state of the electron spin into account for the magnitude of the quadrupole splitting in (25,31-33) for use in calculations of the ESE modulation for nuclei with I = 1, in particular deuterium.

Numerical calculations of the modulation in disordered systems with the help of more rigorous but cumbersome approximate expressions yield closer correspondence to a precise numerical calculation at long times, but the results obtained do not change the above-presented qualitative conclusions and estimates of the effect of weak nqi.

Note that the use of theoretical results has so far been restricted to model calculations and, except for two examples of investigations of systems with known crystalline structure (31-33), these results were not as yet applied to analyze real experimental data. It is therefore useful to employ the simple qualitative analysis of expression [22] to hazard some conjectures on the possible manifestations of weak nqi in experimental ESEEM spectra. For example, for spin I = 1, the product $\cos 2\pi(2\nu_I\tau) \cos 2\pi(2\Upsilon\tau)$ may be represented as a sum of two terms. This implies that for matrix nuclei far removed from the electron spin, for which the second order effects of hfi fail to contribute much to the harmonic frequency, it is expected that the $2\nu_I$ harmonic in the primary ESE modulation spectra will split into two components. One of these components is to be shifted slightly down and the other slightly up from the exact $2\nu_I$ position. The splitting between the maxima of these components is in the order of the nqi magnitude. It should be noted that a negative shift of the combination harmonic cannot be caused by hfi.

We have observed a splitting of the $2\nu_I$ harmonic in the spectra of the primary ESE modulation from the matrix deuterium nuclei in many paramagnetic samples. An example of this splitting is visible in Fig. 10, which represents a spectrum of the primary ESE modulation shown previously in Fig. 9b. Sometimes the quadrupole splitting of the ν_I harmonic also manifests itself in the spectra if for all matrix nuclei contributing to this harmonic the constant of quadrupole coupling exceeds the anisotropic hfi parameter.

It has been noted in (22) that the condition of weak nqi at X-band fre-

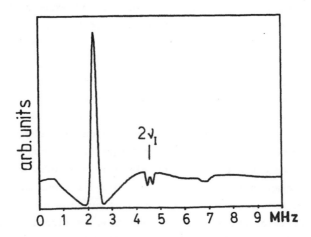

Fig. 10. Modulus Fourier spectrum of primary ESEEM of D atoms in D_2SO_4/D_2O.

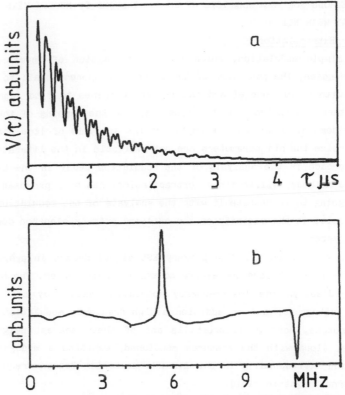

Fig. 11. The decay of the primary ESE signal for F-centers in LiD and the modulus Fourier spectrum of its modulation.

quency is met well enough for many nuclei: ^{7}Li, ^{23}Na, ^{27}Al, ^{133}Cs, which produce ESE modulation. The validity of this approximation is indirectly confirmed by the presence of a simple modulation caused by these nuclei, which is a superposition of Zeeman and double Zeeman harmonics. Fig. 11 shows the decay of the ESE signal of F-centre in lithium deuteride (LiD). The spectrum of this envelope involves the lines ν_I and $2\nu_I$ of ^{7}Li in the applied field B_o = 338 mT. Note that in this example the amplitude of the $2\nu_I$ harmonic decays only slightly, and the modulation spectrum lacks its quadrupole splitting, which for I = 3/2 must, in accordance with [22], represent a triplet with the center at $2\nu_I$. The absence of such a splitting may be caused by a small value of the quadrupole coupling constant (23). Also other attempts to observe a similar multiplet splitting caused by the matrix nuclei of spin I ≥ 3/2 have so far not produced a positive result.

3. SOME METHODS FOR THE ANALYSIS OF ESEEM INDUCED BY NUCLEI WITH $I = \frac{1}{2}$ AND $I \geq 1$ AT WEAK NQI

3.1 Linear Extrapolation

For a simple modulation, which is a superposition of Zeeman and double Zeeman frequencies, the positions of the spectral components fail to yield data on the hfi with matrix nuclei and the necessity arises to find approaches related to either the analysis of the line shape or the damping of the harmonics in the time domain. At present, examples of the analysis of line shapes in order to determine the hfi parameters are not available in the literature. A more fruitful approach is the analysis of the modulation decay in the time domain. One of these methods, called linear extrapolation, has been proposed in (26,34) and we are going to illustrate it with the analysis of the modulation caused by deuterium nuclei in the vicinity of the oxidized primary electron donor P700 of plant photosystem I.

Fig. 12 represents the P700 primary ESE signal decays in preparations of Vicia faba bean leaf tissue containing normal and heavy water. The bottom trace in Fig. 12 displays the low-frequency modulation caused by hfi and nqi of chlorophyll nitrogens, and in addition a high-frequency modulation induced by hfi with protons, both of chlorophylls and of their nearest neighbours. The upper trace, along with the features mentioned, exhibits a modulation with a 440 ns period corresponding to the 2.2 MHz deuterium Zeeman frequency in the 338 mT external magnetic field.

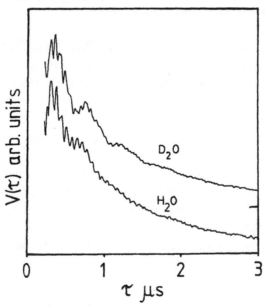

Fig. 12. The decay of the P700$^+$ primary ESE signal in preparations of freeze-dried leaf tissue of <u>Vicia faba</u> with H_2O and D_2O. From (35).

Since the contributions of different nuclei to the primary ESE modulation are multiplicative, the modulation induced by deuterium nuclei may be derived by dividing the upper envelope in Fig. 12 by the lower one. The result (Fig. 13, curve E) may be analytically expressed as $V(\tau) = V_D(\tau)/V_H(\tau)$, where $V_H(\tau)$ is the contribution to the lower modulation envelope of those protons which are substituted by deuterium in the sample giving the upper trace. For the envelopes in Fig. 13 the division of V_D by V_H results in high-frequency beats of small amplitude.

The frequency composition of the observed deuterium modulation is seen from Fig. 13b, representing the cosine Fourier transform of curve E in Fig. 13a. The spectrum has two components, with maxima at 2.2 and 4.5 MHz, the amplitude of the latter being negative. This allows to attribute the 4.5 MHz harmonic to the combination frequency $\nu_\alpha + \nu_\beta$.

The shift in 0.1 MHz of the combination harmonics maximum from the exact $2\nu_I$ position is too small for a correct quantitative determination of the anisotropic hfi value, since it may be largely caused by the nqi effect together with distortions introduced by Fourier-transformation in the presence of dead time.

Under these conditions estimates of the anisotropic hfi and the number of the deuterium nuclei contributing to the ESE modulation may by obtained from an analysis of the damping of the ν_I harmonic amplitude. Prior to the analysis, note that this harmonic decays in a time $\tau < 2$ µs, so that on the basis of the estimates made for deuterium nuclei in section 2.4, it may be assumed that this decay is mainly caused by anisotropic hfi and is only slightly affected by nqi.

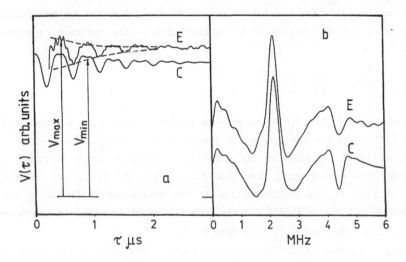

Fig. 13. a) Deuterium ESEEM (E - experiment, C - calculation). b) Cosine Fourier spectra corresponding to the envelopes E and C. From (35).

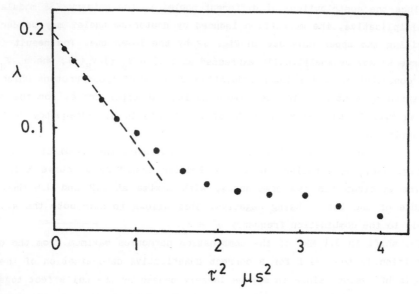

Fig. 14. The dependence of the modulation amplitude λ of the ν_I harmonic on τ^2. From (35).

To describe quantitatively the time dependence of the ν_I harmonic amplitude, the parameter λ is introduced:

$$\lambda = 1 - V_{min}/V_{max} \, , \qquad\qquad [24]$$

where V_{min} and V_{max} are the envelopes depicted by the dashed lines in Fig. 13a. According to (34), the modulation amplitude must decrease in the initial time region as a linear function of τ^2. Analytically, this decrease is expressed as follows:

$$\lambda = \frac{32}{5} \sum_i \frac{T_{\perp i}^2}{\nu_I^2} \left(1 - \frac{4\pi^2}{14} T_{\perp \, i}^2 \, \tau^2\right) . \qquad\qquad [25]$$

Extrapolation of this linear dependence to zero time yields a value proportional to $\sum_i T_{\perp i}^2$, whereas the tangent is proportional to $\sum_i T_{\perp i}^4$.

The linear decrease of the modulation amplitude as a function of τ^2 is observed with satisfactory precision for $\tau^2 < 1$ μs^2 (Fig. 14). From the data given in Fig. 14, the initial value of the modulation amplitude $\lambda_0 = 0.19$ and the slope $\Delta\lambda/\Delta\tau^2 = 0.1$ μs^{-2}. Hence, using equation [25] it is found that $\sum_i T_{\perp i}^2 = 0.144$ MHz^2 and $\sum_i T_{\perp i}^4 = 0.0268$ MHz^4. $\qquad\qquad [26]$
Assuming N equivalent nuclei to contribute to the modulation, we obtain from

[26] $N = 0.77$ and $T_\perp = 0.43$ MHz.

The above result indicates that in the case under consideration, the contribution to the ESE modulation is made by a single nucleus coupled with the unpaired electron with anisotropic hfi $T_\perp \approx 0.4$ MHz. The fact that the number of nuclei is slightly below unity may be related to the approximate character of the analysis or to the partial substitution of protons by deuterons.

Numerical calculation of the modulation for one deuterium nucleus yields good agreement with the experimental amplitude and damping of the harmonic ν_I for $T_\perp = 0.4$ MHz (Fig. 13a,b, trace C). In the calculations for the decrease of the $\nu_\alpha + \nu_\beta$ harmonic amplitude at large time τ, the nqi has to be taken into account. The close correspondence between the experimental envelope and the calculated envelope obtained with the linear extrapolation procedure confirms the validity of the estimates involved.

In conclusion, it should be noted that in the original paper (34) the technique described above has been developed for nuclei with $I = \frac{1}{2}$ and experimentally verified for protons by determining the structure of the nearest environment of hydrogen atoms trapped in a frozen H_2SO_4/H_2O glass. In particular, the dependence of the proton Zeeman harmonic amplitude on τ^2 in (34) had two linear regions, which allowed one to separate the contributions to the modulation and find the structural parameters for the protons of the first environment sphere and for the other matrix protons. However, a comparison of the ESE modulation decay of hydrogen atoms with that of deuterium ones in an analogous, deuterated matrix D_2SO/D_2O (26) showed that in the last case the second linear region at large τ^2 is absent due to the nqi effect, which makes the separation of the contributions from different groups of matrix nuclei impossible. Under these conditions, using the remaining low-time linear region one can only estimate the distance to the nearest nuclei and their number. Such an estimation will yield too large a value for the number of nearest-neighbour nuclei and their distance to the paramagnetic center.

Returning to the example mentioned, it is noteworthy that the application of the technique for deuterium resulted in a reliable estimate due to the presence of only one nucleus with hfi value exceeding the nqi.

The attempts to create an analogous situation in model systems, such as chlorophyll a (Chl a) radical cations in solution of CH_2Cl_2 in CH_3OD or CD_3OD (1 : 3) have failed. In the ESE of monomeric Chl a$^+$ in a solution with CH_3OD the deuterium modulation was practically absent. It has been observed only when using the fully deuterated alcohol (Fig. 15). In this case, however, the slow damping of the deuterium modulation points to a substantial distance between matrix nuclei and the region of spin density localization.

Hence, to produce the deuterium anisotropic hfi observed for P700$^+$, a special structural organization is required, which is absent in the solutions

84

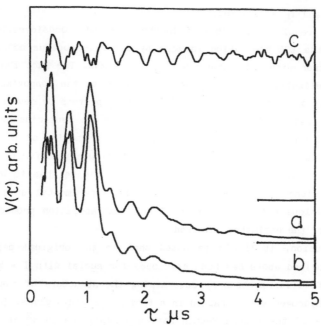

Fig. 15. The decay of the Chl a$^+$ primary ESE signal in solutions of CH_2Cl_2 in CD_3OD (a) and CH_3OH (b). Modulation from deuterium nuclei (c).

of monomeric chlorophyll but is present in reaction centers of photosystem I. On the strength of these data, it was concluded in (35) that the deuterium manifesting itself in the ESE modulation belongs directly to the chlorophyll dimer aggregate and is part of a water molecule or a nucleophilic protein side group that binds two parallel macrocycles.

3.2 Variational Calculations of Modulation Damping

The technique of linear extrapolation allows one to determine the number of nuclei and the distance to them from two parameters resulting from appropriate processing of the experimental envelope. In addition, there exist a number of works, mainly with paramagnetic centers in deuterated matrices, in which analogous data for deuterium nuclei are obtained using an approach based on the variational calculation of the modulating damping (36).

The analysis is carried out by comparing the experimental ratios $\alpha_i = V^i_{min}/V^i_{max}$ representing the modulation decay at different times with those calculated theoretically. The curves V^i_{min} and V^i_{max} are drawn, as previously, through maxima and minima of the ν_I harmonic.

In the approximation of N equivalent nuclei, the experimental ratio may be represented as $\alpha_i = \beta_i^N$, where β_i is the ratio V^i_{min}/V^i_{max} for one nucleus. The optimal parameters are found using a least-squares fit, i.e. the value $F = \sum_i (\alpha_i - \beta_i^n)^2$ is minimized.

The formula for F may be transformed in the following way:

$$\alpha_i^2 [(\beta_i^n / \alpha_i - 1)]^2 \approx [\ln(\beta_i^n / \alpha_i)]^2, \qquad [27]$$

then:

$$F = \sum_i [n \ln \beta_i - \ln \alpha_i]^2 . \qquad [28]$$

The number of parameters varied (two or three, depending on whether one allows for isotropic hfi in the calculations or not) may be reduced by one, since:

$$n = (\sum_i \ln \alpha_i \ln \beta_i) / (\sum_i (\ln \beta_i)^2 . \qquad [29]$$

This technique was first applied in (36) to determine the nearest environment of the electrons stabilized in a methyltetrahydrofuran glass. Furthermore, it has been used to analyze the modulation induced by the deuterium nuclei in the ESE of trapped electrons, atoms, radicals and ions of transition metals in frozen solutions and in various heterogeneous systems. The results obtained are described in reviews (7,8,37,38). It should be noted, however, that this technique possesses two disadvantages that may cause errors in determining the structural parameters, namely that the entire contribution is ascribed to the nearest-neighbour nuclei and the nqi is not taken into account.

3.3 Variational Calculation of Divided Envelopes

Time domain analysis of ESE modulation is often associated with the problem how to take into account the relaxation decay of the ESE signal. Comparing calculation with experiment, one must either extract the relaxation decay from the experimental envelope, approximating it with some monotonously decreasing function, or multiply the calculated curve by the decaying function to obtain visual correspondence. Since the modulation damping is accompanied by an overall decrease of the ESE signal amplitude, there is a real possibility of distorting the modulation upon taking into account the relaxation decay.

To overcome this problem a procedure has been proposed in (39,40) consisting in the removal of the decay from the stimulated ESE envelopes by means of dividing them by the stimulated ESE envelope recorded at a different τ value. For example, the envelopes recorded at different τ_i are divided by one obtained at the shortest time τ_0. The decay functions do not depend on τ, so that the results of the division $V(\tau_i, T)/V(\tau_0, T)$ are curves parallel to the base line after the modulation is completely damped.

On the next step calculations are carried out to search for a best agreement with the envelopes resulted from the division. To automatize the fit, a non-linear least-squares procedure may be used. In the course of this procedure the stimulated ESEEM in disordered sample are calculated for different τ_i values. Furthermore, the results of the division $V(\tau_i,T)/V(\tau_0,T)$ are compared with the experimental values for all τ_i simultaneously.

The procedure described was used to determine the hfi parameters of ^{15}N nuclei in the bacteriochlorophyll a radical cation (BChl a^+) and the oxidized primary electron donor P860$^+$ of bacterial photosynthesis (39-41). For BChl a^+, e.g., a set of 23 curves $V(\tau_i,T)/V(300\ ns,\ T)$, with τ varying from 320 to 800 ns, was used in the calculations. The results are analyzed in detail in section 5.1.

Note in conclusion that the efficiency of the above procedure depends dramatically on the quality of the initial approximation to the parameters that are varied, which is especially true for systems containing a large number of nonequivalent magnetic nuclei. To choose these parameters, it is necessary to use additional available information about the systems under study. For the paramagnetic centres indicated, the initial values of the variable hfi parameters were chosen on the basis of the results of ENDOR experiments on these systems.

3.4 Filtering of Matrix Contribution

If the environment of an unpaired electron includes close-lying nuclei having isotropic and anisotropic hfi well above those for other matrix nuclei, the hfi parameters and the number of the former nuclei may be determined using the technique described in (42). We shall illustrate it using as an example the analysis of the ESEEM induced by deuterium nuclei of CH_2OD radicals trapped in irradiated frozen glassy solution of perdeuterated methanol, CH_3OD.

Fig. 16 presents the primary ESE signal decay of CH_2OD radicals. The low-frequency parts of the modulus and cosine Fourier transforms of the primary decay envelope are shown in Fig. 17. They explicitly display lines with maxima at frequencies of 2.2, 3.4 and 4.9 MHz, the amplitude of the latter being negative as seen from the cosine spectrum. This harmonic is absent in the stimulated ESE modulation (42).

These spectra are interpreted as follows. The 2.2 MHz frequency coincides with the deuterium Zeeman frequency, and an intense line at this frequency is contributed by the matrix deuterons, coupled to electron by a weak anisotropic hfi. The matrix nuclei must also cause the appearance of small-amplitude lines near the double Zeeman frequency which are likely to be masked by more intense lines at 3.4 and 4.9 MHz.

The 3.4 and 4.9 MHz lines correspond to the frequencies $\nu_{\alpha(\beta)}$ and $\nu_\alpha + \nu_\beta$ of a hydroxyl deuterium nucleus with non-zero electron spin density. Using the value 0.5 MHz for the shift of the $\nu_\alpha + \nu_\beta$ harmonic from the exact $2\nu_I$ position and assuming 3.4 MHz to correspond to $\nu_{\alpha(\beta)}$, one obtains the estimate: $a = \pm 0.9$ MHz and $T_\perp = \pm 1.5$ MHz from the spectrum (another possible estimate: $a = \mp 3.9$

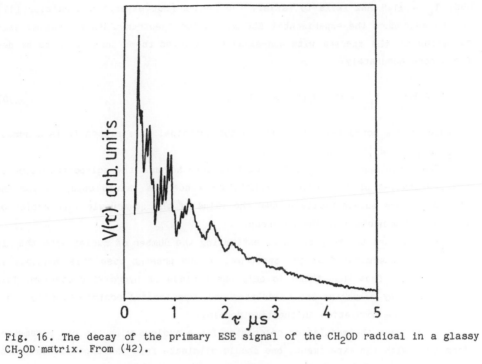

Fig. 16. The decay of the primary ESE signal of the CH$_2$OD radical in a glassy CH$_3$OD matrix. From (42).

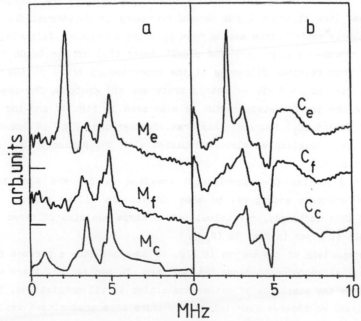

Fig. 17. Modulus (a) and cosine (b) Fourier transforms of the primary ESEEM envelopes: M$_e$,C$_e$ - experimental, M$_f$,C$_f$ - experimental after rejector filtering of the suppression frequencies of 2.2 and 14.4 MHz, M$_c$,C$_c$ - calculated for one deuteron with the hfi parameters of [30]. From (42).

MHz, $T_{\perp}$ = ±1.5 MHz fails to satisfy the square root expansion condition [17] and to reproduce the experimental ESE modulation spectrum). More detailed calculations of the spectra with non-axial hfi allowed these parameters to be defined more accurately:

$$a = \pm 1.1 \text{ MHz}, \quad T_{11} = \pm 1.7 \text{ MHz}, \quad \alpha = 0.8 , \tag{30}$$

in which for a non-axial hfi tensor in the principal axes system it is assumed: $|T_{33}| \geq |T_{11}| \geq |T_{22}|$, $\alpha = T_{22}/T_{11}$.

The above considerations did not take into account the nqi contribution to the frequencies of the nuclear transitions since the values obtained for the hfi parameters substantially exceed the value of the quadrupole interaction of the deuterium nucleus in the OD group.

We now turn to the problem of estimating the number of nuclei with the hfi parameters determined from the spectrum. In the present case this consists in verifying the fact that there is only one nucleus of hydroxyl deuterium. This may be done by comparing the calculated and experimental contributions of this nucleus to the modulation in the time domain.

To compare the modulation calculated for a nucleus with the hfi parameters from [30] with the experiment, one should eliminate from the experimental envelope the contribution of the matrix nuclei, displayed in the spectra mainly as an intense line at the 2.2 MHz Zeeman frequency of deuterons. The elimination of the matrix contribution may be done by using a rejector filtering technique. Fig. 18 represents a primary ESE signal decay (P_f) of the CH_2OD radical that resulted from rejector filtering of the experimental trace at the Zeeman frequencies 2.2 and 14.4 MHz of matrix deuterons and protons. The result of filtering in the low-frequency region is also seen in Fig. 17 showing the modulus (M_f) and cosine (C_f) Fourier transforms. Furthermore, Fig. 18 demonstrates the primary ESE modulation envelope calculated with the parameter set of [30] for one nucleus (P_c).

When filtering the experimental envelope at 2.2 MHz also the hydroxyl deuterium spectrum undergoes, to some extent, the filter effect. To take account of this influence, the calculated envelope was also filtered at 2.2 MHz. The result is given in Fig. 18 (P_{cf}).

A comparison of the curves in Fig. 18 does not show a perfect fit but just some qualitative correspondence between them. In particular, there exist a convergence in the positions of maxima and minima in all modulations. Furthermore, in all cases we observe four initial periodes of a predominant amplitude after which, at $\tau > 1.1$ μs, the amplitude of modulation decreases. But this decrease in the calculated envelopes is not so sharp as in the experimental one. Such a difference in modulation damping is known to be caused by the nqi effect and by

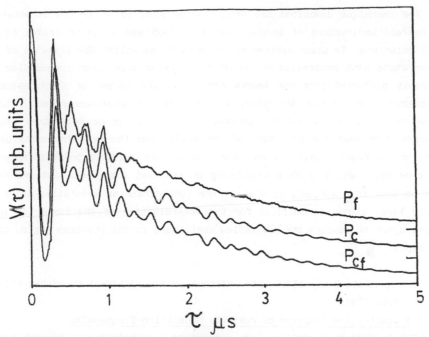

Fig. 18. Primary ESE modulation envelopes: P_f - resulting from rejector filtering of the experimental one at the suppression frequencies of 2.2 and 14.4 MHz,
P_c - calculated for one deuteron with hfi parameters of [30], P_{cf} - resulting
from filtering of the envelope P_c at the suppression frequency of 2.2 MHz. From
(42).

some statistical scattering of the hfi parameters, which were neglected in the
calculations.

Among other reasons for the difference in the reported curves one could
mention the distortions caused by the filter for the harmonics with frequencies
close to that of the suppression. An example of this influence is seen in the
spectra in Fig. 17, where the line shape of the 3.4 MHz harmonic, which is
fairly close to the 2.2 MHz suppression frequency, is somewhat changed by the
filtering procedure.

It should be noted also that filtering of modulation envelopes with and
without an intense harmonic at the Zeeman frequency yields curves in the time
domain that are somewhat different in shape in the low-time region. Thus, it is
believed that with the above approach a perfect fit of the calculated and experimental envelopes, even in the low-time region, which is fairly free from the
nqi effect, should not be expected. Nevertheless the available correspondence
in the shape and amplitude of this region reliably indicates that only one nucleus contributes to the modulation. For example, the calculation for two equivalent nuclei leads to a considerable (about two-fold) excess in the calculated
modulation amplitude compared to the experimental one.

The technique described has been applied also to study the structure of the nearest environment of trapped electrons (43) and atoms in frozen aqueous (D_2O) solutions. In these systems we happened to establish the presence of deuterium atoms with noticeable hfi constants, giving modulation frequencies considerably different from the Zeeman frequency. The number of these atoms per paramagnetic center was determined by means of the above-described filtering procedure. To do this, the hfi parameters need only to be determined with an accuracy such that the positions of the maxima and the peak intensity of the lines in the Fourier spectrum are reproduced. Since nqi or a possible spread in hfi constants, which cause a broadening of spectral components and are difficult to take into account, show up in the time domain as modulation damping at fairly large times, it suffices for the determination of the number of interacting atoms to compare the modulation amplitudes in the low-time region of the envelopes.

4.   ESE MODULATION IN DISORDERED SYSTEMS IN THE PRESENCE OF STRONG NQI (FOR EXAMPLE ^{14}N)

4.1 A Qualitative Analysis of Possible Modulation Frequencies

The condition of small nuclear quadrupole interaction as compared with the Zeeman frequency is not fulfilled for ^{14}N nuclei (I = 1). The values of their quadrupole coupling constants vary in the range of 1 - 5 MHz, depending on the type of nitrogen coordination with other nuclei (23,44). On the other hand, in the most frequently used microwave X-band range, the Zeeman frequency of ^{14}N is about 1 MHz (e.g. for B_o = 339 mT, ν_I = 1.04 MHz). Hence, the splitting of energy levels caused by nqi is comparable to the Zeeman splitting or may even exceed it.

At the same time, there is experimental evidence that disordered samples often show a deep modulation induced by ^{14}N nuclei in the range of several microseconds up to several tens of microseconds, yielding modulation spectra with narrow lines. Most often this modulation is stimulated by ^{14}N nuclei that are separated from the region of unpaired electron density by several bonds, and therefore coupled to it by a weak anisotropic hfi of the order of several tenths of MHz. Simultaneously these nuclei can possess a noticeable isotropic hfi constant, up to several MHz.

Based on the above observations a qualitative analysis was performed in (45,56) allowing for only Zeeman, nuclear quadrupole and isotropic hyperfine interactions between electron and nucleus. It was shown that the manifestation of narrow lines corresponding to the frequencies of nuclear transitions in the ESE modulation spectra is governed by the ratio of effective frequencies ν_{ef} = $|\nu_I \pm a/2|$ corresponding to various electron spin manifolds to the value of the quadrupole coupling constant.

If $\nu_{ef}/K < 1$ ($K = e^2Qq/4h$, one-fourth of the quadrupole coupling constant), the frequencies of the ^{14}N nuclear transitions will approach the zero-field nqi frequencies:

$$\nu_+ = K(3 + \eta), \qquad \nu_- = K(3 - \eta), \qquad \nu_o = 2K\eta . \qquad [31]$$
$$(\eta = \text{asymmetry parameter})$$

Alternatively, for $\nu_{ef}/K \geq 1$ the appearance of a narrow peak should be expected near the frequency:

$$\nu_d = 2[(\nu_I \mp {}^a/2)^2 + K^2(3 + \eta^2)]^{\frac{1}{2}} , \qquad [32]$$

though the existence of a wide smeared-out line centered at $\nu_d/2$ is also possible. The manifestation of these frequencies in the spectrum may be illustrated by using a schematic diagram of the ^{14}N energy levels versus the external magnetic field (Fig. 19).

In zero field, the ^{14}N nucleus has three nqi-split energy levels between which three transitions with frequencies [31] are possible. Then, if the value of the isotropic hfi is such that any one of the ν_{ef}-frequencies is small, a situation close to a zero-field NQR is realized and the frequencies of nuclear transitions will be close to those in [31]. They are independent of the external magnetic field direction thus leading to the appearance of narrow peaks in the modulation spectra.

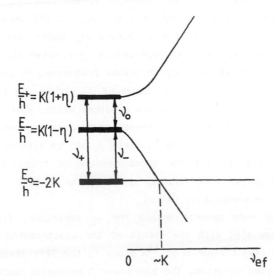

Fig. 19. The energy levels of the nucleus with I = 1 (^{14}N) as a function of external magnetic field intensity.

With increasing magnetic field, the inversion of the second and third levels occur at $\nu_{ef} \sim K$, which roughly accounts for the use of the condition $\nu_{ef}/K < 1$ for the presence of three NQR frequencies in the region of small effective fields. This scheme also indicates that with increasing ν_{ef}/K, the component ν_+ will change its position by far the least, which is confirmed by the results of precise numerical calculations (47).

At $\nu_{ef} \geq K$, after the inversion of the second and third levels, the system also has three transitions. But two of them - one-quantum transitions - show a strong dependence on the magnetic field orientation, while for the third - a two-quantum transition - this dependence is weak. Therefore, the narrow peak in the ESE spectrum is possible only for the last transition.

As follows from this consideration, the maximum possible number of narrow lines originating from one ^{14}N nucleus in ESE modulation spectra equals four.

The above qualitative regularities have been verified in (48) by numerical calculations of spectra, using general expressions for modulation amplitude factors and a spin hamiltonian of a $S = \frac{1}{2}$, $I = 1$ system, neglecting anisotropic hfi. The widths, amplitudes and positions of the line maxima of the nuclear transitions in disordered samples, and their connection with the nqi and hfi parameters in the case of exact cancellation of the Zeeman frequency by hyperfine interaction (i.e. when one of the ν_{ef} frequencies equals zero), have been analyzed. Simultaneously the same data have been obtained for the combination harmonics in the primary ESE. In some cases, the spectral changes in the absence of exact cancellation, which leads to shifts of the maxima positions of the spectral lines, have also been numerically analyzed.

From the above, it is evident that the nqi and hfi parameters from modulation spectra may be determined most accurately under the condition of exact cancellation. The only variable experimental parameter that could be used to attain this condition is the nuclear Zeeman frequency. It may be changed in two ways. The first is related to the change of ν_I within the excitation region of the EPR spectrum and is limited by the spectral width, which for radicals typically amounts to several mT and increases up to tens or even hundreds of mT for transition metal ions. Hence, for radicals ν_I may be altered in this way by only a few per cent, but for the broad spectra of transition ions it may be changed by tens of per cent and may approach the condition of exact cancellation. This will be exemplified below.

Another and more general method for ν_I variation, first implemented in (49,50), is associated with the change of the spectrometer carrier frequency. For example, in (49) the ESEEM of ^{14}N atoms in the DPPH stable radical has been studied in a frozen solution. The microwave frequencies used in this work were 4.45, 4.94, 5.47, 6.68 and 9.10 HHz, and the change in modulation spectra with frequency was analyzed with the aim to find the condition of full compensation.

This kind of ESE experiments makes it possible to eliminate the ambiguities arising in the interpretation of modulation spectra of nuclei with considerable quadrupole interaction that are obtained at only one value of the carrier frequency.

4.2 Experimental Examples of Nitrogen ESEEM in Model Systems

Since ^{14}N NQR frequencies are characteristic of the type of nitrogen coordination with other atoms, it is of interest to determine regularities in the spectra of ^{14}N ESE modulation observed in various model molecular groups, specifically comparing the modulation frequencies with the published data on NQR frequencies in the same groups. In this section we shall consider as model systems some stable radicals in which nitrogens are coordinated with different types of other atoms.

4.2.1 The NOH Group. Fig. 20 presents the decay of the stimulated ESE signal of the radical:

(R1)

showing a slightly damped modulation of a fairly large amplitude. In the ESE of radicals of analogous structure lacking the NOH group, there are only high-frequency modulations from the radical and the matrix protons. The nitrogen nuclei in the radical cycle do not contribute to the modulation. Moreover, the ESE modulation of radical R1 changes when substituting ^{14}N by ^{15}N. These facts unambiguously testify that the modulation in Fig. 20 (R1) is induced by the nitrogen nucleus in the =NOH, oxime group. Its modulus Fourier transform (Fig. 20) exhibits three well-resolved, narrow components with maxima at 1.3, 3.5 and 4.8 MHz, which frequencies are in fair agreement with the NQR frequencies available in the literature for ^{14}N involved in the =NOH group in various compounds (44). Now, using relations [31] we obtain the quadrupole parameters K = 1.4 MHz and η = 0.47 for the ^{14}N nucleus in the =NOH group.

The stimulated ESE modulation spectra in the case considered lack the line corresponding to a double quantum transition, which, for the condition $\nu_{ef}/K < 1$ and using expression [32] is estimated to be in the range 5.7 to 8.6 MHz.

The authors of [48], on the basis of the results of numerical calculations, have shown that for isotropic hyperfine coupling a double quantum frequency must always be present in the spectra. In some cases its absence in

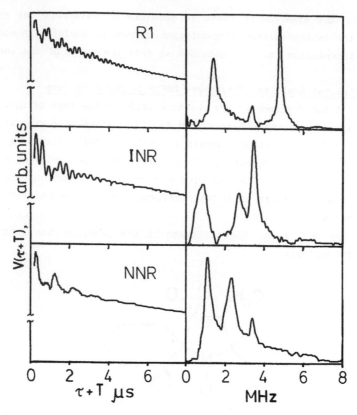

Fig. 20. The decay of the stimulated ESE signal for radicals R1 (τ = 180 ns), INR and NNR with the NH_2 group (τ = 200 ns) and corresponding modulus Fourier spectra.

stimulated ESE could be explained by incorrectly chosen values of the time interval τ. However, the peak corresponding to the double quantum transition is not observed in the primary ESE modulation spectrum either, although here the observable frequencies do not depend on the choice of τ. Hence, the absence of this line in the spectra must have a different origin. It may be related to the effect of anisotropic hfi, which, as we have shown by numerical calculations, may considerably broaden the spectrum components. In addition, in the given case the absence of the double quantum peak may be associated with the existence of different conformations of the cyclohexane ring, for which the nitrogen nucleus has different hfi constants. The latter is possibly favoured by experiments with R1 radicals in which the =NOH group contains [15]N (40).

4.2.2 <u>The NH$_2$ group</u>. The shape of the ESR spectra of nitronylnitroxyl:

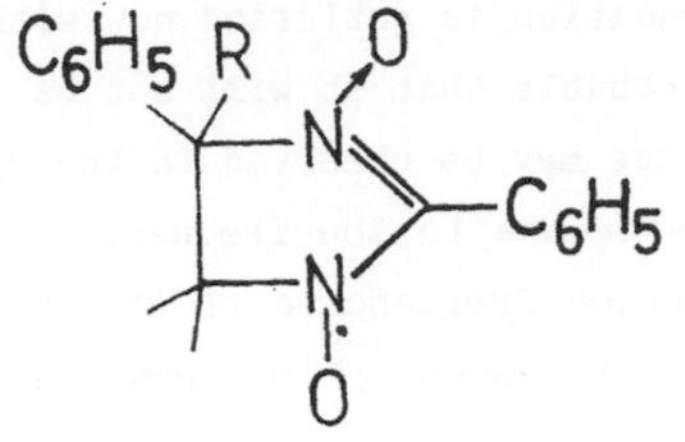

(NNR)

and iminonitroxyl:

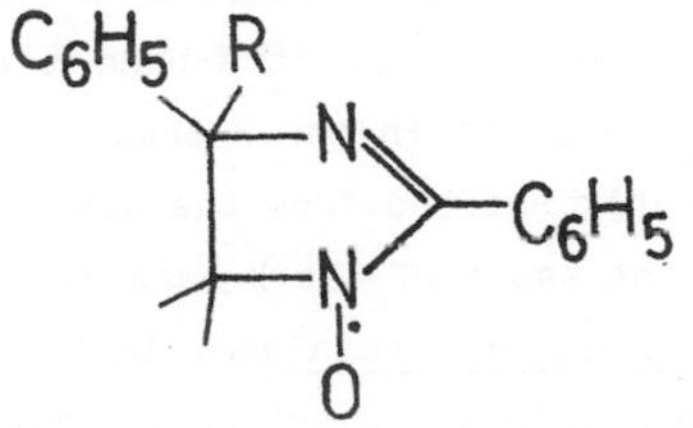

(INR)

radicals in frozen solution is determined by hfi of both nitrogen atoms. For example, in NNR the largest principal components of the hfi tensor are equal and amount to 1.8 mT, while those in INR are different and are equal to 0.8 and 2.2 mT.

The ESEEM of these radicals with substituents of the -CH$_3$, -OCH$_3$, -C$_6$H$_5$ types exhibit only the high-frequency proton modulation, but after the introduction of the -NH$_2$ group as a substituent the low-frequency modulation is also present.

The modulus Fourier transformation of the NH$_2$-containing NNR and INR stimulated ESE envelopes yield spectra with three components at 1.1, 2.4, 3.5 and 0.9, 2.7, 3.6($\pm$ 0.03) MHz, respectively, which frequencies are obviously the NQR ones. Now, using [31], one finds K = 0.98 MHz, η= 0.56 and K = 1.05 MHz, η = 0.43. The published data (44) indicate that the K value for ^{14}N in the NH$_2$ group in the primary aliphatic amines ranges from 0.995 to 1.02 MHz, and in the primary aromatic amines from 0.9 to 1.0 MHz. Thus, our values of K are in rather good agreement with those obtained by NQR measurements. Unfortunately, in the ESE modulation spectra of these radicals we have also failed to observe the peak corresponding to the double quantum transition, which if present could have aided in estimating the value of the nitrogen isotropic hfi constant.

4.2.3 The NO_2 group. From the data available in the literature (44) it follows that the nitrogen nucleus in the NO_2 group has a considerably smaller quadrupole coupling constant (K = 0.3 MHz) than, for example, in the NOH or NH_2 groups. The $\nu_{ef}/K < 1$ condition is satisfied now within a small interval of ν_{ef} values and it is quite probable that it will not be satisfied when working only in the X-band. Hence, lines may be observed in the spectra, which cannot be interpreted by a simple reference to the frequencies of [31,32], and it will be useful to change the carrier frequencies of the microwave pulses, allowing to increase or decrease ν_{ef} by means of a corresponding change of the nuclear Zeeman frequency.

The situation described above prevails in the modulation spectra induced by nitrogens in the NO_2-groups of the DPPH radical (49-51). The three components: 0.207, 0.736 and 0.943 MHz, close to the NQR frequencies of NO_2 groups available from the literature, were observed in the modulation spectra only when using the microwave C-band (49). At X-band a similar triplet is not displayed and the prominent harmonic in the X-band spectra has a frequency of 1.3 MHz, i.e. it is considerably shifted from the NQR frequency ν_+ [31] to which it was assigned in (51) on the basis of the X-band data alone.

4.2.4 A Biological Example: T-catalase. In the last few years a new class of non-haem bacterial catalases was discovered, which, unlike the haem containing enzymes originating from animals, are not inhibited by cyanides. Studies of the non-haem T-catalase from the extremely thermophilic bacterium Thermus thermophilis HB-8 have shown that this protein of a molecular weight of 210,000 D consists of six identical subunits and every subunit contains a pair of closely situated manganese atoms (52,53). The manganese in the T-catalase active center forms a binuclear cluster. In the reduced form of the protein, the excited state $Mn^{2+}.Mn^{2+}$ with an integer spin is paramagnetic. In the completely oxidized state $Mn^{3+}.Mn^{4+}$ the manganese ions coupled by a strongly antiferromagnetic interaction, form a cluster with net electron spin $S = \frac{1}{2}$.

The nature of the ligands in the coordination sphere of the manganese in T-catalase has so far not been established. The EPR spectrum of the oxidized T-catalase is about 130 mT wide due to multicomponent hyperfine structure of the two manganese atoms.

Fig. 21 shows as an example the stimulated ESE envelopes of T-catalase recorded at two values of the external magnetic field: B_o = 295 and 370 mT with τ = 300 ns. The modulus Fourier transform envelopes recorded at B_o = 295, 309, 329.3 and 370 mT are also depicted in Fig. 22. All the spectra in Fig. 22 have three components with maxima positions at 0.54, 1.56 and 2.1 MHz, accurate to 0.05 MHz. In addition, in the spectra at B_o = 329.3 and 370 mT another peak is explicitly displayed. Its intensity is smaller and its width is larger than the corresponding values of the other peaks, and the position of its maximum in-

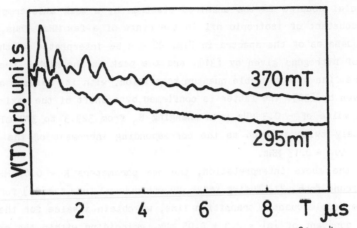

Fig. 21. The decay of the stimulated ESE signal of the $Mn^{3+}.Mn^4$ cluster in T-catalase for τ = 300 ns and B_o = 295 and 370 mT.

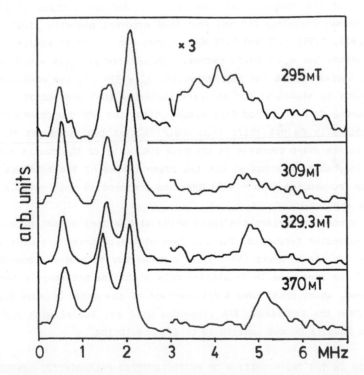

Fig. 22. Modulus Fourier spectra of the stimulated ESE modulation of the $Mn^{3+}.Mn^{4+}$ cluster in T-catalase for τ = 300 ns and B_o = 295, 309, 329.3 and 370 mT, respectively.

creases with increasing external magnetic field.

The spectra observed exhibit the above-described features of the spectra of ^{14}N nuclei coupled to the unpaired electron by a weak anisotropic hfi and having a constant of isotropic hfi in the range of a few MHz. Thus, the three lower frequencies of the spectra in Fig. 22 can be interpreted as the NQR frequencies of nitrogens given by [31], and the peak in the higher frequency region as the line of the double quantum transition, with its maximum at the frequency given by [32]. The latter is confirmed by a shift of the maximum of this line by a value of ~ 0.3 MHz when changing B_O from 329.3 to 370 mT, which is approximately twice as much as the corresponding increment of the ^{14}N Zeeman frequency: $\Delta \nu_I$ = 0.13 MHz.

From the above interpretation, the nqi parameters K = 0.61 MHz and η = 0.44 are found from [31]. Using these parameters and formula [32] for the maximum of the double quantum transition line, we obtain a value for the isotropic hyperfine constant of $|a|$ = 2.3 ± 0.05 MHz, coinciding within the experimental accuracy for both values of the magnetic field, B_O = 329.3 and 370 mT. The value for the isotropic hfi constant thus obtained allows us to compare the fulfilment of the compensation condition for different types of transitions. The ^{14}N Zeeman frequency for the indicated external magnetic field intensities equals 0.908, 0.95, 1.01 and 1.14 MHz. Hence, for the 'NQR' region of the modulation spectra the ν_{ef}/K ratio decreases successively: ν_{ef}/K = 0.4, 0.33, 0.22, 0, as B_O increases. At the highest value, B_O = 370 mT, the condition of exact cancellation is attained. Our analysis indicates that even under conditions of fairly strong deviation from full compensation, the NQR components of the spectrum practically do not shift from their initial positions. At the same time there occurs a sharp decrease in the peak intensity of the double quantum transition line, whose maximum at the two lower values of the external field must be at the frequencies 4.65 - 4.75 MHz; its presence is only revealed by magnifying the spectra.

The protein comprises the amino acids with a limited set of nitrogen-containing molecular fragments. The NQR frequencies observed are closest to those described for the NH group (44). Then, a model may be proposed assuming coordination of the manganese in T-catalase to a histidine heterocycle through a nitrogen atom, which could have a hfi constant of the order of some 0.5 mT as estimated from the individual EPR linewidth of 2 mT. Nuclei with such hyperfine constants usually do not contribute to ESE modulation.

5. ESEEM IN THE INVESTIGATION OF PHOTOSYNTHETIC PARAMAGNETIC CENTERS

By now we can point to two types of real biological objects actively investigated with the help of ESEEM spectroscopy. The first type contains metalproteins, where the problem of the ligand environment of the paramagnetic metal ions was solved using ESE. The main results here have been obtained by Mims and Peisach et al. and are considered in chapter 1 of this volume.

Another type of biological systems actively studied by the ESE technique by several groups, contains the primary electron donors in the electron transport chain of plant and bacterial photosynthesis. In a number of studies, the ESE modulation induced by ^{15}N and ^{14}N nuclei in radical cations of bacteriochlorophyll a (BChl a) and chlorophyll a (Chl a) have been investigated. This modulation was compared further with that observed for the oxidized primary electron donors P860 and P700 of the photosystem of purple bacteria and photosystem I of higher plants, respectively, to obtain information on the delocalization of spin density in the primary donors and on their structure. Below we shall consider briefly the experimental results for these systems, which simultaneously will allow us to illustrate the various alternative approaches to the analysis of ESE modulation in the time and frequency domains.

5.1 Paramagnetic Centers with ^{15}N

BChl a$^+$. Most of the data on the hfi of nitrogens have been obtained for this radical-cation, which has been studied by ENDOR (54) and, by different groups of authors, by the ESE technique (20,39,40,55).

The isotropic hfi constants for the four ^{15}N of BChl a$^+$ in liquid solution, evaluated from the ENDOR spectrum, equal 3.15, 3.25, 4.05, 4.4 MHz. The accuracy achieved in the ENDOR spectra of BChl a$^+$ in solid samples has allowed the determination of two sets of hfi parameters for two pairs of (approximately) equivalent ^{15}N nuclei (see Table 1).

Fig. 23 demonstrates the decay of the primary and stimulated ESE signal of BChl a$^+$-^{15}N. The deep low-frequency modulation at these decays differs substantially from that observed for BChl a$^+$-^{14}N (Fig. 23). This indicates that its origin is due to the interaction of the unpaired electron with the nitrogen nuclei. In (39) a set of hfi parameters, different for all four nitrogens, has been obtained by searching a best agreement between the calculated and experimental stimulated ESE envelope when varying the hfi parameters of the ^{15}N nuclei and using the ENDOR data as an initial approximation (Table 1). An axial hfi was used in the calculations; however, speculations were made on the presence of some non-axiality.

A series of lines is reliably distinguished in the ESE modulation spectra of BChl a$^+$-^{15}N (Fig. 24). In the low-frequency region there is a broad, strong peak with the maximum position ranging from 0.5 to 0.9 MHz, depending on the time interval τ. Besides, one may distinguish narrow lines with maxima at 2.5 and 2.9 MHz and – at small τ values – a wide peak with a maximum at 4.6 MHz.

Taking into account the dead time, the lines with the maxima at 2.5 and 2.9 MHz have been assigned to ν_α transitions for two groups of approximately equivalent nuclei. A weak 4.6 MHz line has been interpreted as an overlap of the $\nu_\alpha^{//}$ components for different nuclei, and a strong peak in the low-fre-

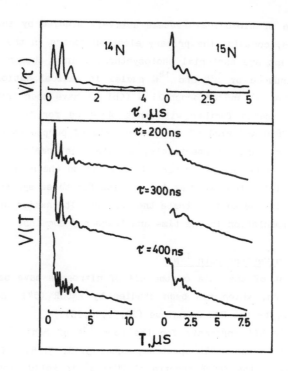

Fig. 23. The decay of the primary and stimulated ESE signals of BChl a$^+$ with ^{14}N and ^{15}N nuclei. From (55).

TABLE 1

Hfi parameters for ^{15}N nuclei in BChl a$^+$ (MHz). From (20).

Nucleus	ENDOR in liquid state (54)	ENDOR in solid state (54)			ESE (39)			ESE (20)		
	a	a	T_{11}	T_{22}/T_{11}	a	T_{11}	T_{22}/T_{11}	a	T_{11}	T_{22}/T_{11}
1	3.15	3.2	-1.2	1	3.16	-1.215	1	3.2	-1.2	0.9
2	3.25	3.2	-1.2	1	3.41	-1.305	1	3.6	-1.4	0.5
3	4.05	4.3	-1.4	1	4.10	-1.335	1	4.3	-1.4	0.9
4	4.40	4.3	-1.4	1	4.45	-1.455	1	4.6	-1.7	0.5

quency region as a superposition of contributions of the $\nu_\beta^{\perp}$ and $\nu_\beta^{//}$ components, and in addition of difference combinations of the transition frequencies of different nuclei.

The ESE modulation spectra calculated with hfi parameters determined by the ENDOR (54) and ESE (39) methods are in qualitative agreement with the ex-

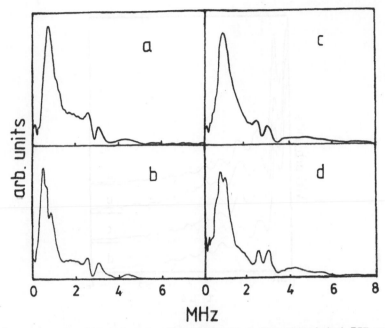

Fig. 24. Experimental modulus Fourier spectra of the stimulated ESE modulation obtained in (55) for BChl a^+-^{15}N (a,c) and calculated spectra (b,d) with the hfi parameters from (20) (Table 1). For spectra (a,b) τ = 300 ns, T_d = τ + T_o = 480 ns; for spectra (c,d) τ = 500 ns, T_d = 680 ns. From (20).

perimental ones. They contain a strong low-frequency line and two lines with maxima at 2.5 and 2.9 MHz. At the same time, a primary ESE modulation calculated for these hfi parameters is distorted at τ > 1.5 μs by a high-frequency (~ 5.6 MHz) modulation which is absent in the experimental envelope (Fig. 25) (20). An analysis performed in (20) has shown that the variation of the hfi parameters within allowed limits determined by the position of components in the ESE and ENDOR spectra does not permit the elimination of the mentioned discrepancy, which arises due to the correlation in the orientation of hfi tensors of nitrogens with about equal hfi constants. This discrepancy could be overcome only by using in the calculations non-axial tensors, which have allowed us to obtain good agreement with the experiment in the time and frequency domains (Figs. 24,25). The parameters determined in this way are also listed in Table 1. The difference between all listed sets of isotropic hfi constants amounts to ~ 10 %, but in the last case there is a qualitative difference related to the non-axiality of hfi tensors.

In conclusion it might be noted that the published data of various authors on the ^{15}N hfi parameters in monomeric BChl a^+ are in satisfactory mutual agreement within the resolution limits of the techniques used.

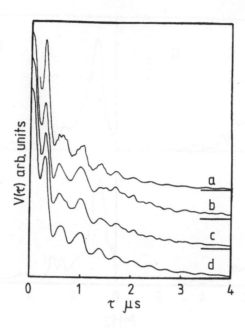

Fig. 25. Experimental primary ESE envelope for BChl a-^{15}N (55) and calculated with hfi parameters from (54) (b), (39) (c) and (20) (d). The calculated envelopes are multiplied by an exponent with a decay time of 1.3 μs. From (20).

<u>P860$^+$</u>. In the ENDOR spectrum of P860$^+$-^{15}N in a liquid sample, investigated in (54) simultaneously with BChl a$^+$, resonance lines corresponding to four isotropic hfi constants: 1.05, 1.61, 2.22, 2.6 MHz were observed. These constants are approximately two times less than those determined for BChl a$^+$. In the ENDOR spectra of these centers in a frozen sample in the frequency region > 2 MHz, a series of resonances were displayed that were attributed to two sets of hfi parameters: a = 2.23 MHz, $T_\perp$ = -0.3 MHz, and a = 2.56 MHz, $T_\perp$ = -0.26 MHz.

The stimulated ESE modulation spectra of P860$^+$-^{15}N presented in (56) contain a strong line in the 0.4 - 0.7 MHz region, a peak of lesser intensity with its maximum at 2.6 MHz and a component close to the Zeeman frequency of the ^{15}N nucleus, ν_I = 1.45 MHz.

Using the isotropic hfi constants for P860$^+$ in a liquid, a numerical calculation of the stimulated ESE modulation has been performed in (41) to find a best fit with experiment. This provided the anisotropic hfi parameters: $T_\perp$ = -0.28, -0.67, 0.88, -0.98 MHz. It has been noted in (41) that the decrease of the obtained anisotropic hfi parameters vs. BChl a$^+$ is on the average close to two, which again testifies in favour of the dimeric model for P860. However, in their next paper (40) the authors, analyzing the results obtained in more detail, claimed that the latter are not in line with the experimental ENDOR spectrum.

Unfortunately, employing the data of (41) we failed to carry out the simulation of the modulation spectra of $P860^+$ and to compare them with those given in (56), since the latter were obtained by the method of autoregression (AR) and the parameters of the AR model used in (56) were not indicated. Our calculation of the stimulated ESE Fourier transforms with the parameters of (41) yielded, irrespective of the P860 model, spectra that were drastically different from those given in (56). In particular, the 2.6 MHz component was not observed in our simulations. In addition, our calculation of the primary ESE modulation for dimeric and monomeric models with the parameters mentioned has shown that the amplitude of the calculated modulation substantially exceeds the experimental one presented in (56). All these discrepancies allow some doubt as to the anisotropic hfi values of the ^{15}N obtained with the aid of the time domain variation procedure for $P865^+$. These parameters are, therefore, subject to refinement.

$Chl\ a^+$. The decay of the primary and stimulated ESE signals of Chl a^+ are also accompanied by a modulation of considerable amplitude, which changes when substituting ^{14}N by ^{15}N (57). In the stimulated ESE modulation spectra of Chl a^{+}-^{15}N there are components with maxima at 0.65, 2.1 - 2.2 and 2.75 MHz, and in addition a broad line with its maximum at 4.6 MHz (58).

The Chl a^{+}-^{15}N ESE modulation envelopes and spectra as well as the dependence on τ of the relative intensity of the low- and high-frequency spectral components in the stimulated ESE are close to those observed for BChl a^{+}-^{15}N.

Taking into account the closely related BChl a and Chl a molecular structures and the similarity in the envelopes and ESE modulation spectra of their radical cations, the Chl a^+ spectra were explained qualitatively in the same way as for BChl a^+. On the basis of this interpretation, using the spectral components at 2.1 - 2.2, 2.75 and 4.6 MHz, the hfi parameters have been estimated for two pairs of nitrogen nuclei with two equivalent nuclei in each pair, in the approximation of axial anisotropic hfi (58):

$$a = 2.9\ \text{MHz} \qquad T_\perp = -1.5\ \text{MHz}$$
$$a = 4.0\ \text{MHz} \qquad T_\perp = -1.4\ \text{MHz} . \qquad\qquad [33]$$

However, the calculated primary ESE modulation for two pairs of equivalent nuclei with coinciding directions of the hfi tensors' principal axes displayed combination frequencies that were absent in the experimental envelope. These frequencies could be eliminated, as in the case of BChl a^+, using non-axial hfi. Only under these conditions has a satisfactory agreement between calculated and experimental envelopes and spectra been obtained. Fig. 26 exemplifies

104

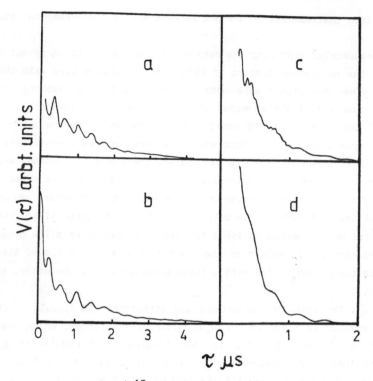

Fig. 26. Experimental Chl a^+-^{15}N (a) and P700$^+$-^{15}N (c) decays of the primary ESE signal; primary ESE envelopes calculated for hfi parameter sets of [34] (b) and [35] (d). For convenient comparison with the experiment, the calculated envelope (b) is multiplied by an exponent with decay time 1.2 μs and the envelope (d) by an exponent with decay time 0.3 μs. From (58).

the primary ESE modulation envelope calculated for the following set of hfi parameters:

a = 3.1 MHz	T_{11} = -1.8 MHz	T_{22}/T_{11} = 1
a = 3.5 MHz	T_{11} = -1.2 MHz	T_{22}/T_{11} = 0.5
a = 4.3 MHz	T_{11} = -1.7 MHz	T_{22}/T_{11} = 1
a = 4.3 MHz	T_{11} = -2.0 MHz	T_{22}/T_{11} = 0.8 . [34]

The sets of the parameters [33,34] show an accuracy of their determination from the ESE data (for a given system and in the absence of additional information) of the order of 0.5 MHz. Unfortunately, the data on the hfi constants of ^{15}N in Chl a^+ in [34] are, at present, the only experimental ones since all efforts to measure them using the ENDOR technique failed (59).

P700$^+$. The modulation effects observed in the ESE of P700$^+$ centers as well as in the above system change when substituting ^{14}N by ^{15}N, demonstrating that they are caused by the interaction with nitrogen nuclei (58,60). The P700[+] ESEEM exhibits a substantially smaller amplitude as compared with that observed for Chl a[+] (Fig. 26) (58). The P700[+] stimulated ESEEM spectra show a strong low-frequency line at 0.55 MHz, a peak at 2.3 - 2.4 MHz and a number of poorly resolved lines in the intermediate region, amongst which the 1.47 MHz harmonic that is close to the Zeeman frequency for the ^{15}N nucleus, is clearly visible (58). It is noteworthy that the spectral structure obtained is similar to that observed in (56) for the P860[+] centers. Taking the similarity in the ESEEM spectra of P700[+] and P860[+] into account, it is expected that the ^{15}N hfi parameters in these centres are close and that their structures are similar.

The calculation of the P700[+] modulation envelopes and spectra to obtain information on the hfi with the nitrogen nuclei is hampered by the small number of characteristic lines in the spectra, by the great number of variable hfi parameters (12 for monomeric and 24 for dimeric models even assuming the same direction of principal axes of all hfi tensors), and by the absence of any information on the hfi parameters extracted by other methods. In these conditions, using the above-noted similarity of the spectral compositions of P700[+] and P860[+], one may attempt to calculate the ESE modulation envelopes and spectra orientating oneself on the ENDOR data for P860[+], which allows one to limit the arbitrariness in varying the hfi parameters. Therefore, in the calculations of the spectra in (58), the existence of two types of ^{15}N nuclei with isotropic hfi constants of 2 MHz and two types of same with lesser hfi constants has been assumed. The anisotropic hfi was considered to be axial and the halves of the dimer in the dimer model equivalent. Under these assumptions we managed to reach qualitative agreement between the calculated spectra and envelopes and the experimental ones. Fig. 26 exemplifies the result of the primary ESEEM calculation with the following hfi parameters:

$$a = 0.2 \text{ MHz} \qquad T_\perp = -0.2 \text{ MHz}$$
$$a = 0.5 \text{ MHz} \qquad T_\perp = -0.5 \text{ MHz}$$
$$a = 2.0 \text{ MHz} \qquad T_\perp = -0.3 \text{ MHz}$$
$$a = 2.1 \text{ MHz} \qquad T_\perp = -03. \text{ MHz} . \qquad\qquad [35]$$

It has been noted in (58) that the shapes of the modulation spectra do not depend substantially on which model (dimer or monomer) is used for the calculations. Besides, variation of the anisotropic hfi values within 0.1 MHz leads to changes of the calculated modulation amplitude of the same order of magnitude as the changes resulting from the use of the monomer model (four nuclei) instead of the dimer model (eight pairwise equivalent nuclei). Therefore, the calculation of the modulation amplitude failed to provide direct evidence in favour of a definite P700[+] model (dimer or monomer) (58).

The two lower values of the isotropic hfi constants in set [35] are considerably less than the values 1.05 and 1.61 MHz derived in (54) for $P865^+$. The ν_α frequency should lie in the range 1.8 - 2.1 MHz for nuclei with the $P865^+$ constants for the anisotropic hfi, $T_\perp$ = -0.2 - -0.3 MHz. But such frequencies were not found in the $P865^+$ and $P700^+$ modulation spectra. A consequence of this fact is that for nuclei with lower hfi constants the magnitude of the anisotropic hfi is close to that of the isotropic one: $|T_\perp| \sim |a|$. Then ν_α will shift to the neighbourhood of the ^{15}N Zeeman frequency, thus explaining the appearance of such a line in the modulation spectra in contrast to the suggestion in (56) that it appears as a result of weak hfi between the unpaired electron and ^{15}N nuclei of the reaction center protein.

Model calculations confirmed the correctness of the supposition that a line appears close to the Zeeman frequency when $|T_\perp| \sim |a|$, but calculated modulation amplitudes in the primary and stimulated ESE exceeded considerably the experimental values when hfi constants of these nitrogens were equal to 1 - 1.5 MHz. Correspondence was achieved only when these hfi constants were reduced to those given in set [35] (58).

In conclusion it should be noted that the results of the above calculation for $P700^+$ should be considered as qualitative. They demonstrate the necessity of further radiospectroscopic experiments on $P700^+$ to obtain additional data on the hfi with nitrogen nuclei that will allow us to refine these parameters.

5.2 Paramagnetic Centers with Nitrogen ^{14}N

We turn now to the results obtained when studying samples with the natural content of nitrogen isotopes (99.63 % ^{14}N). In the Chl a^+-^{14}N modulation spectra there are five components with maxima at 0.8, 1.6, 2, 2.4 and 2.8 MHz and in addition a series of lines of small intensity in a higher frequency region (60). For the five lines indicated we distinguish two groups, three lines in each, with maximal positions satisfying the condition $\nu_- + \nu_o = \nu_+$: 0.8, 1.6, 2.4 MHz and 0.8, 2, 2.8 MHz (component 0.8 MHz is common for both groups), which obviously are the NQR frequencies at I = 1. These two groups of lines were ascribed to NQR frequencies of ^{14}N with different electron orbital configuration: $\overset{|}{N}\diagdown$ and $\overset{..}{N}\diagdown$. The nqi parameters for these nuclei are the following: K = 0.67 MHz, η = 0.6 and K = 0.8 MHz, η = 0.5.

The stimulated ESE spectra of BChl a^+ have also exhibited five lines with maxima at 0.9, 1.2, 1.6, 2.5 and 2.8 MHz, which also have yielded two sets of nqi parameters: K = 0.73 MHz, η = 0.8 and K = 0.68 MHz, η = 0.67 (55). The first set refers to the lines of 1.2, 1.6, 2.8 MHz and the second one to those of 0.9, 1.6, 2.5 MHz.

The observation of modulation frequencies corresponding to the NQR frequencies in Chl a^+ and BCh a^+ is somewhat unexpected since the experimental results on the centers with ^{15}N nuclei are indicative of the presence of a substantially anisotropic hfi. However, the numerical estimates of the spectra have confirmed the validity of the given interpretation. The NQR lines obsered in these spectra proved to correspond to the statistically most probable orientations of the hfi tensor axes perpendicular to the direction of the external magnetic field, for which orientation the compensation condition is satisfied. The data on the nqi parameters are of interest because they reflect the distribution of electron density around the nucleus. Unfortunately, we failed to establish a correspondence between the derived groups of hfi and nqi parameters.

The P700^{+}-^{14}N samples have been studied in (60). Due to the weakness of the echo signals observed, the stimulated ESE signal decay was recorded in (60) only at one value of τ = 200 ns. In the modulation spectrum at this τ there were two lines with maxima at 1.3 and 2.6 MHz. Later, using the data accumulation technique, the decays of the stimulated ESE signal were obtained at various τ and the presence of three broad lines with maxima at 1.05, 1.55 and 2.6 MHz (Fig. 27) was registered in the corresponding spectra.

The simplest way to interpret the spectra is, presently, to correlate the lines with the NQR frequencies with the corresponding parameters K = 0.69 MHz and η = 0.76, and to accept that the nqi parameters of P700^{+} and Chl a^{+} are different. However, a more careful look at the three prominent lines of the P700^{+} spectra allows one to notice that they have a complex structure. For

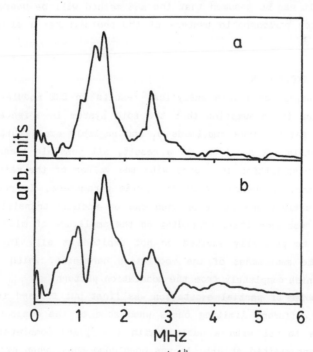

Fig. 27. Modulus Fourier spectra of P700^{+}-^{14}N stimulated ESE modulation at τ = 280 ns (a) and 350 ns (b).

example, the line with the maximum at 1.05 MHz has a shoulder stretching towards low frequencies down to 0.45 MHz, and the central line with the main maximum at 1.55 MHz has a side maximum at 1.3 MHz. The observed complication of the spectra may be related to the contribution to the low frequency region of lines arising from nuclei with smaller hfi constants, for which the effective magnetic field compensation condition does not hold. As a result, for these nuclei the narrow NQR lines will be broadened, thus decreasing the spectral resolution and complicating the structure.

Experiments with the $P865^+$-^{14}N centers have been performed in (56). In the AR spectra of the stimulated ESE modulation components at 0.9, 1.6 and 2.5 MHz were observed (56). In addition, a line at 0.5 MHz was detected, but taking into account the results of the Fourier analysis for $P700^+$ and the properties of the AR method it is probable that the appearance of this line is associated with the asymmetry of the 0.9 MHz low-frequency component. Not taking this 0.5 MHz line taken into account, the spectra of $P700^+$ and $P865^+$ are quite similar.

The data presented show that the use of the ESE method for the investigation of cation-radicals of the primary electron donors of bacterial and plant photosynthesis has considerably extended the possibility of EPR spectroscopy, allowing one to pursue the study of their structure on a qualitatively new level, based on a comparison between the hfi constants rather than the EPR linewidths. It can be assumed that the ESE method will be useful also for the study of other paramagnetic centers of the photosynthetic electron transport chains.

6. PARTIAL EXCITATION

The presently available analytical results in ESE modulation theory are obtained under the assumption that the spin system is affected by microwave pulses of a fairly large amplitude ν_1 and neglibly small duration t_p, which excite the whole EPR spectrum. As a result, all possible transitions are induced with probabilities in accord with the values of the matrix elements of the $\hat{S}_x$ operator, independently of the pulse parameters. In real experiments, however, situations are possible when one or another transition will be far enough from resonance that, depending on the amplitude of microwave field, it will either be partially excited or not excited at all. This will cause a change in the amplitudes of the modulation harmonics, which in the weak ν_1 limit may vanish completely from the modulation pattern.

The problem of partial excitation was first put forward in (13) and considered for different limiting cases when some of the components of the EPR spectrum due to hfi with a nucleus with $I = \frac{1}{2}$ are completely excited, and others are not excited at all. It was concluded that, when exciting only part of all possible spectrum components, a necessary condition for the appearance of ESE modulation is that the corresponding transitions should have a common energy level. Each level is connected with a set of modulation frequencies made up of the differences of the frequencies of the electronic transitions from this common level. The experimental verification of the regularities formulated in (13) were presented in (61,62). Although this approach allows one to explain the absence of some frequencies under the condition of partial excitation, it does not allow to quantitatively describe the degree of excitation of the transition that is at $\Delta\nu = \nu - \nu_o$ off resonance. For a long time the dominant opinion in the literature, based on qualitative estimates and experiments, was that at an amplitude of microwave pulses ν_1 those transitions will be efficiently excited that have their frequencies within the interval from $\nu_o - \nu_1$ to $\nu_o + \nu_1$ (under the condition $\nu_1 t_p \sim 1$). For instance, for the efficient excitation of the matrix protons the necessary ν_1 value must be equal to about 14 MHz. The example of the experimental observation of the proton modulation at $\nu_1 = 26.6$ MHz and of its absence at $\nu_1 = 8.4$ MHz is described in (63).

However, a number of new works have recently appeared in which further efforts have been made to study the effect of partial excitation on ESE experimentally and to describe it analytically. It was shown in (64) that, when studying the proton modulation in a frozen solution of Cu^{2+}, the complete excitation effect of the inhomogeneously broadened EPR lines that is realized for short powerful microwave pulses, remains for a small amplitude of the first pulse with a duration satisfying the condition $\nu_1 t_p < \frac{1}{4}$.

The effect of partial excitation on the stimulated ESE modulation was analyzed in (65). To this end a general expression for the stimulated ESE modulation, modified for the partial excitation case, was derived. The following conclusions were made using this expression. In the one-dimensional spectrum of stimulated ESE modulation (the spectrum obtained by Fourier transform of the envelope at fixed τ value) the frequencies of single nuclear transitions will be displayed, which change their amplitude under conditions of partial excitation. But in the two-dimensional spectrum (the spectrum obtained by Fourier integration of a set of 1D spectra over τ) of a partially excited system, additional peaks arise being combinations of three nuclear transition frequencies. Two of these frequencies belong to one electron spin manifold, and the third to another one. The theoretical conclusions obtained were confirmed in experiments with F-centers in KCl single crystals. In addition, a numerical calculation was performed in (65) demonstrating the effect of the value of ν_1 on the amplitude of several modulation harmonics at one orientation of the single crystal.

An analytical calculation of the primary ESE modulation of a spin system $S = \frac{1}{2}$, $I = \frac{1}{2}$ in the weak hfi limit was recently carried out in (66). The calculation yielded a rather cumbersome expression for the amplitudes of the ν_I,

$2\nu_I$ and T_{zz} modulation harmonics as a function of the microwave pulse parameters. Using this expression one may calculate the change of the amplitudes of
these harmonics in disordered samples. To verify the theoretical expressions
derived for the amplitudes of the ν_I and $2\nu_I$ harmonics, experiments were performed in (66) to study their dependence on the pulse parameters for proton
modulations observed in ESE of the stable nitroxide radical:

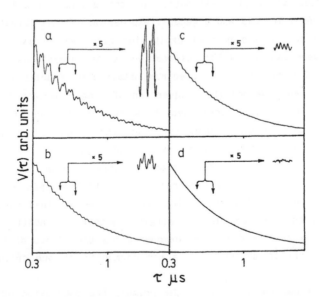

in frozen methanol glass. Fig. 28 shows the primary ESE modulation envelopes
recorded at different durations of the 90° and 180° pulses. The pulse durations
varied from 15 and 30 ns up to 90 and 180 ns, respectively. The microwave field
amplitude ν_1 varied correspondingly from 16.7 down to 2.8 MHz. It follows from
the figure that at short pulses, when ν_1 is maximal, the amplitude of the ν_I
harmonic is several times higher than that of the $2\nu_I$ harmonic. The T_{zz} harmonic is not observed in this experiment due to its low frequency and comparatively strong dependence on orientation. When increasing the duration of the

Fig. 28. Experimental decays of the primary ESE signal for different durations
of the 90° and 180° pulses: a) 15, 30 ns, b) 70, 140 ns, c) 80, 160 ns, d) 90,
180 ns. From (66).

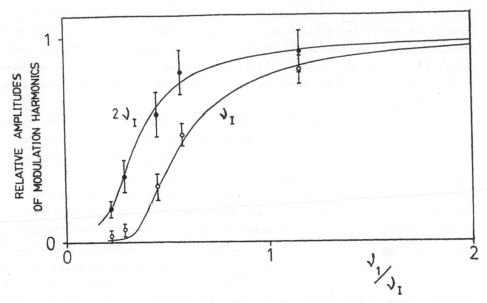

Fig. 29. Comparison of the experimental and theoretical dependences of the amplitude of the primary ESE modulation on the microwave field intensity for a 90°, 180° pulse sequence. The amplitude of every harmonic in the strong ν_1 limit is assumed to be unity. From (66).

pulses and decreasing ν_1, a decrease of the amplitude of all harmonics, and simultaneously a decrease of the relative amplitude of the ν_I harmonic is observed. For example, in Fig. 28b the amplitude of the ν_I harmonic is approximately equal to that of the $2\nu_I$ harmonic. Finally, the modulation vanishes completely (Fig. 28d).

Fig. 29 demonstrates the dependence of the amplitudes of the ν_I and $2\nu_I$ harmonics on ν_1, obtained from the analysis of the experimental envelopes. The appropriately normalized theoretical curves are also drawn in this figure. The theoretically predicted changes of the amplitudes of the ν_I and $2\nu_I$ harmonics agree well with those obtained experimentally.

7.   ESEEM IN SYSTEMS WITH HIGH g-FACTOR ANISOTROPY

In all the above-presented considerations of ESE results for disordered systems a model of spherical averaging was used. According to this model, in disordered samples all orientations of the paramagnetic center contribute to the modulation. There are, however, a number of systems with high anisotropy of the electronic g-factor that display EPR spectra, different parts of which correspond to different orientations of the paramagnetic center. Then, one can expect the appearance of an orientation dependence of the modulation spectra when exciting different parts of such an anisotropic EPR spectrum by microwave pulses. The observation of this orientation dependence provides a possibility to directly determine the hfi and nqi constants of magnetic nuclei in different orientations. As an example, we shall analyze the orientation-dependent modulation spectra of $VO^{2+}(acac)_2$ in frozen solutions (67). These systems were investigated also with the ENDOR method (68), which allows us to compare the results of both techniques.

The EPR spectrum of $VO^{2+}(acac)_2$ exhibits a pronounced g-factor anisotropy and regions corresponding to $g_{//}$ and $g_{\perp}$ may be distinguished. When exciting the EPR spectrum regions corresponding to either $g_{//}$ or $g_{\perp}$, those orientations of the complex having the V = 0 axis parallel or perpendicular to the direction of the external magnetic field contribute to the ESE modulation correspondingly. The ESE modulation of $VO^{2+}(acac)_2$ in toluene exhibits only the high-frequency proton Zeeman and double Zeeman harmonics, since protons are the only type of magnetic nuclei in this system. After adding CH_3OD to the solution, modulations with lower frequencies arise, induced by the interaction of the unpaired electron with the alcohol deuterons (Fig. 30a,b). Fig. 30c depicts the stimulated ESEEM spectrum of $VO^{2+}(acac)_2$ in toluene with admixture of CH_3OD, recorded when exciting the $g_{//}(-7/2)$ EPR spectrum component at B_0 = 282.3 mT. The modulation

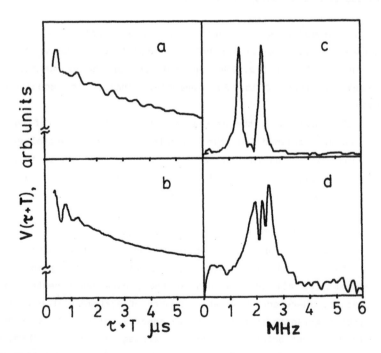

Fig. 30. Stimulated ESEEM of the adduct of $VO^{2+}(acac)_2$ with CH_3OD at τ = 200 ns under excitation by microwave pulses of the ESE spectrum components $g_{//}(-7/2)$ (a) and $g_{\perp}(3/2)$ (b) and the corresponding modulus Fourier spectra (c,d).

spectrum exhibits two lines at 1.42 and 2.25 MHz. The relative intensity of these lines changes with changing τ.

The Zeeman frequency of deuterons, ν_I, equals 1.845 MHz at B_0 = 282.3 mT. The two lines observed are situated symmetrically around ν_I and the splitting between them is 0.83 MHz. These two lines may be assigned to the transition frequencies $\nu_{\alpha(\beta)}^{//} = |\nu_I \pm A_{//}/2|$ of the deuteron in a CH_3OD molecule coordinated with $VO^{2+}(acac)_2$. Hence, $A_{//}$ = 0.83 MHz.

The stimulated ESEEM spectrum obtained when exciting the $g_\perp(3/2)$ component of the EPR spectrum at B_0 = 341.7 mT is shown in Fig. 30d. In contrast to the spectrum with $g_{//}$, it contains three components. The frequency of the middle component is ν_I = 2.23 MHz and its relative intensity decreases with decreasing the alcohol concentration. This component is induced by distant matrix deuterons. The relative intensity of the other two lines does not change and they may be assigned to the frequencies $\nu_{\alpha(\beta)} = |\nu_I \pm A_\perp/2|$. The splitting between these two lines gives $A_\perp$ = 0.6 MHz.

From the constants obtained for deuterons, one can calculate for protons: $A_{//}$ = 5.4 MHz, $A_\perp$ = 3.9 MHz. The corresponding values measured by ENDOR are $A_{//}$ = 6.0 MHz, $A_\perp$ = 3.12 MHz (68). It has been shown in (68) that these experimental values are approximately equal to the proton dipolar hfi calculated for a model in which one OH group is axially coordinated to the $VO^{2+}(acac)_2$. It follows from the comparison between the ESE and the ENDOR results that the difference between the $A_{//}$ values determined by these two methods is about 10 % and is probably governed by the accuracy of the determination of the line maxima. The difference between the $A_\perp$ values is somewhat larger, about 20 %. This difference may be attributed to the overlap of the hydroxyl proton signal with other proton signals in the ENDOR spectra and to a weak nqi effect in the ESE spectra.

The quadrupole splitting of the deuteron transition lines was inobservable in the ESEEM spectra, which is especially surprising for the spectrum for $g_{//}$, since this one corresponds to a definitely ordered system. To explain its absence, we note that it depends on the angle between the quadrupole tensor axis and the external magnetic field direction as $(e^2qQ)/8 \, (3 \cos^2\theta' - 1)$. Thus, this splitting approaches zero at θ' = 55°, which angle therefore probably gives the direction of the OD band upon coordination to the sixth ligand position of $VO^{2+}(acac)_2$. We did observe the additional quadrupole splitting in the ESEEM spectra when exciting the $g_{//}$ EPR spectrum component of the $VO^{2+}(acac)_2$ adduct with CH_3COOD acetic acid.

Analogous experiments were performed also with other aliphatic alcohols with the aim to elucidate the effect of the aliphatic chain structure on adduct formation. Modulation spectra identical to those of CH_3OD were recorded for CH_3CH_2OD, $CH_3CH_2CH_2OD$ and $(CH_3)_2CHOD$. Hence, these alcohols coordinate with

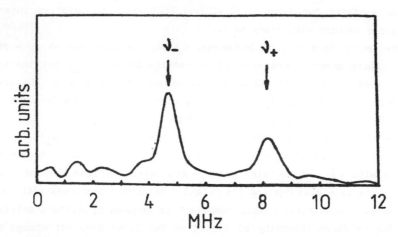

Fig. 31. Modulus Fourier spectrum of stimulated ESEEM of the adduct of $VO^{2+}(acac)_2$ with pyridine at τ = 280 ns under excitation of the $g_{//}(-5/2)$ EPR spectrum component.

$VO^{2+}(acac)_2$ and their alkyl chain geometry does not affect the position of the deuterium atom in the adduct and its interaction with the unpaired electron. In contrast, for $(CH_3)_3COD$ deuterium ESE modulation was not observed, which points to the absence of adduct formation, probably due to sterical factors.

When some pyridine is added to the $VO^{2+}(acac)_2$ solution in toluene, one also observes ESE modulation with frequencies lower than those of protons. These low frequencies are induced by the nitrogen nucleus of pyridine (69).

Fig. 31 demonstrates the spectrum of the stimulated ESE modulation of the adduct of $VO^{2+}(acac)_2$ with pyridine recorded when exciting the $g_{//}(-5/2)$ component of the EPR spectrum at B_0= 303.8 mT. Two intense lines are visible in the spectrum. Analogous two-component spectra were obtained also for other EPR spectrum components for $g_{//}$ and $g_\perp$. To figure out the correspondence between these lines and the ^{14}N nuclear transitions, the dependence of their maxima on the external magnetic field intensity was studied (the change of the field here means the excitation of another EPR spectrum component). It was found that the change of the field intensity results in shifts of the maxima of these lines approximately twice as big as the change of the ^{14}N Zeeman frequency. This means that these two lines in the modulation spectrum correspond to double quantum transitions of ^{14}N for different electron spin manifolds. From the frequencies observed it was calculated in (69), using [32]: $A_{//}$ = 6.0 MHz, $A_\perp$ = 5.6 MHz, which gives an idea of the value of the isotropic and anisotropic parts of the ^{14}N hfi in the adduct.

The $VO^{2+}(acac)_2$ adducts with pyridine were studied also by the ENDOR method in (68). In the $g_{//}$ spectra four transitions were observed, which were ascribed to one-quantum transitions. Using this interpretation, the values A_z = 6.5 MHz, Q_z = 0.85 MHz were obtained. The $g_\perp$ spectra in (68) were of poor quality and no parameters were estimated for this orientation. Unfortunately, the field dependence of the ENDOR spectra was not studied in (68).

The comparison of both methods shows that, although they give close estimates of the hfi constants, they display different types of ^{14}N transitions in the spectra. This result is confirmed by the work in (70), where the ^{14}N transitions in the Ni(II) bis(N,N-diethyldithiocarbamate) with 0.3 % Cu were observed by ENDOR and ESE methods and numerical calculations of the spectra were carried out.

In conclusion we note that the experimental results presented above show a significant orientation dependence of ESEEM spectra in systems with high g-factor anisotropy. It provides a possibility to measure hfi and nqi constants directly for different orientations in random samples.

ACKNOWLEDGEMENT

We thank Prof. A.J. Hoff for critically reading the manuscript and for useful comments, and Mrs. T. Veldhuyzen-van der Meer for help in the preparation of the manuscript. SD thanks the Netherlands Organization for Scientific Research (NWO) for financial support and the Department of Biophysics, University of Leiden, for its hospitality.

REFERENCES

1 W.B. Mims, in: S. Geschwindt (Ed.), Electron Paramagnetic Resonance, Plenum Press, New York, 1972, pp. 263-351.
2 K.M. Salikhov, A.G. Semenov and Yu.D. Tsvetkov, Electron Spin Echo and Its Applications, Science, Novosibirsk, 1976.
3 L. Kevan: in: L.Kevan and R.N. Schwartz (Eds.), Time Domain Electron Spin Resonance, Wiley-Interscience, New York, 1979, pp. 279-341.
4 W.B. Mims and J. Peisach, in: L.J. Berliner and J. Reuben (Eds.), Biological Magnetic Resonance, Vol. 3, Plenum Press, New York, 1981, pp. 213-263.
5 J.R. Norris, M.C. Thurnauer and M.K. Bowman, in: J.H. Lawrence, J.M. Gofman and T.L. Hayes (Eds.), Advances in Biological and Medical Physics, Vol. 17, Academic Press, New York, 1980, pp. 365-416.
6 W.B. Mims and J. Peisach, in: R.G. Shulman (Ed.), Biological Applications of Magnetic Resonance, Academic Press, New York, 1979, pp. 221-269.
7 P.A. Narayana and L.Kevan, Magn. Res. Rev., 1 (1983) 234-274.
8 S.A. Dikanov and Yu.D. Tsvetkov, Zh. Strukt. Khim., 26, N 5 (1985) 136-167 (J. Struct. Chem. 26 (1985) 766-803).
9 Yu.D. Tsvetkov and S.A. Dikanov, in: H. Sigel (Ed.), Metal Ions in Biological Systems, Vol. 22, M. Dekker, New York, 1987, pp. 207-263.
10 W.B. Mims, Phys. Rev. B, 5 (1972) 2409-2419.
11 W.B. Mims, Phys. Rev. B, 6 (1972) 3543-3545.
12 L.G. Rowan, E.L. Hahn and W.B. Mims, Phys. Rev. A, 137(1965) 61-71.
13 G.M. Zhidomirov and K.M. Salikhov, Teor.Eksp. Khim., 4 (1968) 514-519 (Theor. Exp. Chem., 4 (1968) 332-334).
14 S.A. Dikanov, V.F. Yudanov and Yu.D. Tsvetkov, J. Magn. Reson., 34 (1979) 631-645.

116

15 S.A. Dikanov, A.A. Shubin and V.N. Parmon, J. Magn. Reson., 42 (1981) 474-487.

16 P.A. Narayana, M.K. Bowman, L. Kevan, V.F. Yudanov and Yu.D. Tsvetkov, J. Chem. Phys., 63 (1975) 3365-3371.

17 A.V. Astashkin, S.A. Dikanov and Yu.D. Tsvetkov, Chem. Phys. Lett., 122 (1985) 259-263.

18 A. de Groot, R. Evelo and A.J. Hoff, J. Magn. Reson., 66 (1986) 331-343.

19 C.P. Lin, M.K. Bowman and J.R. Norris, J. Chem. Phys., 85 (1986) 52-62.

20 A.V. Astashkin, S.A. Dikanov and Yu.D. Tsvetkov, Chem. Phys. Lett., 130 (1986) 337-341.

21 A.V. Astashkin, S.A. Dikanov and Yu.D. Tsvetkov, Chem. Phys. Lett., 136 (1987) 204-208.

22 A.A. Shubin and S.A. Dikanov, J. Magn. Reson., 52 (1983) 1-12.

23 G.K. Semin, T.A. Babushkina and G.G. Yakobson, Applications of Nuclear Quadrupole Resonance in Chemistry, Chemistry, Leningrad branch, 1972.

24 J. Kowalewski and H. Kovacs, Z. Phys. Chem., Neue Folge, 149 (1986) 49-63.

25 A.A. Shubin and S.A. Dikanov, J. Magn. Reson., 64 (1985) 185-193.

26 S.A. Dikanov, Yu.D. Tsvetkov, A.V. Astashkin and A.A. Shubin, J. Chem. Phys., 79 (1983) 5785-5795.

27 M. Heming, M. Narayana and L. Kevan, J. Chem. Phys., 83 (1985) 1478-1484.

28 T. Ichikawa, J. Chem. Phys., 83 (1985) 3790-3797.

29 M. Romanelly, M. Narayana and L. Kevan, J. Chem. Phys., 83 (1985) 4395-4399.

30 M. Romanelly and L. Kevan, J. Chem. Phys., 86 (1987) 4369-4374.

31 M. Iwasaki and K. Toriyama, J. Chem. Phys., 82 (1985) 5415-5423.

32 M. Iwasaki, K. Toriyama and K. Nunome, J. Chem. Phys., 86 (1987) 5971-5982.

33 M. Iwasaki and K. Toriyama, in: J.A. Weil (Ed.), Electronic Magnetic Resonance of the Solid State, CSC, Ottawa, 1987, pp. 545-570.

34 S.A. Dikanov, Yu.D. Tsvetkov and A.V. Astashkin, Chem. Phys. Lett., 91 (1982) 515-519.

35 S.A. Dikanov, A.V. Astashkin, Yu.D. Tsvetkov, M.G. Goldfeld and A.G. Chetverikov, FEBS Lett., 224 (1987) 75-78.

36 T. Ichikawa, L. Kevan, M.K. Bowman, S.A. Dikanov and Yu.D. Tsvetkov, J. Chem. Phys., 71 (1979) 1167-1174.

37 L. Kevan, J. Phys. Chem., 85 (1981) 1628-1636.

38 L. Kevan, in: J.A. Weil (Ed.), Electronic Magnetic Resonance of the Solid State, CSC, Ottawa, 1987, pp. 281-294.

39 J. Tang, C.P. Lin and J.R. Norris, J. Chem. Phys., 83 (1985) 4917-4919.

40 C.P. Lin, M.K. Bowman and J.R. Norris, J. Chem. Phys., 85 (1986) 56-62.

41 C.P. Lin and J.R. Norris, FEBS Lett., 197 (1986) 281-284.

42 S.A. Dikanov, A.V. Astashkin and Yu.D. Tsvetkov, Chem. Phys. Lett., 144 (1988) 251-257.

43 A.V. Astashkin, S.A. Dikanov and Yu.D. Tsvetkov, Chem. Phys. Lett., 144 (1988) 258-26 .

44 I.A. Safin and D.Ya. Osokin, Nuclear Quadrupole Resonance in Compounds of Nitrogen, Science, Moscow, 1977.

45 S.A. Dikanov, Yu.D. Tsvetkov, M.K. Bowman and A.V. Astashkin, Chem. Phys. Lett., 90 (1982) 149-153.

46 A.V. Astashkin, S.A. Dikanov and Yu.D. Tsvetkov, Zh. Strukt. Khim., 25, N 1 (1984) 53-64.

47 E.J. Reijerse and C.P. Keijzers, J. Magn. Reson., 71 (1987) 83-96.

48 H.L. Flanagan and D.J. Singel, J. Chem. Phys., 87 (1987) 5606-5616.

49 H.L. Flanagan, G.J. Gerfen and D.J. Singel, J. Chem. Phys., 88 (1988) 20-24.

50 H.L. Flanagan and D.J. Singel, Chem. Phys. Lett., 137 (1987) 391-397.

51 S.A. Dikanov, A.V. Astashkin and Yu.D. Tsvetkov, Zh. Strukt. Khim., 25, N 2 (1984) 35-39 (J. Struct. Chem. 25 (1984) 200-203).

52 S.V. Khangulov, V.V. Barynin, V.R. Melik-Adamian et al., Bioorgan. Khim., 12 (1986) 741-748.

53 S.V. Khangulov, N.V. Voevodskaya, V.V. Barynin, A.I. Grebenko, V.R. Melik-

Adamian, Biofizika, 32 (1987) 960-966.

54   W. Lubitz, R.A. Isaacson, E.C. Abrech and G. Feher, Proc. Natl. Acad. Sci. USA, 81 (1984) 7792-7796.

55   A.J. Hoff, A. de Groot, S.A. Dikanov, A.V. Astashkin and Yu.D. Tsvetkov, Chem. Phys. Lett., 83 (1985) 40-47.

56   A. de Groot, A.J. Hoff, R. de Beer and H. Scheer, Chem. Phys. Lett., 113 (1985) 286-290.

57   M.K. Bowman, J.R. Norris, M.C. Thurnauer, J. Warden, S.A. Dikanov and Yu.D. Tsvetkov, Chem. Phys. Lett., 55 (1978) 570-574.

58   A.V. Astashkin, S.A. Dikanov, Yu.D. Tsvetkov and M.G. Goldfeld, Chem. Phys. Lett., 134 (1987) 438-443.

59   M. Huber, F. Lendzian, W. Lubitz, E. Tränkle, K. Möbius and M.R. Wasielewski, Chem. Phys. Lett., 132 (1986) 467-473.

60   S.A. Dikanov, A.V. Astashkin, Yu.D. Tsvetkov and M.G. Goldfeld, Chem. Phys. Lett., 101 (1983) 206-209.

61   V.F. Yudanov, A.M. Raitsimring and Yu.D. Tsvetkov, Teor.Eksp. Khim., 4 (1968) 520-526.

62   V.F. Yudanov, V.P. Soldatov and Yu.D. Tsvetkov, Zh. Strukt. Khim., 15 (1974) 600-606 (J. Struct. Chem., 15 (1974) 510-515).

63   L. Kevan, M.K. Bowman, P.A. Narayana, R.K. Boeckman, V.F. Yudanov and Yu.D. Tsvetkov, J. Chem. Phys., 63 (1975) 409-416.

64   L. Braunschweiler, A. Schweiger, J.M. Fauth and R.R. Ernst, J. Magn. Reson., 64 (1985) 100-106.

65   H. Barkhuijsen, R. de Beer, B.J. Pronk and D. van Ormondt, J. Magn. Reson., 61 (1985) 284-293.

66   A.V. Astashkin, S.A. Dikanov, V.V. Kurshev and Yu.D. Tsvetkov, Chem. Phys. Lett., 136 (1987) 335-341.

67   S.A. Dikanov, R.G. Evelo, A.J. Hoff and A.M. Tyryskhin, Chem. Phys. Lett., in preparation.

68   B. Kirste and H.J. van Willigen, J. Phys. Chem., 86 (1982) 4743-4749.

69   A.V. Astashkin, S.A. Dikanov and Yu.D. Tsvetkov, Zh. Strukt. Khim., 26, N 3 (1985) 53-58.

70   E.J. Reijerse, N.A.J.M. van Aerle, C.P. Keijzers, R. Böttcher, R. Kirmse and J. Stach, J. Magn. Reson., 67 (1986) 114-124.

'HAPTER 3

MULTIFREQUENCY ESEEM: PERSPECTIVES AND APPLICATIONS

DAVID J. SINGEL

1. INTRODUCTION

A broad objective of the application of ESEEM to the study of paramagnetic centers in orientationally disordered solids is to ascertain, as fully as possible, the values of the various spin couplings - the magnetic interactions - of the nuclei in the nearest environment of the center. Properly interpreted, the spin couplings provide a wealth of detailed information about the chemical, electronic, and molecular structure of the system under investigation. This lucrative reward has inspired a variety of enterprises designed to facilitate the detailed quantitaive analysis of ESEEM patterns and their corresponding spectra.

In considering experimental strategies through which this goal may be attained, it is instructive to recall the analogous challenge faced in NMR spectroscopy of diamagnetic solids. This challenge has been met in NMR (1) through the realization that the properties (the spin Hamiltonian parameters) of a complex spin system can be *actively manipulated* in such a way as to generate simplified spectra with enhanced sensitivity, resolution, and spectral "clarity" - simplified connections between the spectral frequencies and the values of the spin coupling constants. Ingenious selective averaging techniques (1) comprise the operational means of manipulation in solid-state NMR. Such methods are unfortunately not easily transferable to electron-nuclear spin systems, because of their substantially larger frequency dispersions and relaxation rates.

If attention is focussed, not on the *internal* spin interactions, however, but on their interactions with the *external field* a simple and readily applicable means of manipulating the properties of an electron-nuclear spin system becomes apparent: *variation of the external field strength*. In this footnote to the chapter by Dikanov and Astashkin (we retain their notation throughout), we discuss the manner in which the parametric variation and judicious selection of spectrometer frequency - and, concomitantly, external field strength - facilitate the analysis of ESEEM spectra. We also discuss recent applications of "multifrequency" ESEEM.

2. IMPACT OF EXTERNAL FIELD STRENGTH ON ESEEM SPECTRA

As indicated in Fig. 1, we identify three, salient properties of ESEEM spectral features that are strongly, and characteristically, influenced by the strength of the external field. Manipulation of these properties can have a variety of favorable consequences, a number of which are also noted in the Fig 1. In this section, we describe the behavior of each of these properties with variation of external field strength, and the manner in which this behavior can be exploited in analysis of ESEEM patterns.

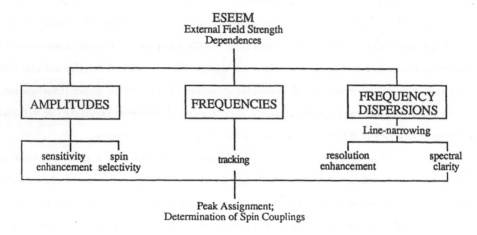

Fig. 1. Relationships between ESEEM spectral properties influenced by variation of external field strength and the impact of this influence on ESEEM spectroscopy.

2.1 Frequency Dispersion

Arguably the important feature of ESEEM spectra that is subject to manipulation by variation of the external field is the frequency dispersion - the spread of the spectral peaks arising from anisotropic spin couplings. Line-narrowing, the removal of this dispersion, can be of great importance in enhancement of spectral resolution, and in mitigating dead-time artifacts. Of even greater importance, however, are its consequences regarding spectral clarity. Through line-narrowing, a more direct link can be forged between the ESEEM frequency components and the values of the spin couplings that we seek to measure. This strategy has decided advantanges over efforts to extract the parameter values by fitting them to complex powder line-shapes or modulation patterns. The *generally* obscure relationship between the parameter values and the experimental data raises questions about the uniqueness of the fitted parameters - particularly since the number of parameters is (potentially) very large. Moreover, failure to specifically include every magnetic interaction in the fitting procedure corrupts, in obscure ways, those values that are fitted. Line-narrowing strategies provide a means to factorize ESEEM spectra, according the influence of each parameter, so that their values may be reliably measured directly from the ESEEM frequencies.

This principle, albeit in passive form, has underlain much of the quantitative analysis of monofrequency ESEEM spectra: the measurement of spin couplings has generally relied on the fortuitous occurrence of clarified, line-narrowed spectral features. The numerous studies (2,3) of ^{14}N nuclei that sustain a nearly isotropic hyperfine interaction, but a substantial quadrupole coupling, provide an outstanding example of this situation. When the spectrometer frequency is selected such that the nuclear Zeeman interaction matches the hyperfine interaction, then, in one of the nuclear spin manifolds (α), the two interactions cancel: the magnetic field, and thus the dispersion associated with its orientation within the quadrupole principal axis system, is removed; narrow lines appear at the pure quadrupole frequencies. From the frequencies of these peaks observed at "exact cancellation" the quadrupole coupling constants can be directly evaluated; the hyperfine coupling constant can be determined from the value of the external field strength at which the line-narrrowing occurs.

An analagous, though less obvious, situation occurs for spin 1/2 nuclei with axially symmetric hyperfine interaction matrices(4). When the nuclear Lamor frequency assumes the particular value $T_\perp/4 + a/2$, the dispersion associated with the secular and non-secular portions of the hyperfine interaction precisely cancel in the α manifold, yielding a dispersionless peak, as illustrated in Fig. 2. The frequency of this peak is $3T_\perp/4$, thus it provides a direct measure of this coupling constant. More subtle line-narrowing phenomena involve the constriction, rather than the removal, of frequency dispersions in conjunction with a skewing of intensity to a restricted region of the line - through the joint effects of ESEEM amplitude anisotropy and the usual orientation weighting factors. The β manifold peak in Fig. 2 shows line-narrowing of this type (4). A further example is provided by the combination peaks of spin 1/2 nuclei (5). The $\nu_\sigma = \nu_\alpha + \nu_\beta$ peak is narrowed by the approximate cancellation of the dispersions of the constituent fundamental peaks when the Zeeman interaction is large compared to the hyperfine interaction. The peak position of this combination is also influenced by amplitude weighting factors. The frequency of this narrowed peak gives a *selective* measure of $T_\perp$ (5). Another example of narrow lines that can appear in ESEEM spectra are the double "quantum" peaks associated with I=1 systems with isotropic hyperfine interactions (2,3).

Table I: Studied Dependences of ESEEM Linewidths and Frequencies on External Field Strength

Orientation Selectivity	Nuclear Spin	Restrictions	ESEEM Peak	External Field for Narrowing	Frequency
None	1/2	axial hyperfine interaction $(a > 0;\ T_\perp > 0)$	ν_α ν_β ν_σ	$4\nu_I = 2a + T_\perp$ $2\nu_I = a - T_\perp$ $\|\nu_I \pm \|\langle \underline{S}\rangle_{\alpha,\beta}\|A\| \geq \|\langle \underline{S}\rangle\|B\|$	$3T_\perp/4$ $a - T_\perp$ $2\nu_I + T_\perp^2(\|\langle \underline{S}\rangle_{\alpha,\beta}\|/\nu_I)$
	1	isotropic hyperfine interaction $(a > 0)$	ν_α $\nu_{\alpha(dq)}$ $\nu_{\beta(dq)}$	$\nu_I \approx \|\langle \underline{S}\rangle_\alpha\|a$ $\nu_I \not\approx \|\langle \underline{S}\rangle_\alpha\|a$ unrestricted	$2K\eta;\ K(3-\eta);\ K(3+\eta)$ $2[(\nu_I + \xi\|\langle \underline{S}\rangle_\alpha\|a)^2 + K^2(3+\eta^2)]^{1/2}$ $2[(\nu_I + \xi\|\langle \underline{S}\rangle_\beta\|a)^2 + K^2(3+\eta^2)]^{1/2}$
Complete	1/2	$\langle \underline{S}\rangle_{\alpha,\beta} \parallel \underline{B_0}$	ν_α ν_β	unrestricted unrestricted	$[(\nu_I + \xi\|\langle \underline{S}\rangle_\alpha\|A)^2 + \|\langle \underline{S}\rangle_\alpha\|^2 B^2]^{1/2}$ $[(\nu_I + \xi\|\langle \underline{S}\rangle_\beta\|A)^2 + \|\langle \underline{S}\rangle_\beta\|^2 B^2]^{1/2}$

Symbols are defined in chapter X, with the exception of ξ, which indicates the alignment of $\|\langle \underline{S}\rangle\|$ along or against the external field ($\xi = +1$ for the β manifold and -1 for the α manifold.)

From this discussion of line-narrowing phenomena, it is evident that sharp peaks can appear in ESEEM spectra under numerous circumstances. With the exception of the last example, however, specific spectrometer frequencies or frequency ranges are required for the onset of narrowing. The regular appearance of narrowed peaks in monofrequency ESEEM spectra illustrates the fundamental importance of line-narrowing in ESEEM spectrosocpy: quantitatively intelligible spectra are precisely those in which line-narrowing has occurred. Demonstrating that the specific conditions for a particular type of line-narrowing have in fact been met, however, lies outside the scope of monofrequency ESEEM. The multifrequency approach provides a means by which the satisfying of line-narrowing conditions can be verifiably demonstrated, and the origin of the peak accordingly determined.

Another, entirely separate means by which ESEEM peaks can be narrowed involves orientation-selection (6). An inherent consequence of the selective-excitation that is typical in ESE spectroscopy (Rabi frequency $\gg$ EPR dispersion), is the selection of a limited range of orientations within the moleulcar frame, whenever the EPR line is spread by anisotropic interactions. When orientation-selection is "complete," that is , when a single orientation is selected, the line-broadening effects of powder averaging are wholly circumvented. When only partial orientation-selection is possible, as in systems with an axially symmetric electron g-matrix, it may still be possible to remove residual frequency dispersions, by combining the orientation-selection with multifrequency line-narrowing. We have recently found (7), for example, that, the "in-plane" hyperfine coupling constants associated with a rhombic hyperfine interaction matrix whose principal axes coincide with those of the (axial) g-matrix, can be determined in a manner analogous to that described in Ref. 4.

2.2 Frequencies

As we have indicated in the preceding section, multifrequency ESEEM provides a valuable tool for recognizing a particular type of spin system through the characteristic behavior of its ESEEM line-widths as a function of external field strength. The spin system may thus be identified, and its spin coupling constants directly measured. The dependence of the ESEEM frequencies themselves on field strength is similarly useful, particularly for those peaks which remain narrow over an appreciable range of field strengths: any peak generated in conjunction with complete orientation-selection; the double quantum peaks in $I=1$ systems with isotropic hyperfine interactions; and the ν_σ peak in $I=1/2$ systems with Zeeman interactions appreciably greater than hyperfine interactions. Through multifrequency ESEEM. the frequencies of these peaks may readily be "tracked" as a function of external field strength. Their functional dependence on field strength provides further detailed information about the type of spin system and thus aids in the peak assignments. Moreover, the spin couplings can be precisely measured through the analysis of these tracks (8-10). We discuss, in Sec. 3, two specific examples of this tracking technique, for $I=1/2$ and $I=1$ nuclei. We present a summary of the field-dependent characteristics of ESEEM peaks thus far discussed, in Table 1.

2.3 Amplitudes

From the time that ESEEM was first observed and theoretically explained (11), it has been realized that the appearance of nuclear modulations in an electron spin echo decay requires the nuclear spin states of the α manifold to be "scrambled" relative to those of the β manifold, and that this scrambling depends sensitively on the strength of the external field. Mims had presented, already in 1972, a simulation which shows the effect of a change in spectrometer frequency on the amplitudes of nuclear modulations (12). Further work has made it clear that "matching" the magnitudes of the nuclear Zeeman and electron-nuclear hyperfine interactions leads to the maximum modulation amplitudes theoretically possible in $I=1/2$ systems (4), and to the maximum amplitudes possible for a given quadrupole coupling

124

Fig. 2. Simulated multifrequency ES-
EEM spectra (fundamental components only) for a spin $1/2$ system
with axially symmetric hyperfine coupling matrix (after Ref. 4). The parameter values employed in the simulation are: $T_\perp = 0.2$ MHz; $a = 1.0$
MHz; Larmor frequencies (ν_n) 0.3250,
0.3625, 0.4010, 0.4375, 0.4750, 0.5125,
0.5490, 0.5875, 0.6250, 0.6625, 0.6990,
0.7375, and 0.7750 MHz. The nar-
rowed ν_α and ν_β peaks (at $\nu_n = 0.5490$
and 0.3990, respectively) are trun-
cated. Horizontal dashed lines mark
values of ν_n - $(a - T_\perp)/2$, $(2a + T_\perp)/4$, and $(a + 2T_\perp)/2$ - that lead
to line-narrowing as noted in Table
1. The angled, dashed lines mark
the associated frequencies: $a - T_\perp$;
$3T_\perp/4$; and $a + 2T_\perp$.

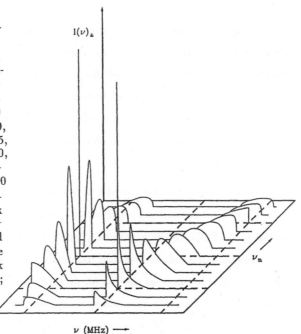

ν (MHz) $\longrightarrow$

I=1 systems (3). As deviations from these matching conditions occur, the modulation
amplitudes rapidly diminish. In monofrequency ESEEM, the very observation of nuclear
modulations (let alone their analysis) is contingent on the chance existence of an appropriate
balance of interactions, whereas in multifrequency ESEEM this balance can (in principle)
be deliberately struck. Multifrequency ESEEM thus provides a means toward sensitivity
enhancement. Moreover, in a spin system comprised of several nuclei with distinct hyperfine
couplings, the external field could chosen to selectively accentuate the contributions of some,
but attenuate the contributions of other spins to the modulation pattern. This hyperfine
coupling-based method of spin-filtration is of obvious utility for the simplification of complex
modulation patterns. Finally, it may be possible to substantiate a tentative assignment of
ESEEM peaks on the basis of the variation of their amplitudes with field strength. It may
also prove possible to determine the values of magnetic interactions through the analysis
of this behavior. This strategy would be particularly useful for systems, such as solvent
complexes of paramagnetic species, in which the the hyperfine interactions are typically too
small to give rise to resolvable splittings in their ESEEM spectra. We have also shown that
the *anisotropy* of ESEEM amplitudes is valuable source of information about spin couplings
(3). This information can be tapped through orientation-selection experiments in which the
ESEEM amplitudes are monitored (13).

3. EXEMPLARY APPLICATIONS

3.1 Multifrequency ESEEM of DPPH: ^{14}N Near Exact Cancellation

Aspects of our study (14) of the nitro groups in the free-radical DPPH (diphenylpicryl-hydrazyl) are presented in the chapter by Dikanov and Astashkin. We mention the study again, however, because the multifrequency ESEEM patterns and associated spectra of DPPH - shown in Figs. 3 and 4 - provide a comprehensive illustration of the attributes of multifrequency ESEEM that were discussed in the previous section. Moreover, they show how these various properties combine to create a spectroscopic signature which reliably reveals the "authorship" of the spectral peaks. The two-pulse ESEEM patterns in Fig. 3 exhibit a systematic reduction in modulation depth as the EPR frequency is varied from 4.45 to 9.10 GHz. The spectra shown in Fig. 4 (from three-pulse ESEEM patterns) additionally reveal systematic changes in the frequencies and in the widths of the ESEEM peaks as the microwave frequency is varied. The spectral properties change rapidly at the highest frequency surveyed, but are approximately stationary at the lower EPR frequencies.

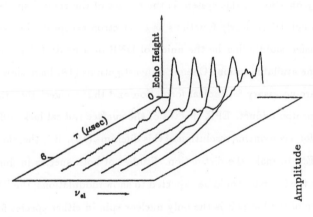

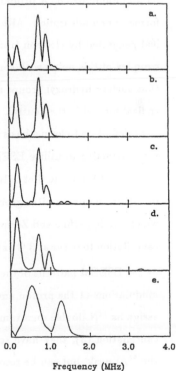

Fig. 3 (above). Two-pulse ESEEM patterns of 10^{-2} M DPPH in toluene at 77 K. Electron spin excitation frequencies (ν_{el}) from left to right: 4.45; 4.94; 5.47; 6.68; and 9.10 GHz.

Fig. 4 (right). Three-pulse ESEEM spectra of 10^{-2} M DPPH in toluene. Excitation frequency and delay (τ) between the first and second microwave pulses: a. 4.45 GHz ($\tau = 1.020$ μsec); b. 4.94 GHz ($\tau = 1.140$ μsec); c. 5.47 GHz ($\tau = 0.780$ μsec); d. 6.68 GHz ($\tau = 1.140$ μsec); e. 9.10 GHz ($\tau = 0.650$ μsec).

This total behavior constitutes the signature of exact cancellation in I=1 spin systems with approximately isotropic hyperfine interactions: deep modulations that give rise to narrrow lines at the quadrupolar frequencies; the stationarity of these properties for deviations from exact cancellation that are smaller than 2K/3; and a reduction of modulation depths, an increase in line-widths, and a shift of the observed frequencies as the deviation exceeds K. Detailed analysis of the ESEEM spectra yields precise values for the hyperfine and quadrupole coupling constants. These values correspond extremely well to those obtained in NMR (15) and neutron diffraction (16) studies of DPPH, and in pure NQR studies of substituted nitrobenzenes (18).

3.2 *m*-Dinitrobenzene on Aluminum Oxide: [14]N Far from Exact Cancellation

In ongoing work (9), we are applying ESEEM, as a site-specific probe of surface-adsorbate interactions, in the study of structural and dynamical aspects of the interfacial electron transfer which occurs upon adsorption of electron donor-acceptor molecules onto dispersed metal oxides. We have initiated this work with the study of DNB (*m*-dinitrobenzene) on aluminum oxide. A primary question in this system is the nature of the radical species formed upon adsorption. Although DNB clearly functions as an electron acceptor, the radical generated by electron transfer could either be the anion of DNB or a neutral, ion-pair complex of this anion with some available cation. Chemical investigations (18) have shown that surface hydroxyl groups are necessary for radical formation and thus suggest that the radical formed is the neutral, proton adduct. EPR spectra of the surface radical lack sufficient definition to form a basis for a convincing resolution of this dilemma (19). It is therefore very attractive to utilize ESEEM to make the distinction. Studies of both species in fluid solution (20) indicate that only the ion-pair would be expected to show modulations; the ^{14}N in the uncomplexed nitro group of the ion-pair is the only nuclear spin in either species for which the hyperfine and Zeeman interactions can be approximately counterbalanced (exact cancellation to occur at 3 GHz) at EPR frequncies in the range spanned by our spectrometer.

ESEEM patterns of the radical, registered at EPR frequencies from 6 - 11 GHz, show modulations at the proton and [27]Al Larmor frequencies, as well as two lower ones that we assign as ^{14}N double quantum frequencies. As expected for a single nucleus far from exact cancellation (3), they have scant modulation amplitude. They are partially obscured by the [27]Al peak, but can be resolved in three-pulse ESEEM by judicious selection of the time between the first pair of pulses (suppression-effect) (11,12). In the process of seeking optimal suppression, we observed that the low-frequency peaks are linked exclusively to each other

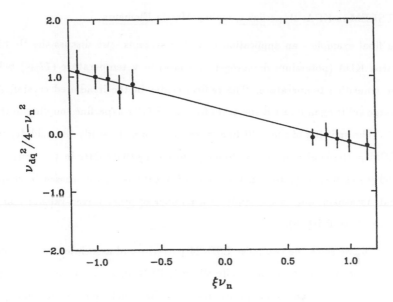

Fig. 5. Frequency-tracking of ^{14}N double quantum peaks of the radical produced by adsorption of DNB on dispersed aluminum oxide. ESEEM results obtained at spectrometer frequencies of 6.48, 7.37, 8.20, 9.08, and 10.34 GHz are included in the plot. The plotted function of observed ESEEM and nuclear Larmor frequencies (ν and ν_n) is linear in field strength for ^{14}N double quantum peaks. The best linear fit to the data (parameter values noted in the text) is also plotted.

in their suppression behavior; this observation further supports their assignment (21). From the dependence of the ESEEM frequencies of these peaks on external field strength, shown in Fig. 5, we conclude that the assignment is indeed correct and, moreover, evaluate the coupling constants:

$|a| = 0.57$ (0.02) MHz; $K^2(3+\eta^2) = 0.32$ (0.03) MHz^2.

The quadrupole coupling is typical of a nitrobenzene (17); the hyperfine coupling is in excellent agreement with the value observed for the proton adduct of the DNB radical anion in fluid solution (18). As a check, we have registered the multifrequency ESEEM spectra of frozen solution of the proton complex of of the DNB radical anion, prepared by photolysis of a solution of DNB (18). Again we observe low-frequency modulation components and assign them as ^{14}N double quantum frequencies. By multifrequency tracking, we derive coupling constants that correspond extremely well to those of the surface-bound radical:

$|a| = 0.58$ (0.05) MHz; $K^2(3+\eta^2) = 0.32$ (0.06) MHz^2,

and thus confirm the identification of the adsorbed radical as the ion-pair species.

3.3 ^{1}H ESEEM in KDA: Hyperfine Coupling Matrix Elements

As a final example - an application to I=1/2 systems - we discuss the ^{1}H ESEEM of γ-irradiated KDA (potassium dihydrogen arsenate) at a temperature (77 K) below its ferroelectric transition temperature. The radical center in the irradiated crystal, AsO_4^{4-}, exhibits a large splitting in its EPR spectrum from a 2.8 GHz hyperfine coupling to the ^{75}As nucleus (22). The splitting of the EPR line makes it possible to study the field dependence of the ESEEM patterns at a *single spectrometer frequency* (9.09 GHz in the present case). This possiblity has been exploited in the study of T-catalase, as discussed in chapter by Dikanov and Astashkin, and in the study of a number of other mixed-valence manganese compounds by R. D. Britt (8).

In Fig. 6, we show the two-pulse ESEEM patterns and associated spectra obtained by excitation of each of the four (nominally allowed) EPR lines - with the field orientation selected to be parallel to the unique principal axis of the axially symmetric ^{75}As hyperfine interaction matrix. In addition to the low-frequency peaks arising from coupling to ^{39}K nuclei (23), we observe peaks assignable, on the basis of prior ENDOR studies,[22] to protons attached (proximate) and hydrogen-bonded (distant) to the AsO_4^{4-} center. The solid and dotted lines superimposed on the spectra track the field dependence of the proximate and distant proton ESEEM peaks, respectively. The relationship between ESEEM frequency and field strength is decidedly non-linear; the non-secular portion of the hyperfine interaction contributes substantially to the observed frequency. This situation is by no means accidental. If the nuclear Larmor frequency were substantially greater than the hyperfine coupling, and a first-order treatment of the ESEEM frequencies were justified, then the modulaion amplitudes would necessarily be very small, precisely because of this mismatch between the Zeeman and hyperfine interactions (4). The variation in amplitude of the proton ESEEM peaks in Fig. 6 is indeed striking. The peaks arising from the proximate protons reach maximum amplitude when the field is 318.6 mT, as clearly evidenced in the low-frequency peak (its high-frequency partner suffers a loss of intensity because of incomplete excitation). When the lowest-field EPR line is excited, however, the contribution of the proximate protons to the modulation pattern is quenched. For the more weakly-coupled, distant protons, the reverse is true: the hyperfine and Zeeman interactions are nearly matched when the low field EPR line is excited; the ESEEM spectrum so obtained is clearly dominated by the fundamental and combination frequencies of the distant protons. These spectra provide a nice illustration of the nuclear selectivity possible in ESEEM spectroscopy.

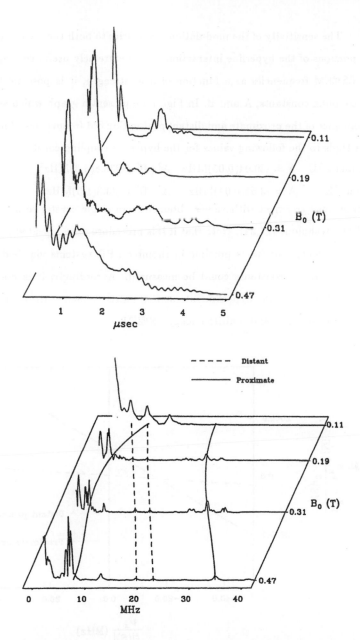

Fig. 6. Two-pulse ESEEM patterns (top) and associated spectra (bottom) of the arsenate radical in γ-irradiated KDA obtained at 9.09 GHz and at the indicated external field strengths, which correspond to the four (nominally) allowed EPR transitions. The orientation of the external field was selected to lie along the unique principal axis of the [75]A hyperfine interaction matrix (by single-crystal alignment). The solid and dashed lines superimposed on the bottom panel track the frequencies of the proximate and distant protons.

The sensitivity of the modulation frequencies to both the secular (A) and non-secular (B) portions of the hyperfine interaction has an extremely useful consequence. By tracking the ESEEM frequencies as a function of field strength, it is possible to evaluate both of the coupling constants, A and B. In Fig. 7 we present a graph which summarizes the field dependence of the proximate and distant proton ESEEM frequencies. Analysis of the plotted data leads to the following values for the hyperfine coupling constants:

proximate ^{1}H   A = -30.04(0.05)MHz A^2+B^2 =938.8(3.3)MHz^2

distant ^{1}H   A = -3.55(0.04)MHz A^2+B^2 = 33.8(1.5)MHz^2,

in excellent agreement with values obtained from single crystal rotation studies using the ENDOR technique (22,24). Note that if this procedure were repeated at three independent, selected orientations (as is possible in rhombic EPR systems via field jumping (25)), six distinct coupling constants could be measured. Accordingly, it is possible to obtain the magnitude of all of the hyperfine interaction matrix elements, through this combination of orientation-selective and multifrequency ESEEM.

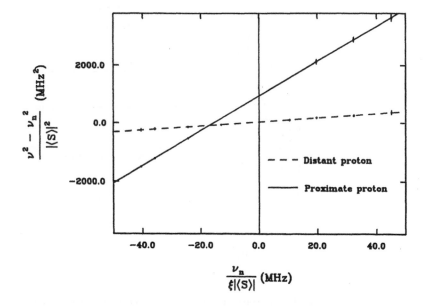

Fig. 7. Frequency-tracking of the ^{1}H fundamentals exhibited in the ESEEM spectra of γ-irradiated KDA. Results for both the proximate and distant protons are included in the plot. The plotted function of observed ESEEM and nuclear Larmor frequencies (ν and ν_n) is linear in field strength for fundamental peaks of spin 1/2 nulcei . The best linear fits to the data (parameter values noted in the text) are also plotted.

It is worthwhile to note that we have deliberately avoided the explicit or implicit incorporation of M_S in our expressions for ESEEM frequencies (Table 1). Instead, we write the magnitude of the expectation value of the electron spin angular momentum, together with the parameter ξ, which specifies the alignment of $\langle \underline{S} \rangle$ along (+1) or against (-1) the external field. When an EPR line is split by a large hyperfine interaction, it is important to account for deviations between the nominal values, M_S, and the actual values of $|\langle \underline{S} \rangle|$, which arise from the mixing of the electron Zeeman states by the large hyperfine interaction. In KDA, the splitting is large enough to make these deviations as large as 40 per cent. When the graph shown in Fig. 7 was generated with the use of the nominal, M_S, values the linearity of the plot was spoiled and *imaginary* hyperfine coupling constants were implied by the fitted parameter values! In more typical situations (DPPH and the DNB radical, for example) *all* hyperfine interactions as so small that the mixing is negligible. (Hence the M_S values are used in Fig. 5.) A useful rule of thumb (10) is that whenever a line is split by (not necessarily resolved) hyperfine interactions to such an extent that it appears sufficient to employ a *field-sampling* rather than a *multifrequency* approach to register the field dependence of ESEEM spectral properties, non-negligible deviations between $|\langle \underline{S} \rangle|$ and M_S can be expected to occur.

4. CONCLUSIONS

In this article, we have presented our perspective on the role that multifrequency ESEEM can play among efforts to facilitate the quantitative analysis of the modulation patterns and their corresponding spectra. We regard this approach as perhaps the simplest means by which ESEEM spectral properties - specifically, the modulation amplitudes, frequencies, and frequency dispersions - may be brought under the control of the spectroscopist. Strategic manipulation of these properties leads to the enhancement of sensitivity, resolution, and clarity in ESEEM spectra. Through the characteristic response of a spin system to such manipulations, spectral assignments can be made with improved confidence, and spin coupling parameters determined with improved precision. We have described a number of possible tactics and have demonstrated several of them - most notably ESEEM frequency tracking - them with exemplary results. In closing, rather than merely recapitulating these first applications of multifrequency ESEEM, we turn, instead to the consideration of possible future developments.

More striking in Table 1 than the experiments that it suggests, are the glaring absences of experimental strategies for a number of important spin systems - systems that simply have not yet been analysed from a multifrequency ESEEM perspective. For example, all studied dependences of ESEEM spectra on external field strength pertain to $I \leq 1$ nuclei. We have, as mentioned (9), observed multifrequency ESEEM from ^{27}Al ($I=5/2$). Further, detailed study of this system is certainly warranted, not in the least because of the great potential of ^{17}O ESEEM investigations in biological applications. A second area that we believe to be prime for further development is the combination of orientation-selection and multifrequency ESEEM (7, 13). Most intriquing in this vein are line-narrowing experiments performed in conjunction with partial orientation-selection (7). In effect, such methods incorporate two echelons of selectivity, and thus provide information reminiscent, in detail and clarity, of that formerly obtained only in single-crystal studies - but now in orientationally-disordered solids, as well.

REFERENCES

1   J. S. Waugh, in M. M. Pintar (Ed.), NMR Basic Principles and Progress. Vol. 13, Springer Verlag, New York, 1976, pp. 23-30; U. Haeberlen, High Resolution NMR in Solids: Selective Averaging, in: J. S. Waugh (Ed.), Advances in Magnetic Resonance, Supplement 1, Academic, New York, 1976.

2   W. B. Mims and J. Peisach, J. Chem. Phys., 69 (1978) 4921-4930; A. V. Astashkin, S. A. Dikanov, and Yu. D. Tsvetkov, J. Struct. Chem., 25 (1984) 45-55.

3   H. L. Flanagan and D. J. Singel, J. Chem. Phys., 87 (1987) 5606-5616.

4   Albert Lai, H. L. Flanagan and D. J. Singel, J. Chem. Phys., 89 (1988) 7161-7166.

5   S. A. Dikanov, A. V. Astashkin and Yu. D. Tsvetkov, Chem. Phys. Lett., 144 (1988) 281-285.

6   B. M. Hoffman, J. Martinsen and R. A. Venters, J. Magn. Res., 59 (1984) 110-123; B. M. Hoffman, R. A. Venters and J. Martinsen, J. Magn. Res., 62 (1985) 537-542; G. C. Hurst, T. A. Henderson and R. W. Kreilick, J. Am. Chem. Soc., 107, (1985) 7294-7299; T. A. Henderson, G. C. Hurst and R. W. Kreilick, J. Am. Chem. Soc., 107, (1985) 7299-7303.

7   D. Lentini and D. J. Singel, in preparation.

8   R. D. Britt, Thesis, University of Califonia at Berkeley, (1988).

9   S. A. Cosgrove and D. J. Singel, in preparation.

10 G. J. Gerfen and D. J. Singel, submitted for publication.

11 L. G. Rowan, E. L. Hahn and W. B. Mims, Phys. Rev. A, 137 (1965) 61-74.

12 W. B. Mims, in: S. Geschwind (Ed.), Electron Paramagnetic Resonance, Plenum, New York, 1972, pp. 263-351.

13 H. L. Flanagan, G. J. Gerfen and D. J. Singel, J. Chem. Phys. 88 (1988) 2162-2168.

14 H. L. Flanagan and D. J. Singel, Chem. Phys. Lett., 137 (1987) 391-397; H. L. Flanagan, G. J. Gerfen and D. J. Singel, J. Chem. Phys., 88 (1988) 20-24.

15 N. S. Dalal, J. A. Ripmeester and A. H. Reddoch, J. Magn. Res., 31 (1978) 471-477.

16 J. X. Boucherle, B. Gillon, J. Maruani and J. Schweizer, Mol. Phys., 60 (1987) 1121-1142.

17   H. D. Hess, A. Bauder and Hs. H. Gunthard, J. Mol. Spectrosc., 22 (1967) 208-220; C. P. Cheng and T. L. Brown, J. Magn. Res., 28 (1977) 391-395.

18   B. D. Flockhart, I. R. Lieth and R. C. Pink, Trans. Farad. Soc. (1970) 469-476.

19   B. D. Flockhart, I. R. Lieth and R. C. Pink, J. Chem. Soc. Chem. Comm., (1966) 885-886.

20   R. L. Ward, J. Chem. Phys., 36 (1962) 1405-1406; P. Rieger and G. Fraenkel, J. Chem. Phys. 39, (1963) 609-629; W. M. Fox, J. M. Gross, and M. C. R. Symons, J. Chem. Soc. (A), (1966) 448-451.

21   H. L. Flanagan and D. J. Singel, J. Chem. Phys., 89 (1988) 2585-2586.

22   N. S. Dalal, C. A. McDowell and R. Srinivasan, Mol. Phys., 24, (1972) 417-439.

23   H. Barkhuijsen, R. de Beer, A. F. Deutz, D. van Ormondt and G. Völkel, Solid State Commun., 49 (1984) 679-684.

24   J. R. Coope, N. S. Dalal, C. A. McDowell and R. Srinivasan, Mol. Phys. 24 (1972) 403-415.

25   A. Schweiger, personal communication.

CHAPTER 4

RESOLUTION ENHANCEMENT IN TIME DOMAIN ANALYSIS OF ESEEM

RON DE BEER and DIRK VAN ORMONDT

1. INTRODUCTION

1.1 Discrete and continuous frequency distributions.

This contribution deals with processing of pulsed magnetic resonance signals, recorded in the *time* domain. A distinction is made between two fundamentally different classes of signals, denoted by I and II. Class I comprises those signals that possess a *discrete* distribution of frequencies, yielding a peaky spectrum in the frequency domain. Model functions that adequately describe class I signals, are made up of damped sinusoids, the model parameters of physical interest being frequencies, damping factors (linewidths), amplitudes, and phases. Physical systems producing signals in class I are e.g. liquids and single crystals. Class II comprises those signals that possess a *continuous* distribution of frequencies, and consequently yield predominantly smooth spectra. Model functions pertaining to class II signals are preferably set up in terms of spin Hamiltonian parameters, such as the static magnetic field, isotropic hyperfine interactions, effective distances between a paramagnetic centre and surrounding nuclei, etc. Typically, class II signals will be obtained from disordered solid systems. An example (1,2) of a signal in each class is given in Fig.1.

In this Chapter, spin Hamiltonians are not considered. Consequently, the contents is concerned mainly with processing of signals made up of damped sinusoids. Sec.2 deals with Maximum Entropy (3,4,5), which is meant for *producing a spectrum*, with or without resolution enhancement. Sec.3 addresses the aspect of *quantitative* analysis, in terms of physical time domain model parameters. The approach chosen by us is to make use of Singular Value Decomposition (SVD) (6,7,8), combined with Linear Prediction (3,9,10) and the State Space formalism (11), which both pertain to the *exponentially* damped sinusoid model. If more *general* model functions (see e.g. 1.4) are needed to quantify the data, we advocate the use of Maximum Likelihood (14,15,16) methods (MLM). An advanced implementation of MLM is treated in Sec. 4.

For recent examples of processing of class II signals, see e.g. (12,13) and references therein.

1.2 Time domain versus frequency domain

As already mentioned above, pulsed Magnetic Resonance phenomena are recorded in the time domain. Should one desire to have a *spectrum*, then the obvious thing to do is to perform a Fast Fourier Transformation (FFT). Subse-

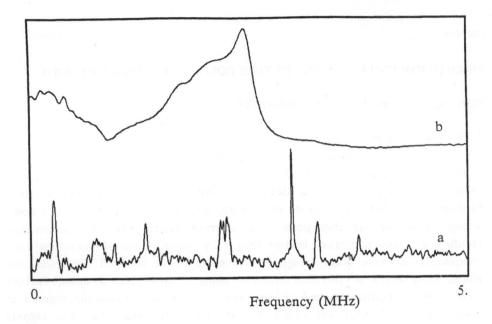

Fig. 1: *Discrete* versus *continuous* frequency distributions. (a) Discrete frequency spectrum (cosine FFT) of three-pulse ESEEM of nitrogen in a free-base porphin in an n-octane single crystal (1). (b) Continuous frequency distribution (modulus FFT) of three-pulse ESEEM of nitrogen in a powder of a quinoprotein (2).

quent quantitative analysis in the frequency domain will yield values of the parameters of (the Fourier transformed version of) the model function. An advantage of analysis in the frequency domain is that it is often easy to concentrate on only a limited spectral region and thus attain a reduction of the computational burden. However, a problem is encountered when the data are incomplete or non-uniformly sampled, and transformation artefacts appear as a result. Should this happen there are two options to choose from: Either one applies a technique to deconvolve the spectrum, or one avoids the frequency domain altogether and performs the analysis directly in the time domain. Various techniques are available for executing the first option. These are, a deconvolution procedure called CLEAN (17, and references therein), the Maximum Entropy Method (MEM), and a computationally efficient implementation of linear prediction, indicated by LPZ (18). Within MEM one can distinguish two versions, indicated by MEM1 (3,9,10) and MEM2 (4,5). It should be pointed out that the extent to which the 'true' spectrum can be recovered by CLEAN, MEM, and LPZ is yet to be rigorously assessed (see 19) for the case of MEM2). When using the second option, the phenomenon of transformation artefacts is of course non-existent and therefore the time domain approach is to be preferred in those cases where precision of parameter quantification (estimation) is of paramount importance. In addition, if the *exponentially* damped sinusoid model is valid, then the time domain offers the possibility of significant improvement of computational efficiency, and/or reduction of user involvement. It should be noted that the feasibility of restricting a time domain analysis to only a few isolated spectral components, while ignoring the rest, has not yet been well investigated.

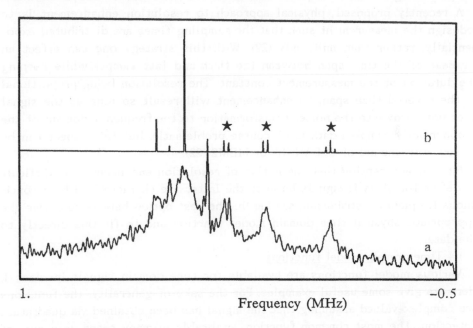

Fig. 2. *Superficial* resolution enhancement via stick-diagram representation of parameters obtained with model function fitting. (a) Cosine FFT spectrum of an *in vivo* 31P NMR signal of a human brain. (b) Stick-diagram representation of peak positions and amplitudes of the same signal, as obtained with time-domain model function fitting with the VARiable PROjection method (VAR-PRO is treated in section 4.2). The multiplets marked by a star could only be resolved by imposing the known ATP multiplet structure (two doublets and a triplet) (20).

1.3 Resolution Enhancement

An important aspect of signal processing is to bring to light splittings of spectral components that are difficult to discern in a spectrum obtained by FFT of a time domain signal. Possible causes of poor resolution are 1) the natural linewidths of the constituent spectral components are larger than the peak separations, 2) the time domain data have been prematurely trunca-ted, 3) the signal-to-noise ratio (SNR) is low.

One way to achieve resolution enhancement, applicable in all three cases just mentioned, is to least squares fit a model function (in either domain), preferably guided by prior knowledge. Since the fit yields values of the model parameters involved, a stick diagram can then be made which suggests superresolution. See Fig.2 for an example (20) of this procedure.

Alternatively, one may apply MEM (21), CLEAN (17), or LPZ (18), depending on the prevailing cause of the poor resolution, and the available computer time.

A recently proposed, physical approach to resolution enhancement is to redesign the measurement such that the sampling times are distributed exponentially, rather than uniformly (22). With this strategy one can effect an increase of the time span between the first and last sample, while keeping the duration of the measurement constant. The resolution being proportional to the covered time span, an enhancement will result so long as the signal does not decay into the noise. Transformation to the frequency domain of the exponentially sampled data is of course problematic, but this aspect can be solved with the aid of CLEAN (17) or MEM2 (22).

Finally, we mention that the notion of resolution enhancement is difficult to define for class II signals because the latter are characterized by a continuous frequency distribution, see Fig.1b. The best option here is to derive the appropriate physical time domain model function and to fit this directly to the data.

1.4 Time domain model functions

Various model functions are available for time domain signals in class I. Here we give some useful examples. For the sake of generality, the functions are complex-valued assuming that the signal has been obtained via quadrature detection. The most common function, applicable in many cases, is a sum of *exponentially* damped sinusoids, i.e.

$$\hat{x}_n = \sum_{k=1}^{K} c_k \exp\left[(\alpha_k + i2\pi\nu_k)(n+\delta)\Delta t\right] , \qquad n = 0, \ldots , N-1 , \tag{1}$$

in which c_k, α_k, ν_k, $k=1, \ldots , K$, are the amplitudes (complex-valued), damping factors and frequencies of the sinusoids, Δt is the sampling interval, $\delta \times \Delta t$ is the point of time of the first sample to be included in the analysis, and N is the number of samples (data points). Exponentially damped sinusoids have favourable mathematical properties enabling one to analytically work out an otherwise computationally expensive step of least squares (LS) model fitting. To be more specific, one can rewrite [1] as

$$\hat{x}_n = \sum_{k=1}^{K} c_k z_k^{n+\delta} \tag{2}$$

where $z_k \stackrel{def}{=} \exp[(\alpha_k + i2\pi\nu_k)\Delta t]$. As a consequence of the simple structure of

[2], one time consuming stage of the LS procedure can be reduced to evaluating summations of the form $\sum_n n^\ell (z_k^* z_{k'})^n$, $\ell = 0,1,2$, for which analytical expressions can be derived. (* denotes complex conjugation.)

Another important model function is that comprising *mixed* damping terms, i.e. similar to [1], but with $\exp[\alpha_k(n+\delta)\Delta t]$ replaced by $\exp\{\alpha'_k(n+\delta)\Delta t + \alpha''_k[(n+\delta)\Delta t]^2\}$. In the frequency domain, this function corresponds to a Voigt line shape, or, if α'_k is put to zero, to a Gaussian line shape. Furthermore, it is easy to accommodate the time domain equivalent of *asymmetric* line shapes such as a ramp function. It can be shown that the latter function, expressed in the time domain is

$$\hat{x}_n = \frac{[1 - i\,2\pi\Delta\nu\,n\Delta t - \exp(-i\,2\pi\Delta\nu n\Delta t)]\,\exp(i\,2\pi\nu n\Delta t)}{(2\pi\Delta\nu n\Delta t)^2}, \quad n = 0, \dots, N-1, \qquad [3]$$

where ν is the frequency of the sharp edge of the ramp, and $\Delta\nu$ is the width of the ramp at zero height. Note that Eqn.[3] can easily be combined with Voigt-type damping.

A very different approach to time domain model functions is Linear Prediction (also referred to as Autoregressive Modelling, or as MEM1). In this approach one expresses a data point as a linear combination of past or future data points, i.e. according to

$$\hat{x}_n = a_1 x_{n-1} + a_2 x_{n-2} + \dots + a_M x_{n-M}, \quad n = M, \dots, N-1, \qquad [4]$$

in the case of forward prediction. In Eqn.[4], $a_1, \dots, a_M$, are the so-called prediction coefficients, which are independent of n. In absence of noise, and provided the signal is made up of exponentially damped sinusoids, linear prediction is *exact*. In other cases, Eqn.[4] is only approximate, but the error can be reduced by increasing the number of prediction coefficients. The usefulness of the linear prediction approach stems from the fact that it can be used to *extrapolate the signal*, so as to alleviate the ill effects of missing data points on the spectrum as obtained from FFT. Alternatively, it appears possible to derive either a power spectrum (3) or an absorption spectrum (18), *directly* from the prediction coefficients. Finally, it should be pointed out that a relation exists between the linear prediction coefficients and the quanties z_k in Eqn.[2]; see 3.4.

In summary, the time domain offers a variety of model functions which enable one to deal with many possible situations occurring in class I. For class II it is hard to present a unified approach.

1.5 Parameter estimation, non-iterative versus iterative procedures

It is important to quantify a signal in terms of *physical* parameters of a model function. Therefore, various techniques have been developed to least squares fit a model function to the data. In this Chapter two main groups of methods are distinguished, namely non-iterative and iterative ones. Properties of non-iterative techniques are 1) The model parameters are evaluated in a single cycle. 2) Starting values of the model parameters are not required. 3) User involvement can be minimized. 4) The model function is restricted to exponentially damped sinusoids and the linear prediction equation. 5) Exploitation of prior knowledge of model parameters is possible only on a limited scale. A popular example of a model function that can be fitted non-iteratively, in various alternative ways, is Eqn.[4].

Properties of iterative techniques are 1) The model parameters are obtained only after a number cycles. 2) Starting values of the model parameters entering nonlinearly in the model function must be provided in the first cycle. 3) There are *no restrictions* on the model function. 4) Exploitation of prior knowledge of model parameters is possible. 5) When choosing a so-called Maximum Likelihood method, the statistical and systematic errors approach theoretical lower bounds. 6) The computational cost is high if the model function is not made up of exponentially damped sinusoids. An example of a model function that can only be fitted by iterative methods is a sinusoid damped by a Voigt-type function. Also, model functions for class II signals are usually so complicated that iterative methods are unavoidable.

1.6 Precision of model parameters, prior knowledge, and experimental design

When an adequate model function of the data is available, and the statistical properties of the noise contaminating the data are known, it is possible to derive lower bounds (Cramér-Rao lower bounds) on the standard deviations of the model parameters to be quantified (estimated). Assuming that the quantification method (estimator) at hand can indeed attain the lower bounds, one can then analyse the precision of a quantification experiment for any chosen set of parameter values. In fact, one may even predict in advance whether or not a particular parameter will be quantifiable with sufficient precision, given the noise level of the spectrometer and the available measurement time. Alternatively, it is feasible to adapt instrumental parameters like the sampling interval and the number of scans used for each sample, so as to attain optimal precision. This constitutes a powerful tool for quantification experiments, called *experimental design*. In addition, it appears often possible to dramatically enhance the precision of certain parameters by fixing the (relative) values of other parameters to *a priori* known values. The Cramér-Rao lower bounds enable one to study this systematically.

2. MAXIMUM ENTROPY METHODS

2.1 Introduction

In this Section we treat two Maximum Entropy Methods (MEM), indicated by MEM1 and MEM2 (24), that are currently in use in Magnetic Resonance signal processing. The primary property of MEM is that it can construct an optimal spectrum from *incomplete* time domain data. (By incomplete we mean, for instance, that data points may be missing because of dead time of the spectrometer.) The importance of this property should be clear when one realizes that the number of spectra that are compatible with incomplete data is *infinite*. From this infinity of possible spectra, MEM selects the one that is 'maximally noncommittal with respect to missing information'. Other properties of MEM are that it can bring about resolution enhancement and/or noise suppression.

A treatment of the *notion* of entropy in the context of signal processing (3,4) is beyond the scope of this Chapter. In the remainder of this subsection we shall briefly introduce two alternative entropy expressions.

A possible approach is to define the entropy, H_1, of a band-limited signal in the *time* domain, as follows (3)

$$H_1 = - \int_{-\infty}^{\infty} \rho(\bar{x}) \, \log(\rho(\bar{x})) \, d\bar{x} , \tag{5}$$

where $\rho(\bar{x})$ stands for the *joint probability density function* (see 4.1) of the sampled time domain signal, $x_0, x_1, \dots , x_{N-1}$. Assuming then that the time domain signal represents a stationary, stochastic zero-mean Gaussian process, and $N \to \infty$, it is possible to show ((3), Chapters 2 and 6) that, apart from a constant factor which is omitted here, Eqn.[5] can be converted to an entropy *rate*, h_1, given by

$$h_1 = \frac{1}{4B} \int_{-B}^{B} \log(S(\nu)) \, d\nu , \tag{6}$$

where $S(\nu)$ is the power spectrum of the signal, and B the band limit. The subscript 1 is added to indicate that Eqn.[6] relates to so-called MEM1, i.e. the maximum entropy method first devised by Burg, in the field of geophysical signal processing (25). In actual practice, the various available implementations of MEM1 are applied to all sorts of signals that do not nececcessarily satisfy the assumptions made above. An alternative approach, devised by Gull and Daniell (26) in the field of astronomy, is to define the entropy *directly* in the domain to which the data are to be transformed. Sibisi (27) and others (28,29,30) have shown that in NMR this leads to the expression

$$H_2 = - \int_{-B}^{B} S(\nu) \, \log(S(\nu)) \, d\nu , \tag{7}$$

where $S(\nu)$ now stands for the absorption spectrum rather than the power spectrum (special measures are required for negative absorption peaks).

In both MEM1 and MEM2, the task is to maximize h_1 and H_2 respectively, subject to constraints that are derived from the *available* time domain data. In the following two subsections the two versions of MEM will be treated, the emphasis being on MEM1 because this technique has already been applied to Electron Spin Echo Envelope Modulation (ESEEM) in the past (31-35), while application of MEM2 to MR has so far been limited to NMR.

Finally we end this introduction by mentioning some contrasting properties of the two MEM approaches: Most MEM1 implementations are non-iterative and computationally efficient. However, so far it is not possible to incorporate prior knowledge of the spectrum other than a rough band shape into the algorithm (36). Also, a constant sampling interval is required. MEM2, on the other hand, is iterative and therefore computationally more demanding. However, all sorts of spectral prior knowledge can be accommodated in the algorithm, see e.g. (37). As a rule, imposition of spectral knowledge results in a rather significant improvement of the result. The sampling times need not be uniformly distributed (22).

2.2 MEM1

2.2.1 Theory

It has been proved by van den Bos that for a stochastic zero-mean Gaussian signal, the spectrum resulting from MEM1 is equivalent to that obtained from so-called AutoRegressive (AR) modelling (38). AR modelling in turn, is equivalent to Linear Prediction (3). Since the notion of Linear Prediction (LP) has recently gained some ground among magnetic resonance spectroscopists, we shall adhere to the latter formalism in what follows. We point out that the philosophy of MEM, namely maximization of the entropy subject to constraints derived from the data, is not easily discernible in the LP implementation of MEM1. For a discussion of the various implementations the reader is referred to (3).

With LP, one expresses a data point as a linear combination of previous or future data points, i.e. with *forward* prediction one writes

$$\hat{x}_n = a_1 x_{n-1} + a_2 x_{n-2} + \ldots + a_M x_{n-M} \quad , \; n=M, \ldots ,N-1. \tag{4}$$

or, with *backward* prediction,

$$\hat{x}_{n-M} = a'_1 x_{n-M+1} + a'_2 x_{n-M+2} + \ldots + a'_M x_n \quad , \; n=M, \ldots , N-1. \tag{8}$$

The caret on x_n in Eqns. [4] and [8] indicates that the quantity on the left-hand side is the predicted value rather than the actual value. We point out that the prediction coefficients, a_m, are independent of n. In addition, when the signal is stationary, we have $a'_m = a^*_m$, $*$ denoting complex conjugation. The prediction coefficients can be evaluated by minimizing the sum of squared residues, $\sum |x_n - \hat{x}_n|^2$, using either Eqn.[4] or Eqn.[8], or a combination

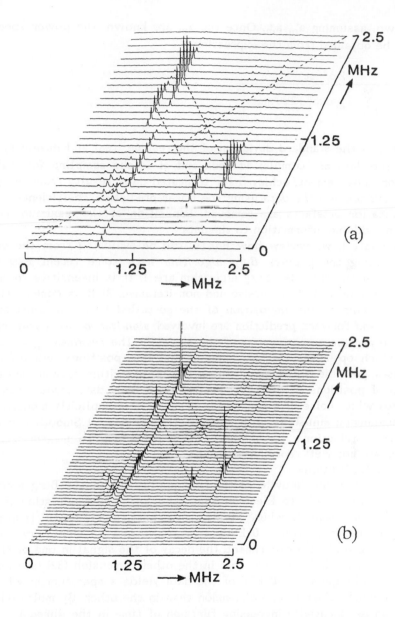

Fig. 3. 2-D three-pulse ESEEM from a single crystal of KCl with F centres (32). (a) 2-D FFT spectrum. (b) Combined FFT-MEM1 spectrum. The symmetry property of the 2-D three-pulse ESEEM spectrum is more clearly demonstrated by resolution enhancement in the dimension processed by MEM1 (pointing 'backward' in the figure).

of the two, assuming $a'_m = a^*_m$. Once the a_m are known, the power spectrum, $S(\nu)$, can be obtained from the expression

$$S(\nu) = \frac{2\,\sigma^2\;\Delta t}{\left|\,1 - \sum_{m=1}^{M} a_m \exp(-i\,2\pi\nu\, m\Delta t)\,\right|^2} \quad , \qquad\qquad [9]$$

in which σ is the standard deviation of the residue (real and imaginary parts each, if the data are complex-valued), and Δt is the sampling interval. The resolution enhancement, achieved for truncated data, results from the possibility to *extrapolate the data* using the linear prediction coefficients. In this way truncation artefacts are removed. Imposition of the maximum entropy constraint adds no information to the data.

In Appendix I, we review how Eqn.[9] can be evaluated from the data in practice, using the popular *Burg algorithm* with some recent extensions. Properties of the extended Burg algorithm are: 1) It is insensitive to truncation of the data. 2) It is recursive and non-iterative. 3) It is computationally efficient, owing to the imposition of the so-called Levinson constraint. 4) Backward *and* forward prediction are involved *simultaneously,* assuming $a'_m = a^*_m$, which presumes *stationarity* of the signal 5) The *residues* $x_n - \hat{x}_n$ are tapered, which optimizes the precision of the line positions and effectively suppresses possible splitting of lines. 6) The algorithm is well suited for retrieval of positions of discrete spectral lines from data made up of sinusoids plus white noise. If a sinusoid is damped, it is implicitly modelled as a set of undamped sinusoids with adjacent frequencies. 7) Smooth continuous frequency distributions, for instance those pertaining to measurements on powders, are not well reproduced, see e.g. (39).

2.2.2 Applications

To the best of our knowledge the first application of the Burg algorithm, including a taper (40), to Electron Spin Echo Envelope Modulation (ESEEM) was in (31). One example where MEM1 proves particularly useful is in producing a 2-D spectrum from 2-D ESEEM measurements on single crystals (32). In a typical 2-D ESEEM experiment, the decay of the signal in one of the two dimensions is much faster than that in the other dimension (32) (see however (41)). As a consequence, 2-D FFT of the data yields a spectrum in which the peaks are much wider in one dimension than in the other. By multiplying the data by an exponentially increasing function of time in the dimension where the decay is fastest, followed by application of MEM1, this disparity can be significantly alleviated. An example of this, taken from (32), is given in Fig. 3.

Another example of the usefulness of MEM1 is a 1-D ESEEM measurement on a single crystal as a function of the angle between the static magnetic field and the crystalline symmetry axes (32,42), or as a function of the temperature near a phase transition (42). The resolution enhancement (line sharpening) readily attainable with MEM1, using the same procedure as in the previous 2-D example, enables one to observe changes of the peak positions with increased precision.

So far, the examples concerned signals from class I. As explained in 1.3, resolution enhancement is not readily applicable to class II signals (powdered samples) because the latter are largely smooth in appearance. However, in some cases peaky features carrying particularly useful information appear which can be enhanced with the aid of MEM1, see e.g. (43) and Fig. 4. It

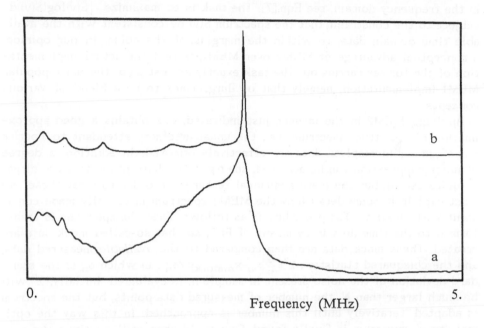

Fig. 4: Resolution enhancement via MEM1. (a) FFT as described in Fig. 1b. (b) MEM1 spectrum of the same signal. The number of data points, N, is 256 and the MEM1 model order, M, equals 0.5N. The signal was made stationary by multiplying the time-domain signal by an exponentially increasing function (time constant 2 ms).

should be realized that the smooth parts of the spectrum may be severely distorted under the influence of the enhancement procedure. We repeat that the ultimate method for class II signals is to fit an appropriate physical model function, based on the spin Hamiltonian parameters.

In conclusion, we can state that MEM1 is a computationally efficient and simple to operate procedure for bringing about resolution enhancement (line sharpening). If optimally accurate quantitative information about each parameter (frequency, damping factor, amplitude, phase) of a spectral component is desired, we recommend the use of the methods treated in 3. and 4.

2.3 MEM2

So far, the application of MEM2 to MR is limited to NMR, the first contribution in that field being by Sibisi (27). Consequently, the following treatment is necessarily restricted to NMR. However, we hope to convey the message that application of MEM2 to ESEEM will become useful.

As pointed out in 2.1, the entropy expression of MEM2 is defined directly in the frequency domain, see Eqn.[7]. The task is to maximize $-\int S(\nu)\log(S(\nu))d\nu$ subject to the constraint that the spectrum $S(\nu)$ be consistent with the available time domain data, to within the margins of the noise. In our opinion, a conceptual advantage of MEM2 over MEM1, is that the actual implementation of the former carries out the task exactly as just said; the most popular MEM1 implementation, namely that of Burg, seems to be a blend of various concepts.

Applying MEM2 in the manner just indicated, one obtains a good approximation of the 'true' spectrum, i.e., the usual artefacts attendant to Fourier transforming incomplete data are effectively removed. In addition, a degree of noise suppression can be achieved, see e.g. (21). These properties are already quite useful, but there are additional features. In order to treat these, we must explain in some detail how the MEM2 spectrum is actually made consistent with the data. The procedure is as follows. First the spectrum is transformed to the time domain by means of FFT, so that so-called mock data are created. These mock data are then compared to the *available* measured data, and the chisquared statistic, $\chi^2 = \sum_n |x_n - x_{n,mock}|^2/\sigma_n^2$, in which σ_n is the standard deviation of the noise present in sample n, is evaluated. Initially, χ^2 will be much larger than N, the number of measured data points, but the spectrum is adapted iteratively until this number is approached. In this way the optimal, 'true', spectrum is finally found. One could then still continue the procedure, but henceforth multiplying the mock data by an exponentially decaying function, before comparison with the measured data. As a result, χ^2 increases, but the program will compensate for this effect by iteratively reducing the line widths in the spectrum, until χ^2 again approaches N. Apparently, the effect of modifying the mock data as mentioned, is to bring about a sharpening of the lines, and thus resolution enhancement. (It is realized, of course, that the modification of the mock data may take place from the start, and not after first retrieving the unsharpened spectrum.) An example of this resolution enhancement is shown in Fig.5.

Note that there is an important difference with MEM1 line sharpening: With MEM1, one multiplies the *measured* data with an exponentially increasing function which magnifies not only the 'pure' signal, but also the noise superimposed on it. With MEM2, on the other hand, the *data remain unaltered*.

Enlarging on the possibility of modifying the mock data, we point out that one may multiply the latter with any function that is known to 'blur' the measured data (37). Examples of this are the inhomogeneity distribution of the magnetic field, or line splitting due to coupling with some nucleus

(NMR). Introducing the correct blurring function will in general remove the blur from the spectrum.

In conclusion, we point out that although MEM2 requires more computation than MEM1, it is much more flexible and conceptually clearer. See the last paragraph of 2.1 for a comparison of other aspects. As for optimal accuracy in quantitatitave applications, we remark that the methods described in 3. and even more so in 4. are generally superior, especially with regard to the amplitudes (intensities) of the individual spectral components.

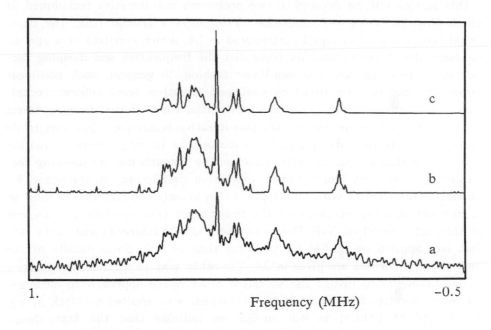

Fig. 5: Resolution enhancement via MEM2. (a) FFT as described in Fig. 2a. (b) MEM2 spectrum of the same signal, when comparing the available measured data directly with the mock data (see text). (c) MEM2 spectrum when comparing the available measured data with exponentially damped mock data (time constant 20 ms). (R. de Beer and D. van Ormondt, unpublished results).

3. SVD-BASED METHODS FOR QUANTITATIVE ANALYSIS

3.1 Introduction

The previous section was concerned with producing a *spectrum* from time domain data. The emphasis was put on reducing spectral artefacts associated with missing data points, and on enhancement of the spectral resolution. No attempt was made to use the spectrum for accurately quantifying physical parameters such as the frequencies, damping factors (linewidths), amplitudes, and phases, including the related statistical errors, of the sinusoids present in the data. In order to carry out such a quantification one would have to introduce a physical frequency domain model function, and fit this to the data. However, since it has yet to be established how close the 'true' spectrum is approximated by MEM (and CLEAN and LPZ, for that matter), we advocate to avoid the frequency domain altogether and instead fit a physical time domain model function (see 1.4) *directly* to the data. As mentioned in 1.5, we distinguish non-iterative and iterative methods, each with specific advantages and disadvantages.

This section will be devoted to two prominent non-iterative techniques, in which Singular Value Decomposition (SVD) plays a crucial role. The basic model function used is Eqn.[1], introduced in 1.4, which consists of a sum of exponentially damped sinusoids. Note that the frequencies and damping factors enter the equation in a *non*-linear fashion. In general, such nonlinear parameters can only be fitted by nonlinear, iterative, least squares techniques, but it can be shown that the *exponential* form of the *damping* terms (i.e., $\exp(\alpha t)$) enables one to still use non-iterative techniques. One way to do this is to invoke the fact that Eqn.[1] satisfies the linear prediction equation exactly, and that a relation exists between the frequencies and damping factors on the one hand, and the linear prediction coefficients on the other. We emphasize that, unlike with MEM1, prediction in *only one* direction is used so as to avoid imposing stationarity. The linear prediction equation, in turn, can be accurately fitted via SVD. This is the method of Kumaresan and Tufts (44), that was applied to ESEEM for the first time in (45). Some details of the theory, and examples are given in 3.4. The other way to circumvent iterative model fitting is to invoke the so-called State Space approach to harmonic retrieval, developed by Kung et al. (11), which was applied to NMR in e.g. (46,47) and to ESEEM in (63). In 3.5 we indicate that the State Space approach, as applied to magnetic resonance, can be formulated in a simple manner with the aid of elementary matrix manipulations. Also, an application to ESEEM is given.

A common feature of the methods of Kumaresan and Tufts, and of Kung et al. is that N consecutive data points are arranged in the form of a matrix, called the *data matrix*. In 3.2 we treat a special property of data matrices formed from data that can be described by the exponentially damped sinusoid model of Eqn.[1]. In both methods, the data matrix is subjected to SVD, which amounts to factorization into a product of two unitary matrices and a diagonal matrix. SVD, and its usefulness for quantification and/or enhancement of time domain signals are treated in 3.3.

3.2 The Data Matrix

Given the data points x_n, n=0, ... , N-1, a data matrix, X, can be formed as follows

$$X \stackrel{\text{def}}{=} \begin{bmatrix} x_0 & x_1 & x_2 & \cdots & & x_M \\ x_1 & x_2 & \cdot & & & \cdot \\ x_2 & \cdot & & & & \cdot \\ \cdot & & & & & \cdot \\ \cdot & & & & & \\ \cdot & & & & & \\ x_L & \cdot & & \cdot & & x_{N-1} \end{bmatrix} ,$$

where M and L are arbitrary, but should both be greater than the number of sinusoids K in Eqn.[1], while L+M+1 =N. Note that X possesses Hankel structure, i.e. all elements on an anti-diagonal are equal. This property is important in the sequel.

We consider first a noiseless signal that comprises only a single, exponentially damped sinusoid and can therefore be exactly represented by Eqn.[2], with K=1. For this case the data matrix has the following simple form

$$X_1 = c_1 z_1^\delta \begin{bmatrix} z_1^0 & z_1^1 & z_1^2 & \cdots & & z_1^M \\ z_1^1 & z_1^2 & \cdot & & & \cdot \\ z_1^2 & \cdot & & & & \cdot \\ \cdot & & & & & \cdot \\ \cdot & & & & & \\ \cdot & & & & & \\ z_1^L & \cdot & & \cdot & & z_1^{N-1} \end{bmatrix} , \qquad [10]$$

in which the subscript 1 of X_1 indicates that the signal comprises only one sinusoid, with number 1. Recall now that the rank of a matrix is equal to the number of linearly independent rows (or columns, which amounts to the same). It can then be seen immediately that the rank of X_1 equals just one. If the signal comprises K exponentially damped signals, then the data matrix can be written as

$$X = X_1 + X_2 + \ldots + X_K . \qquad [11]$$

Adding the matrices of the individual sinusoids, as in Eqn.[11], one obtains K linearly independent rows (or columns). Apparently the rank of a data matrix composed of K *noiseless*, exponentially damped sinusoids is equal to K. If noise is added to the signal, all linear dependence between rows (or columns) is destroyed, and the rank of X becomes full, i.e. min[L+1,M+1]. However, so long as the SNR is not too low, one can still define an approximate rank which equals K. It should be clear at this stage that a rank analysis of the data matrix yields the number of sinusoids. One way to do this is to perform an SVD.

3.3 SVD

Any rectangular matrix can be converted to diagonal form with the aid of SVD (6,7,8), which amounts to the following factorization

$$
\boxed{X} = \boxed{U}\ \boxed{\Lambda}\ \boxed{V^{\dagger}} \qquad [12]
$$

where U and V are unitary matrices of size $(L+1)\times(L+1)$ and $(M+1)\times(M+1)$ respectively, $\dagger$ denotes Hermitian conjugation, and Λ is an $(L+1)\times(M+1)$ diagonal matrix whose diagonal entries, the *singular values* $\lambda_\ell \overset{\text{def}}{=} \Lambda_{\ell\ell}$, $\ell = 1, \dots , L+1$, are greater than or equal to zero. The sizes of the boxes in Eqn. [12] reflect a possible choice of the values of L and M (here we chose L<M).

Once the decomposition (factorization) according to Eqn. [12] has been achieved, the rank of X can be inferred immediately. This can be seen as follows. Since U and V are unitary (full rank), the rank of the product $U\Lambda V^{\dagger}$ is determined by the rank of Λ, which in turn is equal to the number of non-zero singular values. It follows then from the theory of the previous subsection that for a noiseless signal comprising K exponentially damped sinusoids, the number of non-zero singular values equals K. As already mentioned above, the rank of X becomes full when noise is added to the signal; this in turn implies that all singular values then become greater than zero. However, provided the SNR is not too low, it will be possible to discern a jump in the singular values at number K, after ordering them according to descending magnitude. A clear example of such a jump is shown in Fig. 6, which is related to the decay of a stimulated electron spin echo experiment as a function of the time lapse between the second and third microwave pulse. Perusing the singular value plot from right to left, a clear magnitude jump is observed when passing from number 4 to number 3. It follows that in this case the decay comprises three exponentials (with frequency equal to zero) plus noise. The decay curve itself and the fit are displayed in the inset of Fig. 6 (no difference between the two is discernible with the scale factor used). Other examples of singular value plots have been given in e.g. (45,48).

It is important to realize that SVD can be used to 'enhance' the signal (49,50). To achieve this effect Eqn. [12] is 'cleaned', i.e. the noise-related singular values are put exactly equal to zero while the remaining K signal-related singular values are left unchanged. (One may yet apply a first order noise correction to the signal singular values (44,45), but this aspect is ignored for the sake of simplicity.) After this, Λ, U, and $V^{\dagger}$ are stripped of the rows and columns that cease to contribute as a result of the cleaning

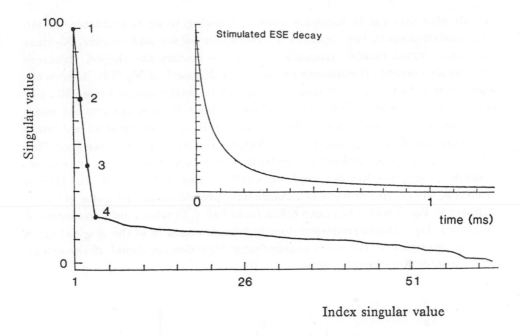

Fig. 6. Plot of the singular values (on a logarithmic scale) related to a three-pulse ESE decay from F centres in KCl (32). It can be seen that, when coming from the right, the first clear jump in the singular values occurs when passing from index 4 to index 3. This demonstrates that the ESE decay in question, shown in the inset, can be adequately modelled by three exponentials. In fact, the difference between the data and the fit is smaller than the random noise at the detector.

procedure, resulting in a truncated version of Eqn. [12]

$$X_C = U_K \Lambda_K V_K^\dagger \qquad [13]$$

where X_C is the cleaned data matrix, Λ_K is a $K \times K$ diagonal matrix, U_K and V_K are respectively $(L+1) \times K$ and $(M+1) \times K$ matrices comprising the K singular vectors related to the retained K signal singular values. X_C is the unique matrix whose rank is *exactly* equal to K, while the sum of squared differences $\Sigma \, |X_{\ell m} - X_{c,\ell m}|^2$

over all elements ℓ,m is *minimum*. However, owing to an unavoidable (small) noise contribution to the signal-related singular values and vectors, X_c does not possess exact Hankel structure. One way to restore the Hankel structure is to simply average all elements on each anti-diagonal of X_c (49). A more advanced way is to fit a Hankel matrix to X_c in the least squares sense (50). Unfortunately, these so-called Hankelization methods in turn destroy the exact rank. This is a 'circular' problem that can be resolved by an iterative procedure comprising the following steps (50). 1) Perform the SVD and truncate according to [13]. 2) Calculate X_c. 3) Restore the Hankel structure. These steps are repeated until convergence has occurred, i.e. until X_c has both rank K *and* Hankel structure. The results of such an enhancement procedure can be quite good, as is shown in Fig. 7 which has been taken from Ref. (50) with kind permission of the author. Fig.7 shows frequency domain presentations of a) the original clean time domain signal, b) the noise-contaminated time domain signal, c) the enhanced (restored) time domain signal.

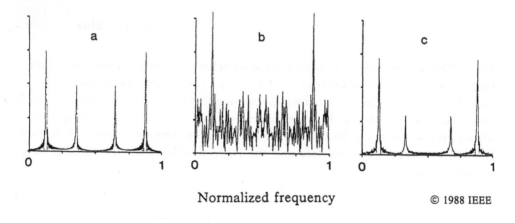

Normalized frequency

© 1988 IEEE

Fig. 7: Signal enhancement via an iterative time domain processing algorithm of Cadzow (50), based on eigencharacterization (i.e. truncated SVD) and structure (i.e. Hankel) of the related data matrix. Fourier transforms of (a) simulated noiseless signal (b) noise contaminated signal (c) enhanced signal. It can be seen for this noise realization, that the peak positions are truly retrieved but not the peak intensities.

For a good understanding of the method the following remarks are in order. 1) The enhancement has taken place entirely in the time domain. 2) Enhancement should not be confused with quantification: The model parameters have yet to be evaluated. In passing, we remark that in actual quantification algorithms (see e.g. Secs 3.4 and 3.5) one usually performs only one round of truncated SVD. However, when the SNR is low, a few iterative enhancement cycles could be useful. 3) The *validity of the model function* is not important in this stage of the signal processing. Those portions of the signal not satifying Eqn. [1] are implicitly expanded into exponentially decaying sinusoids, and that part of the expansion contained in the signal singular values and vectors is salvaged automatically. (See (51) for SVD of a single, *Gaussian*-damped, sinusoid as an example of this aspect.) The main involvement of the person operating the program is to decide where to truncate the set of singular values. If the true jump in the singular values is difficult to locate, one usually retains a few extra singular values to be on the safe side. It may of course happen that the number of sinusoids is known a *priori* from spectroscopic knowledge. However, when the SNR is low, it is advisable to then still take some additional singular values into account.

Finally, it should be mentioned that the available algorithms for performing an SVD are applicable to any real- or complex-valued rectangular matrix. On the one hand this generality is favourable in that there are no restrictions to the use of SVD, but on the other hand the attendant computational load is proportional to the third power of the size of the matrix. By consequence SVD is expensive when the data set is large. For this reason, mathematicians are searching for a way to exploit the Hankel structure present in the data matrix of a 1-D signal, when performing an SVD. This appears to be a difficult problem.

3.4 Linear Prediction and SVD (LPSVD) (44,45)

3.4.1 Theory

As already noted before, the frequencies and damping factors enter the model function in a *non*-linear fashion. This property is undesirable if one wants to quantify *non*-iteratively. The problem can be circumvented by temporarily abandoning Eqn. [1], and switching to an alternative model function that depends linearly on the model parameters. A necessary condition is of course that a relation exists between the parameters of the two alternative model functions. Such is the case. The temporary model function is the *linear prediction* equation which amounts to expressing each sample as a linear combination of past or future samples, i.e., in the case of *backward* prediction

$$\hat{x}_n = a_1 x_{n+1} + a_2 x_{n+2} + \ldots + a_M x_{n+M} \quad , n=0, \ldots , N-M-1, \quad M \geq K, \quad [14]$$

where K is the number of exponentially damped sinusoids.

Note that the prediction coefficients a_m, $m=1, \ldots , M$, are independent of n, and therefore cannot carry information about the amplitudes and phases. In absence of noise and provided the signal comprises only exponentially damped sinusoids, Eqn. [14] is exact. Otherwise, Eqn. [14] is only approximate. However, the error can be made quite small by choosing $M \gg K$, thus incorporating noise (44) and 'higher order expansion terms' (loosely speaking) of *non*-exponentially damped sinusoids (52). In matrix form, the combined Eqns. [14] for all n are written as

$$
\begin{bmatrix}
x_1 & x_2 & x_3 & \cdots & & x_M \\
x_2 & x_3 & & & & \cdot \\
x_3 & \cdot & & & & \cdot \\
\cdot & & & & & \cdot \\
\cdot & & & & & \\
\cdot & & & & & \\
x_{N-M} & & \cdots & & & x_{N-1}
\end{bmatrix}
\begin{bmatrix}
a_1 \\ a_2 \\ a_3 \\ \cdot \\ \cdot \\ \cdot \\ a_M
\end{bmatrix}
\cong
\begin{bmatrix}
x_0 \\ x_1 \\ x_2 \\ \cdot \\ \cdot \\ \cdot \\ x_{N-M-1}
\end{bmatrix} ,
\qquad [15]
$$

which can be solved for a_m, m=1, ... , M, using a standard linear least squares procedure which involves SVD and subsequent truncation so as to reduce the influence of noise. The fact that the data matrix of Eqn. [15] is one row and one column smaller than that of Eqn. [9] is unimportant; details, including an optional correction of the signal singular values for the presence of noise, can be found in e.g.(44,45).

The next task is to establish the relationship between the prediction coefficients of Eqn. [14] and the frequencies and damping factors of Eqn. [1]. The theory behind this has been well established, see (44) and references therein, and therefore it suffices to quote here the result without further comment. The wanted quantities z_k, k=1, ... , K, which contain the frequencies and damping factors, are equal to the inverses of the K largest (in terms of absolute values) roots of the prediction *polynomial* (M»K)

$$
z^M - a_1 z^{M-1} - \ldots - a_M z^0 = 0 , \qquad [16]
$$

in which z is a complex-valued variable. Once the z_k are known, the complex-valued amplitudes c_k can be determined by a standard, non-iterative, linear least squares fit of the original model of Eqn. [1] to the data.

3.4.2 Applications

An example of LPSVD applied to a (class I) three-pulse ESEEM signal of an X-irradiated single crystal of KH_2AsO_4 (42) is given in Figs. 8, 9, and 10. Fig. 8 shows the time domain signal and the residue (magnified by a factor of 50) of the fit. The portion left of the dotted line could not be measured because of experimental conditions. However, using the fitted model parameters one can actually reconstruct the missing part. An indication of the truthfulness of the reconstruction is that all sinusoids are found in phase at only 8 ns from the time origin. Fig. 9 shows the singular values of the data matrix formed from the signal. Apparently there is no clear transition from signal-

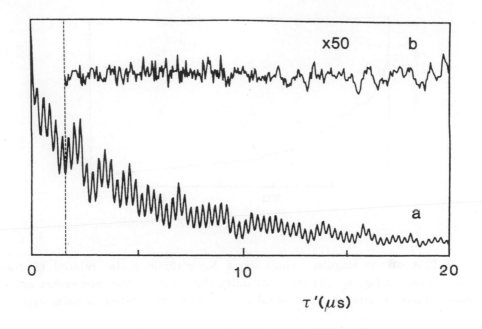

Fig. 8: Quantification via LPSVD. (a) Three-pulse ESEEM from a single crystal of X-irradiated KH_2AsO_4 (42). The portion left of the dotted line represents the 'dead time' of the experiment. It was reconstructed from the parameters obtained by LPSVD. Constructive interference of the spectral components is correctly reproduced at only 8 ns away from the 'true' time origin of the experiment. (b) Time-domain residue of the LPSVD fit, multiplied by a factor of 50. The change in the structure of the residue, when going from left to right, indicates some degree of inadequacy of the exponential decay model (Lorentz line).

related to noise-related singular values. Somewhat arbitrarily, the transition was chosen at K=110. This choice is not critical. Fig. 10 shows the FFT of the data, the difference (magnified by a factor of 50) between the latter and the FFT of the time domain fit shown in Fig. 8, and a spectrum whose resolution is enhanced by artificially reducing the fitted damping factors. For a theory of the statistical errors in the fitted parameters, see 4.4 and (45c).

Other applications of LPSVD are: 1-D NMR (52), 2-D NMR (48,53,54), 2-D ESEEM (45a), 2-D ESE (55,56), a simulated Gaussian-damped 1-D sinusoid (51), and 1-D NMR of *large* data sets (57,58). As for Ref. (58) by Gesmar and Led, it should be noted that these authors succeeded in rooting a polynomial of order M=2400. Finally, we mention that a variation of the Kumaresan-Tufts method, using the so-called QR Decomposition (QRD) rather than SVD, has been devised by Tang et al. (59).

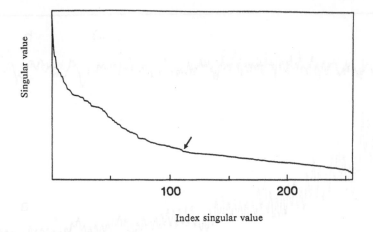

Fig. 9: Plot of the singular values (on a logarithmic scale) related to the
signal shown in Fig. 8a. The arrow at index 110 indicates the (somewhat arbitrarily) chosen transition from signal-related to noise-related singular values.

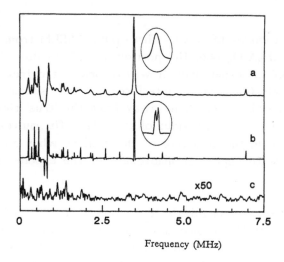

Fig. 10: Fourier transforms of (a) the signal shown in Fig. 8a. A (known)
splitting of the large peak near 3.5 MHz is not resolved (see inset). (b) a
noiseless simulated signal, reconstructed from the parameters obtained by
LPSVD. In order to achieve resolution enhancement, the reconstruction was
performed with reduced damping. As a result the peak mentioned above
shows a splitting (see inset). (c) the residue of the LPSVD fit, multiplied by a
factor of 50. Like the residue in the time domain (see Fig. 8b), a varying
structure in the frequency domain (below 2 MHz) indicates some departure
from Lorentzian lineshape.

3.5 State Space Formalism and SVD (HSVD) (11,46,47)

3.5.1 Theory

The so-called State Space formalism, combined with SVD, is an alternative to LPSVD. Among other things, it circumvents the steps of polynomial rooting and root selection. The method has been developed by Kung et al. (11) for retrieval of sinusoids from 1-D and 2-D signals and was applied to Magnetic Resonance for the first time in Ref. (46), under the name HSVD where H stands for Hankel. Subsequently, it was shown that, at least for Magnetic Resonance, the State Space formalism as such is unnecessary ballast and can be replaced by more generally known matrix algebra(47). Up to and including SVD of the data matrix and subsequent truncation of the noise contribution, LPSVD and HSVD are identical (apart from the exact size of the data matrix, which is an irrelevant detail).

The basic idea behind HSVD is that a noiseless data matrix comprising K exponentially damped sinusoids can be decomposed (factorized) as follows

$$
X = \begin{bmatrix} 1 & . & . & . & 1 \\ z_1^1 . & & & & z_K^1 \\ . & & & & . \\ . & & & & . \\ . & & & & . \\ z_1^L & . & . & . & z_K^L \end{bmatrix} \begin{bmatrix} c_1' & & & \\ & . & & 0 \\ & & . & \\ 0 & & & c_K' \end{bmatrix} \begin{bmatrix} 1 & z_1^1 & . & . & . & z_1^M \\ . & . & & & & . \\ . & . & & & & . \\ 1 & z_K^1 & . & . & . & z_K^M \end{bmatrix}, \qquad [17]
$$

in which the middle matrix is diagonal with entries $c_k' = c_k z_k^\delta$, k=1, ... , K, the matrix on the left is the well-known Vandermonde matrix which we shall indicate by the symbol $\zeta_K^{(L)}$, and the matrix on the right is the transpose of $\zeta_K^{(M)}$. The validity of Eqn. [17] can easily be verified by carrying out the successive matrix multiplications.

As an important aside we mention briefly that extension of Eqn. [17] to 2-D signals is straightforward(11). In fact, the formalism seems even more natural in the 2-D case because the data are in the form of a matrix from the start. The factorization has the same form as Eqn.[17], but the left Vandermonde matrix contains the signal poles of one dimension, and the right one those of the other dimension; the amplitude matrix is in general not diagonal. Under certain conditions, a *single* SVD of the full $N_1 \times N_2$ data matrix will suffice to perform the quantification. This approach seems superior to consecutive, separate treatments of the two dimensions, and in the opinion of the authors merits thorough investigation. See also Appendix II.

After this brief digression on 2-D signals, we proceed by treating the relation between Eqn. [17] and the truncated SVD equation, Eqn. [13]. First we note

158

that U_K and $\zeta_K^{(L)}$ have the same size *and* full rank. Since both matrices pertain to the *same* space (the signal space), it follows then that a nonsingular K×K matrix Q exists that transforms U_K into $\zeta_K^{(L)}$. However, given the fact that U_K is in reality slightly disturbed by noise, Q can be calculated only in the least squares (LS) sense. Fortunately, a simple linear LS procedure, requiring little computational time, is available (44). In symbolic notation, the steps executed so far are

$$X \xrightarrow[\text{truncation}]{\text{SVD}} U_K \Lambda_K V_K^\dagger = U_K Q Q^{-1} \Lambda_K V_K^\dagger \xrightarrow{\text{LS}} \zeta_K^{(L)} Q^{-1} \Lambda_K V_K^\dagger \quad . \qquad [18]$$

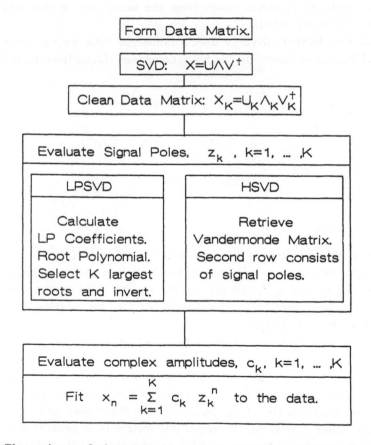

Fig. 11: Flow chart of the various steps in LPSVD and HSVD. The difference between the two methods lies in the evaluation of the signal poles: LPSVD involves an extended order linear prediction polynomial from which the correct roots are to be selected. HSVD involves transformation of the signal singular vectors into a Vandermonde matrix.

Once $\zeta_K^{(L)}$ has been obtained, its second row immediately yields the wanted signal poles; see Eqn. [17]. Also, it can easily be seen by multiplying the middle matrix and right matrix of the right-hand side of Eqn. [17], that the amplitudes are contained in the first column of $Q^{-1}\Lambda_K V_K^\dagger$. An alternative way to arrive at the amplitudes is to fit Eqn. [1] to the data, treating the signal poles resulting from Eqn.[18] as fixed values, similar to the procedure of LPSVD.

For comparison, a flow chart of the various steps in LPSVD and HSVD is displayed in Fig. 11. A theoretical and numerical study, comparing the performance of the two methods under the influence of noise, has recently been made by Tufts et al. for a single sinusoid (60). The result was that HSVD is more precise than LPSVD, but the difference is not dramatic. However, we repeat that apart from this, HSVD circumvents polynomial rooting and root selection, as indicated in Fig. 11. Note that prior knowledge of phases and amplitudes can be imposed in the last step. However, the values of the frequencies and damping factors do not benefit from this. When using iterative techniques, they will.

Very recently, Mayrargue has shown (61) that under certain conditions, HSVD is equivalent to so-called ESPRIT (62), another new non-iterative method gaining popularity.

3.5.2 Applications

In this subsection we treat an application of HSVD to a two-pulse ESEEM signal (class I) of the photo-excited state of pyridine (63). The time domain signal, displayed in Fig. 12, was superimposed on a large background signal which does not favour FFT if preprocessing is to be avoided for reasons of precision. In addition, initial data points were lost because of the dead time of the spectrometer. The residue of the time domain fit is also shown. From the fact that this residue resembles white noise, we conclude that HSVD can very well fit components of the signal that do not satisfy the exponentially damped sinusoid model. In this case the background signal has been effectively expanded into exponentially damped (or augmented) sinusoids. Fig. 13a shows the frequency domain representation (FFT) of a preprocessed version of the signal. Fig. 13b shows the FFT of the HSVD fit obtained with 32 sinusoids, and Fig. 13c the FFT of the time domain residue. For the sake of comparison, the processing prior to FFT was the same in Figs 13 a, b, and c. The spectrum of Fig. 13d is the FFT of the twelve main sinusoids of the HSVD fit, extrapolated to the time origin; preprocessing was omitted. Among other things, it turned out that HSVD correctly reproduces the inverted lines common in two-pulse ESEEM.

160

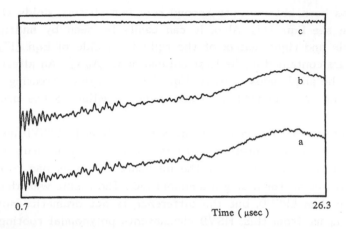

0.7   26.3

Time (µsec)

Fig. 12: Quantification via HSVD. (a) Two-pulse ESEEM from pyridine-d5 in a single crystal of benzene-d6 (63). (b) Noiseless simulated signal, reconstructed from the parameters obtained by HSVD. (c) Time domain residue of the HSVD fit (same scaling factor as (a)). The flat, noisy character of the residue indicates that the exponentially decaying sinusoid model reasonably describes the ESEEM signal. Note also that the increasing background signal has satisfactorily been fitted by a series of exponentials.

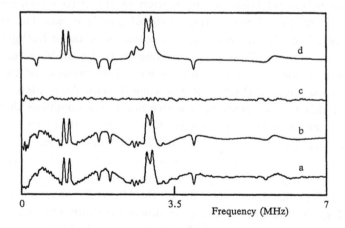

0   3.5   7

Frequency (MHz)

Fig. 13: Cosine Fourier transforms of (a) the signal shown in Fig. 12a. (b) a noiseless simulated signal, reconstructed from the parameters obtained by HSVD. In the fit 32 sinusoids were used. (c) the time domain residue. (d) a noiseless simulated signal, reconstructed from the 12 main sinusoids of the HSVD fit. In order to avoid phase correction, the missing 'dead time' of the signal was also reconstructed. It can be seen that HSVD correctly reproduces inverted lines common in two-pulse ESEEM.

4. MAXIMUM LIKELIHOOD METHODS

4.1 Introduction

In Sec.2 we showed that MEM can be used to avoid spectral artefacts associated with missing data points, and also to bring about resolution enhancement. Subsequently, it was advocated in Sec.3 to rather fit an appropriate time domain model function directly to the data if precise quantitative results are desired. The model function chosen in Sec.3 was based on the exponentially damped sinusoid, see Eqn.[1], which enables one to execute the fitting in a non-iterative manner, using SVD-based methods like LPSVD and HSVD. Although exponential damping applies to many situations, more complicated damping functions occur also, especially when investigating solids. Should this be the case, one is forced to use a more general model function (see e.g. 1.4), and by consequence, a more general fit procedure.

In this section, we advocate to apply the so-called Maximum Likelihood approach to parameter estimation, a well-established method, proposed by R.A. Fisher as early as 1922 (see e.g. (64), p. 436). The Maximum Likelihood Method (MLM) should not be confused with the Maximum Entropy Method (MEM). The essence of MLM is well described in (14,15,16); here, only a brief review of the broad principles is given.

A prerequisite of MLM is that a *joint probability distribution function* of the data be available. Expressed more simply, the requirements are that 1) the probability distribution function of the noise added to each 'true' (i.e. noise-free) datum be known, and 2) a model function which can exactly reproduce the noise-free data, given the 'true' model parameters, be available. An example of a joint probability function, valid for real-valued data corrupted by white (uncorrelated) Gaussian noise is (14,15,16)

$$\varrho(\bar{x},\bar{p},\bar{\sigma}) = \prod_{n=0}^{N-1} \left(\frac{1}{2\pi\sigma_n^2}\right)^{1/2} \exp\left[\frac{1}{2\sigma_n^2}\left(x_n - \hat{x}_n(\bar{p})\right)^2\right] , \qquad [19]$$

in which $\bar{x}$ represents the data, x_n, n=0, ... ,N-1, $\bar{p}$ represents the set of model parameters in the model function $\hat{x}_n(\bar{p})$, n=0, ... ,N-1, and $\bar{\sigma}$ represents the set of standard deviations σ_n, n=0, ... , N-1 of the noise contribution to datum no. n=0, ... ,N-1.

Given the values of the model parameters, one can calculate from [19] the probability that a certain set of data values occurs. Conversely, *given a set of measured data*, one can determine the values of the model parameters that maximize the value of the joint probability distribution function pertaining to that particular set of data. The latter principle is the essence of the maximum likelihood approach.

The maximum likelihood method can be worked out for various alternative noise distribution functions. An important and useful result is that if the noise distribution is Gaussian, MLM is *identical to least squares fitting* of the model function. In signal processing, one therefore speaks of MLM when a least squares fit has been performed to data corrupted by Gaussian noise. An advantage of knowing that the method used satisfies the maximum likelihood conditions is that the favourable statistical properties pertaining to MLM are in force. A notable property of MLM is that for a sufficient number of data points, the standard deviations of the statistical errors in the estimated parameters approach a fundamental lower limit (see 4.4), while the systematic errors approach zero. In addition, if $g(\bar{p})$ is an invertible function defined for all estimated parameters $\bar{p}$, then $\mathbf{g}(\bar{p})$ is just $g(\bar{\mathbf{p}})$ (invariance property), where the bold script indicates that the quantity in question is an MLM estimate. The latter property is very useful, for instance for the following: Should one estimate linear prediction coefficients by means of some MLM procedure (65,66,67), then the signal poles obtained by subsequently rooting the prediction polynomial, Eqn. [16], are also MLM estimates (67).

In the sequel we treat an advanced, nonlinear, iterative, least squares model fitting procedure, called the Variable Projection (VARPRO) method (69), see 4.2. The main characteristic of VARPRO is that the linear model parameters, i.e. the (complex-valued) amplitudes, are temporarily eliminated from the fit procedure, and treated later in a separate non-iterative round. One advantage of this approach is that starting values of the linear parameters need not be provided. An application of VARPRO to a simulated signal is treated in 4.3. An important aspect of 4.3 is that VARPRO can easily accommodate *prior knowledge* of spectral parameters, which results in a significant reduction of the statistical errors in the parameters to be fitted (estimated). Finally, 4.4 is devoted to a method of finding *lower bounds* on the statistical errors of the fitted model parameters, and on the notion of *experimental design*.

4.2 The Variable Projection Method.

The general model function for quantification in the time domain can be written as

$$\hat{x}_n = \sum_{k=1}^{K} c_k \, f_k(\gamma_k, t_n), \qquad n = 0, \dots, N\text{-}1, \tag{20}$$

where the c_k are complex-valued amplitudes, the $f_k(\gamma_k, t_n)$ are the constituent model functions in which each γ_k represents a *set* of *non*-linearly entering model parameters such as the frequency and the damping factor, and t_n is the sampling time. A function $f_k(\gamma_k, .)$ may in turn comprise a set of functions with fixed relative amplitudes and interrelated frequencies and damping factors, together forming a multiplet. Note that Eqn.[20] applies not only to damped sinusoids, but to virtually any 'physical' function, which *includes signals in class II* (defined in 1.1.).

The general model function of Eqn. [20] can be compounded into a matrix equation spanning all sampling times. We write

$$
\hat{x} =
\begin{bmatrix}
\hat{x}_1 \\
\hat{x}_2 \\
\cdot \\
\cdot \\
\cdot \\
\hat{x}_{N-1}
\end{bmatrix}
=
\begin{bmatrix}
f_1(\gamma_1,t_0) & f_2(\gamma_2,t_0) & \cdots & f_K(\gamma_K,t_0) \\
f_1(\gamma_1,t_1) & f_2(\gamma_2,t_1) & \cdots & f_K(\gamma_K,t_1) \\
\cdot & & & \cdot \\
\cdot & & & \cdot \\
\cdot & & & \cdot \\
f_1(\gamma_1,t_{N-1}) & f_2(\gamma_2,t_{N-1}) & \cdots & f_K(\gamma_K,t_{N-1})
\end{bmatrix}
\begin{bmatrix}
c_1 \\
c_2 \\
\cdot \\
\cdot \\
\cdot \\
c_K
\end{bmatrix}
= F\bar{c}, \qquad [21]
$$

which at once defines the vectors $\hat{x}$ and $\bar{c}$, and the matrix F. In this notation, the sum of squared residues of the actual data minus the model function, R, becomes

$$
R = \|\bar{x} - \hat{x}\|^2 = \|\bar{x} - F\bar{c}\|^2 , \qquad [22]
$$

in which the column vector $\bar{x}$ contains the data. R is to be minimized as a function of the variable parameters contained in both F and $\bar{c}$. With VARPRO one now proceeds by first eliminating the amplitudes from the expression for R. This can be effected by momentarily assuming F to be known, and minimizing R as a function of the amplitudes only, using well-known linear least squares techniques. This leads to the following solution for $\bar{c}$

$$
\bar{c} \cong (FF^\dagger)^{-1} F^\dagger \bar{x} . \qquad [23]
$$

It can be seen by multiplying both sides of Eqn. [23] from the left by F, that the operator $F(FF^\dagger)^{-1}F^\dagger$ projects $\bar{x}$ onto the column space of F; hence the name variable projection. Substituting now Eqn. [23] into Eqn. [22], one obtains a new expression for the sum of squared residues, R',

$$
R' = \|\bar{x} - F(FF^\dagger)^{-1} F^\dagger \bar{x}\|^2 . \qquad [24]
$$

Obviously, the amplitudes have now been eliminated from the sum of squared residues. R' is to be minimized only with respect to the nonlinear parameters, which constitutes an attractive feature. The price to pay for this is that the dependence on the nonlinear parameters has become more complicated. However, this aspect has been solved in (68). As usual, starting values of the nonlinear parameters have to be provided in the first iteration cycle. Once the nonlinear parameters have been evaluated, the amplitudes are obtained from the non-iterative solution, Eqn. [23], which does not require starting values. It has been proved in (68) that the (partly iterative) two-step solution

164

is as precise as the one that would be obtained in the single (fully iterative) step of directly minimizing R. Note that *known* amplitudes and phases are *not* eliminated from R. Therefore the fitting of the nonlinear parameters *benefits from such prior knowledge*. When using the non-iterative techniques of 3., this is not the case.

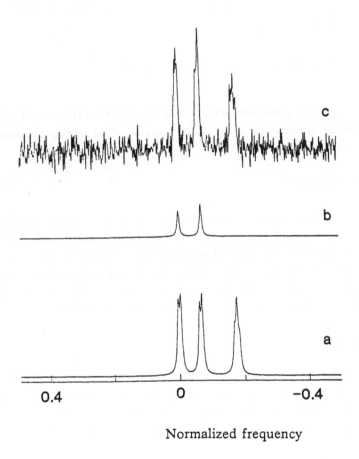

Normalized frequency

Fig. 14: Study of the importance of imposing prior knowledge. Fourier transforms of (a) a seven-sinusoid signal with known relations between frequencies, damping factors and amplitudes (together forming two doublets and a triplet). (b) two additional sinusoids whose frequencies coincide with those of the doublets. (c) the sum of (a), (b) and a Gaussian noise realization.

Another important point is that VARPRO, because of its universal applicability, does *not* exploit the favourable mathematical properties of exponentially damped sinusoids when the use of the latter is warranted. As a result, VARPRO is not as computationally efficient as possible in the case of exponentially damped sinusoids. Various fast versions of VARPRO, capable of exploiting Linear Prediction, have been devised for this particular case (65,66,67). See (23) for yet another fast MLM estimator for exponentially damped sinusoids, not based on VARPRO, but more suitable for imposing prior knowledge.

Finally, we point out that the non-iterative methods treated in Sec.3 are not MLM estimators.

4.3 <u>Application</u> to a <u>simulated</u> signal. The <u>importance</u> of <u>imposing prior knowledge</u>

As pointed out in 4.2, Eqn.[20] can accommodate all sorts of prior knowledge. This property is important because imposing prior knowledge of known parameters improves the precision of the remaining unknown parameters. The signal selected for illustrating this is a simulation comprising nine damped sinusoids, consecutively contaminated with 200 different realisations of Gaussian white noise, each with the same standard deviation. The quantification having to be executed 200 times (Monte Carlo experiment), computational efficiency was required. Therefore, the damping of the sinusoids was made exponential, and one of the modified fast versions, (23), mentioned in 4.2 was applied. The example is focused on quantifying two weak sinusoids whose amplitudes are of the order of the noise and whose frequencies coincide with those of other, stronger lines. The stronger lines are members of a set of seven related lines originating from the same magnetic centre.

A frequency domain display of the situation, including one of the 200 noise realizations is given in Fig. 14. The underlying idea here is to treat the seven related lines as a single entity, with just one variable 'centre' frequency, one variable overall amplitude, and one variable damping factor. As a result of this strategy, the overlap of those lines whose frequencies coincide is reduced to such an extent that quantification becomes feasible. The averaged result of the 200 quantifications is shown for the amplitudes in Table 1.

TABLE 1.
Exploitation of prior knowledge in the quantification of weak, overlapping sinusoids. See Fig.14. The standard deviation of the noise is 0.3 (real and imaginary part, each); 200 noise realizations were used. Result for the *amplitudes* of the two small, single sinusoids (183 successes, 17 failures, see text).

true value	average value	standard deviation	Cramér-Rao lower bound
0.400	0.396	0.128	0.128
0.500	0.524	0.150	0.129

Failure, i.e. no convergence, occurred 17 times. In the remaining 183 quantifications, the results were satisfying: the average values of the amplitudes were near the true ones, while the standard deviation of one amplitude equalled the theoretical lower bound (see next section) and that of the other amplitude was 16% above this lower bound.

4.4 Cramér-Rao Lower Bounds and Experimental Design

In this subsection we review some aspects of the statistical errors in the fitted model parameters caused by noise superimposed on the 'pure' signal. We assume that the model function adequately describes the signal, and that the properties of the added noise are known. For details, the reader is referred to (14,15,16). An important feature of the subject is the existence of theoretical lower bounds to the errors in question, the Cramér-Rao (CR) lower bounds. In the following we first give a recipe for deriving the CR lower bounds pertaining to white Gaussian noise, then we explain that the lower bounds can serve to test the efficiency of a parameter estimation method and to design an optimal measurement. Finally, we comment on the question of whether or not 'signal enhancement' yields increased precision of fitted model parameters.

In the case of general noise characteristics, the formulae for deriving the CR lower bounds may be complicated. However, if the noise is white and Gaussian, matters become much simpler. Denoting the set of all variable model parameters to be fitted by p_j, j=1, ... , J, then the CR lower bounds $\sigma_{p_j,CR}$ are equal to the square roots of the diagonal elements of the inverse of a matrix M, i.e.

$$\sigma_{p_j,CR} = \sqrt{(M^{-1})_{jj}} \qquad\qquad [25]$$

the elements of M being given by

$$(M)_{jj'} = \mathrm{Re}\left[\sum_{n=0}^{N-1} \frac{1}{\sigma_n{}^2} \frac{\partial \hat{x}_n{}^*}{\partial p_j} \frac{\partial \hat{x}_n}{\partial p_{j'}} \right] \quad , \quad j, j' = 1, ... , J \qquad\qquad [26]$$

In Eqn. [26], σ_n is the standard deviation of the real and the imaginary parts of the noise present in sample no. n, while values of all parameters and sampling times are to be substituted in the model function $\hat{x}_n$. * denotes complex conjugation. One way to establish σ_n is to do a blank measurement, i.e. one in which the signal itself is suppressed. Eqn. [26] is easy to program, especially in the case of exponentially damped sinusoids. It should be added that the non-diagonal elements of M^{-1}, if properly normalized, give valuable insight in correlations existing between the various parameters (14).

Suppose now that approximate knowledge of the model parameters of a certain case is available, from previous experiments or otherwise. If the noise characteristics of the spectrometer at hand are also known, one can then calculate the lower bounds on the standard deviations of the model parameters to be fitted, as a function of the SNR. Thus, if some parameter is to be evaluated with an error of, say, 10%, the minimum SNR required for this precision can be predicted. This is an important tool in the preparation of a quantification experiment. We point out that the calculation poses no restrictions to the choice of the sampling times or to the way the samples are averaged within the available measurement time. A possible application of the latter feature is resolution enhancement brought about by non uniform (e.g. exponential) sampling, as recently proposed by Barna et al. (22). Alternatively, non-uniform averaging may be considered, resulting in a non-uniform SNR. Before executing such measurements, one can actually design an optimal strategy that minimizes the lower bounds, and thus save spectrometer time. Of course, approximate prior knowledge of the model parameters must be available.

Apart from predicting the maximum possible precision under the physical conditions in force, the CR lower bounds also enable one to test the quality of the various quantification methods. The latter is done by setting up a Monte Carlo (MC) experiment as indicated in 4.3. If it is found that $\sigma_{P_j,MC}$ approaches $\sigma_{P_j,CR}$ for each j, then the method is optimal at the given SNR. At high SNR, good agreement is usually found between the CR lower bounds and the actual standard deviations. However, as the SNR decreases, all methods eventually fail to produce optimal results. In general, Maximum Likelihood methods remain optimal at lower SNR, than do the non-iterative methods, see e.g. (67). Moreover, the systematic deviations from the true parameter values (bias), which can be inferred also from an MC experiment, are usually lower when using the Maximum Likelihood methods.

The SNR at which a particular quantification method ceases to function optimally is called the *threshold SNR*. Assuming that the threshold has not yet been reached, one may quote the CR lower bounds rather than the actual quantification errors (45c). This is because the latter are convenient to derive.

Finally, we comment briefly on the notion of signal enhancement, see 2. and 3.3, in the light of the theory of the present Section. Suppose one enhances each of the 200 noise-contaminated signals mentioned in 4.3. Most of the 200 resulting low-noise signals (spectra) will reveal part of the structure that was lost by adding the noise. However, in our opinion, differences between the 200 enhanced signals (spectra) will be apparent, which in turn should reflect the magnitudes of the CR lower bounds corresponding to the original noise level. It follows then that enhancement is useful, or perhaps sometimes indispensable for obtaining a qualitative picture of what was previously hidden, but that the precision of subsequent quantification of these signals remains subjected to the original CR lower bounds.

5. SUMMARY and CONCLUSIONS

In this Chapter we have treated some aspects of processing time domain signals. First of all a distinction was made between two classes of time domain signals, namely those made up of a finite number of damped sinusoids (class I), and those containing a continuous distribution of sinusoids (class II). Signals in class I are readily amenable to general signal processing techniques such as MEM, CLEAN, LPZ, LPSVD, HSVD, etc., that do not necessarily require knowledge of the particular system under investigation. Signals in class II, on the other hand, are to be analysed preferably in terms of a specific spin Hamiltonian and the physical parameters therein. The latter approach is beyond the scope of this contribution.

Three different general signal processing techniques were treated.

In 2. we reviewed two Maximum Entropy methods, MEM1 and MEM2, which are meant to produce a well-resolved spectrum from incomplete time domain data. MEM1 is computationally efficient, but not very flexible. MEM2, on the other hand requires significantly more computing time, but is very powerful. In fact, the possibilities of MEM2 have not nearly been fully explored. Application of MEM2 to ESE and ESEEM has not yet been effected, but is recommended. An aspect of MEM which has not yet received sufficient attention is how well the true spectrum is approached. Therefore one has to be careful when using MEM for a comprehensive *quantitative* analysis of spectral parameters.

In order to circumvent worries about the quality of the spectrum, we advocate to fit an appropriate model function directly to the time domain data (should parameter quantification be desired). In 3., this strategy is worked out for a subclass of class I, namely *exponentially* damped sinusoids. A data matrix formed from signals containing such functions has convenient mathematical properties enabling one to fit all model parameters in a non-iterative way, and this without having to provide starting values. Keywords in this context are Linear Prediction, State Space formalism, and Singular Value Decomposition. An advantage of non-iterative quantification is that user involvement can be minimized, which in turn is favourable for automation. However, it should be kept in mind that the methods in question can hardly accommodate spectroscopic prior knowledge. This constitutes a disadvantage at low signal-to-noise ratio.

If the damping of the sinusoids is *non*-exponential, the SVD-based methods will usually continue to function properly. However, many of the quantified parameters cease to have a physical meaning. In such cases the field of application is reduced to 'editing' of the signal; however, this may still be quite useful at times. Nevertheless, whenever the correct model function is available (this is a problem area!), we advocate that it be used. This entails resorting to a nonlinear, iterative, least squares model fitting procedure such as the Variable Projection method, VARPRO. If the noise contained in the signal is Gaussian, VARPRO is a Maximum Likelihood estimator, with the attendant favourable statistical properties. In addition, VARPRO easily accommodates prior knowledge of spectral parameters which can result in a significant improvement of the wanted quantification.

We have indicated that the precision of the quantified model parameters can be approximated by evaluating the Cramér-Rao (CR) lower bounds, assuming that the threshold SNR has not yet been reached and that systematic errors are not significant. Moreover, it was pointed out that the theory of the CR lower bounds enables one to design optimal experiments if approximate values of the model parameters and the properties of the noise are known.

Finally, a matrix decomposition method for quantification of 2-D MR signals satisfying the 2-D exponentially damped sinusoid model was presented. In this method *all data are processed simultaneously*, rather than row- or column-wise.

ACKNOWLEDGMENT

This work was performed as part of the research program of the "Stichting voor Fundamenteel Onderzoek der Materie" (FOM) and the "Stichting voor Technische Wetenschappen" (STW), with financial aid from the "Nederlandse Organisatie voor Wetenschappelijk Onderzoek" (NWO).

APPENDIX I. The Tapered Burg Algorithm

As mentioned in 2.2.1, the Burg algorithm is recursive. In fact, the number of prediction coefficients, M, is initially put to one, and is subsequently augmented by steps of one. Consequently, the prediction coefficients and various other quantities involved must be given a label to keep track of the number recursions. Thus, we write the forward and backward *prediction errors*, $\varepsilon_{f,n}^{(M)}$ and $\varepsilon_{b,n}^{(M)}$, and the *prediction error power*, P_M, as follows.

$$\varepsilon_{f,n}^{(M)} = x_n - a_1^{(M)} x_{n-1} - a_2^{(M)} x_{n-2} - \cdots - a_M^{(M)} x_{n-M} , \qquad M \geq 1, \qquad [AI.1]$$

$$\varepsilon_{b,n}^{(M)} = x_{n-M} - a_1^{*(M)} x_{n-M-1} - a_2^{*(M)} x_{n-M-2} - \cdots - a_M^{*(M)} x_n, \quad M \geq 1, \qquad [AI.2]$$

$$P_M = \sum_{n=M}^{N-1} \frac{|\varepsilon_{f,n}^{(M)}|^2 + |\varepsilon_{b,n}^{(M)}|^2}{2N} \quad , \qquad [AI.3]$$

The procedure is initiated by writing

$$\varepsilon_{f,n}^{(0)} = \varepsilon_{b,n}^{(0)} = x_n \quad , \qquad\qquad [AI.4]$$

and

$$P_o = \frac{1}{N} \sum_{n=0}^{N-1} x_n x_n^* \qquad\qquad [AI.5]$$

Next, P_M is minimized recursively, i.e. for $M=1, 2, \ldots$, etc. The minimization is executed with respect to the 'end coefficient' $a_M^{(M)}$ only, while the prediction errors are tapered for reasons given above. It can be shown that this results in the following expression

$$a_M^{(M)} = 2 \frac{\displaystyle\sum_{n=M}^{N-1} \varepsilon_{f,n}^{(M-1)} \varepsilon_{b,n-1}^{*(M-1)} w_n^{(M)}}{\displaystyle\sum_{n=M}^{N-1} \left[|\varepsilon_{f,n}^{(M-1)}|^2 + |\varepsilon_{b,n-1}^{(M-1)}|^2 \right] w_n^{(M)}} \quad , \quad M \geq 1 \quad , \qquad [AI.6]$$

where $w_n^{(M)}$ is the taper function, which will be briefly specified later on. The recursion expressions for the remaining quantities are

$$P_M = P_{M-1} \left(1 - |a_M^{(M)}|^2 \right) \qquad\qquad M \geq 1 \quad , \qquad [AI.7]$$

$$
\begin{bmatrix}
1 \\
-a_1^{(M)} \\
\cdot \\
\cdot \\
\cdot \\
-a_M^{(M)}
\end{bmatrix}
=
\begin{bmatrix}
1 \\
-a_1^{(M-1)} \\
\vdots \\
\\
-a_M^{(M-1)} \\
0
\end{bmatrix}
- a_M^{(M)}
\begin{bmatrix}
0 \\
-a_{M-1}^{*(M-1)} \\
\vdots \\
\\
-a_1^{*(M-1)} \\
1
\end{bmatrix}
\quad , \quad M > 1 \quad , \qquad [AI.8]
$$

$$\varepsilon_{f,n}^{(M)} = \varepsilon_{f,n}^{(M-1)} - a_M^{(M)} \varepsilon_{b,n-1}^{(M-1)} \qquad\qquad M \geq 1 \quad , \qquad [AI.9]$$

$$\varepsilon_{b,n}^{(M)} = \varepsilon_{b,n-1}^{(M-1)} - a_M^{*(M)} \varepsilon_{f,n}^{(M-1)} \qquad\qquad M \geq 1 \quad , \qquad [AI.10]$$

For sinusoids in white noise, the recursion is continued until M is in the region of $N/3 \leq M \leq N/2$. Finally, the spectrum is given by the expression

$$S(\nu) = \frac{P_M \Delta t}{\left| 1 - \sum_{m=1}^{M} a_m^{(M)} \exp(-i\, 2\pi\nu m\Delta t) \right|^2} \qquad [\text{AI.11}]$$

the denominator of which can be evaluated by means of FFT.

Various alternative taper functions have been devised to remove the phenomenon of line shifting and line splitting that occasionally marred a Burg spectrum in the past. These tapers are the Hamming window (40), an optimum parabolic window (69), and a data adaptive window (39). See also (70). Another important improvement could yet be achieved by modifying the expression for the first 'end coefficient' $a_1^{(1)}$ (71,39,72). For a discussion of the deficiencies of the original, untapered, Burg algorithm the reader is referred to e.g. (73,74).

APPENDIX II. 2-D Time domain model fitting

In this appendix we treat quantification of a 2-D Magnetic Resonance signal, $x_{nn'}$, $n=0, \dots , N-1$, $n'=0, \dots , N'-1$, that can be adequately described by a model function comprising a sum of exponentially damped sinusoids, i.e.

$$\hat{x}_{nn'} = \sum_{k=1}^{K} \sum_{k'=1}^{K'} c_{kk'} z_k^{n} z_{k'}^{'\,n'} \quad , \; n=0 \dots , N-1, \quad n'=0, \dots , N'-1 \quad , \qquad [\text{AII.1}]$$

where a primed and an unprimed dimension are distinguished, $c_{kk'}$ is a complex-valued amplitude, and the product $z_k^{n} z_{k'}^{'\,n'}$ a 2-D sinusoid. See 1.4 for a further explanation of the symbols. (The 2-D analogue of the quantity δ is omitted here for simplicity.) From the data we form a matrix X, according to

$$X \stackrel{\text{def}}{=} \begin{bmatrix} x_{00} & x_{01} & x_{02} & \cdots & & x_{0\,N'-1} \\ x_{10} & x_{11} & \cdot & & & \cdot \\ x_{20} & \cdot & & & & \cdot \\ \cdot & & & & & \cdot \\ \cdot & & & & & \\ \cdot & & & & & \\ x_{N-1\,0} & & & \cdots & & x_{N-1\,N'-1} \end{bmatrix} \qquad [\text{AII.2}]$$

In absence of noise, X can be decomposed as follows

$$X = \begin{bmatrix} 1 & . & . & . & 1 \\ z_1^1 & . & . & . & z_K^1 \\ . & & & & . \\ . & & & & . \\ . & & & & \\ . & & & & \\ z_1^{N-1} & . & . & . & z_K^{N-1} \end{bmatrix} \begin{bmatrix} c_{11} & \cdots & c_{1K'} \\ \vdots & & \vdots \\ c_{K1} & \cdots & c_{KK'} \end{bmatrix} \begin{bmatrix} 1 & z_1'^1 & \cdots & z_1'^{N'-1} \\ . & . & & . \\ . & . & & . \\ 1 & z_{K'}'^1 & \cdots & z_{K'}'^{N'-1} \end{bmatrix} , \qquad [AII.3]$$

which can be checked by working out the successive matrix multiplications. Note that *all* data points are involved simultaneously. When noise is added, Eqn. [AII.3] is an approximation, as pointed out in 3.5 for the 1-D case.

If K=K', then the above decomposition can be achieved via a single SVD, applying the same approach as in 3.5, but this time using both the left *and* right singular vectors (11). However, the restriction to a diagonal amplitude matrix, made in (11), is too strict. In fact, all that is needed is that the amplitude matrix have full rank. If K≠K', a complication arises because the rank of X equals min[K,K']. One may then proceed as follows. Using the modified fast VARPRO method of (76), X is decomposed according to

$$X = \begin{bmatrix} 1 & . & . & . & 1 \\ z_1^1 & . & . & . & z_K^1 \\ . & & & & . \\ . & & & & . \\ . & & & & \\ . & & & & \\ z_1^{N-1} & . & . & . & z_K^{N-1} \end{bmatrix} \begin{bmatrix} A_{11} & \cdots & A_{1N'} \\ \vdots & & \vdots \\ A_{K1} & \cdots & A_{KN'} \end{bmatrix} . \qquad [AII.4]$$

Subsequently, the K×N' matrix A on the right-hand side of Eq. [AII.4] is in turn decomposed, using once more the method of (76), which results in

$$A = \begin{bmatrix} c_{11} & \cdots & c_{1K'} \\ \vdots & & \vdots \\ c_{K1} & \cdots & c_{KK'} \end{bmatrix} \begin{bmatrix} 1 & z_1'^1 & \cdots & z_1'^{N'-1} \\ . & . & & . \\ . & . & & . \\ 1 & z_{K'}'^1 & \cdots & z_{K'}'^{N'-1} \end{bmatrix} . \qquad [AII.5]$$

The method just described was successfully tested on a simulated 2-D signal with K=3, K'=4, N=N'=50, adding fifty different noise realizations. Two of the twelve amplitudes were equal to half the noise standard deviation (75). It turned out that the standard deviations of all parameters approached the Cramér-Rao lower bounds.

REFERENCES

1   D.J. Singel, W.A.J.A. van der Poel, J. Schmidt, J.H. van der Waals, and R. de Beer, J. Chem. Phys., **81** (1984) 5453-5461.

2   R. de Beer, J.A. Duine, D.van Ormondt, and J.W.C. van der Veen, unpublished results.

3   S. Haykin (Ed.), Nonlinear Methods of Spectral Analysis, 2nd Edition, Springer, Berlin, 1983.

4   C.R. Smith and W.T. Grandy (Eds.), Maximum Entropy and Bayesian Methods In Inverse Problems, Reidel, Dordrecht, 1985.

5   J. Skilling (Ed.), Proc. 8th MAXENT Workshop (1-5 Aug., 1988, Cambridge, UK), Kluwer, Dordrecht, 1989.

6   C.L. Lawson and R.J. Hanson, Solving Least Squares Problems, Prentice-Hall, Englewood Cliffs, 1974.

7   V.C. Klema and A.J. Laub, IEEE Trans. Aut. Control, AC-**25** (1980) 164-176.

8   G.H. Golub and C.F. Van Loan, Matrix Computations, The Johns Hopkins University Press, Baltimore, 1983.

9   S.L. Marple, Digital Spectral Analysis with Applications, Prentice-Hall, Englewood Cliffs, 1987.

10   S.M.Kay, Modern Spectral Estimation, Prentice-Hall, Englewood Cliffs, 1988.

11   S.Y. Kung, K.S. Arun, and D.V. Bhaskar Rao, J. Opt. Soc. Am., **73** (1983) 1799-1811.

12   W.B. Mims, J. Magn. Reson., **59** (1984) 291-306.

13   P.A. Snetsinger, J.B. Cornelius, R.B. Clarkson, M.K. Bowman, and R.L. Bedford, J. Phys. Chem., **92** (1988) 3696-3699.

14   A. van den Bos, Ch.8 in: Handbook of Measurement Science, P.H. Sydenham (Ed.), Wiley, London, 1982.

15   M.B. Priestley, Spectral Analysis and Time Series, Vols. 1 and 2, Academic Press, London, 1981.

16   J.P. Norton, An Introduction to Identification, Academic Press, London, 1986.

17   J.C.J. Barna, S.M. Tan, and E.D. Laue, J. Magn. Reson., **78** (1988) 327-332.

18   J. Tang and J.R. Norris, J. Magn. Reson., **78** (1988) 23-30.

19   M.L Waller and P.S. Tofts, Magn. Reson. Med., **4** (1987) 385-392.

20   J.W.C. van der Veen, R. de Beer, P.R. Luyten, and D. van Ormondt, Magn. Reson. Med., **6** (1988) 92-98.

21   F. Ni, G.C. Levy, and H.A. Scheraga, J. Magn. Reson., **66** (1986) 385-390.

22   J.C.J. Barna, E.D. Laue, M.R. Mayger, J. Skilling, and S.J.P. Worrall, J. Magn. Reson., **73** (1987) 69-77.

23   R. de Beer, D. van Ormondt, and W.W.F. Pijnappel, in: R. Lacoume (Ed.),

Proc. Fourth European Conference on Signal Processing, Elsevier, Amsterdam, 1988.

24 N. Wu, The Maximum Entropy Method and its Applications in Radio Astronomy, Ph. D. Thesis, Sydney, Australia, 1985. See abstract in IEEE ASSP Magazine, October 1986, p.33.

25 J.P. Burg, Maximum Entropy Spectral Analysis, 37th Ann. Intern. Meeting, Soc. Exploration Geophysicists, Oklahoma City, 1967.

26 S.F. Gull and G.J. Daniell, Nature, **272** (1978) 686-690.

27 S. Sibisi, Nature, **301** (1983) 134-136.

28 E.D. Laue, J. Skilling, J. Staunton, S. Sibisi, and R.G. Brereton, J. Magn. Reson., **62** (1985) 437-452.

29 P.J. Hore, J. Magn. Reson., **62** (1985) 561-567.

30 J.F. Martin, J.Magn. Reson., **65** (1985) 291-297.

31 D. van Ormondt and K. Nederveen, Chem. Phys. Lett., **82** (1981) 443-446.

32 H. Barkhuysen, R. de Beer, E.L. de Wild, and D. van Ormondt, J. Magn. Reson., **50** (1982) 299-315.

33 S.A. Dikanov, A.V. Astashkin, Yu. D. Tsvetkov and M.G. Goldfeld, Chem. Phys. Lett., **101** (1983) 206-210.

34 S.A. Dikanov, A.V. Astashkin and Yu. D. Tsvetkov, Chem. Phys. Lett., **105** (1984) 451-455.

35 A.V. Astashkin, S.A. Dikanov and Yu. D. Tsvetkov, Chem. Phys. Lett., **136** (1987) 204-208.

36 C.L. Byrne and R.M. Fitzgerald, IEEE Trans. Acoust., Speech, Signal Processing, ASSP-**32** (1984) 914-916.

37 M.A. Delsuc and G.C. Levy, J. Magn. Reson., **76** (1988) 306-315.

38 A. van den Bos, IEEE Trans. Information Theory, IT-**17** (1971) 493-494.

39 B.I. Helme and C.L. Nikias, IEEE Trans. Acoust., Speech, Signal Processing, ASSP-**33** (1985) 903-910.

40 D. N. Swingler, Proc. IEEE, **67** (1979) 1368-1369.

41 P. Höfer, A. Grupp, H. Nebenführ, and M. Mehring, Chem. Phys. Lett., **132** (1986) 279-282

42 H. Barkhuysen, R. de Beer, A.F. Deutz, D. van Ormondt, and G. Völkel, Sol. State Commun., **49** (1984) 679-684.

43 A. de Groot, A.J. Hoff, R. de Beer, and H. Scheer, Chem. Phys. Lett., **113** (1985) 286-290.

44 R. Kumaresan and D.W. Tufts, IEEE Trans. Acoust., Speech, Signal Processing, ASSP-**30** (1982) 833-840.

45 a) H. Barkhuysen, R. de Beer, W.M.M.J. Bovée, and D. van Ormondt, J. Magn. Reson., **61** (1985) 465-481. b) H. Barkhuysen, R. de Beer, and D. van Ormondt, J. Magn. Reson., **64** (1985) 343-346. c) H. Barkhuysen, R. de Beer, and D. van Ormondt, J. Magn. Reson., **67** (1986) 371-375.

46 H. Barkhuysen, R. de Beer, W.M.M.J. Bovée, A.M. van den Brink, A.C.

Drogendijk, D. van Ormondt, and J.W.C. van der Veen, in: I.T. Young et al. (Eds), Signal Processing III: Theory and Applications, Elsevier, Amsterdam, 1986, pp. 1359-1362.

47   H. Barkhuysen, R. de Beer, and D. van Ormondt, J. Magn. Reson., **73** (1987) 553-557.

48   B.A. Johnson, J.A. Malikayil, I.M. Armitage, J. Magn. Reson., **76** (1988) 352-357.

49   D.W. Tufts, R. Kumaresan, and I. Kirsteins, Proc. IEEE, **70** (1982) 684-685.

50   J.A. Cadzow, IEEE Trans. Acoust., Speech, Signal Processing, ASSP-**36** (1988) 49-62.

51   H. Barkhuysen, R. de Beer, A.C. Drogendijk, D. van Ormondt, and J.W.C. van der Veen, in B. Maraviglia (Ed.), Proc. International School of Physics "Enrico Fermi" on The Physics of NMR Spectroscopy in Biology and Medicine (1986), Italian Physical Society, 1988, pp. 313-344.

52   T. Dumelow, P.C. Riedi, J.S. Abell, and O. Prakash, J. Phys. F, **18** (1988) 307-322.

53   A.E. Schussheim and D. Cowburn, J. Magn. Reson., **71** (1987) 371-378.

54   H. Gesmar and J.J. Led, J. Magn. Reson., **76** (1988) 575-586.

55   G.L. Millhauser and J.H. Freed, J. Chem. Phys., **85** (1986) 63-67 .

56   J. Gorcester and J.H. Freed, J.Magn. Reson., **78** (1988) 292-301.

57   H. Barkhuysen, R. de Beer, A.C. Drogendijk, D. van Ormondt, and J.W.C. van der Veen, Proc. XXIII CONGRES AMPERE on Magn. Reson., 1986, pp. 30-35.

58   H. Gesmar and J.J. Led, J. Magn. Reson., **76** (1988) 183-192.

59   J. Tang, C.P. Lin, M.K. Bowman, and J.R. Norris, J. Magn. Reson., **62** (1985) 167-171.

60   A.C. Kot, S. Parthasarathy, D.W. Tufts, and R.J. Vaccaro, Proc. ICASSP, 1987, pp. 1549-1552.

61   S. Mayrargue, Proc. ICASSP, 1988, pp. 2456-2459.

62   R. Roy, A. Paulray, T. Kailath, IEEE Trans. Acoust., Speech, Signal Processing, ASSP-**34** (1986) 1340-1342.

63   J. Schmidt, W.J. Buma, E.J.J. Groenen, and R. de Beer, J. Mol. Struct., **173** (1988) 249-253.

64   M. Dwass, Probability and Statistics, Benjamin, New York, 1970.

65   R. Kumaresan, L.L. Scharf, and A.K. Shaw, IEEE Trans. Acoust., Speech, Signal Processing, ASSP-**34** (1986) 637-640.

66   Y. Bresler and A. Macovski, IEEE Trans. Acoust., Speech, Signal Processing, ASSP-**34** (1986) 1081-1089.

67   S. W. Park and J.T. Cordaro, Proc. ICASSP, 1987, pp. 1501-1504.

68   G.H. Golub and V. Pereyra, SIAM J. Numer. Anal., **10** (1973) 413-432.

69   M. Kaveh and G.A. Lippert, IEEE Trans. Acoust., Speech, Signal Processing, ASSP-**31** (1983) 438-444.

70   K.K. Paliwal, Indian J. Technol., **24** (1986) 677–681.

71   P.F. Fougere, J. Geophys, Res., 1977, pp. 1051–1054.

72   M.K. Ibrahim, IEEE Trans. Acoust., Speech, Signal Processing, ASSP-**35** (1987) 1474–1476.

73   R.W. Herring, IEEE Trans. Acoust., Speech, Signal Processing, ASSP-**35** (1980) 692–701.

74   M.K. Ibrahim, IEEE Trans. Acoust., Speech, Signal Processing, ASSP-**35** (1987) 1476–1479.

75   R. de Beer, D. van Ormondt, and W.W.F. Pijnappel, Signal Processing, **15** (1988) 293–302.

76   R. Kumaresan and A.K. Shaw, Proc. ICASSP, 1985, pp. 576–579.

CHAPTER 5

TWO DIMENSIONAL AND FOURIER TRANSFORM EPR

JEFF GORCESTER, GLENN L. MILLHAUSER, AND JACK H. FREED

1. INTRODUCTION

In the past decade, an extensive range of modern time domain two-dimensional EPR techniques for the study of spin relaxation and motional dynamics of small and macromolecules in fluids, including membranes, have been developed based on electron-spin-echo and on Fourier Transform (FT) techniques. A primary concern has been with the spectra of nitroxide spin labels and spin probes.

The nitroxide spin-label technique has, in the last 20 years, permeated many areas of biophysics (1-4). The principal reliance has been on cw-EPR methods despite the somewhat limited features of the characteristic three-line hyperfine (hf) pattern from the ^{14}N nucleus. Of course, as rotational motions slow down, spectra take on a greater range of shapes in the "slow-motional" regime. The analysis of such spectra is more complicated (5,6), but rewarding, because it can supply more detailed information on microscopic dynamics and ordering in membrane systems (6-11). Recent developments in computational methodology have now lead to a convenient but powerful computer program to simulate slow motional spectra that is readily available (12).

But the fact remains that these spectra exhibit rather low resolution to the details of the dynamics, a matter which is exacerbated by the proton (and/or deuteron) super-hyperfine (shf) splittings. In a significant sense, EPR lagged behind NMR in its applications to the study of dynamic molecular structure, despite its virtues of greater sensitivity to fewer spins and the shorter time scale over which it examines motional dynamics. In NMR, after all, one can use spin echoes to measure homogeneous T_2's; pulse techniques give T_1's directly, and modern two-dimensional (2D) methods based on FT-NMR can provide detailed mappings of complicated cross-relaxation pathways.

But now it is fair to say that the new, though sophisticated, time-domain EPR techniques have been able to come to the aid of the EPR spin-labeling method. Modern electron-spin-echo methods can provide a display, in a two-dimensional format, of the homogeneous lineshapes across an inhomogeneous EPR spectrum. Alternatively, they can provide the cross-relaxation rates from each point in the spectrum. Such 2D spectra are found to provide much enhanced sensitivity to motional dynamics, and they even place greater demands on the simulation algorithms because of this enhanced sensitivity (13,14).

Perhaps an even more dramatic development has been the recent introduction of two-dimensional FT-EPR spectroscopy for nitroxides. Instead of looking at a simple three-line EPR spectrum, one may now observe a two-dimensional display of "auto-peaks" and of "cross-peaks" whose intensities relate directly to cross-relaxation phenomena such as Heisenberg spin exchange and/or dipole-dipole interactions modulated by the diffusion processes. Extension to the slow-motional regime would lead to the complete 2D mapping of the (integrated) transition rates between each and every point in the spectrum. This can only enhance the sensitivity, reliability, and accuracy with which EPR may be applied to questions of dynamic molecular structure.

In this chapter we review these recent developments, explaining the background theory and the methods, as well as the applications which have already been realized, and the potential for further applications. Other recent reviews focus on related matters, including the application of EPR time-domain methods for the study of surfaces (15), studies of rotational dynamics in fluids and membranes (16), and the theoretical and computational methods for extracting the molecular dynamics from the EPR spectra (14).

In Section 2, the relevance of electron-spin-echoes (ESE) to studies of molecular motions is reviewed. Then in Section 3, the 2D-ESE method for obtaining T_2 variations across the spectrum is described in some detail. Section 4 provides the description of techniques to study magnetization transfer (e.g. cross-relaxation) variations across the spectrum. This includes inversion recovery, stimulated echoes, and spin-echo electron-electron double resonance (ELDOR). The effects of electron-spin echo envelope modulations on motional studies are outlined in Section 5. The new 2D-FT-EPR techniques are discussed in Section 6, which includes sub-sections on spin-echo correlated spectroscopy (SECSY), correlation spectroscopy (COSY), and 2D-ELDOR. The initial applications of these methods and their future potential are also discussed in this Section.

2.   ELECTRON-SPIN ECHOES AND MOTIONAL DYNAMICS

The principal motivations for applying electron-spin-echo (ESE) techniques to spin relaxation in model membranes include: 1) the ability to separate homogeneous from inhomogeneous contributions to the linewidths (or T_2), as well as to profit from the resulting increase in resolution; 2) the ease of simultaneously performing T_1 measurements; 3) the possibility that special ESE techniques could provide information on motional dynamics in addition to that from cw studies; and 4) the possibility of extending the range of study to slower motions. Fig. 1 summarizes the most typically used ESE pulse sequences.

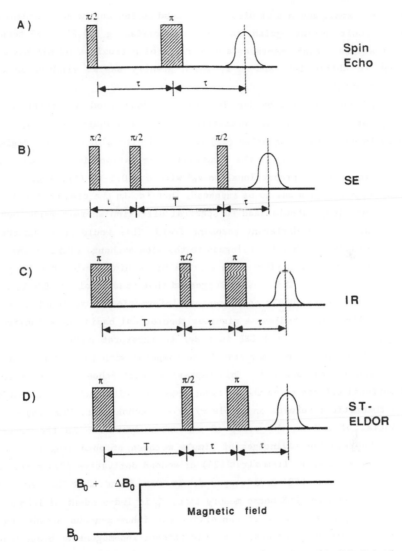

Fig. 1. Typical spin echo pulse sequences: (a) Hahn spin echo (b) Stimulated Echo (c) Inversion Recovery (d) ST-ELDOR echoes.

When an echo technique (17-21) is applied to spin labels with inhomogeneously-broadened cw-EPR lines (e.g. due to unresolved--or partially resolved--proton shf interactions in the case of nitroxides) the measured phase-memory time T_M is equal to T_2, the homogeneous linewidth of a single spin packet (21-24) (cf. Fig. 2). Past attempts at removing, or at least reducing, the inhomogeneous broadening were to utilize deuterated spin probes such as PD-Tempone for this purpose, but it is very inconvenient to have to perdeuterate all the various spin labels. A more serious problem occurs in the case of oriented model membrane samples, where small amounts of disorder can lead to inhomogeneous broadening that is very difficult to distinguish from motional broadening (10,25). ESE techniques now permit accurate experiments on a wide range of nitroxides of different sizes and shapes to obtain T_2's and T_1's, which greatly assist studies on dynamic molecular structure.

The ESE work on nitroxides in liquids shows good agreement with the motionally-narrowed linewidths extracted by cw techniques (cf. Fig. 2). An especially interesting observation (22) was that in the slow motional regime, for $\tau_R \approx 10^{-7}$ to 10^{-6} sec, where τ_R is the rotational correlation time, the phase memory time T_M was found to be proportional to τ_R^{α} with $\alpha \approx 0.5\text{-}1$, (cf. Fig. 2). Simple arguments suggested this was to be expected. That is, in the slow motional regime, reorientational jumps should lead to spectral diffusion wherein each jump takes place between sites of different resonant field. This would be an uncertainty-in-lifetime broadening that is analogous to the slow exchange limit in the classic two-site case, and it contributes to T_M. The broadening would then be given by τ_R^{-1}, the jump frequency. This result suggested that studies of the ESE T_M in slow-motional spectra would supply complementary information on motional dynamics to that from cw lineshape studies. A rigorous theoretical basis was established for the analysis of slow-motional ESE in order to interpret such experiments with confidence (23). The theory may readily be computed with the newer cw-EPR slow-motional computer programs (12). This emphasizes that echoes relate to the same type of motional effects as do the cw-lineshapes, but with much better resolution.

The theoretical results on simple $\pi/2\text{-}\tau\text{-}\pi\text{-}\tau$ echoes (cf. Fig. 1a) clarified the potential of ESE in the study of molecular dynamics. In the theoretical analysis, the relative advantages of simple methods of obtaining τ_R from slow-motional cw spectra (e.g. from first (26) or second derivative (27) outer extreme linewidths) vs. from ESE were compared. It was found that ESE methods are preferable, because the ESE phase memory time, T_M is independent of inhomogeneous broadening. Simple approaches at cancelling out inhomogeneous broadening in cw studies (26,27) are not rigorous, since significant inhomogeneous broadening will have a non-linear effect on the observed cw widths (23,26). It was concluded that careful study of the variation of T_M (i.e. T_2) across the spectrum would be a useful way to proceed in applying ESE to this problem. This led to the two-dimensional spectroscopy approach outlined below.

Let us first summarize the theoretical results on simple $\pi/2\text{-}\tau\text{-}\pi\text{-}\tau$ echoes. It has been demonstrated that (23): 1) the ESE decay envelopes show a short-time behavior with an $e^{-c\tau^3}$ dependence on τ and a long-time behavior of $\exp[-\tau/T_M^{\infty}]$. 2) The asymptotic phase-memory decay constant T_M^{∞} shows a significant dependence on models, and this was traced to the different mechanisms of spectral diffusion induced by these models. 3) The T_M^{∞} obtained from selective echoes on different parts of the (nitroxide) spectrum show significant differences. 4) The short-time

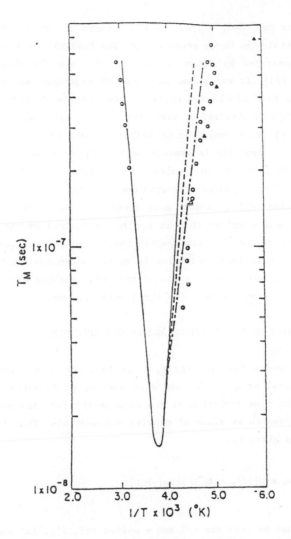

Fig. 2. Plot of T_2 (or T_M) vs T^{-1} for the nitroxide probe Tempone in 85% glycerol/H_2O. The circles show data collected from spin echo experiments; the triangles show T_2 as measured from the center maximum of 2D ESE spectra. The different lines show T_2 calculations from the same spectral region for the different models of jump diffusion (solid line), free diffusion (dashed line) and Brownian diffusion (dashed-dotted line). The calculations employed the values of τ_R extrapolated from the fast-motional regime. [From Ref. 29].

behavior yields a $c \propto \tau_R^{-1}$, and c is independent of diffusional model. The experimental results, such as those in Fig. 2 show that in actual ESE experiments one is primarily sensitive to the T_M^∞.

There is an important variant on the simple spin echo. It is the Carr-Purcell (CP) sequence: $\pi/2 - (\tau - \pi - \tau)_n$, where the π pulse is applied n times giving rise to n echoes at times $2m\tau$, m=1, 2...n, and one detects the decay of the echo train for given τ to obtain the decay constant T_M^C. The feasibility of such experiments by ESE was demonstrated by Eliav and Freed (28), and the theory is given by Schwartz et al. (23). It was shown that such CP sequences are potentially more informative than just simple two-pulse echoes. In the latter, usually only a single T_M (e.g. T_M^∞) is obtained at each spectral position, while in the former a whole set of $T_M^C(\tau)$ vs. τ would, in principle, be obtained, which could be used to explore motional dynamics in greater detail. This is particularly true as τ becomes comparable to τ_R, which results in $T_M^C > T_M^\infty$ (23). In fact, as $\tau/\tau_R \to 0$, the successive π pulses repeatedly reverse the electron-spin precession so rapidly that the hyperfine and g-tensor anisotropy is (coherently) averaged more effectively by these π pulses than it can be randomized by the slow tumbling. [This is analogous to the NMR case wherein the CP sequence is used to remove the effect on T_2 of translational diffusion in an inhomogeneous applied field].

We now inquire how a two-dimensional technique enables us to obtain maximum information from T_2 measurements. (We shall refer to the T_M as T_2 below)

3.   2D-ESE STUDIES OF T_2 VARIATIONS ACROSS THE SPECTRUM

3.1 The Method

The 2D-ESE method for studying T_2 variations across the spectrum is based upon the theoretical study of ESE and slow motions by Schwartz et al. (23). It was shown that the time evolution of the echo height for slow motions could, in general, be represented as a sum of complex exponentials. That is, the observed echo signal, s is given by:

$$s(2\tau+t) \propto \mathrm{Re} \sum_{j,k} a_{kj} \exp[-(\Lambda_k + \Lambda_j^*)\tau]\exp[-\Lambda_k t] \tag{1}$$

where τ is the time between the $\pi/2$ and π pulses (cf. Fig. 1a) and where $2\tau+t$ is the time period measured from the initial pulse. The echo maximum is expected at time 2τ. In Eqn. [1], Re represents the real part of the complex function to its right and Λ_j and Λ_k are complex eigenvalues discussed below. Eqn. [1] can best be understood in terms of the concept of the "dynamic spin packet" (DSP). That is, in the rigid limit, the inhomogeneously – broadened EPR spectrum from a polycrystalline (or glassy) sample is made up of the broad envelope of contributions from many, many (actually a continuum of) spin packets, each one resonating at a definite frequency and associated with the spins on molecules oriented at the appropriate angle with respect to the static magnetic field. The rigid limit spectrum is determined in the usual way in terms of the g-tensor and hyperfine tensor anisotropies.

As the rotational motion increases, each spin-packet in the continuum will "interact" with nearby spin-packets, since the motion will transform molecules contributing to one spin-packet into contributors to another at a different orientation. As a result of this coupled behavior of the original (i.e., rigid-limit) spin packets, there will be new "normal-mode" solutions, the DSP's, which will be linear combinations of the original spin packets. The physics of the problem is described by the stochastic Liouville equation, (5,14), which simultaneously includes the reversible quantum-mechanical spin dynamics and the irreversible molecular dynamics (usually treated classically). Typical spin Hamiltonians and dynamical models are summarized elsewhere (5,14). The eigenvectors of the stochastic Liouville operator are the DSP's, and the Λ_j in Eqn. [1] are the corresponding eigenvalues in the rotating frame. The imaginary parts $Im(\Lambda_j)$ represent the resonance frequencies of the associated DSP [i.e. $\omega_j = Im(\Lambda_j)$], while the real parts $Re(\Lambda_j)$ represent the corresponding natural or homogeneous widths and are associated with the observed $T_{2j} = [Re(\Lambda_j)]^{-1}$. (Note that the asterisk in Eqn. [1] represents a complex conjugation, which requires a simple generalization (23) if echo envelope modulation is to be considered.) In general, the relative weights of the DSP's given by the "diagonal" coefficients a_{kk} in Eqn. [1] and the "beats" between pairs of DSP's given by the "off-diagonal" coefficients a_{kj} ($k\neq j$) have a complicated dependence upon the form of these "normal-modes" and of the spin transition moments as averaged over the equilibrium ensemble of molecular orientations. However, near the rigid limit, the a_{kj} simplify to:

$$a_{kj} = (O^T U)_j^2 \, \delta_{kj} \qquad\qquad\qquad [2]$$

where O is the (nearly) real orthogonal matrix associated with the transformation that diagonalizes the stochastic Liouville operator, while U is the vector of (appropriately averaged) spin transition moments.

When Eqn. [2] applies, Eqn. [1] becomes:

$$s(2\tau+t) \propto \sum_j a_{jj} \exp[-2Re(\Lambda_j \tau)] \exp[-Re(\Lambda_j t)] \cos[Im(\Lambda_j t)] . \qquad [3]$$

Thus, in a "two-dimensional" plot of s vs. the two independent variables τ and t, the t dependence includes both the resonance frequency of each DSP and its $T_{2,j}$. If we set t=0 and step out τ, then we obtain the echo envelope as a superposition of exponential decays corresponding to the different $T_{2,j}$ values from each DSP contributing appreciably to the signal. However, such a result would be difficult to disentangle. Instead, let us take advantage of the additional resolution provided by two-dimensional spectroscopy. Let us perform a double Fourier transform of Eqn. [3], recognizing however that in the near-rigid limit the linewidth for each DSP will largely be masked by various sources of inhomogeneous broadening such as unresolved shf interactions and site variations in the spin-Hamiltonian parameters. We shall therefore assume a Gaussian inhomogeneous width $\Delta \gg T_{2,j}^{-1}$. This yields the two-dimensional spectrum given by:

$$S(\omega,\omega') \propto \sum_j a_{jj} \frac{T_{2,j}}{1+\omega^2 T_{2,j}^2} \exp[-(\omega'-\omega_j)^2/\Delta^2] \qquad [4]$$

which is a sum of Lorentzians along the ω-axis (the FT of τ) and the sum of Gaussians along the ω'-axis (the FT of t). But more importantly, we have separated out the role of the relaxation of each spin packet given by its $T_{2,j}$, which is plotted along the ω axis, from its resonance position in the spectrum given along the ω' axis. In fact, for $\omega=0$ we almost recover the expression for the cw lineshape (with Gaussian inhomogeneous broadening) along the ω'-axis. Thus this separation allows us to <u>observe how the $T_{2,j}$ vary across the whole spectrum</u>. In other words we have succeeded in resolving the different dynamic behavior of the various spin packets, subject however, to the resolution limit due to the inhomogeneous broadening, Δ. Examples of such 2D-ESE spectra appear in Fig. 3.

The actual experimental technique that has been utilized by Millhauser and Freed (MF) (29) is to sweep through the static magnetic field B_0 very slowly, but with a microwave-field intensity B_1 small enough that only DSP's associated with widths well within the inhomogeneous width Δ are effectively rotated by the pulses. MF show that Eqn. [4] applies to their experiment provided that the following set of inequalities apply:

$$\gamma B_s > \Delta \gg \gamma B_1 \gg T_{2,j}^{-1} \qquad [5]$$

where γB_s is the full extent of the spectrum. In this format, one sweeps through B_0 and observes the echo height at 2τ; i.e. one obtains an "echo-induced EPR spectrum" for each value of τ. This provides an $S(\tau,\omega')$ such that the FT in t is automatically obtained. Then an FT in τ is performed after the experiment is repeated for a sufficient number of values of τ to obtain the form of Eqn. [4]. This technique has the advantage of simplicity, a minimum of experimental artifacts, and only low-power microwave pulses are required. However, it is time consuming, typically requiring about 5-10 hours for a complete set of 2D data. One may use a standard ESE spectrometer such as the one illustrated in Fig. 4. We shall show in Sect. 6 that modern techniques now permit such an experiment to be performed with large enough B_1 pulses such that the whole spectrum can be

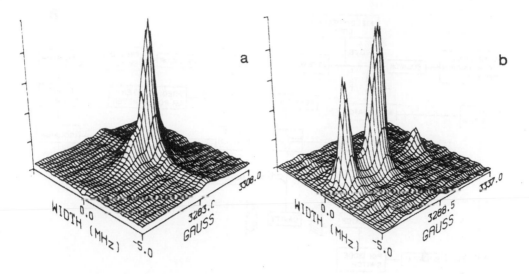

Fig. 3. Experimental 2D-ESE spectra from CSL spin probe in oriented multilayers of low water content DPPC: a) $\theta=0°$, T=-20°C b) $\theta=90°$, T=-20°C where θ denotes the orientation of the director with respect to the external field B_0. The "width" axis provides the homogeneous lineshape, whereas the "Gauss" axis (in units of 10^{-4} T) supplies the inhomogeneous EPR lineshape. [From Ref. 30].

irradiated. For this newest technique, the relevant inequalities would be:

$$\gamma B_1 \geq \gamma B_s > \Delta \gg T_{2,j}^{-1} \tag{6}$$

in order that Eqn. [4] be obtained after a double FT in τ and t. This newest technique is more difficult. Nevertheless, by removing the need to sweep slowly through B_0, but instead to gather the whole spectrum after each echo sequence, at the least there is an order-of-magnitude savings in time.

3.2 Examples

The actual data are most usefully displayed not by the full 2D representation, but by normalized contours. These are produced by dividing $S(\omega,\omega')$ by the "zero MHz slice" [i.e., $S(0,\omega')$] to normalize and then to display the constant contour lines (and also the zero MHz slice). A set of horizontal lines imply that there is no T_2 variation across the spectrum, whereas contour lines with curvature indicate the presence of at least some variation. One finds that these contours are very sensitive not only to the rate of reorientation but also to the model of molecular reorientation, (e.g. whether it is by jumps, free diffusion, or Brownian motion) with different characteristic patterns for each! We show in Fig. 5 an actual experimental demonstration of the sensitivity to motional anisotropy by comparing the results for Tempone, which tumbles nearly

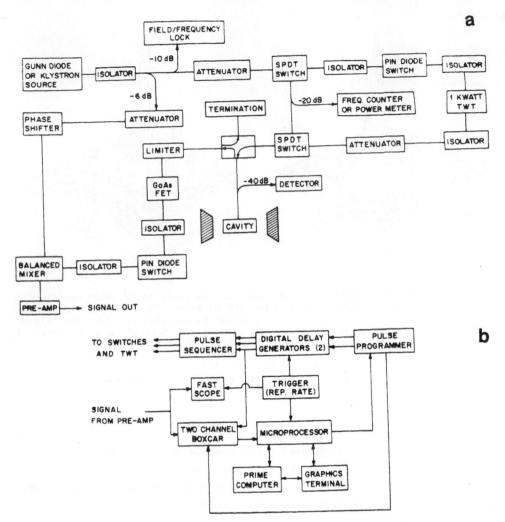

Fig. 4. Block diagram of conventional electron-spin-echo spectrometer that was used for field-swept 2D-ESE experiments. a) Spin-Echo Bridge; b) Digital Electronics.

isotropically vs. those for CSL, whose motion is anisotropic. While the T_2's are comparable, the contour shapes are significantly different, emphasizing the large anisotropy for CSL.

It should also be emphasized that Fig. 5 shows patterns that are consistent with a Brownian reorientation model. In the general theoretical analysis of this experiment, MF show that Brownian reorientation, which occurs by infinitesimal steps, will lead to a T_2 variation, because of the different sensitivity of different spectral regions to a small change in molecular orientation [i.e. $dS(\omega')/d\theta$ varies across the spectrum]. On the other hand, reorientation by substantial jumps would not show any T_2 variation.

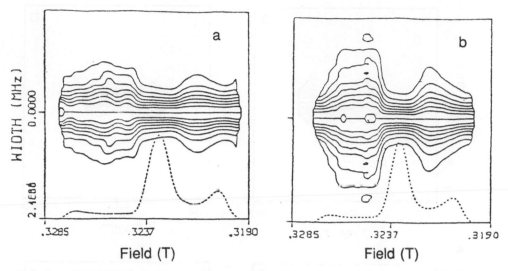

Fig. 5. Normalized contours and zero MHz slices from spectra of two different nitroxides. (a) shows the spectrum of Tempone in 85% glycerol/H$_2$O at -75°C. (b) shows the spectrum of CSL in n-butylbenzene at -135°C. The T$_M$'s for these spectra, under these conditions, are approximately the same.

We have applied the 2D-ESE technique to oriented phospholipid samples (30) (cf. Fig. 3). In Fig. 6, we show a sequence of experimental contours and zero MHz slices from oriented multilayers of low water-content dipalmitoyl-phosphatidyl choline (DPPC) doped with cholestane spin label (CSL) for different temperatures and angle of tilt θ, and in Fig. 7 we show typical simulations which relate to these results (in particular, Fig. 6a) showing specific sensitivities to the orienting potential as well as details of the dynamics.

We wish to emphasize the importance of this sensitivity. Our studies with CSL in oriented lipid samples have shown that even in the slow-motional region, where cw-spectral simulations are only slightly sensitive to motion, it is very difficult to obtain a unique set of parameters characterizing the system under study (8-10). In fact, temperature-dependent inhomogeneous broadening may dominate the cw-EPR line shapes in the very slow motional region. In the cw line shape analysis, there is a danger of misinterpreting this effect as due to motion! The 2D-ESE results are much more sensitive to these matters as illustrated in the simulations of Fig. 8. In Fig. 8a we show a cw-EPR simulation for high ordering (S=0.87) and very slow motion R $\approx$ 10^4 sec^{-1}. We superimpose the results for isotropic (N=1) and very anisotropic (N=100) motions to demonstrate that they are almost indistinguishable. However, in Figs. 8b and 8c we show the 2D-ESE contours and 0 MHz slices for the same parameters. They clearly differ both in magnitude and shape and are very easily distinguishable!

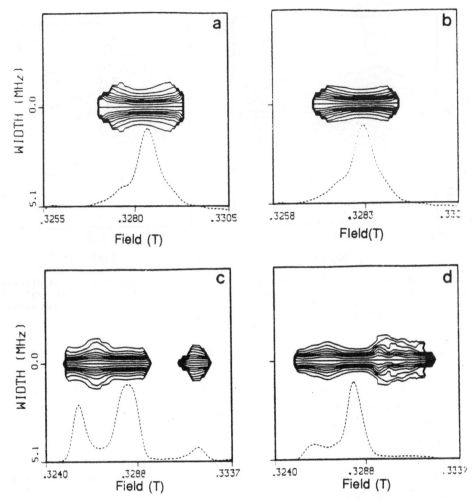

Fig. 6. Normalized contours of experimental spectra from oriented multilayers of low water content DPPC doped with CSL spin probe. (a) $\theta=0°$, T=0°C; (b) $\theta=0°$, T=-20°C; (c) $\theta=90°$, T=-20°C; (d) $\theta=45°$, T=-20°C. [Note b and c correspond to the spectra shown in Fig. 3a and b respectively]. [From Ref. 30].

It is this sensitivity to dynamics and ordering that may be exploited in many biophysical applications. For example, we have obtained well-aligned 2D-ESE spectra for higher-water-content samples (20 wt. % H_2O) prepared by a combined evaporation and annealing method. Several plate samples are stacked together to increase signal strength. Typical results (31) are shown in Fig. 9. These are interesting because they show significant variation of T_2 across the spectrum, more than previously obtained for lower water content (cf. Fig. 7), and this should enhance the ability to distinguish structure and dynamics.

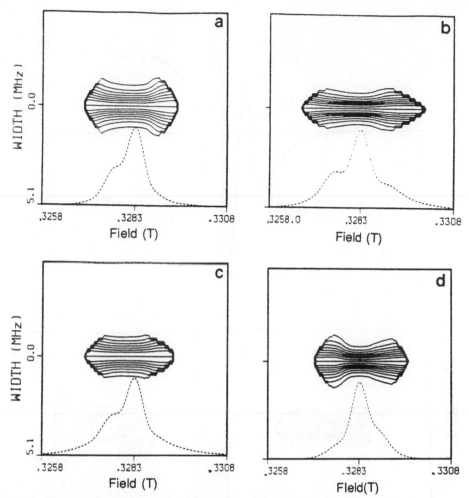

Fig. 7. Normalized contours and zero MHz slices of simulated DPPC/CSL oriented spectra (θ-0°) to illustrate the sensitivity to motion and ordering. (a) $R_\parallel$=4×10^4 sec^{-1}, $R_\perp$1=0.8×10^4 sec^{-1}, or N=5 , S=0.85; (b) $R_\parallel$=2×10^4 sec^{-1}, $R_\perp$=0.5×10^4 sec^{-1}, or N=4, S=0.85; (c) $R_\parallel$=$R_\perp$=1×10^4 sec^{-1}, S=0.7; (d) $R_\parallel$=$R_\perp$=1×10^4 sec^{-1}, S=0.91. Here $R_\parallel$ and $R_\perp$ are the parallel and perpendicular components of the rotational diffusion tensor, N = $R_\parallel/R_\perp$, and S is the order parameter. The effects of an intrinsic (solid-state) T_2^{ss} (0.7 μs) and inhomogeneous broadening (0.32 mT) have been included.

Before closing the summary of this technique, we wish to point out that an important experimental artifact has not been included in Eqn. [4], viz. the effect of a non-zero dead-time τ_d after the second pulse (partly due to cavity ringing), so that one is restricted to $\tau \geq \tau_d$. Eqn. [4] may be corrected for this effect by multiplying by a factor: exp[-2τ_d/T$_{2,j}$] inside the summation. However, Millhauser and Freed (32) have shown that a modern technique of data analysis, viz. linear prediction with singular-value decomposition (LPSVD), enables one to back-

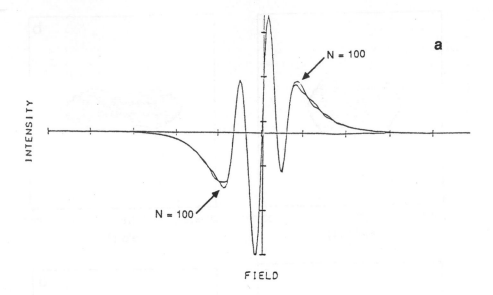

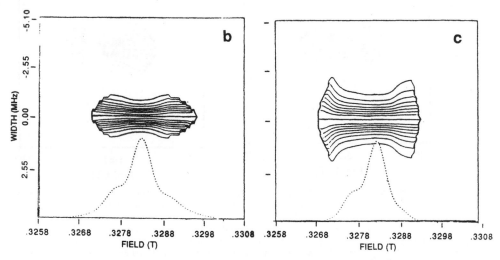

Fig. 8. A comparison of the relative sensitivity of cw vs. 2D-ESE to motional anisotropy. (a) Two superimposed spectral simulations where one spectrum has $R_\perp = 10^{-4}$ s^{-1} and N=1, and the other has the same $R_\perp$ but with N=100. The markers on the x-axis are 9.77×10^{-4} T apart. The normalized contours are simulated from the same parameters with (b) N=100 and (c) N=1. Case of high ordering, with order parameter S=0.87 and $\theta = 0°$.

extrapolate the 2D-ESE data set to estimate the signal in the range $0 < \tau \leq \tau_d$. The LPSVD method also leads both to significantly improved resolution enhancement of the complex 2D line shapes as well as to the least-squares values

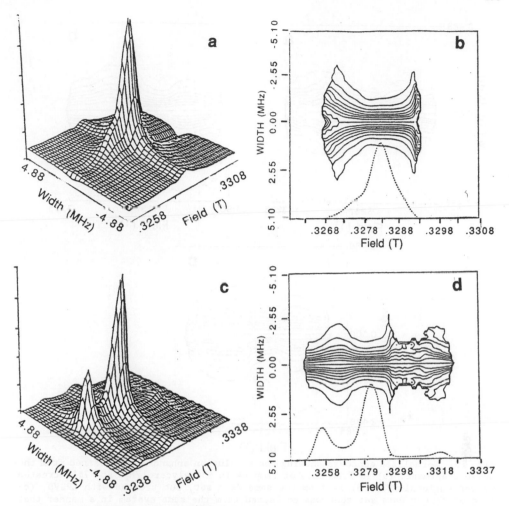

Fig. 9. 2D-ESE spectra of DPPC containing a high water content (20% by weight) and doped with CSL spin probe at -40°C. (a) and (b) show the spectrum and associated normalized contours for $\theta=0°$. (c) and (d) are for $\theta=90°$. [From ref. 30].

characterizing the exponential decays consistent with Eqns. [3] and [4] (32). Furthermore, it removes a difficulty with Fast Fourier Transform (FFT) methods. That is, to avoid so-called FFT window effects, it is necessary to collect data over a considerable time range, before performing an FFT. This means that a considerable amount of time is spent collecting data when the signal-to-noise ratio is low, and, hence, the spectral resolution is low. Instead, with LPSVD, one need only collect over those time ranges for which there is a significant

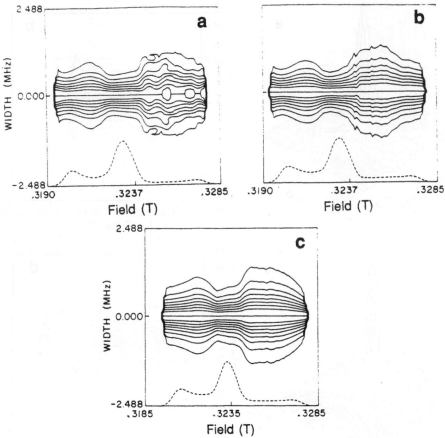

Fig. 10. Normalized contours showing the resolution enhancement obtained from the LPSVD treatment. (a) is from data of Tempone in 85% glycerol/H_2O at -75°C treated by conventional FFT. (b) is from the same data set, but treated with LPSVD. (c) is a different data set that was collected from the same system in a manner that maximizes the efficiency of the LPSVD algorithm. [From Ref. 32].

signal, thereby greatly increasing the efficiency of the data acquisition. These features are illustrated in Fig. 10.

Thus, we have found that the use of LPSVD in the data analysis leads to a much more powerful and useful 2D-ESE method. It means much better discrimination of detail in the 2D contours which translates into a much better analysis of molecular dynamic structure. Its ability to recover signal from noise means, that as long as one is able to obtain echoes yielding an estimate to T_M by usual procedures, then useful 2D-ESE contours may be obtained with the use of LPSVD. This is an important development for model membrane studies. In the case of low-water-content samples (10), it was possible to obtain estimates of T_M from ESE over most of the temperature range from +140° C to -150° C for both CSL and the phospholipid spin label 16-PC, as shown in Figs. 11. However, only a restricted

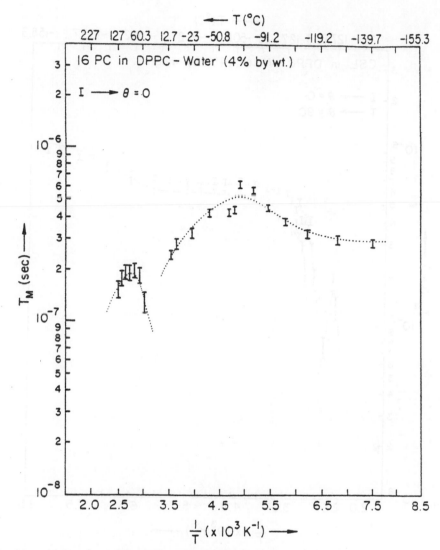

Fig. 11a. Variation of T_M with temperature for CSL in low water content DPPC. The height of the symbols (I,T) marking the data points give an idea of the error associated with each data point.[From ref. 10].

range (e.g. cf. Fig. 6) could yield sufficient S/N by conventional FFT methods to recover meaningful 2D-ESE contours from experiments of reasonable duration (ca. 5 hours). With LPSVD and the improved data collection it permits, the whole range of Figs. 11 is now available to 2D-ESE.

Given the fact that this 2D-ESE technique is very sensitive to slow motions, i.e. $\tau_R < 10^{-3}$ sec, it also has applications to the study of slow motional reorientation of spin labeled proteins. In Fig. 12, we show the 2D-ESE spectra

194

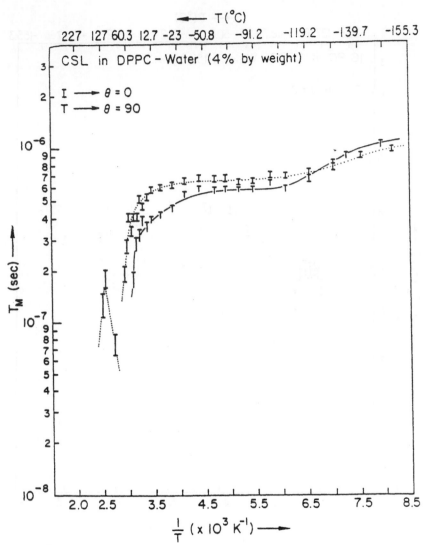

Fig. 11b. Variation of T_M with temperature for 16-PC in low water content DPPC. [From ref. 10].

that we have obtained from spin labeled α-chymotrypsin immobilized on CNBr-Sepharose (33). The effects of the rotational motions are still visible in the 2D contours. A more extensive 2D-ESE study of slow tumbling for an immobilized protein has been performed by Kar *et al*. (34).

Finally, we note that saturation transfer (35) is a cw-EPR technique frequently used in a more qualitative fashion to study very slow motions. While relatively easy to perform, the analysis is quite complex. The analysis of 2D-ESE, as discussed above, is more straightforward.

195

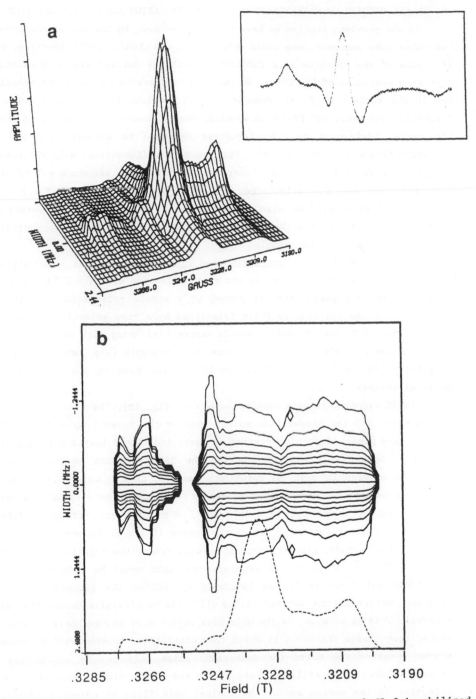

Fig. 12 (a) 2D-ESE Spectrum of α-Chymotrypsin spin labeled with SL-2 immobilized on CNBr-Sepharose Recorded at -55°C. Inset shows derivative of cw-EPR spectrum. (b) Contours for (a). [From Ref. 33].

4. 2D-ESE STUDIES OF MAGNETIZATION-TRANSFER VARIATION ACROSS THE SPECTRUM

In the previous section we have demonstrated how, by the use of the standard two-pulse echo sequence, one could obtain a slow-motional 2D-ESE spectrum, which is a plot of the spectrum as a function of both the natural line widths and the resonance positions of the DSP's. We now wish to describe an analogous experiment but in the context of T_1 measurements, which provides a "map" of the rates of "magnetization transfer" from each spectral region. However, we must first provide the needed background on echo sequences designed to measure T_1's and more generally "magnetization transfer". First we review (Subsection 4.1) the standard T_1-type measurements, i.e., the inversion recovery (IR) sequence π-T-$\pi/2$-τ-π-τ (cf. Fig. 1c) and the stimulated echo (SE) sequence $\pi/2$-τ-$\pi/2$-T-$\pi/2$-τ (cf. Fig. 1b) (17,21,24,36). We also summarize their theoretical interpretation in terms of motional dynamics. We then describe (Subsection 4.2) the appropriate 2D techniques and provide an example.

4.1 Inversion Recovery and Stimulated Echo Sequences: Full Irradiation

The IR sequence first inverts the z-magnetization with a π pulse, and after a time T, the z-magnetization is probed by a simple $\pi/2$-π echo sequence (cf. Fig. 1c). If the spectrum is fully irradiated by a "non-selective" pulse, then by stepping out T, one observes a simple exponential decay with decay constant T_1 equal to $(2W_e)^{-1}$, where W_e is the pure electron-spin-flip rate, assumed for simplicity, to be isotropic. This result is true both in the fast-and slow-motional regimes.

The SE sequence is more sophisticated (cf. Fig. 1b). The first pulse of this sequence (called the preparation pulse) nutates the dynamic spin packets first into the x-y plane, where they precess at their individual Larmor frequencies for a time τ. During this evolution period the DSP's get out of phase, and the projection of each DSP onto the rotating y'-axis at time τ, which will be nutated onto the z-axis by the second pulse (taken as being about the x'-axis) will depend upon its Larmor frequency. In this manner, the DSP's are "frequency-labeled" during this first τ interval. During the second (i.e., T) interval (referred to as the mixing period), electron-spin flips will reduce the negative magnetization component M_z just as in the IR sequence, and this would be independent of the individual DSP labeling for an isotropic W_e. Unlike the situation in the IR sequence, motion and nuclear-spin flips will also be effective during this mixing interval. This is because, as the molecules reorient or change their nuclear-spin state, they change the DSP's to which they contribute. An echo will be formed at a time τ after the third (or detection) pulse only to the extent that the frequency labeling is still accurate at the end of the mixing interval. The more effective is the motion and/or the nuclear spin flips at changing (during the mixing period) the DSP's to which a molecule contributes, the weaker will be the echo after the detection pulse.

At this point the reader may be concerned by the fact that DSP's are being "interconverted" by the molecular dynamics. Are they not the normal-modes? The answer to this question lies in the fact that the DSP's as introduced above refer to normal-mode solutions for spin-packets precessing in the rotating x'-y' plane, hence the designation of T_2-type behavior. When the spins are rotated to the (negative) z-axis, then the effective stochastic Liouville operator (in particular the spin part) is different, so that the normal-mode solutions are different, <u>even though the molecular dynamics is unchanged.</u> These are the T_1-type normal modes (24,36). For example, for the case of a simple g-tensor spin-Hamiltonian describing ultra-slow motion near the rigid limit, the DSP's are to a first approximation the rigid-limit spin packets (i.e., a continuum with each member representing a particular orientation), while the T_1-type normal modes are found to be the spherical harmonics (i.e., the eigenfunctions of the simple rotational diffusion operator which span the Hilbert space defined by the molecular orientations). This is why it is possible for the T_2-type DSP's to be interconverted during the period when T_1-type relaxation is occurring. The general theory (24,36-38) allows one to consider the appropriate normal modes at each stage of the pulse sequence and their interconversion by each pulse.

The theoretical results therefore predict that the T_1-type relaxation in slow motional "magnetization-transfer" (MT) experiments may be expressed as a weighted sum of exponential decays, whose time constants represent the T_1-type normal modes. Let us call them the MT modes. The weighting factors depend, in part, on τ, which measures how long the DSP's dephase by their precession in the x-y plane after the first $\pi/2$ pulse.

As discussed above, the case of the IR sequence with full irradiation reduces to just a single exponential form:

$$s(T+2\tau) = A[1-2\exp(-2W_e T)] \qquad [7]$$

where $s(T+2\tau)$, the signal at time $T+2\tau$, is that of the echo maximum (cf. Fig. 1c). For SE with full irradiation, the result is written as:

$$s(T+2\tau) = \sum_p b_p(\tau)\exp(-\Lambda_{d,p} T) \qquad [8]$$

where $\Lambda_{d,p}$ is the decay rate for the pth MT mode. While it would be difficult to detect many superimposed exponentials, we have found that in the limit of $W_e \tau_R \ll 1$, this expression is fit to a good approximation by a sum of only two exponential decays, i.e., as:

$$s(T+2\tau) = A + B \exp(-2W_e T) + C \exp[-T/T_A] \qquad [9]$$

where $T_A(\tau)$ is the MT time, which decreases with τ. One can show that at $\tau=0$, $C=0$, so there is only a single exponential decay in $2W_e$, since there is insufficient evolution time for the spins to precess appreciably in the x-y plane, and the first two $\pi/2$ pulses thus add to a single π pulse. The more interesting limit is for $\tau > T_2^*$, the effective decay time of the free-induction decay (FID), (while also $\tau < T_2$), for which T_A approaches an asymptotic value such that $T_A < (2W_e)^{-1}$. Also B and C take on asymptotic values independent of τ that are in general significant. In this case the frequency labeling during the evolution period is complete, so that MT during the mixing period can have its maximum effectiveness. We find this asymptotic value to be $T_A \sim \tau_R/b$ where $b \approx 2.5$ for Brownian motion and $b \approx 1$ for jump motion (24,36). In the limit $W_e\tau_R \gg 1$, the slow-motional MT is too weak, so the results are again simply described by Eqn. [7].

4.2 2D-ESE and Magnetization Transfer: Partial Irradiation

By analogy to the 2D technique providing T_2 variation across the spectrum, we can devise an experiment to provide the MT or T_A variation across the spectrum by employing partial irradiation. However, in the case of partial irradiation, both IR and SE sequences yield signals of the form of Eqn. [8] which are well represented by Eqn. [9] for all positions in the spectrum. There is now an extra MT mechanism, viz. MT shifts spins that are initially irradiated to frequencies outside the irradiated region. When this occurs, the spin is no longer detectable, and it is as though it has relaxed back to equilibrium. More precisely, spins rotated into the x-y plane by the (frequency) selective preparation pulse will not even be affected by the second or mixing pulse and/or the detection pulse if they are transferred out of the irradiation "window". This mechanism is more important for IR than for SE, since in the latter case the basic spin-delabeling mechanism, described above, already supplies a closely related effect on T_A.

In summary, the relative effectiveness of MT out of the different spectral regions can be studied by such a partial irradiation technique, but first the two decays in $(2W_e)^{-1}$ and T_A^{-1} must be separated in the data processing. This is again accomplished by applying LPSVD.

We now consider mechanisms whereby T_A can vary across the spectrum. Suppose a molecule whose principal magnetic axis is parallel to the applied field ($\theta=0°$) makes a small Brownian jump by $\Delta\theta$. Its new spectral frequency will hardly change, because the orientation-dependent part of the resonance frequency mainly goes as $3\cos^2\theta-1$, which for $\theta=0°$ or $90°$ gives a change of $\Delta\omega \propto (\Delta\theta)^2$. However, for a molecule oriented at $\theta \sim 45°$, that same Brownian jump will cause a corresponding spectral frequency change given by $\Delta\omega \propto \Delta\theta$. Thus, the spin packet at the $\theta=45°$ orientation experiences the more effective MT out of an irradiated region. Similarly if there is anisotropy of the diffusion tensor (e.g., let there be rapid rotation only about the molecular y-axis) then DSP's associated with the perpendicular (i.e. x and z) orientations will experience more rapid MT than DSP's associated with the y-orientation, thus yielding a greater T_A^{-1} for those DSP's associated with the perpendicular orientations.

The more general form of Eqn. [8] for SE with partial irradiation is (24,36,37,38):

$$s(T+2\tau) = A' \exp(-2W_e T) \sum_p b_p^\omega(\tau) \, \exp(-T/\tau_p) \tag{10}$$

where

$$\tau_p^{-1} = \Lambda_{d,p}^{-1} - 2W_e \tag{11}$$

and

$$b_p^\omega(\tau) = \sum_{n',\ell',j'}^{\omega} \sum_{m,k,i} \sum_{s,q} O_{o,qn'} O_{o,mn'} O_{d,mp} O_{d,k,p} O_{o,k\ell'} [O_{o,i\ell'} O_{o,ij'}^{*}] O_{o,sj'}^{*} \tag{12}$$
$$\times \exp[-(\Lambda_{o,n'} + \Lambda_{o,j'}^{*})\tau] .$$

(For a nitroxide with 3 allowed transitions, s and q take on values -1, 0, and 1.) Here O_o is the orthogonal transformation introduced in the previous section, which diagonalizes the stochastic Liouville operator associated with the DSP's and O_d is the equivalent orthogonal transformation for the MT modes. The $\Lambda_{o,n'}$ are the eigenvalues of the DSP's while the $\Lambda_{d,r}$ are those for the MT modes. The primes appearing on the indices n', ℓ', and j' imply that these are summed only over those DSP's which are within the irradiation window of the selective pulse, (i.e. they depend upon $\omega_j-\omega$). This gives an implicit dependence upon ω as indicated. We have actually used this form for the theoretical predictions given below (24,36). (In the near rigid limit, when Eqn. [2] applies, we may let $\sum_i O_{o,i\ell'} O_{o,ij'}^{*} \approx \delta_{\ell'j'}$ to achieve some simplification of Eqn. [11].) Fortunately, the results are approximately fit by Eqn. [9] at each position in the spectrum! Also, W_e is found to be constant, within experimental error, across the spectrum. (But we shall encounter an expression like Eqns. [10]-[12] in Sect. 6.) Thus, after extracting the term in $\exp(-T/T_A)$ by LPSVD, and after FT, one obtains a plot of $T_A/(1+\omega^2 T_A^2)$ across the spectrum with normalized contours showing the variation of T_A.

We illustrate this 2D-MT experiment for the NO_2/vycor system in which the NO_2 is physisorbed on the vycor surface. We display in Fig. 13 the 2D normalized contours for T_A obtained at 35 K. Focussing on the $M_I=0$ region (the central region), we see that the broadening that corresponds to the x and z molecular axes

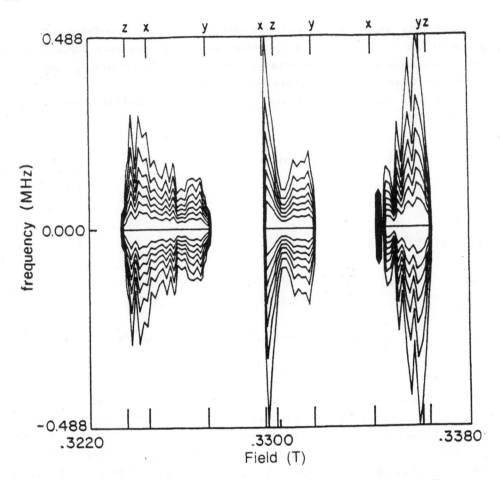

Fig. 13. Two-dimensional ESE contours from stimulated-echo sequence for NO_2 on vycor at 35 K showing rates of magnetization transfer. It shows relatively rapid rotation about the molecular y axis (i.e. the axis parallel to oxygen-oxygen internuclear vector). [From Ref. 36].

being parallel to the field is quite dramatic. This is clear evidence for more rapid rotation about the y-axis, which is the axis parallel to the line connecting the two oxygen atoms. The contours in the $M_I = \pm 1$ regions indicate less clear trends, probably due to inhomogeneous broadening from site variation in the hf-matrix. This example indicates the potential utility of the 2D-MT experiment, which can provide useful information about molecular dynamics even prior to a thorough quantitative spectral analysis. Furthermore, it appears that rotational motion is detectable at lower temperatures than with the T_2-type 2D-ESE experiment described in Sect. 3. The relationship to the new 2D-FT techniques will be discussed in Sect. 6.

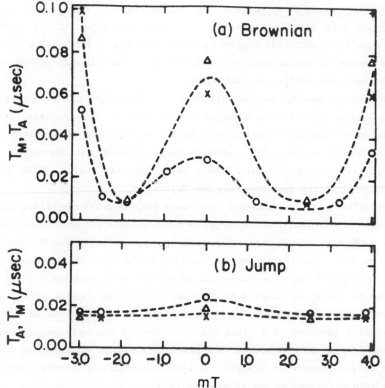

Fig. 14. Calculated time constants T_A and T_M versus dc field across the nitroxide spectrum for partial irradiation. The irradiation window is 4×10^{-4} T. The T_A from the IR sequence are ($\times$), those from the SE sequence are (Δ), while the T_M from a spin-echo sequence are (0). (a) Brownian motion and τ_R=170 nsec, W_e=0.05/τ_R; (b) Moderate jump model, τ_R=17 nsec, W_e=0.05/τ_R [From Ref. 36].

We conclude with some theoretical predictions (24,36), summarized in Fig. 14, which show that this technique is very sensitive to the model of molecular motion.

4.3 Spin-Echo ELDOR

A spin-echo ELDOR technique has also been developed and has been applied to the study of slow motions (39). While ELDOR has typically been applied in the past as a cw technique (40), pulsed ELDOR is also possible (41). The use of cw-ELDOR for the study of slow motions was suggested some time ago by Bruno and Freed (42), and detailed cw studies and analysis were performed by Hyde and Dalton (43). The use of spin echoes in such studies was suggested previously (38). The advantages of an echo technique for ELDOR are 1) the absence of the radiation fields (as well as any dc field modulation) during the evolution time of the spins and the (rotational) diffusion of the molecules, 2) the cancellation of effects of inhomogeneous broadening, and 3) the direct measurement of relaxation rates rather than just their ratios, such as obtained by cw ELDOR. This latter advantage also exists for pulsed ELDOR based on saturation recovery techniques (44). Thus (1) permits a much simpler theoretical analysis (in terms of the stochastic Liouville equation), while (2) suggests greater accuracy in data analysis (without having to resort to deuteration of spin labels as we already noted) and (3) removes the need for additional techniques.

ELDOR may be regarded as the epitome of magnetization transfer experiments by EPR. One first inverts the z-magnetization of some spin packets, and at a later time, one "observes" whether this spin inversion has been transmitted to another part of the spectrum. Thus, whereas the previous MT methods merely observe the rate of loss of magnetization from a given spectral region, ELDOR shows to which regions(s) of the spectrum they are delivered.

The precise technique that was first used (39) is a stepped-field method in conjunction with a conventional three-pulse π-T-$\pi/2$-τ-π-τ sequence used to measure T_1's by IR (cf. Fig. 1d). We call this technique spin-echo ST-ELDOR. The first π pulse inverts the magnetization at one resonant field B_0 in the spectrum. Then B_0 is rapidly stepped by ΔB_0 to a new region of the spectrum, and the $\pi/2$-τ-π pulses are applied to yield the usual echo at time τ after the last π pulse. In this experiment T is varied, while τ is maintained at a fixed value. The echo amplitude, as a function of T gives the variation of the spin magnetization at the spectral region $B_0 + \Delta B_0$ due to the initial inversion of the spins at B_0. It is thus a direct measure of transfer of spin polarization. Similar experiments have also been performed by Soviet workers (45).

Stepped magnetic fields in the local region of the EPR sample are produced by a coil wound on the inside tube of a variable temperature dewar cavity insert. A specially-constructed pulsed current supply provided the current for driving the coil. It is driven by a variable width and amplitude pulse from a pulse generator triggered by the EPR spectrometer. Current pulses with a rise time of 100 nsec could be produced, but the characteristics of the turn-off are much poorer. Thus the field-step is turned on after the first π pulse and maintained during the $\pi/2$-τ-π-τ sequence.

The characteristic feature of the results is the initial decrease in signal vs. T to a minimum followed by a slower return to an equilibrium value. (cf. Fig. 15a). The shape of s(T) is consistent with a model in which a portion of the spins inverted in one region is transferred to another and then relaxes back to equilibrium. The transfer of spin polarization can be due to rotational reorientation, while W_e would then restore the spins to equilibrium.

A simplified analysis of this experiment, based on the model of strong jumps leads to the following discussion (39). The orientation-dependent echo signal, $S(\Omega_i,T)$ at Euler angles specified by Ω_i due to initially irradiating the spectrum corresponding to Ω_j is given by:

$$S(\Omega_i,T) \propto \frac{-1}{8\pi^2} \{e^{-T/T_1} + e^{-\omega'T}[8\pi^2\delta(\Omega_i-\Omega_j) - 1]\} \qquad\qquad [13]$$

where $\omega'=T_1^{-1} + \tau_R^{-1}$, and $\delta(\Omega_i-\Omega_j)$ is the Dirac delta function. Here $T_1=2W_e$ which is assumed to be largely independent of orientation. Thus, if $\Omega_i \neq \Omega_j$ one has:

$$S(\Omega_i,t) \propto \frac{-1}{8\pi^2} [e^{-T/T_1} - e^{-\omega'T}] \qquad \Omega_i \neq \Omega_j \cdot \qquad\qquad [14]$$

Eqn. [14] clearly shows that there will be a non-negligible effect only if τ_R is not much longer than T_1. In actual fact, a range of Ω_j are affected by the first π pulse, while the detecting pulses "observe" a range of Ω_i. Thus we may write:

$$S(\Omega_i,T,\tau) = S_o(\Omega_i,\tau) \{1-C(\Omega_i,\Omega_j)[e^{-T/T_1}-e^{-\omega'T}]\} \qquad \Omega_i \neq \Omega_j \qquad [15]$$

where $S_o(\Omega_i,\tau)$ measures the echo signal from a conventional $\pi/2$-τ_2-π-τ_2 sequence that arises from the dynamic spin packets in the appropriate range of orientations centered about Ω_i. Also, $C(\Omega_i,\Omega_j)$ is a factor determined by the range of orientations centered about Ω_i, whose spins are initially inverted, and the range about Ω_j, whose spins are "observed." It is both an instrumental factor via B_1 and a function of the spectral lineshape. This factor can be removed from the analysis by solving for T_{min} such that $S(\Omega_i,T,\tau)$ is a minimum. One easily obtains:

$$\ln(1 + T_1/\tau_R) = T_{min}/\tau_R \cdot \qquad\qquad [16]$$

However, the "integrating effect" over a range of orientations still remains in the technique. (2D-FT-ELDOR described in Sect. 6 would not suffer from such integrating effects and would permit Ω_i - Ω_j to be arbitrarily small.)

An extensive theoretical analysis of ELDOR spin-echoes based upon the stochastic Liouville equation has been developed (24). This analysis confirms the semi-quantitative validity of the simple analysis for jump diffusion, but it also rigorously extends the theory to all types of motions. The general results can again be approximated by Eqn. [15] with specific values of S_o and C given as a function of the resonant frequencies initially inverted and then observed. Also, the precise definitions of T_1 and ω' are modified somewhat. However, a mechanism of orientation-independent nuclear spin flips was found to be potentially important in generating ELDOR effects (cf. Fig. 15b).

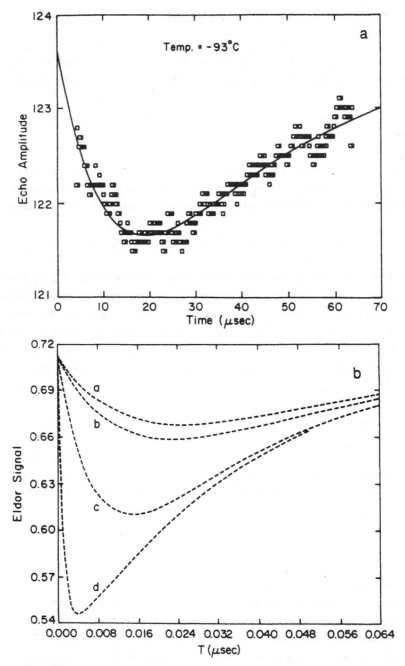

Fig. 15. (a) Experimental stepped-field ELDOR echo amplitude versus time between first π pulse and $\pi/2$ pulse for PD-Tempone in 85% glycerol/H_2O at -93°C. Solid line: Best two-exponential fit to data. [from Ref. 39] (b) Calculated ELDOR curves for slowly tumbling nitroxide. In sequence a-d orientation-independent nuclear spin flip rate increased, thereby increasing the ELDOR effect. [From Ref. 24].

The primary limitation of the original experiments (39) is a 1-2 μsec delay after stepping the field. We have redesigned this experiment so that two frequency sources (a klystron and a Gunn diode) feed separately pulsed signals to the travelling wave tube. This configuration allows one to pulse the pumping and observing regions with time delays of as little as 10-20 nsec. This arrangement is used in conjunction with a low Q cavity with an absorption half-width of 125 MHz which permits one to irradiate spectral regions separated by ΔH < 4.5 mT. Such an experimental arrangement could be used to study possible exchange between ordered and disordered lipids or between fluid and immobilized lipids in oriented samples containing polypeptides or proteins (46). It can also be effectively used for studies of motional dynamics in the regions corresponding to fast motions (e.g. high water content liquid crystalline phases) where one may directly measure the electron spin, and cross-relaxation rates (35,38,47). This has been done with cw-ELDOR (48,49). However, the 2D-FT methods such as those that are described in Section 6 below are more powerful.

5.   ELECTRON-SPIN ECHO ENVELOPE MODULATION (ESEEM)

An important application of ESE techniques enables one to obtain more detailed information about the location of spin probes. This may be achieved by studying the electron-spin-echo modulations (ESEEM) resulting from the dipolar interaction of the unpaired electron of the nitroxide group with deuterated (or protonated) sites. In particular, the use of these ESEEM for the study of site geometries has been utilized by Kevan and co-workers (50), and was applied by them to the study of doxylstearic acid spin labels in frozen micellar solutions of surfactants (51) with deuterated head groups, deuterated counterions, or deuterated terminal methyl groups. By measuring the ESEEM as a function of the location of the nitroxide group on the hydrocarbon chain, they were able to assess the location of the spin probe nitroxide group relative to the head group region. Also, by measuring the deuterium interactions from D_2O with the different labeled doxylstearic acids, they could investigate the degree of water penetration into the micellar systems.

The theory for ESEEM has been reasonably well developed and is summarized elsewhere in the volume. It enables the measurement of extremely weak electron-nuclear dipolar interactions such as exist between the unpaired electron of the nitroxide label and the deuterons (protons) on the nearby surfactants or D_2O (H_2O) molecules. Aside from measurements of modulation frequency which can be sensitive to the dipolar interaction, the deuterium (proton) modulation depth depends on the number of interacting deuterons and their average distance from the spin probe (or more precisely, their average dipolar interaction). Kevan and co-workers (51) used their estimate of the average numbers of surfactant head groups or counterions a spin probe interacts with to extract information on the relative distances of

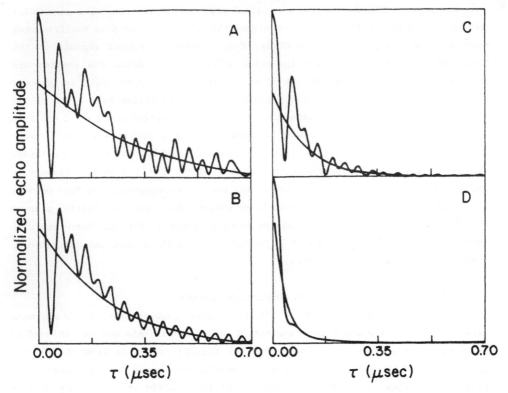

Fig. 16. Theoretical simulations of ESEEM patterns for a spin probe containing a single proton undergoing Brownian diffusion with rotational correlational times of (A) 3.6×10^{-5} sec, (B) 3.6×10^{-6} sec, (C) 3.6×10^{-7} sec and (D) 3.6×10^{-8} sec. Here $A_{\perp}=0.325$ mT, $A_{\parallel}=0.65$ mT, $B_0=0.22$ T. Superimposed are the best fits to $\exp[-2\tau/T_M]$. [From Ref. 23].

the spin-probe from the head-group region. They found that deuteron ESEEM is generally detectable for determining interaction distances less than 6 Å.

While ESEEM has primarily been used for structural studies, it is also possible to utilize them in the study of motional dynamics. In general, motion averages out the ESEEM. A general theoretical approach to the problem of how motion averages ESEEM has been developed (23,24). We show in Fig. 16 a theoretical computation illustrating how they become averaged out as the rotational motion speeds up from rigid-limit values.

We find in model experiments (24) that for PD-Tempone, where the ESEEM are due to the weak deuteron shf splittings, that the patterns change significantly for τ_R values ranging from about 10^{-4} sec to 10^{-7} sec, although residual effects may be seen for somewhat faster and slower motion (cf. Fig. 17). Note that the deuteron shfs is significantly smaller (e.g. $A_z \approx 100$ mG) than the ^{14}N hf term ($A_z \approx 3.0$ mT). Thus, it will be averaged out by motions that are considerably slower

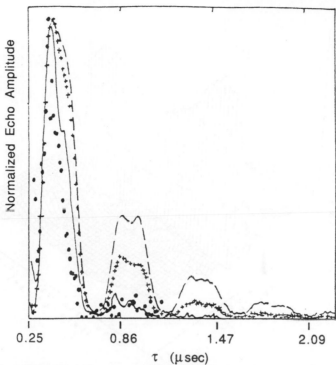

τ (μsec)

Fig. 17. Experimental $\pi/2$-τ-π-τ ESEEM patterns as a function of temperature, for PD-Tempone. Dots, -43°C; solid line, -55°C; crosses, -64°C; dashed line, -73°C. Solvent is 85% glycerol/H_2O. [From Ref. 24].

than required for normal ESE spectra. Effects from deuterons directly on PD-Tempone are averaged by any reorientation. Deuterons of the solvent can also cause ESEEM in the rigid limit (52), which would be averaged by the relative translation and rotation of solvent and solute molecules.

Another way of studying the ESEEM is to use a stimulated echo sequence: $\pi/2$-τ-$\pi/2$-T-$\pi/2$-τ and step out T for fixed τ. However, by varying the pulse interval τ, one can markedly affect the shape of the ESEEM. This effect has been called the suppression effect in solids (53). When these ESEEM are affected by motions, then studies as a function of both τ and T could be most useful.

We now address the question as to the relevance of ESEEM for 2D-ESE experiments. We recall that the T_2-type 2D-ESE experiments of MF required a small B_1 such that $\gamma B_1 \ll \Delta$ the inhomogeneous width (cf. Eqn. 5). However, it <u>also</u> <u>requires</u> that $B_1 \ll a_i/\gamma$, where a_i is the shf constant for the ith deuteron (or proton) in order that ESEEM <u>not</u> be excited. In actual fact, it is only practical to achieve this inequality for <u>fully</u> protonated samples, so only such samples were used. On the other hand, if $B_1 > a_i/\gamma$, then the 2D-ESE experiments would include the effects of ESEEM, and in fact, one could map out how the ESEEM varies

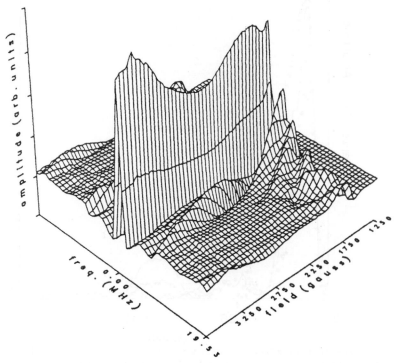

Fig. 18. 2D-ESE spectrum of ovotransferrin-carbonate containing Fe(III). [From Ref. 53].

across the EPR spectrum. This can be expected to enhance the interpretation of these patterns, both for structure and dynamics. We show such a 2D-ESE spectrum in Fig. 18, not for a nitroxide, but for ovotransferrin-carbonate containing Fe(III) (54). It dramatically demonstrates the sensitivity of ESEEM to position in the EPR spectrum (i.e. the field position).

Actually, special 2D techniques have been demonstrated to help one to unravel complicated ESEEM patterns (55-57). In fact, the work by Merks and de Beer (55,56) on two-dimensional echo spectroscopy for improving spectral resolution of ESEEM in solids was the first example of 2D-EPR. LPSVD methods should be useful in the analysis of the deuteron modulations. We show in Fig. 19 1D results obtained from the model system of PD-Tempone in deuterated glycerol/H_2O. In particular, we could distinguish six distinct frequencies with the aid of LPSVD. Thus LPSVD methods are likely to greatly enhance the sensitivity of related 2D studies, (cf. the chapter by de Beer and van Ormondt).

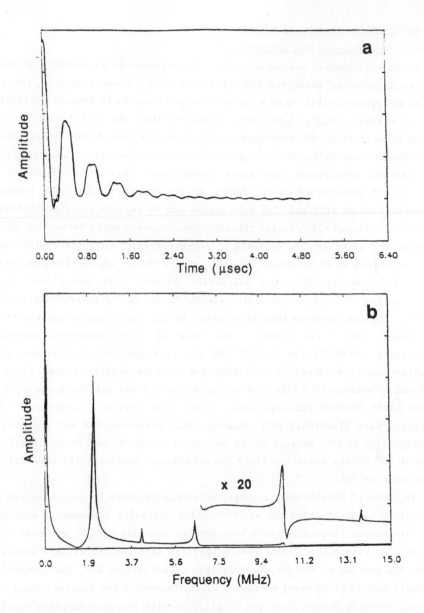

Fig. 19. ESEEM patterns of PD Tempone in 85% glycerol/H_2O at -80°C (a) Experimental data (dashed line) and LPSVD fit (solid line). (b) Fourier transform of LPSVD curve showing six spectral components.

6.   TWO-DIMENSIONAL FOURIER-TRANSFORM EPR

6.1   Introduction and Motivation

Ernst and Anderson demonstrated that the response of an ensemble of nuclear spins to a pulse of polarized radiofrequency (rf) is equivalent to the slow-passage NMR spectrum (58). When a two-level spin system is in thermal equilibrium, the -1/2 state is only slightly more populated than the +1/2 state. The slow-passage (unsaturated) NMR experiment is based on the well-known linear response theory wherein one excites nuclear spin transitions without significantly altering these thermal equilibrium spin state populations. This is accomplished with continuous rf irradiation at low power. In a high-power pulsed NMR experiment the populations of +1/2 and -1/2 spin states can be rapidly equalized before the rf power is switched off. Fourier transform spectroscopy works because in the time following the rf pulse, the ensemble of coherently-excited spins evolves in the complete absence of rf radiation. Linear response theory applies because there is no rf field during this free precession period. This means that Fourier transformation of the free-precession signal (known as the free-induction decay) generates the NMR spectrum that is obtained in the slow-passage experiment.

Jeener (59) first proposed the idea of two-dimensional correlation spectroscopy (2D-COSY). In 2D-COSY NMR the time-correlation function of the magnetization is obtained by utilizing two or more closely spaced rf pulses. Ernst and co-workers (60) first demonstrated this concept utilizing two $\pi/2$ pulses in the basic 2D-COSY NMR experiment. Since that original experiment, 2D-NMR techniques have flourished (61). However, that potential had not been realized experimentally in EPR, because of the requirement that the entire spin Hamiltonian be "rotated" by the radiation field, an experiment substantially more difficult in EPR than in NMR.

In terms of Fourier spectroscopy, other distinctions between electron spins and nuclear spins are that the electron spins typically resonate at much higher (i.e. microwave) frequencies with much greater spectral widths, and they exhibit much shorter relaxation times. These properties have imposed certain limitations on the application of Fourier spectroscopic techniques to EPR. The generation of extremely narrow (5-10 nsec) microwave pulses necessary for Fourier transform EPR has only recently become practical. Digitizers with adequate sampling capability to record free induction decay signals of electron spins have also not been available until very recently. Now that such components are commercially available, the same techniques that were introduced to NMR beginning twenty years ago are being introduced to EPR. Simultaneously, Bowman and co-workers performed FT-EPR experiments of short-lived organic free radicals of spectral bandwidth ~ 40 MHz (62,63), and Gorcester and Freed performed FTEPR experiments on the broader   (~90 MHz) spectra of nitroxide radicals (64) based on earlier considerations of Hornak and Freed (65). In a communication that followed shortly thereafter, Gorcester and Freed first demonstrated 2D-SECSY and 2D-FT exchange EPR experiments on the same nitroxide radical samples (66). Since these initial demonstrations, there have been significant developments and improvements of 2D-FT-ESR instrumentation and methodology (67,68).

The principal application of 2D-FT-EPR demonstrated thus far is to the ELDOR experiment. As we have already discussed in Sect. 4.3, ELDOR requires the use of two excitation fields: one to modify the populations (the pumping field at frequency ω_p) and another to detect the response (the observing field at frequency ω_o) elsewhere in the spectrum. In the slow-passage ELDOR experiment, one typically keeps the two constant at a precise frequency difference, while sweeping the DC magnetic field B_0 through the entire spectrum. One could repeat this experiment in which the difference between pumping and observing frequencies is varied and thereby obtain a two-dimensional spectrum with $\gamma_e B_0$ on one axis and $\omega_o - \omega_p$ on the other. Such a data set can be manipulated to produce the two-dimensional spectrum as a function of $\omega_o - \gamma_e B_0$ and of $\omega_p - \gamma_e B_0$, which are the "natural variables" for interpreting the spectrum. Alternatively, such a procedure can be applied in the time-domain by utilizing the ST-ELDOR-echo method discussed above. Such experiments would be very time consuming and are typically not practical. However, such an experiment may be performed very efficiently with 2D-FT-EPR methodology. With 2D-FT-ELDOR (67), one obtains all of the combinations of pump and probe frequencies simultaneously in a single two-dimensional experiment! In fact, the matrix of 2D-ELDOR peak intensities is rather simply related to the matrix of transitions probabilities. Such an experiment can typically be performed in less than an hour, and analyzed in a few minutes.

The analysis of 2D experiments, whether NMR or EPR, typically involves 2D Fourier transformation of a two-dimensional time series, followed by phase corrections and numerical determination of volume integrals. LPSVD presents an alternative to Fourier analysis (69,70), and we do find that LPSVD may be very usefully applied to 2D-FT-EPR, (71). Also, we have found a variation of this technique that is customized to exploit many of the symmetries in 2D-FT EPR or NMR (71).

We now consider how the newly-developed FT-EPR methods can improve and extend the range of 2D experiments on molecular dynamics. To date, this has been done for cases involving fast motional nitroxide EPR spectra, but the principles and methods can be extended to slow motions.

6.2 Spin-Echo Correlated Spectroscopy: (SECSY)

The 2D-FT analogue of our field-swept T_2-type 2D-ESE experiment may be referred to as spin-echo correlation spectroscopy (66,68) (SECSY). The basic pulse scheme for SECSY is illustrated in Fig. 20. The SECSY experiment involves acquisition of the electron spin echo during the t_2 time period for a series of equally spaced values of t_1. The 2D-SECSY spectrum (Fig. 21) is obtained after

212

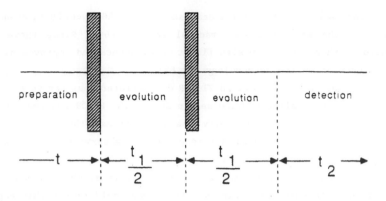

Fig. 20. SECSY pulse sequence. The homogeneous decay of the spin echo occurs during the evolution period t_1; the inhomogeneous decay occurs in the detection period t_2.

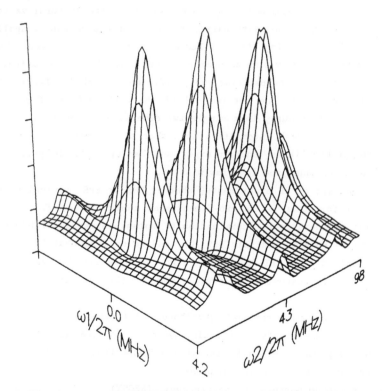

Fig. 21. SECSY spectrum of 1.17×10^{-3} M PD-Tempone in toluene-d_8 at 22°C. $T_2^* \simeq 75$ nsec; $T_2 \simeq 174$ nsec; 0.86 nsec resolution in t_2 providing 256 complex data points, each the average of 2048 transients. Time resolution in t_1 is 12 nsec; pulse width is 15 nsec; acquisition time 60 minutes. [From Ref. 66].

making a small first-order phase correction in ω_2 and a linear amplitude correction. One observes the inhomogeneously-broadened three-line hf pattern along ω_2 and the Lorentzian homogeneous lineshapes along ω_1. [The DC magnetic field homogeneity was destroyed to suppress the FID from the second $\pi/2$ pulse. This would not be necessary for a slow-motional spectrum (29,30,36).]

The advantage of SECSY in comparison to field-swept 2D-ESE is the order-of-magnitude shorter data acquisition time. In addition, SECSY has the potential for observing (weak) cross-correlations which cannot be observed with techniques using narrow band excitation.

The theory for this experiment (67,68) is very similar to that of field-swept 2D-ESE (cf. Sect. 3). One obtains:

$$ s'(t_1,t_2) = \sum_{k,m} c_{km} \exp(-\Lambda_k t_2) \; \text{Re} \sum_{j} b_{mj} \exp\left[(\Lambda_k - \Lambda_j^*)t_1/2\right] \qquad [17] $$

for the pulse sequence in Fig. 20, where

$$ c_{km} = \sum_{n} U_n O_{o,nk} V_m O_{o,mk} \qquad [18] $$

and

$$ b_{mj} = \sum_{l} O_{o,mj}^* V_j^* O_{o,lj}^* U_l . \qquad [19] $$

We have neglected electron spin echo envelope modulation (ESEEM) arising from cross-polarization of nuclei via the hf tensors (but see Sect. 5). These terms are important in the slow-motional regime.

The V_j give the correction for variation of effective B_1 across the spectrum for the "$\pi/2$ pulse," i.e. the jth DSP resonating at angular frequency ω_j experiences an effective rotation that depends on $\omega_j - \omega$ (with ω the irradiating frequency) and on B_1 (65,67,68,72). In comparing Eqn. [17] with Eqn. [1], we note that the a_{kj} of Eqn. [1] is now replaced by $\sum_m c_{km} b_{mj}$. Also U is the vector of (appropriately averaged) spin transition moments.

After Fourier transformation with respect to t_1 and t_2 we obtain

$$ S'(\omega_1,\omega_2) = \sum_{k,m} c_{km} \frac{1}{i\omega_2 - \Lambda_k} \; \text{Re} \sum_{j} b_{mj} \frac{1}{i\omega_1 - \frac{1}{2}(\Lambda_k + \Lambda_j^*)} . \qquad [20] $$

Thus inhomogeneous broadening is removed along the ω_1 axis.

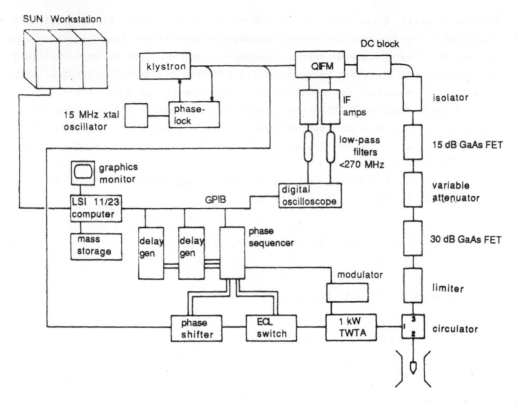

Fig. 22. Block diagram of spectrometer used for 2D-FT-EPR experiments.

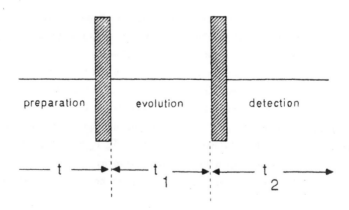

Fig. 23. 2D-COSY pulse sequence.

A comparison of Eqn. [20] with Eqn. [4] shows that this full FT method provides both the in-phase and quadrature components with respect to ω_2. They are obtained from the two arms of the quadrature detector. It is also possible to obtain the out-of-phase part of the signal with respect to ω_1. First, let us refer to the pulse sequence yielding Eqn. [20] as: $(\pi/2)_x - (\pi/2)_x$, where the subscript x implies rotation of the spins by the microwave radiation along the rotating x-axis. Then the out-of-phase part of the signal with respect to ω_1 is obtained by phase shifting the first $\omega/2$ pulse by 90°, which we then write as a $(\pi/2)_y - (\pi/2)_x$ sequence. It yields $S''(\omega_1,\omega_2)$ which is given by Eqn. [20] but with the real part of the expression in curly brackets replaced by its imaginary part. We shall refer to this approach as the 2D phase quadrature scheme.

An illustration of the principal features of our 2D-FT-EPR spectrometer is shown in Fig. 22.

6.3 Correlation Spectroscopy (COSY)

The previous method was based on detection of the echo signal. We now turn to FT-EPR experiments which are based upon detection of the free-induction decay (FID).

Two-dimensional correlation spectroscopy (2D-COSY) in its various forms has gained widespread use in NMR as a method of observing coherence transfer between coupled spin transitions. The pulse sequence $\pi/2$-t_1-$\pi/2$-t_2 constitutes the simplest of the COSY experiments and is used in NMR in the separation of scalar interactions (60,61). A COSY-EPR spectrum is obtained in much the same manner as for NMR; the basic pulse scheme is illustrated in Fig. 23. The preparation period consists of a $\pi/2$ pulse to generate the initial transverse magnetization components. Free precession of the magnetization occurs during the evolution period of duration t_1 during which the components become amplitude encoded according to their precessional frequencies in the rotating frame. The FID is recorded during the detection period of duration t_2, which begins with the final $\pi/2$ pulse. For each t_1 the FID is collected, then the phase of the first pulse is advanced by 90°, and a second FID is collected. These two signals depend on terms oscillatory in t_1 that are in phase quadrature, i.e. the 2D phase quadrature scheme. The oscillatory behavior in t_1 is illustrated in Fig. 24 after Fourier transforming one of these signals with respect to t_2 only. Two-dimensional complex Fourier transformation generates a spectrum over the two frequency variables, ω_1 and ω_2.

This two-step phase alternation sequence yields the required frequency discrimination in ω_1, and it provides the phase information necessary for the pure absorption representation of the 2D spectrum (73). In order to cancel "images" resulting from imperfections in the quadrature detector, (cf. Fig. 22), we combine this two-step procedure with the four-step CYCLOPS image cancellation

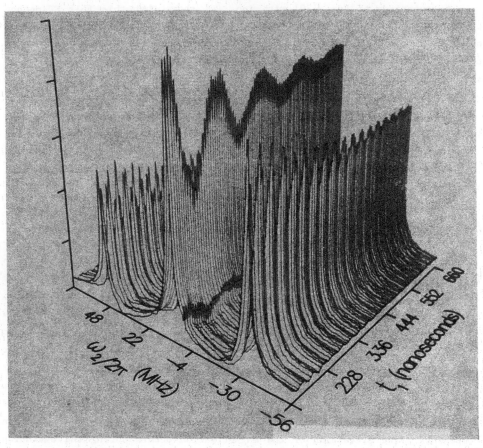

Fig. 24. 2D-COSY-EPR spectrum of 5.1×10^{-4} M PD-Tempone in toluene-d8 prior to FT in the t_1 domain in order to show the oscillatory behavior in t_1 (cf. Fig. 25 for a fully transformed 2D spectrum).

method (74) to obtain a sequence consisting of eight steps as described elsewhere (67,68). A COSY-EPR spectrum of 5×10^{-4} M PD-Tempone in toluene-d_8 obtained in this manner is shown in Fig. 25. Resonances at positions for which $\omega_1 = \omega_2$ will be referred to in the standard fashion as autopeaks, because they represent auto-correlations. Cross-peaks representing cross-correlations are not found in the spectrum of Fig. 25 because the contributions of the electron-electron dipolar and chemical or Heisenberg exchange interactions are too weak. Such mechanisms can induce "off-diagonal" relaxation between the different transitions (47,74). Additional peaks at positions for which $\omega_1 = 0$ arise because of electron spin flips during the evolution period; these peaks are referred to in the standard manner as axial peaks (60,61).

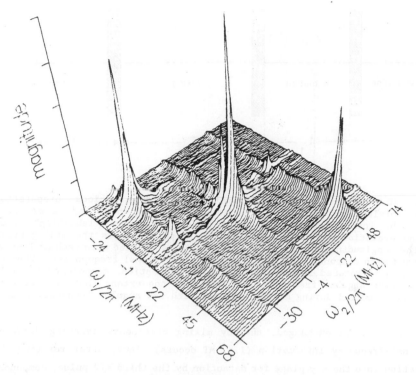

Fig. 25. Absolute value 2D-COSY-EPR spectrum of 5.1×10^{-4} M PD-Tempone in toluene-d_8 at 21°; $\tau_p=15.5$ nsec; $\Delta t_1=3.9$ nsec; 90 samplings in t_1; eight-step phase alternation sequence with 30 averaged FID per step; dead time in t_2 of 100 nsec; dead time in t_1 of 120 nsec; acquisition time 10.6 minutes. [From Ref. 67].

6.4 2D-ELDOR

We now describe the 2D-FT experiment that displays the effects of MT. More precisely, it is a way of performing ELDOR but with only a single frequency source! The spectral band produced by a finite pulse is coherently related, and this is the basis for the power of the technique. 2D-ELDOR involves a procedure similar to that of simple COSY except that three $\pi/2$ pulses are applied in the sequence $\pi/2$-t_1-$\pi/2$-T-$\pi/2$-t_2 with the mixing time T being held constant. The application of this sequence (in NMR) to the study of chemically exchanging species was first illustrated by Jeener et. al. (75). The basic pulse scheme for 2D-ELDOR is illustrated in Fig. 26. The preparation period consists of a $\pi/2$ pulse to generate the initial transverse magnetization. The phase of this pulse determines the phase of the amplitude modulation that results from the "frequency labeling" during the subsequent evolution period of duration t_1 as described above for the COSY-EPR experiment. The second $\pi/2$ pulse marks the beginning of the mixing period wherein the longitudinal magnetization components associated with

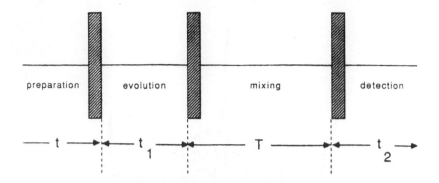

Fig. 26. 2D-ELDOR pulse sequence, involving four time periods: preparation, evolution, mixing, and detection. After the preparation period, a $\pi/2$ pulse generates the initial transverse magnetization. "Frequency labelling" occurs during the evolution period t_1. The second $\pi/2$ pulse starts the mixing period wherein the longitudinal magnetization components associated with each hf line can be exchanged with components having different precessional frequencies. When the magnetization is rotated into the x-y plane for detection, components initially precessing with angular frequency $\omega_1=\omega_a$ will, to the extent that magnetization transfer has occurred during mixing, precess with new angular frequency $\omega_2=\omega_b$.

each hf line can be exchanged, thereby mixing components carrying different precessional-frequency information (i.e. MT occurs). Thus, after rotating this magnetization into the x-y plane for detection by the third $\pi/2$ pulse, components initially precessing with angular frequency $\omega_1=\omega_a$ will (to the extent that MT occurred by exchange curing the mixing period) precess with new frequency $\omega_2=\omega_b$. As in the COSY-EPR experiment, the pulse sequence is repeated for a series of equally-spaced values of t_1, where for each t_1 the FID is collected and then the phase of the preparation pulse is advanced by 90°, followed by collection of a second FID.

A 2D-ELDOR spectrum of 1×10^{-3} M PD-Tempone in toluene-d_8 (67) is shown in Fig. 27 with the corresponding contour map in Fig. 28. Magnetization transfer induced by Heisenberg spin exchange (HE) during the mixing period gives rise to cross-correlations and hence to the appearance of cross-peaks (66-68). The cross-peaks in Fig. 27 have the characteristics predicted for an exchange process, viz. there are comparable cross-peaks between all pairs of auto-peaks. This arises because HE has no nuclear spin selection rules. Measurement of the relative intensities of autopeaks and cross-peaks gives a direct determination of the Heisenberg exchange rate ω_{HE} (cf. Sect. 6.5).

Quantitative determination of exchange rates with 2D-ELDOR is complicated by the presence of unwanted coherences which add intensity only to the autopeaks of a motionally narrowed 2D spectrum, and in the slow-tumbling regime to the

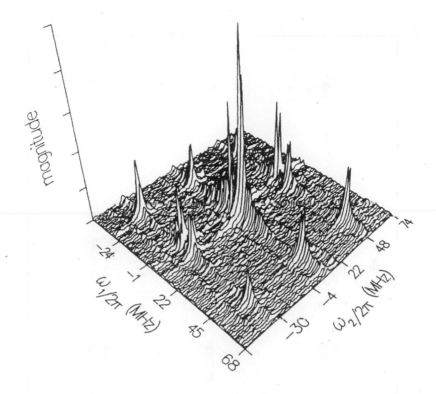

Fig. 27. Absolute value 2D-ELDOR spectrum of 1.17×10^{-3} M PD-Tempone in toluene-d_8 at 21°; τ_p=15.2 nsec; Δt_1=6 nsec; Δt_2=3.9 nsec; 90 in t_1; 16 step phase alternation sequence with 30 averaged FID per step; dead time in t_2 of 100 nsec; dead time in t_1 of 120 nsec; mixing time T=3.10×10^{-7} sec; 256 complex data points per FID extending to 1 μsec; acquisition time 27 minutes. [From Ref. 67].

cross-peaks as well. This phenomenon, known as transverse interference (76), arises from transverse magnetization following the first $\pi/2$ pulse which freely precesses for the rest of the sequence and interferes with the FID recorded during the detection period. A two-step phase alternation sequence has previously been suggested for the cancellation of transverse interference (76). We can combine this sequence with our eight-step 2D image cancellation sequence to obtain a sixteen-step procedure tabulated elsewhere (66). However, we have discovered a more compact phase alternation sequence for 2D-ELDOR which achieves the same result as the sixteen-step sequence but requires only eight steps. It is also tabulated elsewhere (66).

The 2D-ELDOR spectra of 1×10^{-3} M PD-Tempone in toluene-d_8 were recorded with the 16-step phase alternation sequence for four different mixing times. The ω_{HE} were determined by comparison of autopeak and cross-peak magnitudes (e.g. volume)

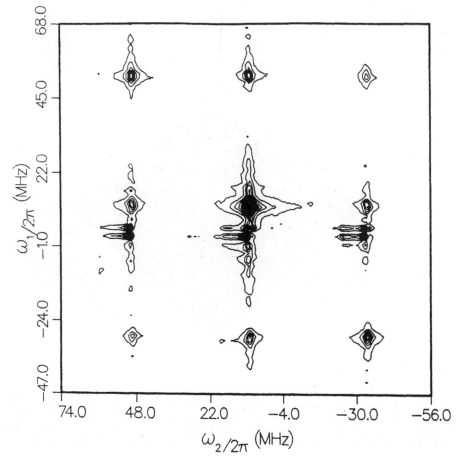

Fig. 28. 2D-ELDOR contour map of spectrum of Fig. 27. Residual axial-peaks appear as doublets centered on the line ω_1; the observed splitting is an artifact of the baseplane correction described in the text.[From Ref. 67].

in each of these spectra according to the theory summarized below. These results could be compared to an ESE study in which six different concentrations of PD-Tempone in toluene-d_8 were used, and T_2 (M_I=0) was determined for each by fitting the ESE envelope to a single exponential (67). The exchange rate determined by ESE corresponds to the rate constant ω_{HE}/C=k_{HE}=3.63 ± 0.63×10^9 M^{-1} sec^{-1}. The result is in agreement, within the experimental uncertainty, with the value determined by 2D-ELDOR (cf. Fig. 28), corresponding to k_{HE}=3.97 ± 0.43×10^9 M^{-1} sec^{-1}.

The 2D-ELDOR technique constitutes a direct observation of Heisenberg exchange on a single sample, whereas the ESE technique is somewhat indirect; that is, one usually fits T_2^{-1} to a linear dependence on concentration, but this is not always valid (77,-79). Electron-electron dipolar (EED) relaxation between probe molecules can also contribute to the observed T_2 in a concentration-dependent fashion. EED is "ELDOR active" (47) but its presence is reflected relatively more substantially in the T_2^{-1}, hence also in the widths with respect to ω_1 and ω_2 of the 2D-ELDOR resonance lines. This allows for discrimination between HE and EED in the 2D-ELDOR experiment; (EED was not important in this experiment in a nonviscous solvent [77,80]). Also, we note that in an ELDOR experiment the sensitivity to ω_{HE} depends upon the ratio ω_{HE}/W_e (where W_e is again the electron spin-flip rate), whereas in a T_2 experiment it depends upon $\omega_{HE}T_2(0)$, where $T_2(0)$ is the concentration-independent T_2, so that when $W_e \ll T_2(0)^{-1}$, as is frequently the case, the ELDOR experiment would be the more sensitive to ω_{HE}. In this connection, one should note that whereas cw ELDOR only yields ratios such as ω_{HE}/W_e, time-domain ELDOR, such as 2D-ELDOR, yields the relaxation rates directly (37,38,47,81).

It is possible to use LPSVD methods to facilitate accurate projection of 2D absorption lineshapes and to suppress certain artifacts that appear in 2D-ELDOR spectra. For this purpose we have developed a new linear predictive technique, based on LPSVD (71), which models two-dimensional time series obtained in COSY type experiments entirely in the time domain, i.e. without Fourier transformation. This new application of complex-valued linear prediction facilitates the projection of 2D absorption lineshapes as well as the rejection of residual axial peaks and much of the noise. We applied 2D linear prediction to the data set leading to Fig. 27. Projection of pure absorption lineshapes was performed in both time domains. Extrapolation in t_1 eliminated artifacts caused by t_1 truncation and enabled a more accurate determination of baseplane offset. In Fig. 29 we illustrate the LPSVD result obtained after eliminating components for which $|\omega_1/2\pi| < 3$ MHz (i.e. narrow reject filtering of axial peaks); note the considerable improvement in signal/noise ratio. The volume integral of each 2D absorption line was measured by numerically integrating in the ω_2 domain and summing the results over the discrete values of ω_1. Estimation of the Heisenberg exchange rate from ratios of volume integrals with the use of Eqn. [44] gives the result $k_{HE}=3.92\times10^9$ M^{-1} sec^{-1}, in good agreement with the result obtained from Fig. 27.

In the experiment of Fig. 27, we have chosen a sample and temperature for which nuclear spin relaxation is very small (80). If the nuclear spin relaxation rate $2W_n$ constitutes a non-negligible source of spin relaxation, then a more general analysis of the 2D-ELDOR experiment is required. Note that W_n arises from rotational modulation of the electron nuclear dipolar interaction involving the ^{14}N nuclear spin, and it is an important parameter in the study of molecular dynamics (47,80). The 2D-ELDOR spectrum of PD-Tempone in the liquid crystal S2 (Fig. 30) is an example of a 2D-ELDOR experiment in which nuclear spin relaxation cannot be neglected. In this case Heisenberg exchange does not play a significant

222

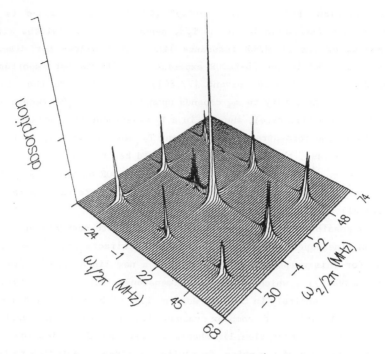

Fig. 29. LPSVD projected pure 2D absorption representation of the spectrum of
Fig. 27. M=24, K=6 in the t_1 domain, M=60, K=12 in the t_2 domain; where M is the
number of linear prediction coefficients and K is twice the number of signal
peaks. The broad peak near the center of the spectrum (at $\omega_1/2\pi=-8$ MHz) has zero
width in ω_2 (hence zero volume) and is apparently an artifact of the computation.
[From ref. 71].

role in the transfer of magnetization. This is apparent by inspection of the
contour map (Fig. 31) associated with the spectrum of Fig. 30 which shows no
significant cross-peaks connecting the outer hf lines. This is consistent with
a nuclear spin relaxation mechanism where $\Delta M_I=\pm 1$, e.g. from the electron-nuclear
dipolar interaction. The spectrum of Fig. 30 yielded the relaxation rate $2W_n=4.25$
$\pm\ 0.25\times 10^5$ sec^{-1}. The quantitative accuracy of this technique for determination of
W_n was demonstrated by comparison with stimulated-echo (SE) results.

The quantitative agreement of 2D-ELDOR and SE methods raises the question:
Why perform 2D-ELDOR experiments, which are significantly more sophisticated than
stimulated echoes, when SE experiments yield the same results? The answer to this
question is two-fold. First, note that the SE experiment required the use of
substantially larger sample volumes in order to obtain adequate sensitivity.

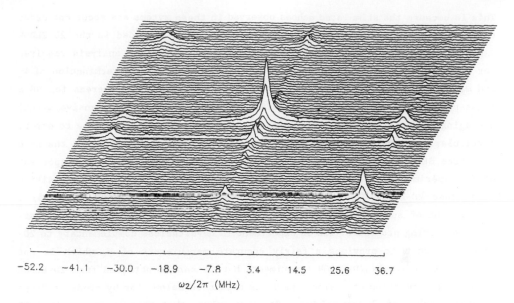

$-52.2 \quad -41.1 \quad -30.0 \quad -18.9 \quad -7.8 \quad 3.4 \quad 14.5 \quad 25.6 \quad 36.7$

$\omega_2/2\pi$ (MHz)

Fig. 30. Absolute value 2D-ELDOR spectrum of 2.05×10^{-3} M PD-Tempone in liquid crystal S2 at $35.5 \pm 0.5°C$ obtained with the director aligned parallel to B_0; $\tau_p=12$ nsec; $\Delta t_1=7$ nsec; $\Delta t_2=5.86$ nsec; 128 samplings in t_1; eight-step phase alternation sequence with 128 averages per step; dead time in t_1 of 66 nsec; dead time in t_2 of 145 nsec; mixing time $T=5.48\times10^{-7}$ sec; 256 complex data points per FID extending to 1.5 μsec; acquisition time 51 minutes. [From Ref. 72].

$\omega_2/2\pi$ (MHz)

$32.9 \qquad 19.2 \qquad 5.5 \qquad -8.2 \qquad -21.8 \qquad -35.5 \qquad -49.2$

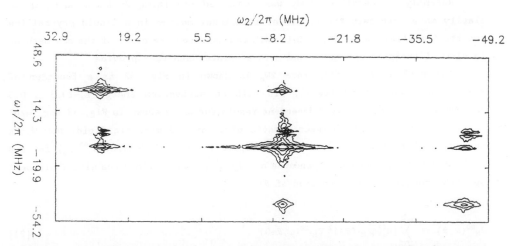

Fig. 31. 2D-ELDOR contour map of the spectrum of Fig. 30. Note the absence of cross-peaks connecting the two outer hf lines, indicative of an ^{14}N nuclear spin relaxation mechanism, [From Ref. 72].

This is because the stimulated echoes observed in these sytems are about one order of magnitude smaller in signal voltage than the FID's recorded in the 2D-ELDOR experiment. Second, it is important to compare the methods of analysis required for the two types of data. In the case of 2D-ELDOR, a direct determination of W_n and ω_{HE} is obtained by comparison of peak volumes (cf. below), whereas for SE a least-squares fit to a sum of exponentials is required. Unless extensive signal averaging is performed, the least-squares procedure will be susceptible to error, particularly in the case of $W_n/W_e \ll 1$ or $\omega_{HE}/W_e \ll 1$. This underscores the need for three SE experiments, each performed at a different hf line. The proper set of SE experiments therefore requires 1) substantially longer data acquisition times than 2D-ELDOR; 2) greater minimum number of spins than 2D-ELDOR; 3) a measurement of spectrometer dead-time; and 4) a five-parameter nonlinear least-squares fitting procedure. The SE experiment is nevertheless complementary to 2D-ELDOR, in that it provides a fairly accurate measurement of W_e which is not obtained in a single 2D-ELDOR experiment, but it can be obtained from 2D-ELDOR results as a function of mixing time T, as discussed below (or by studying "type E" axial peaks, cf. below and Ref. 68). [Note also, that these comments about SE apply also to saturation recovery (SR) discussed by Hyde in this volume. In SE, however, the spins are first prepared in a well-defined state prior to studying their time evolution. Also, in SR one observes not an echo (nor an FID), but a low power cw signal.]

Recently, a detailed study was completed utilizing 2D-ELDOR and ESE to clarify more precisely the nature of rotational motion in a liquid crystalline smectic (lamellar) phase (72). This work demonstrates the power of the new methods in view of the more extensive and more accurate data they provide.

The nuclear spin flip rate $2W_n$ is shown in Fig. 32 as a function of temperature, and it is readily fit with an activation energy $E_a = 10.3 \pm 0.3$ kcal/mole. A typical angular dependent result for W_n is shown in Fig. 33 for T=20° C. Here θ is the angle between nematic director and magnetic field. Now W_n is proportional to the (electron-nuclear dipolar) spectral density $J_1^{DD}(\omega_a)$ for $\theta=0$ (47). However, the angular dependence of $2W_n = J_1^{DD}(\omega_a, \theta)$ provides one with additional spectral densities as a function of θ:

$$J_M^{DD}(\omega, \theta) = \sum_{m'=-2}^{m} |d^{(2)}_{M,M'}(\theta)|^2 \, J_{M'}^{DD}(\omega, \theta=0) \qquad [21]$$

where the $d^{(2)}_{M,M'}(\theta)$ are the reduced Wigner rotation matrix elements of rank two. The use of Eqn. [21] in EPR requires that the spin probe is not highly

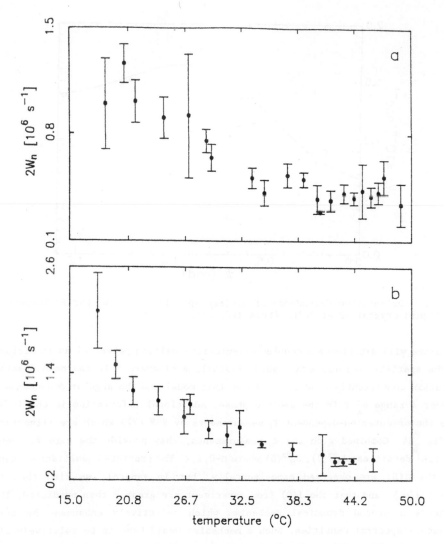

Fig. 32. Temperature dependence of nuclear spin flip rate W_n for PD-Tempone in the liquid crystal S2 at orientation (a) $\theta=0°$; (b) $\theta=90°$. [From Ref. 72].

ordered (82). Here M represents a quantum number for projection of the spin axes in the director frame. Eqn. [21] yields three basic quantities: $J_0^{DD}(\omega_a)$, $J_{\pm1}^{DD}(\omega_a)$ and $J_{\pm2}^{DD}(\omega_a)$, which are obtained from a non-linear least squares fit to the orientation-dependent data. In particular, the data of Fig. 33 lead to $J_1^{DD}(\omega_a) < J_0^{DD}(\omega_a) < J_2^{DD}(\omega_a)$, which immediately rules out a variety of mechanisms proposed for liquid crystalline phases. However, a model due to Moro and Nordio (83) for molecular dynamics in smectic phases was found to be consistent with the results. It is based on the idea that the orientational potential felt by a molecular probe should depend upon the probe location within the bilayer. Thus

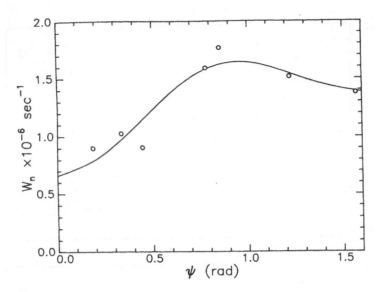

Fig. 33. Orientation dependence of nuclear spin flip rate W_n for PD-Tempone in the liquid crystal S2 at 20°C. [From Ref. 72].

the probe will experience a coupled orientation-position potential as it diffuses in the spatially non-uniform liquid crystalline bilayers. It has been possible to obtain consistently good fits between this model and the angular dependence of W_n over a range of T in the smectic phase. Additional information is obtainable from the orientation-dependent T_2 measurements by ESE (72) which are illustrated in Fig. 34. Combined with the W_n measurements, they provide the zero frequency spectral densities: $J_M^{DD}(0)$, $J_M^{DD}(0)$ where M=0,1,2. The predicted qualitative trends for the $J(0)$ are observed experimentally, but in general, we find that the $J(0) > J(\omega_a)$, and that the $J(0)$ from experiment are greater than predicted. This suggests a second dynamical mechanism which selectively enhances the zero-frequency spectral densities. Such a mechanism would have to be relatively slow in order to affect $J(0)$ predominantly. It was speculated that this may be due to cooperative fluctuations in the hydrocarbon chains sensed by the probe.

Finally, we demonstrate the importance of measuring accurate homogeneous T_2's rather than cw linewidths (i.e. $T_{2,cw}^{-1}$) in ordered media, as discussed in Sect. 2. We show in Fig. 35 how the two differ, especially for the $M_I=+1$ hf line for which $T_{2,cw} < T_2$. Furthermore, we find that the orientation dependence of $T_{2,cw}$ is inconsistent with that for the homogeneous T_2's from ESE. In fact, the observed $T_{2,cw}$ may be obtained by simulating the effects of a static distribution of director orientations which is magnetic-field dependent, such as shown by Lin and Freed (25) earlier.

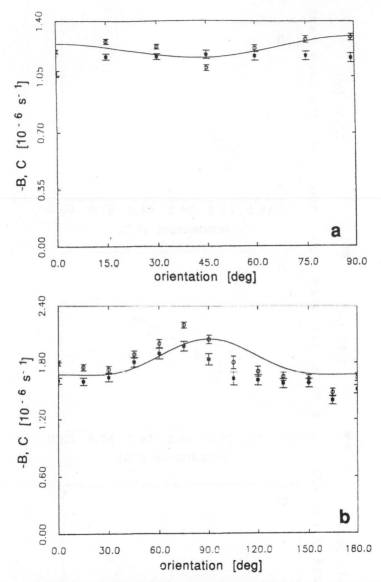

Fig. 34. Orientation dependence of linewidth parameters B and C determined by ESE for PD-Tempone in S2 at (a) 38°C (b) 31°C. B and C were calculated from experimental T_2's according to $T_2(M_I)^{-1} = A + BM_I + CM_I^2$. [From Ref. 72].

In summary, note (i) the extensive and reliable information provided by these new techniques which bear directly on molecular dynamics in multi-bilayers; (ii) their utility in discriminating between models.

228

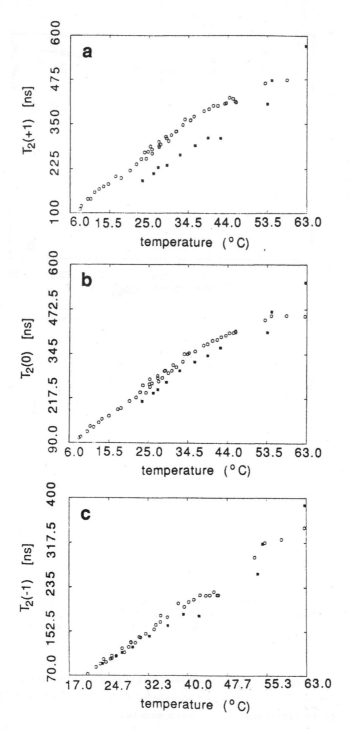

Fig. 35. Temperature dependence of homogeneous (o) T_2 from ESE and (■) $T_{2,cw}$ for the three hyperfine lines of PD-Tempone in S2. [From Ref. 72].

6.5 Theory of 2D-ELDOR and COSY (67,68,72)
The FID following a single pulse is given by the expression

$$s(t) = B' \sum_j a_j \exp(-\Lambda_j t), \tag{22}$$

where the complex coefficients a_j are given by:

$$a_j = \sum_{l,m} U_1 O_{o,1j} O_{o,mj} U_m, \tag{23}$$

according to the notation we have already introduced.

The expression for 2D-ELDOR is:

$$s'_{ELDOR}(T,t_1,t_2) = B' \sum_{l,m,n} c_{lmn} \exp(-\Lambda_n t_2) \exp(-T/\tau_m) \operatorname{Re} \sum_j b_{lj} \exp(-\Lambda_j t_1) \tag{24}$$

where

$$c_{lmn} = \sum_{k,r} U_r O_{o,rn} O_{o,kn} O_{d,km} O_{d,lm} \tag{25}$$

and

$$b_{lj} = \sum_p O_{o,1j} O_{o,pj} U_p \tag{26}$$

where, for simplicity, we have ignored the V_j factors, corresponding to the assumption of uniform spectral rotation by the pulses. The result of Eqns. [24]-[26] is closely related to that of Eqn. [12] for the field-swept MT experiment, except for the fact that an FID (rather than an echo is detected).

Eqns. [24]-[26] represent only one of several terms predicted by the theory; in particular, they represent the contribution to the 2D spectrum which provides the ELDOR information. We may identify four independent contributions to the time-domain spectral function $s(T,t_1,t_2)$,

$$s'_{(x,x,x)}(T,t_1,t_2) = s'_{ELDOR} + s'_{Transverse} + s'_{Axial(E)} + s'_{Axial(M)} \tag{27}$$

where (x,x,x) identifies, for each pulse, the axis of rotation by B_1. s'_{ELDOR} is the contribution given by Eqns. [24]-[26], while $s'_{Transverse}$ is the contribution associated with the transverse interference discussed above, and $s'_{Axial(E)}$ and

$s'_{Axial(M)}$ are the terms associated with axial peaks (60). The quadrature component is written analogously as

$$s_{(y,x,x)}(T,t_1,t_2) = s''_{ELDOR} + s''_{Transverse} + s''_{Axial(E)} + s''_{Axial(M)}. \qquad [28]$$

The phase cycling sequences cancel out $s_{Transverse}$ and $s_{Axial(E)}$. The $s_{Axial(M)}$ are removed by first measuring the constant baseplane offset described below.

In the case of Figs. 25 and 28, corresponding to the motional narrowing regime with three well-separated hf lines, our expressions are greatly simplified by the fact that $O_o=1$, the unit matrix, and we can write them as

$$s'_{ELDOR}(T,t_1,t_2) = B' \sum_{n,m,j} c_{nmj}A_n(t_2) \exp(-T/\tau_m) \, Re[A_j(t_1)] \qquad [29]$$

and

$$s''_{ELDOR}(T,t_1,t_2) = B' \sum_{n,m,j} c_{nmj}A_n(t_2) \exp(-T/\tau_m) \, Im[A_j(t_1)], \qquad [30]$$

where $A_i(t)$ is defined by

$$A_i(t) = \exp(-\Lambda_i t). \qquad [31]$$

The real coefficients c_{nmj} are defined by

$$c_{nmj} = U_n O_{d,nm} O_{d,jm} U_j, \qquad [32]$$

with

$$
O_{d,nm} = \begin{array}{c} \\ -1 \\ 0 \\ +1 \end{array}
\begin{array}{ccc} 1 & 2 & 3 \\ \left[\begin{array}{ccc} 3^{-1/2} & 2^{1/2} & 6^{-1/2} \\ 3^{-1/2} & 0 & -2(6)^{-1/2} \\ 3^{-1/2} & -2^{-1/2} & 6^{-1/2} \end{array} \right] \end{array} \qquad [33]
$$

and with $n=M_I= -1,0,1$ and

$$\tau_1^{-1} = 2W_e, \qquad [34]$$
$$\tau_2^{-1} = 2W_e + 2W_n + \omega_{HE}, \qquad [35]$$
$$\tau_3^{-1} = 2W_e + 6W_n + \omega_{HE}. \qquad [36]$$

[The presence of chemical exchange and/or pseudo-secular EED terms would lead to an effective exchange frequency ω_{HE} replacing ω_{HE} in these expressions (47).]

The U_n in Eqn. [32] are given by $U_n=i\epsilon/2$ where $-\epsilon$ is the equilibrium population difference for an isotropic system, given by $g\beta_e H_o/k_B T$ in the high-field limit. Eqns. [33]-[36] are appropriate when the nonsecular contributions from the g-tensor, hf-tensor, and EED tensor may be neglected. More generally, up to five τ_m would be required for a nitroxide. This case is discussed in detail in (68,72).

We see immediately from Eqns. [29] and [30] that the effect of the 2D quadrature phase alternation scheme is to select the real part of the time-domain spectral function $A_j(t_1)$ in the case of s' and the imaginary part in the case of s".

Upon FT with respect to t_2 we obtain $\hat{s}_{(x,x,x)}(T,t_1,\omega_2)$ and $\hat{s}_{(y,x,x)}(T,t_1,\omega_2)$ at which time we form (66):

$$\hat{s}(T,t_1,\omega_2) = \text{Re}[\hat{s}_{(x,x,x)}] + i\text{Re}[\hat{s}_{(y,x,x)}]. \tag{37}$$

Then FT with respect to t_1 yields the 2D spectrum given by

$$S_{ELDOR} = S'_{ELDOR}(T,\omega_1,\omega_2) + iS''_{ELDOR}(T,\omega_1,\omega_2) \tag{38}$$

with

$$S'_{ELDOR}(T,\omega_1,\omega_2) = B' \sum_{n,m,j} c_{nmj} \frac{1}{i\omega_1 - \Lambda_j} \exp(-T/\tau_m) \text{ Re } \frac{1}{i\omega_2 - \Lambda_n} \tag{39}$$

and with $S''_{ELDOR}(T,\omega_1,\omega_2)$ given by Eqn. [39] but with Re replaced by Im.

As noted above, we are left with the one remaining unwanted $S_{Axial(M)}$ after phase cycling, and it is removed during data analysis. One finds (67,68) that $s'_{Axial(M)}(T,t_2)$ is independent of t_1, and it constitutes the only contribution to $\bar{s}_{(x,x,x)}(T,t_1,t_2)$ with this property. Thus we first estimate a constant baseplane offset in $\bar{s}_{(x,x,x)}(T,t_1,\omega_2)$, which we identify with $\bar{s}'_{Axial(M)}(T,\omega_2)$. Then we subtract this baseplane from $\bar{s}_{(x,x,x)}(T,t_1,\omega_2)$, and repeat this procedure with $\bar{s}_{(y,x,x)}$ prior to FT with respect to t_1. The experimental implementation is not perfect. There is residual amplitude associated with axial peak contributions, and it gives rise to the distortions seen along $\omega_1=0$ in the spectra of Figs. 23, 25, and 28. [Additional distortions near $\omega_1=0$ arise from the extra amplitude modulation in t_1 due to distortions in the second MW pulse (causing a variation of the rotation angle for this pulse) as a function of interpulse delay t_1, which is a problem analogous to that in NMR (84).] Combining Eqns. [32] and [39] and introducing the factors V_j, we obtain

$$\text{ReS}'_{\text{ELDOR}}(T,\omega_1,\omega_2) = B' \sum_{n,m,j} U_n V_n \frac{T_{2,n}}{1+(\omega_2-\omega_n)^2 T_{2,n}^2} U_j V_j^2 \frac{T_{2,j}}{1+(\omega_1-\omega_j)^2 T_{2,j}^2}$$

$$\times \; O_{d,nm} O_{d,jm} \, \exp(-T/\tau_m), \qquad\qquad [40]$$

applicable in the motional narrowing regime with well-separated hf lines. Here, $T_{2,j}$ and ω_j are the T_2 and resonant frequency of the jth hf line. For the purpose of simulation one need only compute Eqn. [40].

We have neglected inhomogeneous broadening in the above analysis, as it does not play a significant role in the interpretation of 2D-ELDOR data in the motional narrowing regime (other than to broaden each resonance line by a small amount). The effect of including significant inhomogeneous broadening in the theory is that there would be several types of spin echo : viz. three spin echoes associated with each pair of pulses, plus a stimulated echo and a twice refocussed echo. [These matters are discussed elsewhere (24,36).] These are the terms which would be important for slow motional 2D-ELDOR spectra, where $T_2^* < \tau_d$, so an FID technique would not be appropriate. Instead, one could observe the stimulated echo, which has a structure very similar to that of Eqns. [24]-[26], whereas the others would be canceled by phase cycling.

Before discussing 2D-ELDOR further, we note that a similar analysis leads to motional narrowing expressions for COSY-EPR:

$$s'_{\text{COSY}}(t_1,t_2) = -B' \sum_n U_n^2 A_n(t_2) \text{Re}[A_n(t_1)] \qquad\qquad [41]$$

and

$$s''_{\text{COSY}}(t_1,t_2) = B' \sum_n U_n^2 A_n(t_2) \text{Im}[A_n(t_1)]. \qquad\qquad [42]$$

Eqns. [41] and [42] predict a COSY spectrum without cross-peaks, consistent with the experimental result of Fig. 25. However, given sufficiently rapid exchange (e.g. ω_{HE} comparable to the hyperfine splittings), O_o is no longer simply the identity operator 1, and cross-peaks are predicted to appear in the COSY spectrum. It is also true in general that $O_o \neq 1$ in the slow-motional regime, hence one finds cross-terms in the appropriate analogues of Eqns. [41] and [42].

We now consider the analysis of the 2D-ELDOR experiment for a motionally-narrowed nitroxide spectrum with three well-separated hf lines characteristic of a nitroxide. We obtain from the experiment the 3×3 matrix Q(T) of volume integrals. [The matrix Q(T) is defined such that Q_{mj} is the volume integral of the cross-peak located at coordinates $(\omega_1,\omega_2)=(\omega_j,\omega_m)$.] According to Eqn. [40] we can write:

$$Q_{nj}(T) = E \sum_m O_{d,nm}O_{d,jm} \exp(-T/\tau_m) \qquad [43]$$

where E is a spectrometer constant. The expressions for the matrix elements Q_{nj} (within the spectrometer constant) are given in Table 1 for the case when Eqns. [33] and [34] are applicable. In the analysis of experiments, one utilizes ratios: $f_{mj} = Q_{nj}/Q_{jj}$ thereby canceling the spectrometer constant, E.

In the very simplest case where W_e is negligible and Heisenberg exchange is the only source of magnetization transfer, one obtains after some algebra:

$$\omega_{HE} = \frac{1}{T} \ln \frac{2f_{mj}V_j r_{2j}+V_m r_{2m}}{V_m r_{2m} - f_{mj}V_j r_{2j}} . \qquad [44]$$

Here the r_{2j} measure the dead-time reduction factors (as well as the filtering effects of the resonator with its finite Q) on the FID's collected as a function of t_2. The products $V_j r_{2j}$ are determined by measuring the normalized peak areas obtained from a single-pulse FID collected under the same conditions (i.e. same dead time, pulse length, etc.) as the 2D-ELDOR spectrum.

Note that Eqn. [44] predicts that all six cross-peaks equivalently reflect the exchange process, and hence that six independent measurements of the exchange rate ω_{HE} are obtained from a single 2D spectrum. The equivalence of the six cross-peaks is a manifestation of the lack of a selection rule for the exchange mechanism (74), i.e. both $\Delta M_I=\pm 1$ and $\Delta M_I=\pm 2$ transitions are equally probable.

TABLE 1. Matrix Elements $E^{-1}Q_{nm}(T)$:[a]

Auto-Peaks[b]	
$E^{-1} Q_{\pm 1,\pm 1}$:	$\frac{1}{3} \exp[-T/\tau_1] + \frac{1}{2} \exp[-T/\tau_2] + \frac{1}{6} \exp[-T/\tau_3]$
$E^{-1}Q_{0,0}$:	$\frac{1}{3} \exp[-T/\tau_1] + \frac{2}{3} \exp[-T/\tau_3]$
Cross-Peaks	
$E^{-1} Q_{0,\mp 1} = E^{-1}Q_{\mp 1,0}$:	$\frac{1}{3} \exp[-T/\tau_1] - \frac{1}{3} \exp[-T/\tau_3]$
$E^{-1} Q_{\pm 1,\mp 1}$:	$\frac{1}{3} \exp[-T/\tau_2] - \frac{1}{2} \exp[-T/\tau_2] + \frac{1}{6} \exp[-T/\tau_3]$

(a) This is the case of negligible non-secular terms.
(b) These apply both to 2D-ELDOR and to Stimulated Echoes from the three hyperfine lines.

The W_n mechanism and the exchange mechanism may be distinguished in these spectra as a result of their different selection rules. Unlike exchange, the W_n mechanism obeys the selection rule $\Delta M_I = \pm 1$ and thus gives rise predominantly to cross-peaks connecting only adjacent hf lines, (cf. Fig. 30), but see below. Thus the geometrical pattern of the spectral contours may be utilized to obtain information regarding the mechanism of magnetization transfer prior to the application of Eqn. [44] or its analogues for $W_n \neq 0$. If we consider, for example, a case where only cross-peaks for which $\Delta M_I = \pm 1$ are significant, we can write by analogy to Eqn. [44],

$$6W_n = -\frac{1}{T} \ln \frac{2\hat{Q}_{mj}V_j r_{2j} + V_m r_{2m}}{V_m r_{2m} - \hat{Q}_{mj}V_j r_{2j}} \qquad [45]$$

which applies to the two cross-peaks for which $m=\pm 1$ and $j=0$. In Eqn. [45], $\hat{Q}_{mj} = Q_{mj}/Q_{oo}$. Thus in the event that $\omega_{HE}=0$ one can obtain W_n directly from two of the four cross-peaks with the application of Eqn. [45]. The expression for the other two cross-peaks (i.e. $m=0, j=\pm 1$), is somewhat more complicated. These cross-peaks nevertheless directly yield W_n provided exchange is absent. If cross-peaks for which $\Delta M_I = \pm 2$ are significant, then the left side of Eqn. [45] requires an additive factor involving ω_{HE}. The full set of linear equations applicable when $\omega_{HE} \neq W_n \neq 0$ has been given elsewhere for all six cross-peaks (68,72).

The overdetermined set of six linear equations may be used to obtain both W_n and ω_{HE}. Note that the presence of cross-peak intensity at $\Delta M_I = \pm 2$ does not always imply the presence of Heisenberg exchange; the W_n mechanism may also contribute intensity to these peaks by a small amount, but only for sufficiently long mixing times T, i.e. when $W_n T > 1$, (cf. Table 1). The absence of $\Delta M_I = \pm 2$ peaks is easily shown by expanding $Q_{\pm 1 \mp 1}$ given in Table 1 to lowest power in $W_n T$.

These expressions for relative peak volumes for a given mixing time, T, do not by any means exhaust the available sources of data from the 2D-ELDOR experiment. There is, after all, a third dimension available, that of mixing time T. From a series of 2D-ELDOR spectra at different values of T, one can map out the T dependence of all nine auto- and cross-peaks. Thus, for example, from Table 1 in the simple special case of $W_n = 0$, we obtain for the auto-peaks

$$Q_{nn}(T) \propto \exp[-2W_e T] + 2 \exp[-(2W_e + \omega_{HE})T] \qquad n=-1,0,1 \qquad [46]$$

and for the cross peaks

$$Q_{nm}(T) \propto \exp[-2W_e T] - \exp[-(2W_e + \omega_{HE})T] \qquad n \neq m \; . \qquad [47]$$

In general, each peak shows a time evolution given by a sum of exponentials which may be fit to provide W_e, W_n, and ω_{HE}. The Q_{nn} vs. T represents equivalent information to that of stimulated echoes, whereas the Q_{nm} vs. T for $n \neq m$ are equivalent to that from ELDOR echoes. We have found that for accurate and reliable estimation of relaxation rates, it is a good idea to obtain 2D-ELDOR at 3 or 4 values of T for which all the peaks are prominent. As 2D-ELDOR techniques improve and data acquisition rates are increased, it should become practical to obtain 2D-ELDOR for significanty more values of T, which would then more truly represent a 3D experiment.

6.6 Future Uses of 2D-FT-EPR

We have illustrated how FT-EPR is capable of routine use in several applications. We plan to develop the technique to be applicable to all studies involving nitroxides. Until very recently, one was limited to the application of FT methods to spectra not exceeding about 100 MHz spectral bandwidth. We have illustrated applications of the FT technique that do not require uniformity of effective rotation by B_1 across the entire spectrum. Such "B_1 uniformity" is not a necessity for the implementation of most two-dimensional techniques. The effects of B_1 non-uniformity can be accounted for in simulations by recognizing that spectral components which are only partially rotated by the pulses have reduced effective transition moments (65,67,68), as we have presented in Eqns. [17]-[19] or [40]. We have illustrated the utility of two-dimensional FT spectroscopy as a double resonance technique which yields quantitative information without necessarily requiring simulation or least-squares fitting of the data. We did find, however, that least-squares fitting in the form of linear prediction is a convenient method for projection of the pure 2D absorption representation of 2D-ELDOR data necessary for determination of volume integrals of the 2D absorption lines. It can also be effectively utilized to remove artifacts, such as residual axial peaks, and to significantly improve the signal-to-noise.

In the application of this technique to slow-tumbling motions of nitroxides one would obtain all of the ELDOR data in a single 2D spectrum. Such spectra would yield information about the couplings between dynamic spin-packets corresponding to different molecular orientations which can provide considerable insight into the microscopic details of the motional process (14,24,36-38,39,85). The implementation of a 2D-ELDOR experiment for the study of slow-tumbling motions would be somewhat different than described for motionally-narrowed systems. FID's in these systems are, in general, rapidly damped as a result of the large inhomogeneous line widths, which are not averaged by the motion as in the motionally-narrowed spectra. The FID will, in general, be lost in the dead-time following a microwave pulse. In these instances, the transverse magnetization can still be refocussed into a spin echo. In the 2D-ELDOR experiment one would have the option of applying an additional refocussing pulse or of recording the stimulated echo following the third pulse. The latter method would have the advantage that axial-peaks will not be refocussed and therefore cannot contribute intensity to the 2D-ELDOR spectrum.

In another application, the implementation of SECSY to slow-tumbling nitroxides would be identical to the method described in Sect. 6.2 for motionally narrowed spectra, except that reduction of the DC magnetic field homogeneity is not required. This SECSY application would be the 2D-FT analogue of the field-swept 2D experiment described in Sect. 3. It would have the advantages of (i) greatly reduced data acquisition times; (ii) pulse widths significantly shorter than the relevant T_2's. It would also be useful for the study of echo-modulation patterns and how they vary across a spectrum (cf. Sect. 5).

Many other possibilities exist for the application of other two-dimensional spectroscopies to EPR. We mention one potential application of COSY-EPR to the study of distances in macromolecules such as proteins. Suppose an ^{14}N and an ^{15}N label are placed at two distinct sites in a protein, and in a rigid-limit spectrum they are coupled by the dipolar interaction between the two electron spins. Then from the COSY cross-peaks between the ^{14}N and ^{15}N spectra, one could, in principle, measure the dipolar interaction from which $1/r^3$ could be obtained.

Access to the full range of nitroxide spectra with FT techniques requires at least a doubling of the spectrometer bandwidth. The main challenge is the generation of sufficiently large B_1 fields at the sample. Based upon our current experience, a 220 MHz spectrum would require at least a 1.4 mT B_1 field at the sample in a rectangular pulse of width about 6 nsec. There is not sufficient power to rotate on-resonant magnetization by $\pi/2$ in a low-Q cavity resonator. However, a loop-gap resonator, (86,87) which has been stabilized against electric breakdown, has enabled the generation of a large B_1 at the sample because of the high conversion efficiency of these resonators. To date we have obtained B_1 field strengths of 1.5 mT in a bridged loop-gap resonator (88) having a Q of 45 (68,89). Such a combination of high B_1 and wide bandwidth facilitates nearly uniform excitation of isotropic motionally-narrowed spectra such as PD-Tempone in toluene-d_8 (cf. Fig. 36 for the "one-dimensional" FT) or PD-Tempone in the nematic liquid crystal phase V (cf. Fig. 37 for the 2D-ELDOR). As a result of these new developments, it is now possible to obtain 2D-ELDOR spectra from lipid dispersions in the liquid crystalline phase containing fatty acid spin labels. These are especially challenging, since the three hyperfine lines are inhomogeneously broadened due to the random orientation of the lipid fragments (6,8) with $T_2^* \approx 20$-30 nsec. Nevertheless, interesting patterns of auto- and cross- peaks may be obtained as illustrated in Fig. 38.

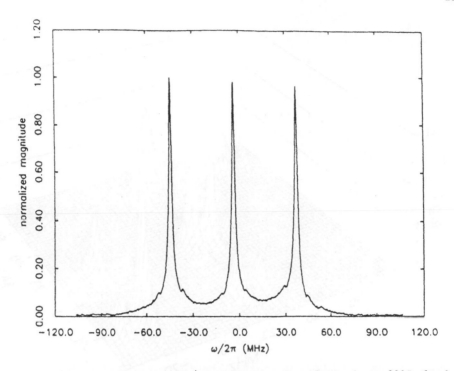

Fig. 36. FT spectrum of 5.0×10^{-4} M PD-Tempone in toluene-d_8 at 22°C obtained with a bridged loop-gap resonator. $\pi/2$ pulse width was 6.7 nsec, resolution in t_2 is 4.69 nsec, 2048 averages, 256 complex data points zero-filled to 1024 prior to Fourier transformation.[from Ref. 68].

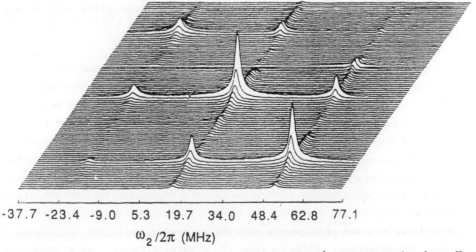

Fig. 37. Absolute value 2D-ELDOR spectrum of 5.0×10^{-4} PD-Tempone in phase V at 24°C obtained with a bridged loop-gap resonator; $\tau_p = 9.5$ nsec; $\Delta t_1 = 7$ nsec; $\Delta t_2 = 3.9$ nsec; 8-step phase alternation sequence; dead time in t_2 of 150 nsec; dead time in t_1 of 100 nsec; mixing time $T = 6.0 \times 10^{-7}$ sec; acquisition time 49 minutes. [From Ref. 68].

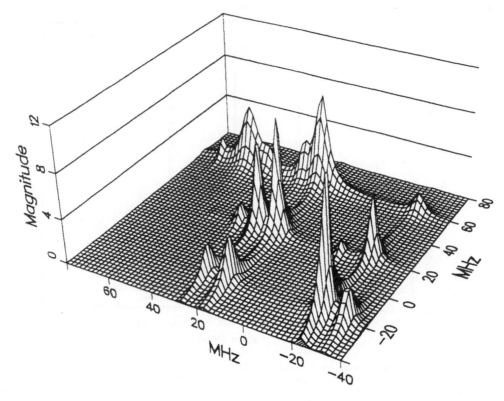

Fig. 38. 2D-ELDOR spectrum of 16-stearic acid spin label in at 0.5 mole percent in DMPC at 74.5°C. The lipid dispersion is in a buffer of 0.01 M KH_2PO_4 (pH=7) at concentration of 0.5 mg phospholipid/μl. The mixing time, T=500 nsec. The data have been treated by 2D-LPSVD to reduce noise, remove axial peaks, and correct for dead-time effects (due to t_1 and t_2 dead-times of 80 and 130 nsec, respectively). [Results obtained by R.H. Crepeau in these labs.]

In order to make the field more uniform across the spectrum, the possibility of using shaped pulses (89,90) is being investigated. A nearly rectangular pulse shape is used only as a matter of convenience; the response of the magnetization to such a pulse is of the form $\sin(\omega t_p)/\omega t_p$ and is far from optimal for FT work because of the points of null-response (65). Shaping of pulses can contain all of the available microwave power within a single band or envelope, thus improving the conversion of microwave power into useful B_1. Other possibilities for improvement of the present spectrometer include the use of a high sweep-rate 200 MHz digitizer which has just become available. We predict a reduction in FID acquisition time by at least an order-of-magnitude with the implementation of such digitizers. Also under investigation is the design of a multi-frequency superheterodyne FT spectrometer taking advantage of new multi-octave stripline devices.

In cases where B_1 field strengths are insufficient for total spectral coverage, variations of the techniques described in this report may be of interest. These are techniques of two-dimensional correlation spectroscopy which require that only a limited region of the spectrum be irradiated (so called "soft COSY" techniques (91)). A "soft" 2D-ELDOR spectra would provide information about the couplings among only those dynamic spin packets rotated by the microwave pulse. In these experiments one could collect several such 2D spectra, each focussing on a different region of the spectrum, and then combine the data during data analysis. Related techniques are already frequently utilized in NMR for 2D nuclear Overhauser spectroscopy of systems of biological interest.

It is clear that full-bandwidth irradiation techniques will be limited to specific suitable applications in EPR. These will probably not, for the foreseeable future, include transition metals in liquid solution. Transition metal spectra have bandwidths of up to several gigahertz requiring B_1's of 10-100 mT, and their T_2's in liquids are too short (typically less than or of the order of nanoseconds). Nevertheless, solid state experiments on inorganic ions by field-swept 2D-ESE such as illustrated in Sect. 5 are quite feasible. The methods described in this section are generally applicable to studies of organic free radicals, because their spectral bandwidths usually do not exceed 100-200 MHz, and their T_2's are usually significantly greater than a nanosecond.

Finally, we believe that the range of applications of 2D-FT-EPR to biophysics will continue to increase as new developments in high-speed electronics, resonators, and microwave devices find their way into the research laboratory.

ACKNOWLEDGEMENTS

We thank David Budil and Betsey Van Sickle for assistance with the final manuscript. This research was supported by NIH Grant #GM25862, NSF Grants CHE87-03014 and DMR 8604200, and the Cornell Materials Science Center.

REFERENCES

1   L.J. Berliner (Ed.), *Spin Labeling: Theory and Applications*, Vol. I, Academic, New York, 1976.
2   L.J. Berliner (Ed.), *Spin Labeling: Theory and Applications*, Vol. II, Academic, New York, 1979.
3   L.J. Berliner and J. Reuben (Eds), *Biological Magnetic Resonance*, Vol. 8, Spin Labeling: Theory and Applications III, Plenum, New York, 1989.
4   J.J. Volwerk and O.H. Griffith, *Mag. Res. Rev.* 13, (1988) 135-178.
5   J.H. Freed in: L.J. Berliner (Ed.), *Spin Labeling: Theory and Applications*, Vol. I, Academic, New York, 1976, Ch. 3.
6   E. Meirovitch, A. Nayeem, and J.H. Freed, *J. Phys. Chem.*, 88 (1984) 3454-3465.
7   E. Meirovitch and J.H. Freed, *J. Phys. Chem.*, 84 (1980) 3281-3295, 3295-3303.
8   H. Tanaka and J.H. Freed, *J. Phys. Chem.*, 88 (1984) 6633-6644.
9   H. Tanaka and J.H. Freed, *J. Phys. Chem.*, 89 (1985) 350-360.
10   L. Kar, E. Ney-Igner, J.H. Freed, *Biophys. J.*, 48 (1985) 569-595.
11   Y.-K. Shin and J.H. Freed, *Biophys. J.* 55, (1989) 537-550.

240

12 D.J. Schneider and J.H. Freed, in: L.J. Berliner and J. Reuben (Eds), *Biological Magnetic Resonance*, Vol. 8, Plenum, New York, 1989, Ch. 1.

13 K.V. Vasavada, D.J. Schneider, and J.H. Freed, *J. Chem. Phys.*, **86** (1987) 647-661.

14 D.J. Schneider and J.H. Freed, in: J.O. Hirschfelder *et al.* (Eds), *Lasers, Molecules, and Methods*, *Adv. Chem. Phys.*, **73** (1989) 387-527.

15 J.H. Freed, in: T. Dorfmüller and R. Pecora (Eds), *Rotational Dynamics of Small and Macromolecules in Liquids*, Springer-Verlag, New York, 1987, pp. 89-142.

16 G.L. Millhauser, J. Gorcester, and J.H. Freed, in: J.A. Weil (Ed.), *Electronic Magnetic Resonance of the Solid State, Can. Soc. for Chemistry*, Ottawa, 1987, pp. 571-597.

17 L. Kevan and R.N. Schwartz (Eds), *Time Domain Electron Spin Resonance*, Wiley-Interscience, New York, 1979.

18 K.M. Salikhov, A.G. Sevenov, and Yu. D. Tsvetkov, *Electron Spin Echoes and Their Applications*, Nauka, Novosibirsk, 1976.

19 W.B. Mims and J. Peisach, in: L.J. Berliner and J. Reuben (Eds), *Biological Magnetic Resonance*, Vol. 3, Plenum, New York, 1981, Ch. 5.

20 J.R. Norris, M.C. Thurnauer, and M.K. Bowman, *Adv. Biol. Med. Phys.*, **17** (1980) 365-415.

21 A.E. Stillman and R.N. Schwartz, *J. Phys. Chem.*, **85** (1981) 3031-3040.

22 A.E. Stillman, L.J. Schwartz, and J.H. Freed, *J. Chem. Phys.*, **73** (1980) 3502-3503.

23 L.J. Schwartz, A.E. Stillman, and J.H. Freed, *J. Chem. Phys.* **77** (1982) 5410-5425.

24 L.J. Schwartz, Ph.D. Thesis, Cornell University, 1984.

25 a) W.J. Lin and J.H. Freed, *J. Phys. Chem.*, **83** (1979) 379-401; b) E. Meirovitch and J.H. Freed, *ibid.*, **84** (1980) 2459-2472.

26 R.P. Mason and J.H. Freed, *J. Phys. Chem.*, **78** (1974) 1324-1329; b) R.P. Mason, E.B. Giavedoni, and A.P. Dalmasso, *Biochemistry*, **16** (1977) 1196-1201.

27 D. Kivelson and S. Lee, *J. Chem. Phys.*, **76** (1982) 5746-5754.

28 U. Eliav and J.H. Freed, *Rev. Sci. Instrum.* **54** (1983) 1416-1417.

29 G.L. Millhauser and J.H. Freed, *J. Chem. Phys.*, **81** (1984) 37-48.

30 L.J. Kar, G.L. Millhauser, and J.H. Freed, *J. Phys. Chem.*, **88** (1984), 3951-3956.

31 Y. Shimoyama, D.J. Schneider, G.L. Millhauser, and J.H. Freed, (to be published).

32 G.L. Millhauser and J.H. Freed, *J. Chem. Phys.*, **85** (1986) 63-67.

33 G.L. Millhauser, D.S. Clark, and J.H. Freed (unpublished); G. A. Marg, G.L. Millhauser, P.S. Skerkev, D.S. Clark, *Ann. NY Acad. Sci.* **469** (1986) 253-258.

34 L. Kar, M.E. Johnson, and M.K. Bowman, *J. Magn. Res.*, **75** (1987) 397-413.

35 J.S. Hyde and L.R. Dalton, in: L.J. Berliner (Ed.), *Spin Labeling: Theory and Applications*, Vol. II, Adacemic, New York, 1979, Ch. 1.

36 L.J. Schwartz, G.L. Millhauser, and J.H. Freed, *Chem. Phys. Lett.*, **127** (1986) 60-66.

37 J.H. Freed, *J. Phys. Chem.*, **78** (1974) 1155-1167.

38 J.H. Freed, in: L. Kevan and R.N. Schwartz (Eds), *Time Domain Electron Spin Resonance*, Wiley Interscience, New York, 1979, Ch. 2.

39 J.P. Hornak and J.H. Freed, *Chem. Phys. Lett.*, **101** (1983) 115-119.

40 M. Dorio and J.H. Freed (Eds), *Multiple Electron Resonance Spectroscopy*, Plenum, New York, 1979.

41 M. Nechtschein and J.S. Hyde, *Phys. Rev. Lett.*, **24** (1970) 672-674.

42 G.V. Bruno and J.H. Freed, *Chem. Phys. Lett.*, **25** (1974) 328-332; b) G.V. Bruno, Ph.D. Thesis, Cornell University, 1972.

43 a) J.S. Hyde, M.D. Smigel, L.R. Dalton, and L.A. Dalton, *J. Chem. Phys.* **62**, (1975) 1655-1667; b) L.R. Dalton, B.H. Robinson, L.A. Dalton, and P. Coffey, *Adv. Mag. Reson.*, **8** (1976) 149-259; c) L.A. Dalton and L.R. Dalton, in: M. Dorio and J.H. Freed (Eds), *Multiple Electron Resonance Spectroscopy*, Plenum, New York, 1979, Ch. 5.

44 J.S. Hyde, W. Froncisz, and C. Mottley, *Chem. Phys. Lett.*, **110** (1984) 621-625.

45 S. A. Dzuba, A.G. Maryasov, K.M. Salikhov, and Yu.D. Tsvetkov, *J. Magn. Reson.* **58** (1984) 95-117.

46 D. Marsh, in: E. Grell (Ed.), *Membrane Spectroscopy*, Springer-Verlag, W. Berlin, 1981, Ch. 2.

47 J.H. Freed, M. Dorio (Eds), *Multiple Electron Resonance Spectroscopy*, Ch. 3.

48 C.A. Popp and J.S. Hyde, *Proc. Nat. Acad. Sci. (USA)* **79** (1982) 2259-2563.

49 a) E. van der Drift and J. Smidt, *J. Phys. Chem.*, **88** (1984) 2275-2284; b) E. van der Drift, Ph.D. Thesis, Delft, 1985.

50 L. Kevan, in: L. Kevan and R.N. Schwartz (Eds), *Time Domain Electron Spin Resonance*, Wiley-Interscience, New York, 1979, Ch. 8.

51 a) E. Szajdzinska-Pietek, R. Maldonado, L. Kevan, R.R.M. Jones, S.B. Berr, *J. Am. Chem. Soc.*, **106** (1984) 4675-4678; b) E. Szajdzinska-Pietek, R. Maldonado, and L. Kevan, *ibid.*, **107** (1985) 6467-6470.

52 a) V.F. Yudanov, S.A. Dikanov, Yu. A. Grishin, and Yu.D. Tsvetkov, *J. Struct. Chem.*, **17** (1976), 387-392; b) A.A. Shubin and S.A. Dikanov, *J. Mag. Reson.*, **52** (1983) 1-12.

53 W.B. Mims, *Phys. Rev.*, **B5** (1972) 2409-2419; *ibid.* **B6** (1972) 3543-3545.

54 D.J. Schneider, Y. Shimoyama, and J.H. Freed, (to be published).

55 R.P.J. Merks and R. deBeer, *J. Phys. Chem.*, **83** (1980) 3319-3322.

56 R.P.J. Merks, Ph.D. Thesis, Delft University, 1979.

57 P. Höfer, A. Grupp, H. Nebenführ, and M. Mehring, *Chem. Phys. Lett.* **132** (1986), 279-282.

58 R. R. Ernst and W.A. Anderson, *Rev. Sci. Instr.*, **37** (1966) 93-102.

59 J. Jeener, Ampère Summer School, Basko Polje, Yugoslavia, 1971.

60 W.P. Aue, E. Bartholdi, and R.R. Ernst, *J. Chem. Phys.*, **64** (1976) 2229-2246.

61 R.R. Ernst, G. Bodenhausen, and A. Wokaun, *Principles of Nuclear Magnetic Resonance in One and Two Dimensions*, Oxford, New York, 1987.

62 M. Bowman, *Bull. Am. Phys. Soc.*, Ser. II, **31** (1986) 524.

63 R.J. Massoth, Ph.D. Thesis, University of Kansas, 1988.

64 J. Gorcester, G.L. Millhauser, and J.H. Freed, *Proc. XXIII Congress Ampère on Magnetic Resonance*, Rome, 1986, pp. 562-563.

65 J.P. Hornak and J.H. Freed, *J. Magn. Reson.*, **67** (1986) 501-518.

66 J. Gorcester and J.H. Freed, *J. Chem. Phys.*, **85** (1986) 5375-5377.

67 J. Gorcester and J.H. Freed, *J. Chem. Phys.*, **88** (1988) 4678-4693.

68 J. Gorcester, Ph.D. Thesis, Cornell University, 1989.

69 R. Kumaresan and D.W. Tufts, *IEEE Trans.* **ASSP-30**, (1982), 671-675.

70 H. Barkhuijsen, R. de Beer, W.M.M. Bovee, and D. van Ormondt, *J. Magn. Reson.*, **61** (1985) 465-481.

71 J. Gorcester and J.H. Freed, *J. Magn. Reson.*, **78** (1988) 291-301.

72 J. Gorcester, S. Rananavare, and J.H. Freed, *J. Chem. Phys.* (1989) in press.

73 D.J. States, R.A. Haberkorn and D.J. Ruben, *J. Magn. Reson.*, **48** (1982) 286-292.

74 J.H. Freed, *J. Phys. Chem.*, **71** (1967) 38-51.

75 J. Jeener, B.H. Meier, P. Bachmann, and R.R. Ernst, *J. Chem. Phys.*, **71**, (1979) 4546-4553.

76 S. Macura and R.R. Ernst, *Mol. Phys.*, **41** (1980) 95-117.

77 M.P. Eastman, R.G. Kooser, M.R. Das, and J.H. Freed, *J. Chem. Phys.*, **51** (1969) 2690-2709.

78 A.E. Stillman and R.N. Schwartz, *J. Magn. Reson.*, **22** (1976) 269-277.

79 A. Nayeem, Ph.D. Thesis, Cornell University, 1986; A. Nayeem, S. Rananavare, and J.H. Freed (to be published).

80 J.S. Hwang, R.P. Mason, L.P. Hwang, and J.H. Freed, *J. Phys. Chem.*, **79** (1975) 489-511.

81 J. Yin, M. Pasenkiewicz-Gierula, and J.S. Hyde, *Proc. Natl. Acad. Sci. (USA)*, **84** (1987) 964-968.

82 G.R. Luckhurst and C. Zannoni, *Proc. Royal Soc. Lond. A.*, **353** (1977) 87-102.

83 G. Moro and P.L. Nordio, *J. Phys. Chem.*, **89** (1985) 997-1001.

84 A.F. Mehlkopf, D. Korbee, T.A. Tiggelman, and R. Freeman, *J. Magn. Reson.*, **58** (1984) 315-323.

85 G.L. Millhauser, Ph.D. Thesis, Cornell University, 1986.

86   W. Froncisz and J.S. Hyde, *J. Magn. Reson.*, **47** (1982) 515-521.

87   J.P. Hornak and J.H. Freed, *J. Magn. Reson.*, **62** (1985) 311-313.

88   S. Pfenninger, J. Forrer, A. Schweiger, and T.H. Weiland, *Rev. Sci. Instrum.*, **59** (1988) 752-760.

89   R.H. Crepeau, A. Dulcić, J. Gorcester, T. Saarinen and J.H. Freed, (*J. Magn. Reson.* submitted).

90   W.S. Warren, *J. Chem. Phys.*, **81** (1984) 5437-5448.

91   R. Brüschweiler, J.C. Madsen, C. Griesinger, O.W. Sørensen, and R.R. Ernst, *J. Magn. Reson.*, **73** (1987) 380-385.

CHAPTER 6

PULSED E P R ON DISORDERED SYSTEMS: NOVEL EXPERIMENTAL SCHEMES

ARTHUR SCHWEIGER

1. INTRODUCTION

The present chapter is devoted to solid state pulsed EPR methodology, including echo–detected EPR, electron spin echo envelope modulation (ESEEM) spectroscopy and pulsed ENDOR. Some of the techniques are especially designed for disordered systems, others are also applicable in single crystal work. The chapter is subdivided into four subjects.

1.1 Echo–detected EPR

Field–swept EPR spectra are often recorded via the change in the observed electron spin echo intensity. This is particularly suitable for the study of systems with broad EPR lines and for short–lived paramagnetic species. Line broadening effects which occur in such echo–detected EPR spectra are discussed in section 2. Section 6 presents a method to improve the resolution of echo–detected EPR. In section 7, a new approach for measuring EPR spectra in a pulsed fashion which is free of line shape distortions due to modulation effects is proposed.

1.2 ESEEM spectroscopy

Echo modulations reveal magnetic interactions which are not resolved in the EPR spectrum (1,2). In particular, hyperfine and nuclear quadrupole frequencies can be obtained conveniently by Fourier transformation (FT) of two– or three–pulse ESEEM patterns of single crystals (3,4). ESEEM spectroscopy is the method of choice for the observation of low–frequency nuclear transitions, since conventional ENDOR suffers from poor enhancement factors at low frequencies.

In disordered systems, magnetic parameters are difficult to evaluate from ESEEM patterns. The reduced information content of such FT–ESEEM spectra is related to the extended spectrometer deadtime τ_D (5,6). For two–pulse echoes and for short pulse delays in three–pulse echo experiments, ESEEM patterns of broad hyperfine lines may fully decay within τ_D. Moreover, in three–pulse sequences, severe line shape distortions occur , since the modulation amplitudes of the nuclear transition frequencies in one m_S state depend on the transition frequencies in the other m_S state, and *vice versa* (so–called blind spots, see Eqn.2). The loss of hyperfine information in FT–ESEEM spectra of randomly oriented solids has so far been a serious drawback of pulsed EPR spectroscopy.

Fig. 1 demonstrates the effect of the deadtime τ_D in a two–pulse experiment for a system with one electron and one proton lying in the plane over which the powder average is carried out (an axial anisotropic **g** tensor is assumed). The undistorted shape of the two hyperfine lines obtained with $\tau_D = 0$ is shown in Fig. 1a. With a delay in acquisition of

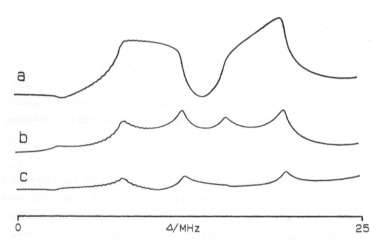

Fig.1. Computed two–pulse ESEEM magnitude spectra for an electron–nuclear distance r = 0.3 nm, $a_{iso}/2\pi = 6$ MHz and $\omega_n/2\pi = 13$ MHz. The sum peak is not shown. (a) Dead-time $\tau_D = 0$, (b) $\tau_D = 100$ ns, (c) $\tau_D = 200$ ns. (Taken from Ref. 5).

$\tau_D = 100$ ns, the two hyperfine lines split into doublets (Fig. 1b) with peaks directly related to the parallel and perpendicular hyperfine components (6). For the data used in the simulation in Fig. 1, a deadtime of $\tau_D = 200$ ns reduces the intensity of the peaks by about a factor of 10 (Fig. 1c). Thus for reasonable τ_D values, the features that allow a direct determination of hyperfine couplings are often buried in the noise.

Methods to overcome the problems due to spectrometer deadtime and blind spot artifacts are presented in section 3, 4 and 7. The evaluation of structural information from the sum peak of disordered systems is discussed in section 5. Finally, ways to improve the orientation selectivity of the ESEEM in powders and to record the whole modulation pattern in a single experiment are presented in section 2 and 8, respectively.

1.3   Pulsed ENDOR

Section 9 is devoted to two pulsed ENDOR schemes recently developed in our laboratory.

1.4   Probe head design

In section 10, a few comments about different types of resonator structures suitable for various sophisticated pulsed EPR experiments are made.

For the convenience of the reader, some of the relevant formulae used in the following sections are briefly summarized.

The echo amplitude of a spin system consisting of one electron and one $I = 1/2$ nucleus is described by (1,2,7)

$$E_{mod}(\tau) = 1 - \frac{k}{2}\left[1 - \cos\omega_\alpha\tau - \cos\omega_\beta\tau + \frac{1}{2}\cos(\omega_\alpha + \omega_\beta)\tau + \frac{1}{2}\cos(\omega_\alpha - \omega_\beta)\tau\right] \tag{1}$$

for the two–pulse sequence, $\pi/2$–τ–π–τ–echo, and by

$$E_{mod}(\tau,\tau') = 1 - \frac{k}{4}\left[(1-\cos\omega_\alpha\tau)(1-\cos\omega_\beta\tau') + (1-\cos\omega_\beta\tau)(1-\cos\omega_\alpha\tau')\right] \tag{2}$$

for the stimulated echo sequence, $\pi/2$–τ–$\pi/2$–T–$\pi/2$–τ–echo, with a fixed τ and a variable $\tau' = T+\tau$.

In [1] and [2], ω_α and ω_β denote the nuclear transition (ENDOR) angular frequencies associated with the electronic spin states $m_s = 1/2$ and $m_s = -1/2$, respectively, with

$$\left|\begin{matrix}\omega_\alpha \\ \omega_\beta\end{matrix}\right| = (\tilde{\ell}\,A_\pm^2\,\ell)^{1/2}, \tag{3}$$

$$A_\pm = \pm A/2 - \omega_n\,E \tag{4}$$

and the nuclear Zeeman angular frequency

$$\omega_n = g_n\,\beta_n\,B_0/\hbar. \tag{5}$$

E is the 3x3–unit matrix. ℓ is the unit vector along the static field B_0 and A is the hyperfine tensor in the molecular frame (in units of rad/s).

The modulation depth parameter k is defined by (8)

$$k = 4\sin^2(\tfrac{\delta}{2})\cos^2(\tfrac{\delta}{2}) \tag{6}$$

with

$$\left|\begin{matrix}\sin^2(\tfrac{\delta}{2}) \\ \cos^2(\tfrac{\delta}{2})\end{matrix}\right| = \frac{|\omega_n^2 - \frac{1}{4}(\omega_\alpha \pm \omega_\beta)^2|}{\omega_\alpha\omega_\beta}. \tag{7}$$

In [6], δ denotes the angle between the two effective magnetic fields at the nucleus with $m_s = 1/2$ and $m_s = -1/2$. The quantities $\cos^2(\frac{\delta}{2})$ and $\sin^2(\frac{\delta}{2})$ are proportional to the transition probabilities of the allowed and the forbidden transitions, respectively, indicating that both types of transitions are required for the observation of an echo modulation.

2. SELECTIVITY IN ECHO–DETECTED E P R AND E S E E M POWDER STUDIES

2.1 Selectivity in echo–detected EPR spectra

Obviously, EPR signals recorded via the echo intensity in a two– or three–pulse experiment are free of modulation broadening effects observed in cw EPR spectra. However, the line shape may be distorted due to power broadening of the short microwave (mw) pulses. In this context, it is interesting to note that the range of excited spin packets contributing to the echo signal is much narrower than the range of spin packets excited by the mw pulses.

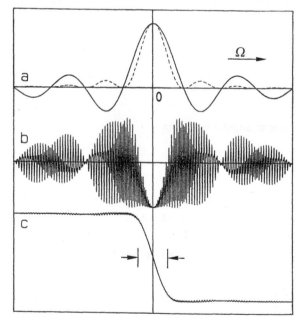

Fig. 2. Magnetization as a function of Ω/ω_1 in an two–pulse echo experiment. (a) M_x after a $\pi/2$ pulse of length t_p (full line), M_z after a π pulse of length $2t_p$ (dashed line), (b) M_x at the echo maximum, (c) Integral of (b); the range between the arrows contributes 90% to the echo intensity.

Consider a standard two–pulse sequence acting on a spin system with an infinitely inhomogeneously broadened line. The transverse magnetization M_x immediately after a $\pi/2$ pulse along y as a function of Ω/ω_1 where Ω is the resonance offset and ω_1 denotes the mw field strength ($\omega_1 = \gamma B_1$), has the well–known shape shown in Fig. 2a. Since M_x decreases slowly with Ω/ω_1, a wide range of off–resonance spin packets is excited.

A more complex situation is found at the time of echo formation (Fig. 2b). Again, many off–resonance spin packets have components along M_x; most of them, however, cancel each other. The integration of M_x over Ω/ω_1 at 2τ (Fig. 2c) shows that only the magnetization in a narrow interval symmetric to $\Omega/\omega_1 = 0$ contributes to the echo signal. Thus, two–pulse echo formation essentially improves selectivity compared to the excitation range of the corresponding $\pi/2$ pulse.

The influence of the pulse widths on a Gaussian line of width Γ in echo–detected EPR and LOD–PEPR (section 7) is shown in Fig. 3a for various Γ/ω_1–ratios.

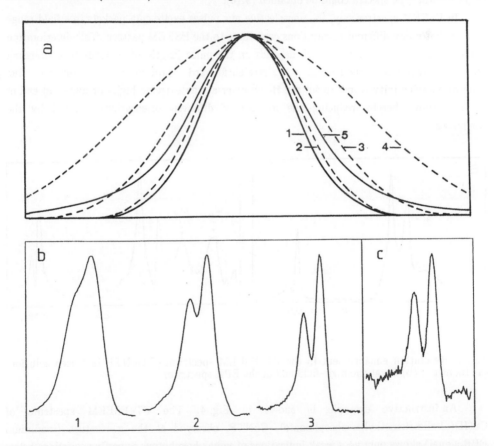

Fig. 3. Line broadening in an echo–detected EPR experiment. (a) Calculated lineshapes for different Γ/ω_1–ratios: 1) undistorted line shape, 2) $\Gamma/\omega_1 = 2$, 3) $\Gamma/\omega_1 = 1$, 4) $\Gamma/\omega_1 = 0.5$, 5) LOD–PEPR, $\Gamma/\omega_1 = 2$, (b) Line broadening of the $m_I = 3/2$ transition (^{63}Cu and ^{65}Cu) of Cu(mnt)$_2$ single crystal, boxcar gate: 10 ns, pulse widths: 1) 10/20ns, 2) 40/80ns, 3) 200/400ns, (c) boxcar gate covers the entire echo, pulse widths: 10/20 ns.

An experimental example for this line broadening in a two–pulse experiment on bis(maleonitriledithiolate)Cu(II), Cu(mnt)$_2$, is given in Fig. 3b for various pulse lengths and a short aperture duration of the boxcar integrator. Integration over the entire echo shape results in an undistorted line shape (Fig. 3c), at the expense of signal–to–noise ratio. A further effect which may lead to strong distortions in echo–detected EPR spectra is discussed in section 7.

2.2   Orientation–selectivity in ESEEM powder studies

In randomly oriented spin systems with sufficiently large anisotropic magnetic interactions, particular orientations may be excited depending on the setting of the B_0–field. This orientation selection has often been used in cw ENDOR where in favourable cases single–crystal type spectra could be obtained (9,10).

In ESEEM spectroscopy, the nonselective mw pulses excite spin packets in a field range ΔB_0 so that many different orientations contribute to the ESEEM pattern. This deterioration of the orientation–selectivity, which depends on the pulse length, often leads to a smearing out of hyperfine frequencies. A straightforward procedure to improve the orientation–selectivity is just to set the B_0 observer at the extreme high– or low–field end of the spectrum, thereby reducing the number of different orientations covered by the mw pulses.

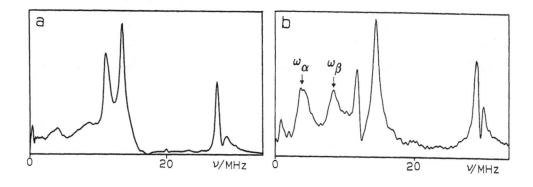

Fig. 4. Resolution enhancement in the FT–ESEEM spectrum of CuNTP in frozen solution. (a) B_0 at $g_\perp$, (b) B_0 at the high–field end of the EPR spectrum.

An instructive example is shown in Fig. 4. The FT–ESEEM spectrum of Cu(II)(nitrilotrismethylenephosphonate), CuNTP, observed at the $g_\perp$–feature (maximum EPR signal) shows only very weak indications of some phosphorous hyperfine couplings below 10 MHz (Fig. 4a). However, if B_0 is set to the extreme high–field end of the EPR spectrum, the number of orientations covered by the mw pulses is reduced, resulting in much stronger phosphorous hyperfine lines (indicated by ω_α und ω_β in Fig. 4b).

249

3. ELIMINATION OF UNWANTED ECHOES IN THREE–PULSE E S E E M

The standard three–pulse experiment (Fig. 5) suffers from various drawbacks. In addition to the stimulated echo (SE) at time $\tau + \tau'$, free induction decays (FID) occur after each pulse, and two–pulse echoes are created at times 2τ, E(I,II), $2\tau' - \tau$, E(II,III), and $2\tau'$, E(I,III) as well as a refocused echo (RE) with opposite phase at time $2\tau' - 2\tau$. For times $\tau' \lesssim 3\tau$, these unwanted echoes may cross the stimulated echo causing glitches to appear in the stimulated echo envelope plotted as a function of τ'. This leads to additional distortions of the modulation pattern, and thus to artifacts in the frequency domain spectrum.

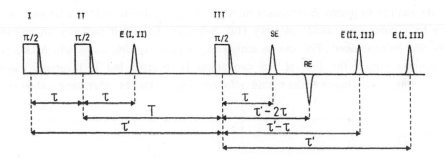

Fig. 5. Free induction decays and spin–echo signals created with a three–pulse sequence with microwave pulses I, II, and III of equal phase. SE: stimulated echo; E(I,II), E(II,III), E(I,III): two–pulse echoes; RE: echo E(I,II) refocused by pulse III. (Taken from Ref. 5).

To avoid the crossing of undesired echoes with the stimulated echo, data acquisition is often not started until $\tau' > 3\tau$. With this expedient, however, 1–1.5 μs of the initial part of the echo trace may be lost. In frozen solutions or powder samples, this amounts to a significant fraction of the whole modulation pattern. Reconstruction of the long truncated part of the envelope function is no longer feasible, making phase sensitive Fourier transforms impossible. The magnitude spectrum must then be calculated, resulting in poorer resolution.

Two remedies to improve the performance of three–pulse echo experiments have been proposed (11): (1) a pulse–swapping technique which allows the measurement of the echo envelope as a function of τ' even for $0 < \tau' \le \tau$ (the short time limit is then exclusively determined by the true spectrometer deadtime); (2) a phase–cycling procedure combining four experiments to eliminate all unwanted echoes and the FIDs after each of the three pulses.

3.1 Extension of sampling to $\tau' < \tau$ by pulse–swapping

The clue for extending the measurement of the echo amplitude to values of $\tau' < \tau$ can be understood by considering [2] for the modulation of a stimulated echo in the case S = 1/2, I = 1/2, which is symmetric with respect to an interchange of τ and τ'. This equation suggests that moving the third pulse right through the second pulse to reduce τ' below the value τ is mathematically equivalent to interchanging the significance of τ and τ' (12). In this manner it is possible to measure the envelope of the stimulated echo (SE) for $\tau' > \tau$ as well as for $0 < \tau' < \tau$. However, it should be mentioned that the relaxation behaviour changes at $\tau' = \tau$: The faster decay of the stimulated echo for $\tau' < \tau$ is affected by both the phase memory time T_M and the spin–lattice relaxation time T_1, while for $\tau' > \tau$ the decay function is described exclusively by T_1 (4).

3.2 Suppression of unwanted echoes by a phase–cycling procedure

By analogy to pulsed experiments in NMR, it is possible to characterize each of the echoes by a coherence–transfer pathway (13). Selecting the proper pathway all unwanted echoes may be suppressed. This can be achieved by a phase–cycling procedure. A two–pulse echo changes sign if the first of the two pulses is reversed but is invariant under a corresponding phase change of the second refocusing pulse. On the other hand a three–pulse

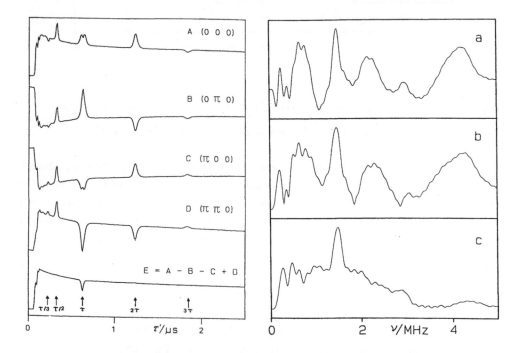

Fig. 6. (left) Demonstration of the phase–cycling technique: Three–pulse echo amplitude of γ–irradiated CaO powder at 8 K as a function of τ' for different phase combinations of the three mw pulses. Arrows indicate echo crossings. (Taken from Ref. 11).

Fig. 7.(right) Fourier transform (0–5 MHz) of the ESEEM of Cu(II)tetraimidazole, $\tau = 300$ ns. (a) Cosine Fourier transform, (b) Magnitude spectrum, (c) Magnitude spectrum of the restricted interval $\tau' > 950$ ns. (Taken from Ref. 11).

echo is sensitive to the phases of all three pulses and changes sign under a phase reversal of each pulse.

In the case at hand all contributions which may disturb the stimulated echo can be eliminated by cycling the first two pulses independently by $\phi_k = 0,\pi$, k = 1,2, leading to a four–step phase cycle with the experiments A, B, C, and D with microwave phases (0 0 0), (0 π 0), (π 0 0) and (π π 0), respectively. By summing the signals of experiments A and D and subtracting B and C, all unwanted echoes and the FIDs of pulses II and III are cancelled; the only remaining signal is the stimulated echo SE. This is valid for both $\tau' > \tau$ and $\tau' < \tau$.

The phase–cycling and pulse–swapping technique is demonstrated on γ–irradiated CaO powder with a negligible concentration of magnetic nuclei. The amplitude of the stimulated echo is then simply described by the relaxation function, possibly distorted by other coinciding echoes. Figures 6a–d show the amplitude of the stimulated echo as a function of τ' for $0 \leq \tau' \leq 4\tau$, $\tau = 600$ ns, and the four phase settings mentioned above. In the combination (A–B–C+D) of the four experiments shown in Fig. 6e, the unwanted echoes are virtually eliminated.

The remaining narrow peak at $\tau' = \tau$ is caused by the interfering overlap of microwave pulse II and III. Some remedies to avoid this feature are discussed in (11).

The effectiveness of this technique is demonstrated in Fig. 7 which shows the Fourier transform of the three–pulse echo modulation of Cu(II)tetraimidazole in frozen solution. Sampling of undesired data points near the pulse crossing region is avoided using time increments $\Delta \tau'$ of 40 ns. The initial points of the echo decay disguised by the short spectrometer deadtime of 140 ns have been reconstructed by a simple straight line extrapolation. Figures 7a,b show the cosine and magnitude Fourier transforms of the ESEEM obtained with the phase–cycling and pulse–swapping technique. The peaks are due to hyperfine couplings to remote nitrogen spins of bound imidazole molecules.

For comparison, Fig. 7c shows the magnitude spectrum of the same modulation function with a starting value of $\tau' = 950$ ns $> 3\tau$, corresponding to the lowest τ' value for which the waveform is not distorted by unwanted echoes in a conventional three–pulse experiment. The lack of data during the first 950 ns distorts the frequency spectrum in an intolerable manner. In particular the peaks at 2.3 and 4.3 MHz are strongly reduced in intensity compared to the corresponding peaks in Figs. 7a or b.

Fig. 7 clearly demonstrates that the recording of the initial time interval is essential for obtaining a faithful spectrum, particularly for disordered systems.

4. RESTORATION OF BROAD HYPERFINE LINES

The technique described in section 3 reduces the number of missing data points in three–pulse ESEEM experiments considerably. Blind spots, however, may not be eliminated with this method. Moreover, as mentioned in the introduction, ESEEM patterns of broad lines may even decay within the remaining instrumental deadtime. Since a fast decay of the

252

modulation amplitudes is due to an interference of modulation frequencies of different orientations and not to a relaxation mechanism, it is possible to *restore* the shape of broad hyperfine lines (14).

The method is again based on the three–pulse stimulated echo sequence. According to [2], an unusually long and fixed delay time τ would increase the density of blind spots, governed by the modulation factors $(1- \cos \omega_\alpha \tau)$ and $(1- \cos \omega_\beta \tau)$. The *envelope* of such a blind spot– or amplitude–modulated spectrum then corresponds to the true shape of the hyperfine line that would be obtained by setting $\omega_\alpha \tau = \pi$ and $\omega_\beta \tau = \pi$.

It is even possible to fully eliminate the blind spots. This is explained by a simple model calculation for a hyperfine line of the form (Fig. 8a)

$$S(\omega) = F(\omega - \omega_0)(1- \cos \omega \tau). \tag{8}$$

$F(\omega)$ represents the undistorted lineshape of the transitions at ω_α or ω_β, whereas $(1- \cos \omega \tau)$ describes the modulation which creates the blind spots.

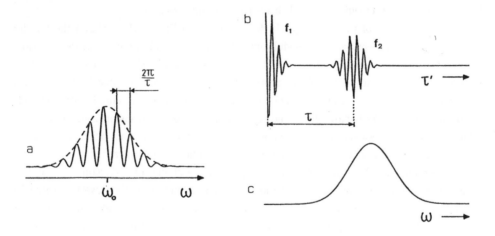

Fig. 8. (a) Undistorted Gaussian lineshape (broken line) modulated by blind spots caused by the function $(1- \cos \omega \tau)$ in Eq. [8] (solid line),(b) Echo modulation pattern as a function of τ' obtained by Fourier transformation of (a), (c) Magnitude spectrum of the echo–modulation echo $f_2(\tau')$. (Taken from Ref. 14).

The real part of the inverse Fourier transform of [8] is given by

$$s(\tau') = f_1(\tau') + f_2(\tau') + f_3(\tau')$$
$$= f(\tau')\cos(\omega_0 \tau') - \tfrac{1}{2}f(\tau'-\tau)\cos[\omega_0(\tau'-\tau)] - \tfrac{1}{2}f(\tau'+\tau)\cos[\omega_0(\tau'+\tau)] , \tag{9}$$

with $f(\tau)$ being the Fourier transform of $F(\omega)$. The signal $s(\tau')$ represents the modulated part of the observed echo intensity. The first term $f_1(\tau')$ is a fast decaying signal for very short τ' values (Fig. 8b) that often decays within the instrumental deadtime. Its Fourier transform represents the original unmodulated hyperfine lineshape $F(\omega-\omega_0)$ indicated by the dashed line in Fig. 8a. The second term $f_2(\tau')$ describes the enhanced modulation pattern near $\tau'\simeq\tau$ in Fig. 8b. We call it the *echo–modulation echo*, since it arises from the refocusing of the blind spot modulation. The third term $f_3(\tau')$ is responsible for an echo–modulation echo in the negative τ' domain. Its extension into the observed positive τ' domain is often negligible.

The Fourier transform of the echo–modulation echo $f_2(\tau')$ *alone* (with time origin at $\tau' = 0$) then yields

$$\mathscr{F}\{f_2(\tau')\} = \int_{-\infty}^{\infty} f_2(\tau')\, e^{-i\omega\tau'}\, d\tau' = \frac{F(\omega-\omega_0)}{4}\, e^{i\omega\tau} . \qquad [10]$$

This function is responsible for the blind spots in Fig. 8a; i.e. for the deviation of the solid curve from the broken line. It is most interesting to note that the magnitude of [10] is just given by the undistorted line shape $F(\omega-\omega_0)/4$. Thus, by computing a *magnitude* spectrum of an echo–modulation echo for a sufficiently long pulse delay τ it is possible to obtain an undistorted ESEEM spectrum as shown in Fig. 8c, irrespective of the instrumental deadtime. Because of the magnitude computation, the position of the time zero for the Fourier transformation is immaterial.

Real spin systems do not follow exactly the idealized modulation pattern described by Eqns. 8–10. This is due to the fact that the dipolar hyperfine interaction is of the same order of magnitude as the nuclear Zeeman frequency and that for orthorhombic hyperfine tensors a partial filling up of the blind spots may occur (14). For most spin systems, however, these effects are not fundamental limitations of the proposed technique.

The utility of the modified three–pulse approach is demonstrated on dibenzene-vanadium, $V(bz)_2$, diluted in ferrocene powder. The original echo modulation amplitude as a function of τ', after elimination of undesired two–pulse echoes at $\tau/3$, $\tau/2$, and τ, is shown in Fig. 9a. The modulation pattern after subtraction of the unmodulated echo decay is plotted in Fig. 9b. The echo–modulation echo $f_2(\tau')$ near $\tau'\simeq\tau$ is clearly evident.

To eliminate the blind spots, the time domain signal was zeroed for $0\leq\tau'\leq1$ μs and apodized near $\tau'= 1$ μs. The corresponding magnitude spectrum, comprised almost exclusively of the contribution of $f_2(\tau')$ is given in Fig. 9c. It shows a narrow matrix proton line at 14.8 MHz (protons of the ferrocene host), flanked by two well developed hyperfine lines due to the twelve $V(bz)_2$ protons.

For comparison, Fig. 9d contains also the two–pulse ESEEM magnitude spectrum of $V(bz)_2$. The two hyperfine lines are indicated by broad featureless humps. Little useful information can be extracted from this spectrum.

The three–pulse ESEEM magnitude spectrum of Fig. 9e, ($\tau= 400$ ns, $\tau'\geq0$), is strongly distorted by blind spots. Again, an unambiguous identification of the hyperfine peaks is

difficult. In comparison to these conventional ESEEM spectra, the spectrum of Fig. 9c
obtained by the novel approach is much more satisfactory and can be analyzed with a higher
degree of certainty. Peaks below 3 MHz are due to improper elimination of the unmodulated
echo decay function.

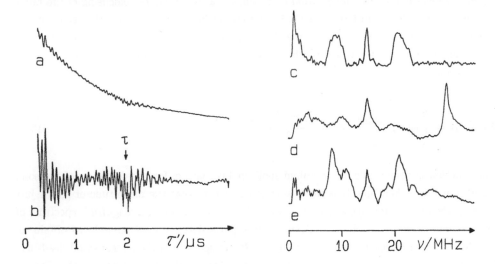

Fig. 9. ESEEM powder pattern of dibenzenevanadium in ferrocene host. (a) Full echo decay
after elimination of unwanted two–pulse echoes, (b) Same data after elimination of the
unmodulated echo decay, (c) Magnitude spectrum computed exclusively from the
echo–modulation echo in Fig. 9b, (d) Magnitude spectrum obtained by a two–pulse ESE
experiment, (e) Magnitude spectrum obtained by a three–pulse ESE experiment with $\tau = 400$
ns, $\tau' \geq 200$ ns. (Taken from Ref. 14).

The experimental example indicates that the improvement of the lineshapes and the
relative intensities of broad hyperfine lines in disordered systems, using large τ values, can be
appreciable.

5. STRUCTURAL INFORMATION FROM THE SUM PEAK OF E S E E M PATTERNS

According to [1], the ESEEM of a two–pulse sequence also contains the sum and the
difference of the ENDOR frequencies ω_α, ω_β. The sum frequency is given by (Eqns. 3–6)

$$(\omega_\alpha + \omega_\beta) = (\tilde{\ell} A_+^2 \, \ell)^{1/2} + (\tilde{\ell} A_-^2 \, \ell)^{1/2} \,, \qquad\qquad [11]$$

which may be approximated by (15)

$$(\omega_\alpha + \omega_\beta) \approx 2\omega_n + [\omega_n^4 - \frac{\omega_n^2}{2}(\omega_\alpha^2 + \omega_\beta^2) + \frac{1}{16}(\omega_\alpha^2 - \omega_\beta^2)^2]/\omega_n^3 .$$ [12]

The second term in [12] is zero for isotropic hyperfine couplings a_{iso} or B_0 parallel to one of the **A** tensor principal axes, and is usually much smaller than $2\omega_n$. This results in a narrow sum peak even in disordered systems where ω_α and ω_β cover a large frequency range. The sum peak can therefore also be observed in systems where a direct observation of hyperfine couplings is not possible. It is particularly pronounced for protons, and contains information about the positions and a_{iso} of local nuclei in paramagnetic species (16,17).

The theoretical shape of the sum peak of a proton in the plane over which the powder average is taken with $r = 0.26$ nm, $a_{iso} = 0$, $\omega_n/2\pi = \nu_n = 13$ MHz and $T_2 = 33$ μs is shown in Fig. 10.

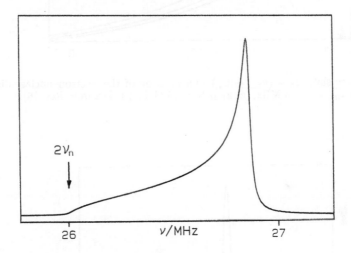

Fig. 10. Typical shape of the sum peak. (Taken from Ref. 16).

For this long relaxation time T_2, only a weak damping of the high frequency components occurs; i.e., the plot in Fig. 10 closely approaches the original shape of the sum peak. The signal intensity is zero at $2\nu_n$ (B_0 parallel or perpendicular to the electron–nuclear direction) and increases smoothly with the angle between B_0 and the electron–nuclear direction. Near the maximum value $\nu_{max} = 2\nu_n + \Delta$, a steep increase in intensity is found, because many orientations contribute to the peak near ν_{max}. It is found theoretically (17) that the sum peak has maximum intensity at ν_{max}. By introducing a deadtime τ_D of $100 - 200$ ns, only the high frequency components survive, resulting in a narrow line very close to ν_{max}.

Plots of the frequency shift Δ as a function of r for three different a_{iso} values are given in Fig. 11. In the important region $0.25 < r < 0.3$ nm, the variation of Δ for $a_{iso}/2\pi$ from 0 to 10 MHz corresponds to a change in r of only 0.01 nm. This small influence can usually be taken into account by using data about a_{iso} from other sources (ENDOR, quantum chemical calculations).

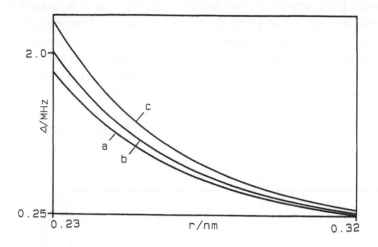

Fig. 11. Frequency shift $\Delta = (\nu_{max} - 2\nu_n)$ as a function of the electron–nuclear distances r. (a) $a_{iso} = 0$, (b) $a_{iso}/2\pi = 6$ MHz, (c) $a_{iso}/2\pi = 10$ MHz. (Taken from Ref. 16).

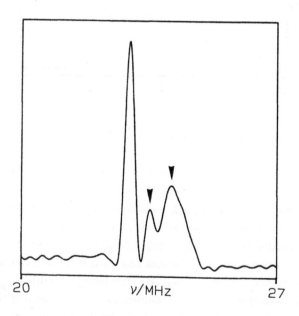

Fig.12. Sum peak region of the cosines FT–ESEEM spectrum of $Cu(H_2O)_6^{2+}$ in frozen solution. Arrows mark peaks due to axial and equatorial water protons. (Taken from Ref. 17).

The experimental example in Fig. 12 shows the cosine FT–ESEEM spectrum of $Cu(H_2O)_6^{2+}$ in frozen aqueous–glycerin solution near $2\nu_n$ recorded at a field position in the EPR spectrum where the $\mathbf{B}_0$ vector makes an angle of $\theta = 40^0$ with the $g_{||}$–direction. The dominant sum peak close to $2\nu_n = 22.97$ MHz is due to ambient water protons. The peak shifted by $\Delta = 0.57$ MHz to higher frequencies presents the protons of the two *axial* water molecules. This feature can only be observed within a small range of the angle θ near 40^0. For θ close to 0^0 or 90^0, the peak is hidden by the one of the ambient water protons. The third transition with $\Delta = 1.14$ MHz originates from the eight *equatorial* water protons. Usually, this peak is also well developed for $\mathbf{B}_0$ near $g_\perp$.

Fig. 12 indicates that the sum frequency spectrum contains important structural information about protons of water or other functional groups near a transition metal ion. This finding is particularly useful in metallo–proteins, where the broad hyperfine lines usually cannot be measured with ESEEM.

A detailed analysis of the sum peak including arbitrary proton positions and $\mathbf{B}_0$–field orientations is given in Ref. 17.

6. ECHO–DETECTED E P R WITH MAGNETIC FIELD VECTOR JUMPS

The interpretation of powder EPR spectra is often impeded by unresolved fine and hyperfine interactions, and by the interference of different paramagnetic species simultaneously present in a sample. The resolution of powder spectra with anisotropic $\mathbf{g}$ tensors can be improved by increasing the microwave frequency. However, this often gives rise to a decrease in the resolution of the hyperfine structure. The resolution of ligand hyperfine interactions may sometimes be improved by using microwave frequencies lower than X–band. This has been demonstrated, for example, for copper complexes where g–strains and copper hyperfine strains cancel each other for particular m_I values of the copper nucleus (18).

In this section a novel pulsed EPR technique to enhance the resolution of echo–detected EPR spectra in disordered systems is introduced (19). This method selects specific crystallite orientations based on sudden changes of the static magnetic field direction $\mathbf{B}_0$ in the time between the microwave pulses in two– or three–pulse ESE experiments. The only prerequisite for the resolution enhancement technique is *anisotropy* either in the electronic Zeeman interaction, the fine structure term, or in one of the hyperfine interactions.

6.1 Basic idea and pulse sequences

To describe the principles we select a spin system with $S = 1/2$ and an axially symmetric $\mathbf{g}$ tensor; hyperfine interactions are disregarded. For a fixed microwave frequency ν, the resonance field strength is given by

258

$$B_{res}(\theta) = h\nu/\beta g(\theta) \, , \tag{13}$$

with the orientation–dependent g factor

$$g(\theta) = (g_{\parallel}^2 \cos^2\theta + g_{\perp}^2 \sin^2\theta)^{\frac{1}{2}}. \tag{14}$$

The resonance field $B_{res}(\theta)$ of a *single* paramagnetic species may be represented geometrically by the surface of an ellipsoid representing the **g** tensor. The principal axes span the molecular frame, as shown in Fig. 13. For a polar angle $\theta=0$, $B_{res}(0)=B_{\parallel}=h\nu/\beta g_{\parallel}$, for $\theta=\pi/2$, $B_{res}(\pi/2)=B_{\perp}=h\nu/\beta g_{\perp}$.

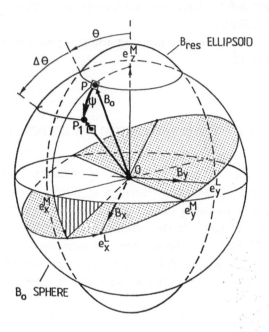

Fig. 13. Ellipsoidal resonance surface B_{res} for an axially symmetric **g** tensor in the molecular principal axes frame. OP and OP_1 denote the field vectors during the microwave pulses in a one–jump two pulse experiment. The penetration points of these vectors through the resonance surface are marked by squares.

A given static magnetic field strength B_0 fulfills the resonance condition $B_0=B_{res}(\theta)$ for all vectors B_0 on a cone with opening angle 2θ.

It is evident that the resolution of a multiline powder spectrum would be considerably improved, if only the features with B_0 along the principal axes $g_{\parallel}$ and $g_{\perp}$ (stationary points on the ellipsoid) could be measured. This holds also for systems with hyperfine interactions and for orthorhombically distorted **g** tensors. With this goal in mind, it can be noted that the principal axis directions correspond to stationary points on the ellipsoid with $dg(\theta)/d\theta = 0$, which show minimum susceptibility to a variation of the magnetic field direction.

Two magnetic field jump ESE experiments are proposed: The one–axis field jump experiment of Fig. 14a uses the standard two–pulse spin echo sequence leading to a primary echo. However, during the application of the refocusing pulse, a dc field pulse is applied transverse to the field B_0. The two–axis field jump experiment of Fig. 14b, based on the three–pulse stimulated echo sequence, employs two transverse dc field pulses applied in orthogonal directions, B_x extending from the second to the third microwave pulse while B_y covers only the third microwave pulse.

We now consider a particular crystallite in a powder distribution and assume that the static field B_0 with $B_0 = B_{res}(\theta)$ points in the direction OP (laboratory z axis), as shown in Fig. 13. The first microwave pulse in Fig. 14a excites all spin packets for all crystallites at angle θ. After the first microwave pulse, the static field is rotated by the small angle ψ about the laboratory y axis in a time short compared to the electron spin relaxation times T_1 and T_2, while keeping the magnitude of the field constant. This changes the resonance condition due to the θ–dependence of the g factor (Eqn. 14).

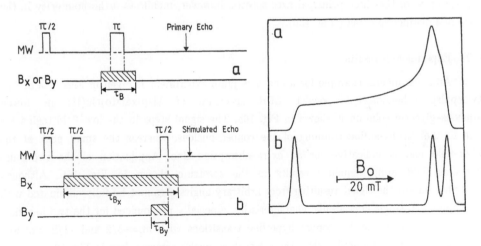

Fig. 14. (left) Pulse sequences with magnetic field vector jumps. (a) Two–pulse one–axis field jump sequence, (b) Three–pulse two–axis field jump sequence.

Fig. 15.(right) Model calculation demonstrating the optimum achievable resolution enhancement. (a) Powder spectrum: Orthorhombic **g** tensor with $g_x = 2.02$, $g_y = 2.05$, $g_z = 2.40$; linewidth $\Gamma = 2$ mT, (b) Superposition of the three single crystal spectra along g_x, g_y and g_z. Same parameters as in (a). (Taken from Ref. 19).

For the principal axes directions, $\theta = 0, \pi/2$, the resonance condition does not change for small tilts and the π pulse in Fig. 14a, applied in the presence of the tilted field will fully refocus the corresponding magnetization. For all other orientations the signal contribution will be *attenuated* but not eliminated as there are always orientations that do not experience a change in the g factor either. This leads to pronounced features in the spectrum at the positions $g_{||}$ and $g_\perp$ and an attenuated signal in between.

In order to eliminate all signal contributions except for $g_\perp$ and $g_{||}$, a second tilt of the magnetic field orthogonal to the first one is required. It can be applied in a stimulated echo sequence (Fig. 14b) during the third pulse. Whether the unwanted signal contributions are completely eliminated by such a two–axes jump experiment depends also on the width and on the selectivity of the pulses relative to the anisotropy of the **g** tensor (20). A more detailed description of the experiment is given in Ref. 19.

To estimate the maximum resolution enhancement that can be obtained with the field jump method, model calculations of conventional (absorption) powder spectra have to be compared with superpositions of three single crystal spectra with B_0 parallel to the principal axes of the **g** tensor. Such plots represent the *upper limit* of attainable resolution and is not realizable experimentally.

A simple example with a slightly orthorhombically distorted **g** tensor is shown in Fig. 15. Hyperfine couplings are not considered. In the powder spectrum shown in Fig. 15a, the orthorhombicity is only indicated by a small hump at the high–field end. The superposition of the three principal axes spectra, however, manifests orthorhombicity in the form of a distinct splitting (Fig. 15b).

6.2 Experimental results

An experimental example for a one–axis jump experiment is given in Figs. 16a,b. The two–pulse echo–detected powder EPR spectrum of bis(oxalato)Cu(II) in frozen aqueous–glycerin solution is shown in Fig. 16a. The signal steps in the low–field region are due to the $A_{||}$–hyperfine coupling of the copper nucleus, whereas the small peak at the high–field end is indicative for an extra–absorption line. Application of the two–pulse one–axis field jump sequence results in the spectrum shown in Fig. 16b. Although contributions to the signal resulting from arbitrary angles θ cannot be fully eliminated with this approach, single crystal–like characteristics, especially pronounced for the two low–field peaks representing the $A_{||}$ copper hyperfine transitions with $m_I=-3/2$ and $-1/2$, are well developed. In the high–field region, the $g_\perp$–peak is much narrower than in Fig. 16a and well separated from the extra–absorption peak.

A two–axis jump experiment which demonstrates the improved resolution of hyperfine splittings is presented in Figs. 16c,d. In the high–field region of the powder spectrum of bis(salicylaldoximato)Cu(II), Cu(sal)$_2$, diluted in the corresponding nickel complex (21), copper and nitrogen hyperfine splittings are only poorly resolved (Fig. 16c). Application of the two–axis jump technique again increases the spectral resolution considerably (Fig. 16d).

The field–jump technique is not restricted to echo–detected field–swept EPR spectroscopy. The two–pulse one–axis jump sequence has also successfully been used to improve the resolution of ESEEM patterns (20). The same procedure is feasible for various pulsed ENDOR schemes, where the sequences can be supplemented by B_0 field jumps.

Finally, it should be mentioned that field jump methods can also be applied to separate overlapping single crystal spectra (20,22).

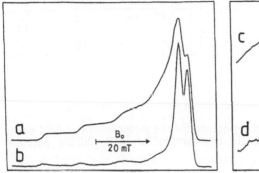 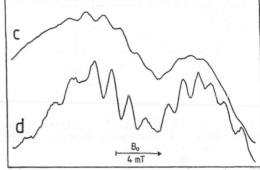

Fig. 16. (a) Two–pulse echo–detected EPR spectrum of bis(oxalato)Cu(II) in frozen H_2O–glycerin solution, temperature: T=10 K, (b) One–axis jump spectrum of (a), (c) High–field region of the three–pulse echo–detected powder EPR spectrum of Cu(sal)$_2$ in the corresponding nickel host, T=10 K, (d) Two–axis jump spectrum of (c). (Taken from Ref. 19).

7. PULSED E P R WITH LONGITUDINAL DETECTION

In this section we discuss a novel recording technique in pulsed EPR (23). The method is based on the measurement of the rapid changes of the M_z magnetization during short microwave pulses of duration $t_P < T_1, T_2$, and flip angles $\beta \leq \pi$. This longitudinal detection technique for pulsed EPR (LOD–PEPR) is particularly well suited for the measurement of short–lived radicals, very short relaxation times, ESEEM patterns and distortion–free field–swept pulsed EPR spectra of disordered systems. Since detection is not perturbed by the microwave pulses, excitation and detection can take place simultaneously. Consequently, there exists no deadtime due to cavity ringing. Thus, by using longitudinal detection, data acquisition of the echo envelope modulation can be initiated after much shorter delays than in a conventional electron spin echo experiment.

7.1 Repetitive one–pulse sequence (field–swept LOD–PEPR)

The on–resonance magnetization component, $M_z(0) = M_0$, of an inhomogeneously broadened EPR line may be rotated by a π pulse in a time t_P along $M_z(t_P) = -M_0$. This corresponds to a hole burned into the EPR line with a depth $2M_z$ and a width $2/t_P$ which recovers with the relaxation time T_1 towards thermal equilibrium.

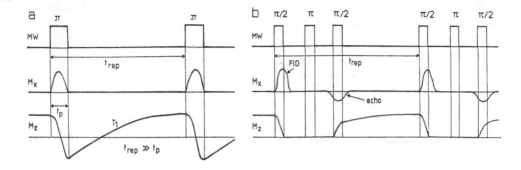

Fig. 17. LOD–PEPR sequences. (a) One–pulse sequence for the measurement of field–swept LOD–PEPR, (b) Three–pulse sequence for the measurement of LOD–ESEEM patterns. (Taken from Ref. 23).

The time derivative of $M_z(t)$ is detected by a pair of pickup coils arranged coaxially with the z axis which form together with a capacitor C a tuned circuit resonating in the frequency range 1 kHz $< \omega_{res}/2\pi <$ 100 kHz. The circuit with its high Q serves as a narrow–band filter that selects one Fourier component of the signal.

For short relaxation times, $T_1 \leq$ 100 μs and correspondingly fast pulsing, one is normally constrained to detect the first harmonic and to set the tuning frequency equal to the repetition rate $\omega_{res}/2\pi = 1/t_{rep}$. When the pulsing rate fulfills the condition $1/t_{rep} > \omega_{res}/Q$, the resonance circuit is thereby continuously stimulated and the output voltage of the circuit is a pure sine wave:

$$U_{out}(t) = U \sin(\omega_{res}t + \varphi). \qquad [15]$$

This signal is then fed through a narrow band–pass filter and recorded by an rms ac–voltmeter.

With the repetitive one–pulse sequence (Fig. 17a), it is possible to record EPR spectra by a slow sweep of the magnetic field. This novel type of EPR experiment is most useful for recording EPR spectra of paramagnetic species in powders and frozen solutions.

Due to the orientation and thus B_0–dependent depth of the echo modulation, powder EPR spectra recorded via the echo intensity may be strongly distorted (24,25). In field–swept LOD–PEPR, however, the magnetization is sampled directly, resulting in a correct absorption–mode EPR spectrum that is only affected by the bandwidth of the pulse (see Fig. 3a, which indicates that the line broadening in a LOD–PEPR experiment is somewhat larger than in an echo–detected EPR for the same Γ/ω_1–ratio). This is demonstrated in Fig. 18 for a frozen solution of Fe(III)tetraphenylporphyrin(imidazole)$_2$.

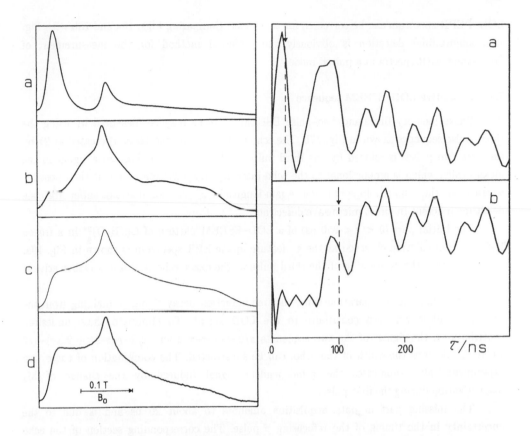

Fig. 18. (left) EPR spectra of Fe(III)tetraphenylporphyrin(imidazole)$_2$ in a frozen solution of CHCl$_3$–CH$_2$Cl$_2$ (1:1); temperature: 5 K. (a) Two–pulse echo–detected EPR spectrum, τ = 180 ns, (b) Two–pulse echo–detected EPR spectrum, τ = 230 ns, (c) cw EPR spectrum (integrated first derivative), (d) Field–swept LOD–PEPR spectrum; pulsewidth: 40 ns. (Taken from Ref. 23).

Fig. 19. (right) ESEEM of Cu(H$_2$O)$_6^{2+}$ in frozen H$_2$O–glycerin solution; temperature: 5 K. (a) LOD–ESEEM sequence: $\pi/2$–τ–π–τ–$\pi/2$; first significant data point at 20 ns is indicated by an arrow, (b) Conventional two–pulse ESEEM sequence; deadtime of $\approx$ 100 ns is indicated by an arrow. (Taken from Ref. 23).

The two–pulse echo–detected EPR spectra in Figs. 18a and 18b are recorded with two only slightly different pulse delays (τ = 180 ns and τ = 230 ns). The lineshapes in both spectra deviate substantially from the undistorted shape obtained by integration of the cw spectrum show in Fig. 18c. Thus, data from powder EPR spectra recorded via the echo intensity should be used very cautiously. The corresponding field–swept LOD–PEPR spectrum shown in Fig. 18d, however, agrees well with the cw EPR spectrum. Some small deviations can be traced back to the linear dependence of the flip angle β on the g value which varies by a factor of 2 across the spectrum, 1.5 < g < 3. The intensities of the

LOD–PEPR spectrum can be corrected for this effect. Comparing Figs. 18a and 18b with Fig. 18d, longitudinal detection is obviously the preferred method for the measurement of field–swept EPR spectra in a pulsed mode.

7.2 Repetitive LOD–ESEEM sequence

Phase memory times and two–pulse ESEEM patterns may be measured by using the three–pulse sequence shown in Fig. 17b. The echo (transverse magnetization) created at 2τ by the first two pulses is rotated by the third pulse to the z axis. This induces a signal in the pickup coils, which is proportional to the echo intensity. By variation of τ it is then possible to trace out the echo envelope function. Again longitudinal detection does not suffer from the deadtime inherent in direct microwave detection.

The initial part ($0 < \tau < 350$ ns) of a LOD–ESEEM pattern of $Cu(H_2O)_6^{2+}$ in a frozen H_2O–glycerin solution observed at the $g_\perp$ feature in the EPR spectrum is shown in Fig. 19a. It is noted that the positioning of the third pulse at the exact echo position is rather critical (± 5 ns).

For a short pulse separation time, the free induction decay signals remaining from the first two pulses will also contribute to the LOD signal. To eliminate these undesired contributions, the signals of the two sequences $\pi/2-\tau-\pi-\tau-\pi/2$ and $\pi/2-\tau-\pi-(\tau + 2\Delta\tau)-\pi/2$ where $\Delta\tau$ denotes the width of the echo, can be substracted. The combination of these two experiments also suppresses the τ–independent signal induced by the change of M_z magnetization during the first pulse.

The missing part in data acquisition amounts to about 20 ns and is due to the uncertainty in the timing of the refocusing π pulse. The corresponding section of the echo modulation obtained by the conventional two–pulse method under the same conditions is shown in Fig. 19b. For $\tau > 100$ ns the modulation pattern agrees well with the LOD–ESEEM features. For $\tau < 100$ ns, however, the instrumental deadtime prevents the detection of the echo.

LOD–PEPR shows a number of advantages compared to the detection of the transverse magnetization. Since the frequency of the detected signal is scaled down from microwave to audio frequencies, a number of expensive microwave devices are not required. From the spectroscopic point of view, LOD–PEPR represents an extremely fast measurement technique and does not suffer from deadtime problems. The method is therefore well suited for the study of short–lived radicals, short longitudinal and transverse relaxation times, and echo modulations at short τ–values. At present, the attainable signal–to–noise ratio for the measurement of field–swept spectra with LOD–PEPR is about the same at liquid helium temperature as found for echo–detected EPR. At room temperature, the sensitivity is reduced by a factor of 2–10.

8. EXTENDED–TIME EXCITATION

The modulation of the echo envelope as a function of the pulse separation in a two– or three–pulse sequence is of major interest in electron spin echo spectroscopy. Since in an ESEEM experiment, the time between the microwave pulses has to be incremented step by step, the measurement of a full echo envelope may be very time–consuming. The use of pulse trains for the simultaneous measurement of the entire echo envelope in analogy to Carr–Purcell experiments (26) is, unfortunately, often not feasible because the typically long deadtime makes it impossible to sample between the pulses.

In this section we present an alternative approach to refocusing which permits the entire echo envelope modulation to be recorded in a single experiment without any pulses during detection (27). In effect the experiment provides a *continuous refocusing* of the interactions by means of a particular extended–time preparation of the spin system prior to detection.

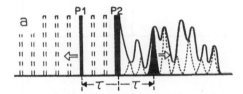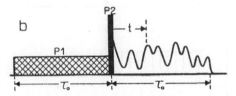

Fig. 20. (a) Conventional two–pulse Hahn–echo experiment. The echo amplitude is recorded pointwise by stepping τ from experiment to experiment, (b) Echo experiment with soft–pulse excitation followed by a short refocusing π pulse producing an entire echo decay in a single experiment. (Taken from Ref. 27).

Instead of applying an initial $\pi/2$ pulse (P1) followed by a π refocusing pulse (P2) as in Fig. 20a, we excite the system by a low–level irradiation V(t) for an extended time τ_0 prior to the π pulse as illustrated in Fig. 20b. In the linear response approximation, each time interval within the excitation period τ_0 causes an echo in the corresponding, symmetric time interval after the π pulse. The superposition of all echoes leads then to a continuous echo envelope which can be measured in a single experiment (Fig. 21).

The irradiation V(t) can either be a continuous wave microwave field of constant amplitude with a frequency placed in the center of an EPR transition, a broadband stochastic noise irradiation, or a burst of small flip angle microwave pulses.

For the basic two–pulse experiment (Fig. 20a), the density operator $\sigma(t)$ at the echo maximum $t = \tau$ is given by

$$\sigma(t=\tau) = -\exp(-i\mathscr{H}\tau)\exp(-i\pi S_x)\exp(-i\mathscr{H}\tau)S_y \exp(i\mathscr{H}\tau)\exp(i\pi S_x)\exp(i\mathscr{H}\tau)$$
$$= \exp(-i\mathscr{H}\tau)\exp(-i\tilde{\mathscr{H}}\tau)S_y \exp(i\tilde{\mathscr{H}}\tau)\exp(i\mathscr{H}\tau) , \qquad [16]$$

with the transformed Hamiltonian $\tilde{\mathscr{H}} = \exp(-i\pi S_x)\mathscr{H}\exp(i\pi S_x)$. The propagator $\exp(-i\mathscr{H}\tau)\exp(-i\tilde{\mathscr{H}}\tau)$ is responsible for the echo amplitude modulation. Of particular importance are the non–commuting parts of $\mathscr{H}$ and $\tilde{\mathscr{H}}$ involving the hyperfine and quadrupole interactions.

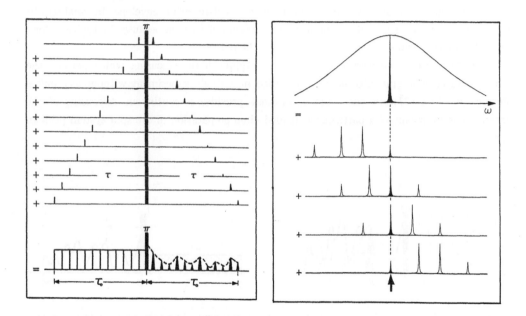

Fig. 21. (left) Formation of the extended–time transient signal.

Fig. 22. (right) Effect of inhomogeneous line broadening: The underlying spectra of the individual spin systems are shifted by different amounts with respect to the center of irradiation (arrow). A selective pulse excites all resonance lines present through only one line in each spin system. (Taken from Ref. 29).

To derive a similar relation for the case of an extended–time low level excitation V(t), we can solve the density operator equation

$$\dot{\sigma}(t) = -i\left[\mathscr{H} + V(t), \sigma(t)\right] \qquad [17]$$

by expanding $\sigma(t)$ in terms of the perturbation V(t)

$$\sigma(t) = \sigma^{(0)}(t) + \sigma^{(1)}(t) + \ldots\ldots \qquad [18]$$

Solving [17] for the linear response density operator $\sigma^{(1)}(t)$ we find after the π pulse

$$\sigma^{(1)}(t) = i\int_{-\tau_0}^{0} \exp(-i\mathscr{H}t)\exp(i\tilde{\mathscr{H}}x)[\tilde{V}(x),\sigma_0]\exp(-i\tilde{\mathscr{H}}x)\exp(i\mathscr{H}t)dx . \qquad [19]$$

For a sufficiently broad inhomogeneous frequency distribution $g(\omega)$ of the EPR line, one can show that only the integrand for $x = -t$ contributes to the integral

$$\sigma^{(1)}(t) = ig(0)\exp(-i\mathscr{H}t)\exp(-i\tilde{\mathscr{H}}t)[\tilde{V}(-t),\sigma_0]\exp(i\tilde{\mathscr{H}}t)\exp(i\mathscr{H}t) \qquad [20]$$

for $0\leq t \leq \tau_0$. Eqn. 20 leads to a refocusing for any $t \leq \tau_0$ and generates a continuous echo envelope which depends on the properties of the spin system and on the excitation $V(-t)$. Schenzle, et al. (28) have shown that the response of an inhomogeneously broadened system to an excitation of length τ_0 can last no longer than for an additional time τ_0.

The response becomes particularly simple for a cw soft–pulse with $V(t) = \omega_1 S_x$ for $-\tau_0\leq t\leq 0$. Full equivalence to [16] is obtained when $\sigma_0 = S_z$ is inserted:

$$\sigma^{(1)}(t) = g(0)\omega_1\exp(-i\mathscr{H}t)\exp(-i\tilde{\mathscr{H}}t)S_y \exp(i\tilde{\mathscr{H}}t)\exp(i\mathscr{H}t) \qquad [21]$$

for $0 \leq t \leq \tau_0$. Thus, the soft–pulse single–experiment echo envelope is *identical* to the one obtained from a series of basic two–pulse experiments as long as the linear approximation is valid for the low–level pulse.

We note here that the survival of the modulation pattern is not affected in this experiment, even though the frequency range of excitation of the long, selective low–level pulse is usually much narrower than the hyperfine couplings. This is in contrast to the usual statement that the frequency range of excitation of all pulses must exceed the maximum envelope modulation frequency to be observed.

The equivalence of selective and non–selective excitation can be proven for inhomogeneously broadened systems where the inhomogeneous broadening is much wider than the spread of the relevant EPR frequencies, as visualized in Fig. 22 for an electron– nuclear two–spin system with S=I= 1/2 (29). The inhomogeneous broadening, in effect, shifts the underlying EPR spectrum by different amounts. A selective perturbation excites all resonance lines, even though only one line in each spin system is resonant with the microwave frequency. The initial density operator for selective excitation, when averaged over the inhomogeneous distribution, becomes proportional to that accomplished by a nonselective pulse.

Fig. 23 shows an experimental verification of the extended–time excitation technique. The echo envelope obtained from a series of two–pulse electron spin echo experiments on N,N'–ethylenebis(acetylacetonatiminato)Co(II), Coacacen, diluted in a single crystal of Niacacen is reproduced in Fig. 23a. The photograph is a multiple exposure of the echoes of 200 τ –values incremented in steps of 10 ns (pulse lengths: 10 ns and 20 ns). The deep modulations are due to the hyperfine and quadrupole couplings of the two nitrogen ligands of Coacacen with nuclear frequencies lying in the range between 1–7 MHz (30).

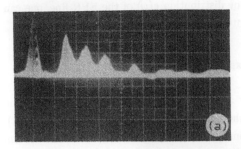 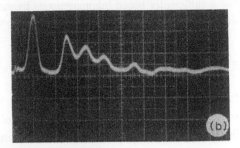

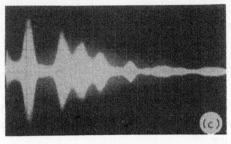

Fig. 23. Echo modulation amplitude of Coacacen diluted in a single crystal of Niacacen (arbitrary crystal orientation). The horizontal deflection is 0.2 μs per division, the vertical axes have been adjusted that all the signals have approximately the same amplitudes. (a) Echo modulation obtained by multiple–exposure to 200 two–pulse experiments, (b) Echo modulation obtained from a *single event* using 2 μs soft pulse excitation, (c) Echo modulation obtained by multiple exposure to 10^4 stochastic response signals of 2 μs duration. (Taken from Ref. 27).

The trace shown in Fig. 23b has been obtained with a single experiment of the type shown in Fig. 20b with a soft excitation pulse of $2\mu s$ length. It is apparent that the echo envelope is faithfully reproduced. It has been found that no visible distortions are encountered as long as the total rotation angle of the soft pulse does not exceed 30^0.

The same echo envelope can also be reproduced by stochastic excitation of equal duration τ_0 prior to the application of the short π pulse. To eliminate the random character of a single response, a series of stochastic response experiments may be combined. Fig. 23c shows the multiple exposure photograph of 10^4 stochastic response experiments with a repetition rate of 1 kHz. Again, a perfect echo envelope matching the one of Fig. 23a is obtained. Experimentally, we found that the amplitude of a single soft–pulse echo decay is approximately 5–10% of a Hahn two–pulse echo while the maximum excursions of the stochastic response can reach up to 30% of a Hahn echo, without causing appreciable envelope distortion. Stochastic excitation leads to enhanced sensitivity as a result of the wider bandwidth of irradiation.

The method may be improved by use of a rapid, repetitive burst of weak microwave pulses for excitation. It may be shown that undesired higher order echoes can partly be eliminated by a proper phase setting and a proper timing between the burst pulses and the refocusing pulse. This allows one to increase the entire flip angle of the weak microwave pulse beyond the linear regime, resulting in an increase of signal intensity.

9. PULSED ENDOR

Pulsed ENDOR is reviewed in this monograph by K.P. Dinse (Chapter 17). In this section we will just make a comment with respect to the blind spots occuring in the Mims–ENDOR scheme and mention two pulsed ENDOR techniques recently developed in our laboratory.

9.1 Blind spots in the Mims–ENDOR experiment

The first pulsed ENDOR scheme, introduced by Mims in 1965 (31) is based on a stimulated echo sequence with an rf pulse between the second and third microwave pulse. The advantage of this type of pulsed ENDOR lies in the fact that nonselective microwave pulses do not prevent the observation of hyperfine splittings covered by the short pulses. This is in contrast to other ENDOR schemes like Davis–ENDOR (32) or coherence transfer ENDOR (33), where selective excitation is a prerequisite for ENDOR detection.

A disadvantage of the Mims–ENDOR is the appearance of blind spots in the spectrum at hyperfine splittings $a = 2\pi n/\tau$, $n = 1,2...$, where τ denotes the fixed time interval between the first and the second microwave pulse. This may be very troublesome, particularly in disordered systems, where the knowledge of the correct ENDOR lineshape is essential for the interpretation of the data. However, as already mentioned by Mims (31), blind spots do not occur for *selective* excitation. Thus, to get rid of the blind spots and to restore the original lineshape, one should use long and selective mw pulses.

An example is given in Fig. 24 (34). The powder ENDOR spectrum of dibenzene-vanadium diluted in ferrocene is recorded at B_0 parallel $g_\perp$ with pulse lengths of 10 ns (Fig. 24ab) and 200 ns (Fig. 24c). The 10 ns pulses are nonselective for both the ferrocene protons (matrix line) and the benzene protons (hyperfine splittings: $9\ \text{MHz} \leq a/2\pi \leq 15\ \text{MHz}$). The 200 ns pulses, however, are selective with respect to the large couplings of the benzene protons, and a line shape free of blind spots equivalent to the cw ENDOR shape is obtained. In contrast to Davis–ENDOR, however, the transitions of the protons of the host compound are still fully developed (not shown in Fig. 24).

Recently the Mims– and Davis–ENDOR sequences have been optimized by supplementing these standard schemes with additional mw and rf pulses (35).

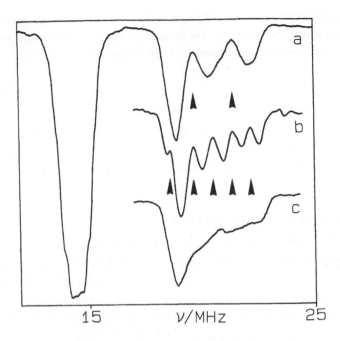

15   ν/MHz   25

Fig. 24. Part of the Mims–ENDOR spectrum of dibenzenevanadium in ferrocene powder. (a) Nonselective excitation, mw pulse lengths 10 ns, $\tau = 300$ ns, (b) $\tau = 600$ ns, Arrows indicate blind spots, (c) Selective excitation, mw pulse lengths 200 ns, $\tau = 600$ ns.

9.2 EPR–detected nuclear transient nutations

Transient nutations (TN) within the nuclear sublevels of an electron–nuclear spin system can be driven by a strong rf field. Such TN may be detected indirectly via the changes in the polarization of the electron spins (36). Several schemes have been developed for the generation and detection of nuclear TN patterns (37). All of them utilize pulsed or continuous–wave mw excitation in order to prepare a large initial nuclear polarization and employ high–sensitivity detection of the electron spin polarization modulated by the nuclear TN.

A scheme where the TN is *continuously* monitored by a weak mw probe pulse together with the resulting nutation pattern is shown in Fig. 25a. With this extended time detection, the entire nuclear TN is obtained in a single experiment (Fig. 25b).

It is well known that in a cw ENDOR spectrum the number of equivalent nuclei contributing to an ENDOR transition is *not* reflected in the intensity; i.e. information about multiplicities in an electron–nuclear spin system is lost (10).

In pulsed ENDOR, the intensities of the transitions are determined by the rf field strength B_{rf}, the enhancement factor E and, possibly, affected by blind spots or insufficient selectivity of the mw pulses. For this reason, line intensities are again not useful for the determination of the number of spins. It has been demonstrated (37) that the nuclear TN pattern of an ENDOR line is characteristic of the number n of equivalent nuclei contributing to that line. The nutation patterns for n≤3 are displayed in Fig. 26a–c. Fig. 26d shows an experimental verification for the case n=2.

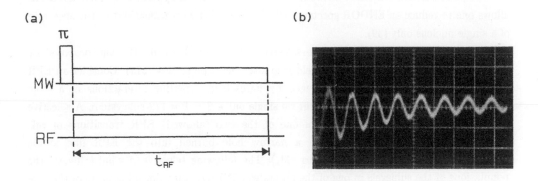

Fig. 25. (a) Pulse scheme for the indirect detection of nuclear TN, (b) Nuclear TN pattern of a copper ENDOR transition in Cu(II)–doped $Mg(NH_4)_2(SO_4)_2 \cdot 6H_2O$, detected with scheme (a). (Taken from Ref. 37).

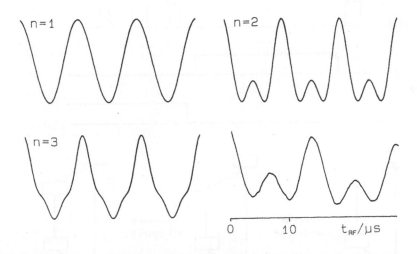

Fig. 26. (a–c) Computed nuclear TN patterns for I = 1/2 nuclei in the case of unresolved hyperfine splittings, (d) Experimental nuclear TN pattern, two equivalent I = 1/2 nuclei (proton line in the ENDOR spectrum of a Cu(II)–doped $Mg(NH_4)_2(SO_4) \cdot 6H_2O$ single crystal). (Taken from Ref. 37).

272

9.3   Hyperfine–selective ENDOR

In spin systems where the unpaired electron interacts with a large number of nuclei, ENDOR spectra may become quite complicated. To simplify and disentangle such complex spectra, various techniques have been developed particularly for cw ENDOR (10,38).

A pulsed ENDOR method introduced very recently, called hyperfine–selective ENDOR, allows one to reduce an ENDOR spectrum which is crowded with transitions to the spectrum of a single nucleus only (39).

The pulse sequence of hyperfine–selective ENDOR is basically the one proposed by Davis (32), supplemented with a B_0–field step after the rf pulse (Fig. 27a). Consider an EPR line which is inhomogeneously broadened by unresolved hyperfine interactions of a large number of nuclei. From this spin system we single out a $S = I = 1/2$ subsystem. A selective mw π pulse inverts the populations of one of the two (allowed) EPR transitions of this subsystem (Fig. 27b–c), resulting in a narrow hole burned into the EPR line at B_0 (z–magnetization changes from $+M_z$ to $-M_z$). The following selective rf π pulse inverts the populations of the sublevels in one of the m_S states (ENDOR transition); i.e. parts of the hole are transferred to a *side hole* at $B_0 + \Delta B_0$ (Fig. 27d), where ΔB_0 corresponds to the hyperfine splitting a of this particular subsystem.

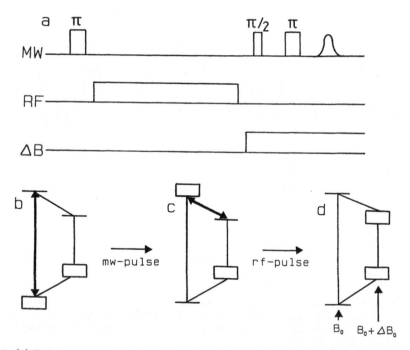

Fig. 27. (a) Pulse sequence for hyperfine–selective ENDOR, (b–d) Change of populations by the mw π pulse and the rf π pulse and position of the observer–field.

In the conventional Davis–ENDOR experiment, the change of electron polarization is monitored at B_0 via a two–pulse echo sequence. Since all the ENDOR transitions in the spin system have one of the two energy levels of the EPR line at B_0 in common, all ENDOR transitions will be observed if the radio frequency is swept through the sublevel spectrum.

If, however, the field is jumped before detection to the side hole at $B_0 + \Delta B_0$, only the ENDOR transitions of a nucleus with a hyperfine coupling $a = \Delta B_0$ will be observed. Thus, the hyperfine splitting may be used as a second dimension in a two–dimensional representation of the data, resulting in an ENDOR spectrum fully disentangled along the hyperfine axis.

Since a B_0–field jump is equivalent to a jump of the mw frequency, the proposed approach represents the first electron–nuclear–electron triple experiment.

A simple experimental example is given in Fig. 28, which compares a section of a single crystal ENDOR spectrum of a copper complex (Fig. 28a) with overlapping proton and nitrogen transitions with the corresponding hyperfine–selective ENDOR four–line spectrum of a pair of magnetically equivalent nitrogen nuclei (Fig. 28b).

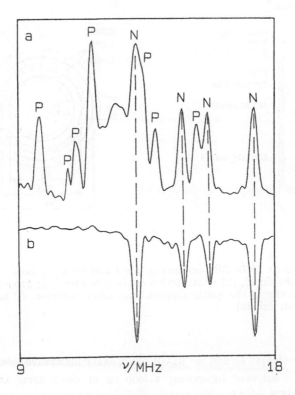

Fig. 28. (a) Section of the single crystal Davis–ENDOR spectrum of bis(picolinate)Cu(II), (b) Hyperfine–selective ENDOR, the intensity variations of the nitrogen side hole is detected. (Taken from Ref. 38).

10. PROBE HEAD DESIGN

All the experiments described in this chapter have been carried out with probe heads developed in our laboratory. The design is based on a new type of a resonant structure for pulsed EPR investigations at X–band frequencies called the bridged loop–gap resonator (BLGR) (40). The BLGR is distinguished by high rf transparency, a property which is especially important in multifrequency experiments like pulsed ENDOR, hyperfine–selective ENDOR, LOD–PEPR and pulsed EPR with jumping magnetic field vectors. The resonator is also well suited for experiments with laser excitation and for Fourier transform EPR where a high filling factor and an adjustable Q value are essential.

The structure of the BLGR is formed by thin gold and/or silver layers on the inner and outer surface of a quartz tube. On the inner surface, the metallic layer is split by two gaps parallel to the tube axis and facing each other. The resulting half loops represent the inductance, the four overlapping regions form the capacitance.

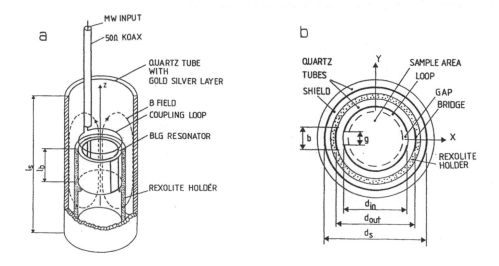

Fig. 29. (a) Arrangement of BLGR, microwave shield and coupling loop. l_b: length of the BLGR, l_s: length of the shield, (b) Cross section of the probe head. g: gap spacing, b: bridge width, d_{in}: inner diameter of the quartz support, d_{out}: outer diameter, d_s: inner diameter of the shield. (Taken from Ref. 40).

A tube of proper length around the BLGR serves as the microwave shield. The coupling to the microwaves is achieved by moving a loop up or down along the sample axis. Illustrations of the BLGR with Rexolite holder, microwave shield and coupling loop are given in Fig. 29. Some of the probe head structures used in recent experiments are schematically shown in Fig. 30.

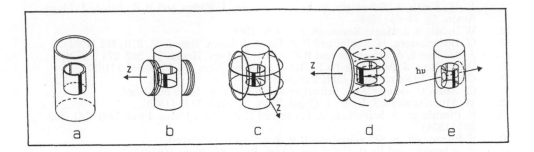

Fig. 30. Probe head structures for various pulsed EPR applications. (a) Standard set up for pulsed EPR, (b) Pulsed EPR with longitudinal detection, (c) Pulsed EPR with magnetic field vector jumps, (d) Hyperfine–selective ENDOR, (e) Experiments with light access.

ACKNOWLEDGMENT

I wish to thank my coworkers who have made the work reviewed in this chapter possible: Gabriele Aebli, Lukas Braunschweiler, Christian Bühlmann, Herman Cho, Jean–Michel Fauth, Jörg Forrer, Claudius Gemperle, Rainer Karthein, Wolfgang Möhl, Herbert Motschi, Marcel Müri, Susanne Pfenninger, René Spycher and Thomas Wacker. I am indebted to Professor Richard R. Ernst for his cooperation. The work has been supported by the Swiss National Science Foundation. The manuscript has been processed by Mrs. Doris Kaufmann and Mrs. Irene Müller.

REFERENCES

1. L. Kevan and R.N. Schwartz, Eds., "Time Domain Electron Spin Resonance", Wiley–Interscience, New York, 1979.
2. S.A. Dikanov and Yu. D. Tsvetkov, J. Struct. Chem. 26, 766–803 (1985).
3. P.A. Narayana and L. Kevan, Magn. Reson. Rev. 7, 239–274 (1983).
4. R.P.J. Merks and R. de Beer, J. Magn. Reson. 37, 305–319 (1980).
5. J.–M. Fauth, Thesis Nr. 8645, ETH Zurich, 1988.
6. A.V. Astashkin, S.A. Dikanov and Yu. D. Tsvetkov, Chem. Phys. Lett. 136, 204–208 (1987).
7. S.A. Dikanov and A.V. Astashkin, Chapter 2.
8. M. Iwasaki and K. Toriyama, J. Chem. Phys. 82, 5415–5423 (1985).
9. G.H. Rist and J.S. Hyde, J. Chem. Phys. 52, 4633–4643 (1970).
10. A. Schweiger, Struct. and Bonding 51, 1–122 (1982).

276

11. J.–M. Fauth, A. Schweiger, L. Braunschweiler, J. Forrer and R.R. Ernst, J. Magn. Reson. 66, 74–85 (1986).
12. W.B. Mims, J. Magn. Reson. 59, 291–306 (1984).
13. G. Bodenhausen, H. Kogler and R.R. Ernst, J. Magn. Reson. 58, 370–388 (1984).
14. J.–M. Fauth, A. Schweiger and R.R. Ernst, J. Magn. Reson. 81, 262–274 (1989).
15. W.B. Mims, J. Peisach and J.L. Davis, J. Chem. Phys. 66, 5536–5550 (1977).
16. G. Aebli, Diploma work, ETH Zurich, 1988.
17. G. Aebli, C. Gemperle, A. Schweiger and R.R. Ernst, to be published.
18. W. Froncisz and J.S. Hyde, J. Chem. Phys. 73, 3123–3131 (1980).
19. S. Pfenninger, A. Schweiger, J. Forrer and R.R. Ernst, Chem. Phys. Lett. 151, 199–204 (1988).
20. S. Pfenninger, A. Schweiger and R.R. Ernst, to be published.
21. A. Schweiger and Hs.H. Günthard, Chem. Phys. 32, 35–61 (1978).
22. S. Pfenninger, A. Schweiger, J. Forrer and R.R. Ernst, Proc. of XXIII Congr. Ampere Magn. Reson., p.568 (1986).
23. A. Schweiger and R.R. Ernst, J. Magn. Reson. 77, 512–523 (1988).
24. W.B. Mims and J.L. Davis, J. Chem. Phys. 64, 4836–4846 (1976).
25. D. Goldfarb and L. Kevan, J. Magn. Reson. 76, 276–286 (1988).
26. H.Y. Carr and F.M. Purcell, Phys. Rev. 94, 630–638 (1954).
27. A. Schweiger, L. Braunschweiler, J.–M. Fauth and R.R. Ernst, Phys. Rev. Lett. 54, 1241–1244 (1985).
28. A. Schenzle, N.C. Wong and R.G. Brewer, Phys. Rev. A22, 635–637 (1980).
29. L. Braunschweiler, A. Schweiger, J.–M. Fauth and R.R. Ernst, J. Magn. Reson. 64, 160–166 (1985).
30. M. Rudin, A. Schweiger and Hs.H. Günthard, Mol. Phys. 46, 1027–1044 (1982).
31. W.B. Mims, Proc. Roy. Soc. A283, 452–457 (1965).
32. E.R. Davis, Phys. Letters A47, 1–2 (1974).
33. P. Höfer, A. Grupp and M. Mehring, Phys. Rev. A33, 3519–3522 (1986).
34. J.–M. Fauth, C. Gemperle and A. Schweiger, unpublished results.
35. C. Gemperle, A. Schweiger, O.W. Sørensen and R.R. Ernst, to be published.
36. M. Mehring, P. Höfer and A. Grupp, Phys. Rev. A33, 3523–3526 (1986).
37. C. Gemperle, A. Schweiger and R.R. Ernst, Chem. Phys. Lett. 145, 1–8 (1988).
38. A. Schweiger in "Electron Spin Resonance", Ed. M.C.R. Symons (Specialist Periodical Reports), The Chemical Society, London 1987, Vol.10B.
39. C. Bühlmann, A. Schweiger and R.R. Ernst, Chem. Phys. Lett. 154, 285–291 (1989).
40. S. Pfenninger, J. Forrer, A. Schweiger and Th. Weiland, Rev. Sci. Instrum. 59, 752–760 (1988).

CHAPTER 7

LOOP GAP RESONATORS

James S. HYDE and Wojciech FRONCISZ

1. INTRODUCTION

1.1 Survey of loop gap resonator geometries

The loop gap resonator (LGR) is based on lumped circuit concepts. In the lumped circuit model, R, L, C are easily defined, and the circuit is small compared with the wavelength. Electric and magnetic fields are independent of each other in the lumped circuit limit.

Cavity resonators are distributed circuits, and characteristic dimensions are of the same order of magnitude as the wavelength with electric and magnetic field vectors inextricably related by Maxwell's equations.

In fact, dimensions of LGRs are characteristically 1/10 to 1/3 of the wavelength, and neither lumped nor distributed circuit vocabulary is fully convenient. The inductance of a simple loop, Fig. 1, is given in the lumped circuit limit by

$$L = \frac{\mu_o \pi r^2}{Z} , \qquad [1]$$

and the capacitance by

$$C = \frac{\epsilon W Z}{tn} . \qquad [2]$$

Here n is the number of gaps. The calculated resonant frequency $(LC)^{-1/2}(2\pi)^{-1}$ is typically in error by about 30% at microwave frequencies because of neglect of the electric field in the loop, the magnetic field in the gap, and both fields at the ends. Our papers (1–3) consider corrections that include these effects in some detail.

Most readers of this article will be familiar with the cylindrical TE_{011} microwave resonator. A cross–section through this resonator is shown in Fig. 2a. The lines of magnetic flux are perpendicular to the page, coming out of the page at the center and going into the page at the walls. Between the center and the wall there is a radial distance where the magnetic field is zero. At this distance the electric field is a maximum and is circumferential as indicated by the dotted line. If the height of the resonator equals the diameter, the latter at X–band is about 4 cm. In some degree all of the loop gap resonators illustrated in cross–section in Fig. 2 are similar to the cylindrical TE_{011} cavity. Similarities and differences are discussed here.

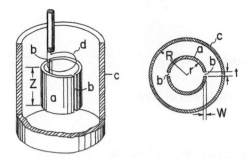

Fig. 1. The loop–gap resonator showing the principal components (a, loop; b, gaps; c, shield; d, inductive coupler) and the critical dimensions (Z, resonator length; r, resonator radius; R, shield radius; t, gap separation; W, gap width). The sample is inserted into the loop a through the coupler, d. The microwave magnetic field in the loop is parallel to the axis. From Ref. 1.

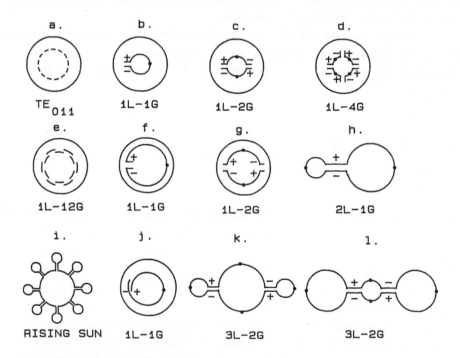

Fig. 2. Cross–sections of various resonators. The microwave magnetic field is perpendicular to the page, and the dc field B_0 lies in the page. The structures are not drawn to scale. The TE_{011} cavity, a, is much larger than the other structures, all of which are loop gap resonators. Solid dots are zero potential loci.

Figures *2b*, *2c*, *2d* illustrate one—loop——one, two and four—gap resonators respectively. The sample goes in the central loop, which is typically in the range of 1 to 3 mm diameter at X—band. A shield of about 2 cm diameter surrounds the structure. One can visualize this geometry as similar to the TE_{011} cavity, but with the electric field shorted out so that the electric field energy is mostly concentrated in the gaps.

The integral of the magnetic induction B_1 over a cross—section of the loop equals the integral of B_1 over the annular cross—section. For a long loop, B_1 is uniform in both loop and annulus, and

$$\frac{B_1(\text{loop})}{B_1(\text{annulus})} = \frac{\text{Area(annulus)}}{\text{Area(loop)}} . \qquad [3]$$

The ratio of filling factors (see Eqn. 6) for a sample in the loop compared to a sample in the annulus is then

$$\frac{\eta(\text{loop})}{\eta(\text{annulus})} = \frac{\int_{\text{loop}} B_1^2 dA}{\int_{\text{annulus}} B_1^2 dA} = \frac{\text{Area(annulus)}}{\text{Area(loop)}} . \qquad [4]$$

Since the loop is generally very small, the filling factor is nearly unity for a sample in the loop. Such a high filling factor is essentially impossible to obtain with the TE_{011} cavity. Typically the Q can be 10 times lower but the filling factor 100 times higher with the LGR. In principle, the structure of *2f* can be made, which would have a high filling factor for an annular sample.

In all of the structures shown, the field is uniform in the loop in the lumped circuit limit if the length is large compared to the diameter. In this limit the electric field is totally concentrated in the gap. In fact there must be an electric potential around the loop, with the points that are indicated in Fig. 2 by solid dots being midway between the plates of the gap and therefore always at zero potential. The spatial variation of the electric field in the loop as determined from the potential around the loop must, by Maxwell's equations, result in some loss of magnetic field homogeneity.

Electric field losses are not negligible for aqueous samples in LGR's. A one—loop——one—gap resonator will have minimum losses when the sample is off—center near the dot. A one—loop——two—gap resonator, Fig. 2c, has a nodal plane extending from dot to dot where the electric field is zero. A sample—containing "flat—cell" can be placed in this plane in analogy to the flat—cell used in the rectangular TE_{102} multipurpose cavity.

Consider now the structure of Fig. *2l*. The central loop of this three—loop——two—gap resonator is identical to that of Fig. 2c. The only difference between 2c and 2l LGR's is the way in which the return flux is managed. In 2c it returns in the annular region and in $2l$ it returns in the two outside loops. Similarly, structures $2f$ and $2k$ can be compared: they have a large central loop, and flux returns in either the small annular region of $2f$ or the two small outer holes of $2k$.

Figure $2h$ is a two—loop——one—gap structure, and one has a free choice of which is the sample loop and which is the return—flux loop. The choice will be based, according to the discussion in the next section, on the properties of the sample.

The dimensions of that portion of the structure that contains the highest energy density determine the resonant frequency. Thus the dimensions of the shield in $2b$ and $2d$ affect the frequency rather little. Similarly the dimensions of the larger inner loop are not critical in $2f$, $2g$, $2k$; the annular region, $2f$, $2j$, or the small outer loops, $2k$, largely determine the frequency.

As the number of gaps increases, say to a one—loop——four—gap structure, $2d$, one has a mechanical problem of how to support the structure. One solution is the Rising Sun resonator, $2i$. We arrived at this structure quite independently and only subsequently learned that it is the basis of the magnetron and serves as the logo for the MIT Radiation Laboratory series of books. This structure has been made at X—band with a 1 cm diameter inner loop and 16 outer loops, with the sum of the outer loop areas equal to the inner loop area. Thus the filling factor is 1/2. With 16 loops, the approximation to full cylindrical symmetry of the TE_{011} cavity resonator is very good.

An alternative geometry employs flexible two—sided printed circuit board material with Teflon substrate, Fig. $2e$ (Cuflon from Polyflon Corp., New Rochelle, NY). Offset strips made on opposite sides of the board using printed circuit techniques provide the gaps. The structure is obviously similar to the Rising Sun, but the return flux is in an annular region rather than 16 holes on a bolt circle.

In the lumped circuit limit, one would expect considerable flexibility as to how the gap is oriented. It can go out as for $2b$, in as for $2f$, or overlap as for $2j$. The structure of $2j$ is somewhat worrisome because of perturbations at the end, but we have done some calculations that suggest that the overlap can be as much as 1/3 of the circumference.

Loop gap resonators tend to radiate, and need to be enclosed in shields. The Q's of microwave resonators always go as the volume/surface ratio. LGR's are smaller than cavities and therefore tend to have lower Q's. But because they are smaller, peak energy densities tend to be much higher. Loop gap resonators have been built by us from 10 MHz (for magnetic resonance imaging) to 35 GHz, and the unloaded Q's that have been realized always fall within a factor of 2, one way or the other, of 1000.

The cavity resonator is a special case of the loop gap resonator, made special only by the fact that Maxwell's equations are easy to solve because of the well—defined boundary values. By making resonators from combinations of loops and gaps a much greater freedom in optimizing the coupling of a transmission line to the spins in a sample is obtained. It seems likely that essentially all EPR experiments will be performed in the future with LGR's. In our laboratory the classic cavity resonator is fast becoming obsolete.

1.2 Review of EPR sensitivity considerations

Some years ago (4) an EPR sample classification system was introduced that was helpful in thinking about sensitivity. Three questions were asked: (a) Does the sample saturate with the available microwave power? (b) Is the sample limited in size because of cost or difficulties in preparation? (c) Does the sample exhibit strong dielectric loss, as for water? The idea was to introduce eight sample classes according to the permutations of the yes or no answers to these three questions, and then to find the optimum microwave geometry for each class from the point of view of sensitivity. Independently, Wilmshurst arrived at a similar perspective (5).

In Feher's fundamental paper on EPR sensitivity considerations (6), the following expression for the signal voltage was derived:

$$S = \chi \eta Q P_o^{1/2} . \qquad [5]$$

Here χ is the radio–frequency susceptibility per unit volume and P_o the incident microwave power. Formally, χ also contains the microwave power to allow, according to the Bloch equations, for microwave power saturation. The filling factor η is given

$$\text{by} \quad \eta = \frac{\int_s B_1^2 \sin^2\phi \, dV_s}{\int_c B_1^2 \, dV_c} \qquad [6]$$

where the integrals are over the cavity (c) and sample (s) and ϕ is the angle between the dc polarizing magnetic field and B_1 (it will be set equal to $90°$ in the following). Q is the loaded (i.e. matched) quality factor of the cavity. Equation 5 can conveniently be used to evaluate the relative sensitivities of two resonant structures, subscripts 1,2, for each of the eight classes of samples:

$$\frac{S_1}{S_2} = \frac{(\chi \eta Q P_o^{1/2})_1}{(\chi \eta Q P_o^{1/2})_2} . \qquad [7]$$

Another form of Eqn. 7 is sometimes useful. Since at match

$$Q = \omega \, \frac{\text{Energy Stored}}{\text{Incident Power}} \quad \text{and} \qquad [8]$$

$$\text{Energy Stored} \; \propto \; \int_c B_1^2 \, dV_c \quad , \tag{9}$$

we can write with $\chi_1 = \chi_2$

$$\frac{S_1}{S_2} = \frac{\left(\int_s B_1^2 dV_s \right)_1}{\left(\int_s B_1^2 dV_s \right)_2} \left(\frac{P_{O2}}{P_{O1}} \right)^{1/2} \quad . \tag{10}$$

It is helpful to introduce the resonator efficiency parameter Λ.

$$\Lambda = \frac{B_1}{P_o^{1/2}} \tag{11}$$

where B_1 is the maximum available microwave field for a given incident microwave power. A particularly important situation where Eqn. 10 is appropriate is for a limited sample that is sufficiently small that dielectric loss is negligible. If the sample is non—saturable, P_o is just the maximum available power and it follows that

$$\frac{S_1}{S_2} = \frac{\Lambda_1^2}{\Lambda_2^2} \quad . \tag{12}$$

One wants the highest possible energy density at the sample for the available microwave power.

If the sample is saturable, the incident microwave power must be adjusted to bring B_1 to the same level in each of the resonators under comparison. Eqn. 10 becomes

$$\frac{S_1}{S_2} = \frac{\Lambda_1}{\Lambda_2} \quad . \tag{13}$$

The resonator efficiency parameter is typically 1 for an X—band cavity resonator if P_o is in watts and B_1 is in gauss. X—band loop gap resonators have been made with Λ as high as 10, resulting in much improved performance whenever the sample is limited.

Thus of the eight sample classes, the LGR structure offers truly substantial improvement for the four that carry the "limited" designation. The concept of "limited" can include not only a small sample, but also limited in one dimension only, limited because of diffusion, or in flow experiments a need to limit the amount of sample that flows through the resonator even though quite a lot is available. Much of the thrust of this article is directed towards loop gap resonators for various sample–limited situations.

The two classes of samples that are unlimited and of negligible dielectric loss include EPR in gasses and paramagnetic materials in low–loss fluids such as benzene or hexane. The resonator can be filled, and the filling factor is unity. For nonsaturable samples

$$\frac{S_1}{S_2} = \frac{Q_1}{Q_2} .$$
[14]

For saturable samples, using Eqn. 8 to eliminate P_o from Eqn. 10, one obtains

$$\frac{S_1}{S_2} \propto \left(\frac{Q_1 V_1}{Q_1 V_2} \right)^{1/2} .$$
[15]

The higher Q and larger volume of the cavity always give it a substantial advantage over LGR geometries for these two classes of samples.

There is a biologically important sample solvent, namely ice, that has low dielectric loss, and is often unlimited, but where Eqns. 14, 15 are oversimplifications. The difficulty is that the high dielectric constant (16 at X–band) can substantially affect the microwave characteristics of the resonator. The LGR provides some flexibility in sample–handling for ice samples, particularly at helium temperatures. This advantage can often outweigh, from a practical perspective, the theoretically favorable sensitivity of high Q cavities.

The remaining two classes of samples, unlimited, high dielectric loss, saturable and nonsaturable, include aqueous samples, and are of great importance. Wilmshurst (5) shows in cylindrically symmetric geometry that for nonsaturable samples the capillary diameter should be such as to lower the Q by 1/2, whereas for flat–cells in a rectangular TE_{102} cavity resonator the cell thickness should be such as to lower the Q by 1/3. However, remarkably, there is no optimum capillary diameter in cylindrical symmetry for saturable samples. One should increase the capillary diameter until sample losses $>>$ resonator losses. For saturable samples in flat–cell geometry the cell thickness should be such as to lower the Q by 2/3.

In the geometries considered by Wilmshurst, the sample was in a microwave electric field node. Hyde pointed out (7) that alternative aqueous sample–cell geometries are possible where the microwave electric field is perpendicular to the sample–cell surface.

These aqueous sample—cell considerations were originally developed for cavity resonators, but they apply equally well for loop gap resonators. In our experience, sensitivities tend to be similar, comparing LGR's and cavity resonators, for aqueous samples when expressed on a molarity basis.

To the present point in this short introduction to EPR sensitivity considerations, we have been concerned only with signal heights. Noise must also be considered. In EPR spectroscopy, noise sources can be divided into (a) the oscillator, (b) the detector, and (c) the environment (microphonics, electrical grounds, radio—frequency interference). With increasing use of field—effect transistor (FET) amplifiers, the detector noise has become so low that in many practical situations the other noise sources are dominant. This is a change from the situation that existed prior to the introduction of the FET. Loop gap resonators tend to have lower Q's, and this is a great advantage in minimizing the demodulation of phase noise that originates in the oscillator or of phase variations in the microwave bridge that arise from microphonics.

2.   MULTIFREQUENCY EPR

In 1979 Froncisz *et al.* (8) studied the EPR spectrum of cytochrome c oxidase using an octave bandwidth microwave bridge operating from 2 to 4 GHz and a rectangular TE_{102} microwave cavity designed by scaling the Varian E—231 X—band multipurpose cavity. Although this paper was important in that it fully established the value of S—band in EPR spectroscopy of copper proteins, some technical problems arose: (a) The amount of sample was 1 ml, limited by sample preparation considerations, and the resulting filling factor was poor. (b) The resonator efficiency parameter Λ was low. In order to approach microwave power saturation for best signal—to—noise ratio (S/N), the incident microwave power was fairly high. But then oscillator phase noise became serious. Octave bandwidth oscillators intrinsically tend to have higher phase noise than narrow band oscillators. Furthermore, the cavity Q exceeded 10,000 also enhancing the demodulation of phase noise.   (c) It was not convenient to vary the microwave resonant frequency of the cavity, and the intrinsic utility of the octave bandwidth bridge was negated. These problems were exacerbated when we completed construction of an L—band (1 to 2 GHz) microwave bridge. Our search for solutions led eventually to the loop gap resonator. In the meantime, octave bandwidth microwave bridges have been built at the ESR Center operating from 0.5 to 1 and from 4 to 8 GHz, resulting in continuous coverage from 0.5 to 8 GHz. By multifrequency EPR we mean the capability of examining one and the same sample over a range of microwave frequencies. Our development of this methodology is dependent on the loop gap resonator. Various aspects of this dependence are discussed in this section.

Figure 3 is an outline drawing of the basic low frequency LGR. The inner diameter was set to accommodate standard X—band 3 mm i.d., 4.5 mm o.d. quartz sample tubes. The structure is silver plated brass. By opening up the gap and by cutting down on W, the gap width, it can be made to resonate as high as 8 GHz. By squeezing on the gap and filling it with thin quartz or sapphire plates, it can be made to resonate as low as 0.6 GHz. Thus we find that it is practical to vary the capacitance by more than 2 orders of magnitude. The inductance remains constant, the microwave field distribution remains mainly constant, and the filling factor for an X—band sample tube is independent of frequency over a range of a factor of 10 in frequency. The amount of sample required is 0.07 ml, suitable for a wide range of biological samples.

Fig. 3. Outline drawing of the one—loop——one—gap resonator used for multifrequency EPR in the frequency range of 0.5 to 8 GHz. The horizontal slots permit penetration of the 100 kHz magnetic field modulation.

The goal of continuous capability in the 0.5 to 8 GHz range has remained elusive. We made a serial production of 10 resonators spaced over this range, all identical except for the gap, resulting in good capability at 10 discrete frequencies. Structures where the capacitance is varied continuously by smoothly inserting a dielectric have been built, but have thus far proven to have inadequate stability against microphonic vibrations. Probably solutions to this problem can be found.

The loop gap resonators in this series have Q's in the range of 500 to 1000. Demodulation of phase noise when tuned to the absorption mode varies as Q^2. When tuned to the dispersion, the dependence of phase noise demodulation on Q is linear and phase noise is a significantly greater problem than when observing absorption. Phase noise demodulation is essentially absent in all of our bridges over the full available power (up to 1 W) when tuned to absorption using LGR's. Very high quality narrow—band klystrons are required in order to eliminate phase noise in this frequency range when cavity resonators are used. When tuned to dispersion, phase noise can still be dominant at higher incident microwave powers, but the LGR permits measurement of dispersion signals over a much increased range of conditions.

One can wonder if loss of signal intensity at low microwave frequencies renders multifrequency EPR impractical for biological samples. We find that the Varian pitch sample at 3 GHz yields about the same signals as in a TE_{102} cavity at 9.5 GHz, because of improved filling factor, and that the signal is about 3 times less at 1 GHz. Of course, such comparisons can be misleading because of field—dependent linewidths, spin—lattice relaxation times and spectral dispersion. It is concluded that the LGR compensates for the Boltzmann factor to a considerable degree in the 1 to 10 GHz range if one uses conventional X—band sensitivity experience as a baseline.

Under development at the ESR Center is a family of liquid helium immersion loop gap resonators. It would be difficult if not impossible to immerse L—band cavities, for example, in helium using practical geometries. The small size of the LGR structure makes this familiar X—band cryogenic methodology practical at low microwave frequencies. Den Blanken *et al* (9) describe an immersion LGR operating between 300 and 1000 MHz.

Also under development are resonators at low frequency that accommodate flat—cells. The dielectric loss decreases with frequency but is not always negligible. Additionally, flat—cell geometry is convenient in the study of oriented bilayers. A multifrequency study of spin probes in oriented bilayers can be expected to yield improved information on motional models.

The main motivation for multifrequency EPR lies in spectroscopy of powders and viscous liquids. The interplay of the anisotropic Zeeman and hyperfine interactions allows the extraction of information using a multifrequency approach that cannot be obtained at a single frequency. Additional rationales exist that are dependent on the microwave frequency dependence of relaxation times and on the dependence of elec—tronic and nuclear state mixing (i.e. forbidden transition probabilities are frequency dependent). Our review article, *The Role of Microwave Frequency in EPR Spectroscopy of Copper Complexes* (10) discusses the rationale for multifrequency EPR in more detail. It is the loop gap resonator that makes the approach practical.

3. FET's AND LGR's

The field effect transistor microwave amplifier lowers the existing overall receiver noise figure from about 10dB in 100 kHz homodyne EPR bridges to about 2dB. Dramatic improvement in system performance does not necessarily follow. Noise originating in the source is about equal to detector noise, for example, when using 100 mW at X—band detecting absorption with a conventional TE_{102} rectangular cavity. Improving the receiver noise figure does very little under these conditions: One must additionally reduce source noise in order to improve system performance.

When tuned to the absorption, phase noise from the source in a reference—arm bridge is given by the expression

$$V_{Noise} \propto \frac{O^{1/2}P_o^{1/2}Q^2G^{1/2}}{V_{Reference}} \qquad [16]$$

where O is the phase noise power in a given bandwidth a given distance from the carrier, V is the microwave reference voltage at the mixer, and G is the FET amplifier gain. (The gain G appears here because it affects the square root of the sum of the squares of the reference voltage and quadrature noise voltage.) The microwave bridge of Fig. 4 (11) illustrates such a microwave circuit, together with some more complicated features to be discussed later in this section. If for example, the gain is

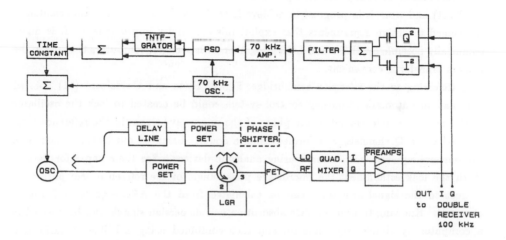

Fig. 4. Reference–arm microwave bridge employing a loop gap resonator (LGR), FET microwave amplifier, balanced quadrature mixer, and new type of automatic frequency control (AFC) circuit. From Ref. 10.

20dB in the example just given, the outstanding performance of the FET would be realized only when P_o was 30dB below 100 mW. However, with a loop gap resonator of Q = 500 the noise voltage is less than that observed with a cavity resonator of assumed Q = 7500 by a factor of 15. The demodulation of noise voltage is reduced by a factor of 225, corresponding to a power ratio of 45dB. Phase noise would never be dominant and the full advantage of the FET would be realized over the full range of microwave powers. Additional freedom from phase noise is achieved by using high–barrier Schottky detector diodes with reference power as high as 50 mW.

When tuned to the dispersion

$$V_{Noise} \propto O^{1/2}P_o^{1/2}Q \; . \qquad [17]$$

Dropping the Q by 15 decreases demodulated phase noise power by 25dB. In a conventional X–band microwave bridge, phase noise equals source noise when $P_o \simeq 1$

mW. Thus even for dispersion, demodulated phase noise is largely absent over the full range of microwave powers in the bridge.

This creates a new opportunity. Dispersion signals can be changed into absorption signals by a Hilbert transform (12), and added. Since the noise floor has already been established, an additional 3dB of signal improvement can be achieved.

Combining this 3dB with the estimated, and now realized, noise figure reduction from 10 to 2dB (i.e. 8dB), the spectroscopist is offered a total of 11dB, or a factor of 3.5 improvement in voltage sensitivity. This improvement has been achieved by us at L-band (11) and work is in progress to achieve it at X-band. The biomedical scientist is challenged to design experiments that exploit this sensitivity improvement. It is made technically possible by FET's and LGR's together with personal computers that make the Hilbert transform convenient.

Referring to the reference arm bridge, Fig. 4, Hyde and Gajdzinski (11) pointed out that an automatic frequency control system could be created to lock the oscillator to the resonator independent of the phase of the microwave signal in the reference arm. Parallel I and Q channels were formed using a double-balanced quadrature mixer, and the outputs squared and summed using analogue devices. See their paper for details. With the lock established, the microwave phase angle of the reference arm path with respect to the signal arm path can be calculated from the AFC signals in I and Q channels. Knowing this angle, pure absorption and dispersion signals can be created in a computer by simple trigonometry, and then combined using a Hilbert transform if desired. Thus the phase shifter in the reference arm can in principle be eliminated. It also is not necessary to use a rotary vane precision attenuator, which was originally employed in microwave bridges because of its minimal phase shift with attenuation.

Thus the cost of a narrowband bridge such as the usual commercial X- and Q-band bridges can be decreased. Perhaps more importantly, broadband bridges become more feasible. A broadband bridge cannot employ waveguide transmission lines or waveguide components. We know of no attenuator with sufficiently low phase shift to be used in general purpose bridges that use semi-rigid coax, microstrip or stripline technology. The FET, LGR, PC and double-balanced quadrature mixer -- all recent technology -- come together to permit not only better signal-to-noise ratio but also greater bridge bandwidth. It is cautioned, however, that further development work is needed to lower phase noise of broadband oscillators. Source phase noise can be seen in the L-band bridge, where these ideas have been implemented, when tuned to the dispersion at powers of about 10 mW and higher.

The argument is summarized.

*   FET's lower the overall receiver noise figure.
*   As a consequence, microwave phase noise becomes the limiting noise source.
*   LGR's reduce demodulation of phase noise.

* With the noise floor established, I and Q channels can be formed without loss of S/N.

* Suitable data processing permits AFC lock and calculation of pure absorption and dispersion signals independent of the reference arm phase.

* Since microwave phase control is not needed, broadboard transmission lines and components can be used.

* Thus octave and greater bandwidth microwave bridges with substantially improved sensitivity and possibly lower cost are feasible.

One wonders what the final limiting noise source will be. The authors suspect that environmental disturbances will become increasingly evident, leading to the necessity for improved isolation techniques.

4. SPECIAL CASES OF LIMITED SAMPLE: FLOW, ELECTROCHEMISTRY, ENDOR, ELECTRIC FIELD

In this section the point is made that the outstanding performance of loop gap resonators for limited samples has a number of consequences that might not initially be apparent. *In situ* electrochemistry can lead to generation of short—lived radicals that essentially lie in a surface around the working electrode. Thus the sample is limited in one dimension by the mean diffusion distance in the mean radical lifetime. Moreover, since capacitances are small because of the small size of electrodes, the time response can be fast and time—domain electrochemical experiments designed. Electric field experiments become increasingly feasible because very high electric fields can be obtained between two closely spaced electrodes using reasonable potentials, and the filling factor for the space between the electrodes can be high. Continuous flow experiments use much reduced sample, essentially a factor of about 100 for comparable flow rates relative to cavity resonators. As will be shown, stop—flow experiments can also be done on a shorter time scale. ENDOR is enhanced because small radio—frequency coils can be used. Since they are very close to a quite small structure, desired rf fields can be achieved with decreased power and the rf field can have very good homogeneity over the sample.

These four topics are discussed here. Other possible applications of LGR's that make use of the good "limited sample" characteristics include (a) EPR imaging where strong magnetic field gradients over the sample can be achieved using modest power in quite small gradient coils; (b) optical irradiation in samples with very high extinction coefficients; (c) improved spatial localization for diffusion measurements (a spin label is envisaged diffusing down a line sample extending through a loop gap resonator of 1 mm length.); (d) temperature jump experiments where reduced energy is required to achieve the desired temperature rise (the thermal recovery is also faster, which may permit more data to be collected per unit time); and (e) low cost EPR spectrometers where the small size of the resonator permits use of smaller, less expensive magnets *etc. etc.* A general instrumental trend towards miniaturization is underway and the loop gap resonator is part of this trend.

4.1 Continuous and Stopped Flow with the LGR

Figure 5 shows the X—band loop gap resonator presently used in our laboratory (Medical Advances, Inc., Wauwatosa, WI). It is a two—loop——one—gap structure. The active sample region is 1 mm diameter, 5 mm long. Quartz sample tubes, 0.6 mm id, are used, resulting in an active sample volume of 1.2 μl. The unloaded Q is 500, and $\Lambda = 6$.

Figure 6 illustrates this structure, together with a mixing chamber for continuous and stopped flow experiments as described by Hubbell *et al.* (13). An important aspect of their experimental configuration was the syringe ram supplied by Update Instruments, Madison, WI, Model 715. This unit is driven by a low inertial mass printed—circuit motor generating linear motion through a screw drive. The ram displacement, direction, and velocity are precisely controlled using input from an optical encoder on the motor shaft. The ram velocities vary between 1 and 8 cm/s. Using 1 ml syringes, this range of velocities is a good match to the geometry of the LGR. Performance of this type of syringe drive is independent of sample viscosity, an important advantage.

Hubbell *et al.* found the following characteristics of their assembly:

* Mixer Volume: 1.6 μl
* Volume from the exit of the mixer to the entrance of the resonator: 4.0 μl
* Continuous flow dead volume (mixer inlet to resonator center): 6.2 μl
* Continuous flow dead time at 8 cm/s: 1.2 ms
* Amount of material required for 100 ms of flow at 8 cm/s: 0.26 ml
* Stop flow dead time (mixing to beginning of data acquisition): 4 ms
* Practical material consumption per stop flow experiment: 33 μl (30 shots per ml).

The resonator was, as mentioned, 5 mm long. Gradients of sample age in the active volume affect the quality of kinetic data. Sample age gradients in the structure used by Hubbell *et al.* are favorable relative to previous work. They can be improved further, as for example, by the use of a 2 mm long resonator (Resonator #1, Table 1, Ref. 1).

Heat of mixing and temperature gradients should also be considered. The short sample minimizes temperature gradients.

An improvement of a factor of two in sensitivity in the stop flow mode was achieved by observing the dispersion signal under partial saturation. In stop flow, one wants to set the instrumental parameters to give the largest possible signal. Spectral

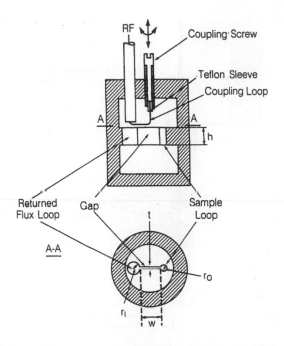

Fig. 5. X−band two−loop−−one−gap resonator showing matching structure. The principle of adjustment of coupling to a coaxial line is illustrated at the vertical cross section. The horizontal cross section at A−A is also diagrammed. From Ref. 13.

distortion will not affect stop flow kinetics. The Varian rapid scan unit was used for continuous flow and it was noted that the small size of the resonator reduced in a meaningful manner the inhomogeneity of the rapid scan magnetic field over the sample.

It is apparent from a study of Fig. 6 that there is some inconsistency between sample volume of 1.2 μl, mixer volume of 1.6 μl, and mixer exit to resonator inlet volume of 4 μl. Future improvements can substantially reduce the 4 μl value to perhaps 1 μl, reducing the dead volume from 6.2 μl to 3.2 μl and the dead time to 0.6 ms. This is probably close to the intrinsic performance limit for EPR mixing experiments.

In summary Hubbell *et al.* achieved a remarkable reduction in the total amount of sample required in continuous flow experiments, of the order of a factor of 50 compared with standard mixing cells (Wilmad WG−804). The sensitivity on a molarity basis for continuous flow was not specified, but probably lower by a factor of 2. For stop flow, it was possible to improve the signal−to−noise ratio by a factor of 2 by observing the dispersion under partial saturation. The time response was improved by a factor of 2 to 3 compared with standard geometries. A number of incidental advantages of miniaturization of the assembly were noted: improved rapid scan and modulation field homogeneity over the sample; decreased sample age gradients in the active volume; and decreased temperature gradients in the sample.

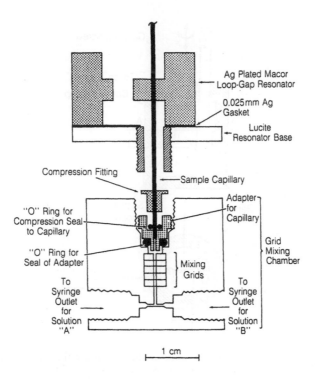

Fig. 6. Adaptation of the grid mixer to the loop gap resonator. The Lucite four—grid mixer is commercially available from Update Instruments, Madison, WI. As the mixer is screwed onto the threaded extension from the resonator base, the compression fitting compresses the small O—ring, making a seal around the sample capillary. The adapter is sealed against the mixer housing by a second O—ring compression seal as shown. Both the compression fitting and adapter were fabricated of brass, although any nonferro—magnetic material will suffice. From Ref. 13.

Yamasaki, Mason and Piette (14) in their seminal paper on reduction of ascorbate by horseradish peroxidase argued that a wide range of one—electron transfer enzymatic reactions should be amenable to study by EPR rapid mixing experiments. The LGR may bring us closer to this goal by greatly deceasing the amount of enzyme required.

4.2 Electrochemistry

Maki and Geske (15,16) first demonstrated *in situ* electrochemical generation of free radicals for electron paramagnetic resonance studies. Shortly thereafter, Piette *et al.* (17) developed an electrochemical cell that remains in very wide usage and is commercially available (Wilmad WG–810). Numerous EPR electrochemical cells have been described since the original Maki–Geske paper. See both the first and second edition of Poole's treatise for citations (18,19).

Allendoerfer *et al.* (20) describe electrochemical generation of free radicals in a loop gap resonator. The overall scheme is shown in Fig. 7 and an expanded view of the sensitive volume in Fig. 8.

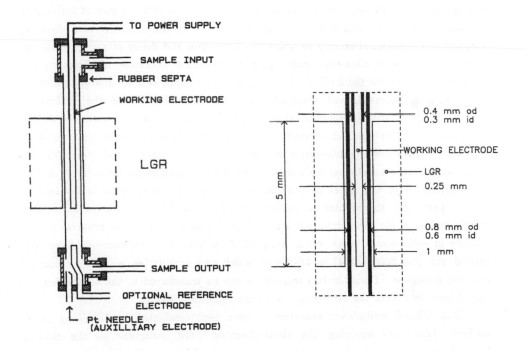

Fig. 7. Schematic overview of an elec–trochemical loop gap resonator assembly with reference electrode. From Ref. 20.

Fig. 8. Enlarged view of the active region of the electrochemical loop–gap resonator geometry. From Ref. 20.

From a microwave engineering perspective, the key finding was that it was possible to insert a metallic working electrode at the center of the one–loop––one–gap resonator. To a good approximation, the Q was unchanged. The microwave frequency was shifted up but remained in the bandwidth of the Bruker Commercial X–band bridge used for the experiment. The microwave field actually increased because the same microwave energy was stored in a space that was reduced by the volume of the electrode.

These workers reported the ESR spectrum of 1 mM p—nitrobenzoic dianion radical in water. Electrochemical preparation of this radical in aqueous solvents had not previously been reported.

J. Gajdzinski, an electrical engineer working in the ESR Center, was asked to develop an equivalent circuit for the cell and to determine the parameters. It was soon apparent that this was a naive request, but he was able to report the following: Using 1 kHz measuring frequency and 1.0 M concentration of the supporting electrolyte (KCl), $|Z| = 2.5 \times 10^3$ Ω, $\phi = -12°$; at 0.1 M, $|Z| = 1.5 \times 10^3$ Ω, $\phi = -6°$. The rationale in making these measurements lay in characterizing the device with the goal of arriving at an electrode geometry with improved time response, with the objective of performing transient electrochemical studies in which the formation and decay of a radical species are on the shortest possible time scale. This is a "kinetics" rationale. Additionally one would like to modulate the EPR signal at the highest possible rate by modulation of the electrochemical potential, with eventual phase—sensitive detection at the modulation frequency. This would be an optimal approach to *spectroscopic* observation of short—lived species. A miniaturized electrochemical cell intrinsically has faster time response, and loop gap resonators intrinsically are superior for small sample volumes. Thus it would appear that a technical basis was created by the work of Allendoerfer *et al.* (20) for examination of very short lived radicals. However, much remains to be done.

In particular, the counter or auxiliary electrode should be in the resonator, thus greatly reducing the electrolyte resistance of the circuit. In work in progress by the authors in collaboration with W. L. Hubbell, dc electric field experiments have been carried out (see Section 4.4) in LGR's in which two conducting wires were inserted into the resonator. Probably this technology can be transferred to the electrochemical experiment to reduce the resistance in a very significant degree.

Unpublished preliminary experiments have been conducted by E. Wong in the authors' laboratory applying the electrochemical LGR structure to the field of bioelectrochemistry. Using methyl viologen at 5 μM concentration as a reducing mediator, the anion flavin radical in glucose oxidase was produced. The enzyme concentration was 0.1 mM in a volume less than 1 μl. This species has been produced previously in this manner and observed using spectrophotometric methods. The electrical potential for the reaction using EPR spectroscopy as the observable was the same as when optical methods were employed.

A principal advantage of the geometry described by Allendoerfer *et al.* is that the electrode geometry is extremely flexible. No glass blowing is required, and the quantities of precious metals used for electrodes is so small that the costs are negligible. The loop gap resonator geometry is itself flexible. It is an interesting thought exercise to design electrochemical cells for the various geometries shown in Fig. 2, and to imagine what the various advantages and disadvantages might be.

4.3   ENDOR

Venters *et al.* (21) have described the use of the loop gap resonator for X–band ENDOR experiments on metalloproteins. Generally the experiments are performed at pumped helium temperatures observing the dispersion using 100 kHz field modulation.

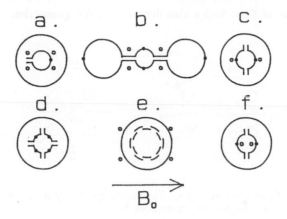

Fig. 9. Loop gap resonator geometries for ENDOR. The small circles represent cross sections of the nuclear radio frequency coil.

Their structure is a one–loop––two–gap resonator with a hairpin rf coil formed from two conducting posts in the annular region at the two zero–potential loci on the loop. See Fig. 9c. They report some difficulty with spurious microwave modes that were observed when the rf structures were introduced. They recognize that microwave currents flow on both inner and outer surfaces of the loop, and that the zero–potential points are particularly suitable for introduction of wires.

Their discussion of sensitivity considerations is similar to that presented here. They suggest that one of the "limited sample" situations where the geometry should be particularly favorable is on ENDOR of small single crystals. In addition, they particularly emphasize the freedom from phase noise when tuned to the dispersion.

Wood *et al.* (2) describe a three–loop––two–gap resonator designed for experiments at S–band. The rf structure was external to the resonator, and they report no microwave complications arising from the presence of the radio frequency coil.

In fact, much remains to be done in development of loop gap resonators for ENDOR. It would appear that there exist five classes of structures: (a) rf coils in the sample loop, Fig. 9f, (b) rf coils in the return–flux region, Fig. 9a,c, (c) rf coils external to the full LGR structure, Fig. 9b,e, (d) multigap resonators in which segments of the microwave resonator carry the rf currents, Fig. 9d, and (e) structures in which the shield carries the rf currents.

In magnetic resonance imaging, the birdcage geometry has become a widely used rf structure. (Ref. 22 and references therein). It is possible that this experience can be transferred to ENDOR, yielding rf fields of improved homogeneity. At the very least, a true saddle geometry should be used (Fig. 10a), which yields better homogeneity over the sample than is the case with 4 conducting elements angled at 90°, Fig. 10b. The rf coil patterns of Figs. 9a,b,e also illustrate saddle geometries.

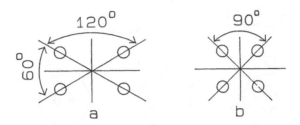

Fig. 10. ENDOR rf coil cross sections. The saddle geometry a yields improved rf field homogeneity relative to the square pattern of b.

Problems that have been identified with ENDOR resonators include the following:

a. If the rf coils are in the microwave field, microwave Q's are lowered. Also pseudo—coaxial lines can be created where the rf inlet leads pass through a resonator wall, and microwave energy can leak out.

b. If the rf coils are external, the skin depth of the microwave resonator must be carefully adjusted, and deleterious heating of the walls can occur.

The two papers that have appeared on ENDOR LGR's do not really address these problems in novel ways, but the LGR does in principle offer some interesting solutions. As mentioned, multigap resonators permit use of resonator segments as rf current elements (Fig. 9d). Also, use of a large number of gaps as in Fig. 9e renders a structure that is effectively transparent to rf field. Note that in the structure of Fig. 9e the shield can be made according to the wirewound principle (23), which has been shown to be transparent to rf (providing it does not go self—resonant).

A theme of this article is miniaturization of EPR sample—containing structures. Obviously the rf coils will also be miniaturized, permitting higher field with lower power. Spurious coupling of rf energy into the EPR receiver is often a problem in ENDOR, and this problem is relieved by miniaturization.

It also occurs to us that ENDOR coherence effects (24) could be studied in great detail because of the high microwave and rf field that can be obtained in miniature resonators. The number of equivalent nuclei contributing to a particular ENDOR transition can, in principle, be obtained by analysis of coherence effects.

Finally, this article emphasizes multifrequency EPR, and multifrequency ENDOR obviously follows.

4.4 Electric field experiments

There is increasing evidence that many biological processes are controlled at the molecular level by electric fields. For example, it has been shown that the conformations of polynucleotides, polypeptides and proteins can be affected by strong electric fields. At a high level of assembly, electric fields can affect cell or liposomal fusion. In work in progress, the authors in cooperation with W. L. Hubbell are studying these kinds of electric field effects using spin label reporter groups in the loop gap resonator geometry. The rationale for this geometry is obvious: Very high electric fields can be obtained with relatively small potentials. The breakdown field for water at pH 7 is about 10^5 V/cm, requiring about 10^4 V for characteristic dimensions of the LGR.

Development of a suitable electrode geometry for aqueous environments is a formidable problem. The ionic conductivity results in temperature rise, hydrogen gas can be readily formed, and polarization charges quickly form on the electrodes, reducing the electric field between the electrodes. It seems likely that the most general solutions will turn out to be pulse electric field approaches, where the time response of the spin label to an electric field pulse is monitored on a time scale as fast as 10^{-7} sec. An alternative time domain approach is the use of sinusoidal electric fields of variable frequency, using phase sensitive detection at the electric field modulation frequency.

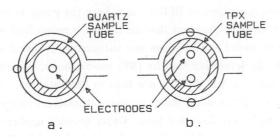

Fig. 11. Geometries for electric field experiments.

Initial experiments have been carried out using the electrode geometries of Fig. 11a,b. In 11a, one electrode is on—axis in contact with the solvent. The counter—electrode is the resonator wall, insulated from the solvent by a quartz sample tube. A peak field of 60 kV/cm was obtained with a 5 kV power supply. This is a simple and appropriate geometry for organic solvents, but in aqueous solvents the polarization charges quickly build up on the quartz wall and the electric field drops in the sample.

Of course the electric field is rather inhomogeneous, varying radially, which is another disadvantage.

The geometry of Fig. 11*b* has both electrodes in water. Here sample heating and difficulties in management of the power supply were encountered. (The current requirements were too great for continuous wave (CW) experiments, and it wasn't configured for pulse electric field experiments.)

In spite of the preliminary and tentative nature of our experience in combining electric fields, spin labels and LGR's it has been decided to report the state of the art to the reader because we are convinced that this is an important opportunity for future research that has been made possible by the loop gap resonator.

5. EXPERIMENTS DEPENDENT ON HIGH B_1 AND LOW Q: PULSE EPR, ELDOR, AND SATURATION TRANSFER.

As previously mentioned, loop gap resonators have a high resonator efficiency factor —— Λ can be as great as 10 G for 1 W of incident power. At the same time, the Q is low. Resonator #1 of Ref. 1 had a Λ value of 8.2 and an unloaded Q of 600. This combination of high Λ and low Q is useful in a number of experiments. In pulse EPR, the desired saturating field can be achieved with much lower power. Expensive microwave power amplifiers may no longer be needed. The energy per pulse is reduced, lessening technical problems of keeping the exciting pulse out of the receiver. Additionally, the lower Q decreases the ringing time. In CW electron—electron double resonance, ELDOR, the low Q permits both pump and observing fields to be incident on the sample at the same time, in the same mode, even though the frequencies are 40 MHz or more apart. A problem in ELDOR is to keep the pump microwave signal out of the receiver. A Λ value of 10 means that for the same pump field at the sample 20dB less pump power is needed and therefore one automatically has an effective increase in isolation of 20dB. In saturation transfer (ST) spectroscopy, the prime consequence of the lower Q and the high Λ is that one can tune to the dispersion out—of—phase signal and detect no excess phase noise under normal saturating conditions for spin labels.

These three topics are discussed here. Other possible applications of LGR's that make use of the combination of high B_1 and low Q include (a) CW saturation studies. With a 10 W TWT amplifier, CW microwave fields as high as 30 G in the rotating frame can be achieved. CW saturation experiments should yield information on T_1 of organo—metallic compounds that has not been previously obtainable. (b) A number of microwave amplitude and phase modulation schemes have been proposed in past years as alter—natives to field modulation. Citations are given in Poole's treatise (19). They generally have inadequate baseline stability because the modulated microwaves find some spurious path to the receiver. Because of the high Λ, the LGR immediately gives a 20dB improvement for any of the published schemes. Additionally, the low Q lessens cavity ringing problems in microwave modulation schemes. (c) Bimodal loop gap resonators have been built and tested by us on the test bench. Cross coupling from one mode to the other is automatically reduced by Λ^2, since so much less power is needed to give a desired microwave field at the sample.

5.1 Pulse EPR

The authors' experience is with saturation recovery, mostly on spin label systems at physiological temperatures. Thus commentary on spin echo EPR in this section will be limited even though we are confident that the LGR is an important structure in that technique also.

In our X−band saturation recovery apparatus, a one−loop−−one−gap resonator is used with $\Lambda = 7$ and sample volume about 2 µl. A 10W TWT is used, but the power output is permanently attenuated to 1 W at the resonator. The full field has never been required in any of our experiments to date.

This is the only LGR that has actually been used thus far for pulse EPR, and it made such a great improvement in system performance compared with a bimodal cylindrical TE_{111} resonator, that we have only recently considered optimization of the geometry. We are generally in a sample unlimited lossy (water) situation. A one−loop−−two−gap resonator that would accommodate a rectangular cross−section capillary with $\Lambda \simeq 3$ would, we judge, give improved signal, and artifacts associated with microwave pulsing in our particular apparatus can, we think, still be managed.

An important aspect of pulse EPR experiments is often the ability to deliver a pulse of uniform intensity over the sample. As the ratio of Length/Diameter (L/D) increases in the LGR, the uniformity gets better and better.

The resonator ringing time is given by

$$\tau = \frac{Q_L}{2\pi\nu} \, . \tag{18}$$

For the resonator used in our saturation recovery apparatus, $\tau = 3.3 \times 10^{-9}$ s in the absence of sample losses. Froncisz and Hyde derived an expression for the Q of an LGR:

$$\frac{1}{Q} = \frac{1}{Q_L} + \frac{1}{Q_C} \, , \tag{19}$$

where the inductive and capacitive Q's are

$$Q_L = \frac{r}{\delta} \left[1 + \frac{r^2}{R^2 - (r + W)^2} \right] \left[1 + \left(1 + \frac{W}{r} + \frac{R}{r} \right) \left(\frac{r^2}{R^2 - (r + W)^2} \right)^2 \right] \tag{20}$$

$$Q_C = 1.7 \times 10^5 t \left[\nu^{3/2} \epsilon WZ(1 + 2.5(t/W)) \right]^{-1} \qquad [21]$$

From Eqn. 20, it is apparent that if a lower Q is desired in order to achieve a shorter ringing time, the resonator must be made smaller. According to EPR sensitivity considerations discussed here, this can be an advantage or a disadvantage, depending on the circumstances. If on balance for a certain experiment it is found desirable artificially to spoil the Q, it is attractive to consider a two—loop——one—gap geometry in which a lossy element is introduced into the unused loop.

In work—in—progress a group under the direction of L. Belford, University of Illinois, has successfully performed spin—echo envelope modulation experiments at S—band using a loop gap resonator. In our own laboratory, saturation—recovery experiments at S—band have been initiated. For this latter experiment, a one—loop——one—gap resonator with length 5 mm, diameter 2.5 mm has been built. The resonator efficiency parameter $\Lambda = 7$. The rationale for S—band spin—echo is largely to exploit the magnetic field dependence of nuclear state mixing. The rationale for S—band saturation recovery is to study dipole—dipole interactions that depend on magnetic field. Multifrequency pulse EPR using loop gap resonator technology is a natural development.

5.2 ELDOR

Hyde *et al.* describe electron—electron double resonance (ELDOR) experiments at X—band using a loop gap resonator (25). The Q was 300 corresponding to 30 MHz between 3dB points. Both pump and observing microwave powers were introduced into the same resonator mode. The separation of hyperfine lines of an ^{14}N—tagged spin label in the motional narrowing limit is 44 MHz, and the characteristic separation of hyperfine lines of ^{14}N and ^{15}N isotopes is 26 MHz. With separations of these magnitudes, the resonator will support both frequencies, although obviously with some compromise. The paper discusses the tradeoffs to be considered between having the pump on—resonance and the observe microwave field off—resonance, or *vice versa*, or spacing the frequencies symmetrically about the resonance frequency of the LGR.

The spin system transfers saturation from a pumped transition to an observed transition. It then becomes necessary to remove the pump microwave power from the signals reflected from the resonator prior to detection of the observe signal. This is accomplished with a microwave trap—filter (25).

Quite remarkably, the signal—to—noise ratio was improved by a factor of 20, expressed on a molarity basis, relative to that achievable with the bimodal cavity geometry that had previously been used. It became possible for the first time to carry out ELDOR experiments on cells.

The key points from the perspective of resonator design are these:

a. Because of the low Q, two fairly well separated frequencies can simultaneously be incident on the sample.

b. Because of the high ηQ product, the amount of pump power required to achieve a given degree of saturation is $\Lambda^2(1)/\Lambda^2(2)$ greater, comparing an LGR (resonator 1) with a cavity (resonator 2). This ratio is between 10 and 100, and automatically the isolation problems −− ie., keeping the pump power out of the observing microwave bridge −− are lessened by this amount.

These were field−swept ELDOR experiments. The two frequencies were set a hyperfine separation apart and the magnetic field was swept. Frequency sweep has not yet been accomplished in LGR geometry. Numerous technical solutions can be proposed. One that is particularly interesting is the bimodal LGR, Fig. 12, which has been built and evaluated on a microwave test bench, but is not yet part of a finished resonator assembly.

Fig. 12. Bimodal loop gap resonator. The sample enters the structure from the right.

Bimodal cavities have been developed by numerous investigators, but have proven difficult to use. It seems likely that bimodal LGR's, because of their low Q and intrinsic simplicity, will prove to be more reliable, convenient, and less microphonic.

The reader is referred to the section of this article on pulse EPR and also to the section on multifrequency EPR. Pulse ELDOR and ELDOR at L−, S− and other bands are obvious technical opportunities that have been made possible by the LGR.

In accordance with a general theme of this article, one can also inquire about the opportunities created in miniaturization using the LGR for ELDOR. The small sample size can make it possible to perfuse cellular systems with carefully adjusted amounts of oxygen −− sufficient to keep respiration above K_M (the Michaelis−Menten constant) but not so high as to relax the spin−labels. Approximately, in our experience, a uniform oxygen concentration of about 2×10^{-6} M would meet these twin objectives.

In addition, small sample size can facilitate heat transfer. Sample heating can be a problem when high pump powers are used in aqueous solvents.

There exist possible ELDOR experiments where the two frequencies are very far apart, and neither the single mode nor the bimodal geometries are appropriate.

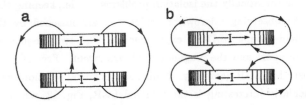

Fig. 13.   The two modes of an axial pair of loop gap resonators.

Multimode LGR's can, in principle, be used (Fig. 13). In the fundamental mode of this co−axial two loop resonator, currents in the two loops are parallel; in the high frequency node they are antiparallel. The separation of the two modes depends on the mutual inductance. (These modes would be somewhat analogous to the cylindrical TE_{011} and TE_{012} modes of a cavity resonator.) An axial sample would experience both fields if sources at the proper microwave frequencies were coupled to the structure. The number of design parameters, L, C for each loop plus the mutual inductance, is large and gives considerable flexibility in tailoring the geometry to the experiment.

The senior author has been working for many years towards the goal of making ELDOR a truly practical technique. With the LGR, it is felt that the goal has been reached.

5.3 Saturation Transfer Spectroscopy

The technique of saturation transfer spectroscopy was introduced by Hyde and Dalton (26) to measure the rotational correlation time of spin−labeled biomolecules in the range of 10^{-7} to 10^{-3} s. This paper used the first harmonic out−of−phase dispersion signal (100 kHz field modulation frequency) as the spectroscopic display. Excess phase noise when tuned to the dispersion led to the introduction of the second harmonic out−of−phase technique by Hyde and Thomas (27). This has become by far the most widely used display.

The introduction of the loop gap resonator permits a reconsideration of the original display, and this has been done in unpublished work at the ESR Center. The magnitude of the signal is about one order of magnitude greater in the first harmonic dispersion display compared with the second harmonic absorption display. Since phase noise is virtually eliminated, the sensitivity is significantly increased when expressed on a molarity basis: a factor of 2 to 5 depending on whether the comparison is with a capillary or a flat–cell in the rectangular TE_{102} multipurpose cavity. On a number of spins basis (i.e. sample–limited), the sensitivity is increased by a factor of about 100. Thus saturation transfer spectroscopy becomes technically feasible in a greatly increased range of experiments.

Utilizing the approach of Hyde and Gajdzinski (11), simultaneously detecting I and Q, it is possible at one and the same time to obtain both first harmonic dispersion and second harmonic absorption out–of–phase displays. The latter has attractive aspects because of the greater number of inflections in the display. Having both displays at hand should be, we suggest, a significant advance. But this aspect has not yet been investigated.

Johnson and Hyde (28) extended the saturation transfer method to 35 GHz, and Froncisz et al. (29) describe a 35 GHz loop gap resonator. The active sample volume is about 100 nl. Thus technical feasibility has been established for saturation transfer experiments on a remarkably small amount of sample. Again, detailed investigations have not yet been carried out.

There are some modest technical advantages in carrying out ST experiments in LGR's. (a) Because of the small size of the LGR, the modulation field amplitude is more homogeneous as is also the the phase of the modulation field. (b) Because of the small size of the sample tube, removal of oxygen, which is essential for ST is facilitated using the gas–exchange method of Popp and Hyde (30). (c) With an aspect ratio of 4:1, height to diameter, which is typical for X–band experiments, the microwave field is significantly more uniform over the sample than is the case with a cavity resonator.

Overall it seems clear that the LGR is a markedly superior structure for saturation transfer spectroscopy compared with cavity resonators.

6. CONCLUDING REMARKS

There are a number of microwave loop gap resonator structures that have been built but that have not yet been introduced into magnetic resonance spectroscopy. Some of these are described in Refs. 31–34. We have also been interested in the development of various microwave devices based on loop gap resonator concepts. See Mehdizadeh et al. (3). In still another line of research, loop gap resonators have been used for in vivo magnetic resonance imaging (MRI) and ^{31}P spectroscopy (35). An industrial application, web thickness measurement (36), has been proposed.

There appears to be a number of additional applications of loop gap resonators to EPR spectroscopy that have not yet been developed. Geometries for photochemical excitation have been made, but much remains to be done. One major feature, the technical capability to introduce very high microwave fields onto a sample, has not been exploited.

Progress continues at a rapid rate in loop gap resonator design. Pfenninger et al. present detailed calculations of radio–frequency field distributions (37). Froncisz et al.

304

describe a resonator suitable for a mouse at 1 GHz (38). And Hyde *et al.* describe a three—loop——two—gap resonator (39) that is, in fact, very similar to the familiar TE_{102} rectangular cavity resonator.

We would like for the reader to accept the notion that our work of the past five years has been not so much the development and application of a series of microwave resonators, but rather the development of a general approach to the design of the optimum resonator for the particular experiment.

Acknowledgement: Preparation of this article was made possible by Grants GM22923, GM27665, and RR01008 from the National Institutes of Health.

REFERENCES

1. W. Froncisz and J. S. Hyde, J. Magn. Reson. **47**, 515—521 (1982).
2. R. L. Wood, W. Froncisz, and J. S. Hyde, J. Magn. Reson. **58**, 243—253 (1984).
3. M. Mehdizadeh, T. K. Ishii, J. S. Hyde, and W. Froncisz, IEEE Trans. Microwave Theory Tech. **MTT—31**, 1059—1064 (1983).
4. J. S. Hyde, Varian Associates Technical Information Bulletin, Fall 1965, p. 10.
5. T. H. Wilmshurst, Electron Spin Resonance Spectrometers, Adam Hilger, London (1967).
6. G. Feher, Bell Syst. Tech. J. **36**, 449—484 (1956).
7. J. S. Hyde, Rev. Sci. Instr. **43**, 629—631 (1972).
8. W. Froncisz, C. P. Scholes, J. S. Hyde, Y.—H. Wei, T. E. King, R. W. Shaw, and H. Beinert, J. Biol. Chem. **254**, 7482—7484 (1979).
9. H. J. Den Blanken, R. F. Meiburg, and A. H. Hoff, Chem. Phys. Letts. **105**, 336—342 (1984).
10. J. S. Hyde and W. Froncisz, in Annual Review of Biophysics and Bioengineering, Vol. 11, Annual Reviews, Inc., Palo Alto, California (1982), pp. 391—417.
11. J. S. Hyde and J. Gajdzinski, Rev. Sci. Instr. **59**, 1352—1356 (1988).
12. R. R. Ernst, J. Magn. Reson. **1**, 7—26 (1969).
13. W. L. Hubbell, W. Froncisz, and J. S. Hyde, Rev. Sci. Instrum. **58**, 1879—1886 (1987).
14. I. Yamazaki, H. S. Mason, and L. Piette, J. Biol. Chem. **235**, 2444—2449 (1960).
15. A. H. Maki and D. H. Geske, J. Chem. Phys. **30**, 1356—1357 (1959).
16. D. H. Geske and A. H. Maki, J. Am. Chem. Soc. **82**, 2671—2676 (1960).
17. L. H. Piette, P. Ludwig, and R. N. Adams, Anal. Chem. **34**, 916—921 (1962).
18. C. P. Poole, Jr., Electron Spin Resonance, J. Wiley, New York (1967).
19. C. P. Poole, Jr., Electron Spin Resonance, Second Edition, J. Wiley, New York (1983).
20. R. L. Allendoerfer, W. Froncisz, C. C. Felix, and J. S. Hyde, J. Magn. Reson. **76**, 100—105 (1988).
21. R. A. Venters, J. R. Anderson, J. F. Cline, and B. M. Hoffman, J. Magn. Reson. **58**, 507—510 (1984).
22. A. Sotgiu and J. S. Hyde, Magn. Reson. Med. **3**, 55—62 (1986).
23. J. S. Hyde, J. Chem. Phys. **43**, 1806—1818 (1965).
24. J. H. Freed, D. S. Leniart and J. S. Hyde, J. Chem. Phys. **47**, 2762—2773 (1967).
25. J. S. Hyde, J.—J. Yin, W. Froncisz, and J. B. Feix, J. Magn. Reson. **63**, 142—150 (1985).
26. J. S. Hyde and L. Dalton, Chem. Phys. Letts. **16**, 568—572 (1972).
27. J. S. Hyde and D. D. Thomas, Ann. N. Y. Acad. Sci. **222**, 680—692 (1973).
28. M. E. Johnson and J. S. Hyde, Biochemistry **20**, 2875—2880 (1981).
29. W. Froncisz, T. Oles, and J. S. Hyde, Rev. Sci. Instrum. **57**, 1095—1099 (1986).
30. C. A. Popp and J. S. Hyde, J. Magn. Reson. **43**, 249—258 (1981).
31. W. Froncisz and J. S. Hyde, U. S. Patent 4,435,680, issued March 6, 1984.
32. W. Froncisz and J. S. Hyde, U. S. Patent 4,446,429, issued May 1, 1984.

33. J. S. Hyde and W. Froncisz, U. S. Patent 4,480,239, issued October 30, 1984.
34. W. Froncisz and J. S. Hyde, U. S. Patent 4,504,788, issued March 12, 1985.
35. J. S. Hyde, in Medical Magnetic Resonance Imaging and Spectroscopy, T.F. Budinger and A. Margulis, eds., Society of Magnetic Resonance in Medicine, Berkeley, CA (1986) pp. 111−120.
36. M. Mehdizadeh, W. Froncisz, and J. S. Hyde, U. S. Patent 4,623,835, issued November 18, 1986.
37. S. Pfenninger, J. Forrer, and A. Schweiger, Rev. Sci. Instrum. 59, 752−760 (1988).
38. W. Froncisz, T. Oles, and J. S. Hyde, J. Magn. Reson. 82, 109−114 (1989).
39. J. S. Hyde, W. Froncisz, and T. Oles, J. Magn. Reson. 82, 223−230 (1989).

CHAPTER 8

ELECTRON PARAMAGNETIC RESONANCE AT 1 MILLIMETER WAVELENGTHS

DAVID E. BUDIL, KEITH A. EARLE, W. BRYAN LYNCH, AND JACK H. FREED

1.   INTRODUCTION

For over two decades, new technical developments in EPR spectroscopy have lagged behind analogous advances in the field of NMR. Techniques such as Fourier-transform and two-dimensional correlation spectroscopy, which have long been standard in NMR, have only relatively recently found application in EPR (1). One of the most important advances in NMR has been the extension to high frequencies, requiring the use of superconducting magnets. This has provided large enhancements in sensitivity and spectral resolution, greatly broadening the utility of NMR techniques, especially for biological applications (2).

Recent years have witnessed a similar trend in EPR spectroscopy. The effort towards higher EPR fields and frequencies has in large part been fostered by an emerging technology in the synthesis and processing of signals in the far-infrared (FIR) and millimeter wave regions (3) for applications in radar, communication, and radioastronomy. Notable in the field of EPR has been the development and application of a spectrometer working at $\lambda = 2$ mm (i.e. 148 GHz) and 5.3 Tesla fields by Lebedev and co-workers (4). They have reported EPR spectra of nitroxide spin labels in liquid and solid phases, emphasizing the significant advantages of high resolution for determining g-tensors and studying molecular dynamics in the slow-motional regime (5). The sensitivity at 148 GHz is also reported to be significantly better than that of conventional EPR spectrometers. More recently Möbius has constructed an EPR spectrometer operating at $\lambda = 3.2$ mm (94 GHz) in a 3.4 T field with similar advantages (6).

Theoretically, even greater enhancements in sensitivity should be realized at still higher frequencies. However, there are two major technical obstacles to extending the frequency range for practical EPR work. First, high resolution EPR at high frequencies would require magnets that can provide very high fields with a homogeneity as good as $1:10^6$. Although EPR has been performed in pulsed magnetic fields of up to 35 T (7), persistent fields with sufficient homogeneity for high resolution EPR work are currently only available up to about 15 T (8). Second, it becomes increasingly difficult to utilize standard microwave technology at higher frequencies, even though this has proved possible at $\lambda = 2$ to 3 mm. Waveguides, cavities, and other components are exceedingly small and lossy, and are often quite difficult to fabricate for shorter wavelengths.

A comfortable compromise is presently accessible at fields of 8-12 T, corresponding to wavelengths of 1.0 - 1.5 mm (200-300 GHz), where superconducting magnets with good homogeneity can be manufactured with relative ease. Even at these wavelengths, however, conventional microwave technologies are no longer adequate. In this chapter we describe a new EPR spectrometer operating at $\lambda = 1.2$ mm (250 GHz) in a field of 9 T that takes advantage of far-infrared technology based upon the principles of Gaussian optics (9). Such quasi-optical techniques greatly simplify the design and performance of the high-frequency EPR spectrometer, and they appear to be the natural way to proceed with submillimeter EPR spectroscopy in the future.

After describing our new design, we present a selection of experiments which illustrate the potential utility of FIR-EPR, particularly with respect to studies of motional dynamics and biologically relevant spin labels. We then discuss currently available components and technology that will make possible future improvements and refinements of EPR spectroscopy at millimeter and sub-millimeter wavelengths.

2. THE 1 MILLIMETER SPECTROMETER

2.1 Overview

Our 1.20 mm EPR spectrometer operates at 249.9 GHz and 8.9 T for $g = 2$. The basic design is a simple transmission cavity configuration, shown schematically in Fig. 1. The cavity is comprised of two mirrors, each of which is attached to a tube that serves as a beam guide for the 1 mm waves. The 1 mm beam travels from the source, located underneath the magnet, through a series of lenses to the cavity, and then through a second series of lenses to a detector which is mounted on top of the magnet Dewar. Since the source delivers only 5 mW of power, optimal alignment of the cavity mirrors and the lenses between source, cavity, and detector is crucial to a successful experiment.

The spectrometer is quite easy to operate; in fact, its operation is in a number of ways very similar to a conventional 9 GHz EPR spectrometer once the main magnet coil is driven up to a field value near resonance. The resonance is found by sweeping the sweep coil. The maximum sweep rate of 3.5 mT/second is quite adequate when compared to a 9 GHz EPR spectrometer, and permits easy initial location of the resonance signal.

From the experimental high-field spectra we have obtained, we estimate a minimum number of observable spins as low as 3×10^{11} spins for a 0.1 mT linewidth with no source attenuation (3.5 mW of incident power), 80 kHz field modulation, and a lock-in amplifier bandwidth of 0.1 Hz. For our resonator and sample holder, this would correspond to a minimum detectable molarity, M_{min} of 8×10^{-9} M in a low-loss solvent. Such a sensitivity is quite comparable to that of a 9 GHz EPR spectrometer for narrow-line spectra.

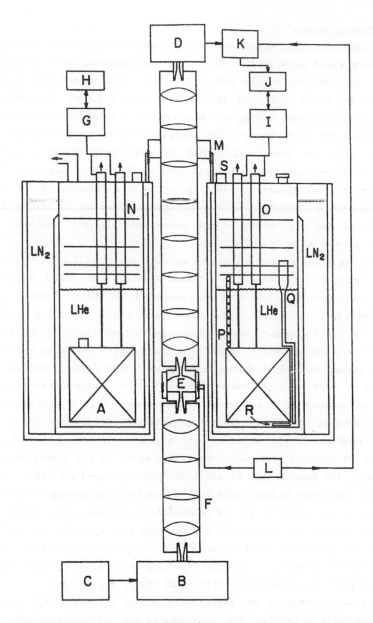

Fig. 1. Block diagram of 249.9 GHz EPR spectrometer. A: 9 T Superconducting solenoid and 57.0 mT sweep coils. B: Phase-locked 250 GHz source. C: 100 MHz reference oscillator. D: Schottky diode detector. E: Semiconfocal Fabry-Perot cavity and field modulation coils. F: 250 GHz quasi-optical beam guide. G: Main coil power supply (100 A). H: Main coil current programmer. I: Sweep coil power supply (50 A). J: Personal computer. K: Lock-in amplifier. L: Field modulation and lock-in reference oscillator. M: Cavity tuning screw. N: Vapor-cooled leads for main solenoid (non-retractable). O. Vapor-cooled leads for sweep coil (non-retractable). P: LHe level indicator. Q: LHe transfer tube. R. Bath temperature and heater resistors. S: LHe blow-off valves. [From Ref. 9].

2.2 Magnet system

The superconducting magnet was built by American Magnetics, Inc. (AMI) and has a maximum field value of 9.2 T at 4.2 K, corresponding to a current of about 72 amps. In addition to the main 9.2 T coil, the magnet has an auxiliary superconducting field-sweep coil, which can be swept ±54.0 mT (38 amps) while the main coil is persistent, making the spectrometer similar in field operation to a conventional EPR spectrometer. The sweep coil is designed to minimize mutual inductance with the main coil, which permits rapid field sweeps without significant interference with the main coil. The magnet measures 33.5 cm in length with an outer diameter of 19.1 cm, and a bore diameter of 6.1 cm. Field homogeneity within a central 1 cm diameter spherical volume is better than 3×10^{-6} without shims for all values of the sweep coil current, and the field is persistent at 9 T to at least 10^{-7}/hour.

The leads supplying current to both main and sweep coils are non-retractable. The main magnet coil is charged by a Hewlett-Packard (HP) 6260B DC power supply which is controlled by an AMI 402A programmer. An HP 6032A DC power supply with a GPIB interface to a PC drives the sweep coil, providing accurate and reproducible field sweeps. The programmable voltage limit of the sweep power supply prevents the buildup of inductive voltage across the sweep coil that would otherwise quench the magnet during fast sweeps. This feature permits a maximum sweep rate of about 3.5 mT/second in either direction, which is useful for initial location of signals or pre-setting the spectrometer to a spectral region of interest. However, the inductive load compensation circuitry of the power supply produces considerable hysteresis at such high sweep rates; to avoid systematic deviations in spectral line positions with respect to the direction or rate of field sweep, the magnet must be swept at 0.03 mT/second or less. Under such conditions, the sweep is reproducible to within the programming resolution of the power supply, or about 19 μT. Reproducibility between runs (that is, after cycling the sweep power supply over a wide range or turning it off) is limited by the programming accuracy of the power supply, and is usually better than 0.1 mT, or 1 part in 10^5 of the applied field. Accurate calibration of the sweep coil was made using the known hyperfine splitting of PDT in decane (10).

The Dewar containing the magnet is a liquid nitrogen (LN$_2$) jacketed warm bore design built by Cryofab, Inc. according to our own specifications. The bore has inner and outer diameters of 4.4 cm and 5.7 cm respectively; the liquid helium (LHe) chamber is 21.6 cm in diameter and 122 cm long. An AMI LHe level detector monitors the liquid level above the magnet. A series of copper baffles reduces the LHe evaporation rate of the empty dewar to about 0.4 ℓ/hr. Boil off caused by the main and sweep coil current leads is minimized by using a design similar to that of Efferson (11). Each lead consists of several brass ribbons (each ribbon is 0.1 mm thick, 6 mm wide and 46 cm long) contained within a fiberglass tube that maintains a flow of cold escaping He vapor over them. The main magnet leads contain 40 ribbons each and are capable of carrying 100 amps; the sweep coil leads can carry 50 amps with 24 ribbons each.

With the magnet and leads in place, the maximum LHe boil off rate is less than 0.75 ℓ/hr. Since our dewar holds 12 ℓ of liquid above the magnet, the time available for experiments between LHe transfers is about 15 hours, which is more than enough time to perform most experiments. All helium gas is collected for reliquefaction. A LHe transfer line extension inside the dewar allows us to transfer liquid efficiently to the bottom of the dewar, where a carbon resistor monitors the temperature during the initial cool down.

2.3 Millimeter-wave source

Our 1 mm wave source is a Millitech Corp. PLS-3F phase-locked solid state source which delivers 5 mW at 249.9 GHz (WR-4 waveguide). The self-locking feature of the source makes it very easy to operate and provides a much less noisy output than other sources of radiation in the 1 mm wavelength region. It is much smaller than an FIR laser system (which can produce 1.222 mm waves), and it does not have the problems of stability and maintenance inherent to a laser (12).

Fig. 2 shows a schematic diagram of the control loop. The heart of the source is a InP Gunn oscillator (Millitech GDM-10T) that is phase locked to a highly stable oven-controlled 100 MHz crystal oscillator (Vectron CO224A59) reference. The Gunn output (40 mW) is locked to 83.3 GHz and then converted to 249.9 GHz using a tripler based on a GaAs diode (13) (Millitech MU3-04T). We estimate (14) that the phase noise is -80 dBc/Hz with respect to 83.3 GHz, and -70 dBc/Hz with respect to 249.9 GHz at an offset frequency of 10 kHz. (At a 100 kHz offset, these figures are approximately -100 dBc/Hz and -90 dBc/Hz at 83.3 GHz and 249.9 GHz, respectively.) The output frequency can be swept within a ± 65 MHz range by sweeping the reference ± 25 kHz. Although the output power is not variable, the power incident on the cavity may be attenuated as needed by inserting calibrated FIR attenuators (made from e.g. carbon black) into the FIR beam.

2.4 Transmission of 1 millimeter waves

Low-loss quasi-optical techniques (15) are used to propagate the millimeter wave beam from the source into the warm bore of the Dewar to the cavity and then on to the detector. Feedhorns "launch" the linearly-polarized beam from a waveguide mode to a free space TEM_{00} mode that can be propagated through a series of lenses. An adequate approximate solution to the wave equation for such a free space mode has a gaussian distribution of the E and H field amplitudes transverse to the direction of propagation.

In our system (cf. Fig. 1) a Millitech scalar feedhorn converts the waveguide mode into a gaussian beam as it exits the source. The base of the feedhorn

312

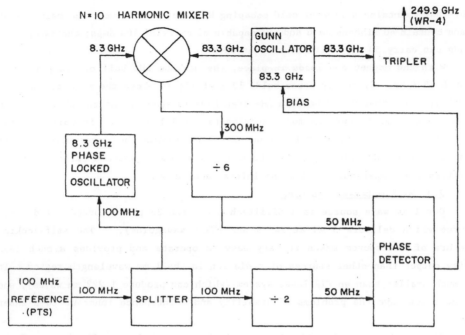

Fig. 2. Block diagram of 249.9 GHz source. [From Ref. 9].

contains a rectangular to circular waveguide transition, so that it may be directly coupled to the source waveguide. The diverging gaussian beam passes through a focusing lens (diameter 2.3 cm, focal length 2.75 cm) and is focussed to a beam waist of radius 1.6 mm. Subsequently a series of longer focal length primary lenses (diameter 3.51 cm, focal length 11.4 cm, beam waist radius 6.6 mm) are used to propagate the beam over longer distances. Both focusing and primary lenses are made of teflon and have anti-reflection surface grooves designed to match the refractive indices of teflon and air at 249.9 GHz (16).

Radiation is coupled into and out of the cavity by using two conical feedhorn/focusing lens pairs. The aperture diameter D of each feedhorn was adjusted to couple to the beam waist radius of the focusing lens, w_0, according to the relationship $D = 3w_0$, which is valid for low-aperture phase error scalar horns (15). Another conical feedhorn (from Custom Microwave)/focusing lens pair couples radiation from the TEM_{00} mode back to a short length of WR-4 rectangular waveguide in front of the detector. Because the FIR radiation maintains its linear polarization during its traverse of the magnet bore, the rectangular waveguide at the detector must be properly aligned relative to the source waveguide for optimal coupling into the detector. Alignment of source and detector waveguides also ensures that the observed EPR signal is in phase (i.e., a pure absorption signal); misalignment of the detector results in a small admixture of dispersion in the EPR spectrum.

This quasi-optical approach provides much more efficient transmission and coupling of 1.2 mm radiation than would conventional waveguide techniques. The loss due to the lenses over a 1.4 m path length is only 2 dB, whereas the theoretical loss of WR-4 waveguide over the same distance is 16 dB. In addition, the gaussian beam is easily coupled into useful waveguide or cavity modes. The conical antennae used in our system are readily fabricated, and do not produce severe losses with gaussian beams. The scalar feedhorn reduces coupling losses even further; its radiation pattern has much smaller side lobes than that of the conical antennae, so that it more closely matches the shape of the fundamental mode of the gaussian beam.

2.5 EPR cavity

The cavity (Fig. 3) is a semi-confocal Fabry-Perot resonator having 2.5 cm diameter mirrors. The radius of curvature of the spherical mirror is also 2.5 cm; thus, the optimum inter-mirror distance for propagating the fundamental mode is one-half the radius of curvature, or 1.25 cm. This corresponds to a mode number, $\nu = 20$ (i.e. the number of half wavelengths between the mirrors). The cavity is coupled to two locally made feedhorns by 1.5 mm diameter coupling holes located at the center of each mirror. To tune the cavity, transmitted power is maximized by raising and lowering the spherical mirror with respect to the flat mirror. The spherical mirror and upper optical guide are translated (and rotated) by a locally made millimeter screw located at the top of the dewar with a vertical travel of 0.635 mm/turn. The spherical mirror slides within a teflon sleeve that is affixed to the lower flat mirror. To measure transmitted 1.20 mm power, we use phase sensitive detection referenced to a homemade (nonmagnetic) beam chopper that operates at 100 Hz. Fig. 4 shows the series of resonances in transmitted power that are observed as a function of mirror translation.

Some performance characteristics of the cavity are apparent from Fig. 4. In addition to the intense, regularly-spaced resonance peaks of the $TEM_{00\nu}$ cavity mode, other less intense peaks corresponding to different cavity modes are observed, often as shoulders of the main peaks. As the lower trace of Fig. 4 shows, the introduction of even low-loss hydrocarbon solvents into the cavity increases such mode conversion and thus cavity losses.

The loaded Q of the cavity, Q_L, is easily determined from Fig. 4 for a given tuning peak. The cavity finesse is given by $f = \lambda/\Delta\lambda$, where λ is the length between successive resonance peaks and $\Delta\lambda$ is the full width at half maximum of a single resonance. Then $Q_L = f\nu$, the product of the finesse and mode number, or about 100. Such a Q-factor is quite low compared with the values of 10^4-10^5 typically obtained in Fabry-Perot type resonators at millimeter wavelengths (17). The major factor limiting Q is the relatively large diameter of the coupling holes

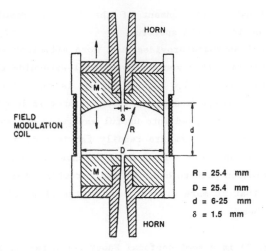

Fig. 3. Fabry-Perot cavity for 250 GHz EPR. M indicates mirror assembly. [From Ref. 9].

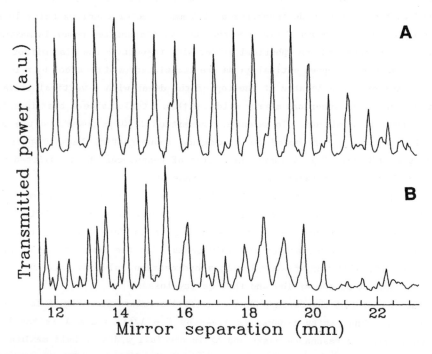

Fig. 4. Transmitted power as a function of mirror separation for the Fabry-Perot cavity under different conditions. (a) Empty teflon sample holder; (b) sample holder plus 50 $\mu\ell$ of toluene (path length 1.0 mm).

compared to the radiation wavelength, leading to a substantially overcoupled cavity. However, a Q of 100 is quite satisfactory for the samples we have examined so far. One advantage of the low Q is that the relatively large width of the cavity resonance peak allows a comfortable tolerance of mechanical misalignment and drift in the cavity dimensions. Thus, cavity tuning is very simple; a resonance is easy to find and maintain at nonambient temperatures.

The flat mirror of the cavity is positioned so that samples placed on it are located within the region of maximum field homogeneity. The field modulation coil is a single solenoid wound around the teflon sleeve (3.2 cm in diameter, 1.25 cm in length, 220 turns of no. 30 magnet wire) and embedded in GE varnish #1202 to reduce mechanical noise at high modulation amplitudes in the 9 T field. The coil is driven by a Wavetek model 197 function generator amplified by a McIntosh audio amplifier, and may be modulated at frequencies from 1 to 100 kHz. The maximum peak-to-peak amplitude currently attainable is only 0.24 mT at the highest frequency; for very broad spectra this limitation somewhat offsets the noise reduction obtained at higher modulation frequencies. At lower frequencies (e.g. 17 kHz) the maximum amplitude is 1.2 mT, which is usually satisfactory for the broad spectra.

2.6 Sample Cells and Temperature Control

The requirement that sample cells be transparent at 1.20 mm places rather strict limitations on the types of materials that can be used in their fabrication. In our initial studies, we have found that polymethylpentene (TPX), teflon, mylar, and Z-cut crystalline quartz (18) have sufficiently low absorbance to be used in sample cells at 250 GHz. However, all these materials have relatively high indices of refraction (e.g., $n \approx 1.4$ for teflon, and $n \approx 2.0$ for crystalline quartz) so that reflection from the sample cell surface is a major concern. It is impractical to machine anti-reflection grooves in the windows of each sample holder; however, reflection can be minimized using techniques adapted from thin-film optics (19).

For millimeter wavelengths, the sample cell wall can be regarded as a "thin film" at the interface of the sample medium with the atmosphere. It can be shown using the Fresnel formulae (19) that the contribution of the film to the reflectance of the interface vanishes when the optical thickness of the film is $m\lambda/2$, where λ is the wavelength of the incident radiation. The only residual effects of the film are a phase shift of $m\pi$ and a slight attenuation by optical absorption. Thus, by adjusting the optical thickness to a convenient integral multiple of $\lambda/2$, many sample cell materials can be made nearly transparent at 1.20 mm.

Fig. 5 shows a sample holder that has proved very successful for hydrocarbons and other low-loss solvents such as thermotropic liquid crystals. The holder is designed to maximize the amount of sample within the cavity at the optimal mirror

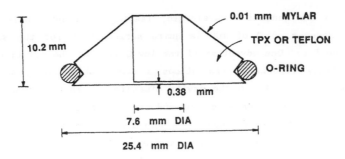

Fig. 5. Sample holder for low-loss liquids in the 249.9 GHz semiconfocal Fabry-Perot cavity. [From Ref. 9].

spacing. The diameter of the sample chamber must be greater than twice the beam diameter at the top of the cell, or at least 5.7 mm, to avoid aperture-limiting the gaussian beam in the cavity. Typical sample volumes are 50-250 $\mu\ell$, which result in a fluid height of several wavelengths.

Samples that require initial degassing to prevent oxygen broadening are degassed on a vacuum line, placed in the sample holder under a nitrogen atmosphere, and then sealed with a 0.01 mm thick mylar film stretched over its top and held in place by an O-ring. During an experiment, the warm bore is swept continuously with nitrogen gas to prevent the dissolution of oxygen into the sample. Nitrogen gas flow also avoids the buildup of frost that occurs when moist room air is trapped in the warm bore by the large diameter of the transmission lens guide tubing. We have found teflon sample cells much more suitable than TPX for working with hydrocarbon solvents and deoxygenated samples.

Initial experiments have also demonstrated Z-cut quartz to a be very useful sample cell material. We estimate its absorbance to be < 0.10 cm^{-1} and the index of refraction to be 2.1 at 1.20 mm (20). The introduction of two 0.28 mm Z-cut quartz windows into the Fabry-Perot cavity has little or no effect on the level or polarization of transmitted power at 250 GHz. The windows are strong enough to permit fabrication of vacuum-sealable sample cells. For our own more specialized purposes, the Z-cut quartz is strong enough be used to prepare well-aligned lipid model membrane samples by a pressure annealing technique (21).

Another material which shows considerable promise for making sample cell windows at 250 GHz is CaF_2, which has very low absorbance at 1.20 mm and a refractive index of about 2.6 (22). Although CaF_2 is very slightly water-soluble, its high strength makes it particularly useful for high pressure applications; furthermore, paramagnetic impurities such as Mn^{2+} may be doped into the cubic CaF_2 lattice sites to serve as field reference markers (23).

Temperature control in the range -170° to +100°C is achieved using a combination of dry nitrogen gas flow through the teflon cavity sleeve and heater resistors located immediately under the bottom mirror. The ambient temperature in the bore is about -20°C; for operation at higher temperatures, the heater resistors are sufficient to provide stable temperature control to within ±1°C, while a slight flow of room temperature nitrogen gas is maintained in the teflon cavity sleeve. For colder temperatures, the nitrogen gas is first passed through a heat exchange coil immersed in a liquid nitrogen bath. Coarse temperature control is obtained by regulating the gas flow, while the heater resistors serve for fine adjustment, again to within ±1°C. Because the tune of the Fabry-Perot cavity is very sensitive to dimensional changes, thorough temperature equilibration of the entire cavity is required before each experiment.

2.7 Millimeter-wave detector

The detector (Millitech DXW-4F) is a Schottky diode coupled to a low noise video amplifier with an upper bandwidth corner frequency of about 80 kHz. The detector is the noise-limiting component of the system, producing 1/f noise with a rms voltage amplitude $V_n \sim 50$ nV/$\sqrt{Hz}$ at 1 kHz. The responsivity, R_0 of the system is 25 V/mW at 1 kHz and an input power of -20 dBm. At input powers near 0 dBm, at which the detector is operated, the responsivity is at least 12 V/mW over the amplifier bandwidth, giving a noise equivalent power (NEP) of $V_n/R_0 < 4.0 \times 10^{-9}$ W/$\sqrt{Hz}$. To reduce the 1/f noise of the system, we currently modulate the field at a selection of frequencies from 17 kHz to 80 kHz. The response of the detector is amplified by a Princeton Applied Research HR-8 lock-in amplifier and then is digitized by a Data Translation DT-2801 board in a PC.

3. APPLICATIONS OF 1 MILLIMETER EPR

3.1 Overview

By analogy with NMR, the extension of EPR spectroscopy to fields requiring superconducting magnets and high frequencies has several advantages. Specifically, these are: 1) increased spectral resolution, permitting very accurate determination of the g tensor components, 2) increased sensitivity to molecular motion on the fast time scales of interest for molecular dynamics studies, 3) increased absolute sensitivity in terms of the number of detectable electron spins per unit field.

The ability of the 250 GHz EPR spectrometer to resolve closely-spaced g-values recommends it particularly to biological applications. Most biologically relevant free radicals such as electron transport cofactors or nitroxide spin labels have g-factors very close to that of the free electron, with very small anisotropies. Moreover, many such radicals have unresolved nuclear hyperfine interactions that produce an EPR linewidth comparable to or larger than the g-anisotropy. Thus, unlike NMR, where the chemical environment of a spin is usually evident simply from its resonance frequency, conventional EPR of biological radicals often gives little information about their chemical structure or identity. In contrast, 250 GHz EPR can separate radicals with close g-values into different spectral regions, providing a better means of identifying a radical or observing changes in its chemical environment.

Increased spectral resolution also improves sensitivity to molecular orientation. At 250 GHz, g-anisotropy will typically be much more important than hyperfine interactions in determining the spectral width. Thus, different orientations of a given radical can be expected to be better resolved into different regions of the spectrum. This feature presents a powerful means of determining the details of molecular structure in samples where the radicals are macroscopically ordered, such as stretched polymer films, well-aligned lipid bilayers or protein crystals. A much more detailed picture of the orienting potentials and ordering tensors is thus available from high-frequency EPR spectroscopy.

An additional advantage of higher frequency is that the range of rotational correlation times that produce slow-motional EPR lineshapes is much higher than that of conventional EPR. Slow-motional spectra from macromolecules are particularly sensitive to details of their motional dynamics. While slow-motional spectra at 9 GHz are observed for rotations with characteristic times longer than a nanosecond, the onset of the slow-motional regime occurs for rotations over an order of magnitude faster at 250 GHz. This permits the 1 mm EPR to reveal details of molecular motion on the time scales typical of biomolecules at physiological temperatures. The sensitivity of the 250 GHz EPR to higher rotational frequencies will be especially useful for identifying ordered and disordered species in biological membranes: motion which appears as rapid averaging at 9 GHz may not be fast enough to lead to spectral averaging at 250 GHz.

An application of 1 millimeter EPR of particular importance for biochemical problems is the study of paramagnetic metal ions, especially those in metallo-proteins. Conventional EPR spectroscopy of such systems is quite limited by the large zero-field splittings (ZFS) that are generally found in the metal ion. Because the ZFS are usually much larger than the Zeeman interaction at 0.3 T spectrometer fields, the only detectable resonances are the $m_s = +\frac{1}{2} \leftrightarrow -\frac{1}{2}$ transitions that are observable because of time-reversal symmetry in ions with an odd number of unpaired electrons. In contrast, the Zeeman interaction at 9 T is comparable to, and in many cases larger than the ZFS of the metal ion. Thus, more EPR transitions should be detectable at 250 GHz, providing much more detailed information about the electronic ground and excited states of the metal ion (24).

3.2 Solid state samples: glasses and polycrystalline powders

The extremely well resolved g-anisotropy of nitroxide spin labels at 250 GHz is most apparent from spectra of randomly-oriented, magnetically dilute samples in the rigid limit. Fig. 6 shows derivative spectra of the PD-Tempone spin probe in a frozen glass of 85%/15% glycerol-$d3$/D_2O taken at both 9.5 and 250 GHz. The 9.5 GHz spectrum exhibits the well-known nitroxide rigid limit lineshape with an overall width of 6.78 mT. At 9 GHz, this width is dominated by the ^{14}N hyperfine interaction (typically 3.5 mT); g-anisotropy (typically only 6.0×10^{-3} for nitroxides) only produces a linewidth of ~1 mT at 9.5 GHz. In contrast, the spectrum at 250 GHz reflects mainly g-anisotropy: the x, y, and z orientations of the molecule are resolved into widely separated spectral regions to produce a lineshape resembling those observed at lower fields for paramagnetic species with much larger g-anisotropy. At each of the canonical molecular orientations, three lines corresponding to the ^{14}N hyperfine splitting appear; thus, the hyperfine tensor elements associated with each molecular axis can be independently and accurately measured. The magnetic parameters obtained from a nonlinear least squares fit (25) of the 250 GHz rigid limit spectrum (dotted line in Fig. 6) are summarized in Table I, together with parameters obtained at 9.5 GHz (36). In addition to the elements of the g and A (hyperfine) tensors, the fit at 250 GHz includes an anisotropic T_2 tensor, to account for variation of the intrinsic inhomogeneous linewidth across the spectrum.

A comparison of the 9.5 GHz and 250 GHz results reveals some interesting aspects of the high frequency spectrum. The most noticeable difference between the magnetic parameters measured at 9.5 and 250 GHz are the values of A_x and A_y. Because peaks from the x and y orientations overlap strongly at 9.5 GHz, there is some uncertainty in the values of the A_x and A_y tensor elements obtained by a least squares procedure. In contrast, the g-factor resolution at 250 GHz affords an unambiguous and more precise determination of the A tensor, as well as greater accuracy in the g-tensor.

Another significant feature of the 250 GHz spectrum is the relatively large variation in linewidth with orientation. Whereas the apparent linewidth is less than 0.3 mT at the z orientation, it almost triples towards the x orientation, nearly obscuring the hyperfine structure on the x peak. Two-dimensional electron spin echo experiments have demonstrated that the homogeneous T_2 of PD-Tempone in a frozen glycerol/H_2O glass is nearly constant across the 9 GHz spectrum (26); this is expected to be the case at 250 GHz as well. Anisotropy in the *inhomogeneous* linewidth might be due to unresolved but anisotropic superhyperfine interactions of the deuterons on the radical; however, the isotropic average of the linewidths reported in Table I is much larger than the intrinsic inhomogeneous linewidths observed at 9.5 GHz.

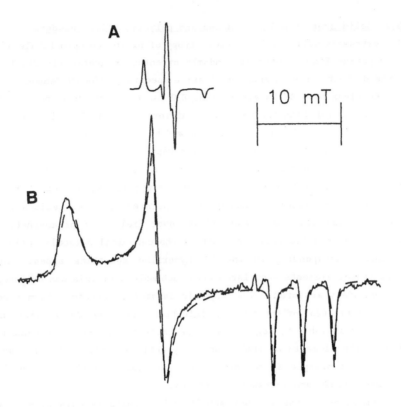

Fig. 6. Rigid limit spectra of 4×10^{-4} M PD-Tempone in a frozen glass of 85%/15% glycerol-$d3$/D_2O (-100°C) taken at (a) 9.5 GHz and (b) 250 GHz.

A more likely explanation of the linewidth variation at 250 GHz is g-strain (27,28), that is, a distribution of g-values arising from inhomogeneity in the local environments of individual spin probes. Since deviations in the effective g from the free electron value g_e = 2.002322 are caused by spin-orbit interactions involving the radical's molecular orbitals (29), effective g-values will be more sensitive to the local environment the further they are from g_e. Thus, inhomogeneities in local spin environments could produce a distribution in g_x values without significantly affecting g_z, which is consistent with the relative order of the linewidths $T_{2x}^{-1} > T_{2y}^{-1} > T_{2z}^{-1}$ reported in Table I.

The phenomenon of g-strain is well known in the case of paramagnetic transition metals (28), where large spin-orbit coupling terms can produce dramatic variations in g with the crystal field. However, such effects are usually neglected for the small g-anisotropies often encountered in organic and biological

Table I. Magnetic parameters of PD-Tempone in 85% glycerol/H_2O (or glycerol-d3/D_2O) obtained at 9.5 GHz and 250 GHz.

	9.5 GHz[a] glycerol-d3/D_2O	250 GHz[b] glycerol-d3/D_2O	250 GHz[b] glycerol/H_2O
g_x	2.0084 ± 0.0002	2.00859	2.00860
g_y	2.0060 ± 0.0002	2.00618	2.00622
g_z	2.0022 ± 0.0001	2.00233	2.00233
A_x (mT)	0.55 ± 0.05	0.62	0.66
A_y	0.57 ± 0.05	0.49	0.52
A_z	3.58 ± 0.03	3.62	3.54
$T_{2x}^{-1(c)}$		0.92	0.96
T_{2y}^{-1}		0.34	0.33
T_{2z}^{-1}		0.23	0.30

(a) Ref. 36.

(b) Uncertainties: g-tensor, approx. 0.00004; A-tensor, approx. 0.03 mT.

(c) Expressed as peak-to-peak Lorentzian linewidth in mT.

radicals. The high resolution of g-values at 250 GHz thus provides a sensitive probe of the chemical environment of organic radicals commonly found in biological systems.

Spectra from a rather more complicated solid state sample, a polycrystalline powder of DPPH, are shown in Fig. 7. The 9.5 GHz spectrum is the well-known exchange-narrowed result with a derivative linewidth of 0.24 mT, but the one at 250 GHz covers over 7.0 mT with four distinct, broad peaks in the first-derivative spectrum. The much smaller peaks are due to individual orientations of microcrystals in the spectrometer field. Because the 1.20 mm beam diameter is small and transmission of the 1.20 mm waves often requires a very thin sample layer, only a limited number of microcrystals can be packed into the active sample volume. We have found that polycrystalline samples must be extremely finely ground to avoid significant deviations from a truly "random" powder pattern.

In most studies of polycrystalline DPPH the g-tensor is taken as isotropic, because its asymmetry is so small (30,31,32). Even at higher frequencies (24-36 GHz) the anisotropy is just barely distinguishable; measurements in this frequency range give $g_\parallel = 2.0028$ and $g_\perp = 2.0039$ (33), in reasonably good agreement with the 5.48 mT splitting of the prominent outer peaks in the 250 GHz spectrum. This demonstrates how small anisotropies in g-tensors virtually unobservable at 9 GHz are dramatically manifested at 250 GHz.

The width of the 250 GHz spectrum is somewhat surprising, since one would expect g-value differences between adjacent molecules to be averaged out by rapid

Heisenberg spin exchange just as the ^{14}N hyperfine tensors are. The exchange interaction has been estimated to be $0.4-1.0 \times 10^{11}$ sec^{-1}, compared with $|g_\parallel - g_\perp|\beta_e B_0/\hbar \approx 10^9$ sec^{-1} at $B_0 = 9$ T (34). However, some features of the 250 GHz spectrum suggest that the exchange may be highly anisotropic. The 5.48 mT splitting shown in Fig. 7 is somewhat narrower than one would anticipate from the g-anisotropy of DPPH in dilute solution measured at 2 mm (4). In addition, the spectrum has a faint broad shoulder on the low-field side (barely detectable in Fig. 7) suggesting a partial averaging of the g-tensor. The additional broad peaks may be due to nearest-neighbor dipolar splittings in the crystal. A useful, detailed interpretation awaits in-depth analysis of the polycrystalline DPPH spectrum; however, our main point here is to emphasize the greatly enhanced spectral sensitivity to both structure and dynamics that is afforded at 250 GHz.

A polycrystalline sample of PD-Tempone similarly produces a single exchange-narrowed line at 9 GHz and a broad spectrum at 250 GHz corresponding to its asymmetric g tensor $g_x = 2.0095$, $g_y = 2.0063$, $g_z = 2.0022$ (35). If the PD-Tempone is prepared in a highly-concentrated smear with $CHCl_3$, then even at 250 GHz only a single exchange-narrowed line is observed, as it should be for strong exchange between randomly-oriented molecules.

3.3 Magnetically dilute liquid solutions

High-frequency EPR can also provide significant new insights into the details of molecular dynamics in the limit of fast motion. At 250 GHz, fast motion corresponds to correlation times $\tau_R < \sim 10^{-10}$ sec, typical of many spin probes in liquid phase solution. In this regime, EPR spectra of nitroxide spin probes consist of three Lorentzian lines from the ^{14}N isotropic hyperfine interaction.

Fig. 8 shows spectra obtained at 250 GHz from 5×10^{-4} M PD-Tempone dissolved in deoxygenated toluene-$d8$ at a series of temperatures from -80° to 70°C. Over this temperature range, the rotational correlation time of the spin probe, τ_R, varies by more than an order of magnitude (36). The spectra at 250 GHz are much broader than those obtained at 9 GHz, and the very sharp lines of the 9 GHz spectra change very little with increasing viscosity (36). In fact, only by studying the small differences in linewidth among the hyperfine lines is it possible to estimate the rate of the rotational tumbling that is averaging out the hyperfine-tensor and the g-tensor (10,36,37). The linewidths at 250 GHz are much more dramatically affected by increasing the viscosity. This is illustrated in Fig. 9 where the linewidth of the central peak is plotted as a function of viscosity for both frequencies. The observed trends are consistent with theory (37), which predicts that the g-tensor contribution to the linewidth should depend quadratically on frequency, so that at 250 GHz it is the predominant source of line broadening. Furthermore, the g-tensor contribution is linear in

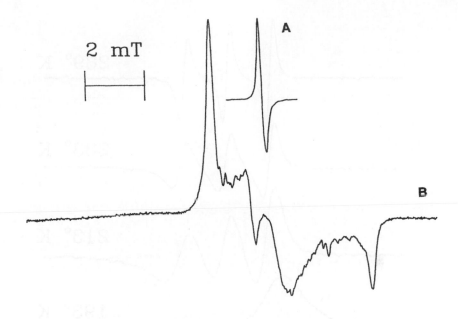

2 mT

A

B

Fig. 7. EPR spectra of polycrystalline DPPH (a) at 9.5 GHz; (b) at 250 GHz at room temperature.

the viscosity for a single solvent, making the 250 GHz linewidths very useful for estimating rotational relaxation times in the motionally-narrowed regime.

In contrast, at 9 GHz both spin rotational relaxation and spin exchange make significant contributions to the linewidth that depend inversely on viscosity (10,37). When these effects are combined with the g-tensor contribution, the viscosity dependence of the linewidth is nonlinear and more difficult to interpret. The fast-motional spectra at 250 GHz thus provide a means of separating the different contributions to the linewidths. There are additional fundamental questions about the frequency dependence of spin relaxation (10,37) which could be addressed by combining studies at 9 and 250 GHz.

At lower temperatures the EPR spectrum at 250 GHz is no longer in the motionally-narrowed regime, and simple relaxation theory no longer applies. This is evidenced by the large overlap of the hyperfine lines as well as the fact that the width is significantly less than that anticipated from the 9 GHz results. At 250 GHz, this slow-motional regime sets in for values of $\tau_R > \sim 10^{-10}$ sec, which is about an order of magnitude faster than for 9 GHz (35).

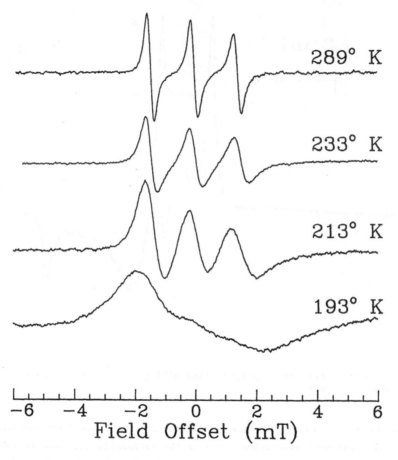

Fig. 8. 250 GHz EPR spectra of 5×10^{-4} M PD-Tempone in toluene-$d8$ at temperatures from -80° to 70°C. All samples are deoxygenated.

3.4 Slow-motional spectra: molecular dynamics

250 GHz studies in the slow-motional regime should provide significantly more detailed information about molecular dynamics than is available from motionally-narrowed spectra. Fig. 10 demonstrates the sensitivity of the 250 GHz EPR to the motion of two different spin probes at equal concentrations of 5×10^{-4} M in the nematic phase of the thermotropic Phase V liquid crystal (Merck). The upper set of spectra were obtained from the PD-Tempone spin probe, and exhibit the narrow three-line spectrum characteristic of fast molecular motions. At the lowest temperature, the hyperfine lines broaden and start to coalesce as τ_R approaches the slow-motional range; however, they remain in the center of the spectrum, indicating nearly isotropic tumbling of the spin probe.

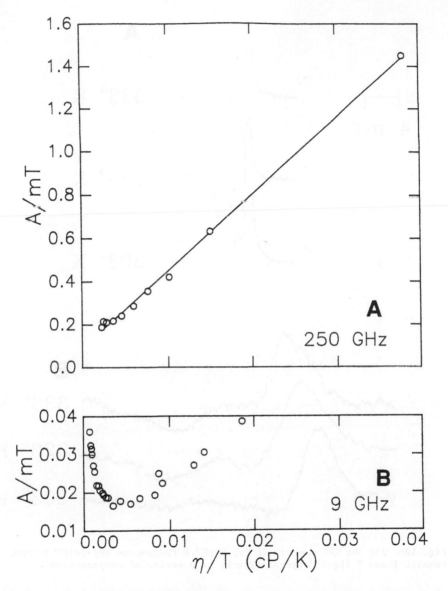

Fig. 9. T_2^{-1} in mT for central hyperfine line of spectra shown in Fig. 8 for 250 GHz and for 9.5 GHz plotted as a function of solvent viscosity, calculated according to reference 36.

The lowest three spectra of Fig. 10 were taken at the same concentration and temperatures using n-octylbenzoyl spin label (OBSL), which is a molecular analog for one of the liquid crystal components. In the nematic phase, the probe aligns

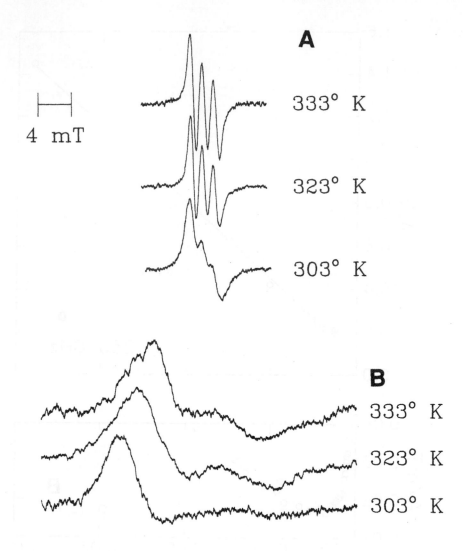

Fig. 10. 250 GHz EPR spectra of (a) 5×10⁻⁴ M PD-Tempone (b) 5×10⁻⁴ M OBSL in nematic Phase V liquid crystal (Merck) at a series of temperatures.

so that the N—O bond of the nitroxide radical is nearly parallel to the liquid crystal director. As the much broader spectra indicate, the bulkier OBSL probe undergoes slower rotation than does PD-Tempone. A strong alignment of OBSL in the spectrometer field is also evident from these spectra: They are most intense at field positions near the x orientation of the spin probe, which is with the N—O

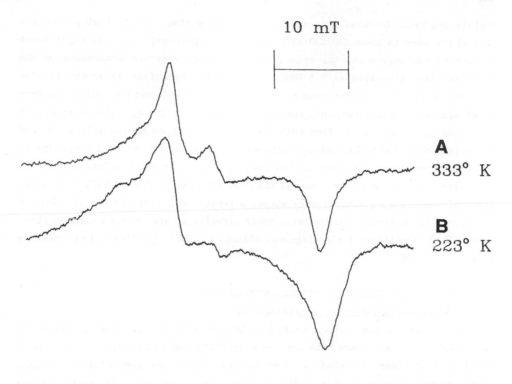

10 mT

A
333° K

B
223° K

Fig. 11. 250 GHz EPR spectra of CSL (concentration 5.7 mole percent) in an isotropic dispersion of DPPC at (a) 60°C (b) -50° C.

bond, (x axis) along the liquid crystal director and the spectrometer field. When the sample is heated, the intensity shifts towards the center of the spectrum as the spin probe motions become more isotropic. This illustrates how the combined sensitivity of 250 GHz EPR to spin motion and molecular orientation can characterize the directional dependence of the spin interaction.

Studies of slow motion should be particularly useful in the investigation of spin-labeled biological membranes and membrane models. Fig. 11 shows initial results from a sample of the spin label CSL, a doxyl derivative of cholestan-3-one, in a randomly-oriented dispersion of the lipid dipalmitoylphosphatidylcholine (DPPC) deposited on a 0.3 mm thickness of Z-cut quartz. These spectra also exhibit the broad features characteristic of slow motion. At the high concentration of CSL used in these initial studies (5.7 mole percent), there is significant Heisenberg exchange among the spin labels, leading to severe broadening of the hyperfine line structure at 9.5 GHz. Although the hyperfine structure is also averaged at 250 GHz, the spectrum still contains much information, extending over 40 mT with three peaks corresponding roughly to g_x, g_y, and g_z. The lineshape of the upper spectrum results from both the molecular motion of the spin probe and the spin motion due to Heisenberg exchange. The extent of spectral averaging by Heisenberg exchange is evident in the lower spectrum, in which molecular motions have been largely frozen out. As was the case for OBSL in Phase V, CSL in aligned lipid bilayers should exhibit very strong orientation effects in the 250 GHz EPR spectrum. For bilayers oriented with their director at some angle to the spectrometer field, additional field alignment effects should be particularly noticeable at 9 T.

4.   FUTURE 1 MILLIMETER EPR SPECTROMETER DESIGNS

4.1 General sensitivity considerations

The transmission cavity configuration of the 250 GHz EPR spectrometer described above was chosen both for its simplicity and to minimize the number of quasi-optical elements needed, at some expense in absolute sensitivity. In many ways, the design is comparable to the early microwave-frequency EPR spectrometers, which were also based on transmission cavities. Similarly, past experience at lower frequencies suggests many refinements that can be expected to enhance absolute sensitivity at 250 GHz. Upgrading the existing spectrometer will require such elements as directional couplers, bridges, switches, mixers, and ultra-sensitive detectors, many of which have only recently become available at 1 mm wavelengths. In this section, we will discuss some of the available components for 1 mm waves in anticipation of progress in EPR spectroscopy at these wavelengths in the near future.

To assess the contributions of each component to the overall sensitivity of the present 250 GHz spectrometer, we examine an expression for the minimum detectable number of spins in a transmission cavity (38,39):

$$N_{min} \propto (\eta Q_u \omega_0)^{-1} \left[(Q_u + Q_r)/Q_L\right] (\Delta B_{pp}/B_0)(\Delta B_{pp}/B_{mod})$$
$$\times (3kT_d\Delta f)^{\frac{1}{2}} \left[F_k - 1 + (t + F_{amp} - 1)L\right]^{\frac{1}{2}} P_w^{-\frac{1}{2}}. \tag{1}$$

Here Q_u, Q_r and Q_L, are the unloaded, radiation, and loaded Q-factors of the cavity (in general, $Q_L^{-1} = Q_u^{-1} + Q_r^{-1}$), ω_0 and B_0 are the FIR frequency and the dc field strengths, ΔB_{pp} is the derivative peak-to-peak EPR linewidth, and B_{mod} is the modulation field amplitude, assumed to be small compared with B_{pp}. Also, F_k is the noise figure of all components before the detector, T_d is the detection temperature, $t = T_d/290$ K is the dimensionless noise temperature of the detector, L is the insertion loss of the detector, F_{amp} is the preamplifier noise figure, and Δf is the effective bandwidth of the lock-in amplifier. Finally, P_w is the mm wave power, which is assumed to be nonsaturating. Strictly speaking, the frequency scaling in equation 1 only applies to resonators of equivalent design (38); hence, we will restrict ourselves to modifications that maintain the present Fabry-Perot design at 250 GHz.

Equation 1 summarizes the factors limiting absolute sensitivity in the current spectrometer design. Specifically, these are (1) the cavity Q_L, (2) the field modulation amplitude ΔB_{mod}, (3) the detector/preamplifier noise $t+F_{amp}$, and (4) the 1 mm wave power P_w. A fifth limitation implicit in Equation 1 is the transmission cavity configuration; comparison with a similar expression for a reflection cavity (38) suggests that sensitivity could be improved at least a factor of 3 over a transmission cavity with a similar voltage standing wave ratio (VWSR).

4.2 Cavity coupling at 1 millimeter

A very significant enhancement of sensitivity could be achieved by increasing the Q_L of the Fabry-Perot resonator. Values of 10^4–10^5 are, in principle, possible with loss-free samples (17). For the resonator shown in Fig. 3, the large coupling holes allow substantial radiative losses from the cavity, so that $Q_r \ll Q_u$ and $Q_L \approx Q_r \approx 100$. According to Equation 1, a critically-coupled resonator with $Q_r = Q_u \approx 10^4$ would increase sensitivity by a factor of 400. A convenient approximation (40) for the hole diameter d at critical coupling is $d \approx \lambda/3$, where λ is the radiation wavelength. For $\lambda = 1.20$ mm, d should be approximately 0.4 mm, whereas $d = 1.5$ mm for the resonator in Fig. 3, resulting in substantial overcoupling. In fact, since the beam waist diameter at the flat mirror is 2.2 mm (41), most of the radiation illuminates the coupling holes, and not the mirrors of our present resonator.

The high Q-value attainable with smaller coupling holes is not expected to be an unmitigated advantage at 1.20 mm, since it will require more exacting mechanical tuning of the cavity. To avoid rotating the upper mirror during tuning, as is necessary in the present micrometer screw arrangement, the cavity could be tuned by a precision translation stage. Since small temperature fluctuations and vibrations in a high-Q cavity might easily cause significant drift from resonance, it may be necessary to construct a feedback loop using a motorized stage to maintain cavity tune during an experiment.

Another problem with a high-Q resonator is that it would require a means of adjusting the coupling for samples of different dielectric loss properties. Sample losses affect the Q_u of the cavity; however, since $Q_r \ll Q_u$ for the resonator shown in Fig. 3, Q_L is quite insensitive to variations in Q_u. Thus, it is not necessary to vary the coupling in our present cavity. However, maintaining the matching condition $Q_r = Q_u$ for different samples in a high-Q resonator will require that the coupling of the 1.20 mm waves be adjustable.

At conventional microwave frequencies, the most common means of adjusting coupling is by mechanically changing the effective size of the coupling hole. Although a modification of this scheme is practical at 3 mm (42), the fabrication requirements for the coupling elements at 1 mm are too severe. For the present Fabry-Perot cavity, some variability in the coupling may be achieved by substituting flat mirrors with holes of different diameters. The easiest way to fabricate mirrors for this purpose would be to evaporate metal directly onto a sample cell window of quartz or CaF_2.

More continuous adjustability would be possible with a variation on the Bethe coupler that finds use at 2-3 mm (43). This scheme takes advantage of the linear polarization of the radiation incident on the cavity: By placing a short length of rectangular waveguide in front of the cavity and rotating its E-plane relative to the polarization plane of the incident beam, the coupling can be varied monotonically with rotation angle.

A different coupling design is required to operate a reflection cavity in an EPR bridge. The quasi-optical diplexers that are presently available for use in a 1 mm bridge (cf. Section 4.6) require the polarization planes of incident and reflected radiation from the cavity to be orthogonal. Since the Bethe tuning arrangement would result in a variable angle between the polarization planes of incident and reflected radiation, it would be unsuitable for this application. Reflection cavities may therefore require substitution of different iris diameters for coarse tuning; fine-tuning might by achieved by partially inserting a low-loss dielectric in the section of waveguide between the horn to the cavity (44).

The coupling schemes discussed may also require scalar feedhorns to couple the gaussian beam into the cavity instead of the conical antennae currently used. In addition to reducing losses, the feedhorns would match the E and H radiation fields, which would be necessary for any coupling schemes requiring polarization independence, circularly-polarized 1 mm waves, or any other superposition of orthogonal polarizations.

Once it is possible to tune the cavity to near critical coupling, it will be necessary to maximize Q_u by minimizing dielectric losses from the sample (45). Lossy samples such as liquid water will require a different holder that can contain the fluid within regions of low E-field in the cavity. For the semiconfocal Fabry-Perot resonator, E = 0 at the flat mirror; if a window of optical thickness $m\lambda/2$ is placed on the bottom mirror, E will also vanish near the surfaces of the window. Thus, excessive dielectric losses could be avoided by restricting the liquid layer to within 0.1 mm of the lower mirror or sample cell window. A more practical scheme may be to make the window thickness slightly smaller than $m\lambda/2$, in order to increase the cavity filling factor at some cost in reflected power. For extremely low spin concentrations, such as might be found in biological samples, alternate "thin layers" of $\lambda/2$ windows and lossy sample could be stacked to maximize the sample volume. An ideal candidate for the windows in this application would be the Z-cut quartz mentioned in Section 2.6.

4.3 Field modulation

Our experience with field modulation in the present spectrometer suggests some simple but useful improvements to the system described in Section 2.5. In EPR spectrometers employing field modulation, coupling between the modulation field and the dc spectrometer field typically produces microphonic vibrations of the modulation coil. Such effects are exaggerated in our 250 GHz spectrometer, where the coil must be placed in a 9 T field. The problems of vibrations are compounded in the present design because the coil is mounted on one of the mirrors. At very high modulation amplitudes, the vibrations modulate the cavity tune, which can produce large baseline offsets in the EPR spectrum. One might expect to avoid such mechanical resonances at higher frequencies; however, the effect does not disappear entirely, presumably because of the large forces involved and the high sensitivity of the cavity to very small dimensional changes. Future designs should mechanically isolate the modulation coil from the Fabry-Perot cavity.

A second difficulty with the current apparatus is the substantial attenuation of the modulation field amplitude at higher frequencies resulting from skin depth effects in the bulk metal of the cavity mirrors. Some mirror heating is also observed at the highest modulation amplitudes as a result of eddy currents in the mirror metal. Such effects will be greatly reduced in the future by constructing the resonator with a minimum amount of metal, preferably only on the surface of each mirror.

4.4 Millimeter-wave sources

There are several reasons to increase the source power at 250 GHz. First, increased power would improve the spectrometer sensitivity (cf. Equation 1) at least until power saturation of the EPR transitions sets in. At the present maximum power of 5 mW, we have never observed any indication of spectral saturation. Second, the availability of saturating powers considerably augments the range of experiments that can be performed, e.g. saturation studies of relaxation times. Finally, higher source powers would provide the groundwork for constructing a pulsed spectrometer at 250 GHz.

For many years there has existed a "power gap" in the frequency range 100 GHz to 1 THz. Most of the progress towards higher powers in this frequency range has come from extensions of microwave technology to the near millimeter region. The available sources fall into three categories: electron tubes, FIR lasers, and solid state oscillators. Table II summarizes the output power and FM noise characteristics of different representative sources that may be viable for EPR at

250 GHz. As Table II shows, higher powers are in general only available at a considerable reduction of noise performance. This is not a major concern for the present spectrometer, which is noise-limited by the detector.

The most powerful millimeter-wave source available are the electron tube sources, which include reflex klystrons (46), gyrotrons (47), ledatrons (48), backward wave oscillators (BWOs) (49), and extended interaction oscillators (EIOs) (50). Such devices are often bulky, require their own superconducting magnets to operate, and may have short tube lifetimes, especially at the higher frequencies. In addition, they are quite noisy even when phase and frequency locked (49), and some produce output in higher waveguide modes that are difficult to transduce efficiently into a gaussian beam mode.

Lasers also present problems of noise, fragility, bulkiness, and high maintenance overhead. In addition, the lasing transition nearest 250 GHz is a rotation of optically pumped $^{13}CH_3F$, a rather expensive medium. One advantage is that it can provide pulses with peak powers of ~1 MW mm and durations of the order $\leq$ 100 ns (51), which would be useful in time-domain EPR experiments. Other laser systems include the free electron laser (FEL), which is the subject of intense investigation and may become a useable laboratory source in the future.

In contrast, solid state sources, primarily Gunn and IMPATT diodes (52-54), have very low phase noise, and their ease of maintenance and low cost make them very attractive in many applications. A major limitation of solid state sources is the low available output powers; moreover, since they only provide useful

Table II. Performance of selected mm-wave sources for use in an EPR spectrometer.

Device	Output Power	Phase Noise dBc/Hz at 100 kHz	Reference
GUNN diode (GaAs)	3 mW	-70	(52)
GUNN diode (InP)	5 mW	-65	(52)
GUNN diode (InP) phase-locked	5 mW	-90	(54)
reflex klystron	8 mW	(a)	(46)
IMPATT diode (280 GHz)	100 mW	-60	(53)
BWO (carcinotron)	500 mW	-80	(49)
EIO (220 GHz)	1 W	(a)	(50)
gyrotron	1 kW	(a)	(47)

(a) Phase noise figure unavailable for this source.

output at fundamental frequencies < ~100 GHz, it is necessary to use frequency doublers or triplers for output at 1 millimeter wavelengths, further reducing the available power and also increasing the phase noise. Because the phase noise of free-running solid state oscillators has approximately a $1/f^2$ frequency dependence between ~10 kHz and ~100 MHz (55), phase-locking considerably improves the noise performance at practical field modulation frequencies (56).

At the modest power levels required to do cw EPR below saturation, the phase-locked, frequency-tripled InP Gunn oscillator used in the present spectrometer provides the best phase noise performance for the available rf power of any available solid state source. The low phase noise at high offset frequencies will also make this source useful as the local oscillator (LO) in a heterodyne detection scheme, so that a relatively noisy but more powerful rf source may be used to irradiate the sample. In principle, a similar heterodyne scheme could also be used with a high-power source in a pulsed spectrometer.

4.5 Detectors

Significant progress in sensitivity at 250 GHz will require a detection scheme with a much lower noise figure than the present direct detection arrangement. Most of the available millimeter wave detectors, such as diode crystals and bolometers, are similar to devices used at conventional EPR frequencies, differing only in the nature of the semiconductor used to make them. The simplest improvement to the present spectrometer would be to use a more sensitive direct detector with the existing transmission cavity; however, more significant noise reduction can be accomplished using coherent detectors (i.e., mixers).

Table III summarizes some performance characteristics for a representative selection of both incoherent and coherent detectors suitable for use in a mm-wave EPR spectrometer. The two most useful characteristics for comparing detectors are the detector noise, which limits sensitivity, and the response time, which determines what sort of noise reduction techniques may be used. For incoherent detectors, these characteristics are specified in Table III as noise equivalent power (NEP) and response time constant, τ; for mixers, they are given as noise temperature T_d and intermediate frequency (IF).

As Table III indicates, detector sensitivity is available at a cost in response time or IF bandwidth. At the most sensitive end of the scale are semiconductor bolometers, which perform optimally at or below LHe temperatures. The extreme is the ^{3}He cooled (300 mK) Ge bolometer (57), which is exquisitely sensitive, but has a minimum response time of about 5 ms, severely limiting field modulation frequency and making it particularly sensitive to acoustical noise. The InSb bolometer has the feature that it may be operated as a "hot electron bolometer" (58), taking advantage of the strong dependence of its conductivity on the temperature of its conduction electrons. In contrast, Schottky diode

Table III.   Performance of selected mm-wave detectors for use in an EPR spectrometer.

Detector	NEP $W/\sqrt{Hz}$	Response Time, τ	Reference
Schottky diode (290 K)	2×10^{-12}	1 nsec	(59)
InSb hot electron bolometer	10^{-13}	100 nsec	(58)
Ge bolometer (^{3}He cooled)	10^{-15}	5 msec	(57)

Mixer	SSB Noise Temperature, T_d	Intermediate Frequency	Reference
Schottky diode (cooled to 20 K)	800 K	1.4 GHz	(62)
InSb hot electron bolometer	360 K	2 MHz	(63)
SIS detector	305 K	1.3 GHz	(64)

detectors are less sensitive, but they provide very fast response times (59), and offer the advantage that they do not require cryogenic cooling. For the present direct detection spectrometer, a hot-electron bolometer/pre-amplifier system (60,61) would provide the optimum sensitivity. This device does require cooling, but offers excellent sensitivity at response times that will accommodate useful field modulation frequencies.

As Table III shows, the performance of mm-wave detectors with very fast reponse times such as Schottky diodes can be considerably enhanced when they are used as mixers (62). The improvements are mostly due to limitation of the input bandwith, and multiplication of the signal by the local oscillator (LO) frequency. For example, the input bandwidth of a waveguide-mounted incoherent Schottky diode detector is roughly half the center frequency of the waveguide, or 100 GHz for WR-4 waveguide. In comparison, the input bandwidth B of a Schottky diode mixer is about 100 MHz, assuming a typical IF of 1 GHz. The NEP for such a mixer is given approximately by $k_B T_d \sqrt{B} \approx 4\times10^{-16}$ $W/\sqrt{Hz}$, over three orders of magnitude smaller than the NEP for a typical incoherent detector (cf. Table III). The performance of InSb hot-electron bolometers may also be considerably improved by using them as mixers (63), although their longer response time results in a somewhat narrow IF bandwidth.

The major difficulty with using a mixer, for example in a heterodyne detection scheme, is that it would considerably complicate spectrometer design. One would require additional quasi-optical elements and two millimeter wave sources for the LO and the rf irradiating the sample. Finally, we note that there has been recent progress (64) towards developing a mm-wave detector based upon superconducting effects at cryogenic temperatures (often called superconductor-insulator-superconductor, or SIS detectors), which may prove to be even more sensitive than Ge bolometers, yet have short enough response times to be operated as mixers.

4.6 Quasi-optical bridges and mixers

Ultimately, it will be necessary to employ more sophisticated detection schemes for EPR at 250 GHz. In this section, we consider the millimeter-wave analogues of the standard waveguide elements used to construct EPR bridges at conventional wavelengths. The necessary quasi-optical devices are all variations on the Martin-Puplett (MP), or polarizing interferometer (65), shown in Fig. 12, which uses a polarizing wire grid to divide an incident beam of plane-polarized light. The roof-top reflectors m_1 and m_2 rotate the plane of polarization by $\pi/2$, so that both the light initially reflected and the light initially *transmitted* by the grid will be directed towards the output. One of the mirrors is moveable, permitting variation of the relative phase of the two arms and thus the output polarization. Viewed as a rf bridge, the MP interferometer is a four-port device with two inputs (orthogonal polarizations I_1 and I_2 incident on the grid) and two outputs (orthogonal polarizations O_1 and O_2, cf. Fig. 12). Division of input power between the two arms of the interferometer can be controlled by rotating the linear polarization of the source, which can readily be accomplished using a K-mirror (66).

For an EPR bridge, the fixed mirror would be replaced with the Fabry-Perot cavity preceded by a quarter-wave plate (QWP), in order to replace the $\pi/2$ rotation of the roof-top mirror. Plane polarized light incident on the QWP will be converted to a circular polarization before it is coupled into the cavity. Since the polarization maintains its "handedness" upon reflection from the cavity, the QWP converts the reflected radiation back into a planar polarization orthogonal to the incident radiation polarization. One consequence of this arrangement is that the cavity coupling elements, including the waveguide between the coupling horn and the cavity, must have cylindrical symmetry to accommodate the circularly-polarized light.

Depending upon the input polarization, the MP interferometer can be operated either as a circulator or as a "magic T" junction. Thus, it will be possible to implement detection methods quite similar to those used at conventional EPR wavelengths; for example, "rf-bucking" to improve the sensitivity of video crystal detectors, balanced-mixer arrangements based on the magic T junction, and

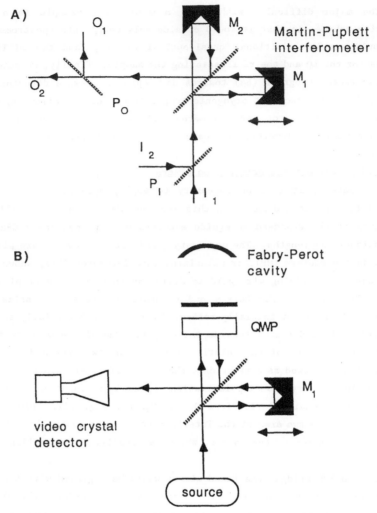

Fig. 12. (a) A Martin-Puplett polarizing interferometer consisting of a polarizing grid and two roof-top mirrors, M_1 and M_2. (b) A quasi-optical EPR bridge (circulator) constructed from a polarizing interferometer.

heterodyne detection using the MP interferometer as an injector for the the local oscillator. Experience with these methods will indicate which is the most advantageous for application at near-millimeter wavelengths; however, based on the sensitivity of the various mixers currently available, it would appear that a heterodyne detection scheme should provide the best sensitivity for 1 mm EPR work in the very near future.

5. OUTLOOK

Although EPR spectrometers at near-millimeter wavelengths have been known for a number of years, such spectrometers have often required specialized apparatus or specific types of samples, and thus have not been available for more widespread application to problems of general chemical and biological interest. However, with the advent of readily available sources and quasi-optical technology at millimeter and sub-millimeter wavelengths, we anticipate that high-frequency, high-resolution EPR will be much more fully utilized in the chemical and biochemical laboratory.

The common availability of high-resolution EPR spectroscopy could prove to be a significant asset for EPR studies of biological systems, similar to the enhancements brought to biological NMR work with the introduction of superconducting magnets. Of particular importance for biological studies are the increased spectral resolution for organic free radicals such as electron transport cofactors or nitroxide spin labels, the enhanced sensitivity to molecular dynamics typical of biomolecules at physiological temperatures, and increased absolute sensitivity to the small number of detectable electron spins often found in biological material. The improved sensitivity to molecular orientation is also quite important, since macroscopic order may be imposed on many biological systems using the techniques of stretched polymer films, lipid bilayer alignment, or protein crystallization. Finally, millimeter and sub-millimeter wave EPR can provide a more detailed picture of many biologically relevant metal ions, for which only limited information is available at standard EPR frequencies.

In addition to the advantages of 1 mm EPR demonstrated in this chapter, it may be possible to exploit its potential further by combining it with other spectroscopic methods, similar to methods employed at lower EPR frequencies. The high g-factor resolution afforded at this wavelength particularly suggests 1 mm EPR for techniques that require resolvable magnetic anisotropies such as magneto-photoselection (67). Many interesting chemical species which could not previously be studied by such EPR methods would be approachable at millimeter wavelengths. High g-resolution may also extend the range of samples exhibiting detectable g-strain (27), permitting studies of site selection in a matrix with narrow band laser excitation (68) and the effects of applied electric fields on the g-tensor (69). No doubt further interesting applications of millimeter and sub-millimeter waves will emerge as high-resolution EPR takes its place as a standard laboratory research tool.

ACKNOWLEDGEMENTS

We thank Dave Schneider, Dan Igner, and Dave Bazell for their help during the early stages of spectrometer design, and Eric Smith and Prof. David M. Lee for their advice on the design of the magnet Dewar. Drs. Paul Goldsmith and Ellen Moore of Millitech Corp. are gratefully acknowledged for many useful discussions about quasi-optical devices.

This work was supported by NIH Grant No. GM-25862, NIH National Research Service Award No. 1GM-12924, and NSF Grants Nos. CHE-8703014 and DMR-8616727.

REFERENCES

1   J. Gorcester, G. Millhauser, and J. H. Freed, this volume.

2   S.J. Opella and P. Lu (Eds.), *NMR and Biochemistry*, Marcel Dekker, New York, 1979.

3   A useful text on recent advances in this field is the series by K. J. Button *et al.* (ed.) *Infrared and Millimeter Waves*, Academic Press, New York.

4   O. Ya. Grinberg, A.A. Dubinski, and Ya. S. Lebedev, *Russian Chem. Revs.*, **52** (1983) 850-865.

5   E. V. Lubashevskaya, L. I. Antsiferova, Ya. S. Lebedev, *Teor. Eskp. Khim.* **23** (1987) 46-53.

6   E. Haindl, K. Möbius, and H. Oloff, *Z. Naturforsch.* **40a** (1985) 169-172.

7   P. deGroot, P. Janssen, F. Herlach, G. DeVos, and J. Witters, *Intl. J. Infrared Millimeter Waves* **5** (1984) 135-145.

8   F. Herlach (ed.), *Strong and Ultrastrong Magnetic Fields and their Applications*, Vol. 57 in *Topics in Applied Physics*, Springer-Verlag, Berlin, 1985.

9   W. B. Lynch, K. A. Earle, and J. H. Freed, *Rev. Sci. Instr.* **59** (1988) 1345-1351.

10   S.A. Zager and J.H. Freed, *J. Chem. Phys.* **77** (1982) 3344-3375.

11   K.R. Efferson, *Rev. Sci. Instr.* **38** (1967) 1776-1779.

12   For example, see: W.A. Peebles, N.C. Luhmann, Jr., A. Mase, H. Park, and A. Semet, *Rev. Sci. Instrum.* **52** (1981) 360-370.

13   N. R. Erickson, *IEEE Trans.* **MTT-29** (1981) 557-561.

14   A. Vickery, Millitech Corp., private communication (1989).

15   P.F. Goldsmith in K. Button (Ed.) *Infrared and Millimeter Waves*, Academic Press, New York, vol. 6 (1982), p. 277.

16   S. B. Cohn, Chap. 14 in: H. Jasik (Ed.) *Antenna Engineering Handbook*, McGraw-Hill, New York, 1961.

17   W. Culshaw, *IRE Trans.* **MTT-9** (1961) 135-144.

18   "Z-cut" windows are cut in the basal plane of the quartz crystal, and are available from Valpey Fisher Corp., Hopkinton, Massachusetts.

19   H. A. Macleod, *Thin-Film Optical Filters*, Adam Hilger Ltd., London, 1969, chapter 2.

20   *Handbook of Physics*, American Institute of Physics, Washington, 1972, p. 295.

21   H. Tanaka and J. H. Freed, *J. Phys. Chem.* **88** (1984) 6633-6644.

22   W. Kaiser, W. G. Spitzer, R. H. Kaiser, and L. E. Howarth, *Phys. Rev.* **127** (1962) 1950-1954.

23   B. Bleaney, P. M. Llewellyn, and D. A. Jones, *Phys. Soc. Proc. Ser. B* **69** (1956) 858-860.

24   R. E. Coffman, *J. Phys. Chem.* **79** (1975) 1129-1136.

25   R.H. Crepeau, S. Rananavare, and J. H. Freed, *Proc. 29th Rocky Mtn. Conference*, Denver, CO, Aug. 2-6, 1987.

26   G. L. Millhauser and J. H. Freed, *J. Chem. Phys.* **81** (1984) 37-48.

27   W.R. Hagen, D. O. Hearshen, R.H. Sands, and W.R. Dunham, *J. Mag. Reson.* **61** (1985) 220-232.

28   W. Froncisz and J.S. Hyde, *J. Chem. Phys.* **73** (1980) 3121-3131.

29   C.P. Slichter, *Principles of Magnetic Resonance*, Springer-Verlag, New York, 1980, chap. 10.

30   R.W. Holmberg, R. Livingston, and W.T. Smith, Jr., *J. Chem. Phys.* 33 (1960) 541-546; Y.K. Kim and J.S. Chalmers, *ibid.* 44 (1966) 3591-3597.

31   N.W. Lord and S.M. Blinder, *J. Chem. Phys.* 34 (1961) 1693-1708.

32   J.S. Hyde, R.C. Sneed, Jr., G.H. Rist, *J. Chem. Phys.* 51 (1969) 1404-1416.

33   P.P. Yodzis and W.S. Koski, *J. Chem. Phys.* 38 (1963) 2313-2314; B . M . Kozyrev, Yu V. Yablokov, R.O. Matevosjan, M.A. Ikrina, A.V. Iljasov, Yu.M. Ryshmanov, L.I. Stashkov, and L.F. Shatrukov, *Opt. Spect.* 15 (1963) 340-345.

34   G. Pake and T. Tuttle, *Phys. Rev. Letters* 3, (1959) 423-425.

35   J.H. Freed in: L. J. Berliner (Ed.) *Spin Labeling: Theory and Applications*, Academic, New York, 1976, chapter 3.

36   J. S. Hwang, R. P. Mason, L.-P. Hwang, and J. H. Freed, *J. Phys. Chem.* 79 (1975) 489-511.

37   S A Goldman, G.V. Bruno, C.F. Polnaszek, and J.H. Freed, *J. Chem. Phys.* 56 (1972) 716-735; S.A. Goldman, G.V. Bruno, and J.H. Freed, *J. Chem. Phys.* 59 (1973) 3071-3091.

38   C.F. Poole, *Electron Spin Resonance*, Interscience, New York, 1967, p. 546.

39   G. Feher, *Bell System Tech. J.* 36 (1957) 449-484; G.K. Fraenkel in: A. Weisberger (Ed.) *Technique of Organic Chemistry: Physical Methods*, Interscience New York, 1960, vol. I, part IV.

40   P. Goy, in: K. Button (Ed.), *Infrared and Millimeter Waves*, Vol. 8, Academic, New York, 1983, p. 352.

41   G.W. Chantry, *Long Wave Optics*, Academic Press, New York, 1984, vol. 1, p. 68.

42   K. Möbius, personal communication, 1989.

43   R. Clarkson, personal communication, 1989.

44   R. Chedester, Millitech Corp. personal communication, 1989.

45   Note that the dielectric loss term ϵ'' for water is only 5.25 at 250 GHz compared to 32.6 at 9.35 GHz; P.R. Mason, J.B. Hasted, and L. Moore, *Adv. Mol. Relax. Proc.* 6 (1974) 217-233; A.M. Bottreau, J.M. Moreau, J.M. Laurent, and C. Marzat, *J. Chem. Phys.* 62 (1975) 360-365.

46   D. Boilard, Varian Corp. (USA), personal communication, 1989.

47   A. V. Gapanov *et al.*, *Int. J. Electron.*, 51 (1981) 277-287.

48   G. Kantorowicz and P. Palluel, in: K. Button (Ed.), *Infrared and Millimeters Waves*, Vol. 1, Academic, New York, 1979, Ch. 4.

49   A. van Ardenne et al., *Rev. Sci. Instrum.* 57 (1986) 2547-2553.

50   Varian Research Corp. of Canada.

51   M.P. Hacker, Z. Drozdowicz, D.R. Cohn, K. Isobe, and R.J. Temkin, *Phys. Lett.* 57A (1976) 328-330.

52   I. G. Eddison and I. Davies, *Radio Electron. Eng.* 52 (1982) 529-533.

53   H. J. Kuno: in: K. Button (Ed.), *Infrared and Millimeter Waves*, Vol 2., Academic, New York, 1980, Ch. 2.

54   A. Vickery, Millitech. Corp., personal communication, 1989.

55   I. G. Eddison in: K. Button (Ed.) *Infrared and Millimeter Waves*, Vol. 11, Academic, New York, 1984, p. 29.

56   W. P. Robins, *Phase Noise in Signal Sources*, vol. 9 of *IEE Telecommunications*, Peter Peregrinus, London, 1982.

57   H.D. Drew and A.J. Sievers, *Appl. Opt.* 8 (1969) 2067-2071.

58   F. Arams, C. Allen, B. Peyton, and E. Sard, *IEEE Proc.* 54 (1966) 612-622.

59   R. B. Erickson, *Microwaves and RF*, 26 (Dec., 1987) 154-157.

60   J.W. Archer and M.T. Faher, *Microwave, J.* 27 (1984) 135-142.

61   A.T. Wijeratne *et al.*, *Phys. Rev. B* 37 (1988) 615-618.

62   Millitech Corp., 1989.

63   T.G. Phillips, P.J. Huggins, G. Neugebauer, and M.W. Werner, *Astrophys. J.* 217 (1977) L161.

64   G.J. Dolan, T.G. Phillips, and D.P. Woody, *Appl. Phys. Lett.* 34 (1979) 347-349.

65   D. H. Martin, in: K. Button (Ed.), *Infrared and Millimeter Waves*, vol. 6, Academic, New York, 1982, chap. 3.

66   P. Goldsmith, Millitech Corp., personal communication, 1989.

340

67   P. Kottis and R. Lefebvre, *J. Chem. Phys.* **41** (1964) 3660-3661; H.S. Judeikis and S. Siegel, *J. Phys. Chem.* **74** (1970) 1228-1235.
68   R.I. Personov, E.I. Alshits, L.A. Bykovskaya, and B.M. Kharlamov, *JETP* **38** (1973) 912-917.
69   W. B. Mims, *The Linear Electric Field Effect in Paramagnetic Resonance*, Clarendon Press, Oxford, 1976.

CHAPTER 9

OBSERVATION OF EPR LINES USING TEMPERATURE MODULATION*

G. FEHER, R.A. ISAACSON and J.D. McELROY

EDITORIAL INTRODUCTION

In cw-EPR sensitivity is commonly enhanced by modulation of the magnetic field and lock-in detection, which yields the first derivative of the EPR absorption line. For very broad lines with little structure such as those of metalloenzymes the first derivative is very weak and one can only detect a signal at the turning points. For line shape analysis it is often important to be able to record the signal between these points (mostly, though not necessarily, the principal g-values). For this purpose a technique must be used that allows the recording of the absorptive line shape itself, while retaining the sensitivity of field-modulated EPR. Such a technique was devised by Feher and coworkers in 1969. Field modulation is replaced by a modulation of the Boltzmann equilibrium through modulation of the temperature of a sample that is equilibrated at a few Kelvin. This modulation is achieved by illumination with a light source whose intensity is modulated at about a few tens of Hz (the low frequency is necessary because of the slowness of heat diffusion). The EPR signal is now intensity modulated and can be sensitively detected using superheterodyne detection.

The temperature modulation technique is only rarely used, which is a pity in view of its unique potential especially for biochemical applications of EPR. It seemed therefore useful to again call attention to this method by devoting a short chapter to it. It seemed unnecessary, however, to write a new piece, since the original article is complete, giving all essential experimental detail and a clear example. I am therefore most grateful to the authors and the copyright holder (the American Institute of Physics) for giving permission to reproduce the article integrally (the notation has been changed to conform to this volume).

Most EPR spectrometers employ external magnetic field modulation followed by phase sensitive detection. In this note we wish to point out and demonstrate some advantages of using temperature modulation instead of field modulation. This technique is particularly useful for broad lines at low temperatures.

The output signal of spectrometers using field modulation is proportional to

$$S \propto (d\chi/dB)B_m \tag{1}$$

where χ is the rf susceptibility (real or imaginary), B the external dc magnetic field, and B_m the peak-to-peak amplitude of the field modulation. Since the derivative in Eqn. (1) is inversely proportional to the effective linewidth ΔB of the resonance line[1], the maximum signal obtained

---

* Reproduced with permission from *The Review of Scientific Instruments*, 40 (1969) 1640–1641.

with field modulation is proportional to $\chi(B_m/\Delta B)$. Although field modulation has the appealing advantage of simplicity, it suffers from the following disadvantages:

(1) For broad lines, limited field modulation amplitude[2] (i.e., $B_m/\Delta B \ll 1$) causes a severe loss of signal.

(2) For broad lines that have structures with steep slopes, B_m has to be kept small resulting in a loss of the broad features.

(3) In order to obtain absolute susceptibilities, a double integration of the signal is required. This causes a large error.

(4) Undesired cavity background signals (e.g., from impurities or cavity wall vibrations) are likely to be observed since all parts of the cavity are being modulated.

(5) Distortion of the line due to different passage effects may occur[3].

The above disadvantages can be eliminated by modulating the temperature of the paramagnetic sample between the limits of T' and T'' [4]. The signal under these conditions will be proportional to

$$S \propto \chi(T') - \chi(T'') \tag{2}$$

where $\chi(T')$ and $\chi(T'')$ are the susceptibilities at temperature T' and T''. For samples obeying the Curie law[5] (i.e., $\chi \propto 1/T$), Eqn. (2) reduces to

$$S \propto \chi(T')\Delta T/T'' \tag{3}$$

where $\Delta T = T'' - T'$. At an operating temperature of $T' = 1.4\,\mathrm{K}$, the "modulation index" $\Delta T/T''$ easily be made to assume values of $0.2 - 0.5$. In the case of magnetic field modulation, for a 100 mT line the equivalent modulation index $B_m/\Delta B$ would be approximately an order of magnitude smaller.

The T-modulation was conveniently accomplished in our experiments by illuminating the sample with IR radiation which passed through the Pyrex Dewars and through light slits at the end of the TE_{102} cavity[6]. The IR source ($1.4\mu < \lambda < 2.7\mu$) was a 600 W tungsten-iodine lamp with Corning Glass color filters CS 7-56 and CS 4-96. A mechanical shutter chopped the light at frequencies up to 30 Hz. A photocell behind the shutter provided the reference voltage for the phase-sensitive detector. A standard 9 GHz superheterodyne EPR spectrometer was used[7]. Special care was taken to minimize the low frequency noise (e.g., due to microphonics) of the spectrometer.

The samples, contained in a 3 mm i.d., 4 mm o.d. quartz tube, were either dissolved in or surrounded by water. This insured the absorption of the IR radiation by the sample without appreciably raising the temperature of the helium bath, cavity, etc. The low specific heat of the samples at low temperatures enabled one to work at relatively high frequencies of light modulation. At 30 Hz and 1.4 K, the average temperature excursion ΔT was only reduced by 50% from the value obtained with steady state illumination.

In Fig. 1 we present the results obtained on a sample of 10^{-3} M Ferrimyoglobin[8] in H_2O at pH $=5.7$ (sperm whale skeletal muscle myoglobin from Nutritional Biochemicals Corporation). Trace 1A was the result of field modulation and 1B was obtained with temperature modulation. The broad features of the line between 120 and 320 mT which are clearly seen in Fig. 1B are lost with the field modulation technique. Note also that the cavity background signals observed with field modulation around $g = 4$ are completely eliminated by the T-modulation technique.

Fig. 2 shows the results of T-modulation of an MgO sample at 1.4 K. The broad resonance line at $g = 3.44 \pm 0.03$ has been attributed to Fe^{2+}.[9,10] With field modulation, this line was barely detectable in this sample.

The T-modulation technique described above is an outgrowth of a search for EPR signals in photosynthetic materials. It is a pleasure to acknowledge the collaboration of Dr. D. Mauzerall in these early experiments in which great care had to be taken to separate out the photochemically induced signals from the "T-modulated" signals[11].

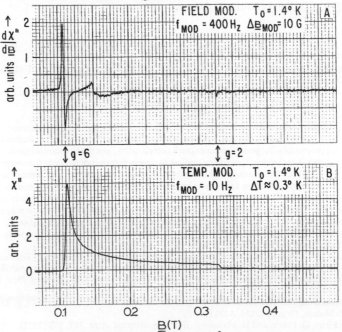

Fig. 1. Comparison of signals obtained from Ferrimyoglobin (10^{-3} M, pH $=5.7$) by field modulation (A) and temperature modulation (B) at 9.4 GHz. Note in the field modulation case the cavity background signals and the loss of the broad features of the line between 120 and 320 mT. The ordinates in Fig. 1A and 1B are in arbitrary units but have been normalized to the same gain settings. The detector time constant in 1A was 0.1 sec and in 1B 0.6 sec.

344

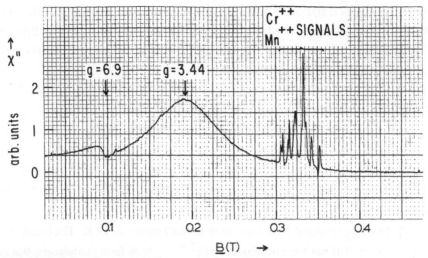

Fig. 2. EPR signal obtained from MgO by the T-modulation technique. $T' = 1.4$ K; $\Delta T = 0.3$ K, $f_{mod} = 10$ Hz; $\nu_{elec} = 9.26$ GHz. The broad resonance line at $g = 3.44$ due to Fe^{2+} was barely detectable with field modulation.

Acknowledgement

This work was supported by the National Science Foundation and Public Health Service Grant GM13191, National Institutes of Health.

References

1   For an up-to-date discussion, see C.P. Poole, *Electron Spin Resonance* (Interscience Publishers, Inc., New York, 1967), Chap. 10, Sec. D.
2   For practical reasons it is difficult to exceed a peak-to-peak amplitude of 10 mT. In the commercial Varian V-4502 spectrometer the maximum B_m is approximately 3 mT.
3   M. Weger, *Bell System Tech. J.* 39, 1013 (1960).
4   "Passage effects" may also occur in the T-modulation technique. The temperature dependence of the spin-lattice relation time T_1 can lead to thermal desaturation of the resonance. Reduction of the signal will occur when the modulation period becomes small compared to T_1 or to the thermal response time of the sample.
5   The T-modulation technique is, of course, useless for temperature-independent paramagnets (e.g., metals). It can, however, be used as a sensitive tool in the investigation of the temperature dependence of susceptibilities. Another interesting application arises by using degenerate p-doped silicon as a g marker [E.A. Gere (unpublished) referred to by G. Feher in *Paramagnetic Resonance II*, W. Low, Ed. (Academic Press Inc., New York, 1963), p. 715]. With field modulation, the g marker produces a signal; however, it may occasionally obscure the desired signal; with T-modulation the signal from the g marker is practically absent.
6   Other means of modulating the temperature (e.g., electrical heating) would, of course, be equally effective.
7   G. Feher, *Bell System Tech. J.* 36, 449 (1957).
8   The first resonances from Ferrimyoglobin were obtained on single crystals by J.E. Bennett and D.J.E. Ingram [*Nature* 177, 275 (1956)]. They found a perpendicular electronic g value of 6 and a parallel g of 2. The traces in Fig. 1 represent essentially the "powder pattern" of ferrimyoglobin [B. Bleaney, *Proc. Phys. Soc. (London)* 75, 621 (1960)].
9   W. Low, *Paramagnetic Resonance in Solids* (Academic Press Inc., New York, 1960), p. 85.
10 D.H. McMahon, *Phys. Rev.* 134, A128 (1964).
11 J.D. McElroy, G. Feher, and D. Mauzerall, *Biochem. Biophys. Acta*, 267, 363 (1972).

CHAPTER 10

TIME—RESOLVED EPR

KEITH A. McLAUCHLAN

1. INTRODUCTION

Optical spectroscopy remains the most widely used method for the detection of free radicals produced by flash photolysis. However it suffers from the limitation of low spectral resolution associated with broad optical lines in solution, and cannot yield positive species identification.   In consequence electron spin resonance methods have been developed in which the linewidth is fundamentally narrow and with which the radical may be identified from observation of the hyperfine structure of its spectrum. This has entailed considerable modification in the technology normally used to observe the EPR spectra of stable radicals. Three major methods are in use, continuous wave (cw) (1,2), spin—echo (3,4) and, very recently, Fourier transform ones (5). For historic reasons most of the development work in the field and in its application to chemical systems has been performed using cw methods. These are also the methods for which conventional EPR spectrometers can be adapted quite easily to give observation times starting from about 30 ns. after the photolysis flash. They form the subject of this chapter.

From the earliest experiments the spectrometers were designed to observe free radicals as soon as possible after their creation, within the first microsecond of their existence, a time shorter than electron spin—lattice relaxation times in most instances. A lower limit exists due to loss of spectral resolution through uncertainty broadening.   A surprising observation was that whilst the hyperfine structure of the radical observed on this timescale was exactly as expected, the intensities of the lines were not. Whole spectra appeared in emission, strongly in absorption or with both emissive and absorptive features. It was apparent that the populations of the hyperfine states of the radicals differed from those expected for species at thermal equilibrium with their surroundings, a state of spin disequilibrium known as Chemically Induced Dynamic Electron Polarization (CIDEP). This phenomenon had been observed earlier, but not recognised as due to spin polarization, in a study of H· atoms generated continuously by radiolysis (6).   CIDEP affects both the appearance of the spectrum at any point in time before the radical becomes relaxed completely and the time—evolution of the signals. It also affects lineshapes, which vary in time.

CIDEP has profound significance to the photochemist and photobiologist in that the phases of the signals disclose directly, without need for further experimentation, the spin multiplicity of the molecular state whose reaction produced the radicals (7). It provides a direct connection between the photophysics of the system and its photochemistry, and gives the information for each radical species present. However the intensity patterns it yields must be understood for these deductions to be made.

This article is consequently in several major parts, a description of cw methods, a summary of polarization phenomena, and an analysis of the time–dependent signals and lineshapes observed. The latter has led to the use of CIDEP in studies of electron, proton and site–exchange processes.

2.   CONTINUOUS WAVE METHODS

Normal EPR spectrometers operate with the frequency of the microwave radiation maintained constant and the spectrum displayed by sweeping the magnetic field of the instrument. At a resonant frequency of 9.6 GHz resonance occurs at about 0.34 T, and it is important to realise that the spectra of e.g. C–centred radicals can extend over 0.02 T whilst e.g. P–centred ones extend to 0.12 T. These are appreciable portions of the main field. As a result the sweep is usually generated by applying a varying current either to separate sweep coils or to the windings of an electromagnet itself. This is a major source of difficulty in observing time–resolved spectra from highly–transient radicals for induction prevents change of the current, and therefore the field, in a time short compared to a microsecond. In consequence an alternative strategy is adopted in which the radicals are created not once but several times, and each photolysis flash occurs at a different setting of the applied field, which is however maintained constant whilst the information following each pulse is gathered. The spectrum is subsequently reconstructed at any time after the flash from samples taken from each individual decay curve (8). The method is exemplified by the two–dimensional (2D) spectrum (9) shown in Fig. 1 in which the decay curves resulting from separate photolysis flashes have been stacked side–by–side in the order of the magnetic fields at which they were obtained. It depicts the early life–history of the radical, with time plotted on one axis and field on the other, and the effects of CIDEP are apparent in the opposite phases of some of the lines. On the time–axis some baseline was recorded before the time–zero of each photolysis flash. The effects of spin relaxation and interaction of the spin system with the microwave field of the spectrometer dominate the early time–dependence of the signals, typically for several microseconds following radical creation. The much slower chemical process which reduces the radical concentration by recombination controlled by diffusion in the bulk solution is observed at later times at the concentrations used (ca. 10^{-4} M/pulse) and is detected largely by its effect on the equilibrium signal to which the initial spin–polarized one has relaxed. However, spin polarization is in fact generated throughout the lifetime of radicals in the system.

The spectrum at any particular time after radical creation can be obtained by taking a cross–section of the surface perpendicular to the time axis. One surface, which takes about an hour to obtain in the signal–to–noise (S/N) ratio shown, may contain the entire information on the radical from the moment it is created to the time it disappears by reaction, and no further experimentation at the temperature of interest is required. The S/N of each spectrum extracted from it can be improved by summation of the signal

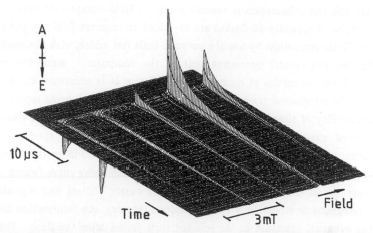

Fig. 1. A two–dimensional spectrum of cycloheptan–1–olyl radicals created as a symmetric pair on reaction of the triplet state of cycloheptanone with cycloheptanol (16). The low–field half of the spectrum is in emission and the high–field half in absorption, an E/A pattern.

obtained over a small period, typically 1 μs, of time. This is the so–called time–integration spectroscopy (TIS) technique (10): summing 100 samples taken every 10 ns improves the S/N by a factor of about 10. The normal mode of operation of our spectrometer is in fact not to store the entire decay curve at each field position, and not to create 2D surfaces, but rather to sample specific chosen parts of it, with digital summation performed on–line and the integrand stored before the field is changed and the process repeated. TIS has the further advantage over point–sampling techniques that it was designed to eliminate misleading sidebands in the spectrum which result from the instantaneous creation of magnetization inside a resonant microwave field (see below) and are not molecular properties of the radical concerned.

Time–integration spectroscopy has been evolved as a digital technique for performing these tasks but similar results can be achieved very simply using analogue sampling with a boxcar averager (11), although when this has been used the spectrometer magnetic field has been swept slowly rather than incremented discretely as in TIS. The field consequently changes as the signal is sampled and unless very slow sweep rates are used the theory specific to TIS given below can only be applied with caution to the interpretation of results. The digital technique also allows facile data–handling, and application of resolution – or S/N – improving mathematical methods (10).

Radicals are created in pulsed laser flashes, of 10–20 ns duration at 308 nm, inside the microwave cavity of a modified Bruker ER200D spectrometer and are detected, after amplification of the microwaves using a low–noise GaAs FET, with a low–noise crystal (Fig. 2). The signals undergo further wide–band amplification before being applied to a Biomation 8100 transient recorder which acts as the primary data store. Following each flash the signal is sampled at 100 MHz maximum as it decays, and this store contains a

decay curve which corresponds to a specific magnetic field value. When the 1024 channels of
the store are full the information is transferred to an IBM–compatible backing computer
where the results of typically 16 flashes are co–added to improve S/N. A pulse repetition
rate of about 20 Hz is attained by use of a purpose–built fast adder, with the resultant signal
only applied to the central processing unit of the computer. Since the spectrum is
reconstructed from the results of many photolysis flashes it is necessary for the sample to
flow through the irradiation region, and this limits the practical pulse repetition rate, as
does the desirability of allowing complete radical decay between successive flashes. In the
two–dimensional experiment the averaged curve is then stored in specific addresses in
memory corresponding to the field value. However the on–line TIS method makes further
use of the period between photolysis pulses. A section of the decay curve (which is displayed
in its entirety on an oscilloscope) is selected under cursor control and digitally summed
before the integrand is stored in a specific address, turning the information into a single
point on the eventual spectrum at the selected time period after the flash. Following the
data gathering and manipulation the computer steps to the next address and outputs signals
to cancel the contents of the transient recorder and to re–arm it, and also to step the applied
field. On the next flash from the laser the whole process is repeated, and this continues until
a pre–selected field–range has been investigated.

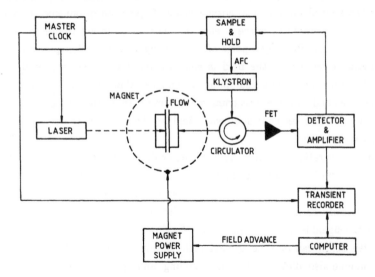

Fig. 2. A block–diagram of a time–resolved continuous–wave EPR spectrometer for use in
two dimensional and TIS applications. The MISTI and MISTIS variants require a
microwave switch, controlled from the master clock, to be placed between the klystron and
the oscilator

The laser firing and the whole timing of the experiment is controlled from a master
clock, which performs another vital function. When a high energy (ca. 50 mJ/pulse) laser

beam is incident on the microwave cavity it produces two major effects apart from creating radicals. In the first place it generates an impulse signal, due to instantaneous mis–matching, which is transmitted via the automatic frequency control (AFC) circuitry to the klystron and causes time–dependent phase changes in the output signal from the spectrometer. This artifact is removed conveniently by use of a sample–and–hold circuit, on the AFC output, which is actuated by a square wave from the clock unit and which temporarily holds the AFC level to that immediately preceding the laser pulse. Secondly the pulse causes heating in the sample and cavity which detunes the latter and causes a comparatively slowly decaying signal to appear even when the spectrometer field is set off–resonance. By careful control of laser output to a constant level this signal can be made completely reproducible, and it is subtracted from each of the experimental decay curves under software control.

An important aspect of a time–resolved spectrometer is its time–response which is determined by the bandwidth of the apparatus, and each individual component of the detection system must be considered. It is convenient in photolysis experiments to use a TE102 cavity which allows irradiation of large areas of sample tubes and this may be detuned to give a quality factor of about 800, corresponding to a response time of ca. 10 ns. The bandwidth of the GaAs FET amplifier is wholly adequate and in practice the device that limits the overall system–response is the detector crystal, with a typical bandwidth of only ca. 6.5 MHz. This implies that the subsequent amplification stages should be designed with this in mind, although in practice band–pass filters are often used to optimise S/N empirically whilst their effect on the signal is checked. It is implicit in this discussion that the normal field–modulation methods associated with EPR spectroscopy are best not used (although high frequency modulation at 2 MHz was used in the earliest experiments), for they limit the bandwidth. In consequence the lines in the spectra are not produced in their normal derivative mode and the pure absorption signal (in the magnetic resonance sense) is obtained. Any variations in its phase are the result of CIDEP processes.

Two variations of the basic experiment are of some interest. In the first place it is tempting to attempt to extract a value of T_1, the spin–lattice relaxation time, from short–time decay curves obtained from radicals created with spin–polarization. As will be shown below, at their simplest these curves are the sum of two exponentials and measurement of T_1 from them is not straightforward. However, if the microwave field is applied somewhat after the photolysis flash, but still in cw mode, the time–integral of the ensuing signal varies in a single exponential manner with the time constant T_1 as the time between the laser pulse and the application of microwaves increases. This is known as the microwave–switched time integration (MISTI) experiment (12). An interesting variant is to maintain the time interval between radical formation and microwave application constant and to perform time–integration on a selected part of the ensuing signal as before, but now to step the magnetic field so as to obtain a spectrum. This is the MISTIS method (9). Its effect, on radicals formed with initial spin polarization, is to yield an appreciable increase in the S/N of the spectrum obtained as compared with a normal TIS spectrum corresponding to the same sampling period after the photolysis flash. This results because T_2 in normal solution is much shorter than T_1 and loss of phase coherence is greater the longer after the microwave field is applied. A consequence, so far untested experimentally, is that the MISTIS method should discriminate against radicals which acquire their spin polarization after their initial creation (i.e. as a result of F–pair encounters, see below).

All the experimental modes, 2D, TIS, mode 1 (to obtain a decay curve at a selected magnetic field value), MISTI and MISTIS are under complete software control implemented in machine code occupying roughly 10 K of RAM on a Zenith 158 IBM–compatible microcomputer. Analysis of the results, and data manipulation to obtain 2D surfaces, is performed on a separate minicomputer. This has an extensive suite of software written to calculate CIDEP patterns and to yield calculated decay curves and complete spectra for comparison with observed ones. Both systems are under continuous development.

3. CHEMICALLY INDUCED DYNAMIC ELECTRON POLARIZATION (CIDEP)

Electron spin polarization arises in two independent processes which often simultaneously affect observations on the timescale of the fastest experiments (13–16). One, the triplet mechanism (TM), yields a contribution to the spectrum wholly in one phase and causes no distortion of the relative line intensities from those observed for radicals at thermal equilibrium; the phase may be absorptive (A) or emissive (E) depending upon the precise system involved. The EPR spectra of transient radical anions often exhibit apparently pure TM behaviour, a result of electron exchange averaging out the inherent contribution from the second, radical pair, mechanism (RPM) of spin polarization (17). The RPM mechanism in its most common (ST_0) form yields a contribution to the spectrum in mixed phase and the sense of the phase behaviour displays the spin multiplicity of the molecular precursor to the radicals. In a less common (ST_{-1}) form, observed in viscous solution, micelles and when nuclear hyperfine interactions are unusually large (18–21), a single–phase contribution is produced which differs from the TM one in showing hyperfine–dependent polarization, with distortion of the normal EPR relative intensities.

These mechanisms have been discussed in detail in the literature (21–25), and the RPM one in the chapter by P.J. Hore in this book, and will be outlined here only in sufficient detail to understand their influence on observations. The TM is the process long–known in the solid state in which the spin polarization arises in the inter–system crossing (ISC) between the excited singlet state, first formed when a ground–state molecule absorbs a photon, and the triplet state which results. The ISC occurs at different rates into the zero–field split levels of the triplet under symmetry–driven molecular selection rules. Coupling to the Zeeman field of the spectrometer produces spin polarization in the triplet in the laboratory frame and this is carried over to the radicals produced by its rapid reaction, which has to compete with rapid triplet spin–lattice relaxation. The radicals are produced with the same magnitude and phase of polarization (in practice typically 20 x the Boltzmann polarization). The spin polarization observed in the radicals is decreased by rapid tumbling of the triplet or by its too rapid dissociation into radicals. TM polarization normally occurs before observations commence although a recent very fast study has observed it to grow in in the first few ns of the radicals' existence (5); this corresponds to the reaction time of the triplet state.

In photochemical and thermal processes radicals are inevitably produced in pairs and the dominant RPM polarization at early times arises from interactions within these geminate pairs. The radical pair is formed with spin–conservation from its molecular precursor and its electrons are consequently spin–correlated. In the comparatively high external magnetic field of an EPR spectrometer a triplet radical pair exists in three non–degenerate Zeeman states, $T_{0,\pm 1}$. Spin polarization arises from the mixing of one or more of these states with the singlet (S) state of the radical pair together with a crucial influence from the electron exchange interaction $J(r)$. The mixing can only occur when two levels are approximately degenerate, a situation attained for long enough only by the S and T_0 levels, as $J(r)$ tends to zero as the the radicals separate by diffusion in liquids of normal viscosity. It then occurs under the influence of the different magnetic fields at the two electrons, and is normally dominated by the hyperfine interaction. For normal organic radicals complete ST_0 mixing takes a few ns, which implies that the radicals diffuse appreciably apart before much spin mixing occurs. Since $J(r)$ falls rapidly with inter–radical separation efficient spin polarization requires the radicals to re–encounter after their first separation so as to experience a non–zero value of $J(r)$ after mixing has occurred. In consequence the relative diffusive motion of the radicals has a strong influence on the magnitude and form of the polarization produced. However provided that the hyperfine interactions are strong enough to cause rapid spin mixing, some polarization may also arise when the initially–formed pair diffuses apart for the first time whilst $J(r)$ is non zero. The polarization produced per pair is slight compared with that in the re–encounter process but its effect is weighted in importance by the fact that all pairs of radicals undergo this initial separation whilst only a few re–encounter.

Overall the polarization produced in the radical pair can be expressed as (26)

$$P \propto [Q^{\frac{1}{2}} - cQ] \ , \tag{1}$$

where c is a constant depending upon the precise radical, solution and temperature involved and Q is a mixing parameter given by

$$Q = \frac{1}{2}(g_1 - g_2)\mu_B B + \frac{1}{2}\sum_n a_{1n} m_{1n}^{(a)} - \frac{1}{2}\sum_m a_{2m} m_{2m}^{(b)} \ . \tag{2}$$

The symbols have their usual significance with e.g. $m_{1n}^{(a)}$ being the magnetic quantum number of the nth nucleus of radical 1 which exists in the overall nuclear spin state (a). This definition is consequently appropriate to a pair of radicals in the specific overall nuclear spin states (a) and (b). However the observed polarization of a single line in the spectrum of one of the radicals results from the radical in the state forming pairs with counter–radicals in all of their possible hyperfine states:

$$P_{1,a} = \frac{1}{x_2} \sum_n P_{1,ab_n} \qquad\qquad\qquad [3]$$

where b_n is the nuclear spin and Zeeman state of nucleus n in radical 2 and x_2 is the total number of such states.

In eqn. [1] the $Q^{\frac{1}{2}}$ term arises in the separation/re–encounter process in normal solution (the exponent changes with viscosity (21)) whilst the second is usually important only in alkyl radicals amongst C–centred ones; it is usually neglected.

The full derivation of ST_0 RPM theory discloses that the polarization is expected to be of mixed phase, half in A and half in E (the overall spin polarization is zero), and that the sense of the polarization depends upon the spin multiplicity of the molecular precursor. An A at low field/E at high field (A/E) pattern results from the reaction of a singlet and an E/A one from that of a triplet. If $g_1 = g_2$ the centre of inversion is the mid–point of the spectrum but otherwise the spectra of the radicals contain unequal amounts of A and E phase. From a triplet precursor one exhibits E^*/A behaviour and the other E/A^*, where the asterisk denotes higher intensity; the E/A sense remains the same for each.

These features are illustrated in Fig. 3 which also demonstrates two further important features of RPM behaviour. Lines near the centre of the spectrum are of low intensity (at exact centre zero) whilst the outer lines are enhanced in their relative intensities as compared with the spectra of equilibrated radicals. Neutral radicals created from triplet precursors usually show both TM and RPM contributions and the overall spectrum has an appearance which depends critically on their relative contributions and phases (Fig. 4). Analysis of such spectra is empirical with the relative contributions of TM and RPM polarization adjusted until agreement is found with experiment. The effect of a TM contribution is to cause the overall spectrum to exhibit a net polarization, which is simply identified provided that both radicals of the pair are observed; if only one is an, e.g. E (from TM) + E/A (from RPM) pattern can only be distinguished from an E^*/A RPM pattern by careful analysis. Although the nomenclature is not elegant, designation of the two different ways in which a qualitatively similar effect can be produced is useful.

When the radicals in the geminate pair are inhibited in their rapid separation, in viscous media or inside micelles, the J(r) interaction does not rapidly become zero and its effect is to produce extra splittings (27), and a new polarization phenomenon, in the early–time spectrum, which is now of the radical pair itself rather than of the separated radicals. This is discussed below. In these media also, or when spin–mixing is accelerated by high hyperfine interactions, the S state of the RPM may be mixed with the T_{-1} state under the influence of the non–secular part of the hyperfine coupling; this assumes, as above, that the S level lies below the T_0 one. Such mixing causes true spin polarization due to the admixture of a pure electron eigenstate of the radical pair with the mixed S one and does not require action of the exchange interaction or diffusive re–encounters.

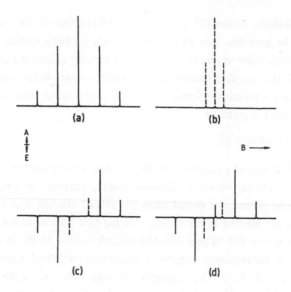

Fig. 3. Calculated unpolarized spectra of (a) a radical with four equivalent protons, and a hyperfine coupling of 2.0 mT, and (b) one with two equivalent protons with a coupling of 0.9 mT. (c) and (d) show the ST_0 RPM patterns expected when the two are formed together from a triplet precursor, (c) with $g_a = g_b$ and (d) with $(g_b - g_a) = 0.001$; the radicals having identical concentrations. Note that in (c) the central line is missing from the spectrum of each, that the lines near the centre are of comparatively low intensity and that each radical possesses E/A polarization. When the g–values of the two are not equal that with the higher g–value (the radical (b)) acquires E*/A polarization and the other E/A*.

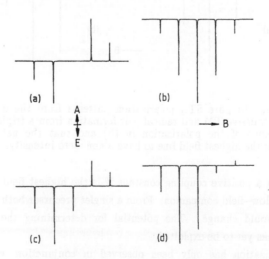

Fig. 4. (a) The calculated ST_0 RPM pattern from the quintet shown in Fig. 3 and formed as a symmetric pair by reaction of a triplet state. (b) The T.M. contribution for the same radical, taken to be emissive. (c) and (d) represent various arbitrary admixtures of the two and demonstrate how much the observed spectrum can vary between the two pure forms.

ST$_{-1}$ polarization, unlike TM polarization, is hyperfine—dependent. It originates in the $I..S_{+}$ part of the hyperfine spin Hamiltonian, which connects nuclear sub—levels in the T$_{-1}$ and S states which differ by one in their nuclear magnetic quantum numbers. Depending upon the sign of the coupling constant, this implies that either the lowest— or the highest—field line in the polarized spectrum has a zero contribution from this mechanism.

The polarization is given by

$$P_{ST_{-1}} \propto a^2[I(I + 1) - m(m + 1)] \qquad\qquad [4]$$

where I is the nuclear spin of a particular nucleus, m the nuclear magnetic quantum number and a the hyperfine coupling constant. However there is another, less obvious, contribution (21): the spin mixing occurs in a period when the overall electron spin of the radical pair may be coupled to any nucleus in the system. During this time what eventually becomes a single radical may form a radical pair with the counter—radical to be, in any of its possible hyperfine states. It consequently acquires a degeneracy—weighted hyperfine—independent polarization, calculated from the analogue of eqn. [4], in addition to its own hyperfine—dependent one. The relative contributions from these two depend upon the specific radicals involved and vary greatly from system to system. The overall effect is shown for a pair of identical radicals in Fig. 5. From a triplet precursor the spectrum is in

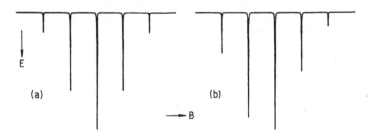

Fig. 5. (a) TM and (b) pure ST$_{-1}$ polarization patterns from the quintet of the previous figures, assuming emissive TM and radical pair formation from a triplet precursor. Note the hyperfine—dependence of the polarization in (b) and that the net contribution to ST$_{-1}$ polarization causes the highest field line to have a non—zero intensity.

emission and with a positive coupling constant it is the highest field line which is of lower intensity than its low—field companion. From a singlet precursor both the phase and the line intensity order would change. The potential for determining the sign of the coupling constant (20,21) has yet to be exploited.

ST$_{-1}$ polarization has only been observed in conjunction with ST$_0$ effects, with sometimes TM ones as well. Furthermore all initially—polarized spectra show progressively increasing single—phase contributions from thermally—equilibrated radicals as time evolves after radical creation. Spectral intensities for comparison with experimental ones then have to be calculated with four contributions, the polarized ones being empirically adjusted as before. The difficulty of obtaining an unique combination implies that studies of ST$_{-1}$ polarization *per se* requires careful choice of experimental systems to accentuate it.

A recent short comment has been published on the extent to which TM and geminate RPM theory is consistent with observation (28).

3.1 CIDEP from F—pairs

So far we have dealt with polarization produced either at creation or in geminate processes within a few tens of ns. thereafter. However, the radicals which escape geminate recombination (including those observed to have geminate polarization) diffuse through the solution and create radical pairs as a result of random encounters for as long as reactive radicals persist in the solution. In these pairs too the spins are correlated, singlet pairs tending to react and triplet ones to undergo spin evolution and polarization development just as they would if created geminately. For an observable polarization to result this underlying difference in reaction probability is essential or else the polarizations created in singlet and triplet encounters would be equal and opposite. With the assumption made above, these pairs produced by free diffusion (F—pairs) are consequently predicted to result in similar polarization in the radicals which escape them to that of geminately—produced triplet pairs, E/A patterns with relative intensities given by eqns. [1–3]. This prediction has been largely borne out in practice in pairs created by a chemistry which apparently requires them to be of this nature (29).

It should be realised that because radical reactions are fast the F—pair contribution to observed polarization often maximises early in time when it is usually, however, swamped by geminate effects. However specific experiments performed deliberately to distinguish between the two (30–33) showed that for alkyl radicals the effect of F—pair polarization was to produce A/E patterns, whilst ketyl radicals showed the expected E/A ones. Furthermore, since initial polarization decays by relaxation typically within 10–20 μs of radical formation, any later behaviour would be expected to be dominated by F—pair processes despite the overall signal having become low. In all cases including alkyl and ketyl radicals in which it has been possible to observe long—term signals initial E/A patterns have apparently inverted to A/E ones in time (26,34,35). Different senses of polarization have even been observed from identical radical pairs in different experiments, all thought to be observing F—pair effects, but this anomaly has been removed by the discovery that the behaviour is concentration—dependent (36).

Possible explanations of these effects have focused on the possibility that J(r) is of different sign in different radical pairs (30), that J(r) changes sign as radicals diffuse apart (37) and that the time—dependence of the phase may be due to electron—nuclear cross—relaxation and not to polarization effects (35). None has proved tenable in general and the most inviting possibility until recently seemed to be that singlet F—pairs have a less than unity probability of reaction, with the further necessity that the distance of closest approach without reaction occurring differs between singlet and triplet pairs (38). In this situation F–pair re–encounters could yield A/E polarization. It is now apparent however that some of the experimental results are false. The inversion of the sense of polarization in time observed in our experiments is, as often speculated upon in our papers, an experimental artifact with origins of some complexity but originating in a coupling capacitor inside the transient recorder used in the experiments (39). This implies that the explanations suggested deserve careful re–examination to consider whether they predict verifiable experimental behaviour, at least in some systems. It is stressed that it is only some of the phase–inversion results which have proved inaccurate and that puzzles in polarization behaviour remain.

3.2 Spin–Correlated Radical Pairs

The spin–correlated radical pair is the intermediate species which exists briefly in solution between the existence of the molecular precursor and the subsequent freely–diffusing free radicals. It is the origin of RPM electron and nuclear polarization, the reaction of geminately–formed radicals and the effects of magnetic and microwave fields on chemical reactions. It has been observed directly in flash–photolysis EPR experiments and in radical–yield detected magnetic resonance (RYDMR) whilst its presence can be inferred from studying the effects of static magnetic fields on radical yields (MARY) (40). The EPR spectrum of the radical pair can be observed as described above and differs from the sum of the spectra of the constituent radicals by exhibiting extra splittings due to the exchange interaction $J(r)$, about which rather little is known in solution. Such spectra are observed for a brief period before the radicals finally separate and they exhibit a novel form of CIDEP (27). This is that the two components of each line split by the $J(r)$ interaction appear in antiphase, that is each normal esr line appears to have an E/A or A/E nature. This contrasts with normal RPM effects in which one field half of the spectrum is in each phase.

The origin of this behaviour lies in the fact that the $T_{\pm 1}$ states of the radical pair remain eigenstates of the system whilst the S and T_0 ones are mixed by the magnetic interactions; a close analogy exists to 'AB' behaviour between strongly–coupled protons in nmr. Four transitions are predicted for the radical pair in each of its nuclear hyperfine levels and their intensities and phases are determined by the relative populations of the states connected by the transitions. On the assumption that a triplet radical pair is formed with equal populations in the $T_{0,\pm 1}$ states an E/A pattern is predicted for alternate lines; a singlet pair would yield an A/E one. The splittings are usually small due to a small average value of $J(r)$ and each apparently single EPR line appears in antiphase (41,42).

4. THEORY OF TIME RESOLVED CONTINUOUS WAVE EPR

In a photolysis flash free radicals are created essentially instantaneously to yield an initial magnetization $M_z(0)$ in the direction of the applied magnetic field of the spectrometer. The observed signal results from its rotation under the influence of a resonant radiation field so as to produce a component M_y in an orthogonal direction. The signal initially increases as this happens but then, if initially due to spin–polarized radicals, decays by fast relaxation processes and, on a somewhat slower timescale, by reaction. Bimolecular encounters also produce F–pair polarization in radicals which escape from them. The esr signal is consequently time–dependent in ways in which that of a stable radical is not, and this also implies time–dependent lineshapes. Here we calculate these effects using modified Bloch equations written in the rotating frame (14,43–47). Our treatment is for the homogenously–broadened lines observed in solution; the inhomogeneously–broadened ones observed in the solid state are considered in the chapter by D. Stehlik, C. H. Bock, and M. C. Thurnauer.

In the absence of spin polarization these may be written for reacting radicals in the matrix form (14),

$$\dot{M}(t) = LM(t) + T_1^{-1}M_{eq}(t) - T_c^{-1}(t)M(t) ,$$ [5]

where it is assumed that the instantaneous value $M_z(t)$ relaxes towards its corresponding equilibrium value. $T_c(t)$ is the instantaneous lifetime of the radical, of concentration $n(t)$, at this time:

$$T_c^{-1}(t) = -\dot{n}(t)/n(t)$$ [6]

and

$$\dot{n}(t) = -k_1 n(t) - k_2 (n(t))^2 ,$$ [7]

where k_1 and k_2 are pseudo first–order and second–order rate constants. The matrix L is given by

$$L = \begin{bmatrix} -T_2^{-1} & \Delta\omega & 0 \\ -\Delta\omega & -T_2^{-1} & \omega_1 \\ 0 & -\Delta\omega & -T_1^{-1} \end{bmatrix} ,$$ [8]

with T_1 and T_2 the usual relaxation times, $\Delta\omega$ the off–set from resonance and ω_1 the microwave field strength, both measured in angular frequency units. Otherwise

$$M(t) = \begin{bmatrix} M_x(t) \\ M_y(t) \\ M_z(t) \end{bmatrix} \quad \text{and} \quad M_{eq}(t) = n(t)P_{eq} \begin{bmatrix} 0 \\ 0 \\ 1 \end{bmatrix} ,$$ [9]

where P_{eq} is the spin polarization at thermal equilibrium. In general the polarization of a particular hyperfine state is defined, with obvious nomenclature, as

$$P_i = (n_\beta - n_\alpha)_i / (n_\beta + n_\alpha)_i .$$ [10]

We shall be concerned also with P_I, the initial value due to TM and geminate RPM spin polarization, and P_F, that created in F–pair encounters.

By defining $L^y(t) = [L - \dot{n}(t)/n(t) \, I]$, with I the unit matrix, and $Q(t) = M_{eq}(t) \, T_1^{-1}$, eqn [5] becomes

$$\dot{M}(t) = L^\dagger M(t) + Q(t) .$$ [11]

This can be solved for the normal experimental condition that initial radical formation and polarization is fast compared with spin lattice relaxation and radical reaction. This implies that the initial magnetization $M(o)$ can be treated as a boundary condition:

$$M(o) = n(o) P_I \begin{bmatrix} 0 \\ 0 \\ 1 \end{bmatrix} .$$ [12]

Eqn [11] then has the general solution

$$M(t) = \exp\left[\int_0^t L^\dagger(t)dt \right] . \left\{ M(o) + \int_0^t \exp\left[-\int_0^t L^\dagger(t)dt \right] . Q(t)dt \right\} .$$ [13]

In our experiments the radical concentration is deliberately kept low at ca. 10^{-4} M per flash (to minimise early–time effects of F–pair polarization) and radical decay is much slower than relaxation. Provided that the flash repetition rate is chosen so that n(t) nonetheless decays to zero between flashes eqn [13] gives (43)

$$M_y(t) = n(t)[P_I g_y(t) + P_{eq} G_y(t)]$$ [14]

where

$$G_y(t) = T_1^{-1} \int_0^t g_y(t)dt ,$$ [15]

and $g_y(t)$ is a function of T_1, T_2, $\Delta\omega$ and ω_1. These are the seminal equations of time–resolved esr with initially–polarized radicals.

The effects of F–pair polarization, which will not be discussed in detail here, are accounted for by a simple addition to the scheme: a bimolecular process which creates spin polarization is added to the original Bloch equation. It results in an additional term in eqn [14] which becomes

$$M_y(t) = n(t)[P_I g_y(t) + P_{eq} G_y(t)(1 + RP_F T_1 k_2 n(t))] ,$$ [16]

where R is the reaction probability per collision of the F–pairs.

The majority of observations have been dominated by initial polarization effects with comparatively little attention given to F–pair ones. We shall consequently use eqns [14] and [15] as the bases of our discussions. Analytical expressions for $g_y(t)$ exist only for certain limiting cases which are however of some interest. In each case we obtain (46,47)

$$g_y(t) = (\omega_1/b) \exp(at) \sin(bt)$$ [17]

where a and b have the following forms:

(i)   At resonance, with $\Delta\omega = 0$,

$$a = -\tfrac{1}{2}(T_1 + T_2) \quad \text{and} \quad b = [\omega_1^2 - \tfrac{1}{4}(T_2^{-1} - T_1^{-1})^2]^{\frac{1}{2}}$$ [18]

(ii)   At resonance, and with $T_1 = T_2$ (which is rare),

$$a = -T_2^{-2} \quad \text{and} \quad b = \omega_1$$ [19]

(iii) Off resonance, and with $T_1 = T_2$

$$a = -T_2^{-2} \quad \text{and} \quad b = (\omega_2^2 + \Delta\omega^2)^{\frac{1}{2}} \ . \tag{20}$$

In early–time experiments it is the $g_y(t)$ term in eqn [14] which dominates the observations and the results are useful in disclosing two important features which hold in the general case also. Firstly the decay of the signal does not occur in a single exponential manner with the time constant T_1. Secondly the response of the signal is in general oscillatory, with the oscillations being at their lowest frequency exactly on resonance. These features are illustrated in Fig. 6 and their implications will be discussed in detail below. If $T_1 = T_2$ and the experiment is performed exactly on resonance the microwave field strength ω_1 can be measured directly from the oscillation frequency; this has proved useful in practice.

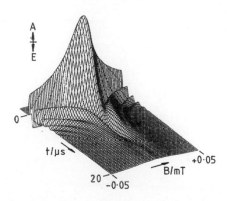

Fig. 6. A calculated two–dimensional spectrum from a single line showing oscillations in the time domain (even at resonance) which increase in frequency as the off–set from resonance increases. These cause undesirable sidebands in the field domain spectrum, obtained by taking a cross–section perpendicular to the time axis. The surface was calculated with $T_1 = 5.0\ \mu s$, $T_2 = 0.8\ \mu s$, $\omega_1 = 1.0$ rad. MHz and an initial polarization $P_I = 10\ P_{eq}$.

When, as is usually the case in solutions of normal viscosity, $T_1 \gg T_2$, it can be shown that at moderate microwave powers the signal does decay in a single exponential manner but with a time constant T_1^{eff} given by

$$1/T_1^{eff} = 1/T_1 + \omega_1^2 T_2/(1 + \Delta\omega^2 T_2^2) \ , \tag{21}$$

and even on resonance the measured values of T_1^{eff} must be extrapolated to zero microwave power to yield the true T_1. In all other cases the signal decay is at best bi–exponential although it does become mono–exponential on resonance approximately $5T_1$ s after the photolysis flash; here again extrapolation to zero microwave power is necessary to obtain T_1. This method has not been found useful in practice since observations have to be made at times when the S/N ratio is poor. A better method is to use the MISTI technique, described above.

In practice the solutions to the Bloch equations used are the general ones obtained under the conditions of slow radical decay by semi–analytical methods with fast numerical techniques. These methods give (47)

$$g_y(t) = \sum_{i=1}^{3} A_{iy} \exp(\lambda_i t) \qquad [22]$$

where

$$A_{1y} = \omega_1(\lambda_1 + T_2^{-2})/(\lambda_1 - \lambda_2)(\lambda_1 - \lambda_3) , \qquad [23]$$

and similarly for i = 2,3 by cyclic permutation. Here

$$\lambda_1 = \alpha_+ + \alpha_- - \theta , \qquad [24]$$

$$\lambda_2 = \beta\alpha_+ + \beta^*\alpha_- - \theta , \qquad [25]$$

$$\lambda_3 = \beta^*\alpha_+ + \beta\alpha_- - \theta , \qquad [26]$$

$$\alpha_\pm = \left[\frac{-q \pm (q^2 + (4/27)p^3)^{\frac{1}{2}}}{2}\right]^{2/3} \qquad [27]$$

$$\beta = -\frac{1}{2}(1 - 3^{\frac{1}{2}}i) \qquad [28]$$

$$\theta = \frac{1}{3}(T_1^{-1} + 2T_2^{-1}) \qquad [29]$$

$$p = -\frac{\delta^2}{3} + \omega_1^2 + \Delta\omega^2 \qquad [30]$$

$$q = -\frac{2}{27}\delta^3 + \frac{1}{3}\omega_1^2\delta - \frac{2}{3}\Delta\omega^2\delta \qquad [31]$$

and

$$\delta = (T_2^{-1} - T_1^{-1}). \qquad [32]$$

These apparently formidable equations are straightforward functions of T_1, T_2, $\Delta\omega$ and ω_1 and are easily treated numerically. Values of all four parameters are required to calculate decay curves and lineshapes, or some may be obtained by curve–fitting. At low ω_1 levels the linewidth is, for example, controlled at early times after radical creation almost entirely by T_2, which may be obtained quite simply.

The predictions of these equations are most clearly seen in the 2D surface computed for a single line and shown in Fig. 6; they seem entirely consistent with (unpublished) single–line experimental observations. The exact appearance of the surface depends critically on the values of the parameters used but in general form there exist oscillations in the signal, which increase in frequency as the distance from resonance increases, but which occur even on resonance if ω_1 is sufficiently high. Cross–sections parallel to the time axis yield the form of the decay curves observed experimentally. A set, at resonance, is shown in Fig. 7b and clearly displays their general non–exponentiality besides the onset of oscillations as ω_1 is increased. Cross–sections perpendicular to the time axis yields the instantaneous lineshape at the time selected. Here the effect of oscillatory behaviour in the time domain is to produce sidebands in the field domain. These effects are regularly observed experimentally (48), although the ω_1 value used to obtain Fig. 1 was set to avoid them. The existence of sidebands in spectra can lead to spectral mis–assignment and their occurrence would appear to limit the usefulness of time–resolved esr techniques. Time integration spectroscopy (TIS) was invented specifically to eliminate them (10) (besides improving the S/N of observed spectra) whilst empirical boxcar methods did so fortuitously.

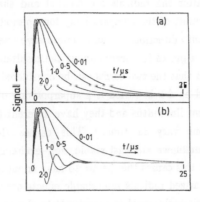

Fig. 7. Calculated decay curves in the presence (a) and absence (b) of electron exchange, with $T_1 = 5$ μs, $T_2 = 1$ μs, $\omega_1 = 0.01\text{--}2.0$ MHz, as indicated. These curves show non–exponential behaviour and the onset of oscillations as ω_1 is increased. These are damped out by electron exchange (see below), here taken to occur at a rate of 0.5 MHz (47).

The principle of TIS is that by digitally summing the signal over a period of time related to the oscillation frequency the oscillations tend to average out. This works well in

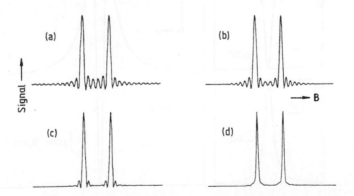

Fig. 8. The effect of the TIS technique on a simulated doublet dominated by initial polarization, calculated with a coupling constant of 0.2 mT, $T_1 = 10$ μs, $T_2 = 1$ μs and $\omega_1 = 0.1$ rad MHz. Sampling starts 1.0 μs after radical formation and continues for (a) 0.01 μs, (b) 0.05 μs, (c) 0.5 μs and (d) 1 μs. The oscillations are progressively averaged out as the integration period increases, resulting in rather little distortion of lineshape. (10).

those usual circumstances where the period of oscillation is not long, an integration period of ca. 1 μs usually being adequate (Fig. 8). The resulting general lineshape is the sum of Lorentzian and squared Lorentzian terms.

Two other features of point–sampled and TIS lineshapes are of interest. In the first place the linewidth is predicted, and observed, to vary in time, being typically 3–4 gauss in the first few tens of ns after the radicals are created and sharpening up thereafter for typically 1 μs. This is a result of the comparatively long period of time required to rotate M_z into the measurement direction at low microwave field–strengths: during the evolutionary period electron precession occurs about the resultant of B_0 and B_1. Secondly, from eqn [14] it is apparent that the two terms can be considered and integrated separately, and the lineshape obtained as $\Delta\omega$ is varied is a superposition of their contributions. The terms correspond to different linewidths and they have different time–dependencies, so that their relative contributions vary in time. This causes the lineshape itself to be time–dependent in a way unknown amongst stable, non–exchanging, radicals. For most of the time that radicals exist this causes the spectra to have no unusual appearance. However, for initially emissively–polarized radicals remarkable effects occur when the polarized term in eqn [14] becomes approximately equal in magnitude to the equilibrium one as a result of spin relaxation (49). The two contributions are then of opposite phase and different width and a lineshape is observed with a broad outer absorptive component and a sharp emissive inner one (Fig. 9).

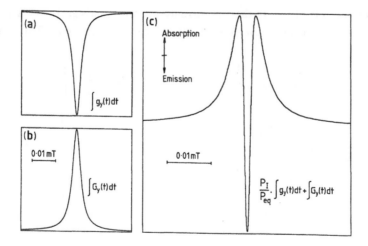

Fig. 9. (a) and (b) show the two contributions to the lineshape (eqn [14]) at a time when relaxation has caused them to have about equal amplitude. (c) shows the resultant characteristic composite lineshape. The parameters used were $T_1 = 18.6$ μs, $T_2 = 2.8$ μs, $\omega_1 = 0.075$ rad. MHz and $P_I = -4.83$ P_{eq}, with TIS applied from 20.0–46.0 μs post radical formation. Reproduced with permission from Molec. Phys. (49).

At slightly higher microwave powers low–frequency oscillations in the decay curves occur which are incompletely averaged out by the TIS method and a second composite lineshape is obtained which is the inverse of the first. Both are sensitive functions of the relative contributions of the two terms in eqn [14] and hence of the polarization ratio (P_I/P_{eq}). They provide sensitive methods for its measurement (49).

5. RATE PROCESSES

Spin–polarized radicals have been observed to undergo reaction to form secondary radicals (50,51), electron (52–54) and proton (54) exchange and site exchange (17), all of which affect decay curves and the appearance of spectra. They may be analysed by adding extra rate terms to the Bloch equations. A result has been to extend exchange studies to include highly transient species.

Provided, as usually occurs, a primary spin–polarized radical reacts within its spin–lattice relaxation period, spin conservation causes any secondary one to be produced in a spin–polarized state. Indeed spin–polarization can be used as a non–invasive label for following the reaction pathway. The secondary signal varies in time in a way which, *inter alia*, allows its formation rate constant to be deduced (47). However the detailed analysis that exists for this situation has not been applied yet in practice and we concentrate instead on the qualitative effect of polarization transfer on observed spectra. If the primary possesses some net polarization (e.g. is E*/A) and transfer occurs into random hyperfine states, only the net contribution is conserved. The secondary then exhibits pure single–phase polarization with no hyperfine distortion (56) and could be confused with a primary radical created with TM polarization. However, if a hyperfine coupling is correlated between two radicals, as happens for example when a P–centred radical adds to an olefin where the electron is coupled to the same P nucleus in the primary and secondary species (57), phase alternations may occur in the spectrum of the secondary (58). The sense of the polarization may alter between the two species if the coupling constant is of opposite sign in the primary and secondary radicals.

A situation almost uninvestigated experimentally is that in which the primary reacts to form the secondary within the initial geminate process. In this case the pair which encounters after initial radical separation differs from that originally formed. A preliminary study has suggested that this would affect the magnitude of the polarization observed in the secondary but would not change the sense of its polarization (59).

Site exchange, as for example between the two interconverting degenerate chair forms of cyclohex–1–olyl radicals, has long been known to cause line–broadening, at appropriate exchange rates, in stable radicals. This results from lines changing their positions between the two forms. If the radicals are spin–polarized this leads to a new phenomenon for then exchange may connect lines of different magnitudes and phases. At exchange rates too low to cause line broadening (i.e. ones much less than spectral splittings) polarization transfer occurs which tends to equalize the populations of the individual Zeeman states of the two sites involved (17). This observation extends EPR rate studies to lower limits than before.

At room temperature cyclohex–1–olyl radicals, created as symmetric pairs by reaction of triplet cyclohexanone with cyclohexanol, exhibit perfect ST_0 RPM geminate polarization in the E/A sense and the one interconnection of lines in equal and opposite phases that occurs from either side of the centre yields a zero average intensity for them. However, as the temperature is lowered a strong ST_{-1} contribution is obtained and the spectrum becomes distorted with more intensity in emission than in absorption. Under these circumstances the effects of population transfer are seen across the spectrum, in parts of which phase alternations of successive lines are observed. This behaviour cannot occur in any normal polarization process (although see above for secondary radicals, and radical pairs, in the second of which all the spectrum shows this characteristic rather than just a few lines). Lines appear in different phases than they would if no exchange occurred.

The phenomenon is simply analysed for a two site exchange process. The signal is proportioned to the y component of the overall magnetization given by the sum of the contributions from $M_A(t)$ and $M_B(t)$ at each site, A and B:

$$M(t) = M_A(t) + M_B(t) .$$ [33]

$M_A(t)$, for example, is obtained from the Bloch equation

$$\dot{M}_A(t) = L\, M_A(t) - k\, M_A(t) + k\, M_B(t) + T_1^{-1}\, M_{A,eq}(t)$$ [34]

where magnetization is transferred between the A and B sites with a rate constant k.

The six resulting coupled differential equations to be solved are then, in matrix form

$$\begin{bmatrix} \dot{M}_A(t) \\ \dot{M}_B(t) \end{bmatrix} = \begin{bmatrix} (L_A - kI) & kI \\ kI & (L_B - kI) \end{bmatrix} \begin{bmatrix} M_A(t) \\ M_B(t) \end{bmatrix} + \begin{bmatrix} M_{A,eq}(t) \\ M_{B,eq}(t) \end{bmatrix} T_2^{-2} .$$ [35]

Geminate polarization is again input as an initial boundary condition, defined as in eqn [12] for each radical. This simple theory seems to reproduce experimental behaviour completely at all exchange rates, accounting for line broadening, exchange narrowing and population transfer.

The fact that exchange connection of lines of opposite phase averages their intensities even at low exchange rates explains why the spectra of radical anions are observed with only net hyperfine–independent spin polarization, usually due to a TM process. Any additional RPM polarization is averaged out even if a very slow electron transfer process occurs to e.g. the parent molecule. For example the electron may leave the radical anion from a hyperfine state which corresponded to emission in the spectrum and enter another, in the new radical formed, which corresponded to absorption. Unlike the site exchange case discussed above, in this bimolecular process there is no correlation between the hyperfine states involved. Ensemble averaging then suffices to average all symmetric ST_0 effects to zero and only net polarization remains. This is a special case of the general remark made above about polarization in secondary radicals.

The fact that radical anions show single–phase polarized spectra with the relative intensities of their lines as in equilibrated radicals considerably simplifies their applications to electron and proton transfer studies. Both have been subject to detailed study. The effect of electron transfer between an anion and its parent is to take the electron from a radical in which it is in a particular hyperfine state, i, onto a molecule whose hyperfine state, j, is random.

$$A_i^{\bar{}} + A \rightleftharpoons A + A_j^{\bar{}} \; . \tag{36}$$

The overall rate of this process depends upon the degeneracy D_j of a given state and the total nuclear spin degeneracy D of the species, A. The Bloch equation then becomes (44)

$$\dot{M}_i(t) = L\,M_i(t) + T_2^{-2}M_{eq,\,i}(t) + \frac{1}{D\tau}\sum_j D_j\,M_j(t) - \frac{1}{\tau}\,M_i(t)\,, \tag{37}$$

where τ is the mean lifetime of the radical between electron jumps. This equation assumes, *inter alia*, that no hyperfine–dependent electron relaxation occurs and it can be solved, subject to the condition that the lines in the spectrum are well separated. In particular, the hyperfine splitting must be greater than T_2^{-2}, τ^{-2} and ω_1. Initial polarization is introduced once more as a boundary condition and equations similar in form to those given above (eqns [22]–[32]) can be obtained.

As with all rate processes, electron exchange affects both the time dependence of the signals and lineshapes. Some decay curves calculated for comparison with those observed in the absence of exchange are given in Fig. 7a. It has two effects, to tend to eliminate oscillations and to prolong the decay time; deviations from bi–exponential behaviour become more pronounced as either the rate of electron transfer or ω_1 increases.

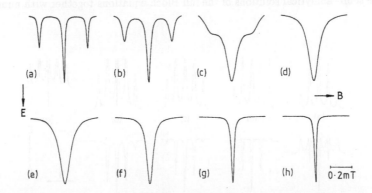

Fig. 10. Calculated spectra for an emissively–polarized radical, with $T_1 = 1\ \mu s$, $T_2 = 0.8\ \mu s$, $\omega_1 = 0.01$ rad. MHz and time–integration from 5–7 μs post the photolysis flash, with τ (a) ∞, (b) 10 μs, (c) 1 μs, (d) 0.1 μs, (e) 0.02 μs, (f) 0.01 μs, (g) 0.001 μs and (h) 0.0001 μs. (55).

In spectra observed away from the cross–over point where lineshapes such as that shown in Fig. 9 result, electron transfer affects polarized spectra in the same ways as it does equilibrated ones. At low exchange rates the lines first broaden and an effective relaxation time is obtained for each hyperfine line given by

$$1/T_{2i}^{eff} = 1/T_2 + (D-D_i)/D\tau .$$ [38]

In practice every radical anion spectrum we have observed has needed the effects of exchange to be considered for accurate reproduction by calculation. At higher rates the lines broaden further, overlap and collapse into a single broad line which eventually exhibits exchange narrowing (Fig. 10). This can be analysed using numerical solutions of eqn [37].

A less obvious effect of exchange is that the effective spin–lattice relaxation rate of the radical becomes dependent upon its specific hyperfine state (47,52,54). In emissively–polarized radicals lines of the highest degeneracy relax through zero first, followed by the others in the order of their degeneracies. This behaviour is detected very directly by the appearance of the composite lineshape (Fig. 9) in successive lines, or sets of lines of the same degeneracy, as time evolves, as shown in Fig. 11. Its origin is that in the slow exchange limit only one transition is excited at a time and the state over–populated in the polarization process loses population as a result of relaxation and microwave pumping. As it does so however it is refilled by electrons hopping from a large reservoir of radicals in different original hyperfine states, and the dependence of the rate of the overall population change process on the degeneracy of the state and the overall degeneracy is obvious. An important implication is that the effect can be tuned by adjustment of ω_1 to occur within the period that the radicals are spin polarized. At low values of ω_1 the effective spin lattice relaxation time of a radical in nuclear hyperfine state i is given by

$$1/T_{1i}^{eff} = 1/T_1 + \omega_1^2 \left\{(1/T_{1i}) - (1/T_{2i}^{eff})\right\}^{-1} ,$$ [39]

where T_{2i}^{eff} is given by eqn [38] and T_1 and T_2 are otherwise assumed hyperfine–independent. In practice semi–analytical solutions of the full Bloch equations together with numerical

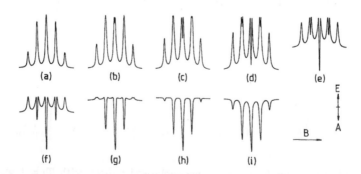

Fig. 11. The variation of an emissive quintet spectrum in time from a radical undergoing slow electron exchange calculated with $T_1 = 5$ μs, $T_2 = 1$ μs, $\tau = 1$ μs, $\omega_1 = 1$ rad. MHz and with a 0.5 μs integration period starting at the following successive times after the photolysis flash: 5 μs, 8 μs, 10 μs, 12 μs, 14 μs, 16 μs, 20 μs, 23 μs and 30 μs. (54).

solutions are always used to analyse observed spectra rather than these derived equations whose application is more limited.

Proton transfer processes have received less attention but can be treated in an analagous way to site exchange ones provided that the nuclear spin states associated with the interconverting transitions are properly accounted for (54). Exchange once more transfers magnetization between radicals with correlated hyperfine interactions and the Bloch equations are simply written. An example would be of the radical trianion from benzene—1,4—dicarboxylic acid interconverting with its protonated dianion form. The former contains 4 equivalent protons with a hyperfine coupling constant of ca. 0.15 mT whilst the latter contains two sets of two equivalent protons, with coupling constants of ca. 0.18 and 0.12 mT; note that the spectral spread is the same in each. On exchange the positions of the outermost lines in the spectrum consequently do not change, as is also true for two—thirds of the components that contribute to the central one. However the positions of the other transitions are modulated by the exchange process. In both unpolarized and polarized radicals this causes an alternating linewidth effect in the observed spectrum over a range of exchange rates. However, the effect is accentuated in the polarized one, and it also affects the time—dependence of the spectrum. Here lines behave differently depending upon the fraction of the overall number of nuclear spin states that contribute to them whose associated transitions change their resonance positions as a result of exchange. The greater this fraction the more effective the exchange process is at transferring magnetization and the faster the decay of the line in time. For example the five lines from the radical trianion above have degeneracies 1, 4, 6, 4 and 1, and the fraction of each line affected by exchange is 0, 1, 1/3, 1 and 0 respectively. The lines of degeneracy 4 consequently decay first in time, followed by the central one, and eventually the outer ones. As with electron exchange the effect is microwave—field dependent, and this field may be adjusted to observe it during the radical's lifetime as a polarized species. Note however that the time—dependence of lines of different degeneracy, and the appearance of the spectrum, differs from the electron exchange case. In systems where both proton and electron exchange may occur experiments at low microwave field strengths tend to expose the former and ones at higher strengths the latter.

6. SUMMARY

In this short account of time resolved continuous wave esr we have sought to introduce both the methods used and the interpretation of the observations. Experiments have been performed within 30 ns of the photolysis flash, although with a 50 ns sampling period, but lack the 1 ns resolution attained in recent Fourier transform pulse experiments (5). However they are unrestricted in their spectral width and it is unlikely that they will be wholly replaced by this new technique. As has been shown here too, cw methods have considerable application to probing rate processes on the few μs timescale.

All transient radicals with normal relaxation times exhibit CIDEP, which affects the appearance of their spectra by affecting line intensities whilst not changing the line positions. It contains important information on the multiplicities of radical processes besides opening up opportunities for study of reaction, relaxation and polarization processes. Here we have sought to give a broad description to illustrate the effect of CIDEP on observed spectra rather than to dwell on the intricacies of the polarization mechanisms themselves. These remain comparatively unexplored in detail by experimentalists.

The creation of spin—polarized radicals instantaneously within the resonant field of a cw spectrometer leads to profound effects on the time—dependence of the signals, on lineshapes and, through reaction or polarization transfer processes, on the actual appearance of spectra. Many of these are without analogy in stable free radicals and attention is drawn in particular to the variation in lineshape with time. In addition geminate RPM and TM polarization contributions to the signal relax in time, whilst both F—pair effects and contributions from equilibrated radicals increase in their relative importance. The intensity patterns in the spectrum consequently change as the relative contributions of mixed phase RPM effects and single phase TM and/or equilibrated radical ones vary in time. The spectra are also sensitive to other rate processes, as described above, and CIDEP spectra must always be analysed with care.

These phenomena however all imply that the study of spin—polarized radicals by cw methods can give light on a wide range of physical, chemical and biological problems and many such applications have been made. Space has not allowed them to be described here. Some examples are given elsewhere in this volume.

REFERENCES

1   P.W. Atkins, K.A. McLauchlan and A.F. Simpson, J. Phys. (E), 3 (1970) 547–551.
2.   B. Smaller, J.R. Remko and E.C. Avery, J. Chem. Phys., 48 (1968) 5174–5181.
3.   A.D. Trifunac and M.C. Thurnauer, J. Chem. Phys., 62 (1975) 4889–4895.
4.   A.D. Trifunac, J.R. Norris and R.G. Lawler, J. Chem. Phys., 71 (1979) 4380–4390.
5.   T. Prisner, O. Dobbert, K.P. Dinse and H. van Willigen, J. Amer. Chem. Soc., 110 (1988) 1622–1623.
6.   R.W. Fessenden and R.H. Schuler, J. Chem. Phys., 39 (1963) 2147–2195.
7.   C.D. Buckley and K.A. McLauchlan, Molec. Phys., 54 (1985) 1–22.
8.   S. Basu, K.A. McLauchlan and G.R. Sealy, J. Phys. (E), 16 (1983) 767–773.
9.   K.A. McLauchlan and D.G. Stevens, Molec. Phys. 57 (1986) 223–239.
10.   S. Basu, K.A. McLauchlan and G.R. Sealy, Molec. Phys., 52 (1984) 431–446.
11.   A.D. Trifunac, M.C. Thurnauer and J.R. Norris, Chem. Phys. Lett., 57 (1978) 471–473.
12.   K.A. McLauchlan and G.R. Sealy, Molec. Phys., 52 (1984) 783–793.
13.   L.T. Muus, P.W. Atkins, K.A. McLauchlan and J.B. Pedersen (Eds), Chemically Induced Magnetic Polarization, Reidel, Dordrecht, 1977.
14.   P.J. Hore, C.G. Joslin and K.A. McLauchlan, J. Chem. Soc. Spec. Period. Rep. ESR, 5 (1979) 1–45.
15.   K.A. McLauchlan, in: J.P. Fouassier and J.F. Rabek (Eds), Applications of lasers in polymer science and technology, C.R.C. Press, in press.
16.   K.A. McLauchlan and D.G. Stevens, Accts. Chem. Res., 21 (1988) 54–59.
17.   K.A. McLauchlan and D.G. Stevens, J. Chem. Phys., 87 (1987) 4399–4405.
18.   A.D. Trifunac, D.J. Nelson and C.T. Mottley, J. Magn. Reson., 30 (1978) 263–272.
19.   A.D. Trifunac, Chem. Phys. Lett., 49 (1977) 457–458.
20.   T.J. Burkey, J. Lusztyk, K.U. Ingold, J.K.S. Wan and F.J. Adrian, J. Phys. Chem., 89 (1985) 4286–4291.
21.   C.D. Buckley and K.A. McLauchlan, Chem. Phys. Lett., 137 (1987) 86–90.
22.   P.W. Atkins and G.T. Evans, Molec. Phys., 27 (1974) 1633–1644.
23.   J.B. Pedersen and J.H. Freed, J. Chem. Phys., 62 (1975) 1706–1711.
24.   L. Monchik and F.J. Adrian, J. Chem. Phys., 68 (1978) 4376–4383.

25. J.B. Pedersen and J.H. Freed, J. Chem. Phys., 59 (1973) 2869–2885.
26. K.A. McLauchlan and D.G. Stevens, J. Magn. Reson., 63 (1985) 473–483.
27. Y. Sakaguchi, H. Hayashi, H. Murai and Y.J. I'Haya, Chem. Phys. Lett., 110 (1984) 275–279.
28. K.A. McLauchlan, in: Proceedings of the 2nd Workshop on Electron Spin Echo Spectroscopy, Amsterdam, 28 – 31 March 1988, Royal Dutch Academy of Sciences, Amsterdam, in press.
29. M.C. Thurnauer, T.–M. Chiu and A.D. Trifunac, Chem. Phys. Lett., 116 (1985) 543–547.
30. I. Carmichael and H. Paul, Chem. Phys. Lett., 67 (1979) 519–523.
31. H. Paul, Chem. Phys., 40 (1979) 265–274.
32. H. Paul, Chem. Phys., 40 (1979) 294.
33. H. Paul, J. Phys. Chem., 86 (1982) 4087–4088.
34. K.A. McLauchlan and D.G. Stevens, J. Chem. Soc. Farad. Trans. 1, 83 (1987) 29–35.
35. V.I. Valyaev, Yu.N. Molin, R.Z. Sagdeev, P.J. Hore, K.A. McLauchlan and N.J.K. Simpson, Molec. Phys., 63 (1988) 891–900.
36. F. Jent, H. Paul, K.A. McLauchlan and D.G. Stevens, Chem. Phys. Lett., 141 (1987) 443–449.
37. A.I. Grant, N.J.B. Green, P.J. Hore and K.A. McLauchlan, Chem. Phys. Lett., 110 (1984) 280–284.
38. D.G. Stevens, D.Phil. thesis, Oxford, 1987.
39. K.A. McLauchlan and N.J.K. Simpson, in preparation.
40. W. Lersch and M.E. Michel–Beyerle, this volume.
41. C.D. Buckley, D.A. Hunter, P.J. Hore and K.A. McLauchlan, Chem. Phys. Lett., 135 (1987) 307–312.
42. G.L. Closs, M.D.E. Forbes and J.R. Norris, J. Phys. Chem., 91 (1987) 3592–3599.
43. J.B. Pedersen, in ref. (13), pp. 169–180.
44. R.W. Fessenden, J. Chem. Phys., 58 (1973) 2489–2500.
45. P.W. Atkins, K.A. McLauchlan and P.W. Percival, Molec. Phys., 25 (1973) 281–296.
46. P.J. Hore and K.A. McLauchlan, Rev. React. Intermed., 3 (1979) 89–105.
47. P.J. Hore and K.A. McLauchlan, Molec. Phys., 42 (1981) 533–550.
48. P.J. Hore, K.A. McLauchlan, S. Frydkjaer and L.T. Muus, Chem. Phys. Lett., 77 (1981) 127–130.
49. K.A. McLauchlan and D.G. Stevens, Molec. Phys., 60 (1987) 1159–1177.
50. K.A. McLauchlan, R.C. Sealy and J.M. Wittman, Molec. Phys., 35 (1978) 51–63.
51. K.A. McLauchlan, R.C. Sealy and J.M. Wittman, Molec. Phys., 36 (1978) 1397–1407.
52. S. Basu, K.A. McLauchlan and A.J.D. Ritchie, Chem. Phys. Lett., 105 (1984) 447–450.
53. P.J. Hore and K.A. McLauchlan, Molec. Phys., 42 (1981) 1009–1026.
54. K.A. McLauchlan and A.J.D. Ritchie, Molec. Phys., 56 (1985) 141–159.
55. K.A. McLauchlan and A.J.D. Ritchie, Molec. Phys., 56 (1985) 1357–1367.
56. J.B. Pedersen, FEBS Letts., 97 (1979) 305–306.
57. C.T. Holder, K.A. McLauchlan and N.J.K. Simpson, in preparation.
58. K.A. McLauchlan and D.G. Stevens, Chem. Phys. Lett., 115 (1985) 108–110.
59. S. Basu, D.Phil. thesis, Oxford, 1983.

CHAPTER 11

TRANSIENT EPR-SPECTROSCOPY OF PHOTOINDUCED ELECTRONIC SPIN STATES
IN RIGID MATRICES

DIETMAR STEHLIK, CHRISTIAN H. BOCK, MARION C. THURNAUER

1. INTRODUCTION

The transient nutation method introduced by Torrey (1) is actually the first transient magnetic resonance technique realized, for a good treatment see Abragam (2). Compared to transient methods with pulsed rf-irradiation it lost practical importance with the advent of modern Fourier-transform spectroscopy in NMR. As outlined in this contribution it is still of considerable interest in EPR.

The descriptive early name "nutation" taken from the physics of symmetric tops makes the phenomenon appear more complex than it is. Today we are used to viewing spin motions in the appropriate interaction frame, especially the so-called rotating frame in the presence of resonant rf-fields. In this reference frame, rotating with the frequency of the rf-field around the static external field, the rf-field becomes a constant field and the Larmor precession around the fixed laboratory field is removed. Viewed in this way the transient nutation is in fact a precession of the magnetization around the effective field in the rotating frame, which at resonance is just the rf-field $\underline{B}_1$.

As pointed out by Torrey, in contrast to steady state methods the transient techniques measure the full magnetic moment $M_z(0)$. This property is of particular importance in transient EPR of photo- or radiation-induced paramagnetic species where large initial spin polarization occurs due to spin selective population processes. Only transient techniques can take full advantage of it. For the Torrey method the possibility of pulsed photoexcitation of the spin species offers in addition a convenient technical simplification, because it eliminates the need of a suddenly turned-on microwave (mw) field and replaces this by a sudden generation of the polarized paramagnetic spin ensemble in a continuous mw-field.

In pioneering work S. I. Weissman has emphasized both the usefulness of the nutation technique as well as the experimental simplicity of the detection of transient electron spin magnetization following photoexcitation (3). In comparison with pulsed EPR-techniques (4) it was pointed out that any transient detection in the presence of the mw-field suffers some drawbacks, limiting the ultimate time resolution. Usually the rise of the transient nutation signal itself is determined by the microwave field, thus, masking any faster kinetics of the paramagnetic species. However, this restriction is limited to cases of homogeneously broadened lines, as in liquid samples, and is not appropriate (5) for the case of an inhomogeneously broadened line, which is the rule rather than the exception in solid samples. Instead, the rise time is determined by the inverse of the inhomogeneous line width and can approach the 15 ns range (6) for a line width of the order of 1 mT. Thus, under these conditions the time resolution of the transient nutation technique (independent of the chosen rf-field strength) competes favorably with pulsed EPR-techniques.

2. BASICS

2.1 The Transient Nutation Experiment

The basic idea of Torrey's experiment (1) is illustrated in Fig. 1. The concept is quite obvious if viewed in the rotating reference frame which rotates around the static magnetic field $B_o \parallel z$ with the resonant mw- or rf-field

$$\omega_{mw} = \omega_0 = (g\beta/\hbar) \, B_o = \gamma B_o$$

set at the Larmor frequency ω_0 of the spin system. In this reference frame the Zeeman-interaction of the static magnetic field B_o is transformed away and the proper circularly polarized component of the usually linearly polarized mw-field $B_{1x}(t) = 2B_1 \cos\omega_{mw} t$ becomes a static field B_1 shown in Fig. 1 along the x-axis of the rotating frame. (The counter-rotating mw-field component can be neglected in high field, e.g. (2)).

When the B_1-field is suddenly turned-on at $t = 0$ the initial magnetization $M_z(0)$ aligned along $B_0 \parallel z$ will precess around the B_1-field. Phase-sensitive detection as commonly applied today can be associated in the rotating frame with a constant detection phase with respect to the B_1-field. Fig. 1 shows the 90° out-of-phase detection along the y-axis (absorption mode). The signal $S(t)$ is determined by the y-component of the magnetization which is readily obtained from the precession motion:

$$S(t) \propto M_y(t) = M_z(0) \, \sin\omega_1 t \, e^{-t/T_2} \tag{1}$$

with $\omega_1 = \gamma B_1$. [1]

Experimentally the precession is always found to be damped by the spin spin relaxation time T_2.

As pointed out first by S. I. Weissman (3) the same result will be obtained with continuous microwave fields when photoinduced paramagnetic species are suddenly brought into existence by an appropriate light pulse (see bottom part of Fig. 1). Obviously, the width of the light pulse should be short compared to the characteristic times of the spin motion ($1/\omega_1$ and T_2). Similarly, the rise time of the spin magnetization should be fast. Finally, the spins should not be generated equally distributed over the spin levels but rather selectively corresponding to a non-zero initial magnetization $M_z(0) \neq 0$ (spin polarization). For this case any standard continuous wave (cw)

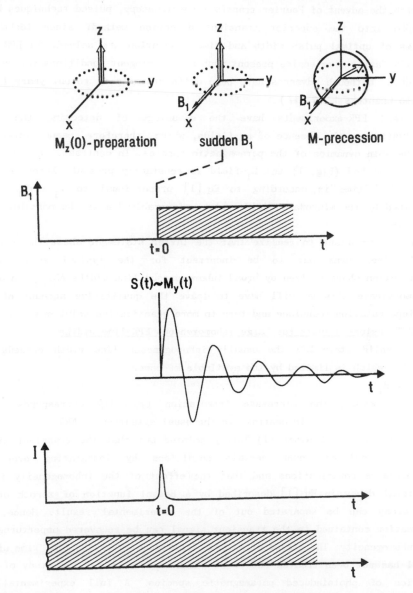

Fig. 1. Scheme of transient nutation experiment. Preparation-, generation- and detection period viewed in the rotating frame. Top: With gated microwave turn-on at t = 0. Bottom: With cw microwave and light pulse at t = 0 for photoinduced paramagnetic states.

EPR-spectrometer will be sufficient to perform this transient EPR-experiment provided the detection bandwidth is adapted to the time resolution needed to observe an undistorted transient signal S(t).

Alternatively, $M_z(t)$ can be observed by pulsed magnetic resonance techniques using the free precession or spin echo signal, see e.g. (2,4). In NMR, with the advent of Fourier transform spectroscopy, pulsed techniques have developed into the superior transient detection method, since technical problems of optimal pulse width and time resolution are solved. In EPR the generally faster time scales present much more stringent conditions for pulsed techniques, although advances have been quite dramatic in recent years (7-9, see also Chapters 1..7, 17).

Pulsed EPR-experiments have the advantage of detecting the spin magnetization in the absence of mw-fields, which, therefore, cannot interfere with the spin dynamics of the paramagnetic species. In contrast, for Torrey's transient method (Fig. 1) the B_1-field is constantly present. Moreover, the signal rise time is according to Eq.[1] proportional to $\omega_1 = \gamma B_1$, i.e. determined by the microwave field, which unfavorably limits the possible time resolution.

It is important to realize that the latter statement about the initial signal rise turns out to be incorrect for the typical solid state EPR-situation characterized by broad inhomogeneous line widths $\Delta B_{1/2}$. In order to demonstrate this we will have to leave the qualitative account of the transient nutation technique and turn to more quantitative solutions.

2.2 <u>Transient Signals for Large Inhomogeneous EPR-line widths</u>

In solid state EPR the usually inhomogeneous line width exceeds the B_1-field, which is limited by the available mw-power:

$$\Delta B_{1/2} \gg B_1 \tag{2}$$

In other words, the microwave irradiation typically corresponds to a narrow-band excitation in contrast to the usual situation in NMR.

In his original paper (1) Torrey pointed out that the transient signal can be calculated under certain conditions by integrating over all off-resonance contributions and that the effect of the inhomogeneity is an additional decay in Eq.[1] described by a Bessel function of zeroth order. This decay can be separated out of the experimental result. Hence, the information contained in the transient signal can be recovered unperturbed by the inhomogeneity. The effect of a large inhomogeneity on the rise time of the signal has not been treated but is of prime importance for the study of fast kinetics of photoinduced paramagnetic species. A full experimental and theoretical account of the transient signals in this case has been given in (5) and the essential aspects will be reviewed here.

We consider an optically generated paramagnetic spin species with the properties mentioned in the previous section. A pair of spin states between which a resonant microwave field induces transitions constitutes an effective two-level system. A high spin polarization is usually associated with photoexcitation corresponding to a large population difference for the two-level system and a large magnetization $M_z(0)$. Its time development is adequately described by the Bloch equations (10,11), which in the rotating frame reduce to the following form:

$$\dot{M}_x = -(1/T_2)\, M_x(t) + \Delta\omega\, M_y(t)$$

$$\dot{M}_y = - \Delta\omega\, M_x(t) - (1/T_2)\, M_y(t) - \omega_1\, M_z(t)$$

$$\dot{M}_z = \omega_1\, M_y(t) - (1/T_1)(M_z(t) - M_{z,eq})\,. \qquad [3]$$

$\Delta\omega = \omega_{mw} - \omega_0$ is the off-resonance setting of the microwave frequency. $\omega_0 = \gamma B_0$ and $\omega_1 = \gamma B_1$ are Larmor frequencies in the respective fields. $M_{z,eq}$ corresponds to the Boltzmann population. In all cases considered here $M_{z,eq}$ can be neglected because the initial optically induced spin polarization is much higher.

After the light pulse the initial condition is assumed to be

$$M_x(0) = M_y(0) = 0;\ M_z(0) \neq 0\,, \qquad [4]$$

i.e. no transversal magnetization or spin coherence is generated. In general, this assumption may not be appropriate. General solutions of the Bloch equations [3] are given in the original work (1) as well as in e.g. (12). We want to use them to handle the narrow-band excitation case described by inequality [2], i.e. we have to consider a superposition of transients, each describing the transient of a spin packet of homogeneous line width. The resonance frequencies of all spin packets constituting the inhomogeneous line are described by a normalized line shape or distribution function $f(\omega_0)$. The normalized signal is then obtained by integration:

$$S(t) = \int_{-\infty}^{+\infty} f(\omega_0)\, M_y(\omega_0, t)\, d\omega_0. \qquad [5]$$

Numerical integration will be appropriate for general line shape functions $f(\omega_0)$. In special cases, analytical results are available, which we want to treat first. We consider three particular aspects of practical interest in the experiments:

a) oscillatory solution for $\omega_1^2 > (1/4)\,(1/T_1 - 1/T_2)^2$ (1), which for $T_2 \ll T_1$ reduces to

$$\omega_1 = \gamma B_1 \gg 1/T_2\,, \qquad [6]$$

b) short time behavior of $S(t)$ in [5] ,

c) case of overdamping with

$$\omega_1 \ll 1/T_1, \ 1/T_2 \ . \tag{7}$$

For case a) we can use the solution $M_y(t)$ given by Torrey (1), which we simplify given the usual solid state situation where $T_1 \gg T_2$. With the notations used for the Bloch equations [3] we obtain for [5]:

$$S(t) = -M_z(0) \int_{-\infty}^{+\infty} f(\omega_0) \frac{\omega_1}{\omega_{eff}} \sin(\omega_{eff} t) \exp\{-\frac{1}{T_2}[1 - \frac{1}{2}\frac{\omega_1^2}{\omega_{eff}^2}]\} d\omega_0 \tag{8}$$

$$\omega_{eff} = \sqrt{\omega_1^2 + (\Delta\omega)^2} = \sqrt{\omega_1^2 + (\omega_{mw} - \omega_0)^2} = \gamma B_{eff}$$

constitutes the rotating frame precession frequency around the effective field $B_{eff} = \sqrt{B_1^2 + [(\omega_{mw}/\gamma) - B_0]^2}$. If we assume a large inhomogeneous line width, [8] can be solved analytically. Specifically, we assume inequality [2] with line width $\Delta B_{1/2} \gg B_1$ and exclude short times (case b) with $\omega_{eff} t > 1$. Then the width of $f(\omega_0)$ is large compared to the width of all factors in [8] and the slowly varying $f(\omega_0)$ can be taken in front of the integral and replaced by $f(\omega_{mw})$, the value of the normalized line shape at the actual microwave frequency, corresponding to a particular field in the magnetic field sweep mode. For each spin packet all factors of the integral are maximal and the integration yields the analytical result:

$$S(t) = -\pi \, M_z(0) \, \omega_1 \, f(\omega_{mw}) \, J_0(\omega_1 t) \, \exp(-\frac{t}{2T_2}) \ . \tag{9}$$

Assuming $T_1 = T_2$ an analogous result has been obtained by Torrey (1), the simpler oscillatory result of Eq.[1] is replaced in [9] by the Bessel function $J_0(\omega_1 t)$ of zeroth order while the relaxation damping remains the same. Eq.[9] predicts a vertical slope at $t = 0$, which is not correct as we have omitted the short time behavior in its derivation. A proper analysis of the short time range is the next case of interest.

For case b) we start with Eq.[8]. For short times the relaxation damping can be ignored and the sine function can be replaced by its argument. We get

$$S(t) = -M_z(0) \int_{-\infty}^{+\infty} f(\omega_0) \, \omega_1 t \, d\omega_0 = -M_z(0) \, \omega_1 t \ . \tag{10}$$

From this we derive a linear slope $dS(t)/dt$. For the rise time of the signal we first introduce a practical definition, the time τ_R needed to reach the maximum signal $S(0)$ of Eq.[8] with the slope of [10] which yields

$$\tau_R = \frac{S(0)}{dS(t)/dt} = \pi \, f(\omega_{mw}) \sim \pi/\Omega \ . \tag{11}$$

For a normalized line shape function the amplitude is about equal to the inverse line width $\Omega = \gamma \, \Delta B_{1/2}$, which gives the last equality in [11]. The rise time is, therefore, determined by the inverse of the inhomogeneous EPR-line width.

For arbitrary line shape functions the signal rise can be obtained from a numerical integration of Eq.[5], or [8] for the oscillatory case. Numerical examples for a Gaussian line profile and a roughly triangular line shape caused by an only partially resolved hyperfine structure are given in Ref.(6). They confirm the above conclusion that the rise time is determined by the inverse of the EPR-line width, see also Section 4.1.

Furthermore, it should be noted that in the analysis of the rise time behavior the microwave field amplitude or ω_1 drops out, see Eq.[11]. Even at very low mw-power the signal rise will be governed by the inverse of the EPR-line width.

For case c) (overdamping) we can adapt the off-resonance solution of Atkins et.al. (12) with $T_1 \gg T_2$ and $M_{z,eq} = 0$. The oscillatory terms of $M_y(t)$ with frequency $\Delta \omega$ will cancel when we integrate over the inhomogeneous line width $\Omega \gg \omega_1$. With low microwave power $(\omega_1^2 \, T_1 T_2 \ll 1)$ the remaining part is damped with $\exp(-t/T_1)$ and integration of Eq.[5] yields

$$S(t) = -\pi \, M_z(0) \, \omega_1 \, f(\omega_{mw}) \, \exp(-t/T_1) \qquad [12]$$

and the rise time $\tau_R \sim \pi/\Omega$ [11] as determined above is independent of ω_1. Note that for a homogeneous line the normal transient signal at low microwave power has a rise time given by T_2, which is replaced here by the inverse line width, a time constant frequently termed T_2^*.

In summary, the oscillatory character with frequency ω_1, the relaxation damping with T_2 in the oscillatory regime and with T_1 in the overdamped case as predicted and demonstrated in the early transient nutation work (1,2), are retained in the case of a large inhomogeneously broadened line characteristic for solid state EPR. However, the rise time of the signals – independent of the chosen microwave power – is determined by the inverse line width. Consequently, this will be the limit of time resolution, which will be typically in the 10 ... 30 ns range and, therefore, much faster than commonly associated with the transient nutation technique.

We have treated here the signal proportional to $M_y(t)$, i.e. 90° out-of-phase to the B_1-field (absorption mode), because noisewise it is the one least affected by the microwave source. The corresponding in-phase signals (dispersion mode) are readily obtained by differentiation with respect to ω_{mw} (2,12) and contain no new information.

2.3 Spectral Information from the Transient Signals S(t)

Both, the oscillatory ($\omega_1 > 1/T_2$) solution [9] and the overdamped ($\omega_1 < 1/T_1$) solution [12] predict the signal amplitude $S(t) \propto f(\omega_{mw})$ independent of time. Therefore, if the microwave frequency ω_{mw} (or at fixed ω_{mw} the magnetic field B_0) is swept across the distribution function $f(\omega_0)$ at any fixed time t_f, a plot of the $S(t_f)$ values will yield the EPR line profile

$$S(t_f, \omega_{mw}) \propto f(\omega_{mw}) \; . \qquad\qquad\qquad [13]$$

This implies that with a sweep of the microwave frequency or the static magnetic field the transient decay function remains unchanged while the amplitude varies with the line shape function $f(\omega_0)$.

At short times after the laser pulse more and more off-resonance contributions are expected to contribute to the line shape function because the pulsed photoexcitation even with continuous microwave field, corresponds at early times to the effect of an intense sharp microwave pulse and broad excitation bandwidth. Consequently, for early times t_f Eq.[13] changes to a broadened line profile unless t_f becomes long compared to the inverse width of the line profile.

For oscillatory conditions [6] the transient decay function bears the full spectral information just like the free induction decay (FID) after an intense microwave pulse and can be displayed on a frequency scale by a Fourier transformation. We do not consider this situation in any further detail because the technical advantages of pulsed microwave methods, i.e. detection in absence of the microwave field, make the latter clearly superior.

In the case of spin species with g-anisotropy, dipolar coupling etc. in isotropic media, i.e. statistical orientations with respect to the magnetic field, the line shape function $f(\omega_0)$ would be a so called powder pattern. In order to maintain an oscillatory character of the transient signal we have to assume that individual spin packets cannot exchange spins during the detection time range. In the dynamic case when exchange of spin packets can occur, Eq.[13] may not hold anymore, because the spins can see different effective fields during the course of the transient signal. The transient decay function will depend on the dynamics of the system and may vary with the spectral position. However, these changes in the decay curve can contain useful information on the molecular dynamics, see Section 4.4.

Reliable line shape data can be expected from [13] for low microwave power, because according to Eq.[12] the time decay function is then determined only by the spin relaxation. Anisotropic spin relaxation may occur. In this case, reliable line shape functions $f(\omega_0)$ can only be obtained at times short compared to the relaxation times.

3. INSTRUMENTAL ASPECTS

The different versions of time resolved EPR techniques are treated in the preceeding chapter of this book. We concentrate here on the transient spectroscopy of photoexcited spin species. As indicated earlier standard cw EPR equipment can be used in this case provided appropriate pulsed light sources are available and the detection part of the spectrometer is adapted to observe the transient response with the chosen time resolution. In the following we will focus on these specific aspects of instrumentation.

Normally, sensitivity problems will be the main concern. In most practical cases they limit the time resolution that can be achieved. For a given application one will always try to use the detection technique that provides the best sensitivity for the required time resolution.

In the Berlin laboratory a standard Varian E-3 spectrometer (X-band, 9 GHz) has been modified for the detection of photoinduced transient signals and in a similar way a home built spectrometer (K-band, 24 GHz). Block diagrams are shown in Fig.2 and 3, respectively. Only in Fig.2 are various detection channels included. In the following, various aspects of general interest are addressed first. They are mostly applicable to either spectrometer set-up. For the K-band spectrometer we comment only the fast-time resolution mode shown in Fig.3. Finally, we add some remarks on the data processing part.

3.1 Light Excitation

Any pulsed light source with a sufficiently narrow pulsewidth and an appropriate wavelength range can be used for the experiments described here. Commonly, pulsed lasers (nitrogen-, excimer-, Nd-YAG-laser, also in combination with dye lasers) are used. Their pulsewidth, in the order of 10 ns, is appropriate for the fastest possible EPR time resolution. The repetition rates in the 1 to 100 Hz range are also suitable because repetitive optical pumping cycles are usually limited by the decay time of the excited states or of other paramagnetic species back to the original ground state. Often this decay rate rather than the available repetition rate of the light source sets the limit for a high averaging rate.

Electrical noise from the laser electronics usually results in unwanted background signals in the EPR broad band detection. Considerable suppression of these signals can be achieved with improved ground connection of the flash lamp switch unit in the laser or additional shielding. Removal of further laser pulse artefacts is mentioned in Section 3.5.

Efficient illumination of the sample inside the cavity may be an

380

important sensitivity aspect. Usually, samples with high optical density are used and distribution of the light intensity over as large a sample area as possible (e.g. with an opalescent glass in front of the cavity's light access window) rather than focusing provides advantages.

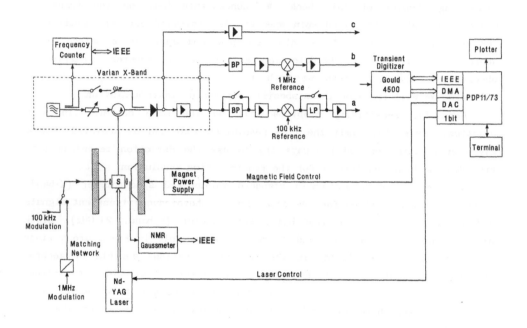

Fig. 2. Block diagram of a modified Varian E-3 spectrometer (X-band, 9 GHz). Three detection channels a,b,c can be used alternatively.

3.2 Modification of Standard X-band Spectrometer

Several groups have reported useful modifications of commercial (X-band) EPR-bridges for the detection of photoinduced transient signals (e.g. 13-17). The modifications installed in the Berlin laboratory are shown in Fig. 2. Three parallel detection channels have been used, corresponding to different requirements in time resolution.

3.2.1 100 kHz Modulation

Usually, the bandwidth of the signal amplifiers is not sufficient to reach the time resolution limit possible with 100 kHz lock-in amplifier. The output low pass filter must be eliminated or modified for higher cut-off frequency (13). Q-reduction of the bandpass input filter by a switchable parallel resistor can further improve time resolution to typically 20 μs.

3.2.2 Higher Modulation Frequency

In order to retain the noise reduction and elimination of baseline drifts as virtues of the modulation technique several systems with 1 or 2 MHz modulation frequency have been introduced with time resolution between 1 and 2 μs (14–17). In addition to a new lock-in amplifier adjustment of the spectrometer's preamplifier to a larger bandwidth is needed. Properly matched field coils have to be designed for the high modulation frequency. Increased losses due to the skin effect result in easier overheating of the cavity. Therefore, the modulation amplitude is limited to < 2 mT.

Compared to the 100 kHz modulation the increased bandwidth for a time resolution of about 2 μs requires a longer averaging time for the same signal-to-noise ratio. As the higher modulation frequency is only useful for the rather narrow 2–20 μs time resolution range, installation of such a detection channel may often not be worth the effort, if direct detection is also available.

3.2.3 Direct Detection

In this mode of operation the signal is directly observed at the output of the mw-diode without field modulation and lock-in technique. Note that EPR spectra taken in this way appear in the direct absorption or dispersion mode not in the usual derivative form associated with effect modulation and phase sensitive detection. The need of very broad bandwidth operation is paid for by reduced signal-to-noise ratio and uncertainty in the baseline. With digitized transient recording the latter problem can be dealt with by suitable methods of baseline correction. The first problem is unavoidable and practically limits fast detection to spin-polarized species, which fortunately are an inherent property of most photogeneration processes.

Sufficient bandwidth must be assured for each of the detection components. For our Varian E-3 spectrometer the signal was derived directly from the mw-diode to a BNC connector with a 0.22 μF capacitor in series and a 22 kΩ resistor in parallel and fed into an Avantek low-noise-amplifier stage (400 MHz bandwidth, 35 dB gain). The high output impedance of the diode mount (600 Ω or higher) was matched with a signal transformer to the 50 Ω input impedance of the amplifier. The internal preamplifier remained connected in parallel to keep the AFC operative. The overall time constant of the detection side was determined to be about 50 ns and was limited by the Q of the standard rectangular cavity (TE 102), which introduces a ringing time $t_R \sim Q/\omega_{mw}$.

The choice of the microwave cavity will depend on the priority requirement of the particular application. Often, in particular with biological samples, sensitivity will be of prime importance, even if some time resolution has to be sacrified. Standard rectangular cavities combine high Q with high aperture for light illumination via grid windows or a waveguide chimney with sufficient length and dimensions smaller than the cut-off size at λ/2. Excellent Q-values can be maintained for aqueous samples by using flat cells. They provide the additional advantages of a large illumination area and small thickness for samples with high optical density. Commonly, the sample cell will be connected to a flow system to prevent overexposure to high light intensity and microwave heating.

If the top priority is fast-time resolution and high B_1-fields, as for pulsed EPR and high nutation frequency experiments, more advanced resonator designs are advisable, like the loop gap (see Chapter 7) or slotted tube resonator (18).

3.3 Direct Detection K-band Spectrometer

Fig.3 shows a block diagram of the K-band (24 GHz) spectrometer. In general, for higher microwave frequencies there are fewer choices of advanced microwave components and they are more expensive. However, the higher spectral resolution (g-anisotropy) may be crucial for particular applications as exemplified in the experimental section.

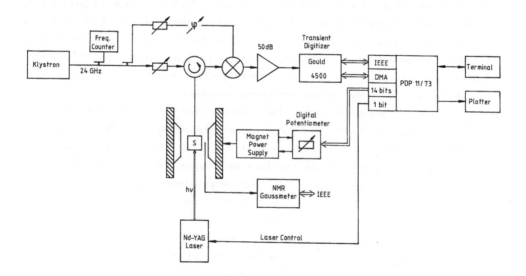

Fig. 3. Block diagram of a home-built K-band spectrometer (24 GHz) in the direct detection mode. Computer control and data processing is similar as that for the X-band spectrometer (Fig. 2).

The output of a cw-klystron (180 mW) reaches the cylindrical TE 011 cavity via an attenuator and circulator. The signal passes back through the circulator and is demodulated in a double balanced mixer providing a bandwidth of 1 GHz. The limited rise times of the available K-band diodes required the use of such a mixer at the cost of some signal-to-noise ratio. A 60 dB gain broad band (400 MHz) amplifier cascade was directly connected to the IF port of the mixer. The total time response reached in this set-up was again mainly limited by the Q of the cylindrical cavity (TE 011) and determined to be about 50 ns.

A phase shifter was introduced in the reference arm with the possibility of selecting an arbitrary detection direction in the xy-plane of the rotating frame (see Fig. 1).

A PIN diode switch (rise time 100 ns, 40 dB attenuation) in the main mw-transmission line was introduced to perform the delayed transient nutation and the driven echo experiments (5), described in Section 4.3.

3.4 Transient EPR-spectroscopy

The transient signal following the light flash at a fixed magnetic field setting is usually stored in a suitably fast transient recorder (in our case: Gould 4500 with a rate of 10 ns/8 bit sample) and averaged over repetitive optical pumping cycles.

In order to obtain transient spectra it is common to use a boxcar integrator. It integrates the transient signal in a time gate of suitable width and with a fixed but adjustable delay time with respect to the excitation flash. The magnetic field may be slowly swept through a predefined scan range. The filtered output of the boxcar integrator then provides a transient EPR-spectrum (14).

When various spin species with different kinetics and spectral patterns contribute to the transient signal, the boxcar technique requires independent knowledge about the system to set the proper time gates. Otherwise the results may be misleading. A more general approach is to start from the complete kinetic and spectral information. This consists in a set of transients taken at a suitable number of magnetic field points and stored in the computer as an original data set. Suitable transient spectra can now be evaluated using digital methods only. Technical requirements for this procedure are a suitable magnetic field control by the computer, proper organization of data acquisition and averaging as well as convenient software for data evaluation and presentation.

3.5 Aspects of Data Acquisition and Processing

In this mode of operation a central computer (PDP 11/73) controls as many parameters as possible. The magnetic field is set via a D/A converter or a digital potentiometer and can be checked by an NMR-probe and Gauss-meter. The

computer controls the sequence of excitation light pulses by activating the
Q-switch in the laser resonator. The transient EPR-signals are digitized in a
transient recorder (Gould 4500). Due to its limited averaging capabilities the
digitized transient data are transferred after each light pulse to the
computer memory via a DMA interface for averaging. A typical transient
contains 500 time points. Data transfer and adding is fast enough to allow a
30 Hz repetition rate, which satisfactorily suits the available laser
repetition rate range and the limiting kinetic rates in virtually all optical
pumping cycles studied. Typically, transients are averaged sufficiently at a
given magnetic field setting before stepping to the next field position.

Full time resolution of the transient signal (500 points) taken at
typically 161 field positions yields data sets of 320 kB. Each such raw data
set is first transferred to a back-up storage for further reference. It may be
subjected to standard procedures like filtering and baseline corrections
carried out in two versions. 1) An average of off-resonance transients is
subtracted, thus eliminating all spurious laser background and shifts due to
photoinduced effects inside the cavity (dynamical correction). 2) The signal
level before the light pulse or after the signal decay is subtracted for each
transient signal in order to eliminate baseline drifts (static correction).

As demonstrated with experimental examples in the next section the best
overall view of the full data set is obtained from a three-dimensional plot of
the signal intensity versus the kinetic (time) and spectral (magnetic field)
axis. From the various methods available we selected the projection of a
surface consisting of a suitable number of principal points using a hidden
line algorithm.

Individual transients at a particular field setting may be examined by
adding all transients in a magnetic field range adjusted to be narrower than
the actual line width studied.

Boxcar type spectra may be obtained by setting an appropriate digital
time gate at a suitable delay after the laser pulse and integrating the
transient signal over this time interval. This procedure has also been termed
time integration spetroscopy (21). Another way of spectra evaluation is very
useful, if more than one paramagnetic species contribute and each follows
different kinetics with possibly multiexponential rise and decay functions.
After the appropriate time functions are fitted, a plot of the coefficients
versus the magnetic field yields the separated EPR-spectra of each of the
different spin species. For an example see Section 4.5 and (49,50).

4.   EXPERIMENTAL EXAMPLES AND APPLICATIONS
Early application of transient EPR involving pulsed radiation excitation

and continuous microwave fields dates back to 1973 (20) and concerned radiation-induced radical reactions in solutions, for a review see (21). Predominant candidates for applications in solids are optically excited molecular triplet states (3,5), which in most cases studied have been well characterized by low temperature kinetic and spectroscopic studies (22). For the transient EPR-spectroscopy described here triplet states have a number of attractive features:

(i)   Following singlet absorption fast Intersystem Crossing (ISC, typically of the order of 10^{-10} s) assures sudden photogeneration on the EPR time scale.

(ii)   Symmetry selection rules provide selective spin level population and, hence, high initial spin polarization ($M_z(0)$ values). Sensitivity, even at high temperatures and fast time resolution, is no serious limit.

(iii)   The high temperature lifetimes are long enough ($\gtrsim 10^{-6}$ s) to fit into the EPR time resolution range. They are also sufficiently short ($< 10^{-2}$ s) to permit reasonable repetition of the optical pumping cycle and thus good averaging efficiency.

Phenazine (3) and acridine (5) doped in single crystal matrices provide typical examples. Both molecules are well characterized with respect to their spectral, molecular and kinetic properties. They are ideal for the demonstration of the various aspects of the transient EPR-technique.

4.1 Characteristic Transient Signals in Solid State EPR (5)

Fig.4 presents the two types of transient signals, the oscillatory one (top) at high mw-power (180 mW, 0 dB) and the overdamped case (bottom) at low power (1.8 μW, -50 dB) for the high field triplet resonance of perdeuterated acridine, oriented with the long molecular in-plane axis along the magnetic field ($B_0 \parallel x$). Because the signals look rather similar to the corresponding ones in solution we emphasize the most important difference: both signals rise very fast within the overall time constant of the detection system (here 100 ns), which is the consequence of the narrow-band excitation within a broad inhomogeneous line width, see Section 2.2.

The solid line in Fig.4 (top) represents a satisfactory fit employing expression [9] with the following parameters, which in part can be evaluated directly from the data: a) The nutation period of 1.4 μs renders the nutation frequency $\omega_1 = 4.5 \times 10^6$ s^{-1} and with $\omega_1 = \sqrt{2}(g\beta/\hbar)B_1$ for triplet states $B_1 = 0.018$ mT, b) The relaxation function can be obtained by eliminating the Bessel function $J_0(\omega_1 t)$. An exponential relaxation with a temperature independent $T_2 = 1.3$ μs was found (5), c) From Eq.[11] with an inhomogeneous line width $\Delta B_{1/2} = 0.5$ mT we obtain a rise time $\tau_R \sim \pi/\Omega = 35$ ns, indeed

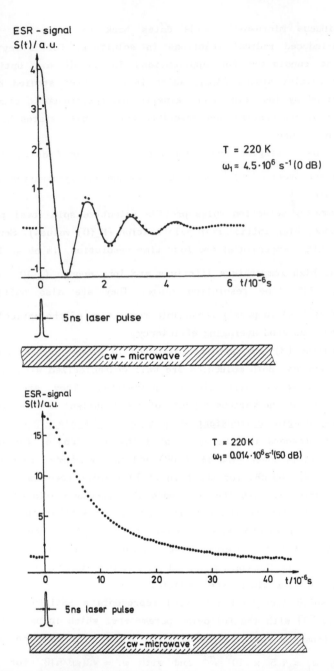

Fig. 4. Transient nutation signal S(t) following laser pulse excitation at t = 0. Sample and experimental details in Ref.(5). Top: Oscillatory case for maximum mw-power 180 mW. Solid line is fit with Eq.[9] and parameters given in text. Bottom: Overdamped case for mw-power reduced by −50 dB. Note the compressed time scale compared to top figure.

smaller than the instrumental rise time (100 ns) and much shorter than the nutation period. Therefore, inequality [2] as a condition for the derivation of [9] is well satisfied. The low power transient (Fig.4 bottom) is in agreement with Eq.[12]. As expected from $T_1 \gg T_2$ a much slower decay is observed. Independent T_1-measurements (see 4.3) show, however, that B_1 is still not small enough to render a reliable T_1-value. We emphasize once more the important aspect that due to the fast rise time the low power spectrum allows reliable transient EPR spectroscopy in the important time range $10^{-8} \ldots 10^{-5}$ s, with modulation techniques available for longer times.

It should be noted that the transient signal decays to zero. Indeed, the Boltzmann signal $M_{z,eq}$ is too small to be detectable as compared to the high initial spin polarization. Assuming complete spin polarization for $M_z(0)$ a signal-to-noise level of $\geq 2 \times 10^3$ would be required to detect $M_{z,eq}$, which is difficult but feasible (23).

With the high polarization the sensitivity is available to probe the initial rise of the signal in more detail. Fig.5 shows the start of transient signals for two line shape functions as taken from Ref.(6). The experimentally observed signal rise can be well reproduced by carrying out the numerical integration [5] over an independently determined line shape function. A Gaussian line with halfwidth $\Gamma = 0.5$ mT for Fig.5 (top) results in a rise time $\tau = 30$ ns as observed experimentally. It corresponds to the fastest time resolution possible for the signals of Fig.4. By rotation of the crystal the line shape can be broadened by increasing ^{14}N hyperfine coupling. With a halfwidth of $\Delta B_{1/2} \sim 1.5$ mT and an approximately triangular line shape, even the finer features and modulations seen in Fig.5 (bottom) can be simulated rather well (6). The approximate rise time of 15 ns approaches the physical limit for time-resolved EPR and is better than or comparable with what can be achieved with pulsed EPR-techniques.

In this context it should be noted that for high mw-power $\gamma B_1 \gg 1/T_2$ the information content of the transient nutation signal becomes equivalent to that of the free induction decay (FID) after a strong microwave pulse. In both cases Fourier transformation will render the spectral line shape provided the excitation bandwidth is available to cover the full width of the spectral line shape. The latter condition, however, is still a serious technical problem for most solid state EPR-applications.

4.2 Complete Kinetic and Spectral Information

As described in Section 3.4 a complete data set is a series of transient signals taken at equidistant magnetic field points covering the total spectral width, i.e. it is a two-dimensional variation of the signal intensity with

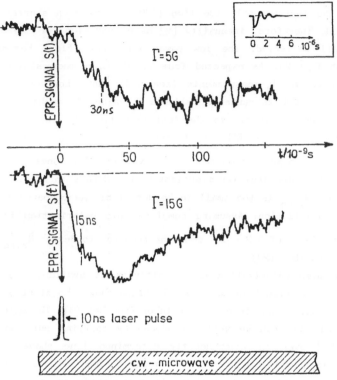

Fig. 5. Rise time of transient signals for two different line widths Γ and line shapes, see text and Ref.(6). The insert shows the full transient signal equivalent to that in Fig. 4 top.

respect to both the magnetic field and the time axis. Such a complete data set can be conveniently presented in a three-dimensional plot as shown in Fig.6 for the high field line of the acridine triplet state with the field $B_0 \parallel z$, the out-of-plane axis with the largest hyperfine splitting due to the ^{14}N-nuclear spin I = 1 (24).

The oscillatory transient decay function of Eq. [9] remains unchanged throughout the whole partially resolved spectrum. Hence, the EPR-spectrum can be obtained at any fixed time t_f after the laser pulse as expressed with Eq.[13]. An exception is only the very early time regime near the rise time $\tau_R \sim \pi/\Omega$ [11], where the line shape is broadened due to fast decaying off-resonance contributions (see Fig.6, bottom).

Fourier transformation of the oscillating transient signal at long times yields a single line for every transient. This indicates clearly that the condition of narrow-band excitation is fulfilled even with respect to the line width of each individual multiplet line. Obviously, for this case the full data set of Fig.6 does not contain any extra information compared to an individual transient as shown in Fig.4 (top) and a boxcar type spectrum at a

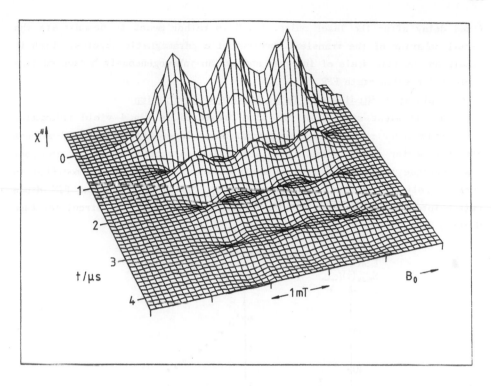

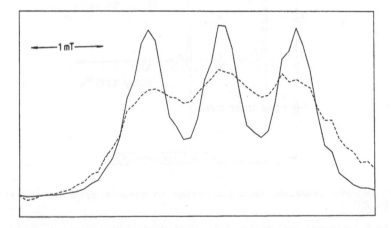

Fig. 6. top: complete kinetic and spectral data set for a triplet signal of acridine split by hyperfine coupling to the nitrogen spin $I(^{14}N) = 1$. The oscillatory pattern extends uniformly over the whole spectrum. bottom: spectra extracted from complete data set (top) within an early time gate after the laser pulse (0 – 100 ns, broken line) and a later gate (150 – 300 ns, solid line). The initial broadening effects quickly die out.

fixed delay after the laser pulse. Fig.6 is rather meant to demonstrate the normal behavior of the transient spectrum of a paramagnetic species, which is stable on the time scale of interest and has an inhomogeneously broadened line typical for solid state EPR.

4.3 Relaxation Studies with Gated Microwave Irradiation

The transient signals of the kind presented in Fig.4 yield relaxation times only under rather restricted conditions. Generally, pulsed EPR-methods are much better suited. However, we would like to mention that the transient EPR-technique based on continuous mw-irradiation can be easily modified to perform reliable relaxation measurements by the insertion of a PIN diode switch into the mw-transmission line between mw-source and the circulator (see Section 3.3).

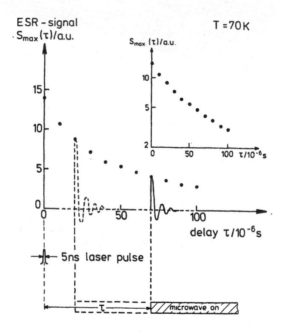

Fig. 7. Delayed transient nutation method to measure spin lattice relaxation rates (5,25).

Fig.7 presents an appropriate method for the determination of spin lattice relaxation times T_1 (5). The mw-field is switched on with a variable delay time τ after the laser flash. The amplitude of the transient nutation signal plotted as a function of τ renders the decay of the spin-polarized initial magnetization $M_z(0)$ towards thermal equilibrium, $M_{z,eq} \sim 0$. For a two-level system the time constant is T_1 as defined in the Bloch equations [3]. For spins $S > 1/2$ the time constant is an effective T_1 for the fictitious spin 1/2 two-level system associated with each resonant mw-irradiation connecting any two spin levels. For the acridine triplet state an extensive spin lattice relaxation study has been carried out with this method (25). The study covered the temperature range between 20 K and room temperature and established the relaxation mechanism to be due to thermal population of vibrationally excited triplet states with fine structure tensors, reoriented compared to the lowest triplet state.

Fig.8 demonstrates the feasability of an echo-type refocusing experiment (5), in which the refocusing pulse is actually a short interruption in the continuous microwave irradiation. This method was first analyzed in broad band solid state NMR and termed "driven echo" (26). It should be distinguished from the "rotary echo" (27), which refocuses by a 180° phase shift of the rf-irradiation. A more rigorous treatment employing the Bloch equations has been given (28) including an experimental comparison with the "rotary echo". The time constant accessible with this technique is essentially $T_{2\rho}$ describing the spin coherence decay in the rotating frame. It should be kept in mind that the driven echo method is of limited use in most relaxation studies because the theoretical interpretation of $T_{2\rho}$ is often aggrevated by additional complications.

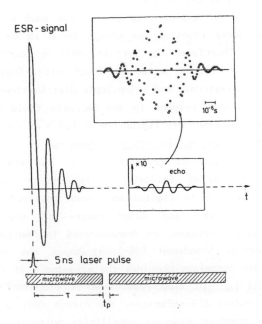

Fig. 8. Transient nutation "driven echo" phenomenon using a microwave turn-off pulse (5).

4.4 <u>Transient Spectra in Glasses and Liquid Crystals</u>

In spite of the large anisotropy the zero-field splitting of molecular triplet states can be observed at high temperatures in disordered solids (glasses) (29-32). Again the high initial spin polarization due to spin selective Intersystem Crossing is crucial to reach the required sensitivity level.

As a particular point, the sensitivity of the transient nutation spectrum to molecular mobility has been noted (31). While the steady state anisotropic triplet spectrum would experience motional averaging only if the motional correlation time τ_c is of the order of or faster than the inverse zero-field splitting (typically $< 10^{-9}$ s) the spin coherence involved in the transient nutation is already affected at much longer correlation times (order of 10^{-6} s).

This approach has recently been used (33) to study the molecular dynamics in a liquid crystal matrix via the transient nutation signal of the photoexcited triplet state of a suitable probe molecule, in this case chlorophyll. On the basis of a spin density matrix formalism a dynamic model has been developed for the simulation of transient EPR-signals, which covers the fast to ultra-slow motional regime and treats the interaction of the spins with the microwave field explicitly.

Although it is not expected in general for such a dynamic regime, the transient signal decay function was always found to be independent of the spectral position. Therefore, the spectra could be constructed according to Eq.[13] for any delay time t_f after the laser pulse. Typical triplet spectra of chlorophyll with isotropic and anisotropic distributions of orientations of the molecular axes with respect to the magnetic field at T = 220 K, both experimental and simulated, are reproduced in Fig.9 as taken from Ref.(33).

4.5 <u>Detection of Transient Paramagnetic Species</u>

The main application of transient EPR is the study of fast decaying paramagnetic spin species following pulsed photo- or radiation-induced generation. Applications in liquids have been reviewed in (4,21). In this context, the advantages and disadvantages of the various transient EPR-techniques were discussed. As demonstrated in Section 4.1 in solids or with inhomogeneously broadened EPR-line shapes the direct detection or transient nutation technique combines the advantages of fast time resolution and good sensitivity. In time resolution it competes well with the most straightforward pulsed EPR-technique, the direct Fourier transform EPR (9), which, however, provides superior sensitivity potential for highly resolved solution spectra, if the overall spectral width does not exceed 2 mT. For solid state EPR the technically much simpler direct detection method

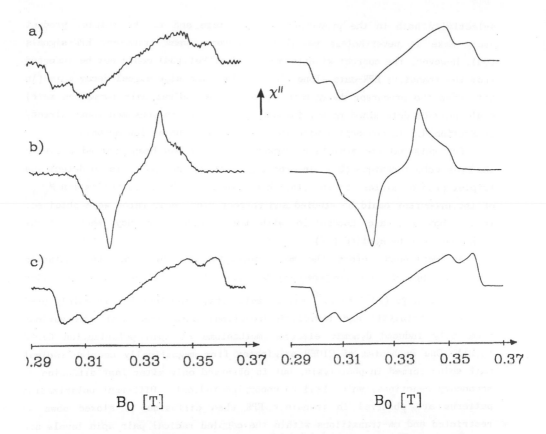

Fig. 9. Transient EPR-spectra of triplet chlorophyll a in a liquid crystalline matrix as taken from Ref. (33). Comparison of experimental (left) and calculated spectra (right). top: isotropic orientational distribution; middle: anisotropic distribution with director oriented parallel to the magnetic field; bottom: anisotropic distribution with director perpendicular to the magnetic field.

described here will probably remain advantageous for some years to come.

A special case of transient spin species detected with transient direct detection EPR (3,34) concerns a photogenerated triplet guest-host complex in doped fluorene crystals, which could be identified as a radical pair product of a reversible H-transfer photoreaction between a guest and a neighboring host molecule (35,36). We will refer to it here only to discuss some aspects of the potential of the transient nutation technique in comparison with pulsed EPR-methods.

The triplet radical pair product at 300 K is formed with a rate (300 µs) slow compared to the triplet decay (0.3 µs) (36). Obviously, such extreme kinetics are nearly prohibitive for a transient detection. High spin selectivity both in the precursor triplet state and in the triplet product decay make it nevertheless possible to observe weak transient EPR-signals (34). However, the correct kinetic model (b) of Ref.(34) could not be inferred from the transient EPR-data alone. In addition, the slow signal decay rate (in this case the precursor decay rate equal to the radical pair formation rate) could only be determined rather inaccurately for the reasons mentioned already in Section 4.1 in connection with the T_1-determination at low mw-power.

For this extreme example a comparative study has been carried out with the spin echo technique (23). Interestingly, it was not possible to detect the triplet product states. On the other hand, the equilibrium magnetization $M_{z,eq}$ of the precursor could be studied and correct slow decay rates were obtained. These, however, are accessible with much higher accuracy when optical techniques can be applied (37).

As mentioned before the main potential of the transient nutation technique applied to inhomogeneously broadened lines will be in the fast time resolution range (10^{-8} to 10^{-5} s). In solutions, transient EPR is mainly used to study photolytic and radiolytic reactions associated with the various chemically induced dynamic electron mechanisms of spin polarization CIDEP (4,21). One mechanism of CIDEP originates from magnetically coupled radical pair spins formed in photolysis, but is observed only after fast diffusion or secondary reactions, which lead to uncoupled radicals. Different polarization patterns are observed in transient EPR when diffusion is slowed down or restricted and mw-transitions within the coupled radical pair spin levels are detected directly (38-41). In this case, transient EPR is able to reveal the geometry of the radical pair complex via the magnetic interactions obtained from the transient spectra as well as dynamical information on the spin polarization mechanism and the reaction kinetics.

It would be particularly interesting to study such weakly coupled transient radical pairs in a rigid matrix where the anisotropic interaction parameters are accessible as well. Radical pairs produced in solid state photoreactions like the ones discussed above (34-37) qualify in principle, but the spin coupling is too strong, so that only pure triplet EPR-transitions have been observed. The charge separated states along the electron transfer chain in photosynthetic reaction centers represent such a solid state-like case, even though it is a rather complicated one. We will focus on the transient EPR-consequences for these systems in the next section.

4.6 <u>Applications to Photosynthetic Reaction Centers</u>

Transient, spin-polarized EPR-spectra from photosynthetic systems have been studied in a number of laboratories, for a review see (42). Recently, the interpretation of the observations has concentrated on the concept of direct detection of EPR-transitions within a coupled radical ion spin pair created with the photoinduced electron transfer in the reaction center (38,43,44). Here we review some recent experimental data supporting this interpretation. We emphasize particular aspects of the transient nutation technique. The light-induced charge separated state of the plant photosystem I provides merely a model system of a weakly coupled electron spin pair in a rigid environment.

As argued in detail in (44), within the fast EPR time range (50 ns ... 5 µs) a charge separated radical ion pair is observed first, in which the electron has moved from the primary donor, a chlorophyll based molecule P_{700} to the acceptor A_1. In the case of bacterial reaction centers, where the structure is known, the magnetic coupling between the spins in each pair can be calculated. This establishes the weakly coupled spin pair between the oxidized primary electron donor and reduced quinone acceptor assumed by analogy in the interpretation of the transient signal found in photosystem I (43,44).

Complete kinetic and spectral data are presented in Fig.10 and 11 for plant photosystem I in deuterated whole cells of the alga *Synechococcus lividus*. The oscillatory behavior at high mw-power (top) and the overdamped case at low mw-power (bottom) are shown in X-band (9 GHz) and K-band (24 GHz), respectively. Assignment to polarized anisotropic powder spectra due to the coupled spin pair $P_{700}^+ A_1^-$ in statistical orientations yields satisfactory simulations (44). Therefore, we base further comments on this assignment.

Inspection of Figs.10 and 11 illustrates the following points: A) At low mw-power there exists a uniformly decaying spectral pattern indicating a single radical spin pair existing in the accessible time range 50 ns ... 5 µs. B) Substantial increase in spectral resolution in going from X- to K-band emphasizes the role of g-anisotropy consistent with the acceptor assignment as a quinone-like molecule (45,46). C) Oscillatory behavior is clearly observed at high microwave power, confirming the anisotropic character of the spectral pattern (47). D) In contrast to the usual case demonstrated in Fig.6 the oscillatory decay function is highly nonuniform and varies considerably with the magnetic field position across the spectral pattern, indicating substantial dynamical effects in the oscillatory spectral pattern.

Neglecting the spectral broadening at early times we can use Eq.[13] to extract the proper spectral pattern from the low power spectrum in Fig.11 (bottom). It is shown as the solid line in Fig.12. In order to demonstrate that detection took place in the absorption mode (χ'', detection phase perpendicular to the B_1-field) Fig.12 includes the broken line spectrum for detection phase parallel to the B_1-field (dispersion mode, χ'). In agreement with expectation (2), after differentiation of the absorption mode spectrum (solid

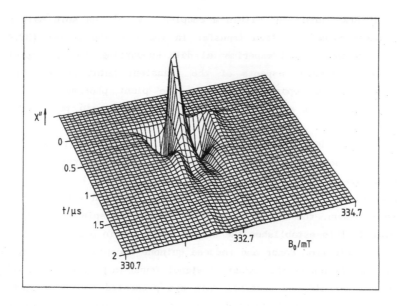

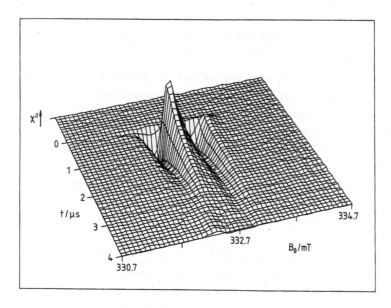

Fig. 10. Complete kinetic and spectral data of the transient EPR-signal of PS I in deuterated whole algae *Synechococcus lividus* at room temperature in X-band (9 GHz). For the presentation the original data set is reduced and averaged to a line spacing of 50 ns at full mw-power (200 mW, top) and of 100 ns at reduced power (2 mW, bottom) and 0.08 mT per line along the field axis.

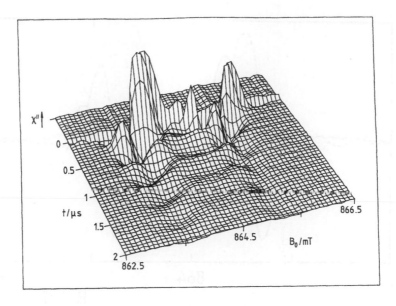

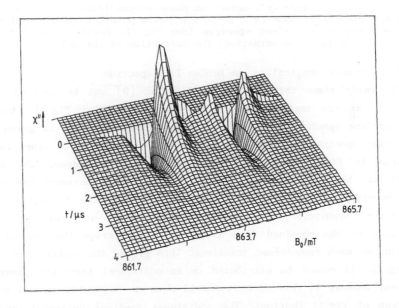

Fig. 11. Same as Fig.10 but in K-band (24 GHz). Note the improved resolution of the g-anisotropy (bottom) and the unusual oscillatory pattern if compared with Fig.6.

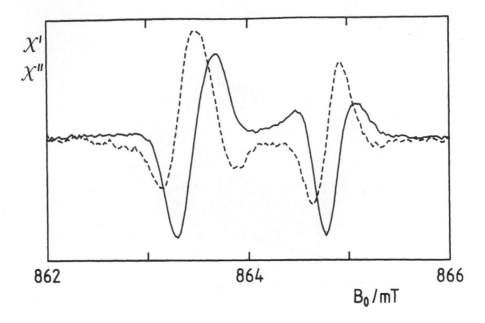

Fig. 12. Absorption mode (χ'', detection phase perpendicular to B_1-field, solid line) and dispersion mode (χ', detection phase parallel to B_1, broken line) for the low-power transient spectrum (see Fig.11, bottom). The broken line spectrum is, in good approximation, the derivation of the solid line spectrum.

line) it becomes identical to the broken line spectrum.

At early times the oscillatory function [9] can be applied to each transient at the various magnetic field positions of Fig.11 (top). The corresponding spectrum as a function of the magnetic field is given as the solid line spectrum in Fig.13. It agrees well with the solid line, low-power spectrum in Fig.12 taken from Fig. 11 (bottom). However, the extended oscillatory behavior observed in Fig.11 (top) cannot be accounted for this way. Spectral information may be extracted from this by subjecting the oscillatory transients to a Fourier transformation and plotting the peak amplitude of the obtained line around $\omega_1 = \gamma B_1$ versus the magnetic field position of each transformed transient. This yields the broken line spectrum in Fig.13. It cannot be attributed to an additional transient paramagnetic species, because in this case it should show up comparably in the low-power spectrum of Fig.11 (bottom). The additional spectral contributions in the high-power spectrum are tentatively taken as indications of dynamical effects, e.g. anisotropic molecular reorientation as demonstrated in liquid crystals (33) or an influence of the microwave field on the spin density matrix development and thus on the spin polarization process.

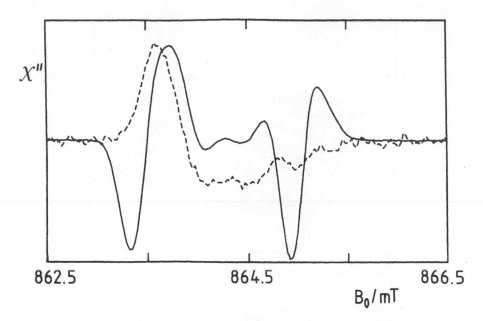

Fig. 13. Transient EPR-spectra extracted from the high-power data set of Fig.11, top. Solid line: spectrum at early times after the laser pulse. Broken line: spectrum obtained from the peak amplitude around $\omega_1 = \gamma B_1$ after Fourier-transformation of the oscillatory behavior.

The fact that only one uniformly decaying low-power spectrum is seen in the spectra of Figs.10 and 11, may be due to the fact that further electron transfer beyond the acceptor A_1 is blocked or significantly slowed down in this particular sample. In other preparations of plant photosystems sequential polarized spectral patterns have been observed in the accessible time range (38,48,49). In Fig. 14 sequential patterns are demonstrated for the case of photosystem I in chloroplasts from spinach (49,50). The early transient spectrum is analogous to the one observed in Fig.10 (bottom). It decays with a time constant of 240 ns while a new transient spectrum rises with the same time constant. Similar behavior has been found in photosystem I preparations of other species (49,50). We conclude that the kinetics of sequential electron transfer steps in photosystem I can be accurately determined by transient EPR-spectroscopy. The spectral characteristics of the charge separated states involved and thus their molecular identity are obtained in addition.

Another characteristic transient EPR-spectrum has been observed mainly in bacterial reaction centers when forward electron transfer is blocked at the quinone acceptor. In this case, the first charge separated state recombines to form the lowest excited triplet state ^3P of the primary donor. The transient

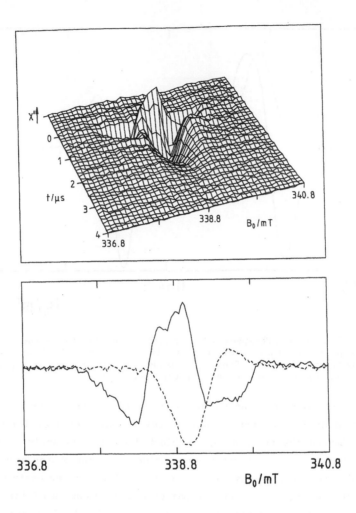

Fig. 14. Top: Complete kinetic and spectral data set of the transient
EPR-signal of photosystem I in spinach chloroplasts. X-band (9.5 GHz),
mw-power 40 mW. Bottom: Digital boxcar-spectrum in the time gate 100 - 250 ns
after laser pulse (solid line) and the spectrum in the time gate 800 - 1000 ns
(broken line).

EPR-spectrum of the characteristic chlorophyll triplet powder pattern is
observable up to room temperature, see e.g. (30) or Fig.15 (32). Due to the
different polarization mechanism via an intermediate radical ion pair the
overall spectral pattern looks quite different compared to ISC populated ones
as in Fig.9, yet the turning points in the powder patterns are quite analogous
to those expected for chlorophyll-like molecules (51).

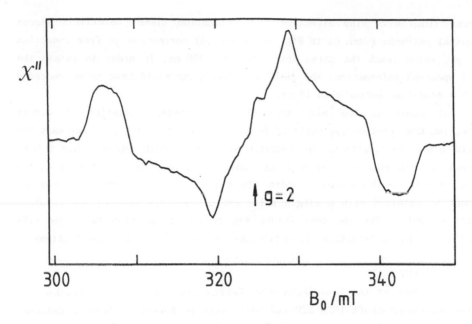

Fig. 15. Primary donor ($^3P_{870}$) triplet spectrum in bacterial reaction centers of Rhodobacter sphaeroides (R26), in which the nonheme iron has been depleted.

In summary, transient EPR spectroscopy can yield useful information about the early charge separated states of photosynthetic reaction centers in a time range as short as 50 ns. Curiously, these complicated reaction centers served as the first model case for the particular spin polarization phenomena observable in photoinduced, weakly coupled spin pairs in a rigid environment.

5.   CONCLUDING REMARKS

As discussed in this and previous chapters of this book, the realistically feasible time resolution in the 10 ns range can be reached by state-of-the-art transient EPR-spectroscopy. Radiation or light-induced processes are the main applications, not only because they concern an important problem in photochemistry and photobiology but because they are connected with a crucial sensitivity advantage due to the inherent spin polarization effects.

Fourier-transform pulsed EPR-techniques (7-9), although the preferable methods and so well established in the NMR-regime, will remain restricted in their applicability due to the existing technical limitations in the faster EPR-regime. The excitation band with, presently of the order of 100 MHz (or 3 mT), is inadequate for most solid state EPR-spectra. Moreover, the detection dead-time (in the 100 ns range) further limits application to reasonably narrow lines with slow effective spin relaxation times. Relatively narrow spectral patterns (such as in Fig. 10,11 and 14) correspond to free induction decays, which reach the noise level in about 150 ns. In order to retain the full spectral information, the free induction decay would have to be recorded with a dead-time better than 10 ns.

In contrast, the advantages of the transient nutation or direct detection EPR-spectroscopy outlined in this contribution are most relevant for solid state applications. An inhomogeneous line width larger than 0.5 mT guarantees the potential of a 10 ns time resolution. No restriction exists with respect to the spectral width. However, the full spectral information cannot be obtained with a single pulse excitation, the spectra are recorded point by point. For the near future the transient nutation technique will remain the only alternative, in particular for solid-state EPR-applications.

ACKNOWLEDGMENT

The work at the Free University Berlin was supported by the Deutsche Forschungsgemeinschaft (Sfb 312 and 337), work performed at Argonne National Laboratory by U.S. Department of Energy, Office of Basic Energy Sciences, Division of Chemical Sciences under contract W-31-109-Eng-38. Part of this work was supported by NATO Research Grant No. 214/84.

REFERENCES

1 H. C. Torrey, Phys. Rev., 76 (1949) 1059-1068
2 A. Abragram, in: "The Principles of Nuclear Magnetism", Chap. III, Part III, Oxford University Press (1961)
3 S. S. Kim and S. I. Weissman, J. Magn. Res., 24 (1976) 167-169, Chem. Phys. Lett., 58 (1978) 326-328, Rev. Chem. Intermed., 3 (1980) 107-120
4 a) A. D. Trifunac and M. C. Thurnauer, in: "Time Domain Electron Spin Resonance Spectroscopy", Eds.: L. Kevan, R N. Schwartz, Wiley Interscience, New York (1979) p. 107-152
   b) A. D. Trifunac, R. G. Lawler, D. M. Bartels, M. C. Thurnauer, Prog. Reaction Kinetics, 14 (1986) 43-156
5 R. Furrer, F. Fujara, C. Lange, D. Stehlik, H.-M. Vieth, W. Vollmann, Chem. Phys. Lett., 75 (1980) 332-339
6 R. Furrer and M. C. Thurnauer, Chem. Phys. Lett., 79 (1981) 28-33
7 J. Gorcester and J. H. Freed, J. Chem. Phys., 85 (1986) 5375-5377 and Proc. XXIII Congr. Amp. on Magnetic Resonance, Ed.: B. Maraviglia et al., Rom (1986) p. 562-567 and Chap. 5 in this book
8 M. Bowman, Bull. Am. Phys. Soc., 31 (1986) 524
9 O. Dobbert, T. Prisner, K. P. Dinse, J. Magn. Res., 70 (1986) 173, J Am. Chem. Soc., 110 (1988) 1622-1623
10 F. Bloch, Phys. Rev., 70 (1946) 460-471
11 R. P. Feynman, F. L. Vernon, R. W. Hellwarth, J. Appl. Phys., 28 (1957) 49-52
12 P. W. Atkins, K. A. McLauchlan and P. W. Percival, Mol. Phys., 25 (1973) 281-296
13 J. R. Norris, J. T. Warden, Varian EPR Newsletter, 1 (1980) 1-3
   A. J. Hoff, P. Gast, J. C. Romijn, FEBS Lett. 73 (1977) 185-190
14 B. Smaller, J. R. Remko, E. C. Avery, J. Chem. Phys., 48 (1968) 5174-5181

15 P. W. Atkins, K. A. McLauchlan, A. F. Simpson, J. Physics E: Sci. Instrum., 3 (1970) 547-551

16 G. E. Smith, R. E. Blankenship, M. P. Klein, Rev. Sci. Instrum., 48 (1976) 282-286

17 P. A. de Jager, F. G. H. van Wijk, Rev. Sci. Instr., 58 (1987) 735-741

18 M. Mehring, F. Freysoldt, J. Phys. E, 13 (1980) 894-895

19 R. W. Quine, G. R. Eaton, S. S. Eaton, A. Kostka, M. K. Bowman, J. Magn. Res., 69 (1986) 371-374

20 N. C. Verma, R. W. Fessenden, J. Chem. Phys., 58 (1973) 2501-2506

21 C. D. Buckley, K. A. McLauchlan, Mol. Phys., 54 (1985) 1-22, K. A. McLauchlan, Chap. 10 in this book

22 "Triplet State ODMR-Spectroscopy", Ed.: R. H. Clarke,Wiley-Intersc.(1982)

23 H. Seidel, M. Mehring, D. Stehlik, Chem. Phys. Lett., 104 (1984) 552-559

24 J. Ph. Grivet, Chem. Phys. Lett., 11 (1971) 267-270

25 F. Fujara, W. Vollmann, Chem. Phys., 90 (1984) 137-146

26 C. Bock, M. Mehring, H. Seidel, H. Weber, ISMAR-AMPERE Internat. Conf. Magn. Resonance, p.421-422, Delft, The Netherlands (1980)

27 I. Solomon, Phys. Rev. Lett., 2 (1959) 301-302

28 R. Gillies, A. M. Ponte Goncalves, Chem. Phys., 78 (1983) 49-55

29 H. Levanon, Rev. Chem. Intermed. 8 (1987) 287-320

30 A. J. Hoff, I. I. Proskuryakov, Chem. Phys. Lett., 115 (1985) 303-310

31 S. S. Kim, F. D. Tsay, A. Gupta, J. Phys. Chem., 91 (1987) 4851-4856

32 C. H. Bock, D. Budil, L. Feezel, M. Thurnauer, D. Stehlik, unpublished

33 J. Fessmann, N. Rösch, E. Ohmes, G. Kothe, Chem. Phys. Lett. 152 (1988) 491-496
     J. Fessmann, doctoral thesis, Univ. Stuttgart (1987)

34 R. Furrer, F. Fujara, C. Lange, D. Stehlik, W. Vollmann, Chem. Phys. Lett., 76 (1980) 383-389

35 D. Stehlik, R. Furrer, V. Macho, J. Phys. Chem., 83 (1979) 3440-3444

36 D. Stehlik, in: "Photoreaktive Festkörper", p.1-18, Ed.: H. Sixl, M. Wahl-Verlag Karlsruhe (1984)
     M. Nack et al., Proc. Congress Ampere, p.357-358, Zürich (1984)

37 B. Prass, F. Fujara, D. Stehlik, Chem. Phys., 81 (1983) 175-184
     B. Prass, J. P. Colpa, D. Stehlik, J. Chem. Phys., 88 (1988) 191-197

38 M. C. Thurnauer, J. R. Norris, Chem. Phys. Lett., 76 (1980) 557-561

39 M. C. Thurnauer, D. Meisel, J. Am. Chem. Soc., 105 (1983) 3729-3731

40 C. D. Buckley, D. A. Hunter, P. J. Hore, K. A. McLauchlan, Chem. Phys. Lett., 135 (1987) 307-312

41 G. L. Closs, M. D. E. Forbes, J. R. Norris, J. Phys. Chem., 91 (1987) 3592-3599
     G. L. Closs, M. D. E. Forbes, J. Am. Chem. Soc., 109 (1987) 6185-6187

42 A. J. Hoff, J. Quart. Rev. Biophys., 7 (1984) 153-282

43 P. J. Hore, D. A. Hunter, C. D. McKie, A. J. Hoff, Chem. Phys. Lett., 137 (1987) 495-500, P. J. Hore, Chap. 12 in this book

44 D. Stehlik, C. H. Bock, J. Petersen, J. Phys. Chem., 93 (1989)1612-1619

45 R. Furrer, M. C. Thurnauer, FEBS Lett. 153 (1983) 399-403

46 M. C. Thurnauer, P. Gast, J. Petersen, D. Stehlik, in Progr. in Photosyn. Res. (Biggins J. ed.) 1 (1987) 237-240, Martinus Nijhoff, Dordrecht
     J. Petersen, D. Stehlik, P. Gast, M. C. Thurnauer, Photosyn. Res. 14 (1987) 15-29

47 C. H. Bock, D. Stehlik, M. C. Thurnauer, Isr. J. Chem., in print

48 a)   H. Manikowski, A. R. McIntosh, J. R. Bolton, Biochim. Biophys. Acta, 765 (1984) 68-73
     b)   Proc. Congr. Ampere, Roma (1986), p.542-543, Ed.: Maraviglia, B. et. al.
     c)   Proc. of the 3rd Int. School, Ustron, Poland (1987)

49 D. Stehlik, C. H. Bock, A. van der Est, Proc. XXIV Congr. Amp., Poznan 1988, Ed.: N. Pislewski, Elsevier 1989

50 C. H. Bock, A. van der Est, K. Brettel, D. Stehlik, FEBS Lett., 247 (1989) 91-96

51 M. C. Thurnauer, Rev. Chem. Intermed. 3 (1979) 197-230

CHAPTER 12

ANALYSIS OF POLARIZED EPR SPECTRA

PETER J. HORE

1. INTRODUCTION

   Chemical reactions often result in electronically, vibrationally or rotationally excited molecules. It is less well known that chemical reactions in magnetic fields often produce paramagnetic species in non-equilibrium electron spin states, that is to say with spin alignments that differ from the Boltzmann distribution at the ambient temperature. These effects go by the name of chemically induced dynamic electron polarization (CIDEP) or, more simply, electron spin polarization (ESP). EPR intensities of polarized species may be enhanced by up to two orders of magnitude with individual resonances appearing in absorption or emission depending on the sense of the polarized population difference.

   Electron spin polarization is generated by two distinct processes: either from interactions within spin-correlated pairs of radicals (the radical pair mechanism) or from spin-selective singlet-triplet intersystem crossing in electronically excited molecules (the triplet mechanism).

   The differences between the populations of the spin states of a paramagnetic molecule are small and easily disturbed. Consequently electron spin polarization is a sensitive probe of a variety of processes including photophysics and photochemistry, molecular motion, the interactions within and between molecules, chemical kinetics and electron spin relaxation. However, the sensitivity to small perturbing influences makes this rich source of information difficult to tap. In what follows I describe some of the techniques that can be used to interpret and learn from electron spin polarization as revealed by EPR spectroscopy. I have concentrated on the radical pair mechanism and have not dealt with the time dependence of polarized spectra, which is covered elsewhere in this volume by McLauchlan and by Stehlik and Bock. Section 2 introduces the radical pair mechanism and explores the origin and properties of radical pair polarization. Sections 3 and 4 present some examples of the interpretation of the polarized EPR spectra of photosynthetic reactants and of free radicals in viscous liquids and micelles.

   Electron spin polarization is the subject of two excellent monographs (1,2) and several reviews (3-5).

2. RADICAL PAIR MECHANISM

The purpose of this section is to show how electron spin polarization arises in spin-correlated radical pairs. The approach adopted is somewhat broader than the conventional treatment of the radical pair mechanism, which deals with the polarization in the free (i.e. non-interacting) radicals formed from the radical pair (6). Here, in addition, we discuss the appearance of the polarized EPR spectrum of the radical pair itself (7-10).

2.1 Background

We consider a prototype radical pair consisting of two radicals (A and B) with a mutual exchange interaction. Within the high field approximation, the various magnetic nuclei to which the two unpaired electrons are coupled need not be treated explicitly except insofar as they affect the EPR frequencies of the two electrons. For simplicity and because the radical pairs of interest are usually short lived, electron spin relaxation will be ignored throughout. The spin Hamiltonian of the radical pair may then be written (in angular frequency units):

$$\hat{H} = \omega_A \hat{S}_{Az} + \omega_B \hat{S}_{Bz} - J(\hat{S}^2 - 1).\qquad\qquad [1]$$

where $\hat{S}_{Az}$ and $\hat{S}_{Bz}$ are the z components of the electron spin angular momentum operators, J is the strength of the (isotropic) exchange interaction and $\underline{\hat{S}}$ is the total electron spin angular momentum operator, $\hat{S}_A + \hat{S}_B$. ω_A and ω_B are the EPR frequencies of the two radicals in the absence of exchange and are determined by Zeeman and hyperfine interactions:

$$\omega_A = g_A \mu_B \hbar^{-1} B_0 + \sum_j A_{aj} M_{Aj}$$

$$\omega_B = g_B \mu_B \hbar^{-1} B_0 + \sum_k A_{bk} M_{Bk}\qquad\qquad [2]$$

in which B_0 is the magnetic field strength, g_A is the g-value of A, and A_{aj} and M_{Aj} are the hyperfine coupling constant and magnetic quantum number of nucleus j in radical A, with similar notation for radical B. This simple Hamiltonian contains all the essential interactions necessary for the generation of polarization. For the sake of clarity, anisotropic interactions will not be introduced until later.

It will be convenient to work initially with the four basis states, S, T_0, T_{+1} and T_{-1}. These are distinguished by their behaviour under $\underline{\hat{S}}$, and its z component, $\hat{S}_z$:

$$\begin{aligned}
\underline{\hat{S}}|S\rangle &= 0 ; & \hat{S}_z|S\rangle &= 0 \\
\underline{\hat{S}}|T_q\rangle &= \sqrt{2}|T_q\rangle ; & \hat{S}_z|T_q\rangle &= q|T_q\rangle , \quad q = +1, 0, -1
\end{aligned}\qquad [3]$$

and may be written in terms of the product states (in which α and β denote the $M_S = +\frac{1}{2}$ and $-\frac{1}{2}$ states of the individual electrons) thus:

$$|S\rangle = 2^{-\frac{1}{2}}(|\alpha_A\beta_B\rangle - |\beta_A\alpha_B\rangle) \qquad |T_{+1}\rangle = |\alpha_A\alpha_B\rangle$$
$$|T_0\rangle = 2^{-\frac{1}{2}}(|\alpha_A\beta_B\rangle + |\beta_A\alpha_B\rangle) \qquad |T_{-1}\rangle = |\beta_A\beta_B\rangle \quad . \qquad\qquad [4]$$

In what follows, T_{+1}, T_0 and T_{-1} will be abbreviated to +, T and – when used as subscripts. The representation of $\hat{H}$ in this basis is:

$$\begin{array}{cccc} |T_{+1}\rangle & |S\rangle & |T_0\rangle & |T_{-1}\rangle \end{array}$$

$$\begin{bmatrix} \omega-J & 0 & 0 & 0 \\ 0 & J & Q & 0 \\ 0 & Q & -J & 0 \\ 0 & 0 & 0 & -\omega-J \end{bmatrix} \qquad\qquad [5]$$

with:

$$\omega = \frac{1}{2}(\omega_A + \omega_B) \qquad \text{and} \qquad Q = \frac{1}{2}(\omega_A - \omega_B) \quad . \qquad\qquad [6]$$

The eigenvectors $|j\rangle$ and eigenvalues (ω_j) of $\hat{H}$ are:

$$\begin{aligned} |1\rangle &= |T_{+1}\rangle & \omega_1 &= \omega - J \\ |2\rangle &= \cos\psi|S\rangle + \sin\psi|T_0\rangle & \omega_2 &= \Omega \\ |3\rangle &= -\sin\psi|S\rangle + \cos\psi|T_0\rangle & \omega_3 &= -\Omega \\ |4\rangle &= |T_{-1}\rangle & \omega_4 &= -\omega - J \end{aligned} \qquad\qquad [7]$$

with :

$$\tan 2\psi = Q/J \qquad \text{and} \qquad \Omega^2 = J^2 + Q^2 \quad . \qquad\qquad [8]$$

Eigenstates 2 and 3 are linear combinations of S and T_0, split in energy by 2Ω. In the weak coupling limit $(|Q| \gg |J|)$ 2 and 3 are respectively $\alpha_A\beta_B$ and $\beta_A\alpha_B$. In the opposite extreme, $(|Q| \ll |J|)$ where the two electrons are magnetically equivalent, 2 and 3 are respectively S and T_0.

2.2 Spin dynamics

In all cases of interest here, the radical pair is generated in either a singlet or a triplet electronic spin state. This initial state may be described by a density operator (11) in the singlet-triplet basis, $\hat{\rho}(0)$. For a singlet radical pair, all elements of $\hat{\rho}(0)$ are equal to zero except $\rho_{SS}(0) = 1$.

For a triplet radical pair all but $\rho_{++}(0)$, $\rho_{TT}(0)$ and $\rho_{--}(0)$ (respectively, the fractional populations of T_{+1}, T_0 and T_{-1} states) are zero. Here it is assumed that no spin selection occurs prior to the formation of the radical pair (e.g. the triplet mechanism), which would cause $\rho_{++}(0)$, $\rho_{TT}(0)$ and $\rho_{--}(0)$ to deviate from 1/3.

Let us consider a radical pair created (at t=0) in a pure singlet state. Being diamagnetic this state has no EPR spectrum. To become polarized, the pair must evolve under the combined action of the exchange and Zeeman/hyperfine interactions. The system at some later time, t, is described by $\hat{\rho}(t)$ which may be obtained by solving the Liouville equation (11):

$$d\hat{\rho}(t)/dt \;=\; -i(\hat{H}\hat{\rho}(t) - \hat{\rho}(t)\hat{H}) \;=\; -i[\hat{H},\hat{\rho}(t)]. \tag{9}$$

This task is straightforward because two of the basis states, T_{+1} and T_{-1}, are eigenstates of $\hat{H}$. Consequently all density matrix elements involving one or both of these states are independent of time. It is therefore only necessary to consider the evolution of the four elements involving S and T_0 namely $\rho_{SS}(t)$, $\rho_{ST}(t)$, $\rho_{TS}(t)$ and $\rho_{TT}(t)$.

Thus [9] becomes:

$$
\begin{bmatrix} d\rho_{SS}/dt \\ d\rho_{ST}/dt \\ d\rho_{TS}/dt \\ d\rho_{TT}/dt \end{bmatrix}
= -i
\begin{bmatrix}
0 & -Q & Q & 0 \\
-Q & 2J & 0 & Q \\
Q & 0 & -2J & -Q \\
0 & Q & -Q & 0
\end{bmatrix}
\begin{bmatrix} \rho_{SS} \\ \rho_{ST} \\ \rho_{TS} \\ \rho_{TT} \end{bmatrix} . \tag{10}
$$

This equation has a simple pictorial interpretation (12), which though giving little extra insight, serves as a useful aide-mémoire. Assembling linear combinations of density matrix elements into a three dimensional vector $\underline{P}$:

$$
\underline{P}(t) \;=\;
\begin{bmatrix} P_x(t) \\ P_y(t) \\ P_z(t) \end{bmatrix}
=
\begin{bmatrix}
\rho_{ST}(t) + \rho_{TS}(t) \\
-i(\rho_{ST}(t) - \rho_{TS}(t)) \\
\rho_{SS}(t) - \rho_{TT}(t)
\end{bmatrix} \tag{11}
$$

one gets:

$$d\underline{P}/dt \;=\; \underline{P}(t) \times (2\underline{\Omega}) \tag{12}$$

where $\underline{\Omega}$ is the vector:

$$
\underline{\Omega} \;=\;
\begin{bmatrix} Q \\ 0 \\ J \end{bmatrix} . \tag{13}
$$

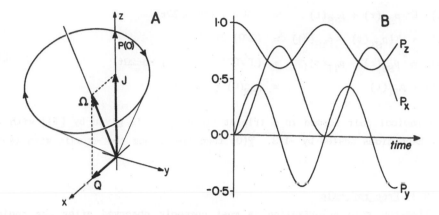

Fig.1   Vector model of the radical pair mechanism. (A): The trajectory of
$\underline{P}(t)$ for an initial singlet radical pair with J > 0 and Q > 0.   (B):   The
corresponding time dependence of the elements of $\underline{P}(t)$.

Eqn [12] describes the precession of $\underline{P}$ about $\underline{\Omega}$ at frequency 2Ω. For a radical
pair formed in a singlet state, $\underline{P}(0)$ points along the positive z axis and has
length unity.   Typical behaviour for J > 0, Q > 0 and an initial singlet state
is shown in Fig. 1. P_z represents the difference in the numbers of S and T_0
radical pairs: P_x, as will emerge later, is closely related to the
polarization.   For now it is sufficient to note that when $|Q| \gg |J|$, $\underline{\Omega}$ is
essentially parallel to the x axis so that $\underline{P}(t)$ precesses in the yz plane.   In
the opposite extreme, $|J| \gg |Q|$, $\underline{\Omega}$ is nearly parallel to the z axis and $\underline{P}(t)$
never deviates much from $\underline{P}(0)$.   Thus the difference in magnetic fields
experienced by the two electrons causes partial interconversion of the radical
pair between S and T_0 states, a process that is inhibited by a strong exchange
interaction.

This behaviour is exactly isomorphous with the vector model based on the
Bloch equations of magnetic resonance (13), $\underline{P}$ being the analogue of the bulk
magnetization and $\underline{\Omega}$ playing the role of the effective magnetic field in the
rotating frame.   Indeed this type of vector model is appropriate for any two
level system (14).

The time dependence of the elements of $\underline{P}(t)$ may now be obtained either
from Fig. 1 by geometry or from [10] by simple algebra:

$$P_x(t) = \rho_{ST}(t) + \rho_{TS}(t) = (JQ/\Omega^2)(1-\cos2\Omega t)$$

$$P_y(t) = -i(\rho_{ST}(t) - \rho_{TS}(t)) = (Q/\Omega)\sin2\Omega t$$

$$P_z(t) = \rho_{SS}(t) - \rho_{TT}(t) = (J^2/\Omega^2) + (Q^2/\Omega^2)\cos2\Omega t \qquad [14]$$

$$\rho_{SS}(t) + \rho_{TT}(t) = \rho_{SS}(0) + \rho_{TT}(0) \ .$$

For a radical pair formed in a triplet state, $\underline{P}(t)$ is given by [14] with the right hand sides scaled by $-1/3$. $\underline{P}(0)$ then lies along the $-z$ axis with length $1/3$.

2.3 Free radicals

Electron spin polarization is most commonly observed after the radical pair has ceased to exist in its original form. In most cases the radicals separate within 100 ns, which is roughly the shortest time needed to record an EPR signal. The detected EPR spectrum is therefore normally that of radicals that are far enough apart that they interact negligibly. In what follows, the polarization of such radicals is calculated.

A radical is polarized when the populations of its α and β spin states differ. More precisely, the polarization of radical A can be defined as the normalised population difference:

$$\mathscr{P}_A = [N(\beta_A) - N(\alpha_A)]/[N(\beta_A) + N(\alpha_A)] \qquad [15]$$

and similarly for B. An excess of spins in the ground state, β, ($\mathscr{P}_A > 0$) gives an absorptive EPR spectrum. Conversely, overpopulation of the excited state, α, ($\mathscr{P}_A < 0$) results in an emissive spectrum.

In terms of the populations of the four radical pair states $\alpha_A\alpha_B$, $\alpha_A\beta_B$, $\beta_A\alpha_B$ and $\beta_A\beta_B$, $\mathscr{P}_A$ and $\mathscr{P}_B$ are given by:

$$\mathscr{P}_A = [N(\beta_A\beta_B) + N(\beta_A\alpha_B) - N(\alpha_A\beta_B) - N(\alpha_A\alpha_B)]/N$$

$$\mathscr{P}_B = [N(\beta_A\beta_B) + N(\alpha_A\beta_B) - N(\beta_A\alpha_B) - N(\alpha_A\alpha_B)]/N \qquad [16]$$

where N is the sum of the four populations. Transforming into the singlet–triplet basis gives:

$$\mathscr{P}_A = (\rho_{--} - \rho_{++}) - (\rho_{ST} + \rho_{TS}) \ .$$

$$\mathscr{P}_B = (\rho_{--} - \rho_{++}) + (\rho_{ST} + \rho_{TS}) \ . \qquad [17]$$

The first term in parentheses in these expressions is the difference in $\beta_A\beta_B$ and $\alpha_A\alpha_B$ populations, which is not affected by the ST_0 interconversion process described above. The second is the difference in $\alpha_A\beta_B$ and $\beta_A\alpha_B$ populations produced by ST_0 mixing.

We may now obtain the polarizations $\mathscr{P}_A$ and $\mathscr{P}_B$ assuming the constituents of the radical pair separate suddenly at time t. From [14] and [17] we have:

$$\mathscr{P}_A = - (JQ/\Omega^2)(1-\cos2\Omega t) \qquad [18]$$

for an initial singlet pair and:

$$\mathscr{P}_A = \tfrac{1}{3}(JQ/\Omega^3)(1-\cos2\Omega t) \qquad [19]$$

for an initial triplet assuming $\rho_{++}(0) = \rho_{TT}(0) = \rho_{--}(0) = 1/3$. In both cases $\mathscr{P}_B = -\mathscr{P}_A$. i.e. the two free radicals are equally and oppositely polarized.

Referring back to the vector model of Fig. 1, it can be seen that the polarization is simply proportional to the projection of $\underline{P}$ onto the x axis. Appreciable polarization only develops when $|Q|$ and $|J|$ are not too dissimilar, i.e. when $\underline{\Omega}$ has significant x and z components.

The simplest, realistic, kinetic model for the separation of the two radicals would be a spin-independent first order process with characteristic time, τ. That is, the exchange interaction would drop instantaneously from J to zero at time t with a probability depending exponentially on t. The polarization in the free radicals at a time much greater than τ would then be (for an initial singlet pair):

$$\mathscr{P}_A = \int_0^\infty \tau^{-1} \exp(-t/\tau) \; (-JQ/\Omega^2) \; (1-\cos2\Omega t) \; dt$$
$$= - 4JQ\tau^2/(1+4\Omega^2\tau^2) \quad . \qquad [20]$$

For a short lived radical pair ($4\Omega^2\tau^2 \ll 1$) this reduces to:

$$\mathscr{P}_A = -4JQ\tau^2 \qquad [21]$$

and $\mathscr{P}_B = -\mathscr{P}_A$, as before. This exponential model has been used for radical pairs in photosynthetic systems where electron transfer causes separation of the paramagnetic centres (section 3.1).

In the liquid state where the radicals can move, the situation is much more complicated. To calculate the polarization, one must sum over all possible trajectories of the radicals taking into account the dependence of J on the separation of the radicals. The most satisfactory way of achieving this end is to solve the appropriate stochastic Liouville equation, that is the Liouville equation [9] for the spin evolution modified to include relative translational motion. This subject has been reviewed by Freed and Pedersen (15).

2.4 Radical Pairs

We now consider the appearance of the EPR spectrum of the radical pair itself (7-10), assuming it to be sufficiently long lived that a spectrum can be recorded. The factor that distinguishes a radical pair from a pair of free radicals is the presence of an appreciable interaction between the two unpaired electrons. We assume that the separation of the two radicals and therefore J remain constant throughout the development and observation of the polarization.

What is actually seen in the spectrum depends crucially on the nature of the EPR excitation and detection. Here we assume the experiment is such that the intensity (I_{jk}) of a resonance between states j and k is proportional to the appropriate population difference scaled by the transition probability. Thus, I_{jk} is assumed to be given by (16):

$$I_{jk} = (p_k - p_j) |(S_y')_{jk}|^2 \qquad [22]$$

where p_k is the fraction of radical pairs in eigenstate k and $\hat{S}_y'$ (the prime denotes the eigenbasis) is the y-component of the total spin angular momentum operator:

$$\hat{S}_y' = (i/\sqrt{2}) \begin{bmatrix} 0 & -\sin\psi & -\cos\psi & 0 \\ \sin\psi & 0 & 0 & -\sin\psi \\ \cos\psi & 0 & 0 & -\cos\psi \\ 0 & \sin\psi & \cos\psi & 0 \end{bmatrix} . \qquad [23]$$

Eqn [22] is valid for slow passage, low power experiments on spin systems for which $\rho(t)$ is diagonal in the eigenbase of the Hamiltonian (17).

Within the high field ($\omega \gg |Q|$ and $|J|$) and high temperature ($\hbar\omega \ll kT$) approximations the normalised populations of the four states at thermal equilibrium are:

$$p_1 = \tfrac{1}{4}(1 - \hbar\omega/kT) \qquad p_2 = \tfrac{1}{4}$$
$$p_3 = \tfrac{1}{4} \qquad p_4 = \tfrac{1}{4}(1 + \hbar\omega/kT) . \qquad [24]$$

Thus there are four transitions with intensities and frequencies ($\omega_{jk} = \omega_j - \omega_k$):

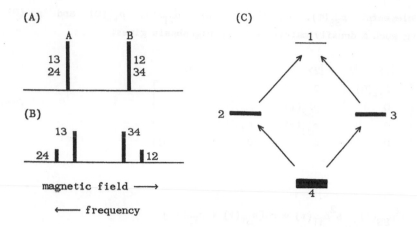

Fig. 2   (A):   Schematic EPR spectrum of two radicals with no hyperfine couplings, J = 0, Q > 0.   (B):   Schematic EPR spectrum of a radical pair with no hyperfine couplings, J > 0 and Q > 0.   When J < 0, the labels 12 and 34 are interchanged as are 13 and 24.   (C):   The corresponding energy level diagram. The thickness of the bars indicates the populations of the states.

$$I_{12} = (\hbar\omega/8kT)\sin^2\psi \qquad\qquad \omega_{12} = \omega - \Omega - J$$
$$I_{34} = (\hbar\omega/8kT)\cos^2\psi \qquad\qquad \omega_{34} = \omega - \Omega + J$$
$$I_{13} = (\hbar\omega/8kT)\cos^2\psi \qquad\qquad \omega_{13} = \omega + \Omega - J \qquad\qquad [25]$$
$$I_{24} = (\hbar\omega/8kT)\sin^2\psi \qquad\qquad \omega_{24} = \omega + \Omega + J \; .$$

A typical spectrum is shown in Fig. 2 together with the corresponding energy level diagram.   The spectrum consists of two pairs of lines centred at $\omega \pm \Omega$ with splitting 2J, the outer lines being weaker than the inner by a factor $\tan^2\psi$.   This pattern is identical to the familiar "AB" NMR spectrum of two strongly coupled protons (18).   The difference in intensity of the inner and outer lines, known as the "roof effect", arises from the different transition probabilities caused by the mixing of states $\alpha_A\beta_B$ and $\beta_A\alpha_B$ in eigenstates 2 and 3.   In the limit $J \longrightarrow 0$ ($\psi \longrightarrow \pi/4$) the spectrum collapses to two lines of equal intensity at frequencies ω_A and ω_B corresponding to the EPR spectrum of the separate, non-interacting radicals.

We are now in a position to calculate the polarized EPR spectrum of the radical pair.   In contrast to the free radical case, it is not helpful to define polarizations $\mathscr{P}_A$ and $\mathscr{P}_B$ because, in general, the mixing of states $\alpha_A\beta_B$ and $\beta_A\alpha_B$ causes all four lines to have both A and B character.   Thus we are forced to consider the population difference for each line scaled by the appropriate transition probability.

The system at any time, t, is described by $\hat{\rho}(t)$ which has, at most, six non-zero elements: $\rho_{SS}(t)$, $\rho_{ST}(t)$, $\rho_{TS}(t)$, $\rho_{TT}(t)$, $\rho_{++}(0)$ and $\rho_{--}(0)$. Transforming such a density matrix into the eigenbasis gives:

$$
\hat{\rho}'(t) \;=\;
\begin{array}{cccc}
|1\rangle & |2\rangle & |3\rangle & |4\rangle \\
\end{array}
\left[
\begin{array}{cccc}
\rho_{++}(0) & 0 & 0 & 0 \\
0 & \rho'_{22}(t) & \rho'_{23}(t) & 0 \\
0 & \rho'_{32}(t) & \rho'_{33}(t) & 0 \\
0 & 0 & 0 & \rho_{--}(0)
\end{array}
\right]
\qquad [26]
$$

where:

$$
\begin{aligned}
\rho'_{22}(t) &= c^2\rho_{SS}(t) + s^2\rho_{TT}(t) + cs[\rho_{ST}(t) + \rho_{TS}(t)] \\
\rho'_{23}(t) &= c^2\rho_{ST}(t) - s^2\rho_{TS}(t) + cs[\rho_{TT}(t) - \rho_{SS}(t)] \\
\rho'_{32}(t) &= c^2\rho_{TS}(t) - s^2\rho_{ST}(t) + cs[\rho_{TT}(t) - \rho_{SS}(t)] \\
\rho'_{33}(t) &= s^2\rho_{SS}(t) + c^2\rho_{TT}(t) - cs[\rho_{ST}(t) + \rho_{TS}(t)]
\end{aligned}
\qquad [27]
$$

and

$$
c = \cos\psi \;\; ; \qquad s = \sin\psi \;\; . \qquad\qquad [28]
$$

$\hat{\rho}'(t)$ differs in two important respects from the density matrix appropriate for thermal equilibrium: the populations of the two central states (ρ'_{22} and ρ'_{33}) have changed and two extra elements, ρ'_{23} and ρ'_{32}, have appeared. The latter represent a coherent superposition of states 2 and 3 – a zero quantum coherence. Unlike the diagonal elements of ρ', which are independent of time for fixed Q and J (see below), ρ'_{23} and ρ'_{32} oscillate at a frequency $\omega_2 - \omega_3 = 2\Omega$:

$$
\rho'_{23}(t) = \rho'_{23}(0)\exp(i2\Omega t) \qquad\qquad [29]
$$

with $\rho'_{23}(0) = -\tfrac{1}{2}Q/\Omega$ and $\rho'_{32}(t) = \rho'_{23}(t)^{*}$ as may be verified by combining [14] and [27]. This is the source of the 2Ω precession frequency noted in connection with the vector model.

Zero quantum coherence is not directly detectable in a magnetic resonance experiment but may be converted into observable transverse magnetization by exciting with microwave radiation one or more of the four allowed transitions. For the moment we ignore this possibility on the grounds that 2Ω is typically in the range 10 – 100 MHz and EPR observations are usually much slower than the inverse of this frequency. For example, if the generation of radical pairs or the acquisition of the EPR signal takes a time that is not short compared to

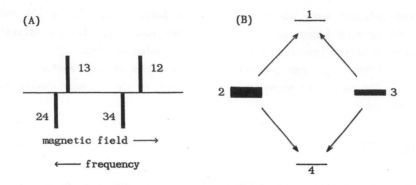

Fig. 3   (A):   Schematic polarized EPR spectrum of an initial singlet radical pair with no hyperfine couplings, J > 0 and Q > 0.   (B):   The corresponding energy level diagram.   The thickness of the bars indicates the populations of the states and the arrows the directions of the transitions.

$(2\Omega)^{-1}$, the net transverse magnetization produced from the rapidly oscillating zero quantum coherence will tend to average to zero.   So, concentrating on the diagonal elements of $\hat{\rho}'$, [22] and [27] give:

$$I_{12} = \tfrac{1}{2}[\; c^2\rho_{SS} + s^2\rho_{TT} + cs(\rho_{ST} + \rho_{TS}) - \rho_{++}]s^2$$
$$I_{13} = \tfrac{1}{2}[\; s^2\rho_{SS} + c^2\rho_{TT} - cs(\rho_{ST} + \rho_{TS}) - \rho_{++}]c^2$$
$$I_{24} = \tfrac{1}{2}[-c^2\rho_{SS} - s^2\rho_{TT} - cs(\rho_{ST} + \rho_{TS}) + \rho_{--}]s^2 \qquad [30]$$
$$I_{34} = \tfrac{1}{2}[-s^2\rho_{SS} - c^2\rho_{TT} + cs(\rho_{ST} + \rho_{TS}) + \rho_{--}]c^2$$

and from [14]:

$$I_{12} = I_{13} = -I_{24} = -I_{34} = Q^2/8\Omega^2. \qquad\qquad [31]$$

The polarized spectrum thus consists of the two doublets in antiphase, Fig. 3. The origin of this spectrum can easily be seen.   For an initial singlet pair, the average populations of 2 and 3 are respectively $\cos^2\psi$ and $\sin^2\psi$ while 1 and 4 remain empty.   Taking transitions 12 and 34, the population differences are respectively $\cos^2\psi$ and $-\sin^2\psi$ and the transition probabilities respectively $\sin^2\psi$ and $\cos^2\psi$ so that the two intensities are proportional to $\pm\sin^2\psi\cos^2\psi$ or $\pm Q^2/4\Omega^2$ as indicated in Fig. 3.   The radical pair spectrum is very weak when $|Q| \ll |J|$ ($\psi \sim 0$) because then transitions 12 and 24 are almost forbidden while 13 and 34 have very little population difference.   The intensity is at a maximum when $|Q| \gg |J|$ provided $|J|$ is not so small that the antiphase doublets cancel one another.   This behaviour contrasts with the situation for the separated radicals where the polarization is largest when $|J| = |Q|$ and small when $|Q|$ is much larger than or much smaller than $|J|$. For an initial triplet pair each of the intensities I_{jk} should be scaled by $-1/3$.

The intensities, I_{jk}, can be interpreted in terms of the radical pair vector model. The polarized signal is proportional to the average change in the z-component of $\underline{P}$:

$$I_{jk} = P_z(0) - \overline{P_z(t)} \qquad\qquad [32]$$

where the overbar indicates a time average. Put another way, the EPR intensity for an initial singlet pair is proportional to the average T_0 population $\overline{\rho_{TT}(t)}$, while that for an initial triplet is proportional to $\overline{\rho_{SS}(t)}$. Significant polarization will develop unless $|J|$ is so large as to prevent ST_0 interconversion.

Eqns [25] and [31] and Fig. 3 show that the phase of the polarization observed (absorption, A or emission, E) depends on several factors. The following rule conveniently summarizes this dependence for both radical pairs and the free radicals formed therefrom (8,10,19):

$$\Gamma = \mu \bullet \text{sign}(J) = \left\{ \begin{array}{cc} - & EA \\ + & AE \end{array} \right. \qquad\qquad [33]$$

The quantity μ has sign "$-$" for an initial singlet and "$+$" for an initial triplet. In the case of radical pairs, EA indicates that the low field component of each doublet is E and the high field component A. For free radicals, EA denotes emissive polarization in the radical that resonates to low field and absorptive polarization in the radical to high field. For example, consider an initial singlet pair ($\mu = -$) with $J < 0$ ($\text{sign}(J) = -$) and $Q > 0$ (i.e. $\omega_A > \omega_B$ which means that A resonates to low field of B): $\Gamma = - \bullet - = +$ (AE). Each of the doublets in the radical pair spectrum has AE polarization and, after they have separated, radicals A and B would be respectively in absorption and emission.

2.5 Anisotropic interactions

Thus far, the constituents of the radical pair have been assumed to interact only by isotropic quantum mechanical exchange. This is an excellent approximation for mobile liquids where anisotropic interactions are averaged to zero by the rapid relative motion of the radicals and, at most, contribute only to the electron spin relaxation of the radical pair. When motion is restricted, as in the solid state, anisotropic interactions affect both the positions of the lines in the EPR spectrum and the polarization, which become functions of the orientation of the sample in the magnetic field.

The exchange interaction is certainly anisotropic, but little is known about its anisotropy. The dipolar interaction, however, is easier to deal with. For an axially symmetric electron-electron dipolar interaction the radical pair Hamiltonian should be supplemented by the term:

$$\hat{H}_D \;=\; \tfrac{1}{2}D_{zz}(3\hat{S}_z^2 - \hat{S}^2) \;=\; \tfrac{1}{2}D(\cos^2\xi - \tfrac{1}{3})(3\hat{S}_z^2 - \hat{S}^2) \qquad [34]$$

in which D is the strength of the interaction and ξ is the angle between the dipolar axis (the vector connecting the two electrons) and the magnetic field direction. $\hat{H}_D$ is diagonal in the singlet-triplet basis; it changes the Hamiltonian to:

$$
\begin{array}{cccc}
|T_{+1}\rangle & |S\rangle & |T_0\rangle & |T_{-1}\rangle
\end{array}
$$

$$
\begin{bmatrix}
\omega - J + \tfrac{1}{2}D_{zz} & 0 & 0 & 0 \\
0 & J & Q & 0 \\
0 & Q & -J - D_{zz} & 0 \\
0 & 0 & 0 & -\omega - J + \tfrac{1}{2}D_{zz}
\end{bmatrix} \qquad [35]
$$

The inclusion of $\hat{H}_D$ changes both the transition frequencies and the polarization of the radical pair (10). The polarization depends on the interaction between the electrons only via the difference in diagonal matrix elements $\langle S|\hat{H}|S\rangle - \langle T_0|\hat{H}|T_0\rangle$, that is $2(J + \tfrac{1}{2}D_{zz})$. Thus the polarized intensities in the presence of a dipolar interaction are obtained simply by replacing J in [31] by $J + \tfrac{1}{2}D_{zz}$. Ω and ψ therefore become:

$$\Omega^2 \;=\; (J + \tfrac{1}{2}D_{zz})^2 + Q^2 \quad \text{and} \quad \tan 2\psi \;=\; Q/(J + \tfrac{1}{2}D_{zz}) \; . \qquad [36]$$

The frequencies of the four allowed transitions in the presence of $\hat{H}_D$ are:

$$
\begin{aligned}
\omega_{12} &= \omega - \Omega - J + D_{zz} \\
\omega_{34} &= \omega - \Omega + J - D_{zz} \\
\omega_{13} &= \omega + \Omega - J + D_{zz} \\
\omega_{24} &= \omega + \Omega + J - D_{zz}
\end{aligned} \qquad [37]
$$

with Ω given by [36]. The doublet splitting, which is 2J in the absence of $\hat{H}_D$ becomes $2(J-D_{zz})$.

A simple example is instructive (Fig. 4). Consider an initial singlet pair with Q > 0, J = 0 and D > 0. When $\xi = 0$, $D_{zz} = 2D/3$ and the spectrum

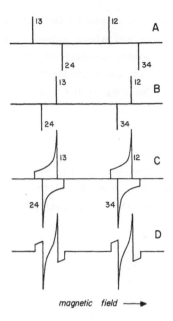

magnetic field ⟶

Fig. 4   Schematic polarized EPR spectra of an initial singlet radical pair with no hyperfine couplings, J = 0 and D > 0.   (A):   ξ = 0.   (B):   ξ = ½π. (C):   Spherical average showing the individual transitions.   (D):   Spherical average showing the total spectrum.

consists of two AE doublets with splitting 4D/3 centred at $\omega \pm \sqrt{(Q^2 + D^2/9)}$. When $\xi = \pi/2$, $D_{zz} = -D/3$, and the two doublets become EA with splitting 2D/3 centred at $\omega \pm \sqrt{(Q^2 + D^2/36)}$.   The spectrum for a randomly oriented radical pair, obtained by spherical averaging, is also shown in Fig. 4.   By analogy with [33] we may write a sign rule for dipolar couplings in the absence of exchange:

$$\Gamma = -\mu \bullet sign(D) \bullet sign(3cos^2\xi - 1) = \begin{cases} - & EA \\ + & AE \end{cases} .$$ [38]

Anisotropic Zeeman interactions can be introduced by making the g-values of the radicals explicitly orientation-dependent (20):

$$g(\theta,\phi)^2 = g(x)^2 sin^2\theta cos^2\phi + g(y)^2 sin^2\theta sin^2\phi + g(z)^2 cos^2\phi$$ [39]

where θ and ϕ are angles defining the orientation of B_0 with respect to the axis system that diagonalises the g-tensor with principal values g(x), g(y) and g(z).   Anisotropic hyperfine couplings could be introduced in a similar way.

2.6 Experimental factors

Everthing that has been said so far has assumed, explicitly or implicitly, that the polarized EPR intensity of a given transition is simply proportional to the difference in populations of the two states involved, scaled by the transition probability. This is not always the case, especially if one follows the EPR signal as a function of time. What in fact is seen depends on the type of experiment performed (continuous wave, Fourier transform, electron spin echo, etc.), on the method of production of the radical pairs (i.e. pulsed or continuous generation), on spin relaxation and on the details of any chemical reactions taking place. For free radicals, the time dependence can usually be understood using the Bloch equations, modified to include the chemistry. This type of analysis and the methods of extracting chemical and physical information has been described extensively elsewhere (21-24). The situation for radical pairs is more complex. As the two electrons cannot be considered independently it is necessary to resort to density matrix methods instead of the Bloch equations. At the time of writing, the level of sophistication of EPR experiments on radical pairs does not seem to warrant such an effort.

However a few general comments on the signals expected from radical pairs can be made by analogy with the corresponding NMR experiments on strongly coupled nuclear spin systems in non-equilibrium states (17,25-30). 1. The transverse magnetization derived from zero-quantum coherences (ρ'_{23}) should be observable when the radical pairs are generated by pulse methods (flash photolysis or pulse radiolysis) and observed by pulsed EPR provided generation and detection are much faster than $(2\Omega)^{-1}$. 2. Simultaneous excitation of more than one radical pair transition may cause mixing of populations for appreciable flip angles. For example, under conditions where selective excitation gives two antiphase EA doublets, a non-selective $90°$ pulse would give an EE in-phase doublet at low field and AA in-phase doublet at high field. 3. Electron spin echoes will exhibit an extra phase modulation even when the radical pair lifetime is long compared to the length of the echo pulse sequence. 4. When the radical pair separates or undergoes a chemical reaction during the course of an electron spin echo or Fourier transform EPR experiment, phase shifts will also be seen (7,31,32).

3. BACTERIAL PHOTOSYNTHESIS

In photosynthetic bacteria, the energy of light is harnessed by a series of chemical reactions:

$$PIX \xrightarrow{h\nu} {}^{1}PIX \longrightarrow P^{+}I^{-}X \longrightarrow P^{+}IX^{-} \longrightarrow \dots \qquad [40]$$

P, the primary electron donor, is a bacteriochlorophyll dimer; I, the first

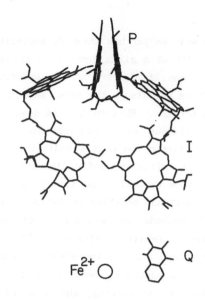

Fig. 5   Arrangement of pigment molecules in the reaction centre of R. viridis determined by X-ray crystallography.   Computed from a coordinate set at 0.3 nm resolution obtained from Dr J. Deisenhofer.

electron acceptor, is a bacteriopheophytin monomer and X, the second electron acceptor, is an iron-quinone complex, $Q Fe^{2+}$. ^1P is an excited singlet state of the primary donor.

These primary reactants are embedded in a protein, the reaction centre (RC) protein, whose three dimensional structure (for Rhodopseudomonas viridis (33) and Rhodobacter sphaeroides (34,35)) has recently been determined by X-ray crystallography (Fig. 5).

The sequence of electron transfers in [40] is linear and essentially irreversible.   The primary radical pair P^+I^- is formed in less than 3 ps and reacts to give the secondary pair P^+X^- in 200 ps.   Subsequent reactions, which cause further charge separation occur on a much slower timescale.   When electron transfer from I^- to X is blocked by removal of X or prior reduction to X^- ($\equiv Q^-Fe^{2+}$), the lifetime of P^+I^- increases to about 15 ns.   Under these conditions P^+I^- either reverts directly to ground state P or recombines to give ^3P, an excited triplet state of P:

$$PI \xrightarrow{\ h\upsilon\ } {}^1PI \longrightarrow P^+I^- \longrightarrow {}^3PI \ . \qquad\qquad [41]$$

Except at low temperatures, P^+I^- can also recombine to give ^1PI.

Electron spin polarization is generated in both primary and secondary radical pairs and has been observed for P⁺, X⁻ and ³P, as reviewed by Hoff (5). The following pages describe some recent attempts to account quantitatively for these polarized signals.

3.1 Polarization in P⁺X⁻

Polarized EPR spectra have been reported for bacterial reaction centres in which the Fe^{2+} has either been removed or uncoupled from the quinone (5,36-38). No polarization is observable when the iron-quinone complex is intact presumably because of rapid spin-lattice relaxation induced by the ferrous ion. At 9 GHz EPR frequency the spectrum consists of a central absorptive feature flanked by emissive wings, as shown in Fig. 6 for randomly oriented reaction centres of Rb. sphaeroides (37). Essentially identical spectra are observed for photosystem I of a variety of plants and algae at 9 GHz (5,7,39-43). It is generally agreed, for bacteria at least, that this EAE spectrum is due to P⁺X⁻; P⁺I⁻ being too short lived to be detectable by EPR. It is also clear from the combination of emission and absorption in the spectrum that the radical pair mechanism rather than the triplet mechanism is responsible for the polarization. The question is, in which radical pair does the polarization originate: P⁺I⁻ or P⁺X⁻ or both.

Initial attempts to account for the EAE spectrum were based on the premise that polarization was generated in P⁺I⁻ and passed by electron transfer to P⁺X⁻ where it was observed (5,37,44). P⁺ and X⁻ were assumed to be sufficiently far apart as to have a negligible interaction so that once electron transfer from I⁻ to X had occurred, the development of polarization came to a halt. The

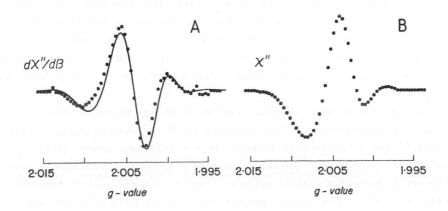

Fig. 6 (A): Experimental (dots) and simulated (solid line) first derivative EPR spectra of P⁺X⁻ radical pairs in Rb. sphaeroides. (B): The integrated form of the experimental spectrum.

electron transfer step was treated as a rapid first order process ($\Omega^2 \tau^2 \ll 1$, Eqn [21]), so that the polarization is linear in Q and therefore in the magnetic field. These approximations greatly simplify the problem, allowing an analytical expression for the polarized lineshape to be derived (44). Using only isotropic interactions and the known g-values of P^+, I^- and X^-, it proved impossible to simulate the observed EAE pattern (5,37). Reasonable agreement with the experimental spectrum could only be obtained by reducing the g-value of I^- from 2.0035 to 2.0007, an exceptionally low value for a pheophytin anion (5,37).

The most serious defect of this approach is the omission of anisotropic interactions. The g-tensor of the semiquinone X^- is appreciably anisotropic (principal values 2.0067, 2.0056 and 2.0024 for ubiquinone (45)) and the dipolar interaction between P^+ and I^- is known to be significant from the anisotropy of magnetic field effects on reaction yields (46). Inclusion of one of these anisotropies on its own has a negligible effect on the simulated spectra for randomly oriented samples. However, because of the correlation between the P^+I^- dipolar axis and the principal axes of the g-tensor of X^-, the two interactions together lead to changes that do not disappear on spherical averaging.

Straightforward extension of the isotropic treatment and inclusion of integration over all orientations of the reaction centre yielded close agreement between simulated and experimental spectra with the literature g-values for P^+, I^- and X^- (47). Two major problems were encountered, however. First, to reproduce the shape of the EAE spectrum, the dipolar interaction between P^+ and I^- had to be nearly 60 times larger than the exchange interaction; other experiments (48) suggest that they are of more comparable size. Second, to account for the observed amplitude of polarization, using [21] and the P^+I^- lifetime appropriate for intact reaction centres, the exchange interaction had to be at least ten times bigger than previous estimates (48). Attempts to use the same approach to account for the much better resolved 35 GHz spectrum of plant photosystem I were even less satisfactory (49).

Much more acceptable results were obtained by taking a diametrically opposed view, namely that P^+I^- is too short lived to generate any polarization and that P^+ and X^- do in fact interact to a significant extent (10). Thus assuming P^+X^- to be formed instantaneously in a spin-correlated singlet state, the theory of polarized radical pairs outlined in 2.4 was used to simulate the spectrum of P^+X^-. The "stick" spectra predicted by [25] were convolved with Gaussian lineshapes to model the effects of unresolved hyperfine couplings and broadening due to spin-spin relaxation. A dipolar interaction between P^+ and X^- and an anisotropic g-tensor for X^- were included and spherical averaging was performed by numerical summation (10).

Fig. 6 shows the "best-fit" simulated spectrum obtained by varying five parameters: the dipolar interaction D_{PX}, the angles α and β that specify the orientation of the P^+X^- dipolar axis relative to the principal axes of the g-tensor of X^-, and Gaussian line-broadenings for P^+ and X^-. The g-tensor components were set at their literature values and J_{PX} was kept equal to zero. The optimum values of the structural parameters so obtained were in broad agreement with estimates (50) based on the unrefined X-ray coordinates of R. viridis (33). In particular, D_{PX} (-0.14 mT) was of the expected sign and similar to the value predicted for two radicals ~ 3 nm apart. That a satisfactory simulation was obtained with $J_{PX} = 0$ also seems reasonable bearing in mind the strong distance dependence of the exchange interaction. Finally, the simulated spectrum is approximately four times more intense at 150 K than the corresponding equilibrium spectrum. This compares favourably with the experimental ratio of about three at this temperature (37).

Although respectable agreement with experiment and reasonable values for D_{PX}, J_{PX}, α and β were found, this hardly constitutes a proof beyond doubt that the proposed mechanism is operative. A more challenging test of the theory has been provided by Stehlik et al. (43) who simulated the much better resolved 24 GHz spectra of deuterated plant photosystem I preparations. Using the same approach as for the 9 GHz bacterial spectrum, they successfully simulated the high frequency experimental spectra by including a small g-tensor anisotropy for P^+. The values of the structural parameters required were also consistent with the X-ray data for R. viridis. A more severe test still, and one that would allow values of the various parameters to be extracted with more confidence, would be provided by spectra of oriented samples where less information would be lost by orientational averaging.

The approach described above can easily be improved in several respects. Hyperfine couplings could be introduced in a more rigorous fashion by averaging ω_A and ω_B separately over Gaussian distributions as done by Tang and Norris (51) in a different context or by using the known hyperfine couplings explicitly. The Gaussian convolution would then represent only the broadening due to spin-spin relaxation. A more serious source of error, perhaps, is the assumption that P^+I^- is too short lived to give rise to significant polarization. This is a reasonable approximation if the lifetime of P^+I^- is 200 ps as measured for native reaction centres (52-54). However if the electron transfer is as slow as the 4 ns measured (55) for Fe^{2+}-depleted Rb. sphaeroides strain R-26.1, substantial polarization would develop in the primary pair. To include this possibility in the simulations, one could simply combine the two extreme approaches described above (10,47). Calculations with reasonable values of D_{PI} and J_{PI} indicate that there is no significant effect on the 9 GHz spectrum for lifetimes of P^+I^- less than about 1 ns (56).

3.2 Polarization in X^-.

Electron spin polarization has also been observed in bacterial reaction centres in which the quinone is both uncoupled from the Fe^{2+} and pre-reduced to Q^- (5,45,57-59). In the dark, the 9 GHz EPR spectrum consists of a single line of ~ 10 G width from X^-. Shortly after flash photolysis or during continuous illumination, the whole line is emissively polarized and suffers an apparent g-value shift (57). At 35 GHz, and especially with deuterated reaction centres, where the g-tensor anisotropy of X^- is resolved, the light spectrum shows a complicated mixture of absorption and emission (45). This effect and the apparent shift in the polarized 9 GHz spectrum indicate that the polarization of X^- is anisotropic.

A tempting but erroneous interpretation of this behaviour is that polarization is generated in the radical pair P^+I^- and passed to X^- via the exchange and/or dipolar interactions between I^- and X^- so that X^- is left in a polarized state when P^+I^- recombines to PI or ^{3}PI. Applying the sign rule [33] with $g_P < g_I$ leads to the conclusion that J_{PI} must be positive in order that I^- and therefore X^- are emissively polarized. If true, this is surprising: the singlet state of a radical pair is normally lower in energy than the triplet i.e. J is usually negative.

Theoretical analysis of the spin dynamics of the three spin system, $P^+I^-X^-$, is substantially more complex than for a radical pair. The three unpaired electrons are coupled by pairwise exchange and dipolar interactions and X^- does not enter the problem in the same way as nuclear spins do in the radical pair mechanism because of its much larger Zeeman interaction. As shown in Fig. 7, the eight energy levels of the three electron spin system are grouped in energy according to the total magnetic quantum number, M_S. Within the high field approximation, only states with the same M_S are coupled by the Hamiltonian. Thus the S state of P^+I^- is mixed not only with T_0 but also with T_{+1} and T_{-1}. This makes the three spin system fundamentally different from the two spin case.

The method for calculating the polarization is basically the same as for a radical pair: one must solve the Liouville equation with an appropriate Hamiltonian and initial condition (15,60). The principal difference is that the Hamiltonian is larger because there are more states involved and the solution must be found numerically rather than analytically. The following discussion follows the treatment by Hoff and Hore (61).

The Hamiltonian has the following general form:

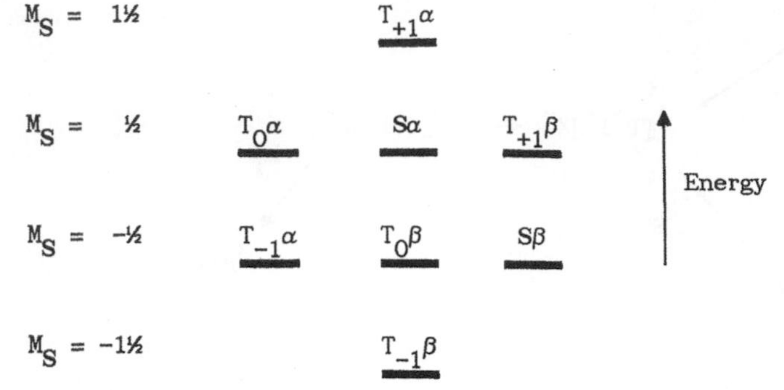

$M_S = 1\frac{1}{2}$ $\underline{T_{+1}\alpha}$

$M_S = \frac{1}{2}$ $\underline{T_0\alpha}$ $\underline{S\alpha}$ $\underline{T_{+1}\beta}$ ↑

                                                                      Energy

$M_S = -\frac{1}{2}$ $\underline{T_{-1}\alpha}$ $\underline{T_0\beta}$ $\underline{S\beta}$

$M_S = -1\frac{1}{2}$ $\underline{T_{-1}\beta}$

Fig. 7   Energy levels of the three electron spin system, $P^+I^-X^-$. S, T_0, T_{+1} and T_{-1} are states of P^+I^-. α and β are states of X^-.

$$
\begin{aligned}
\hat{H} = \; & \omega_P\hat{S}_{Pz} + \omega_I\hat{S}_{Iz} + \omega_X\hat{S}_{Xz} \\
& - J_{PI}[\hat{S}^2_{PI} - 1] - J_{IX}[\hat{S}^2_{IX} - 1] - J_{PX}[\hat{S}^2_{PX} - 1] \\
& + \tfrac{1}{2}D_{PI}[\cos^2\xi_{PI} - \tfrac{1}{3}][3\hat{S}^2_{PIz} - \hat{S}^2_{PI}] \\
& + \tfrac{1}{2}D_{IX}[\cos^2\xi_{IX} - \tfrac{1}{3}][3\hat{S}^2_{IXz} - \hat{S}^2_{IX}] \\
& + \tfrac{1}{2}D_{PX}[\cos^2\xi_{PX} - \tfrac{1}{3}][3\hat{S}^2_{PXz} - \hat{S}^2_{PX}]
\end{aligned}
\tag{42}
$$

where ω_K is the resonance frequency of spin K in the absence of couplings between the radicals, J_{KL} and D_{KL} are the exchange and dipolar couplings between spins K and L, $\hat{S}_{KL}$ is the sum of the spin angular momentum operators of K and L ($\hat{S}_K + \hat{S}_L$) and ξ_{KL} is the angle between the KL dipolar axis and B_0. The exchange interactions are assumed to be isotropic and the dipolar interactions axial. Convenient basis states are the following combinations of the singlet and triplet states of P^+I^- and α and β states of X^-: $T_{+1}\alpha$, $T_{+1}\beta$, $T_0\alpha$, $T_0\beta$, $T_{-1}\alpha$, $T_{-1}\beta$, $S\alpha$, $S\beta$. $\hat{H}$ is block diagonal in this basis with two 3×3 blocks corresponding to $M_S = +\frac{1}{2}$ ($S\alpha$, $T_0\alpha$, $T_{+1}\beta$) and $M_S = -\frac{1}{2}$ ($S\beta$, $T_0\beta$, $T_{-1}\alpha$) and two 1×1 blocks for $M_S = +1\frac{1}{2}$ ($T_{+1}\alpha$) and $M_S = -1\frac{1}{2}$ ($T_{-1}\beta$), see Fig. 7.

It is necessary to include in the Liouville equation the two chemical processes affecting P^+ and I^-, namely the collapse of P^+I^- to give P either in its electronic ground state or its excited triplet state, ^3P (Fig. 8). Both reactions are assumed to conserve electron spin angular momentum. That is, ^3PI is only formed from triplet radical pair states and ground state P is only formed from singlet radical pair states. Thus:

426

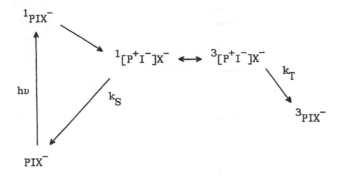

Fig. 8 Simplified reaction scheme for pre-reduced photosynthetic reaction centres.

$$d\hat{\rho}/dt \;=\; -i[\hat{H},\hat{\rho}] \;-\; \tfrac{1}{2}k_T[\hat{P}^T\hat{\rho} + \hat{\rho}\hat{P}^T] \;-\; \tfrac{1}{2}k_S[\hat{P}^S\hat{\rho} + \hat{\rho}\hat{P}^S] \qquad\qquad [43]$$

where k_S and k_T are first order rate constants. The triplet and singlet projection operators are defined by:

$$\hat{P}^T \;=\; \sum_q \{|T_q\alpha\rangle\,\langle T_q\alpha| \,+\, |T_q\beta\rangle\,\langle T_q\beta|\} \qquad q = +1,\ 0,\ -1$$

$$\hat{P}^S \;=\; |S\alpha\rangle\,\langle S\alpha| \,+\, |S\beta\rangle\,\langle S\beta| \quad. \qquad\qquad\qquad\qquad\qquad\qquad [44]$$

Eqn [43] describes the evolution of the density matrix of $P^+I^-X^-$. To calculate the polarization of X^- after P^+I^- has disappeared, one also needs the kinetic equations for the formation of X^- from $P^+I^-X^-$:

$$dp_\alpha/dt \;=\; k_S\,\rho_{S\alpha,S\alpha} \,+\, k_T \sum_q \rho_{q\alpha,q\alpha} \qquad q = +,\ 0,\ - \qquad\qquad [45]$$

(and similarly for p_β) where $p_\alpha(t)$ is the fractional population of the α spin state of X^-. The aim of the calculation is to determine $p_\beta(\infty) - p_\alpha(\infty)$.

One method of achieving this goal is to solve [43] and [45] by Laplace transformation (15,60). The density matrix elements $\rho_{jk}(t)$ and the two X^- populations $p_\alpha(t)$ and $p_\beta(t)$ are first assembled into a vector $\underline{\sigma}(t)$. As there are 8 basis states and therefore 64 density matrix elements, $\sigma(t)$ in general has dimension 66. However, the block diagonal nature of $\hat{H}$ within the high field approximation may be exploited to reduce this size. Rearranging [43] and [45] gives:

$$d\underline{\sigma}/dt \; = \; -\underset{\sim}{W}\underline{\sigma} \tag{46}$$

where $\underset{\sim}{W}$ is a square matrix determined by $\hat{H}$, k_S and k_T. Laplace transformation of [46] gives:

$$(s\underset{\sim}{1} + \underset{\sim}{W})\underset{\sim}{\tilde{\sigma}}(s) \; = \; \underline{\sigma}(0) \tag{47}$$

where $\underset{\sim}{1}$ is the unit matrix with the same dimension as $\underset{\sim}{W}$. $\underline{\sigma}(0)$ describes the initial state of the system (equal population of $S\alpha$ and $S\beta$) and

$$\tilde{\sigma}(s) \; = \; \int_0^\infty e^{-st}\sigma(t) \; dt \quad . \tag{48}$$

The polarization of X^- is then:

$$\mathscr{P}_X \; = \; p_\beta(\infty) - p_\alpha(\infty) \; = \; \lim_{s \to 0} s[\tilde{p}_\beta(s) - \tilde{p}_\alpha(s)] \tag{49}$$

where $\tilde{p}_\beta(s)$ and $\tilde{p}_\alpha(s)$ are the elements of $\underset{\sim}{\tilde{\sigma}}(s)$ corresponding to $p_\beta(t)$ and $p_\alpha(t)$ in $\underline{\sigma}(t)$. Eqn [47] is solved for $\underset{\sim}{\tilde{\sigma}}(s)$ using standard numerical methods.

To date no attempt has been made to simulate the polarized spectrum of X^- using the full Hamiltonian for $P^+I^-X^-$. Although straightforward in principal, such calculations would be expensive in computer time. Instead, the orientationally averaged polarization of X^- has been calculated assuming all g-tensors to be isotropic and the exchange interactions between P^+ and X^- and between ^3P and X^- to be negligible (61).

On the basis of a rather restricted search of the D_{PI}, D_{IX}, J_{PI}, J_{IX} parameter space, Hoff and Hore concluded that in order to get appreciable emissive polarization for X^-, J_{PI} had to be positive in agreement with the simple argument above (61). However more extensive calculations using dipolar parameters calculated from the crystal structure of R. viridis suggest that this need not be the case (62).

Some results of these calculations are summarized in Fig. 9. $\mathscr{P}_X$ is displayed as a function of J_{PI} and J_{IX} for different fixed values of D_{PI} and D_{IX}. In the absence of dipolar interactions ($D_{PI} = D_{IX} = 0$) the sign of $\mathscr{P}_X$ is determined principally by the sign of J_{IX}. This result runs counter to the naive argument above which suggested that J_{PI} should control the sign of $\mathscr{P}_X$. This highlights the dangers of basing predictions for this three spin system on experience with radical pairs.

A more satisfactory and completely different qualitative argument can be

428

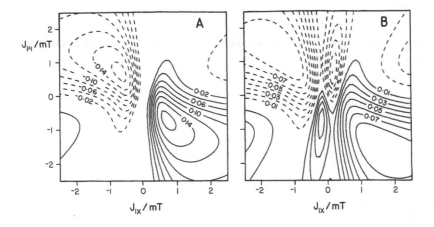

Fig. 9   Contour plots of $\mathcal{P}_X$ as a function of J_{PI} and J_{IX}.   (A): $D_{PI} = D_{IX} = 0$.   (B): $D_{PI} = -1.1$ mT, $D_{IX} = -1.9$ mT.   Solid and dashed contours indicate, respectively, absorptive and emissive polarization.

constructed.   The principal difference between the three electron spin system and a radical pair is that X^- causes the S and T_0 states of P^+I^- to mix with T_{+1} and T_{-1}.   In particular $S\alpha$ and $T_0\alpha$ couple to $T_{+1}\beta$ while $S\beta$ and $T_0\beta$ couple to $T_{-1}\alpha$.   The strength of the coupling is determined by the relative sizes of the off-diagonal matrix element between the two states involved, for example $|\langle S\alpha|\hat{H}|T_{+1}\beta\rangle|$, compared to the corresponding difference in diagonal elements, $|\langle S\alpha|\hat{H}|S\alpha\rangle - \langle T_{+1}\beta|\hat{H}|T_{+1}\beta\rangle|$.   It turns out that, for $J_{PI} = D_{PI} = D_{IX} = 0$, and $J_{IX} > 0$, the difference in diagonal elements is greater for $S\beta/T_{-1}\alpha$ than for $S\alpha/T_{+1}\beta$ while the off-diagonal elements are the same in the two cases.   The same is true of $T_0\beta/T_{-1}\alpha$ and $T_0\alpha/T_{+1}\beta$.   So, $S\alpha$ and $T_0\alpha$ are converted more rapidly into $T_{+1}\beta$ than $S\beta$ and $T_0\beta$ are into $T_{-1}\alpha$.   That is, there will be more X^- radicals with β spins than with α which means absorptive polarization.   For $J_{IX} < 0$, the argument is reversed, α spins are produced to a greater extent than β and the polarization is emissive.   Thus $\mathcal{P}_X$ should be an antisymmetric function of J_{IX} when $J_{PI} = 0$, as shown in Fig. 9.   Presumably this sort of argument could be extended to cope with non-zero values for J_{PI}, D_{PI} and D_{IX}.

$\mathcal{P}_X$ calculated (62) using dipolar parameters estimated (50) from the crystallographic structure of R. viridis ($D_{PI} = -1.1$ mT, $D_{IX} = -1.9$ mT) is shown in Fig. 9.   The principal area of emissive polarization peaks at $J_{PI} \sim +1.2$ mT, $J_{IX} \sim -1.2$ mT with a subsidiary region near $J_{PI} \sim +1.0$ mT, $J_{IX} \sim +0.2$ mT.   As found before (61), the emissive polarization indeed occurs mainly for $J_{PI} > 0$, although weak emission is predicted for $J_{PI} \sim -0.5$ mT, $J_{IX} \sim -1.0$ mT.

On the basis of early RYDMR experiments, Norris et al. inferred that if $D_{PI} <$ 0, then J_{PI} had to be positive. Subsequent RYDMR results combined with observations of static magnetic field effects, led the same authors to conclude that if $D_{PI} < 0$, then either $J_{PI} = +0.7$ mT, $|D_{PI}| < 2.1$ mT or $J_{PI} = -0.7$ mT, $|D_{PI}| < 4.2$ mT. This uncertainty in the sign of J_{PI} is not resolved by calculations of the type presented in Fig. 9. More definite information on the signs and magnitudes of the exchange interactions should be available from simulations of the complete anisotropic spectrum of X⁻ rather than its orientational average.

3.3 Polarization in ^3P

The excited triplet state of P, which is formed in high yield in blocked reaction centres, has a strongly polarized EPR spectrum (5,65-69). This is a natural consequence of the ST_0 radical pair mechanism: the singlet state of P^+I^- couples only to T_0, which reacts with conservation of spin angular momentum to give ^3P principally in its $M_S = 0$ state. The other two ($M_S = \pm 1$) triplet states remain largely empty. Thus the $T_{+1} \leftarrow T_0$ transitions are in enhanced absorption and the $T_0 \rightarrow T_{-1}$ transitions are in emission.

At thermal equilibrium the appearance of the EPR spectrum of ^3P is governed principally by the anisotropic zero-field splitting, that is the dipolar interaction between the two unpaired electron spins that comprise the triplet state (70). As shown in Figs. 10A and 10B, the six main features of the powder spectrum are associated with the transitions occurring when the magnetic field is parallel to one of the principal axes of the zero-field splitting tensor. For each of the principal directions (X, Y and Z) there are two components (+ and -) corresponding to the allowed transitions $T_{+1} \leftarrow T_0$ and $T_0 \leftarrow T_{-1}$, respectively. Thus, for a triplet formed in its T_0 state the spectrum has polarization AEEAAE (Fig. 10C). This pattern is found in numerous photosynthetic systems, plants as well as bacteria at low temperatures.

At higher temperatures, however, the initial polarization of ^3P in pre-reduced (X → X⁻) R. viridis becomes -EAEA- where "-" indicates an absence of detectable signal (71-74). That is, the central portion of the spectrum (Y⁺ and Y⁻) is inverted while the two outer features (Z⁺ and Z⁻) have very little polarization (Fig. 10D). This observation clearly requires that the T_{+1} and T_{-1} states of ^3P are formed more rapidly than the T_0 state when the magnetic field direction coincides with the Y axis of the zero field splitting tensor and that the three states are populated at roughly equal rates when the magnetic field is along the Z axis. Experimentally, this change is observed only when the Q⁻ Fe²⁺ complex is intact and in the presence of a large exchange interaction between I⁻ and Q⁻. It was suggested (71-74) that rapid spin relaxation of the paramagnetic Fe²⁺ ion is transmitted to P^+I^- via the I⁻Q⁻

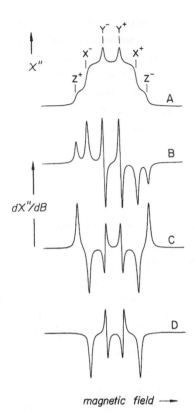

Fig. 10 Schematic EPR spectrum of the triplet state ^{3}P in R. viridis. (A): Unpolarized spectrum. (B): Unpolarized, first derivative spectrum. (C): Low temperature, polarized first derivative spectrum (AEEAAE). (D): High temperature, polarized first derivative spectrum (-EAEA-).

coupling and that this disturbs the development of polarization in some way. This observation is not easy to explain, especially because the presence of X$^-$ renders suspect any qualitative arguments based on the properties of radical pairs (see 3.2).

An attempt to rationalise this behaviour has been made using the technique described in 3.2 (75). Because of the added complexity associated with the Fe^{2+}, drastic simplifications were needed to render the problem computationally tractable. The Fe^{2+} ion was not treated explicitly; dipolar interactions were excluded; only I$^-$ and X$^-$ were assumed to have an (isotropic) exchange interaction; the only anisotropy was the Zeeman interaction of X$^-$; hyperfine couplings and the reaction P$^+$I$^-$ $\longrightarrow$ PI were ignored. Calculations were actually performed for different values of an isotropic X$^-$ g-tensor with the understanding that different g-values corresponded to different (but unknown)

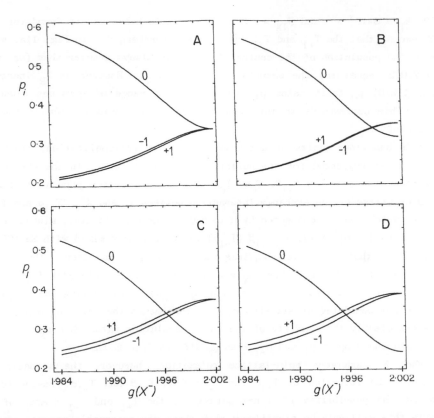

Fig. 11 Calculated fractional populations of ^3P in states T_{+1}, T_0 and T_{-1} as a function of g_X. (A) $k_R = 0$. (B) $k_R = 10^8$ s^{-1}. (C) $k_R = 10^9$ s^{-1}. (D) $k_R = 10^{10}$ s^{-1}. From [75].

orientations of the reaction centre in the magnetic field. Thus the Hamiltonian and the Liouville equation were reduced to:

$$\hat{H} = \omega\hat{S}_{Pz} + \omega\hat{S}_{Iz} + \omega_X\hat{S}_{Xz} - J_{IX}[\hat{S}_{IX}^2 - 1] \qquad [50]$$

$$d\hat{\rho}/dt = -i[\hat{H},\hat{\rho}] - \tfrac{1}{2}k_T[\hat{P}^T\hat{\rho} + \hat{\rho}\hat{P}^T] + \hat{R}\hat{\rho} \qquad [51]$$

The final term in [51] describes the spin-lattice relaxation of X^- by means of a Redfield-type superoperator $\hat{R}$. Eqn [51] was solved as described in 3.2 to give the populations of the three states of ^3P (75).

Fig. 11 shows these populations, p_i (i= = +1, 0, -1) as a function of g_X, the g-value of X^- for four different relaxation rates (k_R) with $g_P = g_I =$

2.002, $k_T = 3 \times 10^8$ s^{-1} and $J_{IX} = -20$ mT. Symmetric behaviour is found for $g_X >$ 2.002 except that the T_{+1} and T_{-1} labels are interchanged. In Fig. 11A, where $k_R = 0$, the population of the central state T_0 is always greater than (or, when $g_X = 2.002$, equal to) the amounts of T_{+1} and T_{-1}. However as k_R increases (Figs. 11B–D) p_0 falls below p_1 and p_{-1} over a range of g-values close to 2.002. This corresponds to inversion of the initial polarization in the EPR spectrum of ^{3}P.

This behaviour admits of a simple qualitative rationalization, along the same lines as suggested for the polarization of X$^-$ in 3.2. In the absence of relaxation of X$^-$, states $T_{+1}\beta$, $T_0\alpha$, $T_0\beta$ and $T_{-1}\alpha$ may become populated from Sα and Sβ by virtue of the exchange interaction between I$^-$ and X$^-$. $T_{+1}\alpha$ and $T_{-1}\beta$, however, which are not connected by $\hat{H}$ to any of the other states, always remain empty. When $|\omega_X - \omega| \gg |J_{IX}|$, $T_{+1}\beta$ ($T_{-1}\alpha$) is not strongly mixed with Sα (Sβ) or $T_0\alpha$ ($T_0\beta$) so that P$^+$I$^-$ never acquires much T_{+1} or T_{-1} character. Thus p_1 and p_{-1} are much less than p_0. When $\omega_X = \omega$, the mixing of Sα (Sβ) with $T_{+1}\beta$ ($T_{-1}\alpha$) is exactly twice as strong as it is with $T_0\alpha$ ($T_0\beta$). Remembering that $T_{+1}\alpha$ and $T_{-1}\beta$ are not populated at all when $k_R = 0$, this means that $p_1 = p_0 = p_{-1}$. Fig 11A shows clearly the equality of the three populating rates when $\omega_X = \omega$ ($g_X = 2.002$) and the drop in p_1 and p_{-1} when g_X differs from 2.002.

When X$^-$ undergoes spin-lattice relaxation, however, the situation is changed. The formation of $T_{+1}\alpha$ and $T_{-1}\beta$ from $T_{+1}\beta$ and $T_{-1}\alpha$, respectively, opens up the possibility of a new pathway to the T_{+1} and T_{-1} states of ^{3}P. Clearly this will only be significant when there are reasonable amounts of $T_{+1}\beta$ and $T_{-1}\alpha$ in the first place, which in turn requires that $|\omega_X - \omega|$ is not large compared to $|J_{IX}|$ as discussed above. Under these circumstances, radical pairs that would have proceeded via $T_0\alpha$ and $T_0\beta$ to give ^{3}P in its T_0 state are diverted along the new routes via $T_{+1}\alpha$ and $T_{-1}\beta$. In this way, the number of radical pairs reacting separately via the T_{+1} and T_{-1} routes may exceed the number going through T_0.

Clearly k_R must be greater than or similar to k_T to get much inversion. When this is not the case, all of the radical pairs will have disappeared before there is time to populate $T_{+1}\alpha$ and $T_{-1}\beta$ by relaxation. Fig. 12 summarizes the important reaction pathways in the absence and presence of spin-lattice relaxation.

In summary, the change from AEEAAE initial polarization to –EAEA– can be rationalised qualitatively if two conditions are satisfied. First the rate of spin-lattice relaxation in X$^-$ must be comparable to k_T. Second, the effective g-value of X$^-$ should be reasonably close to 2.002 when the magnetic field is directed along the Y axis of ^{3}P, very different from 2.002 when the field is along the X axis and of an intermediate value for the Z axis. The EPR frequency of X$^-$ is strongly anisotropic (76,77) covering the range g $\approx$ 0.8 to g

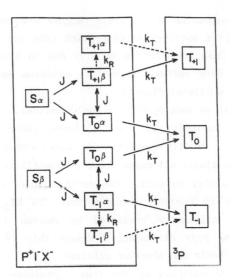

Fig. 12 The important pathways involved in the reactions $^1[P^+I^-]X^- \longrightarrow$ $^3[P^+I^-]X^- \longrightarrow {}^3PIX^-$. The steps indicated by dashed arrows are only important when X^- undergoes rapid spin-lattice relaxation.

≈ 5.0. So, for some orientations of the reaction centre the EPR absorption must occur near $g = 2$. If these orientations are also those for which the magnetic field is directed along the Y axis of ^3P, then inversion of the Y peaks would be expected.

The above discussion rests on the assumption that the observed EPR intensities are strictly proportional to the corresponding population differences. This is a reasonable approximation in the present case, where the microwave field is weak and spin-spin relaxation fast. It would not be appropriate, however, for strong microwave fields which might, inter alia, affect the temperature at which the Y peaks are observed to invert.

4. LIQUIDS

Chemical reactions in the liquid state initiated by thermolysis, photolysis or radiolysis often involve radical pairs as intermediates. As a consequence EPR studies of such processes commonly reveal radical pair polarization. For mobile liquids, one usually observes well separated free radicals that no longer experience an appreciable exchange interaction. The most commonly encountered situation is that of an initial triplet radical pair with a negative exchange interaction J, giving EA polarization, Eqn [33]. When Q is dominated by the difference in Zeeman interactions of the two unpaired electrons, the radical resonating at low field (higher g-value) is entirely in emission while the other is entirely in absorption (the net effect). In the other extreme, when the main contribution to Q is due to hyperfine couplings, the low field lines of both radicals appear in emission and the high field lines in absorption (the multiplet effect).

Quantitative analysis of such spectra is problematical. The difficulty is that the motion of the radicals and their exchange interaction are almost inextricably intertwined. Usually too little is known about one to determine much about the other. In almost all cases there is insufficient information in the polarized EPR intensities to permit the extraction of quantitative data (15). In practice, it is commonly assumed that the ST_0 polarization is proportional to $\sqrt{Q}$ as originally predicted by Adrian (19) from simple considerations of radical pair diffusion, although this is probably a poor approximation in viscous media and micellar solutions.

However, quantitative analysis of the time dependence of the polarized intensities is both feasible and profitable and reveals detailed information on the chemistry, kinetics and spin relaxation of transient free radicals (21-24). This and other aspects of the radical pair mechanism together with the triplet mechanism are discussed by McLauchlan in Chapter 10.

Radicals in viscous liquids or trapped in micelles sometimes exhibit EA polarization for individual lines (78-81). Such "antiphase" lineshapes (low field half emission, high field half absorption) disappear within a few microseconds of the formation of the radicals. These curious effects arise because the high viscosity of the solution or the micelle boundary restricts the diffusion of the radicals such that the exchange interaction has not fallen to zero at the time of observation. Thus, each line in the spectrum is split into an EA doublet by the exchange interaction. Two essentially identical treatments of this effect have appeared (8,9). Both obtain the polarized spectrum by adding spectra calculated for a range of radical separations using an exchange interaction with an exponential distance dependence. Simulated spectra (8) for a prototype radical pair with no hyperfine couplings, formed initially in a triplet state are shown in Fig. 13. For well separated equilibrated radicals the spectrum consists of two Lorentzian lines as expected. Shortly after the creation of the radical pair both lines show the EA antiphase lineshapes observed experimentally and predicted by [33]. As the product of the relative diffusion coefficient of the two radicals and the time after their creation (Dt) increases, the polarized spectra become weaker because faster diffusion and/or longer delays reduce the number of radical pairs with appreciable J. To tailor the calculation to a specific radical pair one has simply to sum spectra of the kind shown in Fig. 13 for all possible nuclear spin configurations of both radicals, using appropriate g-values and

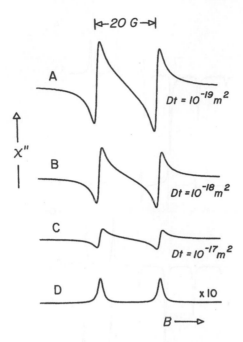

Fig. 13 Simulated EPR spectra of a prototype radical pair averaged over a range of separations. (A)–(C) Polarized spectra. (D) Equilibrium spectrum. From [8].

hyperfine couplings to calculate ω_A and ω_B in Eqn [2]. Fig. 14 shows the results of such simulations for the radical $(CH_3)_2C$-OH formed by flash photolysis of acetone in isopropanol (8). The agreement with experiment once contributions from the conventional radical pair mechanism and triplet mechanism are added is very satisfactory.

The conclusions of these approximate treatments have been confirmed (82) by more rigorous calculations using a stochastic Liouville equation to describe both the spin evolution and diffusive motion of the radical pair. Motion in the body of a sphere or on the surface of a sphere (models of micelles) and in an infinite viscous medium were investigated. As shown in Fig. 15 for a radical pair with no hyperfine couplings, the spectrum shortly after formation of the radical pair, has three distinct contributions. First, there is the conventional polarized spectrum with the low field radical emissive and the high field radical absorptive, Fig. 15A. Second, each line has a superimposed EA antiphase contribution as anticipated, Fig. 15B. The third contribution is an unexpected AE antiphase signal at the mean resonance frequency of the two radicals, Fig 15C. It grows in more slowly as a function of time than do the

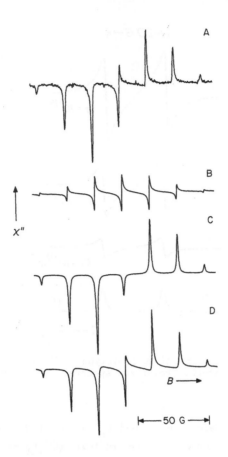

Fig. 14 (A) Experimental EPR spectrum of $(CH_3)_2C-OH$ radicals at 199 K. (B) Simulated EPR spectrum of $[(CH_3)_2C-OH]_2$ radical pairs averaged over a range of separations. (C) EPR spectrum calculated using the conventional radical pair and triplet mechanism theories. (D) Calculated spectrum obtained by adding (B) and (C). From [8].

EA doublets and is more pronounced in micelles than in liquids. This signal arises from pairs of radicals that are re-encountering, having first separated to a point where J is small, as the following argument demonstrates.

Once the radicals reach a separation where $|Q| \gg |J|$, the interconversion of S and T_0 states occurs freely so that when averaged over all radical pairs, with their different diffusive trajectories, the populations of S and T_0 are equalised. Thus for an initial triplet radical pair: $\rho_{SS} = \rho_{TT} = 1/6$ and $\rho_{++} = \rho_{--} = 1/3$ assuming no spin relaxation and no triplet mechanism polarization. Putting these values into Eqn [30] for the intensities of the radical pair spectrum and neglecting the conventional polarization one finds:

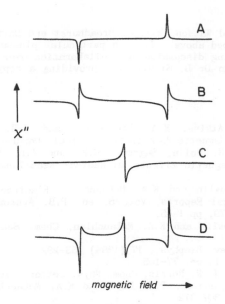

Fig. 15 Simulated EPR spectra of radical pairs in a micelle. (A) The conventional ST_0 polarized spectrum of the separated radicals. (B) Spectrum of separating radical pairs. (C) Spectrum of re-encountering radical pairs. (D) The sum of (A), (B) and (C).

$$I_{12} = -s^2/12 \; ; \quad I_{34} = c^2/12 \; ; \quad I_{13} = -c^2/12 \; ; \quad I_{24} = s^2/12 \; . \qquad [52]$$

Thus, a spectrum consisting of EA doublets is once again predicted but with unequal amplitudes for the two components of each doublet. The spectrum of re-encountering radical pairs is dominated by the two central lines, which are most intense for large $|J|$ (small separations) and which appear close to the mean resonance frequency ω (because $\Omega \simeq |J|$, see Eqn. [25]). The outer lines are almost forbidden for large $|J|$ and appear at $\sim \omega \pm 2J$. The AE enhancement reflects the polarization of the inner components of the two EA doublets.

Not surprisingly, the predicted amplitude of the AE signal is stronger in micelles than in viscous liquids because of the greater re-encounter probability resulting from the micelle boundary. In mobile liquids the central feature is greatly attenuated. Since the EA antiphase signals are due to radical pairs as they first separate while the central AE feature is caused by re-encountering pairs the two together would provide an interesting source of experimental information on the motion of the radicals. The central AE signal has yet to be reported experimentally.

ACKNOWLEDGEMENTS
   I am indebted to Ed Watson, William Broadhurst and David Hunter who did much of the work described above. It is a particular pleasure to thank Keith McLauchlan for stimulating discussion and collaboration over a period of many years. I am grateful to Dr D. Stehlik for providing a copy of reference 43 prior to publication.

REFERENCES

1 L.T. Muus, P.W. Atkins, K.A. McLauchlan and J.B. Pedersen (Eds), Chemically Induced Magnetic Polarization. Reidel, Dordrecht, 1977.
2 K.M. Salikhov, Yu.N. Molin, Sagdeev, R.Z. and A.L. Buchachenko, Spin Polarization and Magnetic Effects in Radical Reactions. Elsevier, New York, 1984.
3 P.J. Hore, C.G. Joslin and K.A. McLauchlan, Electron Spin Resonance, Specialist Periodical Reports, Vol. 5, ed. P.B. Ayscough. The Chemical Society, London, 1979, pp 1–45.
4 P.J. Hore, C.G. Joslin and K.A. McLauchlan, Chem. Soc. Rev., 8 (1979) 29–61.
5 A.J. Hoff, Quart. Rev. Biophys., 17 (1984) 153–282.
6 F.J. Adrian in ref. 1, pp. 77–105.
7 M. C. Thurnauer and J. R. Norris, Chem. Phys. Letters, 76 (1980) 557–561.
8 C.D. Buckley, D.A. Hunter, P.J. Hore and K.A. McLauchlan, Chem. Phys. Letters, 135 (1987) 307–312.
9 G.L. Closs, M.D.E. Forbes and J.R. Norris, J. Phys. Chem., 91 (1987) 3592–3599.
10 P.J. Hore, D.A. Hunter, C.D. McKie and A.J. Hoff, Chem. Phys. Letters, 137 (1987) 495–500.
11 C.P. Slichter, Principles of Magnetic Resonance. Springer, Berlin, 1978, pp. 150–167.
12 L. Monchick and F.J. Adrian, J. Chem. Phys., 68 (1978) 4376–4383.
13 A. Abragam, The Principles of Nuclear Magnetism, Oxford University Press, Oxford, 1961, pp. 44–53.
14 R.P. Feynman, F.L. Vernon and R.W. Hellwarth, J. Appl. Phys., 28 (1957) 49–52.
15 J.H. Freed and J.B. Pedersen, Adv. Magn. Reson., 8 (1976), 1–84.
16 A. Abragam, The Principles of Nuclear Magnetism, Oxford University Press, Oxford, 1961, p. 28.
17 S. Schäublin, A. Höhener and R.R. Ernst, J. Magn. Reson., 13 (1974) 196–216.
18 A. Carrington and A.D. McLachlan, Introduction to Magnetic Resonance. Harper and Row, 1969, pp. 44–47.
19 F.J. Adrian, J. Chem. Phys., 57 (1972) 5107–5113.
20 J.E. Harriman, Theoretical Foundation of Electron Spin Resonance, Academic Press, New York, 1978, pp. 158–162.
21 J.B. Pedersen, J. Chem. Phys., 59 (1973) 2656–2667.
22 P.W.Atkins, K.A. McLauchlan and P.W. Percival, Molec. Phys., 25 (1973) 281–296.
23 P.J. Hore and K.A. McLauchlan, Molec. Phys., 42 (1981) 533–550.
24 K.A. McLauchlan and D.G. Stevens, J. Chem. Phys., 87 (1987) 4399–4405.
25 S Schäublin, A. Wokaun and R.R. Ernst, Chem. Phys., 14 (1967) 285–293.
26 S. Schäublin, A. Wokaun and R.R. Ernst, J. Magn. Reson., 27 (1977) 273–302.
27 C.R. Bowers and D.P. Weitekamp, Phys. Rev. Letters, 57 (1986) 2645–2648.
28 C.R. Bowers and D.P. Weitekamp, J. Amer. Chem. Soc., 109 (1987) 5541–5542.
29 T.C. Eisenschmid, R.U. Kirss, P.P. Deutsch, S.I. Hommeltoft, R. Eisenberg, J. Bargon, R.G. Lawler and A.L. Balch, J. Amer. Chem. Soc., 109 (1987) 8089–8091.
30 M.G.Pravica and D.P. Weitekamp, Chem. Phys. Letters, 145 (1988) 255–258.
31 M.C. Thurnauer and D. Meisel, J. Amer. Chem. Soc., 105 (1983) 3729–3731.
32 T. Prisner, O. Dobbert, K.P. Dinse and H. van Willigen, , J. Amer. Chem. Soc., 110 (1988) 1622–1623.

33 J. Deisenhofer, O. Epp, K. Miki, R. Huber and H. Michel, J. Mol. Biol., 180 (1984) 385–398.
34 C.-H. Chang, D. Tiede, J. Tang, U. Smith, J. Norris and M. Schiffer, FEBS Letters, 205 (1986) 82–86.
35 J.P. Allen, G. Feher, T.O. Yeates, H. Komiya and D.C. Rees, Proc. Natl. Acad. Sci. USA, 84 (1987) 5730–5734; 6162–6166.
36 A.J. Hoff, P. Gast and J.C. Romijn, FEBS Letters, 73 (1977) 185–190.
37 P. Gast, Thesis, University of Leiden, 1982.
38 P. Gast, M.C. Thurnauer, J. Petersen and D. Stehlik, Photosyn. Res. 14 (1987) 15–29.
39 R. Furrer and M.C. Thurnauer, FEBS Letters, 153 (1983) 399–403.
40 J.L. McCracken and K. Sauer, Biochim. Biophys. Acta, 724 (1983) 83–93.
41 P. Gast, T. Swarthoff, F.C.R. Ebskamp and A.J. Hoff, Biochim. Biophys. Acta, 722 (1983) 163–175.
42 M.C. Thurnauer and P. Gast, Photobiochem. Photobiophys. 9 (1985) 29–38.
43 D. Stehlik, C.H. Bock and J. Petersen, J. Phys. Chem., 93 (1989) 1612–1619.
44 J.B. Pedersen, FEBS Letters, 97 (1979) 305–310.
45 P. Gast, A. de Groot and A.J. Hoff, Biochim. Biophys. Acta, 723 (1983) 52–58.
46 S.G. Boxer, C.E.D. Chidsey and M.G.Roelofs, Ann. Rev. Phys. Chem., 34 (1983) 389–417.
47 P.J. Hore, E.T. Watson, J.B. Pedersen and A.J. Hoff, Biochim. Biophys. Acta, 849 (1986) 70–76.
48 A.J. Hoff, Photochem. Photobiol., 43 (1986) 727–745.
49 R.W. Broadhurst, A.J. Hoff and P.J. Hore, Biochim. Biophys. Acta, 852 (1986) 106–111.
50 A. Ogrodnik, W. Lersch, M.E. Michel-Beyerle, J. Deisenhofer and H. Michel, Antennas and Reaction Centres of Photosynthetic Bacteria", ed. M.E. Michel-Beyerle. Springer, Berlin, 1985, pp 198–206.
51 J. Tang and J.R. Norris, Chem. Phys. Letters, 92 (1982) 136–140.
52 K.J. Kaufmann, P.L. Dutton, T.L. Netzel, J.S. Leigh and P.M. Rentzepis, Science, 188 (1975) 1301–1304.
53 M.G. Rockley, M.W. Windsor, R.J. Cogdell and W.W. Parson, Proc. Natl. Acad. Sci. USA, 72 (1975) 2251–2255.
54 C. Kirmaier, D. Holten and W.W. Parson, Biochim. Biophys. Acta, 810 (1985) 33–48.
55 C. Kirmaier, D. Holten, R.J. Debus, G. Feher and M.Y. Okamura, Proc. Natl. Acad. Sci. USA, 83 (1986) 6407–6111.
56 D.A. Hunter and P.J. Hore, unpublished results.
57 P. Gast and A.J. Hoff, Biochim. Biophys. Acta, 548 (1979) 520–535.
58 A.J. Hoff and P. Gast, J. Phys. Chem., 83 (1979) 3355–3358.
59 P. Gast, R.A. Mushlin and A.J. Hoff, J. Phys. Chem., 86 (1982) 2886–2891.
60 F.J.J. de Kanter, J.A. den Hollander, J.A. Huizer and R. Kaptein, Molec. Phys., 34 (1977) 857–874.
61 A.J. Hoff and P.J. Hore, Chem. Phys. Letters, 108 (1984) 104–110.
62 P.J. Hore, unpublished results.
63 J.R. Norris, M.K. Bowman, D.E. Budil, J. Tang, C.A. Wraight and G.L. Closs, Proc. Natl. Acad. Sci. USA, 79 (1982) 5532–5536.
64 J.R. Norris, C.P. Lin and D.E. Budil, J. Chem. Soc., Faraday Trans. I, 83 (1987) 13–27.
65 P.L. Dutton, J.S. Leigh and M. Siebert, Biochem. Biophys. Res. Commun., 46 (1972) 406–413.
66 M.C. Thurnauer, J.J. Katz and J.R. Norris, Proc. Natl. Acad. Sci. USA, 72 (1975) 3270–3274.
67 A.J. Hoff, Phys. Reports, 54 (1979) 75–200.
68 A.J. Hoff, Biophys. Struct. Mech., 8 (1982) 107–150.
69 P. Gast, M.R. Wasielewski, M. Schiffer and J.R. Norris, Nature, 305 (1983) 451–452.
70 J.E. Wertz and J.R. Bolton, Electron Spin Resonance. Chapman and Hall, London, 1986, pp 238–242.
71 F.G.H. van Wijk, P. Gast and T.J. Schaafsma, Photobiochem. Photobiophys. 11 (1986) 95–100.

440

72   F.G.H. van Wijk, P. Gast and T.J. Schaafsma, FEBS Letters, 206 (1986) 238-242.
73   F.G.H. van Wijk, C.B. Beijer, P. Gast and T.J. Schaafsma, Photochem. Photobiol., 46 (1987) 1015-1019.
74   F.G.H. van Wijk and T.J. Schaafsma, Biochim. Biophys. Acta, 936 (1988) 236-248.
75   P.J. Hore, D.A. Hunter, F.G.H. van Wijk, T.J. Schaafsma and A.J. Hoff, Biochim. Biophys. Acta, 936 (1988) 249-258.
76   W.F. Butler, R. Calvo, D.R. Fredkin, R.A. Isaacson, M.Y. Okamura and G. Feher, Biophys. J., 45 (1984) 947-973.
77   G.C. Dismukes, H.A. Frank, R. Friesner and K. Sauer, Biochim. Biophys. Acta, 764 (1984) 253-271.
78   Y. Sakaguchi, H. Hayashi, H. Murai and Y.J. I'Haya, Chem. Phys. Letters, 110 (1984) 275-279.
79   Y. Sakaguchi, H. Hayashi, H. Murai, Y.J. I'Haya and K. Mochida, Chem. Phys. Letters, 120 (1985) 401-405.
80   H. Murai, Y. Sakaguchi, H. Hayashi and Y.J. I'Haya, J. Phys. Chem. 90 (1986) 113-118.
81   A.D. Trifunac and D.J. Nelson, Chem. Phys. Letters, 46 (1977) 346-348.
82   D.A. Hunter, unpublished results.

CHAPTER 13

LIQUID-STATE ENDOR AND TRIPLE RESONANCE [1]

KLAUS MÖBIUS, WOLFGANG LUBITZ AND MARTIN PLATO

1. INTRODUCTION

In many important chemical and biological (photo) reactions in solution free radicals, both charged and uncharged, are involved as intermediates with vastly varying lifetimes. For the elucidation of a structure-function relationship and, thereby, for an understanding of the reaction dynamics it is of key importance to identify and characterize the radical species. Optical spectroscopy, which is frequently used in studies of such systems, is distinguished by its inherent sensitivity and time-resolution. It suffers, however, from relatively poor spectral resolution to the extent that electron-nuclear hyperfine interactions cannot be resolved. This hampers the identification of intermediates, since it is the hyperfine structure (hfs) that carries a wealth of detailed information about the electronic and spatial properties of the radicals. It is, therefore, challenging to design experiments to detect hyperfine couplings (hfc's) from radical intermediates by means of high-resolution, high-sensitive electron spin resonance (EPR) methods. Only for small molecules or molecules of high symmetry straightforward EPR alone allows to resolve the hfs. In many important bio-reactions, however, such as the primary reactions in photosynthesis, the radicals involved are large and asymmetric in their structure so that hfs can no longer be resolved due to inhomogeneous broadening of the EPR lines. Electron-nuclear double or triple resonance techniques have to be applied for extracting the hfc's.

As a general strategy of improving the resolution of inhomogeneously broadened spectra by multiple resonance spectroscopy, one aims at eliminating redundant lines in the spectrum. This is achieved by introducing more than one selection rule, i.e. by applying more than one resonant electromagnetic radiation field to the system, thereby restricting its spectral response.

A typical example of this strategy is ENDOR (electron-nuclear double resonance) where EPR and NMR transitions are driven simultaneously. The pioneering work was done by Feher (1) in the solid state, by Cederquist (2), Hyde and Maki (3) in solution and by Freed (4), who developed the theoretical foundation of

---
[1] Dedicated to the memory of our former collaborator Reinhard Biehl who contributed so much to this field, but died so early (June 6, 1987) at the age of 43 years.

steady-state liquid-phase multiple resonance spectroscopy. In a steady-state ENDOR experiment one adjusts the magnetic field value to an EPR line, saturates it and simultaneously sweeps a saturating NMR rf field through the resonance region of the various nuclei, thereby changing the saturation parameter of the EPR transition. The first ENDOR experiments were performed at low temperature, where all the relaxation times are sufficiently long to obtain saturation easily. For radicals in liquid solution, however, the relaxation times are much shorter - in the order of 10^{-5} to 10^{-7} s - and, consequently, much larger saturating microwave and rf fields have to be applied. This probably explains why the first ENDOR-in-solution experiments required many more years before they were successful.

In our laboratory, ENDOR-in-solution was extended to electron-nuclear-nuclear triple resonance (TRIPLE) experiments with the following goals in mind:

In "Special TRIPLE" (5, 6), which applies to only one group of equivalent nuclei, the spectral sensitivity is considerably improved. Furthermore, the lines are generally narrower and their intensity ratios, in contrast to ENDOR, approximately reflect the number of nuclei involved in the transition. This yields important information for assigning the hfc's to individual molecular positions. In a Special TRIPLE experiment two NMR rf fields are frequency swept symmetrically about the nuclear Larmor frequency while the EPR transition is saturated.

In "General TRIPLE" (6, 7), which applies to more than one group of nuclei, relative signs of hfc's can be directly determined from characteristic intensity changes of the ENDOR lines occurring when one NMR transition is additionally pumped while sweeping the second rf field through the nuclear resonance region. Such sign information is very important in conjunction with theoretical predictions and for assigning hfc's to different types of nuclei.

Some books and earlier review articles devoted to ENDOR spectroscopy should be listed right at the beginning:

The first textbook dealing predominantly with ENDOR has been written by Kevan and Kispert (8). It provides an excellent introduction to the subject and covers experimental aspects as well as representative examples of ENDOR from liquid- and solid-phase samples. In the book "Multiple Electron Resonance Spectroscopy", edited by Dorio and Freed (9), a wide variety of experimental and theoretical ENDOR studies on inorganic, organic, biochemical and polymer systems is thoroughly discussed. It specifically includes a chapter on liquid-state TRIPLE resonance spectroscopy by Möbius and Biehl (6). Recently, a book by Schweiger (10) on ENDOR of solid-state transition metal complexes with organic ligands was published. In the experimental sections, principles and instrumentations of ENDOR spectroscopy are discussed with particular emphasis on ingenious variants that supplement "conventional" ENDOR (see also (11)). Most

recently, a book by Kurreck, Kirste and Lubitz (12) highlighted the applicability of ENDOR spectroscopy of radicals in solution in the fields of organic and
biological chemistry. The book describes the principles of ENDOR and TRIPLE
resonance and illustrates them by a broad survey of examples. It discusses in
detail topics such as proton and non-proton ENDOR, multispin systems, dynamic
effects, and ENDOR-in-liquid-crystal solutions.

An extended review article about radicals in solution studied by ENDOR ad
TRIPLE resonance spectroscopy was published by Möbius et al. (13). More recently, the field of ENDOR spectroscopy in photobiology and biochemistry was reviewed by Möbius and Lubitz (14), the field of ENDOR of photochemically generated radicals by Möbius (15). Schweiger wrote a critical survey of the state of
the art of continuous wave ENDOR methodology (11); Mehring and collaborators
(16, 17) gave general views of pulsed ENDOR and triple resonance schemes and of
the physics behind them. The review articles cited in Ref. 18 cover specific
areas of ENDOR applications that are related to this Chapter.

Our multiple-resonance-in-solution experiments can be grouped into two
blocks: (i) Steady-state ENDOR and TRIPLE on long-lived radicals (lifetime $\tau \gg$
spin-lattice relaxation time T_1), (ii) ENDOR and TRIPLE on transient radicals
($\tau \le T_1$) that show strong CIDEP effects in their spectra (CIDEP = Chemically
Induced Electron Polarization). For both blocks examples will be presented,
which are representative for the recent work of this laboratory. Before doing
so, the theoretical background and the instrumentation will be briefly described.

2. THEORY OF ELECTRON-NUCLEAR MULTIPLE RESONANCE IN SOLUTION

2.1 Stable Radicals

Up to the present time most publications on ENDOR/TRIPLE work have been
concerned with radicals with lifetimes $\tau \gg T_1$ so that the following sketch of
ENDOR/TRIPLE theory will be concerned with such "stable" radicals.

2.1.1 Phenomenological Description

ENDOR

Under steady-state conditions, ENDOR can be described as an NMR-induced
desaturation of a saturated EPR line. As a result, the EPR line is enhanced,
provided EPR and NMR transitions have common energy levels. In the following we
shall first look more closely at the physical principles governing this ENDOR
enhancement, i.e. the actual "ENDOR effect", and then show why ENDOR spectra
are generally better resolved than EPR spectra.

ENDOR Enhancement

Conceptually, the ENDOR enhancement of an EPR line can be best understood for the simplest case S = 1/2, I = 1/2, i.e. in a four-level diagram (see
Fig. 1). The various relaxation transition rates are indicated by wavy lines:

444

W_e describes the relaxation rate of the electron spins, W_n that of the nuclear spins. W_{x1} and W_{x2} are cross-relaxation rates which describe the flip-flop and flop-flop processes of the coupled electron and nuclear spins. For isolated radicals in solution, the relaxation transitions are mainly induced by spin rotational interaction and by modulation of electronic Zeeman and electron-nuclear dipole interactions as a result of Brownian rotational tumbling. The corresponding time-dependent Hamiltonian will be discussed later.

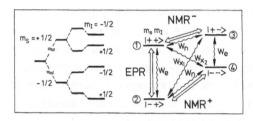

Fig. 1. Four-level diagram of a system with S =1/2, I = 1/2 with relaxation rate $W_{\alpha\beta}$. (A positive sign of the isotropic hyperfine coupling constant is assumed.) Only one EPR, but both allowed NMR induced transitions, NMR^+ and NMR^-, are shown. In ENDOR, NMR^+ or NMR^- are driven consecutively, whereas in Special TRIPLE these two transitions are driven simultaneously.

In a phenomenological description, which neglects all sorts of microwave and rf field induced coherence effects, the ENDOR experiment, in which an EPR and an NMR transition (either NMR^+ or NMR^-, see Fig. 1) are saturated simultaneously, can be visualized as the creation of an alternative relaxation path for the pumped electron spins, which is opened by driving the NMR transition and which passes via $W_e(|+-\rangle\leftrightarrow|--\rangle)$ and $W_n(|--\rangle\leftrightarrow|-+\rangle)$ - or even better via $W_{x1}(|+-\rangle\leftrightarrow|-+\rangle)$ or W_{x2} $(|--\rangle\leftrightarrow|++\rangle)$. The extent to which this relaxation by-pass can compete with the direct W_e route $(|++\rangle\leftrightarrow|-+\rangle)$ determines the degree of desaturation of the EPR line and, therefore, determines the ENDOR signal intensity. The intensity pattern of ENDOR lines, therefore, does not generally reflect the number of contributing nuclei, in contrast to EPR and NMR.

Obviously, good ENDOR signal-to-noise ratios require that this delicate interplay of the various induced rates and relaxation transition rates is carefully optimized. The parameters to be varied by the experimentalists are the radical concentration, the solvent viscosity, the temperature, and the microwave and rf field strengths. In the limit of high NMR saturation and exact "on-resonance" conditions, one immediately finds by analyzing the four-level transition pattern (4) for the simple case $W_{x1} = W_{x2} = 0$ a relative change, E, of the EPR signal amplitude

$$E = \frac{1}{2(2+b+b^{-1})} \qquad [1]$$

where $b = W_n/W_e$. The quantity E is normally called "ENDOR enhancement". The maximum of E is found in the "matching condition" $b = 1$, i.e. for $W_n = W_e$ (see

Fig. 2). This optimum condition yields E_{max} = 1/8, showing that ENDOR signals are weaker than EPR signals by almost an order of magnitude in the absence of efficient cross-relaxation rates.

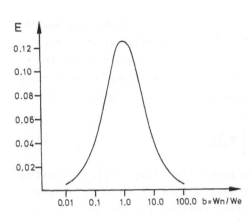

Fig. 2. Dependence of the ENDOR enhancement E on b = W_n/W_e. For details, see text and Eqn.1.

Cross-relaxation becomes operative as soon as W_x exceeds the smaller of the two direct rates, W_n or W_e. W_{x1}, $W_{x2} \neq 0$ always lead to an increased enhancement for both low- and high-frequency ENDOR lines, because these routes by-pass W_e and W_n (see Fig. 1). However, if $W_{x1} \neq W_{x2}$, which is the most frequent case, the now unsymmetrical relaxation network will also produce unsymmetrical ENDOR line patterns. The largest ENDOR effect will obviously be observed for that NMR transition which forms a closed loop with the EPR transition and the larger of the two W_x rates. The observation of such unsymmetrical ENDOR line patterns not only provides clear evidence of considerable cross-relaxation contributions, but also information about the relative signs of different isotropic coupling constants of a molecule (13).

We shall now look more closely at the temperature and solvent dependence of the ENDOR enhancement E by returning to the simpler situation where W_{x1} = W_{x2} = 0. According to Eqn. 1, E depends on the ratio W_n/W_e. It will be shown later in Section 2.1.2 that W_n and W_e usually depend differently on a rotational correlation time τ_R characteristic of the Brownian tumbling motion of the molecule in the liquid solution. In most cases $W_n \propto \tau_R$, whereas $W_e \propto \tau_R^{-1}$, which yields $W_n/W_e \propto \tau_R^2$. Since $\tau_R \propto \eta/T$, where η is the viscosity of the solvent, we obtain the important relation

$$b = W_n/W_e \propto (\eta/T)^2. \qquad [2]$$

In the region b << 1, i.e. W_n << W_e, which is the most frequent situation for free radicals in solution at room temperature, we have from Eqns.1 and 2

$$E(b << 1) \propto (\eta/T)^2. \qquad [3]$$

Since η/T strongly increases with decreasing temperature, mainly due to the strong temperature dependence of η (see Fig. 9, Section 2.1.2), the ENDOR effect can therefore be strongly enhanced by cooling the sample. Fairly seldom,

$W_n \gg W_e$ at room temperature. For such cases, Eqn. 1 yields

$$E(b \gg 1) \propto (\eta/T)^{-2} \qquad [4]$$

implying that the sample has to be heated to maximize the ENDOR effect.

ENDOR Resolution

Now let us explain why ENDOR spectra are better resolved than EPR spectra. For this purpose we refer to Fig. 3 which shows, for simplicity, the energy level scheme of a radical (S = 1/2) in isotropic liquid solution containing four equivalent protons (I_i = 1/2 for each of them) in a strong magnetic field. The interactions responsible for the various splittings are summarized in the following static spin Hamiltonian:

$$\hat{H}_o/h = (g\mu_B/h)\ \underline{B}_o\hat{\underline{S}} - \sum_i (g_{ni}\mu_K/h)\ \underline{B}_o\hat{\underline{I}}_i + \sum_i a_i\hat{\underline{S}}\hat{\underline{I}}_i \qquad [5]$$

The leading term is the electronic Zeeman interaction, $(g\mu_B/h)\underline{B}_o\hat{\underline{S}}$, followed by the nuclear Zeeman and hyperfine interactions, $(g_n\mu_K/h)\underline{B}_o\hat{\underline{I}}$ and $a\cdot\hat{\underline{S}}\hat{\underline{I}}$, respectively, summed over all nuclei. The hyperfine coupling constant (hfc), a, is scalar as long as radicals in isotropic solutions are considered. In the strong field approximation with $\underline{B}_o \parallel$ z axis, the energy eigenvalues, classified by the magnetic spin quantum numbers, m_s and m_I, are given by

$$E_{m_sm_I}/h = (g\mu_B/h)\ B_om_s - \sum_i (g_{ni}\mu_K/h)\ B_om_{Ii} + \sum_i a_im_sm_{Ii} \qquad [6]$$

i.e. in a specific m_s manifold the hyperfine levels are equidistant.

In an EPR experiment, therefore, because of the selection rules, $\Delta m_s = \pm1$, Δm_I = 0, five lines are observed (see Fig. 3) with binomial intensity distribution owing to the transition frequency degeneracies for equivalent nuclei. In an ENDOR experiment, on the other hand, the sample is additionally irradiated with an rf field of varying frequency driving NMR transitions $\Delta m_I = \pm1$ of nuclei coupled to the unpaired electron, while m_s stays unchanged (either +1/2 or -1/2 in doublet radicals). Referring to Eqn. 6, this yields the general (first order) ENDOR resonance condition for a group of equivalent nuclei, all having the same nuclear Larmor frequency $\nu_n = (g_n\mu_K/h)B_o$ and the same hfc $a_i = a$:

$$\nu_{ENDOR}^\pm = |E_{m_sm_I+1} - E_{m_sm_I}|/h = |\nu_n \pm a/2|. \qquad [7]$$

Thus, every group of equivalent nuclei - no matter how many nuclei are involved and what their spin quantum number m_I is - contributes only two ENDOR lines due to the first order degeneracy of the NMR transitions (see Fig. 3). In a doublet radical, these two ENDOR lines are symmetrically displayed about ν_n or $|a|/2$, whichever is larger. Consequently, with increasing number of groups of equivalent nuclei the number of ENDOR lines increases only additively,

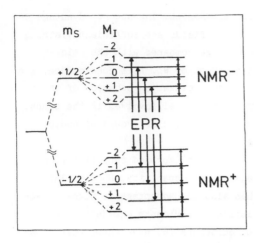

Fig. 3. Energy level diagram of a radical in solution with four equivalent protons in a high magnetic field. The magnetic spin quantum number of the electron is $m_s = \pm 1/2$; the total magnetic quantum number of the nuclei is given by $M_I = \Sigma_i m_I$. According to the selection rules of EPR and NMR, five EPR lines, but only two ENDOR lines, appear.

whereas the number of EPR lines increases in a multiplicative way. Particularly in large low symmetry radicals, tyical for biological systems, the gain in spectral resolution can become very drastic. Examples for such cases will be shown in Section 4 (Applications).

Another important aspect concerning the gain in resolution by ENDOR is the fact that nuclei with different magnetic moments $g_n \mu_K$ appear in different frequency ranges on account of their different Larmor frequencies $\nu_n = (g_n \mu_K/h)B_0$. Thus the ENDOR lines of ^{14}N nuclei ($\nu_n \approx 1$ MHz at $B_0 \approx 0.33$ T) are often well separated from the ENDOR lines of protons ($\nu_n \approx 14$ MHz). This will also be demonstrated by various examples in Section 4.

TRIPLE

Not for all systems an equalization of nuclear and electronic relaxation rates can be achieved by temperature and solvent selection. This is particularly true for biological samples. Then $W_n \ll W_e$, i.e. the slow W_n is the rate-determining step in the relaxation by-pass. Consequently, in cases of vanishing cross-relaxation, W_n acts like a bottle-neck for the EPR desaturation and limits the ENDOR signal intensity to less than one percent of the EPR intensity. There is an obvious way out of this dilemma by short-circuiting the W_n bottle-neck by a second saturating rf field. In this electron-nuclear-nuclear triple resonance experiment, the two rf fields are tuned to drive both NMR transitions, NMR$^+$ at ν^+ and NMR$^-$ at ν^-, of the same nucleus simultaneously (see Fig. 1), thereby enhancing the efficiency of the relaxation by-pass. As a consequence, one gains considerably in signal intensity. Such a triple resonance was theoretically proposed by Freed (19) and experimentally realized by Dinse et al. (5). Since both rf fields are applied at a frequency separation of the hyperfine constant of the same nucleus, this is a special version of triple resonance ("Special TRIPLE").

448

The increased sensitivity is one important aspect. The second advantage of Special TRIPLE over ENDOR is that, when both rf fields are sufficiently strong so that the induced transition rates become large compared with the relaxation rates W_n, the EPR desaturation becomes independent of W_n. As a consequence, the line intensities are no longer determined by the relaxation behavior of the various nuclei, but rather reflect the number of nuclei involved in the transition. TRIPLE lines can therefore be assigned to particular groups of nuclei in the molecule more easily than ENDOR lines.

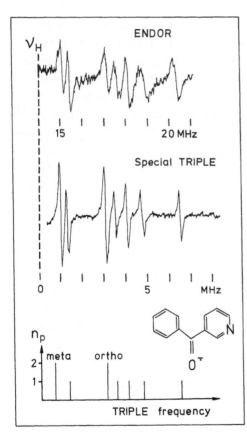

Besides improved sensitivity and assignment capability, Special TRIPLE has also the advantage of higher resolution. Experiments and density matrix calculations (6) have shown that at a given power level, the effective NMR saturation, which determines the observed linewidth, is smaller in TRIPLE than in ENDOR.

In Fig. 4, the ENDOR and Special TRIPLE spectra are shown for the 3-pyridyl-phenyl ketone anion radical as a representative example (20).The spectra were recorded at a deliberately chosen high temperature, where the rings are rotating rapidly on the EPR time scale. At this temperature, however, $W_n \ll W_e$, and consequently the ENDOR signal-to-noise ratio is rather poor. Since the NMR transitions are already saturated at the applied field amplitude of 1 mT_{rot} (rot = rotating frame), all the ENDOR lines show equal intensity within the noise limits. The Special TRIPLE spectrum was obtained at the same total rf power level and hence the rf field amplitudes per side band are reduced to $(1/\sqrt{2})mT_{rot}$. In qualitative accordance with the theoretical predictions, the signal-to-noise is increased in the TRIPLE spectrum. Furthermore, the intensity ratios 2:1:2:1:1:1:1 of the TRIPLE lines reflect nicely the number

Fig. 4. ENDOR half-spectrum and Special TRIPLE resonance spectrum of the 3-pyridylphenyl ketone anion radical (solvent, DME; counterion, Na^+; T = 240 K). The intensity ratios of the stick diagram reflect the number of equivalent protons, n_p, involved in the respective NMR transitions. The Special TRIPLE spectrum was recorded at the same total NMR power level as the ENDOR spectrum. From (20).

of protons involved in the NMR transitions (the ortho and meta protons of the phenyl ring are equivalent). Finally, a reduction of the linewidth in the TRI-PLE spectrum is also clearly visible.

Electron-nuclear-nuclear triple resonance can be generalized ("General TRIPLE") to include more than one nucleus, for example two inequivalent protons (7). The first-order solution of the time-independent spin Hamiltonian of Eqn. 5 for the simplest three-spin system $S = I_1 = I_2 = 1/2$ leads to an eight-energy level scheme which can be characterized in the basis $|m_S m_{I_1} m_{I_2}\rangle$, see Fig. 5. From this figure it is obvious that we now can desaturate a pumped EPR transition by driving the NMR transitions of both nuclei. For the EPR transition shown in Fig. 5, and neglecting cross-relaxation, this can be achieved, for instance, by driving the NMR transitions at ν_1^+ and ν_2^+ of nucleus 1 and 2, the relaxation by-pass being closed via W_e, W_{n1}, W_{n2}. Because all the NMR transitions are doubly degenerate, there principally exist several of such closed relaxation by-passes involving ν_1^+ and ν_2^+ and, hence, a complete picture would be very confusing in such a two-dimensional representation of the energy levels. We, therefore, resort to a three-dimensional representation of the eight-energy level scheme.

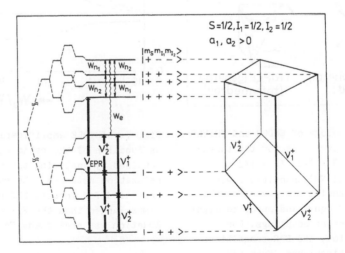

Fig. 5. General TRIPLE resonance on a radical with two inequivalent protons. All nuclear transitions are doubly degenerate in first order. A TRIPLE induced relaxation by-pass for the EPR transition shown, involving the high-frequency NMR transitions at ν_1^+ and ν_2^+ of both nuclei, can be achieved via the routes W_{n2}, W_{n1}, W_e and W_{n1}, W_{n2}, W_e (cross-relaxation rates neglected). This situation can be more clearly visualized by a three-dimensional representation of the energy level scheme in form of a (distorted) cube. From (13).

As is depicted in Fig. 5, the eight energy levels can be arranged to form the corners of a (distorted) cube, in which the EPR transitions occur vertically, the NMR transitions horizontally. In which particular plane of the cube the NMR transitions occur, solely depends on the relative signs of the two hfc's a_i - and it is this sign-information which can additionally be obtained from a General TRIPLE resonance experiment.

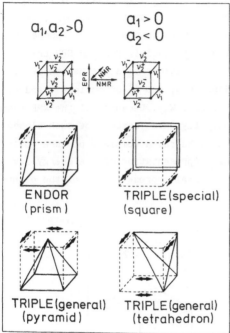

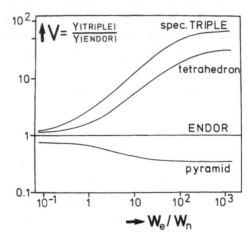

Fig. 6. Topology of ENDOR and TRIPLE resonance experiments for the three-spin system S = 1/2, I_1 = 1/2, I_2 = 1/2. The NMR transitions with their frequencies are given for the two cases of equal and opposite signs of the hyperfine couplings. For simplification, the EPR transitions are not distinguished. From (20).

Fig. 7. TRIPLE amplification factor as function of the ratio W_e/W_n obtained by analyzing the various relaxation networks. The curves shown are valid for induced NMR rates 100 times larger than the W_n; cross-relaxation was neglected. From (20).

This will be explained with the aid of some topological games summarized in Fig. 6. At the top of Fig. 6 the energy level arrangement is depicted for the two different cases, a_1, $a_2 > 0$ and $a_1 > 0$, $a_2 < 0$. In the case that both coupling constants have the same sign, the doubly degenerate low-frequency NMR transitions, ν_1^- and ν_2^-, of both nuclei occur in one plane, the doubly degenerate high-frequency transitions, ν_1^+ and ν_2^+, in the other. If the couplings have opposite signs, low- and high-frequency transitions of both nuclei occur in a mixed fashion in both planes. By now considering the level <u>populations</u>, the different multiple resonance experiments can be visualized by different geometrical figures , which are derived from the cubes by contracting those corners that are connected by induced NMR transitions. This represents the limiting case of highly saturated NMR transitions, where the populations of the connected levels are equalized. In this representation, an ENDOR experiment, where only one of the doubly degenerate NMR transitions is saturated, for instance at ν_1^- (see Fig. 5), forms a <u>prism</u>. Special TRIPLE, where ν_1^- and ν_1^+ are driven simultaneously, forms a <u>square</u>. In General TRIPLE two different cases have to be distinguished: assuming ν_1^- to be the pumped frequency, all low-frequency NMR transitions (ν_1^- and ν_2^-) are saturated in the same plane if the hfc's have the same sign. In this case the four corners are contracted to a single point and a <u>pyramid</u> is formed. If, on the other hand, the hfc's have opposite signs, the low-frequency transitions are saturated in different planes and a <u>tetrahedron</u> is formed.

The impact of these topological games is shown in Fig. 7 giving the result of an analysis of the relaxation networks of the various geometrical figures (6). The TRIPLE amplification factor V, which is plotted versus the ratio W_e/W_n, is defined as the ratio of TRIPLE and ENDOR line amplitudes. The analysis shows that the tetrahedron experiment always yields markedly larger TRIPLE line amplitudes than the pyramid experiment, and the difference between pyramid and tetrahedron becomes more pronounced in those cases, where W_e is much larger than W_n. This is frequently occurring for organic radicals in solution. Relative signs of coupling constants can therefore easily be determined from intensity changes in General TRIPLE spectra.

As an example, Fig. 8 shows ENDOR and General TRIPLE spectra of the fluorenone ketyl/sodium ion-pair in solution (21). The intensity of the low- and high-frequency proton and sodium ENDOR lines is the same, since experimentally almost constant transition moments over the whole frequency range have been provided. This symmetry of

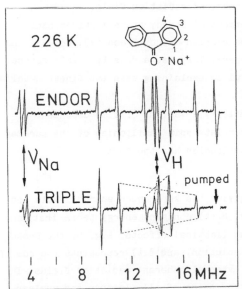

Fig. 8. ENDOR and General TRIPLE resonance spectra of the fluorenone anion radical (solvent, tetrahydrofuran; counterion, Na^+; T = 226 K). From (21).

the ENDOR line intensities is lost in the General TRIPLE spectrum, where the high-frequency proton transition belonging to the largest hfc is additionally pumped. For the protons with the two small couplings, the high-frequency lines are larger than their low-frequency counterparts, thereby demonstrating that for the former a tetrahedron experiment is performed with respect to the pumped transition. The largest and the two small proton hfc's, therefore, must have opposite signs. For the second largest coupling, however, the intensity pattern is clearly reversed, yielding a sign different from that of the two smaller couplings. Also heteronuclear General TRIPLE resonance works: When pumping a proton line, also the sodium line pair changes its intensity ratio yielding the same sign of the sodium hfc as that of the two smaller proton hfc's. It should be pointed out that for heteronuclei with a negative magnetic moment the intensity behavior would be reversed (22, 118).

2.1.2 Relaxation Theory of ENDOR/TRIPLE-in-Solution

The basic principles of ENDOR and TRIPLE experiments have been described in the beginning of this Section in a phenomenological way. This treatment is based on simple rate equations for the level populations. Although such an approach is very helpful in providing a first understanding of the basic ideas underlying these experiments, it is hardly suited to explain the details of an ENDOR spectrum, such as the linewidths, the saturation behavior, and coherence effects caused by the presence of more than one radiation field.

The most detailed theoretical treatment of steady-state multiresonance experiments in the liquid phase has been carried out by Freed (19, 23, 24) and Freed et al. (25, 26) in a series of papers. They use a density matrix method, which has proved to be extremely powerful in explaining even the finest details of multiresonance spectra.

The density matrix method is essentially a variant of time dependent perturbation theory starting from the complete spin Hamiltonian of the paramagnetic molecule. Most generally, this Hamiltonian has the form

$$\hat{H}(t) = \hat{H}_0 + \hat{H}_1(t) + \hat{\varepsilon}(t) \qquad [8]$$

where $\hat{H}_0$ is the time-independent part giving the zero-order energy levels and transition frequencies (compare Eqn. 1), $\hat{H}_1(t)$ contains energy terms that are randomly time-modulated by environmental (lattice) effects, e.g. by the Brownian tumbling motion of the molecule in solution, and $\hat{\varepsilon}(t)$ represents the sum of the interactions of the spins with the external coherent radiation fields. In ENDOR (or TRIPLE), $\hat{\varepsilon}(t)$ contains the interaction of all electron and nuclear spins with one microwave and one (or two) rf radiation fields.

The term of primary interest in Eqn. 8, regarding the behavior of ENDOR and TRIPLE spectra, is $\hat{H}_1(t)$ as it determines the magnitude of all spin-lattice relaxation rates $W_{\alpha\beta}$ and linewidth parameters $(1/T_2)_{\alpha\beta}$ for any two levels α, β of the spin system.

Various types of interactions can contribute to $\hat{H}_1(t)$ for a doublet radical in liquid solution, e.g.

1. the coupling between the electron spin $\underset{\sim}{\hat{S}}$ and the rotational momentum $\underset{\sim}{\hat{J}}$ of the molecule:

$$\hat{H}_{SR}/h = \underset{\sim}{\hat{S}} \cdot \underset{\sim}{B} \cdot \underset{\sim}{\hat{J}}; \qquad [9]$$

2. the anisotropic electron Zeeman interaction between the magnetic moment of the electron $g_e\mu_B\underset{\sim}{\hat{S}}$ and the external magnetic field $\underset{\sim}{B}_0$ due to spin-orbit coupling:

$$\hat{H}_G/h = (\mu_B/h)\underset{\sim}{B}_0 \cdot \underset{\sim}{G}' \cdot \underset{\sim}{\hat{S}}; \qquad [10]$$

3. the anisotropic dipole-dipole hyperfine (END) interaction between the electron spin $\underset{\sim}{\hat{S}}$ and the nuclear spins $\underset{\sim}{\hat{I}}_i$:

$$\hat{H}_A/h = \sum_i \underset{\sim}{\hat{S}} \cdot \underset{\sim}{A}'_i \cdot \underset{\sim}{\hat{I}}_i; \qquad [11]$$

4. the anisotropic quadrupole hyperfine interaction of nuclei with $I_i > 1/2$:

$$\hat{H}_Q/h = \sum_i \underset{\sim}{\hat{I}}_i \cdot \underset{\sim}{Q}_i \cdot \underset{\sim}{\hat{I}}_i; \qquad [12]$$

5. the isotropic electron-nuclear Fermi contact interaction

$$\hat{H}_a/h = \sum_i a_i\underset{\sim}{\hat{S}\hat{I}}_i; \qquad [13]$$

6. the isotropic electron Zeeman interaction

$$\hat{H}_g/h = (g_e\mu_B/h)\underset{\sim}{B}_0 \cdot \underset{\sim}{\hat{S}}. \qquad [14]$$

All interactions 1 to 6 are given in frequency units. The four quantities $\underset{\sim}{B}$, $\underset{\sim}{G}'$, $\underset{\sim}{A}'_i$, and $\underset{\sim}{Q}_i$ are the spin-rotational coupling, anisotropic g, electron-nuclear dipolar, and quadrupolar hyperfine tensors. They are traceless second rank tensors which, in isotropic solutions, do not affect the positions of spectral lines. In order to contribute to $\hat{H}_1(t)$, any of the interactions must be randomly time modulated by the rotational motion of the molecule in solution and/or internal molecular motions, such as rotation of groups of atoms (e.g. CH_3) or vibrations. The last two interactions 5 and 6 can, obviously, only be affected by internal type of motions since they are spatially isotropic.

Proceeding along the ideas of Freed and co-workers by solving the "equation of motion" of the density matrix, it can be shown that certain matrix elements of the "relaxation matrix" R are directly related to unsaturated linewidths and spin lattice relaxation rates, e.g. $R_{\alpha\beta\alpha\beta} = -(1/T_2)_{\alpha\beta}$ and $R_{\alpha\alpha\beta\beta} = W_{\alpha\beta}$. They can be calculated from the spectral density functions $j^{(\mu)}(\omega)$ of the various time dependent terms $\mu = 1, \ldots, 6$ of $\hat{H}_1(t)$ considering the selection rules for the spin operators involved in that particular interaction. An important example is given by the electron-nuclear dipole (END) hyperfine interaction ($\mu = 3$), which, for a system with $S = 1/2$ and a single nucleus with $I = 1/2$, produces the following relaxation matrix elements (see Fig. 1):

$$R_{1122} = W_e = \frac{1}{2} j^{(A')}(\omega_e) \qquad R_{2233} = W_{x1} = \frac{1}{3} j^{(A')}(\omega_e \pm \omega_n)$$

$$R_{2244} = W_n = \frac{1}{2} j^{(A')}(\omega_n) \qquad R_{1212} = -1/T_{2e} = -\frac{2}{3} j^{(A')}(0) \qquad [15]$$

$$R_{1144} = W_{x2} = 2 j^{(A')}(\omega_e \pm \omega_n) \qquad R_{2424} = -1/T_{2n} = -\frac{7}{6} j^{(A')}(0).$$

Here the subscripts 1 to 4 refer to the four states $|++\rangle$, $|-+\rangle$, $|+-\rangle$, and $|--\rangle$, respectively. If the Brownian tumbling motion is considered to be the only significant mechanism causing relaxation, the spectral density function $j^{(A')}(\omega)$ is given by

$$j^{(A')}(\omega) = \frac{\pi^2}{5} (Tr \underset{\sim}{A'}^2) \frac{\tau_R}{1 + \omega^2 \tau_R^2} \qquad [16]$$

where $Tr\underset{\sim}{A'}^2$ is the sum of diagonal elements (trace) of the squared dipolar hyperfine tensor $\underset{\sim}{A'}$ introduced in Eqn. 11, and τ_R is the rotational correlation time. In order to discuss Eqns. 15 on the basis of Eqn. 16, we assume the following conditions that are normally met in ENDOR and TRIPLE studies: $\omega_n \tau_R \ll 1$, i.e. "fast tumbling limit" in the NMR frequency range, but $\omega_e \tau_R > 1$, i.e. "slow tumbling limit" in the EPR frequency range. This yields $W_e^{END} = c_1 \tau_R^{-1}$, $W_n^{END} = c_2 \tau_R$, $W_{x2}^{END} = 6 W_{x1}^{END} = c_3 \tau_R^{-1}$. The proportionality constants c_1, c_2, and c_3 can be calculated from the molecular property $Tr\underset{\sim}{A'}^2$.

After having set up the R-matrix from all relevant interactions contributing to $\hat{H}_1(t)$, one can then work out steady-state solutions of the equation of motion of the density matrix $\rho(t)$ for all non-vanishing diagonal and non-diagonal elements of ρ. This is done on a computer that also puts out the final ENDOR or TRIPLE spectrum for the specific conditions chosen in the experiment (27).

The obvious advantage of such a rigorous theoretical approach to ENDOR and TRIPLE phenomena is that (i) the number of parameters determining the different spin lattice relaxation rates and unsaturated linewidths is reduced to a relatively small number of independent fundamental molecular parameters, e.g. $Tr\underset{\sim}{A'}^2$ and τ_R, and (ii) the influence of the microwave and rf fields is correctly taken account of at any power level. It also leads to a quantitative interpretation of all types of coherence effects typical for multiresonance experiments, which show up as line distortions and sometimes even splittings (28).

Recently, an attempt has been made to find as simple relations as possible between optimum ENDOR conditions, e.g. optimum temperature for a given solvent and microwave and rf fields, and a few of the most relevant molecular proper- ties by employing the rigorous density matrix formalism described above (27). The aim of this study was to find a systematic approach to the ENDOR behavior of different magnetic nuclei in different molecular environments and solvents. The following simplifying assumptions were made which are quite realistic for most organic doublet radicals in solution:

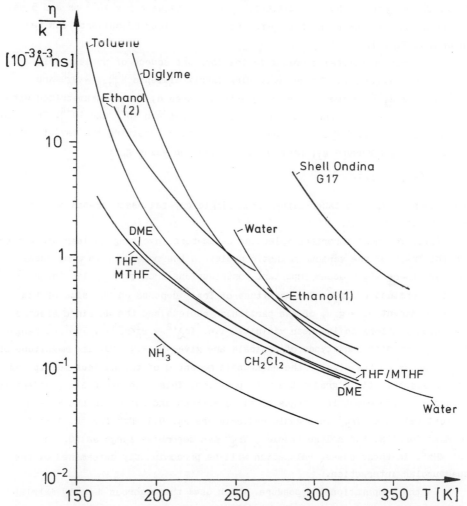

Fig. 9. Dependence of η/kT on T for solvents often used in ENDOR spectroscopy. The rotational correlation time τ_R in nanoseconds for a particular radical- solvent combination is obtained by multiplying η/kT in the given units by the effective rotational radical volume V_{eff} in units of 10^3 Å^3 (DME: 1,2-dimeth- oxyethane; THF: tetrahydrofuran; MTHF: 2-methyl THF; Diglyme: diethylene glycol dimethyl ether; Shell Ondina G17: mineral oil). From (12, 27).

(i) Brownian rotational diffusion is taken to be the dominant source of relaxation and can be described by a single rotational correlation time, τ_R. This correlation time is related to molecular and solvent properties by the well-known Einstein-Debye relation

$$\tau_R = V_{eff}\eta/kT \qquad\qquad [17]$$

in which V_{eff} is the effective volume of the solvated molecule, η the viscosity of the solvent, T the absolute solvent temperature, and k the Boltzmann constant. Under typical ENDOR conditions, τ_R ranges between 5×10^{-3} ns and 5 ns. Values of η/kT as function of temperature for a variety of solvents can be taken from Fig. 9.

(ii) Spin-rotational coupling is the dominant source of nuclear spin independent relaxation. The corresponding matrix elements $R^{SR}_{\alpha\alpha\beta\beta} = W^{SR}_e$ and $-R^{SR}_{\alpha\beta\alpha\beta}/2 = 1/(2T^{SR}_{2e})$ between pairs of electron states α, β are parametrized with respect to τ_R as B/τ_R, where B ranges typically between 10^{-6} and 10^{-4}. The value of B may be estimated from a semiempirical relation involving the deviation of the g tensor components g_{ii} from the free electron value g_e :

$$B \approx \frac{1}{2} \sum_i (g_{ii} - g_e)^2. \qquad\qquad [18]$$

The values of g_{ii} are taken either from single crystal measurements or from theoretical estimates.

(iii) The most important molecular parameters involving nuclear properties are the $\mathrm{Tr}\underset{\sim}{A_i'}^2$ of the various magnetic nuclei in the molecule. In some cases, however, there might occur important contributions from quadrupolar interactions determined by $\mathrm{Tr}\underset{\sim}{Q_i^2}$. The magnitude of $\mathrm{Tr}\underset{\sim}{A'}^2$ depends on the size of the magnetic moment $\mu_n = g_n\mu_K$ of the particular nucleus and the unpaired electron spin distributions in its near vicinity, i.e. $\mathrm{Tr}\underset{\sim}{A'}^2 \propto \mu_n^2 \langle r_n^{-3}\rangle^2$. Typical ranges for $\mathrm{Tr}\underset{\sim}{A'}^2$ in different types of radicals are given in Fig. 10. The magnitude of $\mathrm{Tr}\underset{\sim}{Q}^2$ depends on the size of the quadrupole moment Q of the nucleus multiplied by the electric field gradient at its position. This electric field gradient is produced by the electric charges of all electrons and nuclei in the molecule. Typical values of $\mathrm{Tr}\underset{\sim}{Q}^2$ in organic radicals are ca. 0.1 MHz^2 for $^2\mathrm{H}$, 1 to 10 MHz^2 for $^{14}\mathrm{N}$. For halogen nuclei, $\mathrm{Tr}\underset{\sim}{Q}^2$ can become as large as 10^4 to 10^7 MHz^2. In these cases, relaxation will be predominantly determined by the quadrupolar interaction.

In the computational procedure, which uses the rigorous density matrix approach, the parameters defined in (i) - (iii) are used as input data besides other data, such as the measured strengths B_e and B_n of the microwave and rf fields. Alternatively, in the procedure of optimizing the ENDOR effect, the values of B_e and B_n (including hyperfine enhancement (29)) are calculated so that they correspond to the constant saturation parameters

$$\sigma_e = \gamma_e^2 T_{1e} T_{2e} B_e^2 = 3 \text{ and } \sigma_n = \gamma_n^2 T_{1n} T_{2n} B_n^2 = 1.$$

The output of the computer program are ENDOR or TRIPLE signal amplitudes, linewidths and line shapes, and the "optimum" microwave and rf field strengths, B_e^{opt} and B_n^{opt}, required to fulfill the saturation conditions given above. As an example, calculated ENDOR signal amplitudes are shown in Fig. 11 as a function of the rotational correlation time τ_R. These calculations were performed for a

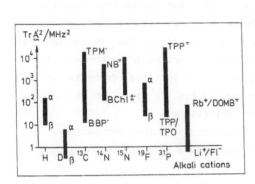

Fig. 10. Typical ranges of $\text{Tr}\underset{\sim}{A}'^2$ for different nuclei. Upper and lower limits are labeled by the names of the corresponding radicals or by the different possible positions α (directly bonded) or β (two bonds away) of the nucleus relative to the π system. The abbreviations stand for: TPM = triphenylmethyl, BBP = di-tert.-butyl-benzoylphenoxyl, NB = nitrobenzene, TPP = triphenyl-phosphabenzene, TPO = triphenylphenoxyl (oxidizing agent), BChl = bacteriochlorophyll, DOMB = di-ortho-mesitoylbenzene, Fl = fluorenone. Other nuclei, so far observed by ENDOR, but not included in the figure, are: ^{10}B, ^{11}B, ^{23}Na, ^{25}Mg, ^{27}Al, ^{29}Si, ^{39}K, ^{117}Sn, ^{119}Sn, ^{113}Cs, ^{203}Tl, ^{205}Tl., see Refs. 12, 13.

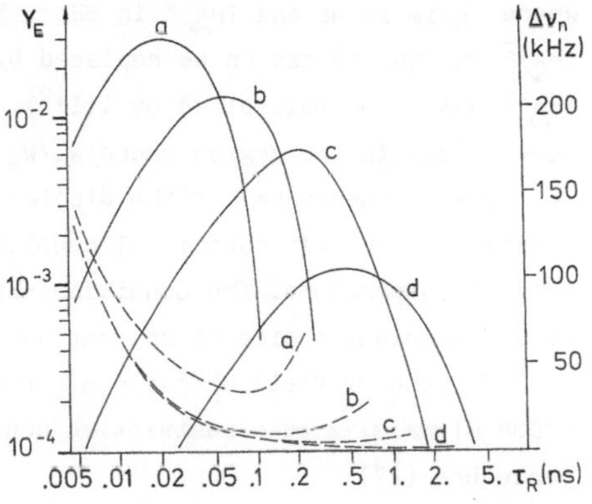

Fig. 11. ENDOR signal amplitude Y_E and saturated ENDOR linewidth $\Delta\nu_n$ (in kHz, broken lines) as functions of the rotational correlation time τ_R (in ns). The hyperfine anisotropy $\text{Tr}\underset{\sim}{A}'^2$ is used as a parameter: $\text{Tr}\underset{\sim}{A}'^2 = 1000$ MHz^2 (curves a), 100 MHz^2 (b), 10 MHz^2 (c), and 1 MHz^2 (d). The calculations were performed for one nucleus with I = 1/2, spin-rotational parameter B = 1×10^{-6}, saturation degrees $\sigma_e = 3$, $\sigma_n = 1$, no exchange. For details, see Ref. 27.

single nucleus with I = 1/2. $\mathrm{Tr}\underset{\sim}{A}'^2$ and the spin-rotational relaxation constant, B, are treated as parameters, the ranges studied cover the majority of nuclei in organic free radicals (27). The curves $Y_E(\tau_R)$ for constant values of $\mathrm{Tr}\underset{\sim}{A}'^2$ and B have well-pronounced maxima at "optimum" values τ_R^{opt}. The steep rise and fall of $Y_E(\tau_R)$ on either side of τ_R^{opt} demonstrate how critically the various relaxation rates have to be balanced out by proper choice of solvent, viscosity, and temperature when performing ENDOR-in-solution work. Evaluation of the computer results gives (27)

$$\tau_R^{opt} \approx 200(B/\mathrm{Tr}\underset{\sim}{A}'^2)^{0.5} \quad , \tag{19}$$

where τ_R is in ns and $\mathrm{Tr}\underset{\sim}{A}'^2$ in MHz^2. For n equivalent nuclei with I = 1/2, $\mathrm{Tr}\underset{\sim}{A}'^2$ in Eqn. 19 has to be replaced by $n \cdot \mathrm{Tr}\underset{\sim}{A}'^2$. For one nucleus with I > 1, $\mathrm{Tr}\underset{\sim}{A}'^2$ has to be multiplied by I(I+1). These modifications are first order corrections in the region where W_n/W_e < 1. Qualitatively, Eqn. 19 implies that an increase (decrease) of the dipolar hyperfine coupling or a decrease (increase) of the spin-rotational coupling require an increase (decrease) of the solvent temperature. The condition expressed by Eqn. 19 can often be fulfilled by a convenient choice of solvent and/or temperature.

For the rf field B_n^{opt} (in mT, rotating frame) required for maximum ENDOR signals, a more restrictive condition follows from the computational procedure (27):

$$\frac{\nu_{ENDOR}}{\nu_H} B_n^{opt} \approx 10(B \cdot \mathrm{Tr}\underset{\sim}{A}'^2)^{0.5} \tag{20}$$

The factor ν_{ENDOR}/ν_H, being the ratio of the observed transition frequency of the studied nucleus and the free proton frequency, takes account of the different magnetic nuclear moments and the enhancement or deenhancement of the rf field by the isotropic hyperfine interaction (29). This hyperfine enhancement factor is of particular importance for nuclei having $\nu_{ENDOR} \approx \nu_H$ in spite of a very small magnetic moment. A typical example is ^{14}N with the ratio of Larmor frequencies $\nu_N/\nu_H = 0.07$, where the isotropic hfc is often so large (>10 MHz) that $\nu_{ENDOR} \approx \nu_H$ at least for the high-frequency ENDOR line. Eqn. [20] also shows that the product $B \cdot \mathrm{Tr}\underset{\sim}{A}'^2$ has to stay within reasonable limits for ENDOR still to be feasible. A practical upper limit for B_n with present technical equipment lies around 3 mT_{rot} which implies a maximum value of $B \cdot \mathrm{Tr}A'^2$ in the order of 0.1 MHz^2 (in the absence of nuclear spin relaxation processes other than the END mechanism).

For the optimum microwave field one obtains B_e^{opt} (mT) $\approx 0.1(B \cdot \mathrm{Tr}\underset{\sim}{A}'^2)^{0.5}$ in the limit of large values of $\mathrm{Tr}\underset{\sim}{A}'^2$. This condition can in some cases be more restrictive than the condition for B_n^{opt}, since the ENDOR effect depends more strongly on B_e than on B_n (27). This is normally not critical for organic radicals, but can easily be fatal for transition metal complexes, which mostly have large values of $B \cdot Tr\underset{\sim}{A}'^2$. For transition metal complexes, therefore, ENDOR-in-solution experiments are only feasible in very few particularly favorable cases (30).

Concluding this Section, some points should be mentioned that are of practical importance in ENDOR-in-solution spectroscopy:

Besides the proper choice of solvent type, viscosity and temperature, the radical concentration is another important factor. In general, the signal-to-noise ratio of ENDOR spectra cannot be improved by increasing the radical concentration well above 10^{-4} M, since then Heisenberg exchange processes become significant and diminish the ENDOR enhancement (4).

It is much more difficult to detect ENDOR responses from non-proton nuclei, which often occur at low rf frequencies. For small isotropic hfc's, $|a| \ll 2\nu_n$, the NMR transitions of such nuclei are difficult to drive into saturation because the hyperfine enhancement of the applied rf field is too weak. Furthermore, when the hfc of a nucleus is so small that it becomes comparable with the homogeneous EPR linewidth, the associated ENDOR lines have vanishing intensity, since the hyperfine-split EPR transitions can no longer be separately coupled by rf irradiation (31).

A major problem in EPR/ENDOR spectroscopy is the assignment of the measured hfc's to specific molecular positions in the radical. Assignment aids can be obtained from (i) intensity ratios in Special TRIPLE spectra, (ii) relative signs of hfc's from General TRIPLE, (iii) temperature dependence of hfc's, (iv) temperature dependence of ENDOR line intensities due to individual nuclear relaxation behavior. Ultimately, such assignments can only be solved by chemical assistance, such as specific isotopic labelling (^{2}H, ^{15}N, ^{13}C, etc.).

2.2 Transient Radicals

Up to this point, ENDOR-in-solution was considered under steady-state conditions with the radicals in Boltzmann equilibrium with the lattice. Realistic ENDOR sensitivity considerations under these conditions require a minimum of 10^{13} radicals to be present in the cavity. As a consequence, steady-state ENDOR is restricted to stable radicals or to relatively long-lived intermediates with lifetimes of at least 10^{-3} s otherwise, even in fast flow-systems, the equilibrium concentration of radicals in the cavity is too small.

However, many reactions of interest in chemistry, biochemistry and in radiolysis involve transient radicals with lifetimes much shorter than 10^{-3} s, which one would like to detect and characterize by EPR and ENDOR. The short lifetime requires fast, i.e. broad-band detection schemes with inherent loss of sensitivity as compared with narrow-band phase-sensitive detection (compare Chapters 10 and 11). Possible ways out of this sensitivity problem are offered by dynamic polarization effects resulting in strong deviations from Boltzmann spin level populations, i.e. by "Chemically Induced Dynamic Electron Polarization" (CIDEP).

The first successful detection of rf pumping of NMR transitions in radical intermediates via changes in the CIDEP intensities of their EPR spectra ("CIDEP-enhanced-ENDOR") was recently performed in this laboratory (32, 33).

2.2.1 Chemically-Induced Dynamic Electron Polarization (CIDEP)

CIDEP effects in transient radicals in solution become observable through pronounced deviations of the EPR line amplitudes from Boltzmann equilibrium values. There are several excellent reviews and books that cover the field of chemically induced magnetic polarization in depth (34-40) so only a brief account of the principles will be given here (see also Chapters 10, 11, 12 for more details).

Spin polarization effects observable in transient radicals in solution arise from two basically different mechanisms: the "triplet mechanism" (TM) and the "radical pair mechanism" (RPM). The TM can only occur in photolytically or radiolytically generated radicals, the polarization originating from selective singlet-triplet intersystem-crossing (ISC) in the molecule. When this molecule eventually reacts from its triplet state with a quencher molecule, doublet state radicals with non-Boltzmann population of the electron spin levels are produced. The RPM can occur in a great variety of chemical reactions involving radical intermediates, the polarization originating from the spin interactions during encounters of pairs of radicals. The non-Boltzmann intensity distribution of the EPR line pattern often reflects contributions from both mechanisms, since TM polarization is related to the process of radical creation and RPM polarization might occur in subsequent radical interactions.

In many photochemical reactions excitation takes place from the singlet ground state of the chromophore to the excited singlet state from which the lowest excited triplet state is populated by ISC. The TM polarization detected in an EPR experiment is generated by an interplay of the electron Zeeman and zero-field interactions and, therefore, the TM CIDEP is hyperfine-independent. This means that the TM creates a "net effect" of the polarization, for which in both radical reaction products the same m_s spin level manifold is overpopulated. The observed EPR spectra, therefore, either show all lines in emission or in enhanced absorption, the mode of polarization depending on the sign of the zero-field splitting parameter and the relative magnitude of the ISC transition probabilities into the zero-field triplet eigenstates (36).

The RPM CIDEP, on the other hand, originates in the electron Zeeman and hyperfine interactions within the radicals and exchange interactions between pairs of radicals and might result in hyperfine-independent ("net effect") as well as hyperfine-dependent ("multiplet effect") polarizations.

Pairs of radicals can either be generated simultaneously as singlet or triplet "geminate" pairs by thermally or photolytically induced radical reactions, or they may be formed by chance encounters of two free radicals in solution ("random" or "F pairs"). In an external magnetic field the two electron spins of a radical pair can orient statistically to form either a singlet state or a triplet state with its three components T_0, T_{+1}, T_{-1}. The amount of bonding singlet character relative to that of the anti-bonding triplet character in the total wavefunction determines the rate of product formation in subsequent radical reactions (see also Chapter 19). EPR detectable polarization might occur because those pairs which can most rapidly acquire the greatest degree of singlet character by virtue of singlet-triplet (S-T) mixing effects will be favored in the reaction products. The crucial point in a diffusion model of the RPM CIDEP (41, 42) is that the radicals undergo a period of diffusive separation (small exchange interaction J), in which the S-T mixing step via Zeeman and hyperfine interactions occurs, followed by a diffusive radical re-encounter dominated by a large J. The polarization is the combined result of the S-T mixing during the time interval between the first and second encounters and of the strong exchange interaction during the re-encounter period giving rise to an excess of "up" spins in one radical (net effect) or an excess of "up" spins in certain states of a radical with a corresponding excess of "down" spins in the other states (multiplet effect).

When both radicals of the pair, $R_1^{\bullet}$ and $R_2^{\bullet}$, separate by diffusion, the distance dependent exchange interaction $J(r)$ becomes so small that the electron spins are decoupled and precess independently under the influence of the different local magnetic fields originating in different g factors or hyperfine interactions. The difference in precession frequencies leads to S-T mixing. In high magnetic fields B_0, only S-T_0 mixing is efficient and, consequently, neither state has a net magnetization with respect to B_0. The amount of S-T_0 mixing is determined by the difference in precession frequencies, δ_{ab}, for radicals 1 and 2 for a particular nuclear spin state (ab). This difference contains contributions from the Zeeman and hyperfine interactions:

$$\delta_{ab} = \nu_{1a} - \nu_{2b} = \frac{1}{h}(g_1 - g_2)\mu_B B_0 + \sum_j a_{1j} \cdot m_{1j}^{(a)} - \sum_k a_{2k} \cdot m_{2k}^{(b)} \qquad [21]$$

where $m_{1j}^{(a)}$ is the magnetic quantum number of the jth nucleus of radical 1 in the overall nuclear-spin state a, and $m_{2k}^{(b)}$ is the magnetic quantum number of the kth nucleus of radical 2 in the overall nuclear-spin state b. In the case of a dominating Δg term, a net polarization effect is observed. Dominating hyperfine terms, on the other hand, lead to the multiplet effect by which different hyperfine levels acquire different degrees of polarization; the resulting EPR spectrum exhibits lines in emission and enhanced absorption (E/A patterns).

As an example of CIDEP-assisted EPR spectroscopy of transient intermediates, photo-induced electron transfer in a mixture of Zn-tetraphenylporphyrin
(ZnTPP) and duroquinone (DQ) in ethanol was studied in this laboratory (43). The
short-lived radical ions were directly detected (without field modulation, see
Section 3) by time-resolved EPR. A similar system has very recently been studied by van Willigen, Dinse and collaborators (44) using FT-EPR and by van
Willigen et al. (45) using time-resolved EPR. After shining a 10 ns laser pulse
at 590 nm selectively into an absorption band of the porphyrin, we observed,
0.3 µs after the laser flash, polarized EPR spectra of two radicals superimposed on each other (Fig. 12). One radical shows resolved hyperfine structure
of equidistant lines (a_{iso} = 0.186 mT) from which, together with the measured
g factor (2.0045), it can be identified as the DQ mono-anion (46). The other
radical exhibits only a 0.65 mT broad Gaussian line at g = 2.0025 which we
identify as the ZnTPP mono-cation (47). Since the EPR spectra of both radicals
have the same polarization phase - enhanced absorption - the TM must be predominant for the CIDEP effect. The observed deviation, however, of the intensity ratios of the hyperfine components from a binomial distribution indicates
a superimposed E/A pattern. This suggests that, at 270 K, there is a minor
contribution from the RPM to the overall polarization with the radical pair
precursor being created in the triplet state. The situation can be explained in

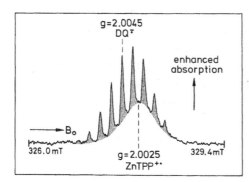

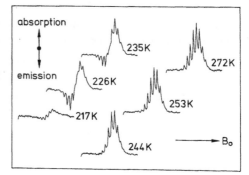

Fig. 12. Photoinduced electron
transfer in Zn-tetraphenyl-porphyrin
(ZnTPP) and duroquinone (DQ) in
ethanol solution. [ZnTPP] ≈ 2 x
10^{-4} M; [DQ] ≈ 2 x 10^{-2} M, T = 270 K,
delay time between laser pulse and
boxcar gate ≈ 0.3 µs. For details,
see text.

Fig.13. Photoinduced electron
transfer in ZnTPP (2 x 10^{-4} M) and DQ
(2 x 10^{-2} M) in ethanol at different
temperatures (delay time ≈ 1 µs). The
emission/ absorption pattern reflects
varying contributions from triplet
and radical pair mechanisms to the
electron polarization. For details,
see text.

the following way: The 590 nm laser flash excites ZnTPP into its first excited
singlet state from which - via selective ISC - spin-polarized triplet state
ZnTPP is generated. Rapid quenching reactions occur via electron transfer to
the DQ and the polarization is carried over to the reaction partners, $ZnTPP^{+\cdot}$
and $DQ^{\bar{\cdot}}$.

Under the experimental conditions (T = 270 K, delay time 0.3 μs) the RPM
cannot yet compete with the TM in terms of CIDEP efficiency. However, the relative importance of both polarization mechanisms can be manipulated by varying,
for instance, the temperature at a fixed delay time of detection after the
laser flash. Fig. 13 demonstrates the switching on and off of the TM with increasing and decreasing temperature. At higher temperatures (above 244 K) all
the lines are in enhanced absorption, i.e. the TM dominates. At lower temperatures (below 235 K), however, some lines appear in emission, some in absorption
showing that the RPM with multiplet effect dominates the polarization kinetics.
Obviously, at higher temperatures the diffusion controlled triplet quenching
electron transfer rate can compete more efficiently with the fast triplet relaxation rate (typically $(10 \text{ ns})^{-1}$). The triplet polarization is then transferred to the doublet radicals, $ZnTPP^+$ and DQ^-, with their relatively long
electronic relaxation times (typically 10 μs). Further experiments with the aim
of elucidating the complex kinetic behavior of this donor-acceptor system and
related ones are in progress.

2.2.2 CIDEP-Enhanced-ENDOR Strategy

The occurrence of electron spin polarization effects makes it possible to
extend EPR to ENDOR with its inherent resolution potential also to short-lived
radical intermediates. The general strategy in such a CIDEP assisted ENDOR
experiment is to detect NMR transitions of short-lived radicals via changes of
the CIDEP efficiency in EPR spectra, the changes being induced by frequency-
swept saturating rf fields.

The critical condition for this experiment is the rf field strength, B_n,
necessary to saturate NMR transitions of transient radicals during their lifetime, τ. The NMR saturation condition for a transient radical is given by

$$\gamma_n^2 B_n^2 T_{1n}^* T_{2n}^* \approx 1 \qquad [22]$$

with $1/T_{1n}^* = 1/\tau + 1/T_{1n}$; $1/T_{2n}^* = 1/\tau + 1/T_{2n}$. An estimate, valid for instance for proton NMR of p-benzosemiquinone radicals, with $T_{1n} = T_{2n} = 4 \cdot 10^{-6}$ s
and $\tau \approx 10^{-3}$ s yields a saturating rf field of $B_n \approx 1.0$ mT_{rot} (rotating frame).
To obtain this field strength with reasonable rf power is not an easy task (32,
33). The required B_n field increases rapidly for $\tau \approx T_{1n}$, T_{2n} and approaches
2.0 mT_{rot} for $\tau = T_{1n} = T_{2n} = 4 \cdot 10^{-6}$ s.

To explain the CIDEP-enhanced-ENDOR detection strategy the simplest model
case of a radical (S =1/2) containing one proton (I = 1/2) will be considered.

464

We assume that spin-lattice relaxation can be ignored, i.e. signals are detected after radical generation in a time shorter than the spin-lattice relaxation time. For the detection strategy the net effect CIDEP behaves differently
from the multiplet effect CIDEP. It should be noted, however, that the specific
mechanism, TM or RPM, leading to CIDEP is irrelevant for the applicability of
the ENDOR method.

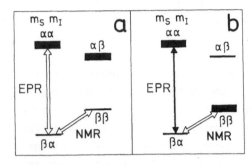

Fig. 14. Spin energy and population
diagram for S = 1/2, I = 1/2 showing
(a) a net effect electron polarization and (b) a multiplet effect
electron polarization. Saturated
transitions are indicated by open
arrows. For details, see text.

In Fig. 14, case (a) represents a
pure net effect polarization, case (b)
a pure multiplet effect polarization.
It is easy to see that in case (a) a
population difference between the nuclear spin levels has to be established
first, for instance by a saturating
microwave field. Driving the NMR transition shown in Fig. 14a, further transfers population from the βα level to
the empty ββ level. Thereby the population difference between the levels
connected by the EPR transition is
increased, resulting in an enhanced EPR
signal (an enhanced emission signal in
this case). For the second NMR transition an enhancement is also obtained.

In the case of a multiplet effect polarization (Fig. 14b) there is already
an initial population difference between the nuclear spin levels, and for the
observed EPR transition saturation must be avoided in order not to reduce this
polarization. Driving the NMR transition decreases the population difference
between the levels connected by the EPR transition which results in a decreased
EPR signal. Net effect polarization can therefore easily be discriminated
against multiplet effect polarization, since in the former case EPR and ENDOR
signals will have the same phase, whereas in the latter case they have opposite
phases.

Simultaneous driving of both NMR transitions, $\alpha\alpha \leftrightarrow \alpha\beta$ and $\beta\alpha \leftrightarrow \beta\beta$, an
experiment analogous to the Special TRIPLE resonance from Section 2.1, will
further increase the ENDOR enhancement. For this case of CIDEP-enhanced Special
TRIPLE, up to 100 % changes in the EPR signal amplitude are expected.

3. INSTRUMENTATION

3.1 Steady-State ENDOR/TRIPLE Resonance

Basic spectrometer designs for continuous wave (cw) ENDOR together with a
variety of cavity constructions are included in the books by Poole (48), Kevan

and Kispert (8), Dorio and Freed (9), and Schweiger (10). The ENDOR review articles (11) and (49) also contain more recent references about instrumentation. Commercial ENDOR spectrometers are available (e.g. from BRUKER, W.Germany; JEOL, Japan) for those with well-stocked research grants.

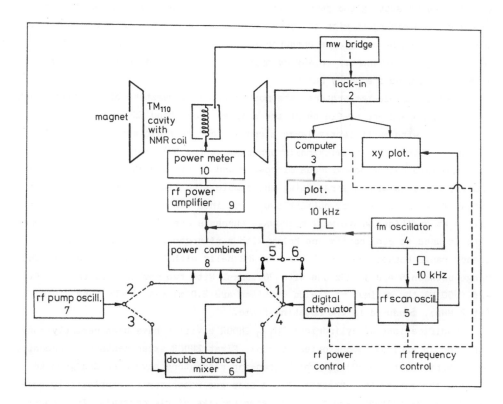

Fig. 15. Block diagram of the steady-state ENDOR/TRIPLE resonance spectrometer built in this laboratory (13). The sample temperature can be varied by an N_2 gas stream. The numbers in the boxes refer to model specifications of instruments actually used: 1 Varian E 101; 2 Ithaco 391 A; 3 Hewlett-Packard 21 MX-E; 4 Wavetek VCG III; 5 Hewlett-Packard 8660 C; 6 Anzac MD 141; 7 Wavetek 7000; 8 Eni PM 400-4; 9 two Eni A-300; 10 Bird 43. The different modes of operation can be achieved by the following connections:

ENDOR:		Special TRIPLE		General TRIPLE	
conn.	1 open	conn.	1 open	conn.	1 closed
	2 open		2 open		2 closed
	3 open		3 closed		3 open
	4 open		4 closed		4 open
	5 open		5 closed		5 open
	6 closed		6 open		6 open

The block-diagram of a self-built X-band ENDOR and TRIPLE resonance spectrometer is shown in Fig. 15. The different modes of operation are explained in the figure caption; conceptual details have been published elsewhere (6). The scan oscillator delivers the frequency modulated rf, whereas the pump oscillator is not modulated. In the Special TRIPLE mode, the double-balanced mixer produces the two sideband fields with the right modulation phase behavior. This is achieved by setting the pump oscillator to the respective nuclear Larmor frequency and mixing it with the scan oscillator frequency. In addition to data acquisition and handling, the computer is used to control the scan oscillator and the digital attenuator to ensure approximately constant NMR transition rates over a wide frequency range.

Besides sample temperature, the most critical parameters of the experiment are the microwave and rf fields at the sample. Broad-band ENDOR and TRIPLE experiments with sufficiently large microwave and rf field amplitudes can be performed with a TM_{110} cavity/helix arrangement (6, 50). This arrangement not only provided rather homogeneous microwave and rf field distributions over the sample, but also allows irradiating the sample with light in photolytic experiments. This is achieved by cutting two slits into the walls of the cylindrical cavity at positions where wall microwave currents are least disturbed. The back slit is used to cool cavity and helix with a nitrogen gas stream. In the frequency range between 0.5 and 5 MHz, an NMR helix with 30 turns is used, in the range between 5 and 30 MHz one with 20 turns. With two combined power amplifiers (600 W) the maximum rf fields obtained are 2.0 mT (at 14 MHz) and 3.5 mT (at 2 MHz), measured in the rotating frame.

Modifications of cylindrical TM_{110} ENDOR cavities have been recently constructed in various laboratories (51-55). First ENDOR experiments have recently been carried out (56-59) for which loop-gap resonators have been designed to replace the cavity (see Chapter 7). A loop-gap resonator is particularly useful for lossy solvents and for small samples like single crystals. In a recent review, Hyde and Froncisz (60) summarize both the theoretical and technical background of loop-gap resonators and their properties and applications. Conceptionally related to the loop-gap resonators, but with different resonance conditions, are the slotted-tube resonators which employ resonant transmission lines (61-63). A folded half-wave resonator, which is particularly suited for light irradiation, has been described recently (64). A particularly promising development is the bridged loop-gap resonator (65). This structure has been thoroughly investigated, both theoretically and experimentally, with respect to electromagnetic field distribution, resonance frequency and Q value.

3.2 Time-Resolved ENDOR/TRIPLE Resonance

Pulsed ENDOR spectroscopy and related techniques are being performed by more and more groups (10, 63, 66-77) (see also Chapter 17). Comprehensive

review articles about electron spin-echo (ESE) ENDOR and ESE envelope modulation, a technique which is particularly useful for determining small hfc's from
inhomogeneously broadened EPR lines, have been published (78-80). For this
situation, cw ENDOR suffers from declining intensity (31, 81). Recently,
Mehring et al. (16, 17) gave a thorough description of various pulsed ENDOR and
triple resonance schemes involving either population or coherence transfer.
Both pulsed and cw time-resolved ENDOR methods have their inherent advantages
and disadvantages (80, 82-84), often making them complementary rather than competing techniques. The most important advantage of ESE-ENDOR lies in the fact
that, unlike cw ENDOR spectroscopy, it does not critically depend on the balance between microwave, rf, and the various relaxation transition rates. In
ESE-ENDOR experiments, changes in the echo intensities up to 100 % have been
reported. In a complex multispin system, however, the evaluation of hfc's from
echo envelope modulations might pose problems of unambiguity.

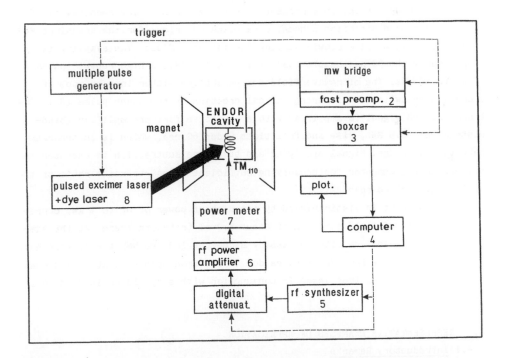

Fig. 16. Block diagram of the time-resolved spectrometer for CIDEP-enhanced
EPR/ENDOR built in this laboratory (33). The sample temperature can be varied by
an N_2 gas stream. For irreversible photoreactions, a temperature controlled
flow-system supplies fresh sample solution to the cavity. The numbers in the
boxes refer to model specifications of instruments actually used: 1 Bruker ER
046 MRP; 2 Bruker ER 047 PH; 3 EG&G 4420; 4 Hewlett-Packard 21 MX-E; 5 Hewlett-
Packard 8660C; 6 Eni A1000; 7 Bird 43; 8 Lambda-Physik EMG 103E and Lambda-
Physik FL 2002.

A time-resolved X-band ENDOR spectrometer employing cw microwave and rf fields and specifically designed for studies of photochemically generated transient radicals has recently been set up in this laboratory (33). The time-delay between radical generating laser pulse and boxcar detection of the ENDOR signal is optimized with respect to the CIDEP enhancement of the ENDOR effect (see Section 2.2.2). For irreversible photoreactions, the spectrometer incorporates a temperature-controlled flow-system supplying fresh solution to the cavity where the photoreaction is started. Fig. 16 shows the block-diagram of the CIDEP-enhanced-ENDOR spectrometer with pulsed laser excitation and time-resolved detection.

The EPR part is rather similar to earlier constructions of the Argonne group (85) and Oxford group (86). In order to catch a large fraction of the initial spin polarization a short response time of 100 ns was striven for by using a TM_{110} ENDOR cavity of low Q (Q = 1000) with internal NMR coil (13). The spectrometer operates without field or frequency modulation but rather with direct boxcar detection. To compensate for the inherently low sensitivity of such a broad-band detection scheme, a sampling technique for the transient EPR signal is applied while slowly varying the magnetic field. The sampling is triggered by the 10 ns tunable laser pulses which have a high repetition rate of up to 100 Hz. The microwave field can be applied either in a cw mode of operation or switched off during radical creation by the laser pulse (87). The transient EPR signal is amplified with a low-noise fast preamplifier (band-width 25 Hz –6.5 MHz, rise and fall times under 60 ns), which is incorporated into the mw bridge. Signal averaging and base line subtraction is executed in the boxcar and computer for preserving sufficient signal-to-noise ratio of the ultimately plotted spectra.

The NMR part is distinguished by its high rf power of up to 1 kW, corresponding to an rf field strength of about 5.0 mT (rotating frame) at the sample. High rf fields are often necessary since the induced NMR transitions have to compete with the fast decay rates due to electron spin relaxation and chemical reactions. An excimer laser in conjunction with a dye laser is used as a light source.

4. REPRESENTATIVE EXAMPLES OF APPLICATION

4.1 Introductory Remarks

The study of structural details of large biomolecules or complexes is particularly difficult because of the various levels of "structure" in biological macromolecules and because of the complex dynamical behavior of such systems. Thus, no single technique is capable of yielding sufficient information about structure and function of such biomolecules, but rather several techniques have to be used in concert (88).

The most powerful method is X-ray diffraction of single crystals, since it can deliver the entire three-dimensional structure of a biopolymer. However, the quality of the data set obtained from X-ray crystallography of biosystems does not always yield sufficiently precise atomic positions. Furthermore, many interesting biosystems, e.g. membrane proteins, are very difficult - if at all - to crystallize. Consequently, other techniques are necessary that either do not require single crystals and/or deliver interesting information that is not directly available from X-ray diffraction, e.g. precise atomic coordinates for light nuclei such as hydrogen.

Magnetic resonance techniques such as advanced NMR and EPR methods often reach or even exceed the precision of X-ray crystallography data. However, generally only certain regions of the entire biomolecule can thoroughly be examined. This is particularly true for EPR spectroscopy which only maps the surrounding of a paramagnetic center. At first sight the restriction to local regions looks like a disadvantage but it can, actually, be an advantage: If paramagnetic species are constituents of larger biological species or if paramagnetic centers can be generated, they may serve as internal spin labels allowing to probe their environment in detail without interfering effects from other groups of the system. Fortunately, the unpaired electron is often located at the most interesting active site. Examples are the light-induced charge separation in photosynthesis in which organic π-radicals of the various constituents of the electron transport chain are created (89) - or the catalytic redox processes mediated by flavoproteins is which flavosemiquinone radicals are prominent species in the active site of the enzyme (90).

Unfortunately, EPR spectroscopy of biological systems has some serious problems:

(i)   Resolution: Owing to the complexity of the paramagnetic centers and their environments the EPR spectra are often poorly resolved due to inhomogeneous line broadenings, e.g. for organic radicals only the g-factors and the linewidths are usually measurable (8, 12, 14).

(ii) Paramagnetic Heterogeneity: Often several paramagnetic species are present in the sample or the same radical species experiences different electronic distributions through interaction with different environments. This problem can, in principle, be approached by EPR studies at much higher frequency bands (91, 92) to select out overlapping resonances of the individual species.

(iii) Immobilization: Since the paramagnetic species are usually quite strongly bound to the biopolymer, they are immobilized. As a result, the anisotropic interactions in the spin Hamiltonian are often not averaged out by rotational molecular tumbling of the whole biopolymer in water at ambient temperature. Consequently, EPR spectra of isotropic fluid solutions do not usually show the high resolution expected for fast tumbling molecules but rather exhibit the typical powder-type patterns (93). One way out of this dilemma would be the elaborate investigation of single crystals (8) which, however, can be grown only for selected biosystems.

These facts in mind, the ENDOR technique is the method of choice to overcome at least some of the aforementioned limitations of EPR spectroscopy. The inherently higher resolution of ENDOR yields detailed information about the hyperfine interactions. The most direct way to obtain isotropic hfc's is to apply liquid phase ENDOR or, even better, TRIPLE resonance. In the fast tumbling limit the hfc's are directly given by line separations in the narrow-line spectra (see Section 2.1.1). Powder-type ENDOR spectra still allow the determination of the anisotropic and isotropic hfc's although much more effort has to be put into the spectrum analysis (8, 12, 14, 93). Under favorable conditions even single-crystal-like ENDOR spectra can be obtained from powder-type EPR spectra taking advantage of the orientational selection achievable with the double resonance techniques. Other advantages of ENDOR and TRIPLE resonance of biological materials are their potential to directly determine the types of interacting nuclei, the relative signs of the hfc's and, under favorable circumstances, the nuclear quadrupole interactions (8-14). In the application of ENDOR spectroscopy to biological systems it is of importance to realize that, in addition to hyperfine interaction with nuclei local to the radical or paramagnetic complex (local ENDOR), also hyperfine interaction with the surrounding medium can be obtained (matrix ENDOR). It is thus apparent that ENDOR spectroscopy has the potential of a very powerful structural probe monitoring the coordination environment of paramagnetic sites in biological systems (14).

In EPR and ENDOR spectroscopy it is useful to divide the paramagnetic species of biological systems into two categories:
(a) <u>Paramagnetic transition metal complexes</u> as found, for instance, in heme, iron-sulfur proteins and other metalloenzymes.
(b) <u>Paramagnetic organic species</u> as occurring, for instance, as doublet state radicals created by electron transfer (redox) reactions or in radiation damaged materials.

Examples from the first category can be found in other chapters of this book. Therefore, we want to confine ourselves to the second category, especially to bioorganic radicals and corresponding model compounds. In order to illustrate the power of the method we have chosen some recent examples from the field of photosynthesis. Advances of a new technique, CIDEP-enhanced-ENDOR, capable of detecting ENDOR of transient radicals (≥ 1 μs) will be summarized. Other applications of liquid state ENDOR and TRIPLE resonance to biomolecules can be found in various monographs (8, 9, 12) and review articles (13-15) and will therefore be only briefly mentioned here.

4.2 Primary Products of Photosynthesis

The primary reactions of photosynthesis have been called a "Garden of Eden" for the EPR spectroscopist (94) since in the process of light-induced charge separation organic π-radicals of the various cofactors involved are created. These radicals - cations of the electron donors and anions of all subsequent electron acceptors - can be directly studied by EPR and ENDOR techniques. From the measured and assigned hyperfine coupling constants a map of the valence electron spin density distribution over the molecule is obtained which forms the basis for a profound theoretical understanding of the investigated paramagnetic species. It should be stressed that the method focuses on the "transferred valence electron", i.e. maps out the frontier orbitals which are important for the electron transfer process. Moreover, additional informations about the spatial structure can be obtained by ENDOR in frozen solution, e.g. about H-bonds between the cofactors and specific protein residues and other environmental effects. The comparison between the in vitro (isolated) radicals and the in-vivo system should ultimately show how the protein alters the electron distribution of the constituents of the electron transfer chain to achieve specific function.

In the following a short introduction to photosynthesis will be given, especially to the processes occurring in the so-called reaction center (RC). For a more extensive treatise the reader is referred to recent monographs and review articles (95-100).

4.2.1 Introduction to Photosynthesis

Photosynthesis is the process by which electromagnetic radiation (sunlight) is converted into chemical free energy in plants, algae and some bacteria. Photosynthetic organisms can be divided into two classes. The species of the first one (algae and green plants) use water as electron donor to reduce CO_2 and produce oxygen and sugar ("oxygenic photosynthesis"). To achieve this goal, two photosystems, PS 1 and PS 2, are needed that work in tandem ("Z-scheme") to transport electrons from a very high to a quite low redox potential. Photosynthetic bacteria have only one PS; they cannot split water ("anoxygenic photosynthesis") but use other reduced materials like H_2S or organic compounds as electron donors. In plants and bacteria, photosynthesis is bound to the presence of pigment molecules, chlorophylls (Chl) and bacteriochlorophylls (BChl), respectively (Fig. 17), which play a multifarious role in the absorption of light, excitation transfer, and charge separation. Since the plant photosystems are fairly complicated (95-98) the following discussion will be confined to bacteria.

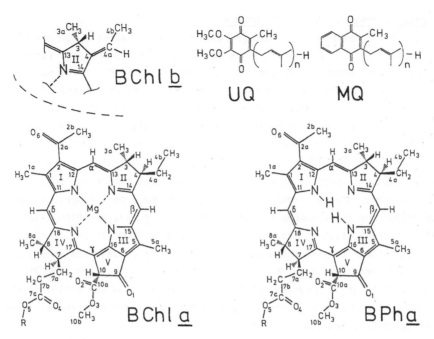

Fig. 17. Molecular structures and numbering schemes of bacteriochlorophyll (BChl) a, BChl b, and bacteriopheophytin (BPh) a, the aliphatic side chain is phytyl (BChl a from Rs. rubrum has R = geranylgeranyl). Two quinones found in bacterial RC's are also shown: UQ = ubiquinone, MQ = menaquinone.

In the primary process of photosynthesis in bacteria light is absorbed by "antenna complexes" (BChl-containing proteins) (98, 99, 101) and funneled to a RC in the photosynthetic membrane where the primary charge separation takes place. RC's have been isolated and purified from several photosynthetic bacteria (102-105). The RC of the purple bacteria Rhodobacter (Rb.) sphaeroides, Rb. capsulatus and Rhodospirillum (Rs.) rubrum, for example, contain three protein subunits (denoted L, M, and H), 4 BChl a, 2 bacteriopheophytin a (BPh a), 1 Fe^{2+} and 2 quinones. Rhodospeudomonas (Rps.) viridis uses modified pigments (BChl b and BPh b, see Fig. 17) and has a fourth protein subunit (cytochrome) attached to the RC. RC's from other phototrophic bacteria have basically similar structures (104, 105).

Reaction centers from Rps. viridis (106) and Rb. sphaeroides (107, 108) were recently crystallized and the spatial arrangements of the prosthetic groups and protein residues were determined by X-ray crystallography. The cofactors are arranged in two branches (Fig. 18), denoted L and M (sometimes also A and B) that are related to each other by an approximate twofold rotation axis. Two of the BChl's form a dimer on the periplasmatic side of the RC which is followed by monomeric BChl's, BPh's and a quinone in each branch. The C_2 axis runs through the BChl-dimer ("special pair") and through the iron (Fe^{2+}) which is located between the primary (Q_A) and secondary (Q_B) quinone.

From optical spectroscopy (109) it is evident that only one branch is photochemically active. After singlet excitation of the primary donor P (BChl-dimer) this species donates an electron to the electron transport chain comprising intermediate I (monomeric BPh) and final quinone acceptors Q_A and Q_B:

$$PIQ \xrightarrow{h\nu} P^*IQ \underset{10ns}{\overset{\lesssim 3ps}{\rightleftharpoons}} P^{\ddagger}I^-Q \underset{0.1s}{\overset{200ps}{\rightleftharpoons}} P^{\ddagger}IQ^-$$

The charge separation and subsequent electron transfer processes have a high efficiency, i.e. possible back reactions are effectively impeded by the fast forward electron transfer and the successively increasing distance of the separated charges.

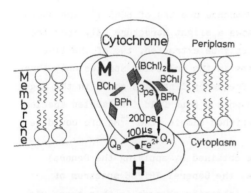

Fig. 18. Schematic drawing of the reaction center of purple bacteria spanning the photosynthetic membrane with subunits L (light), M (medium), H (heavy), attached cytochrome, and the cofactors (see text). The electron transfer path is indicated by arrows and the approximate transfer times are given (for details, see Refs. 106-109).

To understand the charge separation in the RC from first principles, a profound knowledge of both the spatial and electronic structure of the participating molecules in their protein surrounding is necessary. Since radical cations (of the primary donor) and radical anions (of all electron acceptors) are formed in this process, EPR and ENDOR can be used to map out the electron spin density distribution by the agency of the hfc's. These hfc's are sensitive probes of the radical itself and its surrounding.

The approach usually employed is to first investigate isolated species or appropriate model systems in organic solvents, and compare the hf data with those obtained from the same species in the RC. In this way, the existence of the BChl dimer forming the primary donor was first proposed by Norris on the basis of EPR studies (110) that were later corroborated by ENDOR experiments (111-113). Furthermore, the identification of monomeric BPh as intermediate acceptor and (ubi)quinone as first stable electron acceptor in bacterial RC's was achieved in a similar way (see Ref. 102 for references). In the following, some more recent ENDOR experiments performed on the pigment radicals in vitro and in the RC should serve as examples to demonstrate the power of the method.

4.2.2 The Primary Donor in Bacterial Photosynthesis

Studies of the Pigment Radical Cations: The radical cation of BChl $\underline{a}$ can easily be generated in CH_2Cl_2/CH_3OH (ca. 6 : 1) by oxidation with iodine, or with zinc tetraphenylporphyrin perchlorate ($ZnTPP^{+\cdot} ClO_4^-$) (114). BChl $\underline{a}^{+\cdot}$ was first studied by EPR and ENDOR spectroscopy in frozen solutions (111, 112) essentially yielding only the hfc's of the freely rotating methyl groups (positions 1a, 5a, see Fig. 17) and an average hfc for the β-protons at positions 3, 4, 7, and 8. In the first ENDOR study of this system in fluid solution, reported by Borg et al. (114), seven isotropic hfc's could be extracted from the spectra leading to an improved picture of the spin density distribution. Subsequently, the system was studies by ^1H and ^{14}N ENDOR and TRIPLE resonance spectroscopy (115) (a paramagnetic by-product present in the samples, the radical cation of 10-hydroxy bacteriochlorophyll $\underline{a}$, was later identified by similar techniques (116)).

In Fig. 19 EPR, ENDOR, and TRIPLE resonance spectra of BChl $\underline{a}^{+\cdot}$ are shown. The first derivative Gaussian EPR line shows a slight structure (A). From the ENDOR spectrum (B) 11 ^1H hfc's could directly be extracted, two of the line pairs are additionally split (117). Whereas in the unlabeled species only three ^{14}N hfc's were accessible (only the high frequency lines are shown in the figure) all four ^{15}N hfc's could be resolved when a four-fold ^{15}N labeled species was used (118). Furthermore, in a ^{25}Mg enriched sample, the ^{25}Mg hfc could be measured (119).

The relative signs of the hfc's were obtained by applying the General TRIPLE resonance technique. As an example, the General TRIPLE spectrum of BChl $\underline{a}^{+\cdot}$ is shown in Fig. 19(C). From the intensity changes in this trace with respect to the ENDOR spectrum signs were obtained not only for all the proton hfc's, but also for the nitrogen-14 couplings.

Assignment of the ^1H hfc's to their specific molecular positions was achieved by various experimental approaches: by comparison with similar radicals concerning magnitudes and signs of the hfc's, by nuclear relaxation studies, and, in particular, by partial deuteration of the BChl $\underline{a}$ molecule using biosynthetic labeling and H/D exchange experiments (117).

The ENDOR spectrum of selectively deuterated BChl $\underline{a}^{+\cdot}$ - which is known (120) to carry protons only at the methyl groups (position 1a, 5a, Fig. 17), at the acetyl group (position 2b), and at positions more than two bonds away from the π-system - is reproduced in Fig. 19(D). The assignment of the hfc's was performed accordingly. The complete ^2H ENDOR spectrum could be recorded in the low-frequency region. Discrimination of the ^2H from the ^{14}N signals was achieved by taking advantage of the different nuclear relaxation behavior of these nuclei. Whereas optimum ^{14}N ENDOR signals were obtained at ca. 270 K and at

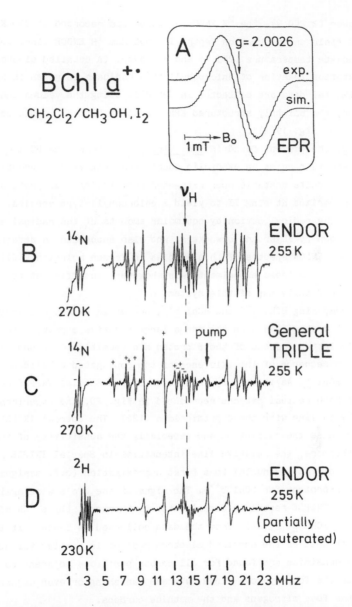

Fig. 19. EPR, ENDOR and TRIPLE spectra of BChl a cation radical (iodine oxidation in CH_2Cl_2 : $CH_3OH \approx 6 : 1$). A: Comparison of experimental and simulated
EPR of BChl a$^{+\cdot}$ at 293 K, for the simulation the ^1H and ^{14}N hfc's and assignments from Table I were used. B and C: ^1H and ^{14}N ENDOR and General TRIPLE
spectra of BChl a$^{+\cdot}$, in C the additionally pumped ENDOR line is indicated and
the measured signs of the hfc's are given (low frequency region of line pattern). D: ^1H and ^2H ENDOR of partially deuterated BChl a$^{+\cdot}$, BChl a isolated
from bacteria grown in D_2O with ^1H-succinate (see text). The respective optimum
temperatures are given.

high RF power ($\geq$ 200 W), the ^{2}H ENDOR spectrum was recorded at 230 K with de-creased rf field and modulation depth. The optimum ^{1}H ENDOR lines were obtained at intermediate temperature (255 K) and rf power. (A detailed discussion of the ENDOR relaxation behavior of chlorophyll-type radicals is given in Ref. 117.) The experimental data are collected in Table I. Using the values and assign-ments given, the slightly structured EPR signal of BChl $\underline{a}^{+\cdot}$ could be perfectly simulated (see Fig. 19(A)).

BChl $\underline{b}$, the pigment found in Rps. viridis, differs from BChl $\underline{a}$ only at ring II where it carries an exocyclic double bond (ethylidene group), see Fig. 17. BChl $\underline{b}$ is quite unstable when extracted from its natural protein environ-ment and isomerizes at ring II to yield a chlorophyll-type species. It can be oxidized to the radical cation by equimolar amounts of the radical salt $ZnTPP^{+\cdot}$ ClO_4^-, as shown by Fajer and coworkers, who also succeeded in detecting the first EPR and ENDOR spectra of this species in frozen solution (121). The reso-lution obtained in these experiments was, however, insufficient to derive a detailed spin density map of this system.

When comparing BChl $\underline{a}^{+\cdot}$ and BChl $\underline{b}^{+\cdot}$, one would expect one additional large methyl proton (position 4b) and a large negative α proton hfc (position 4a) for the latter instead of the β proton hfc (position 4) found in BChl $\underline{a}$. This is a consequence of the existence of the conjugated ethylidene group at ring II in BChl $\underline{b}$. As is seen by inspection of the Special TRIPLE resonance spectra of both radical cations reproduced in Fig. 20, the experimental results are exactly in line with these predictions (122). The Special TRIPLE mode was used to increase the resolution and especially the sensitivity of the experi-ment; furthermore, the relative line intensities in Special TRIPLE reflect the number of contributing nuclei to a first approximation (c.f. assignment of methyl and β-protons for BChl $\underline{b}^{+\cdot}$). The signs of the hfc's were again obtained from General TRIPLE resonance experiments. Furthermore, three ^{14}N hfc's were detected for BChl $\underline{b}^{+\cdot}$ (122). From the data collected in Table I it is obvious that in both species the unpaired electron resides in similar valence orbitals. Large spin densities are found for all carbon positions adjacent to the nitro-gens within the four pyrrole rings, whereas vanishing or even negative values occur at the four nitrogens and the methine carbons.

In Fig. 20 (right) the s spin densities, derived from the isotropic hfc's of BChl $\underline{a}^{+\cdot}$, and BChl $\underline{b}^{+\cdot}$, are compared with those calculated by an all-va-lence-electron SCF method. This method is based on the INDO parametrization (123) and uses a restricted Hartree-Fock approach (half electron method) with subsequent perturbation treatment of spin polarization effects (RHF-INDO/SP), for details, see Refs. 124, 125. The geometry of the molecules was optimized on the basis of an energy minimization procedure, in particular for ring V, for the rotational angle of the acetyl group, and the geometry of the hydrated rings II/IV.

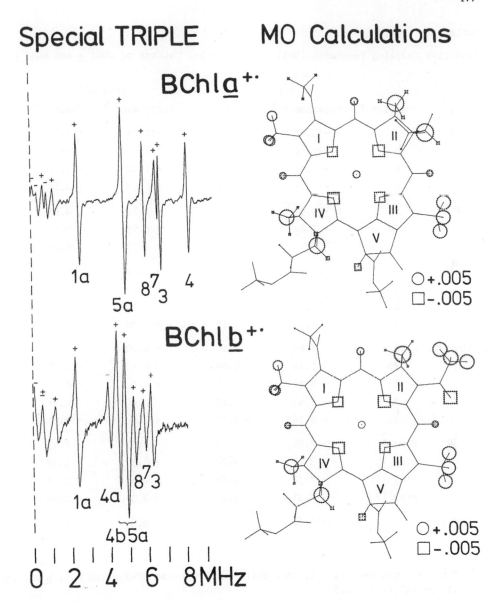

Fig. 20. ^{1}H Special TRIPLE spectra of BChl $\underline{a}^{+\cdot}$ (left, top) and BChl $\underline{b}^{+\cdot}$ (left, bottom), the signs indicated are from General TRIPLE, for the largest hfc's the experimental assignments to molecular positions (see Fig. 17) are given. On the right the experimental hfc's (dotted) are compared with the calculated ones (solid lines), their values are proportional to the area of squares ($\rho < 0$) and circles ($\rho > 0$), respectively. MO calculations performed by the RHF-INDO/SP method (see text). The side chains were replaced by methyl; mostly standard bond lengths and angles were used, except for some critical regions where the optimum geometry was found by energy minimization (see text). Theoretical carbon and oxygen spin densities were omitted for clarity. For more details, see Table I and Refs. 117, 122.

Table I

Hyperfine Coupling Constants (MHz) of the Radical Cations of BChl a and BChl b

Nucleus	Position [a]	BChl a[+•]		BChl b[+•]	
		Experiment [b]	Theory [c]	Experiment [b]	Theory [c]
β H	3	+13.47	+13.22	+12.45	+14.81
	4	+16.35	+15.30	--	--
	7	+13.11	+13.06	+11.60	+11.61
	8	+11.76	+12.08	+10.55	+10.88
CH_3	1a	+ 4.93	+ 5.00	+ 4.70	+ 4.51
	5a	+ 9.62	+ 9.57	+ 8.95	+ 9.08
	4b	--	--	+ 9.70	+10.96
α H	4a	--	--	- 8.02	- 7.23
	α	+ 2.35	+ 3.20	+ 2.50	+ 3.12
	β	+ 1.30	+ 2.48	+ 1.20	+ 2.43
	δ	+ 1.30	+ 2.68	+ 1.20	+ 2.35
Other H	10	- 1.64	- 2.65	- 1.20	- 2.48
	2b	- 0.15	- 0.04	< 0.4	- 0.05
	3a	- 0.36	- 0.31	< 0.4	- 0.16
	7a	- 0.53	- 0.61	- 0.40	- 0.54
	8a	- 0.15	- 0.24	< 0.4	- 0.21
^{14}N	ring I	- 2.25	- 2.10	- 2.3	- 1.95
	ring II	- 3.14	- 3.28	- 3.3	- 3.38
	ring III	- 2.32	- 2.56	- 2.3	- 2.54
	ring IV	- 2.91	- 3.22	- 3.0	- 2.90
^{25}Mg		(-)0.30	- 0.31	n.d.	- 0.31

[a] For positions see Fig. 17, note differences between BChl a and BChl b at ring II (positions 4a, 4b).

[b] Experimental hfc's from ENDOR/TRIPLE at 220 K (BChl b[+•], BChl a[+•]) and at 270 K (^{14}N in BChl a[+•]), n.d. = not determined (117, 122).

[c] Theoretical values from RHF-INDO/SP using Q-factors relating s-spin densities to isotropic hfc's (125): $Q(^1H) = 1420$, $Q(^{14}N) = 650$ and $Q(^{25}Mg) = -168$ MHz.

The agreement between experiment and theory is very satisfactory (see also Table I) indicating that the ground state molecular wave function is calculated reliably by this approach. Thereby it is important to notice that – at least for BChl $\underline{a}^{+\cdot}$ – an almost complete assignment of proton hfc's to specific molecular positions was achieved experimentally.

The Primary Donor Radical Cation $P_{865}^{+\cdot}$

Since various BChl $\underline{a}$ containing bacteria show a light-induced bleaching of the optical band at $\approx$ 865 nm assigned to the primary donor, this species is referred to as P_{865}. The comparison of $P_{865}^{+\cdot}$ with the protein-free (monomeric) BChl $\underline{a}^{+\cdot}$ shows that in the RC's the EPR linewidth is reduced by $\approx 1/\sqrt{2}$, which was believed to be indicative of a delocalization of the unpaired electron over two BChl $\underline{a}$ molecules (110). This "special pair" hypothesis was supported by subsequent ENDOR experiments in frozen solution showing the expected halving of the hfc's of $P_{865}^{+\cdot}$ as compared with BChl $\underline{a}^{+\cdot}$ (111, 112). The information obtained from these experiments in frozen matrices was, however, somewhat limited since anisotropic broadening marked all those signals that do not belong to freely rotating methyl groups (111, 126). In subsequent liquid phase ENDOR work at room temperature the assumption of a dimer forming the primary donor proved to be basically correct:

The first successful liquid state ENDOR experiments on $P_{865}^{+\cdot}$ in RC's of Rb. sphaeroides at room temperatur (127) yielded a highly resolved spectrum, and seven isotropic ^{1}H hfc's could be extracted. Obviously, the rotation of the protein of the reaction center (molecular weight $\approx$ 100,000) was fast enough (rotational correlation time $\tau_R \approx$ 30 ns) to effectively average out the g factor and the hyperfine anisotropies of the various nuclei in this radical. In Fig. 21 the Special TRIPLE resonance spectra in fluid solution of the model compound BChl $\underline{a}^{+\cdot}$ and of light-induced $P_{865}^{+\cdot}$ in RC's from two different bacteria are depicted (113, 128, 129). Although the bacteriochlorophylls $\underline{a}$ of the two bacteria have different side chains (phytyl or geranylgeranyl), the ENDOR spectra of the respective monomeric cation radicals are identical. In contrast to that, the ENDOR spectra of the primary donor radical cations in the RC's show a measurable difference. Interestingly, different strains of the same bacterium (e.g. carotinoidless mutants and wild type) show the same $P_{865}^{+\cdot}$ ENDOR spectra. It is believed that these findings are due to different protein environments of the primary donor in different bacteria.

In $P_{865}^{+\cdot}$ of both bacteria all hfc's - assigned by biosynthetic deuteration - were found to be reduced in magnitude as compared with BChl $\underline{a}^{+\cdot}$, but not by a constant factor of 2 as suggested in the simplest model for the special pair with weakly interacting monomeric π-systems (130). This indicates a perturbation of the wavefunctions of the monomeric bacteriochlorophylls caused by the dimerization process itself and/or by the protein surrounding. The average

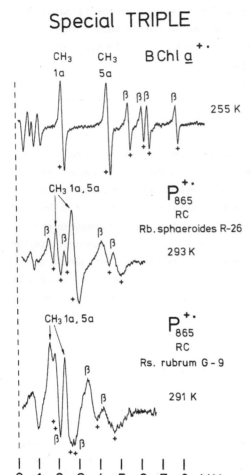

Special TRIPLE

Fig. 21. ^{1}H Special TRIPLE spectra of BChl a$^{+\cdot}$ and of the primary donor cation radical P$_{865}^{+\cdot}$ in RC's of Rb. sphaeroides R-26 and Rs. rubrum G-9. The lines belonging to the larger hfc's were assigned to the β-protons (positions 3, 4, 7, 8; see Fig. 17, hfc's/MHz: +9.50, +8.60, +4.45, +3.30 (for R-26) and +8.50, +7.50, +5.28, +3.95 (for G-9)) and to the methyl protons (position 1a, 5a, hfc's/MHz: +5.60, +4.00 (for R-26) and +4.85, +3.40 (for G-9)) based on deuteration experiments (see Refs. 128, 129). The smallest hfc/MHz, +1.40 (R-26) and +1.60 (G-9), could not be assigned.

reduction factor of spin densities in both bacteria is, however, very close to 2. This shows that the unpaired electron is indeed delocalized over two BChl a molecules and indicates the formation of a supermolecule. No splitting is observed of the ENDOR signals, e.g. of those assigned to the methyl groups in positions 1a or 5a. Hence, it is a sensible assumption that the two moieties of the dimer are (magnetically) equivalent, i.e. the dimer possesses a C_2 symmetry axis.

Additional support for the dimer model came from ENDOR studies in liquid solution performed on biosynthetically ^{15}N labeled P$_{865}^{+\cdot}$ in RC's of Rb. sphaeroides (118). The average reduction factor of the four measured ^{15}N hfc's was again close to 2 as in the case of the ^{1}H hfc's. The simulation of the unresolved Gaussian EPR line by use of the hfc's obtained from ENDOR experiments is also in agreement with the dimeric nature of the P$_{865}^{+\cdot}$ species (128, 129).

The geometry of the dimer can be obtained from the measured isotropic hfc's in conjunction with a reliable MO calculation, e.g. the above-mentioned RHF-INDO/SP method. Calculation on various BChl a dimers in vacuo including a limited geometry optimization by energy minimization (124) yielded a dimer geometry which is quite similar to that found in the recent X-ray crystallographic study of the RC of Rb. sphaeroides (107, 108). Calculations on the X-ray structure of the BChl a dimer (including the protein surrounding) have not been performed so far.

The main result of this combined ENDOR spectroscopic and MO theoretical study is that the $P_{865}^{+\cdot}$ species, structurally and electronically, is a supermolecule formed by two BChl a molecules. The unpaired electron is delocalized in a supermolecular orbital extending over both moieties. From the so far limited experimental data available, there is no indication that the spin density distribution in the BChl a dimer cation radical of Rs. rubrum or Rb. sphaeroides is asymmetrical. However, it cannot be excluded that a refined data set based on different hfc assignments might lead to some structural alterations.

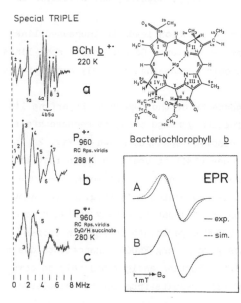

The Primary Donor Radical Cation $P_{960}^{+\cdot}$

In the photosynthetic bacterium Rps. viridis - which contains BChl b instead of BChl a - the primary donor predominantly absorbs at 960 nm. Although the X-ray structure analysis of the RC clearly shows the existence of a BChl b dimer (106), the cation of the primary donor, $P_{960}^{+\cdot}$, does not display the $1/\sqrt{2}$ narrowing of the EPR linewidth nor the 50% reduction of the solid-state ENDOR splittings when compared with BChl $b^{+\cdot}$ (121). Inspection of the Special TRIPLE spectra of BChl $b^{+\cdot}$ and $P_{960}^{+\cdot}$ in fluid solution (Fig. 22 (a, b)) also shows no halving of the hfc's in the latter as would be expected for a symmetric dimer structure. Instead, the spectrum obtained for $P_{960}^{+\cdot}$ closely resembles that of BChl $b^{+\cdot}$ - except for line broadening effects or some shifts of spectral components towards lower frequencies (113, 131).

Fig. 22. ^{1}H Special TRIPLE spectra of BChl $b^{+\cdot}$ (a) and of $P_{960}^{+\cdot}$ in protonated (b) and in partially deuterated RC's (c) of Rps. viridis, see text. In (a) the signs of hfc's and assignments to molecular positions (see structure, right top) are given; in (b) and (c) the lines are numbered sequentially. The hfc's [MHz] for $P_{960}^{+\cdot}$ are (1) +1.4, (2) +2.0, (3) +3.4, (4) +6.0, (5) +7.2 (6) +8.5, and (7) +11.5; lines 3, 4, 5 belong to methyl groups. The experimental EPR (A, B solid line) is compared with two simulations (dashed lines) using the above hfc's with two different assignments to molecular positions, A: symmetric dimer, B: asymmetric dimer (see text).

A thorough comparison of the spectra requires an individual assignment of the measured hfc's to the molecular positions. This could be achieved by various methods for BChl $\underline{b}^{+\cdot}$, (122) in which the largest couplings were assigned to the β-protons in rings II and IV (pos. 3, 7, 8), to the methyl groups (5a, 4b, 1a) and to the α-proton (4a), see above. For $P_{960}^{+\cdot}$, however, the situation is more difficult for the following reasons:

(i)   Slow tumbling of the large protein in solution leads to increased line-widths in the ENDOR spectra even at room temperature. Consequently, some lines might become undetectable due to anisotropic broadening (a likely candidate is the α-proton in pos. 4a);

(ii)   for an asymmetrical dimer one expects twice the number of hyperfine lines;

(iii) these lines appear in a smaller frequency range owing to reduced hfc's.

These complications lead to the poorly resolved spectrum shown in Fig. 22 (b). By use of computer deconvolution techniques (131) it could be shown that lines 3, 4, 5 and 7 consist of several spectral components. Furthermore, the comparison with the spectrum of $P_{960}^{+\cdot}$ in partially deuterated RC's (Fig. 22 (c) shows that the methyl groups (positions 1a, 5a and 4b) can be safely assigned to lines 3, 4 and 5. This is supported by ENDOR experiments on $P_{960}^{+\cdot}$ in frozen solutions (131) where only methyl groups give strong ENDOR enhancements due to favorable relaxation conditions (111, 126). Further assignments rely on comparison with BChl $\underline{b}^{+\cdot}$ (see Table I) and with results from the MO calculations (125).

A crucial test for any assignment is the simulation of the observed EPR spectrum of $P_{960}^{+\cdot}$ (Fig. 22, insert). For such a complex hyperfine pattern there may exist several different assignments which fit the unresolved EPR line equally well within experimental error. Any assignment, however, that fails to reproduce the EPR spectrum must be discarded. In Fig. 22 (bottom right) two attempts are shown (A and B) to simulate the EPR. Both simulations take into account all experimental hfc's and assignments from the partial deuterations. Assignment A relies on a symmetrical dimer model and on a comparison with the monomeric BChl $b^{+\cdot}$. In this case the simulated EPR line is clearly too broad. Assignment B relies mainly on a comparison with the MO calculations using the coordinates of the X-ray dimer structure (106), see below. The simulated EPR line agrees well with the observed one. In this assignment, the distribution of the unpaired electron over the two dimer halves is asymmetrical. The ratio of the sum of the large hfc's (positions 1a, 4a, 4b, 5a, 3, 7 and 8) is 2.0. This is close to the result of the MO calculations which yield 2.2 in favor of the dimer half bound to the L branch of the protein. (For more details of the experimental procedures and assignments refer to Ref. 125).

In Fig. 23 experimental and calculated s-spin densities are compared, which are directly related to the isotropic hfc's. The calculations were performed on the dimer geometry given by the X-ray structure analysis refined to 2.9 Å resolution (106). Furthermore, electric field effects produced by the partial atomic charges of the eight nearest amino acid residues were included in the calculation. The agreement between experimental and theoretical values is quite satisfactory (for details see Ref. 125).

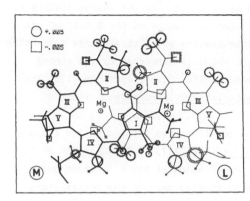

Fig. 23. Theoretical (solid lines) and experimental (dotted lines) s-spin densities for the BChl b dimer cation radical $P_{960}^{+\cdot}$. Calculations performed on the X-ray structure (106) including 8 surrounding amino acids (with backbone). Representation of spin densities as in Fig. 20. The assignment of experimental s-spin densities is the same as that used for the EPR simulation in Fig. 22B (for details, see Table I in Ref. 125).

The result of this work on $P_{960}^{+\cdot}$ is rather surprising with regard to the very near C_2 symmetry of the two pigment/protein branches, L and M. This quasi-symmetrical arrangement is a common feature of both bacteria, Rb. sphaeroides and Rps. viridis, as has been shown by recent X-ray diffraction results (106-108). Obviously, at least in Rps. viridis, there are sufficiently strong local distortions of the dimer and/or its surrounding to produce the observed marked asymmetrical spin density distribution. This breaking of the C_2 symmetry is consistent with the fact that essentially only the L-branch is photoactive in both bacteria. The "unidirectionality" ratio of the electron transfer rates k_L/k_M has been found to be larger than 10 (132, 133). Closer inspection of the X-ray coordinates of the pigments in Rps. viridis shows that the L-half of the dimer is more remote from a nearby "accessory" BChl b monomer in the photoactive branch than the M-half. This accessory monomer is considered to play an important mediating role ("superexchange" mechanism) between the donor P and the intermediate acceptor I (BPh) (132, 133).

The asymmetrical spin density distribution of the donor P in its oxidized state could be an essential requirement for the high efficiency of the primary charge separation PI → $P^{+\cdot}I^{-\cdot}$ in the photosynthetic reaction center. It is evident that the back reaction would be effectively impeded if the unpaired electron (or positive hole) in $P^{+\cdot}$ retreats to that side of the dimer which is more remote from the accessory monomer in the photoactive L-branch (or from some other intermediate species acting as a spacer) as is, in fact, found experimentally. Similarly, the efficiency of the forward reaction would gain by an asymmetrical orbital charge distribution of the "excited" electron in the state $^1P^*$ in favor of the other dimer half facing the photoactive branch (134). Calculations of the spin density distribution of the donor anion radical $P_{960}^{-\cdot}$ in which the unpaired electron occupies the same orbital as the excited electron in $^1P^*$ do, in fact, show a reversal or "switch behavior" of orbital charge distributions (134) leading to concentration of electron charge on the M-half of the dimer facing the photoactive branch. These results point to a charge separation within the dimer upon singlet excitation, which is supported by the strong Stark effect experienced by the dimer band in the optical spectrum (135, 136).

4.2.3 The Electron Acceptors in Bacterial Photosynthesis

The Intermediate Acceptor (I)

The intermediate acceptor radical anion $I^{-\cdot}$ can be trapped at low temperatures in RC preparations of Rb. sphaeroides (102, 137) by illumination at low redox potentials (reduced quinones) in the presence of cyt c^{2+}, which quickly reduces $P^{+\cdot}$. This species can therefore be studied only in the frozen state. The obtained EPR signal is similar to that of BChl $\underline{a}^{-\cdot}$ and BPh $\underline{a}^{-\cdot}$ prepared in vitro. Solid-state ENDOR experiments revealed only one (methyl) hfc that was very similar for all three species (102). It was therefore concluded that $I^{-\cdot}$ is a monomeric anion radical of BChl $\underline{a}$ or BPh $\underline{a}$. Similar observations were made for RC's of Rps. viridis (121).

A clear distinction between BChl $\underline{a}^{-\cdot}$ and BPh $\underline{a}^{-\cdot}$ in vitro is possible by means of their ENDOR-in-solution spectra (138, 139): Up to ten proton and all four nitrogen hfc's could be determined. In Fig. 24 the similar EPR but distinctly different ^{14}N ENDOR and ^1H Special TRIPLE spectra of BChl $\underline{a}^{-\cdot}$ (top) and BPh $\underline{a}^{-\cdot}$ (bottom) are shown. The signs were measured by General TRIPLE resonance and assignments to molecular positions were made by comparison with MO calculations. Furthermore, in ^{25}Mg enriched BChl $\underline{a}$ the ^{25}Mg hfc could be determined by liquid state ENDOR (+0.85 MHz), see Ref. 119.

Recently, BChl $\underline{a}^{-\cdot}$ and BPh $\underline{a}^{-\cdot}$ were measured by solid state ENDOR in different matrices (140) and the results were compared with similar experiments performed on freeze-trapped $I^{-\cdot}$ in different RC preparations of Rb. sphaeroides (140, 141). It could be shown that $I^{-\cdot}$ is indeed a BPh $\underline{a}^{-\cdot}$ species. The spin density is, however, somewhat different in the RC. By MO-theoretical studies the observed changes could be traced back to specific structural effects imposed on the embedded species by the protein. Probably, the protein acts to adjust the electronic structure to optimum function in the electron transfer process. Interestingly, an exchangeable hydrogen was detected in $I^{-\cdot}$ (D_2O exchange) that could belong to a specific H-bond between the protein and the BPh a (141). This might play a structural role and/or could be involved in the stabilization of the electron on the photoactive branch.

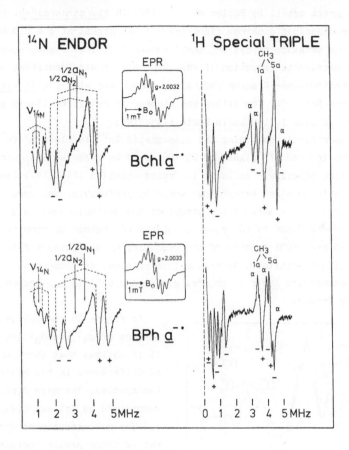

Fig. 24. EPR (293 K), ^{14}N ENDOR (280 K), and ^{1}H Special TRIPLE (255 K) spectra of the radical anions of BChl a (top) and of BPh a (bottom) generated electro-lytically in 1,2-dimethoxyethane. For the largest hfc's the assignment to mo-lecular positions (see Fig. 17) is given, signs are from General TRIPLE; for further details, see Refs. 138, 139. The RHF-INDO/SP calculations (not shown) of the spin density distribution yield very good agreement with the experimen-tal values.

The Quinone Acceptors (Q_A and Q_B)

In the charge separation process of purple bacteria the electron is fi-nally passed to the first (Q_A) and second (Q_B) quinone acceptor (see Fig. 18). Q_A acts as a one-electron gate, whereas Q_B can accept two electrons. In Rb. sphaeroides, for example, both quinones are ubiquinones UQ-10 (Fig. 17). Their different physicochemical properties must therefore result from a different environment in the protein matrix. Both quinones are magnetically coupled to a high-spin Fe^{2+} - located between Q_A and Q_B - that considerably broadens the EPR signal of the respective semiquinones. These ferroquinone complexes have been studied in great detail by Butler et al. (142) in <u>Rb. sphaeroides</u>. In iron-removed RC preparations a narrow light-induced EPR signal at g = 2.0046 was detected in the presence of cyt c^{2+} that reduces $P^{+\cdot}$. This signal could also be created by dithionite reduction of the quinones. It was identified as being due to the ubisemiquinone-10 anion radical by comparison with the <u>in vitro</u> radical (UQ-10$^{\overline{\cdot}}$), see Ref. 102. The latter was studied in detail by liquid-state ENDOR (143, 144) and also in the solid state (145, 146).

In preparations containing the diamagnetic Zn^{2+} in place of Fe^{2+}, the narrow light-induced EPR signal of UQ-10$^{\overline{\cdot}}$ could be observed and detection of ENDOR was also possible - at least in frozen solution (144, 147) (see Fig. 25). ENDOR-in-liquid-solution experiments would be preferable, but were not successful so far due to the large g anisotropy of the semiquinones. This anisotropy cannot be averaged out by the rotation of the RC protein at room temperature.

The EPR and ENDOR spectra of the secondary ubisemiquinone anion radical $Q_B^{\overline{\cdot}}$ could also be obtained. To create this species, Zn^{2+}-containing RC's were given one saturating laser flash ("single turn-over") in the presence of cyt c^{2+} and rapidly frozen.

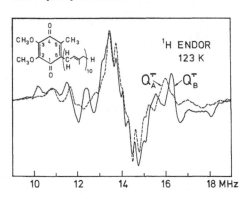

Fig. 25. Comparison of frozen solution ^{1}H ENDOR spectra of the ubisemiquinone anion radicals $Q_A^{\overline{\cdot}}$ and $Q_B^{\overline{\cdot}}$ (broken and solid line, respectively) in RC's of <u>Rb. sphaeroides</u> R-26 containing Zn^{2+} instead of Fe^{2+} (see text and Refs. 144, 147).

In Figure 25 the solid-state ENDOR spectra of $Q_A^{\overline{\cdot}}$ and $Q_B^{\overline{\cdot}}$ are compared. It is obvious that there are pronounced differences in the hfc's of these two species. The main findings can be summarized as follows (144, 147): (i) The well-structured spread-out matrix ENDOR region indicates a very close contact of both semiquinones with the surrounding amino acid residues. This hydrophobic environment is different for $Q_A^{\overline{\cdot}}$ and $Q_B^{\overline{\cdot}}$; the X-ray data of the RC of <u>Rb. sphaeroides</u> support this finding (107). (ii) Both semiquinones are bound to the protein via hydrogen bonds. For $Q_B^{\overline{\cdot}}$ these are probably symmetric and fairly long. ($r_{0\ldots H} \cong 2.0$ Å, obtained from the point-dipole approximation using an oxygen π spin density of $\rho_0^\pi = 0.3$.) For $Q_A^{\overline{\cdot}}$ the ENDOR results clearly indicate the existence of two hydrogen bonds of different strengths, one being fairly short (1.55 Å) and one being longer (1.78 Å). Besides the influence of the divalent metal ion, the asymmetric hydrogen bond situation for the A site – which was also postulated from other studies (148) – might be important for adjusting the redox potential and locking Q_A as a one-electron gate.

(iii) The spin density profiles of $Q_A^{\overline{\cdot}}$ and $Q_B^{\overline{\cdot}}$ in the reaction center are different as was deduced from the ^1H hfc's (Fig. 25) and additional measurements of the ^{17}O hfc in respectively labelled species (144). Whereas $Q_B^{\overline{\cdot}}$ is similar to UQ-10$^{\overline{\cdot}}$ in frozen solvents, $Q_A \cdot$ shows a pronounced asymmetry with respect to the long molecular axis as well as around the in-plane axis perpendicular to it. The latter effect is probably caused by the different hydrogen bond strengths to the two carbonyl oxygens, the former by a close-lying, non-symmetrically placed amino acid residue. (A likely candidate is a tryptophan present in the A site of Rb. sphaeroides, see Ref. 107.)

It has been demonstrated by the extensive ENDOR work on the ubisemiquinones in the RC's of Rb. sphaeroides that valuable information about the spatial and electronic structure of these species in a protein matrix can be obtained which – together with future theoretical considerations and the aid of other spectroscopic and analytical methods – may help to unravel the functional details of protein-bound quinones in membrane processes.

4.2.4 Model Systems

Model reactions and model systems for photosynthetic RC's have attracted much attention in recent years, since they are very important both for a basic understanding of the natural process and for the construction of artificial solar energy conversion and storage devices (149).

EPR and ENDOR have been extensively used to study the radical anions and cations of simple models derived from porphyrins, quinones and related systems occurring in photosynthetic systems. This fields has been thoroughly reviewed by Fajer and Davis (150) and by Pedersen (151). One more recent example of a novel molecular system, the "porphycene", which is a porphyrin isomer, should be mentioned here. A first EPR and ENDOR characterization of anion radicals of porphycenes has just been published (152). In Fig. 26 EPR and ENDOR spectra of the free base porphycene radical anion are shown. One ^{14}N and four ^1H hfc's including signs could be deduced and assigned to molecular positions by comparison with selectively substituted systems. Thereby a detailed map of the spin density distribution was obtained, which is in good agreement with that obtained from advanced MO calculations of the RHF-INDO/SP type (153). A series of porphycenes and metallo-porphycenes (Zn, Ni, Pd, Pt) with various alkyl substituents was studied in a similar way (153). Model systems such as the porphycenes, which are structurally different from porphyrins or even chloro-

488

phylls, but still bear characteristic similarities to the natural molecules, are important for exploring synthetic possibilities for donors and acceptors in future artificial systems for photoinduced electron transfer.

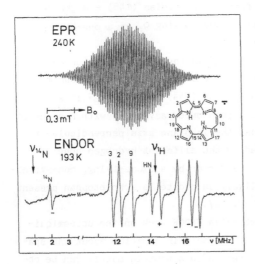

Fig. 26. EPR and $^{14}N/^{1}H$ ENDOR spectra from free-base porphycene radical anion in tetrahydrofuran. The signs of hfc's are from General TRIPLE, assignments to molecular positions were made on the basis of comparison with similar systems and with MO calculations (153), for numbering see molecular diagram. Note that the EPR spectrum could be perfectly simulated by use of the hfc's from ENDOR.

Studies of the paramagnetic states of the isolated cofactors (chlorophylls, pheophytins, quinones etc.) by EPR/ENDOR are required for an understanding of the influence of the protein environment upon the electronic structure of these species. Accordingly, a considerable number of papers has been devoted to these systems (for recent reviews, see (12, 14, 15)).

In a next step, it was attempted to construct in vitro models for RC components which mimic some physical and chemical properties of these in-vivo species. The interest here has mainly focused on the primary donor which is known to be a BChl-dimer in bacterial RC's with sufficiently strong π-π overlap leading to electron spin delocalization. Various dimeric chlorophylls have been synthesized (154), one of which, a "hinged" bis(chlorophyll)cyclophane, was studied by high resolution ENDOR techniques (155). In the radical cation of this system a

localization of the unpaired electron on one dimer half was found. Apparently, the relative large distance (4-6 Å) between the two macrocycles and the position and flexibility of the linking bridges prevented an effective π-π overlap. Other examples of π-stacked radical cations and anions studied by liquid-state ENDOR can be found among the cyclophanes (156, 157), see also Ref. 12 (Chapter 5.1.7). Depending on bridge length, substitution pattern, and counterion/solvent type these systems either show a localization of the unpaired electron on one half of the "dimer" or a delocalization over both halves. In the latter cases the expected halving of the hfc's is observed. Even asymmetrical spin distributions within such systems could sometimes be observed caused by environmental (counterion) effects (156). Furthermore, the cation radical of several covalently-linked porphyrins were studied by ENDOR as models for the primary donor. These dimers were derived from zinc mesotetratolylporphyrin and linked at the

ortho or para position of one phenyl ring via alkoxy groups with varying
bridge lengths (158). In these systems a delocalization of the unpaired electron
over both dimer halves could not be observed contrary to tetraphenylporphyrin
dimers bridged by one of the phenyl rings (159).

Chemically modified chlorophyll-type radicals have also been studied by
ENDOR as models for the natural ones, e.g. the oxidized Chl a enol (stabilized
as silyl ether) (160) or the 10-hydroxy bacteriochlorophyll a (116).

So far, we have only discussed models for isolated donor or acceptor molecules in photosynthesis. For studying electron transfer in artificial systems,
donor-acceptor complexes are necessary, preferentially linked by rigid molecular spacers. This field of vivid research has very recently been reviewed by
Wasielewski (161). The radical cations and anions created in the light-induced
charge separation process are usually very short-lived (ps to ns). Lifetimes as
long as several μs are, however, needed to measure all the relevant electron-
nuclear hfc's (> 100 kHz). In such systems the respective radical cation (from
the donor) or radical anion (from the acceptor) can be created separately and
studied be EPR/ENDOR techniques. In this way, not only the electronic structure
of the isolated donor or acceptor can be obtained, but also that of the molecular spacer or bridge between them, which is of great importance for the electron transfer process (162). Very recently, donor-acceptor tetrads (carotenoid-
porphyrin-diquinone and carotenoid diporphyrin-quinone systems) have been synthesized in which the light-induced charge separation occurs with high quantum
yield ($\geq$ 0.25) and the back-reaction is slow enough to yield radical ion pair
lifetimes of several microseconds (163, 164). In such systems, which mimic the
long-lived charge separation products in RC's, a simultaneous detection of the
created radical ions by time resolved EPR and CIDEP-enhanced-ENDOR should be
feasible.

4.3 Spin-Polarized Transient Radicals

For the first application of CIDEP-enhanced-ENDOR, UV irradiated p-benzoquinone in ethylene glycol solution was chosen (32) because in earlier CIDEP
studies (165) the kinetic data of this system have been analyzed. They show that
the polarization is due to a combination of strong TM and weak RPM effects. UV-
photolysis of p-benzoquinone in solution at pH $\approx$ 7 produces the monoprotonated
p-benzosemiquinone radical (PBQH$\cdot$), which has a lifetime $\tau \approx 10^{-3}$ s and yields
an EPR spectrum with pronounced polarization effects. In the first experiment,
cw mode of operation was chosen for reasons of simplicity and ENDOR was detected by rf induced destruction of the polarization. The signal-to-noise ratio
obtained was, however, rather poor because cw detection catches only a small
integrated value of the polarization. Increased signal-to-noise is obtained by

a time-resolved mode of operation of CIDEP-enhanced-ENDOR (33). Fig. 27 shows the
ENDOR spectrum of PBQH$^\bullet$ detected in the time-resolved mode. All lines appear in
emission as do the EPR lines demonstrating that the net effect dominates the
polarization mechanism. The ENDOR intensity amounts to about 10 % of the EPR
intensity and could be increased by a factor of two when, in a Special TRIPLE
experiment, the two NMR transitions belonging to one hfc were driven simultaneously.

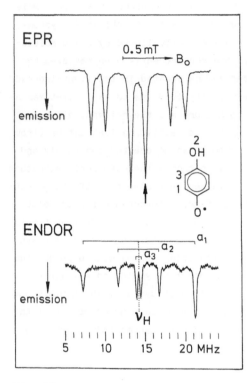

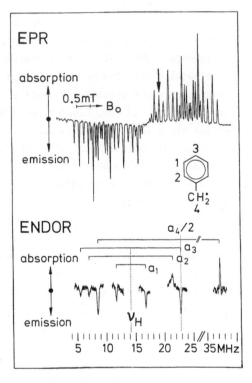

Fig. 27. Time-resolved detection of
EPR and ENDOR of electron spin-polarized p-benzosemiquinone radicals
generated by UV-photolysis of p-benzoquinone in ethylene glycol solution. The arrow indicates the EPR
line saturated for the ENDOR recording. For details, see Ref. 33.

Fig. 28. Time-resolved detection of
EPR and ENDOR of electron spin-polarized benzyl radicals generated by
UV-photolysis of dibenzylketone in
propandiol solution. For the ENDOR
recording the EPR line indicated by
an arrow was chosen without saturating it. For details, see Ref. 33.

The second example of CIDEP-enhanced-ENDOR is the benzyl radical (33) produced by UV-photolysis of dibenzylketone in propandiol (166-168). The polarization arises from the RPM, two identical benzyl radicals forming the radical
pair. Fig. 28 shows the strong multiplet effect, the low-field half of the EPR
spectrum is in emission, the high-field half is in enhanced absorption. Depend-

ing on the magnetic spin quantum number m_I of the chosen EPR line and its emission/absorption polarity, the ENDOR effect was either a decreased absorption or a decreased emission. Up to 50 % changes of the EPR intensity was observed for the largest hfc.

The third example of CIDEP-enhanced-ENDOR published so far is the photolysis of ω,ω-dimethoxy-ω-phenyl-acetophenone (DMPA) in toluene (169). Because of its importance as an efficient initiator for UV-curing of unsaturated polyester and acrylic resins, the photochemistry of DMPA is of particular interest (170). Taking advantage of strong CIDEP effects, three transient paramagnetic species could be detected and identified as benzoyl, dimethoxybenzyl, and methyl radicals by their hyperfine couplings and g-factor. The time-resolved EPR and ENDOR results support earlier conclusions about the primary steps in the UV-photolysis of DMPA (171-173).

Although the number of applications is still small, it is believed that CIDEP-enhanced-ENDOR has future potential for the characterization of photochemically generated radicals, since for short-lived species this method combines high resolution capability with high detection sensitivity (up to 100 % EPR sensitivity is in principle possible). In fact, this method should be applicable to all systems for which pronounced CIDEP effects are observable in their EPR spectra. As an example for a possible field of application, photosynthesis is anticipated, where CIDEP effects have been observed in the primary reaction steps both in green plants and in bacteria (40, 174-177).

5. OUTLOOK

In this Chapter we have described how the problem of inhomogeneously broadened EPR lines typical for radicals in biology and biochemistry can be overcome by extending EPR to ENDOR/TRIPLE techniques. As representative examples of application we have chosen the radical ions occurring as primary photoproducts in reaction centers of photosynthetic bacteria and the corresponding model systems. Similar, but less extensive studies have also been performed for the radical ions occurring in the photosynthetic plant systems, e.g. for the primary donor cation radical $P_{700}^{+\cdot}$ of PS I (reviewed in Ref. 14) or the intermediate acceptor $I^{\bar{\cdot}}$ in PS I and PS II (140, 178, 179). Interestingly, the species mediating between the PS II donor P_{680} and the water-splitting system, called "signal II", has recently been identified as a tyrosine radical (for a short review, see Ref. 180). However, there remain many open questions in the field of primary processes in photosynthesis, e.g. the question whether the structure-function relationship found for the electron transfer chain of Rps. viridis and Rb. sphaeroides is a universal principle and also applies to other photosynthetic bacteria, or even to the plant systems. We are convinced that ENDOR will continue to play a very important role in the investigation of the electronic and spatial structures of radicals involved in photosynthesis research.

Another important class of biological species, but not covered in this Chapter, comprises the flavins which occur in various enzymatic systems ("flavoenzymes") found in almost all biological redox processes (e.g. in the respiratory chain), see Ref. 90. The flavin cofactors can exist in three different redox and various protonation states. For EPR and ENDOR studies the half-reduced semiquinone radical form is of particular interest. It actually occurs in many flavoenzymes mediating "one-electron transfer" reactions. In flavoenzymes strictly involved in two-electron transfer, one can, however, often create the radical state artificially, which can then be used as internal "spin probe" of the system. Owing to the importance of the flavoproteins, these species have been studied relatively early by ENDOR (in the solid state); this work has already been reviewed in the book by Kevan and Kispert (8). Most recently, Kurreck and coworkers considerably extended the studies to a large variety of flavin systems. They also applied liquid-state ENDOR and mapped out the spin density distribution of all the important isolated flavin radicals. Furthermore, many flavoproteins were studied yielding deep insight into electronic and spatial structures. The work has been reviewed by Kurreck, Kirste and Lubitz (12), see also (181).

Many more biologically active organic species have been studied by liquidstate ENDOR, mostly in form of their isolated radical states. These comprise vitamin quinones, such as ubiquinone (coenzyme Q), riboflavin (vitamin B_2), and the K and E vitamins, catechols and catecholamines, and various phenothiazine drugs. For information on such systems, the reader is referred to some recent reviews (12, 14, 15).

To a large extent, we have restricted the discussion to ENDOR on samples in liquid solution because powders or single-crystals are discussed in detail in other chapters of this book. ENDOR-in-solution is distinguished by its high resolving power due to narrow-line spectra and yields a detailed map of the spin density distribution in the molecule. From this map valuable geometrical information can be extracted with the help of advanced MO calculations. On the other hand, in isotropic solvents traceless parts of the static Hamiltonian, e.g. the electron-nuclear dipole and quadrupole interactions with their inherent structure information, are averaged out because of rapid molecular tumbling. The use of liquid crystals as anisotropic solvents can retain information about anisotropic interactions, but still provides narrow-line spectra. In external fields of some hundred mT, nematic solvents and solute molecules can be partially aligned along the field direction. As a consequence, line shifts and splittings occur when going from the isotropic to the nematic phase of the solvent. They depend on the degree of ordering achieved at a particular temperature and on the magnitude of the anisotropic interactions. ENDOR-in-liquid-crystals sounds a bit esoteric, but might be appreciated by gourmets in the EPR community. For example, the first determination of ^{14}N (182) and ^{2}H (183) quadrupole couplings in organic radicals was achieved in this laboratory by ENDOR-in-liquid crystals. During the last few years, Kurreck and collaborators applied this technique to a number of interesting organic systems in nematic liquid crystals. They extended the ENDOR studies to smectic mesophases and showed that even ion radicals can be dissolved and oriented in liquid crystals. These experiments are thoroughly discussed in the book by Kurreck, Kirste and Lubitz (12).

Other unconventional solvents recently used in ENDOR spectroscopy are micelles. Janzen and collaborators (184) performed the first ENDOR study of an inorganic radical (Fremy's salt) in reversed micelles ("water inside"), whereas Kurreck and collaborators (185) were the first to use reversed micelles in their ENDOR experiments on organic radicals (e.g. radical anions of lumiflavins). The ENDOR spectra obtained from micellar solutions generally exhibit good signal-to-noise ratio and narrow lines. This may be attributed to vanishingly small chemical and Heisenberg exchange since each micelle contains, if at all, only one radical molecule. Micellar systems attract increasing attention since they offer the potential (i) to investigate cage effects in thermal and light-induced chemical reactions, and (ii) to study the interaction of encapsulated (bio)molecules with the surrounding micellar environment.

Single-crystal ENDOR and TRIPLE resonance studies, naturally, yield a maximum of structure information, often even supplementing X-ray crystallography. Unfortunately, however, single crystals are difficult to obtain for biological systems. In cases of biomolecules of large molecular weight frequently only rather broad powder-type EPR spectra can be recorded. For large g factor and/or hfs anisotropies, however, single-crystal type ENDOR can be obtained by magneto-selection of specific molecular orientations. For small g factor anisotropy this is not possible for X-band spectroscopy. Instead, the Zeeman term in the Hamiltonian has to be sufficiently enlarged by use of drastically higher magnetic fields than employed in conventional X-band EPR and ENDOR. In this context, our first successful 3 mm cw ENDOR experiments (W-band: 94 GHz) should be mentioned (91). W-band ENDOR is distinguished by its high selectivity in disordered systems, since - due to its ten times higher magnetic field - even for small Zeeman anisotropies canonical orientations can be well-separated with single-crystal like ENDOR spectra.

Concluding, we are convinced that the ENDOR/TRIPLE spectroscopy in the fields of biology and biochemistry has a very promising future.

494

ACKNOWLEDGEMENT

It is a pleasure to acknowledge the numerous valuable contributions from the present members of the Berlin ENDOR group, in particular from Olaf Burghaus, Vera Hamacher, Martina Huber, Petra Jaegermann, Robert Klette, Bernd von Maltzan, Jenny Schlüpmann and Eberhard Tränkle, as well as from the former coworkers Reinhard Biehl (†), Peter Dinse, Friedhelm Lendzian, Wolfgang Möhl and Christopher Winscom. Many ENDOR results on photosynthetic systems and on model compounds have been obtained in collaboration with Hugo Scheer (University of Munich), Arnold Hoff (University of Leiden), Haim Levanon (Hebrew University of Jerusalem) and Hans van Willigen (University of Massachusetts, Boston). Some of the work on photosynthetic radicals has been obtained in collaboration with George Feher and coworkers (University of California, San Diego). We are grateful to Harry Kurreck and his coworkers from the Institute of Organic Chemistry (Free University of Berlin) for a long-standing collaboration. The stimulating discussions with Renad Sagdeev and Kev Salikhov (Academy of Sciences of the USSR, Novosibirsk) regarding the development of CIDEP-enhanced-ENDOR are remembered with gratitude. The work on porphycenes has been performed in collaboration with Emanuel Vogel, Matthias Köcher (University of Cologne) and Moshe Toporowicz (Hebrew University of Jerusalem). We thank Helga Reeck and Brigitte Röttger for their skill in preparing the manuscript. This work was supported by the Deutsche Forschungsgemeinschaft, which is gratefully acknowledged.

REFERENCES

1   G. Feher, Phys.Rev. 103 (1956) 834-837.
2   A. Cederquist, Ph.D. Thesis, Washington University, St. Louis, Missouri, 1963.
3   J.S. Hyde and A.H. Maki, J.Chem.Phys. 40 (1964) 3117-3118.
4   J.H. Freed, in Ref. 9, pp. 73-142.
5   K.P. Dinse, R. Biehl and K. Möbius, J.Chem.Phys. 61 (1974) 4335-4341.
6   K. Möbius and R. Biehl, in Ref. 9, pp. 475-507.
7   R. Biehl, M. Plato and K. Möbius, J.Chem.Phys. 63 (1975) 3515-3522.
8   L. Kevan and L.D. Kispert, Electron Spin Double Resonance Spectroscopy, John Wiley, New York, 1976.
9   M.M. Dorio and J.H. Freed (Eds.), Multiple Electron Resonance Spectroscopy, Plenum Press, New York, 1979.
10 A. Schweiger, Electron Nuclear Double Resonance of Transition Metal Complexes with Organic Ligands, Structure and Bonding, Vol. 51, Springer Verlag, Berlin, 1982.
11 A. Schweiger, in: M.C.R. Symons, (Ed.), Electron Spin Resonance, Spec. Period.Rep. 10B (1986) 138-184.
12 H. Kurreck, B. Kirste and W. Lubitz, Electron Nuclear Double Resonance Spectroscopy of Radicals in Solution, VCH Verlagsgesellschaft, Weinheim, 1988.
13 K. Möbius, M. Plato and W. Lubitz, Phys.Rep. 87 (1982) 171-208.
14 K. Möbius and W. Lubitz, in: L.J. Berliner and J. Reuben (Eds.), Biol. Magn.Res., Plenum Press, New York, 7, (1987) 129-247.
15 K. Möbius, Magn.Reson.Rev. 12 (1987) 285-332.

16   P. Höfer, A. Grupp and M. Mehring, in: J.A. Weil, M.K. Bowman, J.R. Morton
     and K.F. Preston (Eds.), Electron Magnetic Resonance of the Solid State,
     Canadian Society for Chemistry, 1987, pp. 521-533.

17   M. Mehring, P. Höfer and A. Grupp, Ber.Bunsenges.Phys.Chem. 91 (1987)
     1132-1137.

18   see p. 287 in Ref. 15.

19   J.H. Freed, J.Chem.Phys. 50 (1969) 2271-2272.

20   K. Möbius, W. Fröhling, F. Lendzian, W. Lubitz, M. Plato and C.J. Winscom,
     J.Phys.Chem. 86 (1982) 4491-4507.

21   W. Lubitz, R. Biehl and K. Möbius, J.Magn.Reson. 27 (1977) 411-417.

22   H. Bock, B. Hierholzer, H. Kurreck and W. Lubitz, Angew.Chem. 95 (1983)
     817-819; Angew.Chem.Int.Ed.Engl. 22 (1983) 787-788; Angew.Chem.Suppl.
     (1983) 1088-1105. 23 J.H. Freed, J.Chem.Phys. 43 (1965) 2312-2332.

24   J.H. Freed, J.Phys.Chem. 71 (1967) 38-51.

25   J.H. Freed, D.S. Leniart and J.S. Hyde, J.Chem.Phys. 47 (1967) 2762-2773.

26   J.H. Freed, D.S. Leniart, and H.D. Connor, J.Chem.Phys. 58 (1973) 3089-
     3105.

27   M. Plato, W. Lubitz and K. Möbius, J.Phys.Chem. 85 (1981) 1202-1219.

28   J.H. Freed, D.S. Leniart and J.S. Hyde, J.Chem.Phys. 47 (1967) 2762-2773.

29   S. Geschwind, in: A.J. Freeman and R.B. Frankel (Eds.), Hyperfine Interac-
     tions, Academic Press, New York, 1967, pp. 225-286.

30   W. Möhl, C.J. Winscom, M. Plato, K. Möbius and W.Lubitz, J.Phys.Chem. 86
     (1982) 149-152.

31   R.D. Allendoerfer and A.H. Maki, J.Magn.Res. 3 (1970) 396-410.

32   R.Z. Sagdeev, W. Möhl and K. Möbius, J.Phys.Chem. 87 (1983) 3183-3186.

33   F. Lendzian, P. Jaegermann and K. Möbius, Chem.Phys.Lett. 120 (1985) 195-
     200.

34   J.H. Freed and J.B. Pedersen, Adv.Magn.Reson. 8 (1976) 1-84.

35   L.T. Muus, P.W. Atkins, K.A. McLauchlan and J.B. Pedersen (Eds.), Chemi-
     cally Induced Magnetic Polarization, Reidel, Dordrecht, 1977.

36   P.J. Hore, C.G. Joslin and K.A. McLauchlan, in: P.B. Ayscough (Ed.), Elec-
     tron Spin Resonance, Spec.Period.Rep. 5 (1979) 1-45.

37   A.D. Trifunac and M.C. Thurnauer, in: Ref. 78, p. 107-152.

38   M.C. Depew and J.K.S. Wan, Magn.Reson.Rev. 8 (1983) 85-115.

39   K.M. Salikhov, Yu. Molin, R.Z. Sagdeev and A.L. Buchachenko, Spin Polari-
     zation and Magnetic Effects in Radical Reactions, Elsevier, Amsterdam,
     1984.

40   A.J. Hoff, Quart.Rev.Biophys. 17 (1984) 153-282.

41   F.J. Adrian in Ref. 35, p. 77-155; p. 369-381.

42   R. Kaptein, J.Am.Chem.Soc. 94 (1972) 6251-6269.

43   J. Schlüpmann, P. Jaegermann, P. Tian, F. Lendzian and K. Möbius, to be
     published.

44   T. Prisner, O. Dobbert, K.P. Dinse and H. van Willigen, J.Am.Chem.Soc. 110
     (1988) 1622-1623.

45   H. van Willigen, M. Vuolle and K.P. Dinse, J.Phys.Chem., in press.

46   B. Venkataraman, B.G. Segal and G.K. Fraenkel, J.Chem.Phys. 30 (1959)
     1006-1016.

47   J. Fajer and M.S. Davis, in: D. Dolphin (Ed.), The Porphyrins, Academic
     Press, New York, 1979, Vol. 4, p. 197-256.

48   C.P. Poole, Jr., Electron Spin Resonance, a Comprehensive Treatise on
     Experimental Techniques, John Wiley, New York, 1983.

49   K. Möbius, in: P.B. Ayscough (Ed.), Electron Spin Resonance, Spec.Period.
     Rep. 4 (1977) 16-29; 5 (1979) 52-65; 6 (1981) 32-42.

50   F. Lendzian, Ph.D. Thesis, Free University Berlin, 1982.

51   G. Hurst, K. Kraft, R. Schulz and R. Kreilik, J.Magn.Reson. 49 (1982)
     159-160.

52   J. Lopez, J. Yamauchi, K. Okada and Y. Deguchi, Bull.Chem.Soc.Jpn. 57
     (1984) 673-677.

53   W. Lubitz, R.A. Isaacson, E.C. Abresch and G. Feher, Proc.Natl.Acad.Sci.
     USA 81 (1984) 7792-7796.

54   T.C. Christidis and F.W. Heineken, J.Phys. E 18 (1985) 281-283.

55   W. Möhl and E. de Boer, J.Phys. E 18 (1985) 479-481.

496

56 R.L. Wood, F. Froncisz and J.S. Hyde, J.Magn.Reson. 58 (1984) 243-253.
57 R.A. Venters, J.R. Anderson, J.F. Cline and B.M. Hoffman, J.Magn.Reson. 58 (1984) 507-510.
58 J.R. Anderson, R.A. Venters, M.K. Bowman, A.E. True and B.M. Hoffman, J.Magn.Reson. 65 (1985) 165-168.
59 R.A. Isaacson and G. Feher, private communication 1986.
60 J.S. Hyde and W. Froncisz, in: M.C.A. Symons (Ed.), Electron Spin Resonance, Spec.Period.Rep. 10A (1986) 175-184.
61 H.J. Schneider and P. Dullenkopf, Rev.Sci.Instrum. 48 (1977) 68-73.
62 M. Mehring and F. Freysoldt, J.Phys. E 13 (1980) 894-895.
63 W.A.J.A. van der Poel, D.J. Single, J. Schmidt and J.H. van der Waals, Mol. Phys. 49 (1983) 1017-1028.
64 C.P. Lin, M.K. Bowman and J.R. Norris, J.Magn.Reson. 65 (1985) 369-374.
65 S. Pfenninger, J. Forrer, A. Schweiger and T. Weiland, Rev.Sci.Instrum. 59 (1988) 752-760.
66 W.B. Mims, Proc.Roy.Soc. A283 (1965) 452-457; Phys.Rev. B5 (1972) 2409-2419; Phys. Rev. B5 (1972) 3605-3609.
67 L.G. Rowan, E.L. Hahn and W.B. Mims, Phys.Rev. A137 (1965) 61-71.
68 E.R. Davis, Phys.Lett. 47A (1974) 1-2.
69 M.K. Bowman, J.R. Norris, M.C. Thurnauer, J. Warden, S.A. Dikanov and Yu.D. Tsvetkov, Chem.Phys.Lett. 55 (1978) 570-574.
70 R.P.J. Merks, R. de Beer and D. van Ormondt, Chem.Phys.Lett. 61 (1979) 142-144.
71 M. Mehring, P. Höfer, A. Grupp and H. Seidel, Phys.Lett. 106A (1984) 146-148.
72 D. van Ormondt and K. Nederveen, Chem.Phys.Lett. 82 (1981) 443-446.
73 S.A. Dikanov, Yu.D. Tsvetkov, M.K. Bowman and A.V. Astashkin, Chem.Phys. Lett. 90 (1982) 149-153.
74 K.P. Dinse and C.J. Winscom, in: R.H. Clarke (Ed.), Triplet State ODMR Spectroscopy, John Wiley, New York, 1982, pp. 83-136 and references cited therein.
75 S.A. Dikanov, A.V. Astashkin, Yu.D. Tsvetkov and M.G. Goldfeld, Chem. Phys.Lett. 101 (1983) 206-210.
76 A. de Groot, A.J. Hoff, R. de Beer and H. Scheer, Chem.Phys.Lett. 113 (1985) 286-290.
77 M. Iwasaki and K. Toriyama, J.Chem.Phys. 82 (1985) 5415-5423.
78 L.Kevan and R.N. Schwartz (Eds.), Time Domain Electron Spin Resonance, John Wiley, New York, 1979.
79 J.R. Norris, M.C. Thurnauer and M.K. Bowman, Adv.Biol.Med.Phys. 17 (1980) 365-416.
80 W.B. Mims and J. Peisach, in: L.J. Berliner and J. Reuben (Eds.), Biol. Magn.Res., Plenum Press, New York, 3 (1981) 213-263.
81 K.P. Dinse, K. Möbius and R. Biehl, Z.Naturforsch. 28a (1973) 1069-1080.
82 A.E. Stillman and R.N. Schwartz, Mol.Phys. 35 (1978) 301-313.
83 P.A. de Jager and F.G.H. van Wijk, Rev.Sci.Instrum. 58 (1987) 735-741.
84 R. de Beer, H. Backhuijsen, E.L. de Wild and R.P.J. Merks, Bul.Magn.Reson. 2 (1981) 420.
85 A.D. Trifunac, M.C. Thurnauer and J.R. Norris, Chem.Phys.Lett. 57 (1978) 471-473.
86 S. Basu, K.A. McLauchlan and G.R. Sealy, J.Phys. E 16 (1983) 767-773.
87 K.A. McLauchlan and G.R. Sealy, Mol.Phys. 52 (1984) 783-793.
88 C.R. Cantor and P.R. Schimmel, Biophysical Chemistry, Freeman and Company, San Francisco, 1980, Vol. 2.
89 A.J. Hoff, Phys.Rep. 54 (1979) 75-200.
90 D.E. Edmondson and G.Tollin, in Top.Curr.Chem. 108 (1983) 111-138.
91 O. Burghaus, A. Tòth-Kischkat, R. Klette and K. Möbius, J.Magn.Reson. 80 (1988) 383-388.
92 H.C. Box, "Radiation Effects: ESR and ENDOR Analysis," Academic Press, New York, 1977.
93 L. Kevan and P.A.Narayana, in Ref. 9, pp. 229-259.
94 A.J. Hoff in Ref. 97, pp. 97-123.

95   R.K. Clayton, Photosynthesis: Physical Mechanisms and Chemical Patterns, Cambridge University Press, Cambridge, 1980.

96   Govindjee (Ed.), Photosynthesis: Volume 1 and Volume 2 Academic Press, New York, 1982

97   J. Amesz (Ed.) Photosynthesis, New Comprehensive Biochemistry, Vol. 15, Elsevier, Amsterdam, 1987.

98   C.J. Arntzen and L.A. Staehelin, Photosynthesis III. Encyclopedia of Plant Physiology, New Series, Vol. 19, Springer, Berlin, 1986.

99   M.E. Michel-Beyerle (Ed.), Antennas and Reaction Centers of Photosynthetic Bacteria. Structure, Interactions and Dynamics, Springer Series in Chemical Physics, 1985, Volume 42.

100   J. Breton and A. Verméglio (Eds.), The Photosynthetic Bacterial Reaction Center, Plenum, New York 1988.

101   H. Scheer and S. Schneider (Eds.), Photosynthetic Light-Harvesting Systems, de Gruyter, Berlin 1988.

102   G. Feher and M.Y. Okamura, in: R.K. Clayton and W.R. Sistrom, (Eds.), The Photosynthetic Bacteria, Plenum, New York, 1978, pp. 349-386.

103   M.Y. Okamura, G. Feher and N. Nelson, in: Govindjee, (Ed.), Photosynthesis: Energy Conversion by Plants and Bacteria Academic, New York, 1982, Vol. 1, pp. 195-272.

104   R.E. Blankenship, J.T. Trost and L.J. Mancino, in Ref. 100, pp. 119-127; J. Amesz, ibid. pp. 129-138.

105   B.K. Pierson and J.M. Olson, in Ref. 97, pp. 21-42.

106   J. Deisenhofer, O. Epp, K. Miki, R. Huber and H. Michel, J.Mol.Biol. 180 (1984) 385-398; Nature (London) 318 (1985) 618-624; H. Michel, O. Epp and J. Deisenhofer, EMBO J. 5 (1986) 2445-2451.

107   J.P. Allen, G. Feher, T.O. Yeates, H. Komiya and D.C. Rees, Proc.Natl. Acad.Sci. USA 84 (1987) 5730-5734; 84 (1987) 6162-6166; 85 (1988) 8487-8491; T.O. Yeates, H. Komiya, D.C. Rees, J.P. Allen and G. Feher, ibid 84 (1987) 6438-6442; T.O. Yeates, H. Komiya, A. Chirino, D.C. Rees, J.P. Allen and G. Feher, ibid 85 (1988) 7993-7997.

108   C.H. Chang, D. Tiede, J. Tang, U. Smith, J. Norris and M. Schiffer, FEBS Lett. 205 (1986) 82-86.

109   C. Kirmaier and D. Holton, Photosynthesis Res. 13 (1987) 225-260.

110   J.R. Norris, R.A. Uphaus, H.L. Crespi and J.J. Katz, Proc.Natl.Acad.Sci. USA 68 (1971) 625-628.

111   G. Feher, A.J. Hoff, R.A. Isaacson and L.C. Ackerson, Ann. N.Y. Acad.Sci. 244 (1975) 239-259.

112   J.R. Norris, H. Scheer and J.J. Katz, Ann. N.Y. Acad.Sci. 244 (1975) 260-280.

113   W. Lubitz, F. Lendzian, M. Plato, K. Möbius and E. Tränkle, in: Ref. 99, pp. 164-173.

114   C.D. Borg, A. Forman and J. Fajer, J.Am.Chem.Soc. 98 (1976) 6889-6896.

115   A.J. Hoff and K. Möbius, Proc.Natl.Acad.Sci. USA 75 (1978) 2296-2300.

116   W. Lubitz, F. Lendzian and H. Scheer, J.Am.Chem.Soc. 107 (1985) 3341-3343.

117   W. Lubitz, F. Lendzian, H. Scheer, M. Plato and K. Möbius, to be published.

118   W. Lubitz, R.A. Isaacson, E.C. Abresch and G. Feher, Proc.Natl.Acad.Sci. USA 88 (1984) 7792-7796.

119   F. Lendzian, K. Möbius, M. Plato, U.H. Smith, M.C. Thurnauer and W. Lubitz, Chem.Phys.Lett. 111 (1984) 583-588.

120   R.C. Dougherty, H.L. Crespi, H.H. Strain and J.J.Katz, J.Am.Chem.Soc. 88 (1966) 2854-2856; J.J. Katz, R.C. Dougherty, H.L. Crespi and H. Strain, J.Am. Chem.Soc. 88 (1966) 2856-2857.

121   M.S. Davis, A. Forman, L.K. Hanson, J.P. Thornber and J. Fajer, J.Phys. Chem. 83 (1979) 3325-3332.

122   F. Lendzian, W. Lubitz, R. Steiner, E. Tränkle, M. Plato, H. Scheer and K. Möbius, Chem.Phys.Lett. 126 (1986) 290-296.

123   J.A. Pople and D.L. Beveridge, Approximate Molecular Orbital Theory, McGraw-Hill, New York 1970.

124   M. Plato, E. Tränkle, W. Lubitz, F. Lendzian and K. Möbius, Chem.Phys. 107 (1986) 185-196.

498

125 M. Plato, W. Lubitz, F. Lendzian and K. Möbius, Isr.J.Chem. (1989) in press.

126 J.S. Hyde, G.H. Rist and L.E.G. Eriksson, J.Phys.Chem. 72 (1968) 4269-4276.

127 F. Lendzian, W. Lubitz, H. Scheer, C. Bubenzer and K. Möbius, J.Am.Chem. Soc. 103 (1981) 4635-4637.

128 W. Lubitz, F. Lendzian, H. Scheer, J. Gottstein, M. Plato and K. Möbius, Proc.Natl.Acad.Sci. USA 81 (1984) 1401-1405.

129 W. Lubitz, F. Lendzian, H. Scheer, M. Plato and K. Möbius, in: A.H. Zewail, (Ed.), Photochemistry and Photobiology, Proceedings of the International Conference, University of Alexandria, Egypt, Harwood Academic Publishers, New York, 1983, pp. 1057-1069.

130 J.R. Norris and J.J. Katz, in: R.K. Clayton and W.R. Sistrom (Eds.), The Photosynthetic Bacteria, Plenum Press, New York, 1978, pp. 397-418.

131 F. Lendzian, W. Lubitz, H. Scheer, A.J. Hoff, M. Plato, E. Tränkle and K. Möbius, Chem.Phys.Lett. 148 (1988) 377-385.

132 M. Plato, K. Möbius, M.E. Michel-Beyerle, M. Bixon and J. Jortner, J.Am. Chem.Soc. 110 (1988) 7279-7285.

133 M.E. Michel-Beyerle, M. Plato, J. Deisenhofer, H. Michel, M. Bixon and J. Jortner, Biochim.Biophys. Acta 932 (1988) 52-70.

134 M. Plato, F. Lendzian, W. Lubitz, E. Tränkle and K. Möbius, in Ref 100, pp.379-388.

135 M. Lösche, G. Feher and M.Y. Okamura, Proc.Natl.Acad.Sci. USA 84 (1987) 7537-7541.

136 D.J. Lockhardt and S.G. Boxer, Proc.Natl.Acad.Sci. USA 85 (1988) 107-111.

137 M.Y. Okamura, R.A. Issacson and G. Feher, Biochim.Biophys. Acta 546 (1979) 394-417.

138 W. Lubitz, F. Lendzian and K. Möbius, Chem.Phys.Lett. 81 (1981) 235-241.

139 W. Lubitz, F. Lendzian and K. Möbius, Chem.Phys.Lett. 84 (1981) 33-38.

140 W. Lubitz, M. Plato, G. Feher, R.A. Isaacson and M.Y. Okamura, Biophys.J. 53 (1988) 67a; and to be published.

141 G. Feher, R.A. Isaacson, M.Y. Okamura and W. Lubitz, in Ref. 100, pp. 229-235.

142 W.F. Butler, R. Calvo, D.R. Fredkin, R.A. Isaacson, M.Y. Okamura and G. Feher, Biophysical J. 45 (1984) 947-973.

143 M.R. Das, H.D. Connor, D.S. Leniart and J.H. Freed, J.Am.Chem.Soc. 92 (1970) 2258-2268.

144 G. Feher, R.A. Isaacson, M.Y. Okamura and W. Lubitz, in: Ref. 99, pp. 175-189.

145 P.J. O'Malley, T.K. Chandrashekar and G.T. Babcock, In: Ref. 99, pp. 339-344.

146 W. Lubitz and G.T. Babcock, Trends Biochem.Sci. 12 (1987), 96-100.

147 W. Lubitz, E.C. Abresch, R.J. Debus, R.A. Isaacson, M.Y. Okamura and G. Feher, Biochim.Biophys. Acta, 808 (1985) 464-469.

148 M.R. Gunner, B.S. Braun, J.M. Bruce and P.L. Dutton, in Ref. 99, pp. 298-304.

149 Proceedings of the Sixth International Confernce of Photochemical Conversion and Storage of Solar Energy, New Journal of Chemistry 11(2) (1987).

150 J. Fajer and M.S. Davis, in: D. Dolphin, (Ed.), The Porphyrins, Vol.4, Academic Press, New York, 1979, pp. 198-256.

151 J.A. Pedersen, EPR Spectra from Natural and Synthetic Quinones and Quinols, CRC Press, Boca Raton, 1985.

152 J. Schlüpmann, M. Huber, M. Toporowicz, M. Köcher, E. Vogel, H. Levanon and K. Möbius, J.Am.Chem.Soc. 110 (1988) 8566-8567.

153 J. Schlüpmann, M. Huber, M. Plato, M. Toporowicz, M. Köcher, E. Vogel, H. Levanon and K. Möbius, J.Am.Chem.Soc., to be published.

154 S.G. Boxer, Biochim.Biophys. Acta, 726 (1983) 265-292.

155 M. Huber, F. Lendzian, W. Lubitz, E. Tränkle, K. Möbius and M.R. Wasielewski, Chem.Phys.Lett.132 (1986) 467-473.

156 F. Gerson, in: Top Curr.Chem. 115 (1983) 57-105.

157 V. Hamacher, M. Plato and K. Möbius, Chem.Phys.Lett. 125 (1986) 69-73.

158 M. Huber, T. Galili, K. Möbius and H. Levanon, Isr.J.Chem., in press.

159 M. Huber, B. von Maltzan, M. Plato, H. Kurreck and K. Möbius, to be published.

160 M.R. Wasielewski, J.R. Norris, L.L. Shipman, C.P. Lin and W.A. Svec, Proc.Natl.Acad.Sci. USA 78 (1981) 2957-2961.

161 M.R. Wasielewski, Photochem.Photobiol. 47 (1988) 923-929.

162 M. Huber, Ph.D. Thesis, Free University Berlin, 1989.

163 D. Gust, T.A. Moore, A.L. Moore, D. Barrett, L.O. Harding, L.R. Makings, P.A. Liddel, F.C. De Schryver, M. van der Auweraer, R.V. Bensasson and M. Rougée, J.Am.Chem.Soc. 110 (1988) 321-323.

164 D. Gust, T.A. Moore, A.L. Moore, L.R. Makings, G.R. Seely, X. Ma, T.T. Trier and F. Gao, J.Am.Chem.Soc. 110 (1988) 7567-7569.

165 L.T. Muus, S. Frydjaer and B. Nielson, Chem.Phys. 30 (1978) 163-168.

166 H. Paul and H. Fischer, Helv.Chim Acta 56 (1973) 1575-1594.

167 H. Langhals and H. Fischer, Chem.Ber. 111 (1978) 543-553.

168 A.I. Grant and K.A. McLauchlan, Chem.Phys.Lett. 101 (1983) 120-125.

169 P. Jaegermann, F. Lendzian, G. Rist and K. Möbius, Chem.Phys.Lett. 140 (1987) 615-619.

170 F. Jent, H. Paul and H. Fischer, Chem.Phys.Lett. 146 (1988) 315-319.

171 H. Dütsch, Diploma Thesis, University of Zurich (1975), unpublished.

172 A. Borer, R. Kirchmayr and G. Rist, Helv.Chim. Acta 61 (1978) 305-324.

173 J.P. Fouassier and A. Merlin, J. Photochem. 12 (1980) 17-23.

174 C. Dismukes, A. McGuire, R. Friesner and K. Sauer, Rev.Chem.Intermed. 3 (1979) 59-88.

175 A.R. McIntosh and J.R. Bolton, Rev.Chem.Intermed. 3 (1979) 121-129.

176 M.C. Thurnauer, Rev.Chem.Intermed. 3 (1979) 197-230.

177 A.J. Hoff, Quart.Rev.Biophys. 14 (1981) 599; Biophys.Struct.Mech. 8 (1982) 107-150.

178 J. Fajer, M.S. Davis, A. Forman, V.V. Klimov, E. Dolan and B. Ke J.Am. Chem.Soc. 102 (1980) 7143-7145.

179 A. Forman, M.S. Davis, I. Fujita, L.K. Hanson, K.M. Smith and J. Fajer, Isr.J.Chem.21 (1981) 265-269.

180 R.C. Prince, Trends Biochem. Science 13 (1988) 286-288.

181 H. Kurreck, N. Bretz, N. Helle, N. Henzel and E. Weilbacher, J.Chem.Soc. Faraday Trans. 1, 84 (1988) 3293-3306.

182 K.P. Dinse, K. Möbius, M. Plato, R. Biehl and H. Haustein, Chem.Phys.Lett. 14 (1972) 196-200.

183 R. Biehl, W. Lubitz, K. Möbius and M. Plato, J.Chem.Phys. 66 (1977) 2074-2078.

184 E.G. Janzen, Y. Kotake, G.A. Coulter and U.M. Oehler, Chem.Phys.Lett. 126 (1986) 205-208.

185 N. Bretz, I. Mastalsky, M. Elsner and H. Kurreck, Angew.Chem.Int.Ed.Engl. 26 (1987) 345-347.

CHAPTER 14

HEME-GLOBIN INTERACTIONS IN NITROSYL-LIGATED HEMOGLOBIN AS PROBED BY
ENDOR SPECTROSCOPY

REINHARD KAPPL AND JÜRGEN HÜTTERMANN

1   INTRODUCTION

Nitrosyl (NO)-ligated hemoglobin (Hb) is considered a paramagnetic model
very close to the physiologically relevant behaviour of oxyhemoglobin. The electronic
structure of the low-spin FeNO complex is very similar to that of FeO_2 (1, 2). Also,
the NO-radical binds nearly isosterically to O_2 with the heme iron in the protein
(3) and in model complexes (4). Furthermore, addition of inositol hexaphosphate (IHP)
induces those changes in the NO-ligated protein that are usually associated with
the quaternary R-T transition from oxy- to deoxyhemoglobin (1, 5). Since these changes
are also reflected in the Electron Paramagnetic Resonance (EPR) spectra, a number of
investigations using this technique have been performed in addition to those usually
applied for non-paramagnetic hemoglobins. The studies involved frozen solutions
and/or single crystals of model compounds, monomeric myoglobin (Mb), the tetrameric
Hb and some modified derivatives. The results have provided a basic insight into features
like the spin-density distribution and the stereochemistry of the FeNO-complex (6-10)
and the participation of α- and β-chains in the quaternary change of the tetrameric
protein (11).

A more detailed insight into the stereochemistry of the NO ligand and its arrangement in the heme crevice is impeded by the spectral resolution of EPR. This
limitation can be overcome by the application of electron-nuclear double resonance
(ENDOR) spectroscopy, which allows for the observation of proton and nitrogen
interactions of the heme moiety and the proximal and distal amino acid residues within
approximately 5 Å of the paramagnetic centers. From these hyperfine interactions spatial
relationships between the unpaired electron and nearby nuclei can be established at
a high resolution (0.1 to 0.5 Å) (12, 13). In Fig. 1 possible interactions of the proximal
and distal environment below respectively above the heme plane are indicated
employing standard nomenclature. Apart from interactions in the heme plane, the
couplings of protons and nitrogen nuclei in one proximal (histidine at position 8 in
α-helix chain F, His F8) and three distal amino acids (His E7, Valine Val E11 and
Phenylalanine in the intermediate chain segment between α-helices C and D, Phe CD1)

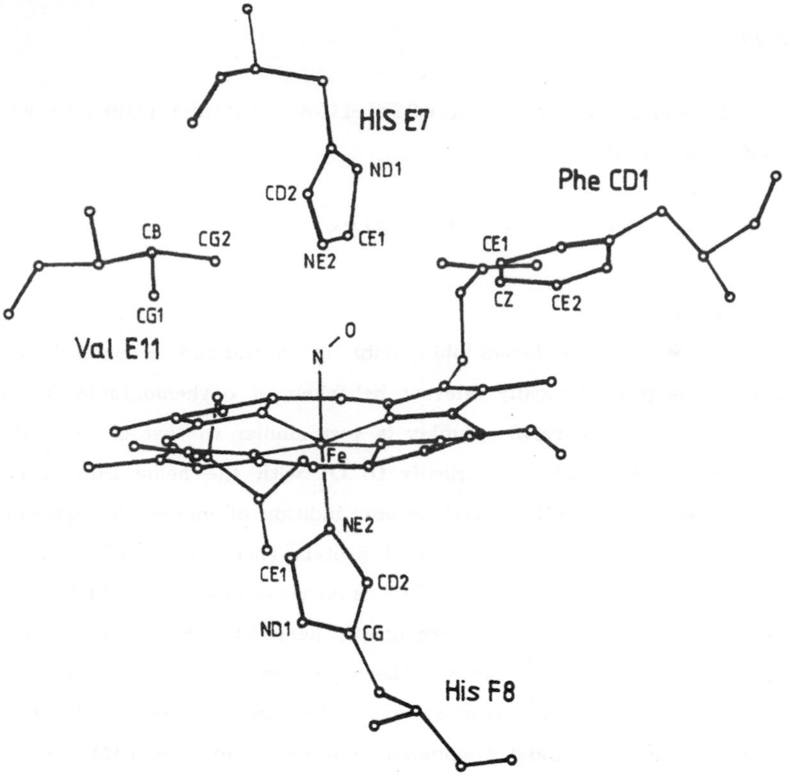

502

Fig. 1. The immediate environment of the heme group in myoglobin

are accessible to study. Essentially, the structural changes of the heme moiety (the allosteric core) induced upon ligand binding are responsible for triggering the R-T transition in hemoglobin (14).

In order to exploit the structural information provided by ENDOR studies on NO-ligated hemoglobin in the R- and T-state as well as possible intermediary quaternary states, a detailed interpretation of hyperfine interactions in frozen solution ENDOR spectra is required, which are easily accessible but exhibit complex powder line patterns. By necessity, a wide variety of systems must be investigated. Within our programme we studied isotopically-substituted model compounds, myoglobin (Mb), the monomeric fraction of Glycera Hb (HbGMNO) that lacks the distal histidine E7 (15), the mutant hemoglobin HbMIwate (in which His F8 is replaced by tyrosine in the α-chains) as well as NO-ligated α- and β-chains and hybrid hemoglobins (13, 16, 17, 18, 19) to arrive at a fairly consistent picture of magnetic interactions. The "model protein" MbNO is particularly useful in this context since it crystallises readily and thus allows for a determination of tensorial parameters of the nitrogen interaction of coordinated histidine (F8) and of some proton interactions of distal and proximal histidine, which have partially been reported previously (13, 18).

In this report we give a survey of the present state of our knowledge about the problem. We include some additional results from the single-crystal ENDOR analysis of MbNO emphasizing the correlation with spectral characteristics of ENDOR powder patterns. From a comparison with powder ENDOR spectra of HbNO in the quaternary T and R states spectral differences will be related to specific structural changes in the heme environment.

2.   ANALYSIS OF ENDOR SPECTRA

The interaction of an unpaired spin in metalloproteins (effective spin 1/2) with the applied magnetic field $\underline{B}$, the nucleus of the central metal ion, the nuclei of coordinated ligands and protons ($\underline{I}$ = 1/2) or nitrogens (^{14}N: $\underline{I}$ = 1, ^{15}N: $\underline{I}$ = 1/2) of the immediate vicinity of the spin center is described by the spin Hamiltonian

$$\hat{H} = \mu_B \, \hat{\underline{S}} \, \underline{g} \, \underline{B} - g_n \, \mu_n \, \hat{\underline{I}} \, \underline{B} + \hat{\underline{S}} \, \underline{A} \, \hat{\underline{I}} + \hat{\underline{I}} \, \underline{Q} \, \hat{\underline{I}} \; . \tag{1}$$

The first and second terms represent the electron (H_{ZE}) and nuclear (H_{ZN}) Zeeman interaction, the third and fourth account for the hyperfine (H_{Hfs}) and the quadrupolar interaction (H_Q). In Eqn. 1 μ_B and μ_n denote the electronic and nuclear magnetic moments, $\hat{\underline{S}}$ and $\hat{\underline{I}}$ the spin operators, $\underline{g}$, $\underline{A}$ and $\underline{Q}$ the g-, hyperfine and quadrupolar tensors and g_n the scalar nuclear g-factor. In the high field limit (i.e. $H_{ZE} \gg H_{Hfs} \gg H_Q$) the spins are quantized along the magnetic field vector $\underline{B}$, m_S and m_I being good quantum numbers. For a $\underline{S}$ = 1/2, $\underline{I}$ = 1/2 spin system (i.e. a proton interacting with a single unpaired electron) and the assumption of isotropic g- and $\underline{A}$-tensors the energy levels are given in first order by

$$E = g_e \, \mu_B \, B \, m_S - g_n \mu_n \, B \, m_I + A \, m_S \, m_I \; . \tag{2}$$

In that simple case the ENDOR frequencies $\nu_{\pm}$ are derived from the selection rules of ENDOR transitions $\Delta m_S = 0$, $\Delta m_I = \pm 1$:

$$\nu_{\pm} = \left| A/2 \pm \nu_n \right| \tag{3}$$

with $\nu_n = g_n \mu_n B$. Depending on the relative magnitudes of ν_n and A/2 ENDOR resonances are either symmetrically located around ν_n and separated by A, or around

A/2 and then separated by 2 ν_n. Eqn. 3 is strictly valid only in solution, but it can be considered a reasonable approximation for analysis of ENDOR spectra of solids in the case of small anisotropy of $\underset{\sim}{g}$- and $\underset{\sim}{A}$-tensors.

More detailed theoretical approaches have been presented by several authors for the analysis of ENDOR spectra in frozen, polycrystalline solutions and single crystals with anisotropic $\underset{\sim}{g}$, $\underset{\sim}{A}$ and $\underset{\sim}{Q}$ tensors of arbitrary, not collinear orientations (12, 20, 21). The ENDOR frequencies for $\underline{S}$ = 1/2 and a single nucleus with spin $\underline{I}$ can be expressed to first order as compact quadratic and bilinear forms (12):

$$\Delta E = c\ (m_S) + 3/2\ Q'\ (m_S)\ (2\ m_I + 1) \qquad \lceil 4 \rceil$$

with $\quad c\ (m_S) = \left| \left(\tilde{\underline{1}}\ \underset{\sim}{C}\ (m_S)\ \tilde{\underset{\sim}{C}}\ (m_S)\ \underline{1} \right)^{1/2} \right|$

and $\quad \underset{\sim}{C}\ (m_S) = (1/g)\ m_S\ g \underset{\sim}{A} - \nu_n\ \underset{\sim}{E}$

$\qquad\qquad g = \left| \left(\tilde{\underline{1}}\ g\ \tilde{\underset{\sim}{g}}\ \underline{1} \right)^{1/2} \right|$

and $\quad Q'\ (m_S) = \tilde{\underline{1}}\ \underset{\sim}{C}\ (m_S)\ \underset{\sim}{Q}\ \tilde{\underset{\sim}{C}}\ (m_S)\ \underline{1}\ /\ c^2\ (m_S)$

where $\underline{1}$ is the unit vector along $\underline{B}$ and $\underset{\sim}{E}$ the unit matrix.

Hence for a nucleus with $\underline{I}$ = 1/2 two first-order frequencies c (1/2), c (-1/2) are observed. In the case of solutions, where anisotropic components average to zero. Eqn. 4 is equivalent to Eqn. 3. For the vector $\underline{B}$ aligned parallel to a principal axis of coaxial $\underset{\sim}{g}$ and $\underset{\sim}{A}$ tensors the hyperfine component A_i can be directly read from the splitting of the ENDOR resonances at $|\ A_i/2 \pm \nu_n\ |$. It has also been shown in solid-state ENDOR on single crystals (20) that the two resonances are, in general, not located symmetrically around ν_n for anisotropic $\underset{\sim}{A}$ tensors, but are always shifted to higher frequencies due to non-negligable off-diagonal elements of $\underset{\sim}{A}$.

For a nucleus with $\underline{I}$ = 1 (e.g. ^{14}N) one obtains four ENDOR energies at frequencies

$$c(1/2) \pm 3/2\ Q'\ (1/2) \quad \text{and} \quad c(-1/2) \pm 3/2\ Q'\ (-1/2). \qquad \lceil 5 \rceil$$

Depending on which m_I-substate (-1, 0, +1) in the EPR-spectrum is used for observing ENDOR transitions, either a two-line ($m_I = \pm 1$) or a four-line ($m_I = 0$) ENDOR spectrum is recorded. Very often, however, the hyperfine structure of ^{14}N-nuclei is buried under broad EPR resonances (in single crystals as well as in powders). The unresolved m_I-substates are then excited simultaneously and a typical quartet line pattern is obtained. If the external magnetic flux $\underline{B}$ is oriented parallel to a principal axis of a coaxial set of $\underset{\sim}{g}$, $\underset{\sim}{A}$ a d $\underset{\sim}{Q}$ tensors, the transition frequencies are given by

$$\left| A_i/2 \pm \nu_n \pm 3/2\ Q_i \right| \qquad \lceil 6 \rceil$$

where A_i and Q_i are the principal tensor values along the axis i, which can then be directly inferred from the ENDOR spectrum. The ENDOR resonances can be grouped into two Zeeman pairs, each line pair being separated by 2 ν_n ($\approx$ 2 MHz for ^{14}N at 350 mT), and shifted from the frequency of the hyperfine component $A_i/2$ by $\pm$ 3/2 Q_i. Second-order corrections to the outlined spectral analysis of interacting protons and nitrogens are discussed in more detail in Refs. 12 and 22. For protons with hyperfine couplings < 20 MHz second-order effects are small and usually can be neglected. They become, however, increasingly important for nitrogen ligands of metal complexes and especially for a correct interpretation of large hyperfine interactions ($\approx$ 100 MHz) of central metal ions.

For transition metal complexes and metalloproteins, EPR and ENDOR experiments usually have to be conducted at low temperatures, due to their fast electronic spin lattice relaxations (T_e) at room temperatures (23). In order to obtain a detectable ENDOR-response in these systems it is necessary to operate at temperatures below 40 K and at high microwave and radiofrequency power levels to satisfy the saturation conditions for EPR and NMR transitions:

$$\text{EPR:} \quad \gamma_e^2 \, B_1^2 \, T_{1e} \, T_{2e} \geq 1 \tag{7}$$

$$\text{NMR:} \quad \gamma_n^2 \, B_2^2 \, T_{1n} \, T_{2n} \geq 1 \ .$$

γ represents the gyromagnetic ratio (e for electron, n for nucleus), T_1 the spin-lattice relaxation times, T_2 the spin-spin relaxation times, B_1 and B_2 the time-dependent magnetic fields of the incident microwave radiation and the radiofrequency, respectively.

As a consequence the ENDOR data of metalloproteins are generally collected in their solid state, from frozen solutions or glasses and from single crystals. Single-crystal EPR-ENDOR analysis yields fundamental information, since the magnitude and the principal directions of the magnetic interactions can be determined and correlated to the molecular structure of the paramagnetic centers. In EPR experiments the g-tensor and, if resolved, hyperfine tensors are obtained, while smaller hyperfine and quadrupole interactions, buried under broad EPR lines, are accessible by ENDOR spectroscopy. The variation of ENDOR frequencies with rotation of the crystal in three orthogonal planes is recorded and the hyperfine and quadrupolar tensors are obtained by least-square fits to the experimental data, for which procedures have been outlined by several authors (21, 24, 25, 26). In general the resulting $\underset{\sim}{g}$, $\underset{\sim}{A}$, $\underset{\sim}{Q}$ tensors are symmetrical and diagonalized to yield the principal coupling parameters and directions without restrictions on the non-coaxial arrangement of principal axes.

In the case of small g-anisotropy, i.e. $(1/g)\,\underset{\sim}{g} \approx \underset{\sim}{E}$ in [4], the component of the hyperfine interaction of a nucleus (e.g. a proton) is then calculated for any arbitrary orientation by

$$\underset{\sim}{A}^2 = \underset{\sim}{l}\,\underset{\sim}{T}\,\underset{\sim}{l} \qquad\qquad [8]$$

where $\underset{\sim}{T}$ is a 3 x 3 matrix and $\underset{\sim}{l}$ the orientation of the magnetic field with respect to the crystal axis. The principal values of the diagonalized $\underset{\sim}{T}_{dia}$ correspond to the squared components of the diagonal hyperfine tensor $\underset{\sim}{A}_{dia}$, which is correlated to the non-diagonal hyperfine tensor in the laboratory frame through the direction cosine matrix $\underset{\sim}{R}$

$$\underset{\sim}{A} = \underset{\sim}{\tilde{R}}\,\underset{\sim}{A}_{dia}\,\underset{\sim}{R} \; . \qquad\qquad [9]$$

The hyperfine interactions with nuclei of ligands in the first coordination sphere and in the immediate environment is constituted from an isotropic and an anisotropic component,

$$\underset{\sim}{A} = \underset{\sim}{A}_{iso} + \underset{\sim}{A}_{dip} \; . \qquad\qquad [10]$$

$\underset{\sim}{A}_{iso}$ arises from non-vanishing s-orbital spin density at the nucleus. The anisotropic interaction $\underset{\sim}{A}_{dip}$ includes the "spin-only" electron-nuclear dipole contribution and a term that takes into account substantial admixture of orbital magnetic moments, which is often found in transition metal complexes and is reflected by considerable g-anisotropy. The spin-only contribution yields an interaction tensor that is symmetric and traceless. The computation of anisotropic hyperfine couplings in terms of an expansion of the molecular orbital as a linear combination of atomic orbitals results in multi-center contributions (12, 27). However, for a sufficiently large distance of an interacting proton from the orbitals containing the unpaired electron the spin density may be considered to be concentrated at a nucleus i and the purely dipolar interaction approximated by the classical point-dipole formula

$$\underset{\sim}{A}_{DD} = g_e\, g_n\, \mu_B\, \mu_n \sum_i \rho_i\, (3\, \underset{\sim}{n_i}\underset{\sim}{\tilde{n}_i} - \underset{\sim}{E}) \cdot r_i^{-3} \qquad\qquad [11]$$

where g_e, g_n are the electron and nuclear g-factors, μ_B, μ_n the magnetic moments of the electron and nucleus, ρ_i is the spin density at nucleus i, r_i the distance between the proton and the i-th spin center and $\underset{\sim}{n_i}$ the direction cosines of $\underset{\sim}{r_i}$ in the molecular frame. If the spin density is not exclusively confined to the central metal but considerably distributed on nuclei of the ligands in the first coordination sphere, a summation over these nuclei is necessary. The presence of strong g-anisotropy leads to an asymmetric proton hyperfine tensor and an anisotropic as well as an isotropic (pseudo-contact) contribution (12, 28), which can be written in the molecular frame (i.e. $\underset{\sim}{g}$ diagonal) and for the spin density located at the central metal as

$$\underset{\sim L}{A} - \Delta \underset{\sim}{g} \underset{\sim}{T} = \Delta \underset{\sim}{g} \, \mu_B \, \mu_n \, g_n \, (3 \, \underline{n} \, \tilde{\underline{n}} - \underset{\sim}{E} \,) \, r^{-3} \qquad [12]$$

so that the hyperfine matrix for a proton is described in the point-dipole approximation as the sum of

$$\underset{\sim}{A} = A_{iso} \underset{\sim}{E} + \underset{\sim DD}{A} + \underset{\sim L}{A} \, . \qquad [13]$$

The classical point-dipole formula (Eqn. 11) is employed to estimate the distance r_i and the location of a purely dipolar-coupled proton in the vicinity of the spin centers at a high spatial resolution ($\sim 1/r_i^3$). The rough approximation of unit spin density at the metal ion gives a first clue for the assignment of a hyperfine tensor to a corresponding proton. This method is, however, not sufficiently precise to simulate the angular variation of a hyperfine coupling measured in single crystals of NO-ligated myoglobin and hemoglobin, due to the large spin density distribution to the axial ligands of the heme group (8, 10, 13, 29). Based on the given coordinates and spin density of the spin-centers of the heme complex (i.e. Fe, NO, NE2 of His F8) it is possible to calculate their individual interaction with a proton at a certain location according to Eqn. 11, thereby neglecting g-anisotropy to first order. The summation of these interaction tensors yields a new tensor that is approximated by an axial dipolar tensor, which is then completely defined by a single center of unit spin density (reduced center of spin density) and the location of the proton. In this way a multi-center spin system is reduced to a single center for which the ENDOR hyperfine splitting are then calculated for any arbitrary orientation of $\underline{B}$ with respect to the crystal axes and compared to the experimental data set (18).

The analysis of ENDOR spectra obtained from polycrystalline or glassy frozen solutions of metalloproteins is usually complicated by the fact, that a large number of molecular orientations are contributing to the resonance pattern at an arbitrary field position selected in the EPR spectrum. It is only for systems with small g- and hyperfine-anisotropy or effective spin diffusion that the EPR transitions of all orientations are saturated simultaneously. For systems with large g-anisotropy (e.g.

metalloproteins) it has been shown by Rist and Hyde (30) that along selected field positions single-crystal-like ENDOR patterns are observed, which may easily be interpreted.

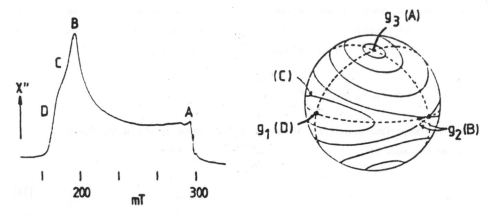

Fig. 2. Correlation between various ENDOR working points across a rhombic EPR spectrum and the selected orientations within the molecular frame (adapted from Ref. 31)

Fig. 2 shows an absorption EPR spectrum with rhombic g-symmetry and the correspondence of various field settings to selected orientations within the molecular frame of the g-tensor. Along the field positions D and A, the extreme turning points, only very few orientations contribute to the ENDOR pattern, the powder character is minimal (i.e. in the ideal case quasi-crystalline spectra are detected) and the resonance pattern is usually well-resolved. For intermediary field settings (C, B) a set of orientations, which is defined by the curves of constant g on the g-tensor elipsoid, is probed simultaneously. As a consequence an anisotropic hyperfine interaction will then result in a distribution of ENDOR-intensities within a certain frequency range for the contributing orientational subset and an ENDOR-powder pattern is recorded, which usually exhibits lower resolution. For an axial g-tensor (i.e. $g_1 = g_2$) one obtains a planar (two-dimensional) powder spectrum along these field orientations. In a similar way for a rhombic system a planar powder spectrum is generated by the orientational distribution for the intermediary g-turning point g_2. If the g- and hyperfine tensors are coaxial one can then infer a hyperfine component for a single-crystal-like orientation directly from the ENDOR pattern, or for a planar powder spectrum the minimal or maximal hyperfine coupling may be observed. However, for an arbitrary arrangement of tensors with respect to the g-tensor the coupling parameters along quasi crystalline orientations do not necessarily represent principal values of hyperfine or quadrupolar tensors. Additional complications in the spectral analysis may arise when large hyperfine interactions are resolved in the EPR spectrum or when extra absorption peaks are present (e.g. in some copper or cobalt complexes). Several theoretical approaches to an analysis of powder ENDOR spectra have been reported in the literature (31 – 34), which describe the ENDOR line pattern in terms of divergencies or accumulation frequencies for the variation of the hyperfine interaction within a subset of an orientational distribution. Simulation of line patterns demonstrate that ENDOR intensities are accumulated at frequencies of absolute or relative extremes of an hyperfine interaction due to the slower change of coupling values with the orientational distribution than for intermediary values.

For considerable g-anisotropy, as found in many transition metal complexes (e.g. Cu^{2+}, Fe^{3+}), Kreilick and coworker (33, 34) have shown the explicit dependence of the proton ENDOR resonances on the effective g-value, which leads to asymmetrical shifts of the ENDOR lines with respect to the free-proton frequency. It is therefore apparent that Eqns. 3 and 6 do not hold for these complex polycrystalline spectra. The authors (34) were able to derive the relative orientation of the hyperfine tensor of an interacting proton in a model complex from the variation of its coupling value at different field settings in the EPR spectrum. As a prerequisite for the application of these methods a clear observation of an ENDOR coupling across the ESR spectrum is necessary, a condition however, that is frequently not fulfilled for the more complex systems of metalloproteins with a large number of interacting nuclei.

It is also worthwhile to consider some additional features imposed on the ENDOR powder patterns of complex biological macromolecules by their intrinsic flexibility and the possible presence of a large number of conformational substates of the protein. In nitrosyl-ligated myoglobin one observes an EPR powder spectrum of rhombic character (see section 4). The spectral separation of the intermediary and minimum g-turning points is, however, not very pronounced. In such a case it may not be expected to obtain single-crystal-like ENDOR spectra along that minimum g-value. This is illustrated for a nearly axial system with small rhombic distortion ($g_1 \approx g_2$) in Fig. 3, from which it can be inferred that the curves of constant g through the intermediary g_2 are closely approaching the extremal g_1-value. Allowing for a small but significant distribution Δg around the constant g_2-curve (i.e. expanding the line to a band) it seems plausible that nearly identical subsets of orientations will contribute to the ENDOR spectra for the adjacent field settings at g_1 in the EPR spectrum. Consequently very similar ENDOR patterns will be detected along g_1 and g_2 due to the admixture of "pure" g_2 orientational distribution for g_1 and vice versa. The origin of the distribution of magnetic parameters g, $\underset{\sim}{A}$ and $\underset{\sim}{Q}$ may be sought in

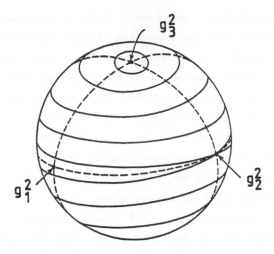

Fig. 3. Curves of constant g-factor for a nearly axial system ($g_1 \approx g_2$). The orientational distribution for g_2 closely approaches g_1.

the flexibility of the complex protein structure. The influence of the variation in magnetic parameters on the EPR and ENDOR spectrum are referred to as g-strain, which is discussed extensively in the chapter by W. R. Hagen in this book.

3.   MATERIALS AND METHODS

The model compound Fe(III)-tetraphenylporphyrin (FeTPP) was obtained from Aldrich. It was dissolved in chloroform and reduced to the ferrous form by rapid mixing with an oxygen-free aqueous solution of sodium dithionite. Addition of a large excess of $NaNO_2$ to the two-phase mixture leads to reduction of nitrite and formation of a five-coordinated NO-Fe(II)-TPP complex. The samples were then filled in quartz tubes in a nitrogen atmosphere and immediately frozen in liquid nitrogen. The completeness of reduction was monitored by the loss of the characteristic g = 6 EPR signal of the ferric complex. The hexacoordinated NO-ligated complex with imidazole (Im) or pyridine (Pyr) as a sixth ligand was produced by adding a fifty-fold molar excess of pyridine or imidazole to the solution followed by rapid mixing. The final concentration of the samples was 7 to 10 mM.

Sperm whale myoglobin was purchased as lyophilized powder from Sigma and dissolved in phosphate buffer at pH 6 to 7.4. Reduction to the ferrous form was achieved by addition of a two- or three-fold excess of sodium dithionite and subsequent purification over a Sephadex G 25 column. Ligation with NO was performed either by exposure of the solution to an atmosphere of nitrosyl gas or by addition of sodium dithionite and sodium nitrite ($NaNO_2$, with ^{14}N or isotopically substituted ^{15}N) in stochiometric amounts. The solutions were filled under oxygen-free conditions into EPR glass tubes and were quickly frozen in liquid nitrogen. Ethylenglycol was used for some samples to encourage formation of glasses (50/50 v/v). Final protein concentrations ranged from 10 % up to 25 %. Single crystals were grown according to the method of Kendrew and Parrish (35) in the aquomet form from solutions in ammonium sulfate ($NH_4 (SO_4)_2$) between pH 6 and 6.4. The crystals exhibited monoclinic shape with a maximal length of 4 to 5 mm along $\underline{b}$. The unit cell contains two molecules (Sites A and B). The crystals were frozen to small teflon holders in a drop of mother liquid or in glass capillaries (for rotation around the elongation axis $\underline{b}$) and mounted to the goniometer to perform rotation in three orthogonal crystal planes. The orientation of the crystals was controlled with a stereomicroscope and by observation of the angular variation of EPR signals from residual aquomet Mb. Misorientation was found to be $\leq 10°$.

Pure hemoglobin (HbA, A = adult) was prepared from concentrated erythrocytes according to the method of Drabkin (36). Separation in α- and β-chains followed a procedure described by Bucci and Fronticelli (37) and modified by Gerraci and Parkhurst (38). β-chains were obtained with high purity, while α-chains were contaminated with residual HbA (20 % to 30 %) as estimated from isoelectric focusing and from EPR spectra of radiation-induced heme-O_2^--complexes at 77 K (39). NO-ligation (with ^{14}NO or ^{15}NO) was identical as described for myoglobin. Inositolhexaphosphate was added prior to NO-ligation to induce transition from the high affinity (R) state to the low affinity (T) state. Sample concentrations ranged up to 30 %.

The g-tensor evaluated from single-crystal EPR-spectra of $Mb^{15}NO$ agreed with the literature (8, 29). ENDOR spectra were recorded with 5° or 10° angular resolution in the frequency range 1 to 30 MHz. The hyperfine and quadrupolar tensors were evaluated to first order according to usual methods (40). The tensorial parameters were then transformed to the heme system of site A for a stereochemical interpretation and compared to X-ray crystallographic data obtained for aquomet-Mb at ambient (41) and liquid nitrogen temperatures (42). In Fig. 4 the conventions used to designate the pyrrole rings and to identify atoms of the porphyrin ring are given. The x- and y-axes are defined by connecting the nitrogens of pyrroles D, B and A, C respectively. The positive z-axis is oriented towards the distal side and points out of plotted plane. All figures involving projections on the heme plane are viewed from the distal side as in Fig. 4.

The ENDOR apparatus employed in our laboratory is based essentially on the components of a Bruker ER 420 EPR spectrometer. The homemade ENDOR probe-

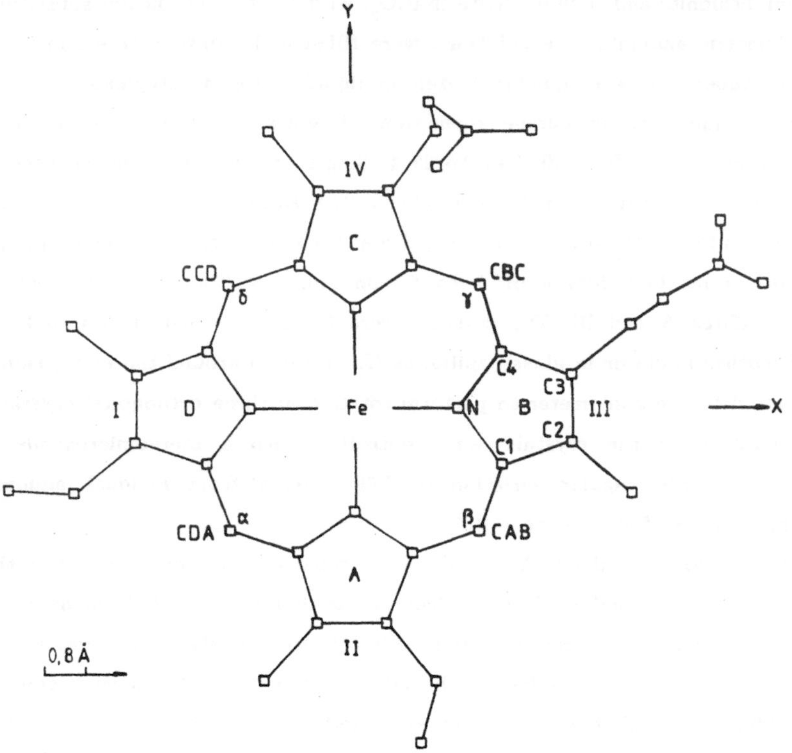

512

Fig. 4. The plot of the heme plane viewed from the distal side designates the pyrrole rings and methine carbons and defines the x- and y-axes of the heme coordinate system.

head consists of an eight turn Helmholtz coil photoprinted on a mylar foil, which is inserted into a Helium flow cryostat (Thor Cryogenics, England) centered in a TE 104 dual cavity. In that arrangement the magnetic components of the radiofrequency and microwave fields are perpendicular to each other and to the static field B. The radio-frequency is generated by a Wavetek 3000 synthesizer and swept repeatedly in the desired frequency range controlled by a computer (Dietz 621/8) that also provides for data acquisition. The rf is frequency modulated with 10 kHZ and the ENDOR signal is recorded as the first derivative. Amplification of the FM modulated rf is achieved with an ENI A150 amplifier to give an rf field B_2 of about 0.1 - 0.2 mT in the Helm-holtz coil dropping at frequencies above 20 MHz. The output of the coil is connected to a dummy load to dissipate the rf energy without reflections. For a more detailed description of the ENDOR set-up see Ref. 13. Typical temperatures for ENDOR measurements varied between 6 and 25 K. Recording a single ENDOR spectrum by multiscan averaging took usually 200 to 600 seconds, in the case of very weak signals up to 2000 seconds.

4.   RESULTS AND DISCUSSION

The well-documented EPR powder spectrum of $Mb^{14}NO$ in frozen solutions is characterized by three rhombic g-components as indicated in Fig. 5, top. Along g_2 the nitrogen hyperfine interactions of the nitroxyl and proximal histidine ligands are poorly resolved. An additional absorption peak is visible at g = 2.041, which is usually related to a minority species with a different ligand-Fe-NO conformation from the major component (43). This absorption peak is also present in the model complex Im-FeTPP-NO (Fig. 5, bottom), which exhibits a more pronounced rhombic g-symmetry and well-resolved hyperfine structures of imidazole and nitrosyl nitrogens (triplet of triplets). ENDOR powder spectra were recorded at various g-factors across the ESR spectrum, which were separated by 0.15 to 0.3 mT depending on the anisotropy of the observed ENDOR line pattern.

In a detailed single-crystal EPR study on $Mb^{15}NO$ Hori et al. (8) determined the g- and the hyperfine tensors of the nitrogen coupling of the NO-ligand. Correlating the direction of the g-component with the least deviation from the free spin value (g_2) to the bond between Fe and the NO-ligand and identifying the minimum hyperfine component (A_{min}) with the N-O bond direction, the authors concluded on a bent "end-on" configuration of the nitrosyl ligand in accordance with X-ray crystal data from HbNO (3) and from NO-ligated Fe(II)TPP model substances (4). An estimate of spin densities yielded a considerable redistribution of spin density onto the heme iron ($\rho_{Fe} \approx 0.8$) away from the ligand ($\rho_{NO} \approx 0.2$). Hori et al. found evidence for a strong temperature-induced reorientation of the NO-ligand between $20°$ C with a more linear arrangement and 77 K with a strongly-tilted configuration and a bond angle of $109°$. No further rearrangements were observed to occur in the temperature range down to 5 K. Nitschke (29) in an independent investigation largely confirmed these findings but gave slightly different directions between $\underset{\sim}{g}$ and $\underset{\sim}{A}$-tensor components at low temperature.

4.1. Nitrogen Interaction in Myoglobin

We have recently completed a single-crystal ENDOR-analysis of $Mb^{15}NO$ that, among others, provided further insight into the orientation of the NO-ligand (13, 18). Although no full tensor could be derived, a definitive discrepancy was found between the estimate given by EPR for the minimum value of the ^{15}N-hyperfine coupling (58.9 MHz) derived from unresolved line width considerations and the respective value of 36.7 MHz derived from partially-resolved ENDOR in the vicinity of the $\underline{c}^*$-axis. The

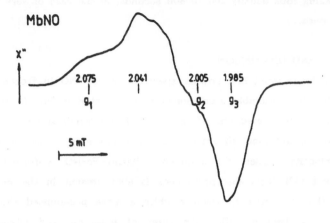

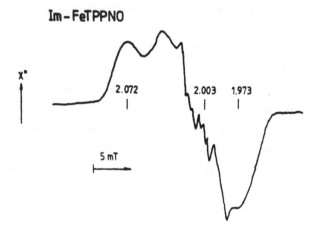

Fig. 5. EPR spectra of MbNO and the model complex Im-FeTPP-NO in frozen solutions (77 K).

latter finding was interpreted in terms of a reorientation of the N-O bond direction towards pyrrole C (IV) in contrast to the previous suggestion involving pyrrole A(II). which was based on the EPR-results (8, 29). With this reorientation another problem is solved that would arise from the previous EPR-assignment, a too short distance (1.1 to 1.25 Å) between O(NO) and H(NE2 His E7) as calculated on the basis of low temperature X-ray data of aquomet Mb (42). It should be noted that in ENDOR-spectra of polycrystalline $Mb^{15}NO$ along g_1, which is the direction close to $\underline{c}^*$ in single crystals, the same small hyperfine coupling value (36.6 MHz) for $^{15}N(NO)$ is obtained, in line with the single-crystal ENDOR observations (13, 44). The discrepancy to EPR single-crystal data must be ascribed to the limited resolution.

The latter factor also prevents the delineation of the ^{14}N-interaction of the bonded nitrogen NE2 of the proximal histidine (F8), which, in single-crystal EPR, can only be observed along very few orientations in the vicinity of $\underline{a}$, indicating the direction of the maximal coupling value, but which otherwise is hidden completely in the broad EPR line width ($\approx$ 3 - 5 mT) (8, 29). The ENDOR-analysis has provided both the hyperfine and the quadrupolar tensors (13, 18). The maximum hyperfine component of 20.4 MHz agrees roughly with the value of 18.8 MHz given from EPR for the $\underline{a}$-direction in single crystals (29). The directions of the maximal $\underline{A}$- and $\underline{Q}$-components were found to be mutually collinear within 6° and within 15° to the Fe-N(NO) bond. Correlating them with the Fe-N(NE2) bond direction yields a tilt of 40° - 50° for that bond. Thus, in two independent measurements, EPR- and ENDOR analysis of Mb[15]NO single crystals have consistently defined the N(NO)-Fe-N(NE2) directions. Both arrive at the same, highly-distorted coordination of proximal and distal ligands prevailing at cryogenic temperatures.

For the direction of the imidazole plane of His F8, the analysis of the quadrupolar tensor has left two alternative orientations, one being parallel to the plane opened by the Fe-N-O bonding scheme derived from previous EPR-single-crystal analysis, which should be discarded on account of the new results on the N(NO) interaction. The other possible orientation coincides with the new Fe-N-O plane resulting from the reorientation of the NO-bond. Both situations are depicted in Fig. 6 as viewed from the side (a) or from a distal position (b). The shaded area depicts the range for the location of O(NO) according to the new results.

Turning now to the powder-type ENDOR-spectra we find that the interpretation of the complex patterns arising from NE2 of proximal histidine (F8) is greatly aided by the single-crystal data analysis. The series of ENDOR spectra in the frequency range 1-11 MHz obtained at various field settings on the powder EPR spectrum of MbNO, is presented in Fig. 7. Along g_1 the prominent ENDOR resonances A and B, which have already been reported earlier in MbNO and HbNO and which were assigned to the proximal histidine nitrogen NE2 (16, 17), dominate the spectrum. The less intense and rather broad signal C was previously shown to be correlated with resonances A and B by a comparison of the resonance patterns in MbNO, HbNO and the monomeric fraction of Glycera dibranchiata hemoglobin (HbGMNO) (13).

In support of this conclusion we have recently observed (18), that the line positions and line shapes of the triplet powder pattern A, B, C are well reproduced by both the summation of single-crystal ENDOR spectra and the projection of line variation onto the frequency scale from an interval of ±30° around the $\underline{c}^*$-axis ($\approx g_1$).

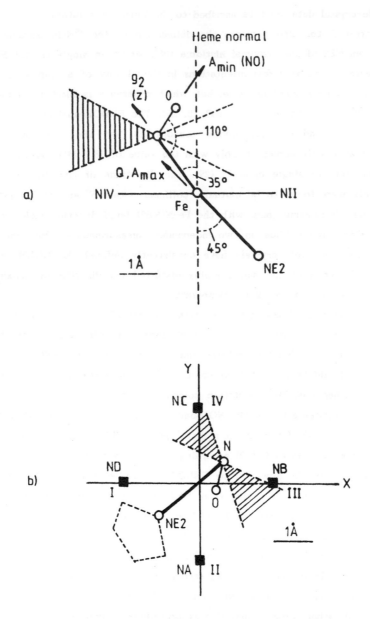

Fig. 6. Bonding geometry of the NE2-Fe-NO complex derived from combined EPR-ENDOR results in a side view cut through pyrroles II and IV (a). The solid lines indicate the orientations of the intermediary g_2-tensor-element (EPR) and of the maximal A- and Q-elements of the NE2-nitrogen interaction (ENDOR). A_{min}(NO) gives the direction of the NO-bond (EPR), while the cone represents possible orientations of the NO-ligand as derived from the minimal observable ENDOR coupling along $\underline{c}^*$-axis. The projection onto the heme plane is indicated in b) showing the imidazole plane in a roughly parallel alignment with respect to the estimated cone of NO-orientation.

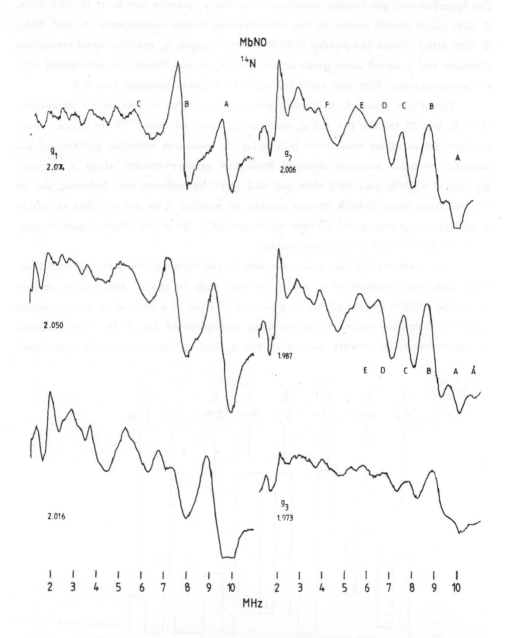

Fig. 7. Sequence of ENDOR spectra along various g-values across the powder EPR spectrum of MbNO. The resonances of the proximal histidine (F8) nitrogen are labeled.

The hyperfine and quadrupolar couplings from the g_1 powder spectrum (A: 15.7 MHz, Q: 0.95 MHz) closely relate to the corresponding tensor components (A: 16.8 MHz, Q: 0.98 MHz). Hence the powder ENDOR spectrum along g_1 exhibits quasi-crystalline character and a set of approximate tensor parameters can directly be determined with sufficient accuracy. This also holds for other hemoglobin specimen (see 4.3).

The triplet group A, B, C evolves to a line pattern of five or six resonances (A' - E, Fig. 7) for the g_2- and g_3-orientation. Since no other ^{14}N-resonances could be identified at higher frequencies (> 11 MHz) the maximum hyperfine splitting of the proximal histidine nitrogen, obtained from EPR powder spectra along g_2 or single-crystal ENDOR data (18.5 MHz and 20.4 MHz) respectively can, however, not be inferred from these ENDOR powder spectra. In addition it is not possible to assign a quartet group among the powder resonances (A' - E) to the third tensor components (A: 15.2 MHz, P: 0.86) unequivocally.

The single-crystal data give some clue to the composition of the powder lines. Since both the directions of g_2 and g_3 are very near to the a- and b-axes, respectively, the ENDOR spectra of the ab-plane can be used in a summation and projection onto the frequency scale. In the resulting histogram of Fig. 8 the line positions of the experimental powder pattern in the g_2 and g_3 region are well reproduced.

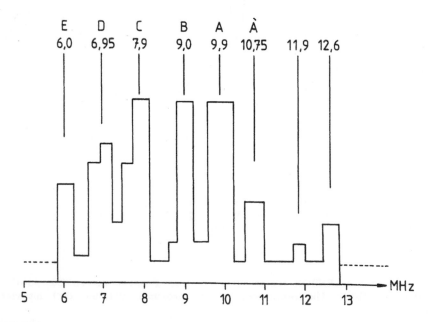

Fig. 8. Projection of line variation in the ab-plane onto the frequency scale. The accumulation frequencies agree well with experimental powder pattern.

With reference to the single crystal the intense accumulation frequencies A to D (Fig. 8) are then inferred to stem from rather isotropic contributions. The anisotropic variations, on the other hand, are distributed in intensity over a wider frequency interval and accumulate significantly only in their terminating regions (e. g. lines E and A'). The accumulated intensity is very low for the most anisotropic variation in the frequency range above 11 MHz, so that their ENDOR response is lost in the "background" and the proton resonances dominate there, at least in the first derivative mode in which broad lines are suppressed. Consequently the resonance patterns for overlapping g_2^- and g_3^-contributions have to be considered phenomonologically just as a "fingerprint" of the interaction of proximal histidine, when attempting to interpret the powder spectra of hemoglobins and hybrids for which no single-crystal data are available.

It is noteworthy that LoBrutto et al. (44) in their ENDOR-study on NO-ligated heme a_3 of cytochrome c oxidase were able to resolve the components of the histidine nitrogen interaction from powder spectra. This probably is due to the employment of absorption mode recording and, more important, to a more collinear arrangement of less anisotropic interaction tensors for that specimen.

4.2 Proton Interactions in Myoglobin

A preliminary account of our investigations of the proton ENDOR-spectra obtained from frozen solution has been first presented some time ago in the context of a descriptive comparison between myo- and hemoglobin (16). An assignment of some of the couplings in the powder system was recently presented on the basis of model system studies and dipolar calculations (13). Among others, an exchangeable interaction due to the distal histidine (E7) NE2-bound proton, a carbon-bound proton on the proximal histidine (F8) and two of the methine coulings could be identified. On some of these couplings, data were also shown previously by LoBrutto and coworkers (44). Very recently, we published the results of a single-crystal analysis of the proton interactions (18). Taking these data together with additional, extensive measurements on model compounds and myoglobin in frozen solutions we now can, for the powder-type system, arrive at a fairly consistent picture of most of the interactions which we wish to delineate here.

A sequence of proton ENDOR spectra along various working points along the powder EPR-spectrum is shown in Fig. 9. The numbering of line pairs along different g-values is according to their relative magnitude and does not imply an assignment to the same proton. The complete set of experimental data from a single crystal is shown in Fig. 10 for the three planes of rotation. Well-resolved lines are indicated by circles, weak lines by the symbol ∩ and broad overlapping resonances by brackets ⊂ ⊃.

Interference of ^{14}N- and ^{15}N-resonances with proton lines is given by the symbol [N.

Let us consider first an H-D exchangeable coupling (XX') of 5.4 MHz along g_1 (2.077.. 2.064). Since this weak, broad line pair cannot be detected in HbGMNO. which lacks the distal His E7 (15), nor in model complexes (19), it is assigned with high confidence to the proton at NE2 nitrogen of His E7 in a strong hydrogen bond to the NO-ligand. By neutron diffraction (45) a hydrogen bond to that proton was shown to exist in O_2-ligated myoglobin. Employing the proton positions derived from room temperature X-ray data (41), it is, however, not possible to correlate the 5.4 MHz coupling to the proton unambiguously. Dipolar calculations indicate that a small but significant shift in proton position of 0.5 Å is required relative to the heme center (13, 19). Just such a dislocation of the proton is provided for in 77 K X-ray data (42) showing, in turn, that the calculations for ENDOR data agree well with X-ray results.

It is interesting to note that the single crystal analysis is only of partial value in this problem. The broken lines of Fig. 10 were simulated for the low temperature position of the His E7 NE2 proton by the reduced spin center method (section 2) and a spin distribution of 70 % and 23 % on Fe and N(NO), respectively. The line variation coincides sufficiently well with the broad extremal resonances at $\underline{c}^*$ ($\approx g_1$) and in the bc^*-plane. However, clearly no match with experimental resonances in the other two planes is obtained. We have shown recently (18, 19) that an orientational distribution of the proton position within ± 0.25 Å around its low temperature coordinates provokes ENDOR-line variations that follow the trend of line width of the extremal resonances in the bc^*-plane, but spread the intensity over a tenfold frequency interval for other orientations (i. e. along the $\underline{a}$ and $\underline{b}$ axes). Consequently, we concluded that the maximal components of this proton interaction are not detectable in the single-crystal **and** powder ENDOR spectra for which only along g_1 ($\approx \underline{c}^*$) sufficient intensity is accumulated.

The estimated range of the proton distribution giving rise to the line width variation is related to the linear root-mean-square (rms) displacement of 0.215 Å, derived from the B-factor of the NE2 nitrogen (42). The rms-value represents the lower limit, since for the ENDOR line width additionally a spatial distribution of the spin centers (i.e. Fe and N(NO)) is superimposed to that of His E7 NE2 proton.

Full tensor evaluations from the single-crystal data for other protons were found to be impeded either by incomplete observation of line variation for larger. more anisotropic couplings or by a severe line overlap for a large number of smaller proton couplings in an interval of ± 2 MHz around the free proton frequency. for which there is too much ambiguity in connecting the line variations in different

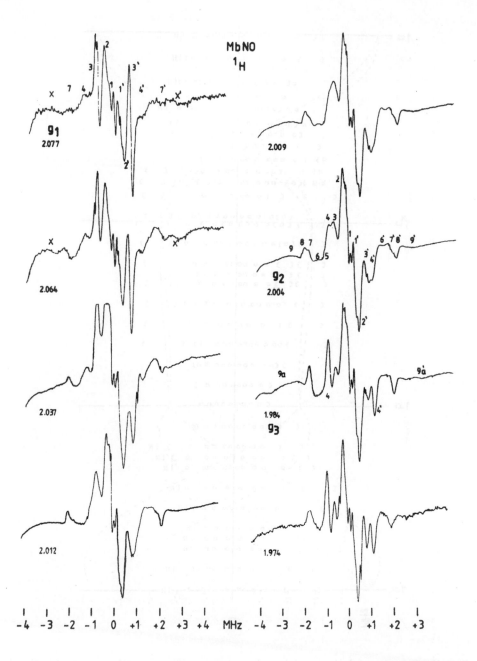

Fig. 9. Sequence of proton ENDOR spectra of MbNO in frozen solutions with line pairs labeled along the principal g-values.

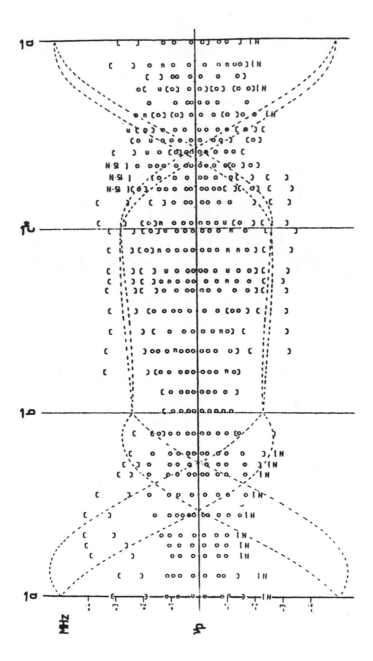

Fig. 10. Experimental data points of proton resonances with respect to the free proton frequency ν_P for the rotation in three crystal planes of Mb^{15}NO single crystals. For definition of symbols see text. The broken lines represent the line variations simulated for the exchangeable proton at NE2 of distal His E7 in a position derived from low temperature X-ray structure (42).

planes. Thus, for identification of other interacting distal protons again reduced spin center calculations were applied, starting from low temperature X-ray data (42). The calculated line variation for one of the methyl protons at CG2 of Val E11 with the lowest z-component is shown here in Fig. 11 to be in good agreement with experimental data points. The calculated maximum coupling of 4.1 MHz is expected to occur at g-factors slightly lower than g_1. This spectral behavior is indeed observed in the powder spectra around g_1 (Fig. 9), where the line pair 77' attains its maximum value at $g = 2.064$; subsequently it disappears and is replaced by the proximal histidine proton couplings 77', 88' along g_2 (see below). The minimal components of 2 MHz expected along g_2- and g_3-orientations are hidden under the intense line pair 44', which, however, can be completely removed in samples extensively equilibrated in D_2O.

In a similar way the line variations for the distal protons at CE1 of His E7 as well as CZ and CE1 of phenylalanine CD1 were determined in single crystals and related to the spectral behaviour of the corresponding powder spectra resonances (18, 19). The simulation parameters are given in Table 1 together with those of the methine protons. Due to the molecular symmetry, these latter protons yield two pairs of inequivalent interactions with maximum couplings of 1.5 to 1.7 MHz. Methine protons HCCD, HCAB acquire their maximal coupling along the $\underline{c}^{*}$ axis, hence indicating that the intense, well-resolved line pair 33' along g_1 of the powder spectra (Fig. 9) is correlated. This is confirmed by the absence of a similar coupling (1.5 MHz) in the six-coordinated model substance Im-FeTPP-NO, which, however, is present in the octaethylporphyrin specimen (44). Besides, dipolar calculations of allowed proton positions for this experimental coupling in powder ENDOR spectra (13) agree very well with the estimated locations of these methine protons from X-ray. For the other pair of methine protons HCDA, HCBC the maximal hyperfine splitting occurs along the b-axis, which deviates from the g_3-orientation by 9°. Therefore line intensities will accumulate along g_3 (and along g_2 due to the powder character) with a hyperfine splitting of ca. 1.5 MHz. Again the associated weak, but well-resolved line pair 33' is seen along g_3 of the powder patterns of MbNO (Fig. 9), which on the other hand were definitely missing in the Im-FeTPP-NO model complex (19). For both proton pairs the minimal tensor components ($\approx$ 0.8 MHz) contribute to the intense central signals (11', 22') and can not be resolved separately. The agreement between the results of dipolar calculations (i.e. simulation of line variations and determination of allowed protons locations) with experimental ENDOR and X-ray data points to a negligable distribution of spin density on the periphery of the porphyrin ring, which therefore should be confined to the orbital system of iron and the axial ligands.

524

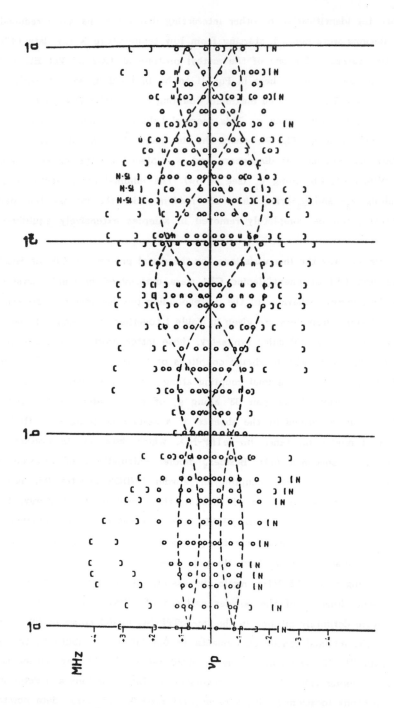

Fig. 11. Simulated line variation for a methyl proton at CG2 of Val E11.

TABLE 1

Simulation parameters of dipolar coupled protons in MbNO

proton at	A_{max} (MHz)	direction[a]	r (Å)[b]
NE2 His E7	10.7	-0.26 0.07 -0.96	2.45
CG2 Val E11	4.3	0.67 -0.48 -0.57	3.32
CE1 Phe CD1	3.05	-0.64 0.26 -0.72	3.73
CZ Phe CD1	1.91	-0.27 0.86 0.43	4.36
CE1 His E7	2.93	-0.57 -0.67 -0.47	3.78
HCDA	1.42	0.71 0.71 0.04	4.80
HCBC	1.75	-0.69 -0.71 0.20	4.49
HCAB	1.68	-0.65 0.68 0.35	4.55
HCCD	1.52	0.73 -0.67 0.14	4.70

[a]with respect to the axes of the heme system as defined in Fig. 4
[b]distance of the protons to the reduced spin-centers

Due to the uncertainties of the imidazole plane orientation in the strongly-tilted bonding configuration compared to the X-ray structure of aquomet Mb, a direct assignment of simulated line variations of carbon-bound CE1, CD2 protons and the exchangeable ND1 proton of proximal His F8 to single-crystal ENDOR data remains ambiguous. There is, however, experimental evidence from model complexes with deuterium-substituted imadazole or pyridine as axial ligands, which relates the

526

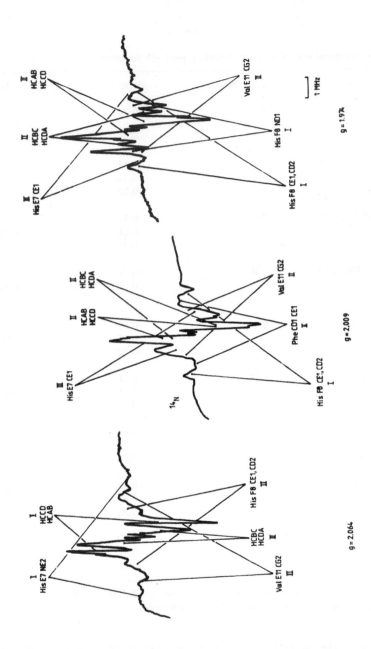

Fig. 12. Summary of assignments for proton couplings in MbNO along three representative g-orientations to various amino acid residues. The roman numbers I – III indicate the quality of assignment (I: certain, highly probable, II: probable, III: tentative).

lines 44` along $g_3$ to the proton at ND1 and line pairs 77`, 88` along g_2 and g_3 (Fig. 9) to the carbon bound protons (13, 19, 44). We recently suggested by reconstructing specific spectral characteristics of powder spectra along g_3 from superimposing single-crystal ENDOR spectra of the ab-plane, that the broad weak line pairs 99', 9a9a' along g_2 and g_3 (Fig. 9) are probably representing the weaker accumulation frequencies of the maximal component from CD2, CE1 protons of His F8 (18). The considerable isotropic components of the tentative tensors derived for these interactions indicate that some spin density is distributed within the imidazole ring, on which a total spin density of 7 % to 10 % should reside (18, 19).

The combination of all the available information for the MbNO system thus allows for a detailed assignment of the proton interactions in frozen solutions, which is summarized in Fig. 12. The couplings are grouped in three classes, I to III, depending on the quality of the individual assignment. Class I contains the lines that were identified by experimental evidence (e.g. H-D exchange, comparison with model complexes, other hemoglobin species and single-crystal data) and for which dipolar calculations yield a good agreement with the coordinates of the low temperature X-ray structure. Their assignment is unequivocal or highly probable. Protons, for which resonances in the powder spectra exhibited a spectral behaviour that could be predicted by single-crystal analysis belong to class II. Class III collects tentative assignments, for which contributions to the powder ENDOR spectra are expected from simulations, but for which no line pairs are resolved separately in the various specimen.

4.3 ENDOR on Hemoglobin

The frozen solution EPR spectrum of HbNO in the "high affinity" (R) state with poorly-resolved hyperfine structure of ^{14}NO and proximal histidine ligand at g = 2.009 (Fig. 13, top) closely resembles that of MbNO (cf. Fig. 5 top). Addition of the allosteric effector inositolhexaphosphate (IHP) forces the hemoglobin to adopt the "low affinity" (T) quaternary state by binding between the two β-chains (4). As first described by Rein et al. (46) the EPR spectrum of HbNO(T) shows a distinct triplet of the ^{14}NO interaction at g = 2.01 and some hyperfine structures at g = 2.075 (Fig. 13, middle tracing) but no nitrogen coupling of proximal histidine. The similarity to EPR spectra of five-coordinated model complexes (9, 47, 48) and the appearance of a second IR stretching mode of the NO-bond in HbNO(T), which was found to be identical in Fe-porphyrin-NO systems, suggested a stretching or breaking of proximal histidine bond for half of the heme centers (2). Optical measurements on hybrids, with proto- and mesoheme-substituted α- and β-chains revealed the structural changes to occur mainly in the α-chains of the tetramer (49). In optical and EPR studies

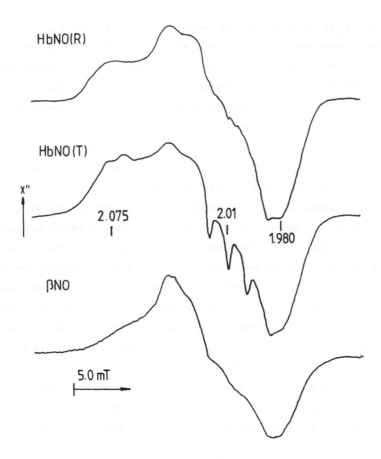

Fig. 13. EPR spectra of frozen solution of HbNO in the quaternary R-state (top), of HbNO in the presence of IHP in the T-state with a characteristic triplet hyperfine structure of a five-coordinated species (middle) and of isolated βNO-chains (bottom).

on hybrids with one type of subunits NO-ligated (i.e. αNOβX and αXβNO with X = deoxy, CO, O_2, F^-, H_2O, N_3^-, CN^-) it was demonstrated that the α-chains are sensitive to the spin state and the tertiary structure of the β-chains but not vice versa (11, 50, 51).

Isolated αNO- and βNO-subunits showed different EPR spectra at physiological pH values (pH 7.4), which could be summed to yield the EPR spectrum of HbNO(R) (52). The spectrum of βNO-chains, given in Fig. 13 (bottom), proved to be insensitive to changes within a pH range from 6.0 to 8.0 and to the presence of IHP, whereas what is termed a "five-coordinated" EPR pattern appeared for αNO-chains at low pH values (pH 6.0) (46).

Information on the heme-ligand configuration in HbNO is available from a low resolution X-ray structure, albeit at room temperature only, which indicates an orientation of N-O towards pyrrole A(II) with a bond angle of 145° (3). In a low temperature (77 K) single-crystal EPR analysis slightly differing NO-bond angles of 137° and 130° for α- and β-chains and deviations of 10° and 8° of the Fe-N(NO)-bond from the heme normal were determined. Spin densities were estimated with values of 20 - 30 % at NO and 80 - 90 % at Fe (10). Again the hyperfine interactions of proximal histidine nitrogens could not be resolved.

Previous ENDOR-studies of NO-ligated hemoglobin and its subunits involved frozen solutions in a wide range of physiological conditions (pH, allosteric effectors) (16, 17). Among others, the loss of nitrogen resonances from proximal histidine was found and taken to be indicative for cleavage of this bond to the heme iron in isolated α-chains at low pH (tertiary t-state). Indirect evidence for a disrupted bond to His F8 in a partially saturated T-state nitrosyl-hemoglobin was provided recently by electron spin echo envelope modulation (ESEEM) techniques (53). For the proton interactions, the hydrogen bond between the NO-ligand and the distal histidine proton could be assigned in the previous study (13). On the whole, however, the interpretation was impeded by the ambiguities arising from the powder character of the samples, so that spectral features of various hemoglobin species had to be compared in a phenomenological way and assignments of most interactions remained tentative. The detailed analysis of powder ENDOR spectra available now for the model protein MbNO together with extended measurements of hemoglobin appears to warrant a new comparison of spectra of R- and T-state HbNO with those of MbNO and with each other with the aim of an extended understanding of the interactions in HbNO.

4.3.1 Nitrogen interactions in hemoglobin

In HbNO(R) the interaction of the proximal histidine NE2 nitrogen exhibits very similar ENDOR line patterns in the 1 - 11 MHz region as in MbNO with the characteristic triplet along g_1 and five resonances along g_2- and g_3-orientations. The line width is slightly increased and the resolution of lines along g_2, g_3-orientations is less pronounced. The positions of lines A and B of the triplet along g_1 yield a hyperfine coupling of 15.85 MHz and a quadrupolar value of 1.0 MHz, which both are somewhat larger than in MbNO as listed in Table 2. The "fingerprint" resonances along g_2 and g_3 are systematically shifted by 0.1 to 0.3 MHz to higher frequencies, the expected maximum lines above 11 MHz are not detectable. These findings indicate that the arrangement of the proximal NE2-interaction tensors relative to the g-tensor is nearly identical in HbNO(R) and MbNO. The concerted shift of lines suggests that anisotropic components are conserved but that the isotropic contribution is slightly increased. This seems to be related to an increased s-spin density on NE2 His F8

TABLE 2

Couplings of the nitrogen NE2 of the proximal His F8 in different hemoglobins

		g_1			g_2		g_3	
		LP[a]	LW[a]	A[a] Q[a]	LP	LW	LP	LW
MbNO	A	9.80	0.4		9.80	0.5	9.85	0.5
	B	7.80	0.4	15.70 ± 0.1	9.00	0.5	9.10	0.6
	C	5.95	0.7	0.95 ± 0.1	7.75	0.4	7.95	0.4
	D				6.80	0.5	7.00	0.5
	E				5.80	0.7	5.90	0.5
HbNO (R)	A	9.95	0.5		9.85	1.2	9.65	0.9
	B	7.95	0.6	15.85 ± 0.1	9.25			
	C	5.90	0.7	1.0 ± 0.1	7.95	0.5	8.20	0.7
	D				7.00	0.6	7.30	0.6
	E				6.00	0.6	6.10	0.6
HbNO (T)	A	9.95	0.7		9.85	1.3	9.50	1.1
	B	7.95	0.7	15.90 ± 0.2	9.30			
	C	6.00	1.2	1.0 ± 0.2	7.95	0.6	8.20	1.4
	D				7.00	0.6	7.20	
	E				6.00	0.5	6.10	0.7
β-NO	A	10.20	0.7		10.00	1.4	10.45	1.0
	B	8.20	0.9	16.45 ± 0.2	9.40		9.35	0.8
	C	6.30	1.1	0.98 ± 0.1	8.05	0.9	8.30	0.8
	D				6.80	0.8	6.90	0.6
	E				6.10	0.6	6.1	0.6

[a]in MHz
LP = line position, LW = line width, A = hyperfine coupling
Q = quadrupole splitting

indicative of a stronger bond to iron in HbNO(R). Compared to MbNO, the less distorted bond geometry of HbNO(R) (10) can be taken to cause increased d_{z^2} orbital contributions leading to the enhanced proximal NE2 spin density. Some minor spectral differences to MbNO (e.g. line width, resolution) at various intermediary ENDOR working points are probably due to the presence of slightly differing tensor arrangements in both α- and β-chains, which otherwise are not discernible.

For the isolated βNO chains similar powder patterns as for HbNO(R) but with less resolution are obtained. It is, however, obvious from the data given in Table 2 that the "tensor component" along g_1 is increased significantly by 0.6 MHz and that there is a concomitant frequency shift of resonances for g_2 and g_3. This implies an increased bond strength in βNO in the tertiary r-state (pH 7). The change in line positions within the g_2, g_3-powder patterns (Table 2) additionally indicates a variant arrangement of hyperfine and quadrupolar tensors relative to the g-tensor in this specimen compared to HbNO(R). These varied spectral parameters may be attributed to the absence of the structural interaction between the subunits of tetrameric HbNO(R), which confirms a previous proposal (16, 17). Remarkably, the spectral characteristics of βNO-chains closely resemble those of Im-FeTPP-NO model complexes, in which restraints from the globin moiety are absent (19). Hence, the proximal ligand seems to adopt a similar relaxed configuration in isolated β-NO-chains due to a modified tertiary structure compared to the chains embedded in the tetramer. As a consequence, results from isolated chains may not be suited to unequivocally identify spectral contributions of subunits in the tetrameric protein. This task has to be tackled using asymmetrically-ligated hybrids.

Considering now the NE2-interactions of proximal histidines in HbNO(T) we find resonance patterns identical to those in HbNO(R) apart from some minor shifts (0.05 MHz) in line positions (Table 2). Since a second nitrogen coupling of reduced hyperfine values is not detected, the NE2-interaction in α-chains is either fully conserved or is completely lost. The latter possibility is in line with experimental evidence from ENDOR on isolated α-chains in tertiary t-state and from ESEEM on partially saturated NO-hemoglobin (16, 53). This is also suggested from preliminary results obtained in αNOβdeoxy-hybrids (in T-state) (54). Consequently the nitrogen resonances in HbNO(T) should arise from β-chain contributions only, which then are insensitive to the change in quaternary conformation induced by the presence of IHP and which are significantly different from isolated β-chains.

The gross changes observed for the NO-interaction in the EPR spectra upon transition from R- to T-state (cf. Fig. 13) are reflected in the corresponding ENDOR spectra. In HbNO(R) only along g_1 broad and weak [14]NO-resonances are visible, yielding a hyperfine coupling of 37 MHz. Compared to the subunit values from single-crystal ESR-results this is larger than the minimal components of 29 MHz (α) and 26 MHz (β) and between the intermediate components of 33 MHz (α) and 44 MHz (β) (10). This discrepancy may either be due to the presence of additional [14]NO-resonances, which are masked by the intense proton signals between 11 and 17 MHz or to a pronounced noncollinearity of this hf-component with respect to the g-tensor (10), which may give rise to accumulation frequencies in the powder spectra that are not related to a tensor component.

In contrast to this for HbNO(T) intense [14]NO-resonances are resolved between g = 2.08 and 2.01 of the EPR-spectrum, which are absent in [15]NO-substituted

532

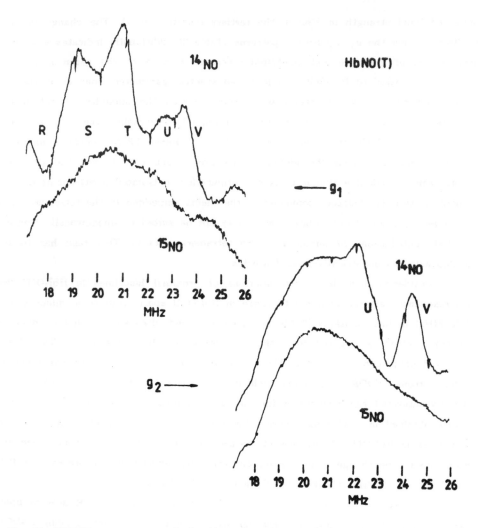

Fig. 14. ENDOR resonances of the ^{14}NO ligand in HbNO(T) along g_1 = 2.075 (top) and g_2 = 2.01 (bottom), which are lost upon substitution with ^{15}NO (lower tracings). The baseline is distorted in this frequency range.

samples as shown in Fig. 14. For g_2 = 2.01 resonances U and V yield a coupling of 47.7 MHz that agrees well with the maximum value of 46.2 ± 1.5 MHz as derived from the resolved triplet hyperfine structure of the EPR spectrum. For the five resonances R – V along g_1 several interpretations appear resonable. First, if all lines are assumed to stem from the five-coordinated species, only an approximate hyper- fine value of 39 or 40 MHz can be estimated due to the distinct powder character. Alternatively, lines R and S may be assigned to the six-coordinated β-chains, since similar lines with A = 37.7 MHz are present in HbNO(R). The resonances T – V should then be ascribed to accumulation frequencies of ^{14}NO-interactions in the α-chains, from which only an approximate hyperfine value of 45 MHz is derived. In any case these values are consistent with a roughly-estimated hyperfine splitting of 38.7 ± 3 MHz for the poorly-resolved structure along g_1 of the EPR spectrum (Fig. 13). Again a more precise assignment of α- and β-chain contributions as well as a more complete determination of hf-couplings can be expected from NO-hybrid studies.

Along the g_2-orientation, which represents the Fe–NO bond direction in the five-coordinated species, obviously the powder character of the ENDOR spectra is minimal due to an approximately coaxial arrangement of interaction tensors, whereas the ENDOR spectra of g_1 are dominated by powder contributions. Since no smaller ^{14}NO-couplings were recorded the line pattern at g_1 contains the minimal hyperfine component (≈ 40 MHz), from which an isotropic part of ca. 44 MHz is estimated. This value is close to the isotropic component of 42 MHz (α) and 44 MHz (β) in HbNO(R) derived from single-crystal ESR-analysis (10). The anisotropy, however, of the hyperfine interactions is reduced from ca. 18 MHz in HbNO(R) to ca. 6 MHz in HbNO(T). Therefore the ENDOR intensity of accumulation frequencies of terminating regions and of isotropic contributions is concentrated within a smaller frequency interval for HbNO(T), so that the broad NO-resonances of the pentacoordinated species are more easily detected in the first-derivative mode of FM-modulated ENDOR.

The estimated s-spin density of 2.9 % in both five- and six-coordinated α-chains remains constant, the p-spin density on the other hand reduces from 24 % in the six-coordinated α-chains (10) to ca. 7 %, yielding a total spin density of 10 % on the NO-ligand in α-chains. This would require a more intense orbital overlap between NO and Fe, which may relax from a position below the porphyrin plane to a position in or above that plane after loss of proximal histidine interaction.

4.3.2 Proton interactions in hemoglobin

The proton ENDOR spectra of HbNO(R), presented in Fig. 15 (left column), exhibit a line pattern very similar to MbNO (cf. Fig. 9). Some resonances are not as well resolved, a feature which is probably related to the superposition of α- and β-chain contributions. The exchangeable proton of the distal histidine E7 shows an identical coupling of 5.4 MHz at g_1 (XX'), but a more anisotropic variation is found on other working points before it vanishes, indicating a slightly different orientation for the hf-tensor relative to the g-tensor. The interaction of one of the methyl protons at CG2 of Val E11 (77') has broadened either by superposition of α- and β-chain couplings or by an increased spatial distribution in HbNO(R). The methine coupling (33') is identical to MbNO. The broad resonances 44', to which the proxi-

534

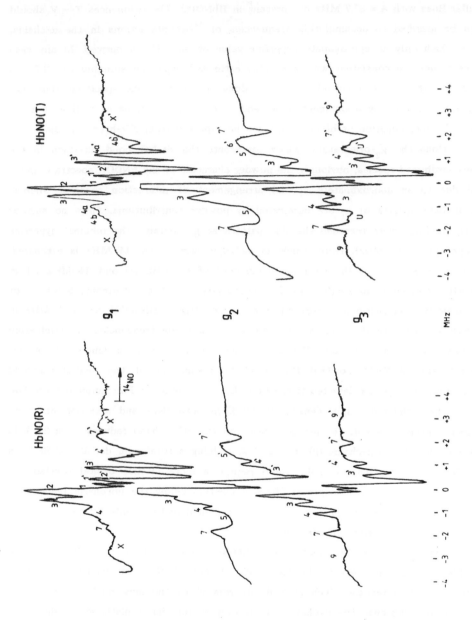

Fig. 15. Proton ENDOR spectra of HbNO(R) (left column) and HbNO(T) (right column) in frozen solutions along the principal g-values.

mal histidine protons at CE1 and CD2 are contributing, are more pronounced. For this interaction also an increased hyperfine coupling (3.8 MHz vs. 3.45 in MbNO) is observed along g_3 (77'). In addition the exchangeable proton of ND1 His F8 is slightly larger (2.10 MHz vs. 2.0 MHz in MbNO). Both results are in accord with the slightly larger NE2 interaction of His F8 in HbNO(R) (section 4.1), which is indicating an increased delocalization of spin density onto the histidine.

The interaction of the CE1 proton of phenylalanine CD1, which is seen along g_2 in MbNO with a hyperfine coupling of 3.0 MHz, is not resolved in HbNO(R). Instead, broad and weak resonances (55', Fig. 15) with 2.5 MHz splitting are observed, for which a distance increased by 0.2 Å to the spin centers is estimated from dipolar calculations. It is known from X-ray analysis of HbO_2 (55), that Phe CD1 is further removed from the heme plane than in MbNO. On the other hand, a comparison of room temperature and low temperature X-ray studies of aquomet Mb (41, 42) reveals an extraordinary large temperature-induced shift of this residue towards the heme. Assuming such effects to occur for HbNO also, it seems plausible that the increased distance is the result of an originally more remote position partially compensated for by the temperature-induced shrinkage of the protein. However, since the minimal components of Val E11 CG2 protons are expected to accumulate ENDOR intensity in the same frequency range as the Phe CD1 coupling, the line pair 55' may be composed from contribution of both these distal residues.

Turning now to the proton ENDOR spectra of HbNO in the quaternary T-state (Fig. 15, right column), we observe along g_1 that the broad resonances of the exchangeable proton at NE2 His E7 are present, exhibiting an identical spectral behaviour as in HbNO(R). It is not clear at present, if these lines are caused both by α- and β-subunits or by one subunit alone. Two well-resolved line pairs, labeled 4a4a' and 4b4b', replace the broad and unstructured resonance 44' in HbNO(R). Probable candidates are protons of Val E11 or Phe CD1 in the pentacoordinated α-chains or carbon-bound protons of the proximal histidines in the β-chains. Exchangeable protons (i.e. at ND1 of His F8, NE2 of His E7) can be excluded from experiments on samples equilibrated in D_2O. The proximal His F8 of the α-NO-chains is ruled out by the presence of these couplings in HbMIwate-NO, in which the proximal His F8 of the α-chains is replaced by a tyrosin (56). A unique identification of these interactions is expected from studies on NO-hybrids, which allow for a separation of α- and β-chain contributions.

A comparison of the spectra along g_2 (Fig. 15) reveals some differences concerning the line pairs 33' and 44'. The hyperfine splitting of 33' (1.30 MHz) in HbNO(R) is 0.2 MHz smaller than the coupling in HbNO(T), so that line pair 44', which originates from the exchangeable ND1 proton of His F8, is not overlapped. The line shape of 33' in HbNO(T) closely resembles that in MbNO (Fig. 12). Since for MbNO distal protons of His E7 and Val E11 were suggested to contribute to this line pair, the arrangement of these residues seems to be nearly identical in HbNO(T), thus implying a more tense conformation of the distal side compared to HbNO(R). This view is supported by the presence of the weak line 6' (Fig. 15), which was assigned to the CE1 proton of Phe CD1, and which is more clearly resolved in HbNO(T) along slightly higher g-factors. Therefore the Phe CD1 residue is presumably located in a similar position with respect to the spin centers in HbNO(T) and MbNO.

Along g_3-orientations a novel coupling UU' of 3.0 MHz is visible, which was not found in MbNO and HbNO(R). Indications for its assignment are provided by a comparison of isolated αNO-chains in their tertiary r- and t-states, which has been reported earlier (16). Those show the substitution of a 3.6 MHz coupling in r by a smaller one of 2.7 MHz in t-state. Combined with the knowledge that the former coupling corresponds to that of a line pair 77' in HbNO(R) which is due to carbon bound protons of proximal His F8 these protons could, in the "five-coordinated" α-subunits in HbNO(T), serve as a source for the UU' interaction. Assuming a cleavage of the bond of Fe to His F8 in the αNO-chains of HbNO(T) no spin density should reside on the imidazole ring, so that the proton interactions must be purely dipolar. Calculations yield a maximal distance of 3.3 Å from the heme iron with 80 % spin density. The anisotropic behaviour of UU', which is only visible along g_3 with a low intensity appears to imply a purely dipolar character of these protons. The difference in coupling values between UU' (3.0 MHz) and the hyperfine splitting of 2.70 MHz in the isolated t-state αNO-chains is probably related to a different F-helix conformation of the α-chains embedded in the tetramer. Additional evidence for the assignment of UU' to protons of His F8 in α-chains of HbNO(T) comes from experiments on HbMIwate-NO, for which this line pair is not observed, probably due to larger distance of protons at the proximal tyrosyl residue, replacing the histidine in the α-chains (56). The identity of the EPR spectra of HbNO(T) and this mutant species is indicative for the prevailance of the low affinity T-state in HbMIwate.

It is also noteworthy, that in isolated αNO-chains the loss of an exchangeable 1.9 MHz coupling was observed upon r-t transition. The interaction was tentatively ascribed to the distal histidine proton (17). However, with the unambiguous assignment of line pair 44' ($\approx$ 2.0 MHz) in MbNO and HbNO(R) to the ND1 proton of His F8 available now, the above finding fits better with the cleavage of the proximal histidine from Fe in the α-chains, since an increased distance to the spin centers and vanishing isotropic interactions due to zero spin density on His F8 should render this interaction very small and undetectable. This interpretation, on the other hand, implies that the line pairs 99', 77' and 44' (g_2, g_3) (Fig. 15) have to be correlated to the proton interactions of the proximal histidine of the β-chains in HbNO(T). Additional evidence for this is provided by the fact that no noticeable spectral changes were observed upon a r-t transition of isolated β-chains (17).

A clear and unique distinction of the spectral contributions from a changed NO-bonding in the α-chains of HbNO(T) cannot be achieved by comparison of the available data from HbNO(R), HbNO(T) and isolated subunits alone. Among others, the interaction of the exchangeable proton at NE2 of distal His E7 in the α-chains remains unidentified, for which a significantly reduced coupling is expected due to shift of spin density to iron away from the NO-ligand in pentacoordinated αNO-chains. On the other hand a small dislocation of the Fe-NO complex towards the distal hemisphere might easily compensate the effect of a smaller spin density on NO. For this reason also the assignment of interactions to other distal protons in HbNO(T) have to be considered as tentative. A more conclusive characterization of proton interactions in α- and β-subunits is expected from NO-hybrids in the quaternary T- and R-state presently under way in our laboratory.

The picture emerging so far from the extensive set of data indicates a very similar arrangement of amino acid residues in the vicinity of the heme in HbNO(R) and MbNO. Minor spectral differences are related to a more intense bonding of the proximal histidine and slightly increased distances of some dipolar-coupled distal protons to the spin centers. This is surprising at first hand considering the less distorted NO-ligand bonding geometry (10) and a less tightly-packed heme crevice (55) in HbNO(R). One has to bear in mind, however, that the temperature-induced shrinkage of the protein may largely compensate structural differences prevailing at ambient temperatures for the proteins. This also implies that the stereochemical interpretation obtained in low temperature experiments may not be directly transferred on the functional protein under physiological conditions. Within the low temperature regime the transition of the tetrameric HbNO to a T-like conformation reveals some major changes both in EPR and ENDOR spectra, which are indicative for a loss of the bond between Fe and the proximal histidine (F8). From the experimental findings it is also apparent that interactions with distal protons are involved in structural changes in the T-like state, which, however, cannot be assigned unambiguously with the present data set.

5. CONCLUSIONS

The application of ENDOR spectroscopy with its capability of high spectral resolution to NO-ligated hemoglobin and myoglobin has greatly expanded the knowledge about structural details of the allosteric core in these paramagnetic models of the oxygenated proteins. ENDOR studies on single crystals of MbNO revealed a similar, low-temperature-induced distortion of the bonding configuration of the proximal histidine (F8) as was suggested previously for the NO-ligand in EPR experiments. Moreover, compared to the latter results a different equilibrium orientation of the NO-bond direction, minimizing sterical interactions with distal histidine (E7) in its low temperature configuration is suggested by ENDOR results. Combined with extensive model compound studies, a mutually consistent picture could be established for the proton interactions on the proximal and the distal side of the heme-NO complex. This led to a fairly confident assignment of the ENDOR-spectra of both nitrogen and proton interactions in frozen solutions of MbNO.

On the basis of these results, the powder-type ENDOR-spectra of tetrameric protein HbNO in the R-state could by and large be identified in its proton and nitrogen interactions. The very important question of their changes upon the quaternary transition to the low affinity T-state still remains unsolved. Only tentative proposals can be presented yet. This is partially due to the uncertainty in the transfer of results from isolated α- and β-chains in their corresponding tertiary states to the tetrameric protein. Moreover, it exemplifies the limitations of ENDOR-spectroscopy on complex macromolecules. In order to fully exploit the high spectroscopic resolution in the elucidation of structural parameters, a whole array of variations in sample parameters is necessary for making the ENDOR-information unequivocal. For the NO-ligated hemoglobin problem we feel that much of the foundation necessary is now available so that asymmetrical hybrids should yield most of the additional information needed to unravel the final assignments.

ACKNOWLEDGEMENTS

Most of the results reported from the authors laboratory were obtained with support by grants from the Deutsche Forschungsgemeinschaft. We gratefully acknowledge the help of Dr. H. Zorn, University of Regensburg, in the preparation of model compounds and of Prof. Dr. Kohne, University of Ulm, for a gift of HbMIwate. We also thank Prof. J. D. Satterlee for stimulating discussions. Valuable help in the preparation of the manuscript came from R. End and J. Marx.

REFERENCES:

1   M. F. Perutz, J. V. Kilmartin, K. Nagai, A. Szabo and S. R. Simon, Biochemistry, 15 (1976) 378-387.
2   J. C. Maxwell and W. S. Caughey, Biochemistry, 15 (1976) 388-396.
3   J. F. Deatherage and K. Moffat, J. Mol. Biol., 134 (1979) 401-417.
4   P. L. Piciulo, G. Rupprecht and W. R. Scheidt, J. Am. Chem. Soc., 96 (1974) 5293-5295.
5   J. J. Hopfield, R. G. Schulman and S. Ogawa, J. Mol. Biol., 61 (1971) 425-443.
6   E. Trittelvitz, K. Gersonde and K. H. Winterhalter, Eur. J. Biochem., 51 (1975) 33-42.
7   B. B. Wayland and L.W. Olson, J. Am. Chem. Soc. 96 (1974) 6037-6041.
8   H. Hori, M. Ikeda-Saito and T. Yonetani, J. Biol. Chem., 256 (1901) 7849-7855.
9   H. Kon, J. Biol. Chem., 243 (1968) 4350-4357.
10 S. G. Utterback, D. C. Doetschman, J. Szumowski and A. K. Rizos, J. Chem. Phys., 78 (1983) 5874-5880.
11 Y. Henry and R. Banerjee, J. Mol. Biol., 73 (1973) 469-482.
12 A. Schweiger, in: Structure and Bonding, 51, Springer-Verlag Berlin, 1982.
13 J. Hüttermann and R. Kappl, in: H. Sigel (Ed.), Metal Ions in Biological Systems, Marcel Dekker Inc, New York, Basel, 1987, Vol. 22, pp 1-80.
14 M. F. Perutz, G. Fermi, B. Luisi, B. Shaanan and R. C. Liddington, Cold Spring Harbor Symposia on Quantitative Biology, Vol. LII, 1987, pp. 555-565.
15 E. A. Padlan and W. E. Love, J. Biol. Chem., 249 (1974) 4067-4068.
16 M. Höhn, J. Hüttermann, J. C. W. Chien and L. C. Dickinson, J. Am. Chem. Soc., 105 (1983) 109-115.
17 M. Höhn, Thesis, Universität Regensburg, 1982.
18 R. Kappl and J. Hüttermann, Israel J. Chem., 1988 (in press).
19 R. Kappl, Thesis, Universität Regensburg, 1988.
20 H. Muto and M. Iwasaki, J. Chem. Phys., 59 (1973) 4821-4829.
21 K. Thuomas and A. Lund, J. Magn. Reson., 18 (1975) 12-21.
22 C. P. Scholes, A. Lapidot, R. Mascarenhas, T. Inubushi, R. A. Isaacson and G. Feher, J. Am Chem. Soc., 104 (1982) 2724-2735.
23 L. Kevin and L. D. Kispert, in: "Electron Spin Double Resonance Spectroscopy", John Wiley and Sons, New York, 1976.
24 D. S. Schonland, Proc. Roy. Soc., 73 (1959) 788-792.
25 S. K. Misra, Physica, 124 B (1984) 53-57.
26 D. W. Marquardt, J. Soc. Ind. Appl. Math., 11 (1963) 431-438.
27 C. P. Keijzers and P. Snaathorst, Chem. Phys. Letters, 69 (1980) 348-353.
28 N. M. Atherton and A. J. Horsewill, J. Chem. Soc. Faraday II, 76 (1980) 660-666.
29 W. Nitschke, Diploma Thesis, Universität Regensburg, 1982.
30 G. H. Rist and J. S. Hyde, J. Chem. Phys., 52 (1970) 4633-4643.
31 B. M. Hoffmann, J. Martinsen and R. A. Venters, J. Magn. Reson., 59 (1984) 110-123.
32 B. M. Hoffmann, R. A. Venters and J. Martinsen, J. Magn. Reson., 62 (1985) 537-542.
33 G. C. Hurst, T. A. Henderson and R. W. Kreilick, J. Am. Chem. Soc., 107 (1985) 7294-7299.
34 T. A. Henderson, G. C. Hurst and R. W. Kreilick, J. Am. Chem. Soc., 107 (1985) 7299-7303.
35 J. C. Kendrew and R. G. Parrish, Proc. Roy. Soc. A., 238 (1956) 305-324.
36 D. L. Drabkin, J. Biol. Chem., 164 (1946) 703-723.
37 E. Bucci and C. J. Fronticelli, J. Biol. Chem., 240 (1965) 551-552.
38 G. Geracci, L. J. Parkhurst and Q. H. Gibson, J. Biol. Chem., 244 (1965) 4664-4667.
39 R. Kappl, M. Höhn-Berlage, J. Hüttermann, N. Bartlett and M. C. R. Symons, BBA, 827 (1985) 327-343.

40   J. E. Wertz and J. R. Bolton, "Electron Spin Resonance: Elementary Theory and Practical Applications", McGraw-Hill, New York, 1972.
41   C. H. Watson, Prog. Stereochem., 4 (1969) 299-333.
42   H. Hartmann, F. Parak, W. Steigemann, G. A. Petsko, D. Ringe Ponzi and H. Frauenfelder, Proc. Nat. Acad. Sci. USA, 79 (1982) 4967-4971.
43   R. H. Morse and S. I. Chan, J. Biol. Chem., 255 (1980) 7876-7882.
44   R. LoBrutto, Y. Wei, R. Mascarenhas, C. P. Scholes and T. E. King, J. Biol. Chem., 258 (1983) 7437-7448.
45   S. E. V. Phillips and B. P. Schoenborn, Nature, 292 (1981) 81-82.
46   H. Rein, O. Ristau and W. Scheler, FEBS Lett., 24 (1972) 24-26.
47   H. Kon, BBA, 379 (1975) 103-113.
48   T. Yoshimura, Inorg. Chim. Acta, 125 (1986) 27-29.
49   K. Nishikura and Y. Sugita, J. Biochem., 80 (1976) 1439-1441.
50   K. Nagai, H. Hori, S. Yoshida, H. Sakamoto and H. Morimoto, BBA, 532 (1978) 17-28.
51   R. Cassoly, J. Mol. Biol., 98 (1975) 581-595.
52   T. Shiga, K. J. Hwang and I. Tyuma, Biochemistry, 8 (1969) 378-382.
53   R. S. Magliozzo, J. McCracken and J. Peisach, Biochemistry, 26 (1987) 7923-7931.
54   A. Kreiter, personal communication.
55   B. Shaanan, J.Mol. Biol., 171 (1983) 31-59.
56   R. E. Dickerson and I. Geis, "Hemoglobin", The Benjamin/Cummings Publishing Company Inc., Menlo Park, California, 1983.

CHAPTER 15

ELECTRON NUCLEAR DOUBLE RESONANCE (ENDOR) OF METALLOENZYMES

BRIAN M. HOFFMAN, RYSZARD J. GURBIEL, MELANIE M. WERST, MOHANRAM SIVARAJA

1. INTRODUCTION

Electron nuclear double resonance (ENDOR) spectroscopy is a multiple magnetic resonance technique in which a nuclear magnetic resonance transition at a site within a paramagnetic center is detected by its effect on the EPR signal of the center (1).

ENDOR spectroscopy can be used to determine the electron-nuclear hyperfine, quadrupole, and Zeeman interaction tensors of a particular site (2,3), but in principle much of this information would appear to be obtainable by single-resonance techniques. Why bother with the additional complexity of this double resonance experiment? The primary advantage of ENDOR spectroscopy is that it is in effect an NMR method whose <u>resolution</u> is orders of magnitude better than that of simple EPR, and this permits the determination of hyperfine interactions too small to observe in EPR. The second advantage is <u>selectivity</u>. Only nuclei that have a hyperfine interaction with the electron-spin system being observed give a resolved ENDOR signal. Thus, for example it is possible to examine ^{57}Fe resonances from one particular metal cluster in the presence of numerous others (4-6).

A third important feature is that ENDOR spectroscopy is inherently <u>broad-banded</u>. Unlike the case of ordinary NMR, it is comparably easy to detect ENDOR resonances from every type of nucleus with $I > 0$. Thus, provided proper isotopic labelling can be achieved, it is possible both in principle and in practice to characterize every atomic site associated with a paramagnetic center. In pursuit of such a goal, we have reported ENDOR studies of biomolecules that describe the resonances from ^{1}H, ^{2}H, ^{13}C, $^{14,15}N$, ^{17}O, ^{33}S, ^{57}Fe, $^{63,65}Cu$, $^{95,97}Mo$ nuclei that are present as constitutive sites of a metal center or as part of a bound ligand or substrate. Finally, although most measurements employ frozen-solution samples, nonetheless <u>angle-selection</u> is achieved in the ENDOR spectrum of a center whose EPR spectrum displays anisotropic magnetic interactions (7), and we have shown that this allows a full determination of nuclear interaction tensors (e.g., 8-10,6).

Although the ENDOR technique is well-established, nevertheless its power to answer fundamental questions about the structure and function of metallobiomolecules only recently has begun to receive wide appreciation. Its broad applicability to the study of metallobiomolecules is illustrated by considering the types of problems that it can address. These can be presented in terms of a series of non-exclusive categories that describe metalloenzyme active sites in terms of a set of oppositions and contrasts (Table 1). Thus a site may be "well-characterized," perhaps with a known crystal structure as with the type-I or blue-copper centers, and the goal of the study may be an increased depth of understanding of its electronic structure and a comparison of family members (11-13). In contrast it might be a "black box," such as the molybdenum-iron (Mo-Fe) cofactor of the nitrogenase MoFe protein (4-6,14, 15); here, not even the composition of the center is agreed upon and ENDOR measurements are of importance even at the level of elemental analysis.

TABLE 1

Representative Metalloenzyme Active Sites

Well-characterized	"Black Boxes"
Blue Copper Proteins (11-13)	Nitrogenase (4-6,14,15)
Hemeproteins (16)	Hydrogenase I,II (28-30)

Resting State	Reaction Intermediates
Nitrogenase	Horseradish peroxidase, Cpd I (17,18)
Sulfite Reductase (31)	Sulfite Reductase, doubly reduced (33)
Cytochrome oxidase, Cu_A (32)	Cytochrome oxidase, Cu_B (34)
	Cytochrome c Peroxidase, ES (35,36)

Metal Center	Substrate Interactions
Cytochrome oxidase, Cu_A	Aconitase (22,23)
Rieske [2Fe-2S] Center (19-21)	Nitrogenase
	Hydrogenase I,II

The resting state of a protein almost invariably is the most accessible to study, as with a variety of heme systems (16), but in favorable cases such as horseradish peroxidase Compound I, ENDOR can provide detailed information about key reaction intermediates as well (17,18). Typically, the first focus of any study will be on the structure of a metal center itself as in our study of Rieske iron-sulfur centers (19-21), but detection of ENDOR signals from a bound substrate can be used to determine the mechanism by which a metal center functions as in our study of aconitase (22,23).

The goal of this review is to describe the application of ENDOR spectroscopy to selected systems representing different categories in Table 1. Discussions elsewhere (24-27), including the chapter by Brustolon and Segre, describe the fundamental issues regarding spin-dynamics (electron and nuclear saturation and relaxation) and we therefore ignore them. We begin with a systematic description of the electron and nuclear spin-Hamiltonians of importance to an ENDOR experiment and develop those expressions that describe observed ENDOR frequencies (See also 24). We then summarize the theory and procedures we have developed for determining magnetic interaction tensors from

ENDOR spectra obtained using frozen solutions. This is followed by discussion of the selected examples mentioned above as representing some of the different areas of Table 1:   The nitrogenase molybdenum-iron protein; horseradish peroxidase Compound I; aconitase; the Rieske iron-sulfur center of phthalate dioxygenase.   References to some of the other systems that have been studied are given in Table 1. The investigations described here have been successful only because of collaborations with the laboratories of the outstanding biochemists whose names are included in the titles of the individual sections.

2.   EPR AND ENDOR ANALYSIS

   2.1 EPR Spectra

      We consider the case of a paramagnetic center with half-integer spin, $S \geq 1/2$. The general spin Hamiltonian for such a system includes terms for the fine-structure and the interaction of the electronic spin with the external field (2):

$$\hat{H}_e = \hat{H}_{zfs} + \hat{H}_z$$

$$= [D(\hat{S}_z^2 - S(S + 1)/3) + E(\hat{S}_x^2 - \hat{S}_y^2)] + \beta \, \underline{\hat{S}} \cdot \underline{g} \cdot \underline{B}. \qquad [1]$$

Here D and E are the axial and rhombic zero-field splitting parameters and the rhombicity is measured by $\lambda = E/D \leq 1/3$; g represents the g-tensor describing the Zeeman interaction and the other symbols have the usual meaning. For $S = 1/2$, $\hat{H}_{zfs}$ is absent; the case $S = 3/2$ will be used as an example of $S > 1/2$. In zero applied field the electron spin manifold of 2S+1 states is split into (2S+1)/2 Kramers' doublets, with the splittings determined by D and E; for $S = 3/2$ there are two doublets separated by the zero-field splitting energy, $\Delta = 2D(1 + 3\lambda^2)^{1/2}$.   We consider explicitly the case in which the zero-field splitting is large compared to $k_B T$ at helium temperatures as well as to the interaction with the external field.   In this case the EPR spectrum is associated solely with the lowest doublet. Typically, in order for such a system to be epr-visible it is required that $\Delta > 0$ in which case the "$m_S = \pm 1/2$" doublet is lowest. We note, but do not explore, systems such as high-spin "rhombic" iron where the splittings are so low that more than one doublet gives measurable EPR intensity (e.g., 37). The epr signal from a single Kramers' doublet can always be described in a representation based on a fictitious spin, $S' = 1/2$.   In this representation the spectrum is characterized by a g'-tensor that is coaxial with the fine-structure interaction (38):

$$\hat{H}'_{ze} = \underline{\hat{S}}' \cdot \underline{g}' \cdot \underline{B} . \qquad [2a]$$

For $S = 1/2$, then $S = S'$ and $g' = g$. As an example, for $S = 3/2$,

$$g_3' = g_{\parallel}$$

$$g_{1,2}' = g_{\perp}[1 + \frac{1 \pm 3\,|\lambda|}{(1 + 3\lambda^2)^{1/2}}] .$$ [2b]

The nitrogenase molbydenum-iron protein Av1 is a representative of $S = 3/2$, with $g_1' = 4.32$, $g_2' = 3.68$, $g_3' = 2.01$; this corresponds to $S = 3/2$ representation parameters, $g_{\parallel} = 2.01$, $g_{\perp} = 2.005$, and $|\lambda| = 0.053$ (5). For $S = 5/2$ and $\lambda = 0$, as in a high-spin aquoferriheme,

$$g_{\shortparallel}' = g_{\parallel} \approx 2$$

$$g_{\perp}' = 3g_{\perp} \approx 6 .$$ [2c]

When the protein is subjected to a magnetic field $\underline{B}$ whose direction within the molecular g'-tensor (fine-structure) reference frame is described by the polar and azimuthal angles (θ, ϕ), then the EPR resonance field for that orientation is described by the angle-dependent g'-value (2),

$$g'^2 = g'^2(\theta, \phi) = \underline{\ell} \cdot \tilde{\underline{g}}' \cdot \underline{g}' \cdot \underline{\ell}$$
$$= \Sigma\ g_i'^2 \ell_i^2 \qquad , \qquad h\nu = g'\beta B,$$ [3]

where $\underline{B} = B\underline{\ell}$ and $\underline{\ell}$ is a unit vector parallel to $\underline{B}$: $\underline{\ell} = (\sin\theta\cos\phi,\ \sin\theta\ \sin\phi,\ \cos\ \theta) = (\ell_1,\ \ell_2,\ \ell_3)$.

The EPR spectrum of a frozen-solution (polycrystalline) protein sample is a superposition of the signals from molecules randomly oriented with respect to the field. The shapes for such powder-patterns are well known (3). Fig. 1 presents the absorption envelope for Av1, which is representative of a center with g'-tensor of rhombic symmetry, $g_1' \neq g_2' \neq g_3'$. When considering ENDOR spectra of polycrystalline samples and their simulations, the observing field is best denoted by the corresponding g'-value. In discussing the experimental results below, we shall omit the primes on g' within subsection 3.1 and of course do so when $S = 1/2$.

2.2 ENDOR Frequencies

The interaction between the electron spin, S, of a paramagnetic center and the nuclear spin, I, of an individual site within the center is described by adding a hyperfine interaction term to the Hamiltonian (2),

$$\hat{H}_{hf,loc} = h\ \hat{\underline{I}} \cdot \underline{A} \cdot \hat{\underline{S}} = h\ \hat{\underline{I}} \cdot (\ ^{A_1}_{\ A_2 \ }_{A_3}\ ) \cdot \hat{\underline{S}}.$$ [4]

The $\underline{A}$ for each site is diagonal within its own local principal axis coordinate frame and the orientation of that frame with respect to the g-tensor reference frame can be expressed in terms of the three Euler angles, (α, β, γ) (Fig. 2)

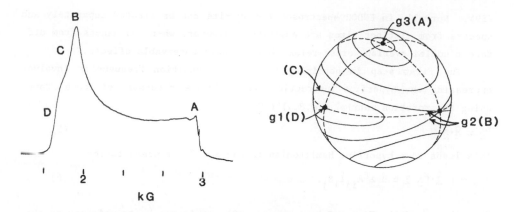

Fig. 1. Correspondences between magnetic field values in an epr spectrum and field orientations within the molecular g-frame. Left: X-Band epr absorption envelope for a nitrogenase MoFe protein, g_1' = 4.32; g_2' = 3.65; g_3' = 2.01. Right: Unit sphere with curves of constant g factor drawn and several correspondences indicated.

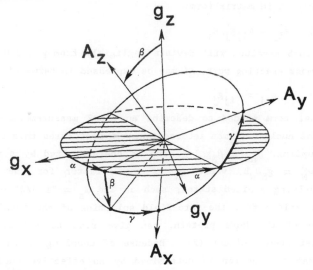

Fig. 2. Definition of the Euler angles (α, β, γ) relating spin-S local (diagonal) nuclear hyperfine tensor axes to g'-tensor (zero-field splitting) axes (Ref. 39).

(39). Normally in ENDOR spectroscopy each site can be treated separately and spectra from distinct sites are additive. However, when resonances from different nuclei overlap, cross-relaxation can have observable effects.

The first step in obtaining the ENDOR transition frequencies involves expressing the hyperfine interaction tensor in the g-tensor reference frame, using the rotation matrix, $M(\alpha, \beta, \gamma)$ (39):

$$^g\underline{A} = \tilde{\underline{M}} \cdot \underline{A} \cdot \underline{M}. \tag{5}$$

This leads to a hyperfine Hamiltonian in the S = 3/2 representation

$$\hat{H}_{hf} = h \; \hat{\underline{I}} \cdot {}^g\underline{A} \cdot \hat{\underline{S}} = h \; \Sigma {}^gA_{ij} \hat{I}_i \hat{S}_j. \tag{6}$$

Through use of the Wigner-Eckart theorem (2), this can be transformed to the effective-spin (S' = 1/2) representation,

$$\hat{H}_{hf} = h \; \hat{\underline{I}} \cdot \underline{A}' \cdot \hat{\underline{S}}' \; ; \quad A'_{ij} = {}^gA_{ij} g'_j / g_j \tag{7a}$$

and can be written in matrix form:

$$\underline{A}' = {}^g\underline{A} \cdot \bar{\underline{g}}' \; ; \quad \bar{g}'_{ij} = \delta_{ij} g'_j / g_j. \tag{7b}$$

Often the spin-S g-values will deviate negligibly from g = 2.0, and then for all but the most exacting work, $\underline{A}'$ may be expressed in terms of the g'-tensor

$$\underline{A}' = {}^g\underline{A} \cdot g'/2 \; ; \quad g'_{ij} = \delta_{ij} g'_j. \tag{7c}$$

The final term needed to describe an ENDOR measurement of an I = 1/2 nucleus is the nuclear Zeeman interaction. In most cases this can be taken as a scalar coupling, $\hat{H}_{nz} = \beta_N g_N \hat{\underline{I}} \cdot \underline{B}$, and is characterized by a nuclear Larmor frequency, $\nu_N^o = g_N \beta_N B$, where g_N is the g-factor for the free nucleus. However, low-lying excited states, such as the m_s = "± 3/2" electronic spin doublet that arises from the zero-field splitting of the total spin S = 3/2 ground state of the MoFe protein, can give rise to a large, anisotropic pseudo-nuclear Zeeman effect (2). Because of coupling to these states, the nuclear Zeeman interaction is determined by an effective nuclear g^N tensor that is coaxial with the zero-field splitting tensor, and thus the g-tensor. For S = 3/2 in the general case of non-coaxial hyperfine and g-tensors, but setting $\lambda = 0$, this takes the form

$$\hat{H}_{nz} = \beta_N \hat{\underline{I}} \cdot g^N \cdot \underline{B} \tag{8a}$$

$$g_{ij}^N = g_N \{ \delta_{ij} + (\frac{3}{2}) (\frac{g_e \beta}{g_N \beta_N}) (\frac{{}^gA_{ij}}{\Delta}) (1 - \delta_{i3}) \} .$$

The same form would hold for other values of S (e.g., see 16), although the numerical factors would differ and there is the possibility of additional contributions from other doublets. Note that in Eqn. 8a, which is correct for

$\lambda \rightarrow 0$, there is no contribution of the pseudo-nuclear Zeeman effect to interactions with the g_3'-component of the external field. Corrections for $\lambda \neq 0$ typically will be small; for the simplest case of coaxial ZFS and hyperfine tensors, $g_{ij}^N = g_i^N \delta_{ij}$ and

$$g_{1,2}^N = g_N [1 + \frac{3}{2} \{ 1 - \frac{2}{3} \sin^2\alpha \ \sin^2\alpha \} (\frac{g_e \beta}{g_N \beta_N}) (\frac{g_{A_{1,2}}}{\Delta})]$$

$$g_3^N = g_N [1 + 3\sin^2\alpha \ \cos^2\alpha \ (\frac{g_e \beta}{g_N \beta_N}) (\frac{g_{A_3}}{\Delta})] \quad ; \quad \tan 2\alpha = \sqrt{3} \ \lambda. \qquad [8b]$$

For nuclei with small g_N, such as ^{57}Fe ($g_{Fe} = 0.18$), the effect can be large and, depending on the magnitude of $g_{A_{ij}}/\Delta$ and the sign of $g_{A_{ij}}$, the observed nuclear Larmor frequency, $|\nu_N|$, may range from much larger than ν_N^0 to ~ 0. Analysis of this effect can yield the signs of the hyperfine couplings as well as a value for Δ; this was done for the ^{57}Fe and ^{95}Mo sites of the FeMo-co cluster of Av1 (5).

For a nucleus with $I = 1/2$, the transition frequencies measured in an ENDOR experiment are determined by the nuclear interaction terms, Eqn. 7 and 8. To first order, these can be derived (40) for a single orientation of the external field, (θ, ϕ), from a nuclear interaction Hamiltonian,

$$\hat{H}_{int} = h \ \hat{\underline{I}} \cdot \underline{V}_\pm \qquad [9]$$

where $\underline{V}_\pm$ is the vector sum of the effective hyperfine and nuclear-Zeeman

$$\underline{V}_\pm = [(\pm\frac{1}{2})\frac{1}{g} \ \underline{A}' \cdot \underline{g}' - \frac{\beta_N}{h} \ \underline{B} \cdot \underline{g}^N] \cdot \underline{\ell}$$

$$\equiv \underline{K}_\pm \cdot \underline{\ell} \qquad [10]$$

fields and the subscript $(\pm)$ refers to the electronic quantum number, $m_s' = \pm 1/2$. Without regard to the relative magnitude of hyperfine and nuclear-Zeeman terms, the first-order, orientation-dependent ENDOR transition frequencies for $I = 1/2$ then are

$$\nu_\pm = K_\pm = [K_\pm^2]^{1/2}$$

$$K_\pm^2 = \hat{\underline{\ell}} \cdot \tilde{\underline{K}}_\pm \cdot \underline{K}_\pm \cdot \underline{\ell} \equiv \hat{\underline{\ell}} \cdot \underline{K}_\pm^2 \cdot \underline{\ell}. \qquad [11a]$$

When either the hyperfine or nuclear-Zeeman term is appreciably larger than the other, it is appropriate to discuss spectra in terms of an approximate equation

$$\nu_\pm = | \pm \frac{A'}{2} + \nu_N | \qquad [11b]$$

where A′, the angle-dependent hyperfine splitting parameter, is given by

$$(g')^2(A')^2 = \tilde{\underline{l}} \cdot \tilde{\underline{g}}' \cdot (\underline{A}')^2 \cdot \underline{g} \cdot \underline{l} \qquad [12]$$

and the nuclear Zeeman splitting is written

$$\nu_N = g^N \beta_N B \qquad [13]$$

where g^N the angle-dependent nuclear g-factor is defined by,

$$g'A'g^N = \tilde{\underline{l}} \cdot \tilde{\underline{g}} \cdot \tilde{\underline{A}}' \cdot g^N \cdot \underline{l} \qquad [13a]$$

which reduces to ν_N^o and g_N when the pseudo-nuclear Zeeman effect is unimportant. For protons in biological systems, $A/2 < \nu_H$ and Eqn. 11b becomes

$$\nu_{\pm} = \nu_H \pm \frac{A'}{2} \qquad [11c]$$

whereas for ^{57}Fe, ^{13}C, and ^{14}N, normally $A/2 > \nu_N$ and Eqn. 11b gives

$$\nu_{\pm} = \frac{A'}{2} \pm \nu_N. \qquad [11d]$$

When the nuclear spin is $I > 1/2$ the ENDOR transition frequencies are modified by the nuclear quadrupole interaction (2,3) The quadrupole splitting tensor, $\underline{P}$, is diagonal in an axis frame that need not be coaxial with the hyperfine tensor frame. Of the three diagonal elements of $\underline{P}$, only two are independent because the tensor is traceless

$$P_1 + P_2 + P_3 = 0 \quad ; \quad P_i = \frac{e^2 q_i Q}{h2I(2I-1)}. \qquad [14]$$

Often, the two independent parameters are taken to be the component of largest magnitude, say P_3 and the anisotropy parameter, $0 \le \eta \le (P_1-P_2)/P_3 \le 1$.

Because we include the possible use of the fictitious spin, S′, the computation of the quadrupole contribution involves expression of this tensor in the g-tensor reference frame using the rotation matrix $M_P(\alpha_P, \beta_P, \gamma_P)$; here the subscripts explicitly admit the use of different Euler angles relating the $\underline{A}$ and $\underline{P}$ tensors to the g-frame:

$${}^g\underline{P} = \tilde{\underline{M}}_P \cdot \underline{P} \cdot \underline{M}_P. \qquad [15]$$

With this transformation, the nuclear-hyperfine, Zeeman, and quadrupole interaction tensors all are expressed in the g′ coordinate frame. The quadrupole coupling contribution to the Hamiltonian has the form

$$\hat{H}_P = h \; \hat{\underline{I}} \cdot {}^g\underline{P} \cdot \hat{\underline{I}} \qquad [16]$$

and its first-order contribution to ENDOR frequencies involves the angle-dependent and electron-spin projection-dependent splitting constant, $P_{\pm}$, given by

$$K_{\pm}^2 P_{\pm} = \underline{\ell} \cdot \underline{\tilde{K}}_{\pm} \cdot {}^g \underline{P} \cdot \underline{K}_{\pm} \cdot \underline{\ell}. \qquad [17a]$$

Higher-order contributions have been discussed (26,41).

The result of these considerations for the ENDOR spectrum is as follows. In a single-crystal ENDOR measurement one expects a set of equivalent nuclei of spin I to give 4I ENDOR transitions (selection rules, $\Delta m_s = 0$ and $\Delta m = \pm 1$). The first-order single-crystal transition frequencies given in Eqn. 11 for $I = 1/2$ are augmented for $I > 1/2$ by an m-dependent quadrupole term and become

$$\nu_{\pm}(m) = \nu_{\pm} + (2m - 1)(\frac{3}{2})P_{\pm} \quad , \qquad -I+1 \le m \le I \quad , \qquad [18a]$$

an equation that is valid for $P_{\pm} < \nu_{\pm}$. The extension of the more approximate equation for transition frequencies, 11a, to nuclei of arbitrary value of I is

$$\nu_{\pm}(m) = \left| \pm \frac{A'}{2} + \nu_N + \frac{3P}{2}(2m - 1) \right| \qquad [18b]$$

where the angle-dependent quadrupole-splitting constant, P, is given by

$$(g')^2(A')^2 P = \underline{\ell} \cdot \underline{\tilde{g}}' \cdot \underline{\tilde{A}}' \cdot {}^g \underline{P} \cdot \underline{A}' \cdot \underline{g}' \cdot \underline{\ell}. \qquad [17b]$$

It is useful to note that the lower Kramers' doublet of an $S > 1/2$ spin system with large zero-field splitting has a strongly anisotropic magnetic moment, which forces the effective hyperfine interaction tensor, $\underline{A}'$ (Eqn. 7), to be highly anisotropic even when the intrinsic site hyperfine interaction is isotropic, $A_1 = A_2 = A_3 = a$ (Eqn. 4). For example, for spin $S = 3/2$ and λ small, by Eqn. 7 the ENDOR frequencies (Eqn. 11 or 18) for a nucleus with isotropic $\underline{A}$ (Eqn. 4) would be determined by a hyperfine tensor that is nonetheless highly anisotropic in the $S' = 1/2$ representation:

$$A'_{1,2} = a \frac{g'_{1,2}}{g_\perp} \sim a(\frac{2S+1}{2}) \quad ; \quad A'_3 = a\frac{g'_3}{g_\parallel} \sim a. \qquad [19a]$$

Thus, for $S = 3/2$ the effective hyperfine couplings associated with g'_1 and g'_2 that are roughly double $((2S+1)/2 = 2)$ the intrinsic values. This doubling spreads out the ENDOR spectra taken at low observing fields and greatly enhances resolution as seen in the ^{57}Fe ENDOR spectra of nitrogenase (Section 3). For high-spin aquoferriheme (16), $(2S+1)/2 = 3$, and the effect is even larger.

2.3 Simulating Polycrystalline ENDOR Spectra

This section briefly presents the elements of the theory (8-10,42), and then describes the procedures that have been developed for its application in determining hyperfine coupling tensors (8-10,6,21,22,31,33). Analysis of quadrupole tensors from polycrystalline spectra is deferred to the discussion of phthalate dioxygenase (21; Section 6).

Consider a frozen solution of a protein with a metal center whose g'-tensor has rhombic symmetry. ENDOR spectra are taken with the external field fixed within the polycrystalline EPR envelope at a selected value, B, which corresponds to a g-value determined by the spectrometer frequency, $g = h\nu/\beta B$. As recognized by Hyde and his coworkers (7), ENDOR spectra taken with the magnetic field set at the extreme edges of the frozen-solution EPR envelope (positions A and D in Fig. 1), near the maximal or minimal g-values, give single-crystal-like patterns from the subset of molecules for which the magnetic field happens to be directed along a g-tensor axis. An ENDOR spectrum obtained using an intermediate field and g-value (e.g., position C) does not arise from a single orientation, but rather from a well-defined subset of molecular orientations. Ignoring for the moment the existence of a finite component EPR linewidth (EPR envelope of a δ-function), the EPR signal intensity at field B, and thus the ENDOR spectrum, arises from those selected molecular orientations associated with the curve on the unit sphere, s_g, comprised of points for which the orientation-dependent spectroscopic split-ting factor (Eqn. 3), satisfies the condition: $g'(\theta,\phi) = g$. However, although g is constant along the curve, s_g, the orientation-dependent ENDOR frequencies, $\nu_{\pm}(m,\theta,\phi)$ (Eqn. 11 or 18), are not. Thus, the ENDOR intensity in a spectrum taken at g occurs in a range of frequencies spanning the values of $\nu_{\pm}$ associated with the selected subset of orientations associated with s_g.

At any observing g-value within the EPR envelope of a polycrystalline (frozen-solution) sample, the intensity of a superposition ENDOR spectrum at radiofrequency, ν, can be written as a sum of convolutions over the ENDOR frequencies (Eqn. 11) that arise on the curve, s_g,

$$^{\delta}I(\nu,g) = \Sigma_m \Sigma_{\pm} \int_{s_g} L(\nu-\nu_{\pm}(m))\; e(\nu_{\pm}(m))\; ds \qquad [20a]$$

$$= \Sigma_m \Sigma_{\pm} \int_{s_g} L(\nu-\nu_{\pm}(m))\; e(\nu_{\pm}(m)) (\frac{\partial s}{\partial \phi})_g\; d\phi \qquad [20b]$$

where

$$(ds)^2 = (d\theta)^2 + \sin^2\theta (d\phi)^2. \qquad [21]$$

L(x) is an ENDOR lineshape function, and $e(\nu)$ is the hyperfine enhancement factor (2,3). For $I = 1/2$ the nuclear transition observed is (m = +1/2 <-> m = -1/2) and the summation involves only the electron spin quantum number, $m_s' = \pm 1/2$; for $I > 1/2$, quadrupole terms must be included (Eqns. 18) and the sum extended over the additional nuclear transitions.

To simulate experimental ENDOR spectra, the restriction to a δ-function EPR pattern must be relaxed. The complete expression for the relative ENDOR intensity at frequency ν, for an applied field set to a value, B, involves the convolution of $^\delta I(\nu,g)$, the EPR envelope function derived by Kneubühl (43), S(B), and a component EPR lineshape function, R(x):

$$I(\nu,B) = \int_{g_{min}}^{g_{max}} dB'S(B') \, ^\delta I(\nu,g') \, R(B-B').$$ [22]

This expression has been implemented as a BASIC program. The component line-shape functions, L(x) and R(x), both were taken as Gaussians and the line-widths as isotropic. In applications such as those discussed below, the hyperfine enhancement factor usually could be ignored ($e(\nu) = 1$; Eqn. 20).

An example of the ENDOR spectra predicted for a polycrystalline sample is presented in Fig. 3. Simulations were performed with Eqn. 20 for a $S' = 1/2$ paramagnetic center that has the rhombic g-tensor of nitrogenase Av1, and an $I = 1/2$ site that has a rhombic hyperfine tensor with $A_1' > A_2' > A_3'$; the $S = 3/2$ tensor components have been chosen as those that characterize site A^1 of Av1 (vide infra), but the pseudo-nuclear Zeeman effect was ignored ($\Delta \to \infty$ in Eqn. 8). Consider first the case of coaxial g' and $\underset{\sim}{A}'$ tensors. The single-crystal like ENDOR spectrum at the low (high)-field $g_1(g_3)$ edge of the EPR envelope is a doublet split by twice the effective nuclear Larmor frequency (Eqn. 8) and centered approximately at $A_1'/2(A_3'/2)$ (Eqn. 11; Fig. 3). For intermediate fields, the pattern shifts and spreads. If the field is set to a value where the maximum and minimum frequencies in an ENDOR pattern do not shift rapidly as g is varied, each of the edges of the pattern appears as a well-defined intensity maximum in the form of a Larmor doublet (when resolved), as seen at $g = 3.69$ in Fig. 3. However, if the observing g is in a range where the ENDOR frequencies vary rapidly with g, then the doublet at an edge can appear as "steps" in ENDOR intensity, rather than peaks, as seen at $g = 3.9$ in Fig. 3.

It is convenient to represent the overall ENDOR response of a particular site in terms of a $(g - \nu)$ plot as in Fig. 3, where each feature in an ENDOR spectrum at a given g is represented by its frequency; for clarity in Fig. 3, a Larmor-split doublet has been given a single point at the frequency corresponding to its center. For the case represented in Fig. 3, with $A_1' > A_2' > A_3'$, and still considering coaxial tensors, the higher frequency doublet, which we label D_{13}, moves smoothly between the frequencies associated with the two extremal, single-crystal-like fields at g_1 and g_3. The other, lower-frequency doublet can best be labelled D_{12} at fields such that $g_1' > g > g_2'$, and D_{23} when $g_2' > g > g_3'$. It shifts to a center at $A_2'/2$ as the field approaches g_2' either from above or below.

If the g' and $\underset{\sim}{A}'$ tensors are not coaxial, then the ENDOR spectra and their field variation become more complex. For example, if the $\underset{\sim}{A}'$ principal-

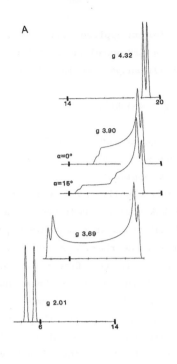

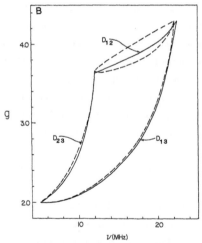

Fig. 3. (Upper) Calculated polycrystalline endor patterns versus observing g—values for the rhombic g—tensor of Av1 (4.32, 3.68, 2.01) and an I = 1/2 hyperfine tensor, (-20.85, -13, -9.8) MHz; Gaussian shape, FWHM = 0.125 MHz; (Δ → ∞) (eq 8); the bottom spectrum has been offset as indicated (Lower). The center-frequency of the resolved features of calculated spectra are given by the intersection(s) of a horizontal with the displayed curves. Collinear A— and g— tensors (—); non-collinear tensors, (α=15°, β=γ=0) (- - -) (See Ref. 6)

axis frame is rotated from the g' frame by a rotation about a single g' axis, say g_i', then the D_{jk} doublet tends to split into two doublets for fields between g_j' and g_k'. In particular, for a rotation about the g_3' axis as illustrated in Fig. 3, the D_{12} step-doublet splits into two such steps in the low-field portion of the EPR spectrum, where $g_1' > g > g_2'$ (Fig. 3). In coaxial cases where the edge of the ENDOR spectrum displays one doublet of peaks, then upon rotation the edge is expected to split into two peak-doublets. Note also that when g and hyperfine tensors are not coaxial the hyperfine values measured at the edges of the EPR spectrum in general do not correspond to a principal-axis value.

More complicated orientations of $\underline{A}'$- and g'-frames, described by two or three Euler angles, can give correspondingly more complex patterns and field variations; conversely, as shown below a set of spectra taken at multiple fields generates a field-frequency pattern that is interpretable in terms of the relative orientation of g'- and $\underline{A}'$-tensors as well as of the principal values of the $\underline{A}'$-tensors.

2.4. Analysis procedure

The process of obtaining hyperfine tensor principal values and orientations begins with the accumulation and indexing of ENDOR spectra at multiple fields across the EPR envelope. Next, a first-approximation to the hyperfine principal values A_3 and A_1 is obtained from the ENDOR frequencies measured in the single-crystal-like spectra obtained, respectively, at the high- and low-field edges of the EPR envelope; A_2 is estimated from the spread of frequencies (when observable) in the spectrum taken at $g_{mid} = g_2'$. Then, the nature of the relative orientation of the g'- and A-tensors is inferred from the development of the ENDOR pattern as the field increases from the low-field, g_1', edge of the EPR spectrum. At this stage, numerous trial and error simulations of selected spectra typically are performed by varying the spin-S hyperfine interaction principal values and the relative orientation of g'-tensor and nuclear coordinate frames. These variations are constrained such that the parameters correctly predict the resonance frequencies measured in the single-crystal-like spectra. The analysis begins with the simplest assumption, namely the minimal noncoaxiality of g' and $\underline{A}$ tensors and is terminated when the entire accessible field-frequency range of ENDOR features had been accommodated.

In all but the ultimate simulations, the ENDOR linewidths are optimized (Eqns. 20) and the possible influence of a finite component EPR linewidth ignored. The final step is to increase the EPR linewidth until this has visible effect, and then to see whether differential broadening alters the positions of the ENDOR features or the shape of the pattern.

For systems with well-resolved ENDOR spectra from a nucleus with $I > 1/2$, it is possible to use these procedures to determine quadrupole and hyperfine tensors simultaneously; an example is the ^{14}N data from the oxidized form of sulfite reductase from Escherichia coli. (31). In the case of nitrogen ENDOR, we show with phthalate dioxygenase that under less favorable conditions it is possible to derive $\underline{A}$ tensors from ^{15}N enriched samples and to use this information to deduce $\underline{P}$ from ^{14}N data (21; Section 6).

3. "BLACK BOX", RESTING STATE: NITROGENASE MoFe PROTEIN (W.H. Orme-Johnson)

The molybdenum-iron (MoFe) protein of nitrogenase is an $\alpha_2\beta_2$ dimer that contains 2 Mo, (30 ± 3) Fe, and approximately 30 labile S atoms (44-46). Mössbauer and EPR measurements indicate that these inorganic components are organized in a minimum of six polynuclear metal clusters (44,47), with the molybdenum-iron cofactor (FeMo-co) cluster, of approximate composition $MoFe_{6-8}S_{9\pm1}$ (48), representing two of these. Extensive efforts have been made to characterize this EPR-active center because there is evidence that it contains the active site of the enzyme. Mössbauer spectroscopy indicated that this cluster probably includes six spin-coupled iron atoms and obtained values for their hyperfine interaction constants (47). EXAFS measurements give average properties of the iron sites and have been used to characterize the local environment of molybdenum (49).

The MoFe protein exhibits an unusual EPR spectrum (Fig. 1) with apparent g-values of ca. $g_1' = 4.3$, $g_2' = 3.6$ and $g_3' = 2$, depending on the species, at temperatures below 25 K. This spectrum is derived from the ground-state Kramers' doublet of the $S = 3/2$ multi-metal enzyme-bound molybdenum-iron cofactor. Our ENDOR measurements of native and isotopically enriched MoFe proteins have provided the most detailed microscopic information available for this remarkable structure. We have observed and analyzed resonances from each type of atom known to be present in the cofactor: ^{57}Fe (4-6), ^{95}Mo (14,15), ^{33}S (5), as well as ^1H and ^{14}N (5). We have further compared MoFe enzymes, Av1, Kp1, and Cp1, isolated from three different organisms, Azotobacter (A.) vinelandii, Klebsiella (K.) pneumoniae and Clostridium (C.) pasteurianum, respectively (5). The observation of an exchangeable proton(s) from each protein source suggests a site on the cluster accessible to solvent and coordinating H_2O or OH^-. The $^{95,97}Mo$ and ^{33}S studies gave the first observed ENDOR signals from these nuclei.

The nifV mutant of K. pneumoniae produces mutant MoFe protein (Kp1 V$^-$) that will reduce C_2H_2 but not N_2 in vivo because the mutant cofactor differs from that of the wild-type (WT) (44,50). The use of ENDOR spectroscopy of ^{57}Fe, ^{95}Mo, and ^{100}Mo enriched samples[10,11] of WT and Kp1 V$^-$ to probe the nature of the alteration in the molybdenum-iron cofactor provided the first observation of physical differences between WT and V⁻ proteins (15).

3.1 ^{57}Fe ENDOR Measurements (4-6)

With the field set to the high-field, $g_3 = 2$ (within subsection 3.1 we delete primes on g-values), edge of the EPR spectrum of ^{57}Fe enriched Av1, the single-crystal-like ^{57}Fe ENDOR spectrum exhibits five resolved ^{57}Fe doublets (Fig, 4A), thereby directly demonstrating that the cluster comprises at least five distinguishable iron sites. The doublets occur in two groups, a lower-frequency trio of resolved doublets centered at $A_3'/2 \sim 5$-6 MHz, and a group of lines centered at $A_3'/2 \sim 10$ MHz consisting of at least two resolved doublets. Because there is no pseudo-nuclear Zeeman effect at g_3 (Eqn. 8), the doublet splittings are simply $2\nu_{Fe}^o$ and they do not give the sign of A_{Fe}.

With the field set to give a single-crystal-like ENDOR spectrum at the low-field, g_1, edge of the EPR envelope, the ^{57}Fe pattern with six well-resolved peaks is quite different from that at $g = 2$ (Fig. 4). For ^{57}Fe($g_{Fe} = 0.18$) the effective nuclear Zeeman splitting at this field (Eqn. 8) may be written in mixed, but convenient units, as

$$2\nu_{Fe} = (1 + 1.02 \ A_1(MHz)/\Delta(cm^{-1}))2\nu_{Fe}^o \qquad [23]$$

where $2\nu_{Fe}^o = 0.433$ MHz at 1575 G, and for clarity only, coaxial g- and A-tensors have been assumed here. This simplified form of Eqn. 8 demonstrates that the effective nuclear-Zeeman splitting can differ drastically from $2\nu_{Fe}^o$. Because of this difference, an ENDOR measurement of $|A_1|$ and ν_{Fe} gives Δ to high precision as well as the sign of the ^{57}Fe hyperfine coupling. Through the use of Eqn. 16 (or 8), the six peaks in Fig. 4B were assigned as resonances from four distinct iron sites. Peaks 1-4, located at lower frequency, comprise two doublets. They correspond to two sites, B^2 and B^1, whose small hyperfine couplings are positive, as shown by the enhanced nuclear Zeeman splittings. Peaks 6 and 5 at higher frequency correspond, respectively, to sites A^1 and A^2; here, the large hyperfine couplings are negative, as shown by the vanishingly small nuclear Zeeman splittings (Eqns. 8,16). Detailed examination further disclosed that peak 4 also derives intensity from an additional distinct site, A^3, with negative hyperfine coupling.

The analysis procedure described above is best applied if the evolution of the ^{57}Fe ENDOR pattern of an individual iron site can be traced as the field increases across the EPR envelope. In general, for Av1 the ^{57}Fe ENDOR peaks can be assigned for fields between $g_1 = 4.3$ and roughly $g \approx 3.5$; poorer resolution at higher field precluded the direct experimental assignment of correspondences with the resonances at g_3.

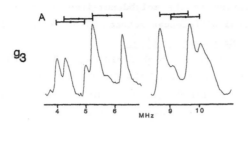

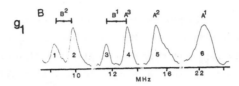

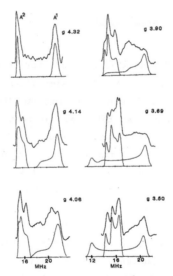

Fig. 4. Single-crystal like ^{57}Fe endor spectra of the Avl protein. The assignment of peaks is indicated. Conditions: T = 2 K; microwave power, 0.63 mW; 100 KHz field modulation, 0.4 mT; rf power 30 W; rf scan rate, 3.5 MHz/s. (g$_3$) B$_o$ = 0.3355 T, 2000 scans; (g$_1$) B$_o$ = 0.1572 T; 5000 scans (Ref. 5).

Fig. 5. ^{57}Fe endor spectra of iron sites A^1 and A^2 of the Avl protein at selected magnetic fields. Computer simulations employ the tabulated hyperfine interaction parameters. Experimental conditions are the same as in Fig. 4. Each spectrum is the average of 1000 scans (Ref. 6).

3.1.1 Analysis of Sites A^1, A^2

Fig. 5 shows ENDOR spectra of iron sites A^1 and A^2 taken at various fields from g_1 to beyond g_2 (6). Both sites have a negative hyperfine coupling and because of the pseudo-nuclear Zeeman effect, Eqn. 8, show a single sharp peak at the low-field, g_1, edge of the EPR envelope (Fig. 4B). The manner in which the pattern for the two sites spreads as the field is increased from g_1 has been analyzed to obtain values for their tensor components and orientation, Table 2, that generate simulations in satisfactory agreement with experiment (Fig. 5). The procedures by which the parameters were obtained serves as an example of this process and are now described.

The A^1 ENDOR pattern broadens rapidly to low frequency as the field is increased, which indicates that the hyperfine tensor is highly anisotropic, with $A_2 < A_1$ (Fig. 5); accompanying this spread is a sharp reduction in intensity that makes detection unreliable by _ca_. g_2. Simulations of the A^1 pattern indicated that the observed spread to low frequency, for example, the breadth for $g \sim 4.14$, cannot be duplicated with coaxial $\underline{A}$ and g tensors, even by setting $A_2 = 0$ MHz. To match these observations requires that the local hyperfine tensor frame for site A^1 departs from the g-tensor frame by a rotation about the g_3 axis. Simulations were performed for various values of the angle α (Fig. 2), and in each case an attempt was made to adjust A_2 and A_1 so that the calculated spectrum fit the low frequency edge of the experimental data at $g = 4.14$ as well as the single crystal peak at $g = 4.32$. As previously noted, when the tensor $\underline{A}$ is rotated about g_3, it is necessary to vary A_1 and A_2 jointly in order to reproduce the single-crystal pattern; for simplicity, we will mention only one of the coupled tensor components. (The same procedure is adopted below when discussing rotation by β about the g_2-axis.) The spectrum at $g \sim 4.14$ could be fit using a range of values, ($\alpha \sim 10\text{-}20°$, $A_2 \sim 12\text{-}14$ MHz). Simulations with α much smaller or A_2 much larger did not reproduce the breadth of this pattern; a larger rotation moved the high frequency edge outside the experimental pattern. An additional rotation of $\underline{A}$ by $\beta \le 10°$ had no influence on the breadth of the pattern simulated. Within the acceptable (α; A_2) range for $g = 4.14$, further calculations showed that the reported values, $\alpha = 15°$, $A_2 = -13$ MHz, best reproduce the spectra at all fields. Simulations of the iron site A^1 with either $A_3 \sim -10$ MHz, as in Fig. 5, or $A_3 \sim -20$ MHz were indistinguishable for the range of g-values that gave resolved spectra ($g > 3.9$). Assignment of the A_3 values of other iron sites led us by elimination to the tentative assignment, $A_3 = \sim -10$ MHz for this site.

TABLE 2

^{57}Fe Hyperfine tensor principal values (MHz) and orientations (degrees)^c relative to fine-structure (g-tensor) principal axes (S = 3/2 representation) for Av1^a

	A^1	A^2	A^3	B^1	B^2
Principal values					
A_1	−20.8	−14.0	−11.6	13.5	8.9
A_2 ^{b}	−13	−18.3	−14	11	11
A_3	−10^c	−19	−10	9	19^c
	(−19)				(10)
Euler angles^d					
α	15	15^e	30	0	12
β	0	10	0	45	0
R^f	0.48 (0.62)	0.72	0.71	0.67	0.47 (0.80)

a) For Av1 the three principal values of the g'-tensor are (4.32, 3.68, 2.01). Uncertainties in the various hyperfine parameters are described in the Discussion. Table taken from Ref 6.
b) The single-crystal-like spectra at g_3 display five doublets in two groups, three with $|A| \sim 10$ MHz (9.7, 9.8, 11.7 MHz) and two with $|A| \sim 19$ MHz (19.35, 19.5 MHz). Only assignments to, not within, a group are possible. (see text.)
c) By elimination, one of Sites A^2 and B^2 must have $|A_3| \sim 10$ and the other, $|A_3| \sim 19$ MHz. The data reported here permit either assignment; the assignment tentatively preferred (without parentheses) is discussed in the text.
d) Euler angles as defined in Fig. 2. In all cases, satisfactory fits were obtained with $\gamma = 0$.
e) In Ref 6 the values for α and β were interchanged.
f) R is defined as the ratio of the smallest to largest components of the $\underset{\sim}{A}$ tensor.

The ^{57}Fe ENDOR pattern for site A^2 remains intense and does not spread very much as the field changes (Fig. 5), which indicates a lesser anisotropy in the hyperfine tensor. The pattern, which starts as a single resonance centered at ~15.5 MHz at g_1, divides into three main peaks (Fig. 5) that shift strongly as the observing field increases. As confirmed by simulations, for all orientations sampled at low observing field, $g > g_2$, the effective nuclear Zeeman splitting for this site is negligible ($2\nu_{Fe} \sim 0$) because of the pseudo-nuclear effect, and the splittings are not due to this interaction. Simulations with collinear $\underset{\sim}{A}$ and g tensor could not reproduce the breadth of the g_2 pattern no matter what value of A_2 was used, nor did these simulations exhibit more than two peaks. Simulations with a rotation of the hyperfine tensor around a single g-tensor axis, g_1 ($\alpha = 90$, $\beta \neq 0$, $\gamma \neq -90$), g_2 ($\beta \neq 0$), or g_3 ($\alpha \neq 0$), could reproduce the breadth of the pattern, but generated only two peaks. Multiple rotations were needed to simulate the ENDOR pattern between g_1 and g_2. Fig. 5 shows that simulations based on the parameters listed in Table 2 reproduce the displayed experimental spectra quite well. Fig. 6 shows in greater detail the agreement between the field dependence of the calculated and observed frequencies for the ENDOR peaks of site A^2.

3.1.2 Sites B^1, B^2, A^3

Fig. 7 shows the ENDOR spectra of iron sites B^1, B^2 and A^3 at various fields along the absorption envelope from g_1 up to and past g_2 (6). Computer simulations using the final parameters for each site (Table 2) as obtained by procedures analogous to those described above for sites A^1 and A^2 are shown below each spectrum. Both sites B^1 and B^2 have positive hyperfine couplings and show a sharp single-crystal-like doublet at g_1 (Fig. 4). The resonance for site A^3 is not distinguishable at this field. However, as the magnetic field is increased (Fig. 7) a shoulder appears on peak 4, the high-frequency partner-peak of site B^1; by g_2, this shoulder has moved to higher frequency and is resolved into a peak we called 4A and assigned as the signal of a distinct iron site, labelled A^3, with negligible Larmor splitting and thus a negative hyperfine coupling.

3.2 ^{1}H ENDOR Results

Setting the magnetic field to the high-field edge of the EPR spectrum of MoFe proteins Av1, Cp1, wild-type or NifV$^-$ Kp1 in H_2O buffer gives single-crystal-like proton spectra (5,51) such as that in Fig. 8. The resonances are described by Eqn. 11b and are associated with many (at least six) sets of magnetically equivalent protons whose coupling constants are in the range of $0.14 \leq A^H \leq 3.3$ MHz. Prior to these studies there had been essentially no information as to whether the cluster of the resting state (as isolated) enzyme has coordination sites open to solvent. To obtain evidence on this point, we examined the proton ENDOR of a number of MoFe proteins subjected to D_2O exchange. When the MoFe protein isolated from wild-type K. pneumoniae is subjected to D_2O exchange there is a reduction in the intensity of the resonance centered at 1.67 MHz (Fig. 8) (51). This difference can be assigned to an exchangeable proton, most likely associated with a H_2O or OH$^-$ group coordinated near the metal cluster. There is also a loss of intensity of the distant ENDOR signal from the solvent.

Computer subtractions of the proton ENDOR patterns for Kp1 WT and V$^-$ proteins, both in H_2O (Fig. 9) (51), show that the NifV$^-$ Kp1 spectrum has an additional proton doublet not present in that of the WT spectrum. Upon D_2O

560

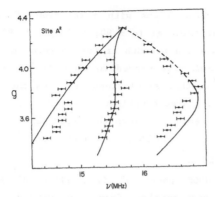

Fig. 6. Endor peak positions versus observing g—values for iron site A^2. The theoretical values, calculated using the tabulated hyperfine tensor parameters, are shown as solid lines for resolved peaks, dotted lines for features visible only as shoulders (Ref. 6).

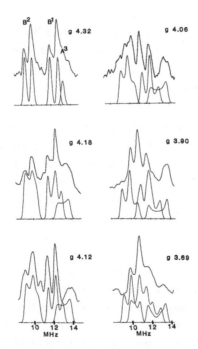

Fig. 7. ^{57}Fe endor spectra of iron sites A^3, B^1 and B^2 of Av1 at selected magnetic fields. The computer simulations using tabulated hyperfine parameters are shown below the corresponding spectrum. Conditions are as in Fig. 4. Each spectrum is the average of 1000 scans (Ref. 6).

561

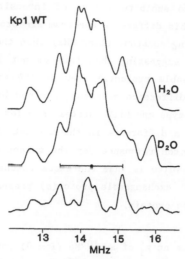

Fig. 8. ^{1}H endor spectra at g_3' of WT Kp1 in H_2O and D_2O. The digital subtraction presented below shows a strongly coupled exchangeable proton as well as a diminished distant endor peak in the D_2O sample. Conditions: B_o = 0.62 T; T = 2 K; microwave power, 0.063 mW; microwave frequency, 9.52 GHz; 100 kHz field modulation, 0.125 mT; rf power, 20 W; rf scan rate, 0.375 MHz/s; 500 scans (Ref. 51).

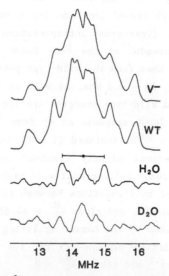

Fig. 9. Comparison of the ^{1}H endor resonances of (A) NifV$^-$ (H_2O buffer) and (B) wild-type Kp1 (H_2O buffer) proteins. The digital subtraction (C) = (A)-(B) shows an extra group of strongly-coupled protons present in the Kp1 V$^-$ protein. The absence of this feature in (D), the digital subtraction of D_2O-exchanged Kp1 WT from D_2O-exchanged Kp1 V$^-$, indicates that the extra proton group is exchangeable. Conditions as in Fig. 8 (Ref. 51).

exchange NifV⁻ Kp1 also exhibits a loss of intensity of a proton doublet. However, the peaks in this difference spectrum are broader and correspond to a slightly smaller coupling constant (1.62 MHz) than the exchangeable proton in the wild-type protein, suggesting that they do not merely correspond to the same group of exchangeable proton(s) seen in the wild-type protein. As a final comparison, the difference spectrum obtained by subtracting the traces for D_2O-exchanged wild-type and NifV⁻ protein samples (Fig. 9D) is essentially featureless except for a difference in the distant ENDOR signal and does not show the additional proton resonance for the mutant protein as in ^1H_2O. The implication of these results is that the extra resonance present in the NifV⁻ protein belongs to (an) exchangeable proton(s) present in the mutant protein but not in the wild-type protein.

3.3 ^{95}Mo ENDOR

When an ENDOR trace at g_3' of an ^{100}Mo (I = 0) enriched Kp1 sample is computer subtracted from the trace of the ^{95}Mo enriched protein, it reveals the rather broad ^{95}Mo resonance in the region of 0-15 MHz (Fig. 10) (4,5,15,51). The resolved peak at ~2.7 MHz on the low-frequency edge of the ENDOR spectra is the ν_- peak of a Larmor doublet associated with the m = ±1/2 nuclear transition of ^{95}Mo, and can be used to obtain the hyperfine coupling in the S = 3/2 representation, A_3^{Mo}. The majority of the ^{95}Mo pattern at g_3' arises from the satellite (m ≠ ±1/2) transitions and for both the wild-type and mutant proteins can be given a first-order interpretation (Eqn. 18) if we use $\nu_{Mo} = \nu_{Mo}^o = 0.9$ MHz. The breadth of the ^{95}Mo ENDOR pattern for the NifV⁻ Kp1 protein is identical to that from the wild-type protein and analysis indicates the quadrupole coupling, P_3 ~ 1.6 MHz, as well as the hyperfine coupling, A_3^{Mo} ~ 1.5 MHz, of the V⁻ and wild-type proteins are indistinguishable.

Comparison of the ENDOR response at g_1' from ^{95}Mo-enriched (I = 5/2) Kp1 protein with that of an ^{100}Mo-enriched (I = 0) sample (Fig. 11) discloses a ^{95}Mo doublet; this becomes clearly resolved as the observing field is increased because the proton pattern shifts to higher frequency (15,51). This ^{95}Mo doublet is assigned to transitions between the m = ± 1/2 level of the I = 5/2 ^{95}Mo nucleus (Eqn. 3), and gives A_1^{Mo} = +4.65 MHz for the WT Kp1. This assignment is corroborated by the Larmor splitting of the doublet (1.07 MHz), which equals the prediction of Eqn. 13 for the pseudo-nuclear effect upon insertion of Δ = 12.5 cm⁻¹ and taking A_1^{Mo} > 0.

The ^{95}Mo ENDOR spectrum at g_1' thus has the appearance expected for a nucleus of I = 1/2, with no hint of the quadrupolar splittings predicted by Eqn. 3, whereas the g_3' pattern is broad, presumably because of such splittings. A full consideration of the field dependence of the ^{95}Mo ENDOR indicated that it is associated with a quadrupole tensor that has P_3(WT) ~1.6 MHz and maximal rhombicity: η~1 (P_1~0, P_2~ $-P_3$). In such a case, the quadrupole

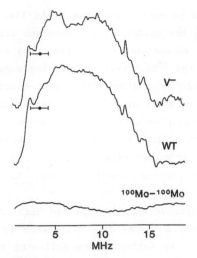

Fig. 10. ^{95}Mo endor of Kp1 WT and Kp1 V$^-$ at g = g$_3$'; each is obtained by subtracting the spectrum of an ^{100}Mo-enriched sample from that of the ^{95}Mo-enriched sample. The hyperfine couplings are described in the text. The bottom trace is the digital subtraction of two separate ^{100}Mo spectra. Conditions: B$_0$ = 0.3360 T; T = 2 K; microwave power, 0.2 mW; microwave frequency, 9.52 GHz; 100 kHz field modulation, 0.4 mT; rf power, 20 W; rf scan rate, 2.5 MHz/s; 700 scans (Ref. 51).

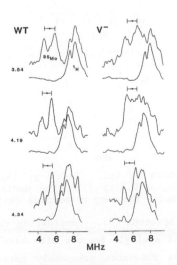

Fig. 11. ^{95}Mo endor of Kp1 WT and Kp1 V$^-$ at g values near g$_1$'. Molybdenum endor spectra of (Left) Kp1 WT and (Right) NifV$^-$ Kp1. The g–values for each set of paired traces are as listed. The upper spectrum of each pair corresponds to the ^{95}Mo-enriched species, the lower to ^{100}Mo-enriched protein. Conditions as in Fig. 8; microwave frequency, 9.52 GHz (Ref. 15).

splittings are predicted to vanish when the field lies near g_1' as observed, while at fields above g_2' the quadrupole interactions are thought to cause the satellite transitions to spread over a broad frequency range.

The small value of the ^{95}Mo hyperfine coupling suggested that the molybdenum is to be classified as a diamagnetic, even-electron ion with an even formal valency, and not as Mo(V) or Mo(III). Further insights into the state of the molybdenum were attempted through considerations of the ^{95}Mo quadrupole coupling parameters inferred from the ENDOR measurements; comparisons involving a series of mononuclear Mo clusters supported an assignment of Mo^{IV} rather than Mo^{VI} for the MoFe protein, consistent with Mo X-ray absorption-edge measurements (49,50).

Comparison of the spectra of wild-type and NifV$^-$ Kp1 at fields near g_1' unambiguously demonstrates a significant difference in the molybdenum resonances at g_1' (Fig. 11). We estimated the following tensor components: WT, $A_1^{Mo} = +4.65$ MHz; NifV$^-$, $A_1^{Mo} = +5.15$ MHz.

3.4 ^{33}S ENDOR

Setting the magnetic field to correspond to the low-field edge (g_1') of the EPR spectrum of a ^{33}S enriched MoFe protein of Cp1 yields a feature that is not present in the natural abundance protein, Fig. 12, and represents the

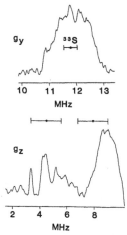

Fig. 12. ^{33}S ENDOR spectrum of Cp1. Left: g_1' ($B_0 = 0.1575$ T); $A^S/2$ is indicated. Right: g_3' ($B_0 = 0.3355$ T). The nuclear Larmor splitting is indicated to help in assessing the many resonances (Ref 5).

first reported ^{33}S ENDOR spectrum (5). Considering that FeMo-co contains 9 ± 1 S^{-2} ions, the pattern of a single feature with barely resolved structure is surprisingly simple. The resonance presumably corresponds to a number of similar sulfur sites, and from the primary sequence it appears that these must primarily, if not exclusively, correspond to the S^{-2} ions that bridge the metal sites of the cluster.

^{33}S has a nuclear spin, $I = 3/2$, and each magnetically distinct site will give a pattern that can contain as many as four resolved lines (Eqn. 18). From the center frequency of the pattern in Fig. 12, $|A_2^S| \approx 11.8$ MHz. Taking into account the pseudo-nuclear Zeeman effect, Eqn. 8, led us to infer that A^S is negative, giving $|2\nu_S| \sim 0.5$ MHz and to suggest that the partially resolved structure reflects small quadrupole couplings and inequivalences among the contributing sulfur sites. It seems plausible that S^{-2} ions would acquire negative spin density through spin polarization and would largely be equivalent as suggested by the spectrum. Furthermore, small quadrupole couplings are suggestive of a closed shell ion whose symmetric charge distribution is not highly distorted by the covalent bonding in the cluster.

3.5 Discussion

The long-range goal in spectroscopic investigations of the FeMo-co cofactor is to elucidate the electronic properties of each individual metal site within the cluster as part of the basis for understanding the molecular mechanism of dinitrogen reduction.

Mössbauer spectroscopic studies originally disclosed that the iron sites of the cofactor-cluster of Av1 fall into two subclasses (44,47). Three sites, labelled here as A^1, A^2, and A^3, have relatively large, negative hyperfine coupling parameters. Another subclass, labelled B, and quantitated at three contributing sites, was found to have smaller, positive coupling constants. The unprecedented application of the ENDOR technique to frozen-solution samples yielded anisotropic hyperfine tensor principal values and orientations relative to the zero-field splitting axes for five distinct iron sites of Av1, A^1, A^2, A^3, B^1, B^2. The results obtained are summarized in Table 1. The ENDOR study dramatically suggests that the cofactor cluster has a remarkably complex structure: Five distinct Fe-sites have been identified, no two of which have equivalent hyperfine tensor components or orientation. Moreover, a variety of arguments strongly suggested that the structural complexity of FeMo-co bound at the active site of Av1 is intrinsic to the cluster, and not imposed by protein constraints.

No previous measurements have given direct evidence regarding sites on the resting state cluster that coordinate small molecules. The observation by ENDOR of exchangeable proton resonances suggests the presence of H_2O or OH^- bound to the cluster in that state.

The most important conclusion drawn from the ENDOR observation of hyperfine couplings to molybdenum was that the molybdenum is indeed electronically integrated into the EPR-visible FeMo-co clusters, because previous studies of CW EPR spectra of MoFe proteins had failed to give evidence of ^{95}Mo hyperfine broadening. The small value of A^{Mo} is consistent with the molybdenum being a non-magnetic, even-electron ion with an even formal valency. This result and a comparison of the ^{95}Mo quadrupole coupling to those estimated from ^{95}Mo NMR studies of mononuclear complexes, supports the assignment of an unsymmetrically coordinated Mo(IV) for molybdenum in the resting state MoFe protein, which focusses interest on synthetic monolybdo-iron-sulfur complexes where Mo is in intermediate oxidation states.

ENDOR spectroscopy further has provided the first physical evidence of differences between the catalytically active centers of the wild-type and NifV⁻mutant Kp1. We speculate (51) that the additional exchangeable proton detected in the ^{1}H ENDOR of nifV⁻ protein may be on a water molecule present in a cavity that owes its existence to the replacement of homocitrate by citrate (52) either as a template ligand during cofactor biosynthesis or as a component (or component precursor) in the finished cofactor entity.

4. REACTION INTERMEDIATE: HORSERADISH PEROXIDASE COMPOUND I (L.P. Hager)

Peroxidases are ferri-hemoproteins that catalyze substrate oxidations by hydrogen peroxide through the action of an enzymic intermediate that is oxidized by 2 equivalents above the ferriheme resting state (53,54). The intermediate in horseradish peroxidase and many other heme peroxidases, called Compound I, is thus at formal oxidation level of Fe(V). Compound I can react with diverse organic, inorganic, or macromolecular substrates in two consecutive one-electron steps. The product of the first step, having the oxidation level of Fe(IV), is called Compound II. We recall key features of Compounds I and II for convenient reference. 1) Their optical spectra are significantly different, HRP I being green and HRP II, red, and differ from that of resting state. 2) Magnetic susceptibility measurements indicate three and two unpaired electrons respectively. Compound I shows an unusual, broad EPR signal that is stoichiometric with the heme and observable only below 30K, but HRP II is epr silent (55). 3) Their ^{57}Fe quadrupole splittings, isomer shifts, and their Mössbauer spectra in high applied magnetic fields are very similar, but their spectra in low applied field differ significantly (55). The characterization of these two Compounds has been a fascinating challenge of long standing and is of even greater importance because analogous states are present in other enzyme classes, such as the catalases and cytochromes P450.

The paramagnetism of HRPI was explained by Schulz et al. (55) in terms of spin coupling between two spin systems, one an iron center that has spin $S_T = 1$ and is subject to a zero field splitting of D ∼ 30 K, and the other, a radical that has spin $S_p = 1/2$. The Compounds I of HRP, catalase and many other heme enzymes as well, are widely believed to contain the second oxidizing equivalent as the π-cation free radical of the porphyrin ring. This identification was proposed by Dolphin, Fajer, Felton and coworkers because the optical spectra of Compounds I resemble those of a series of synthetic porphyrin π-cation radicals (56,57). In addition, they further divided the optical spectra into two classes and assigned them to porphyrin radicals with different ground state symmetries. However, Morishima and Ogawa concluded that their proton NMR spectra of the heme peripheral substituents were inconsistent with the localization of an odd-electron on the ring and suggested the data might be better interpreted in terms of a radical located on a nearby amino acid residue (58,59). On the other hand, La Marr and de Ropp (60) concluded that the observed hyperfine shifts are consistent with a porphyrin cation radical. Thus, the NMR evidence available at the time of our ENDOR investigation did not provide direct evidence for the occurrence of a π-cation radical intermediate.

The task of characterizing the active site as confronted in our ENDOR measurements (17,18) could then can be formulated with three questions: i) What is the identity of the radical center, ii) What is the detailed character of the iron center, and iii) What is the description and origin of any coupling between centers?

4.1 ^{1}H and ^{14}N ENDOR of Compound I.

Fig. 13 presents typical spectra taken at 2K with the magnetic field set to g ~ 2, the peak of the absorption-shaped EPR signal obtained with dispersion-mode EPR under conditions of rapid adiabatic passage (17). The observed resonances cover the frequency range up to 25 MHz; no signal other than those shown were obtained at radio frequencies of up to 100 MHz. Similar patterns are obtained at all values of B_0 within the EPR envelope, thus proving the entire signal is associated with a single center.

In the ENDOR pattern obtained at 9.6 GHz, Fig. 13A, three types of signals may be discerned. The intense resonances located in the frequency range of 11.5-18 MHz occur in pairs whose center falls at the free-proton Larmor frequency of ν_H. The strength and shape of the resonances suggest that they arise from β-protons, those attached to carbon atoms that are in turn bonded to spin-bearing carbon (3,61). Spectra taken at other field positions show the small hyperfine splitting anisotropy expected for such protons, and also suggest that at least some of the resonances in Fig. 13 are superpositions from more than one set of protons having similar coupling constants. Table 3 lists the β-proton coupling constants obtained as the inter-pair spacing from spectra in which the 11-18 MHz region has been expanded.

Two weak features at frequencies above 18.5 MHz (Fig. 13A) do not have obvious, well resolved partners symmetric to ν_H but shift appropriately with the field (Fig. 13B) and thus are protons as well. The partner features are overlapped with the low-frequency peaks in Fig. 13A, and are visible, although barely, in Fig. 13B. The low intensity of the features and the anisotropy of

TABLE 3

Hyperfine coupling constants (MHz) for Horseradish Peroxidase Compound I[a]

	A(g=2)
β-protons	1.30
	2.76
	4.41
	5.80
α-protons	9.26
	11.90
Nitrogen (^{14}N)	7.2
	Tensor Components
Oxygen (^{17}O)	$A_1' \sim 17$
	$A_2' \sim 19$
	$A_3' \sim 0$

[a]Estimated uncertainties: proton couplings ±0.05 MHz; nitrogen coupling, ±10%. Data taken from Refs. 17 and 18.

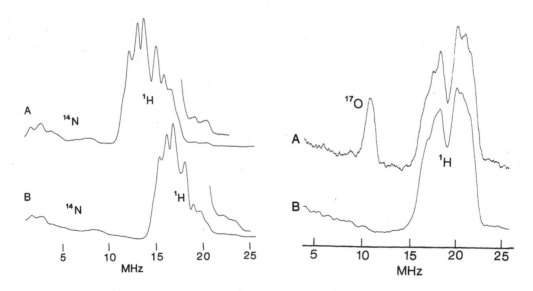

Fig. 13. (Left) ^{1}H and ^{14}N endor spectra of horseradish peroxidase Compound I taken at g ≃ 2.0 with microwave frequencies 9.60 GHz (A) and 11.66 GHz (B). Arrows on the spectra indicate ν_H, the proton Larmor frequency (Ref. 17).

Fig. 14. (Right) Endor spectra of HRPI. (A) HRP(^{16}O) and (B), HRP(^{17}O) at g ≃ 1.5. Conditions: 2-μW microwave power; T = 2 K; field modulation ≃ 0.3 mT, scan rate ~ 2 MHz/s; ~ 300 scans (Ref. 18).

the associated coupling constant suggest that they arise from α-protons, those directly attached to a spin-bearing carbon atom (3,61). Table 3 lists the two coupling constants obtained by doubling the frequency difference between these two peaks and ν_H.

Moderately strong resonances are also observed in the frequency region 0-9.5 MHz. A comparison of Figs. 13A and 13B shows that these signals do not appreciably shift with field. This is even more clearly seen upon comparing expansions of the low frequency portion of the ENDOR spectrum (17). Thus, these signals are not proton resonances and are assigned to ^{14}N. In the range of 1-5 MHz it is possible to roughly discern a four line pattern whose two lower frequency members are particularly apparent, and which is approximately described by the equation for the $\Delta m = \pm 1$ transitions for ^{14}N (see Eqns. 11,18). The center of the pattern is thus associated with $A^N/2$, and the resulting coupling constant is listed in Table 3. The peak at ~ 9.5 MHz is assigned to $\Delta m = \pm 2$ transitions with frequency approximately given by $\nu = A^N + 2\nu_N$; this assignment leads to a value for A^N consistent with that obtained from the $\Delta m = \pm 1$ lines. As discussed above, a frozen-solution EPR signal is a superposition of intensity from molecules with a range of orientations with respect to the field. Although the EPR spectrums of HRPI is not well understood, the well-resolved $\Delta m = \pm 1$ features of the ENDOR response suggest that the signal is dominated by orientations corresponding to a particular component (or two comparable components) of the ^{14}N hyperfine splitting tensor. Spectra taken at other field settings indicate that these features correspond to the largest element(s) of the tensor. The appearance of shoulders on the $\Delta m = \pm 1$ features could be due to slight chemical nonequivalences among a set of similar nitrogens, but also might represent contributions from other orientations.

The spectra presented here clearly show the radical site of horseradish peroxidase Compound I to contain ^{14}N and β-protons, and probably α-protons as well. This immediately suggests that the S = 1/2 portion of the horseradish peroxidase Compound I intermediate is the porphyrin π-cation radical, with ^{14}N resonances arising from the pyrrole nitrogens and/or the proximal histidine, and with ^1H resonances from the α-protons on the meso-positions of the porphyrin ring and/or the vinyl groups and from the β-protons on the peripheral methyl and methylene substituents. ENDOR measurements following selective deuterium substitution on the porphyrin periphery proved this conclusion to be correct. For example, the doublet associated with the β-proton coupling constant A_3 (Table 3) disappears upon deuteration of the methyl groups at ring positions 1 and 3.

Equations that relate the hyperfine coupling constants of β- or α-protons to the π-electron spin density of the nearby spin-bearing carbon atom, and the hyperfine coupling of a ^{14}N atom to its own spin density are well known (3,61). When applied to the ENDOR data the resulting spin densities show a striking similarity with an early extended-Hückel prediction (62) as well as with more recent X-α calculations (63) for an electron hole in the a_{2u} orbital and disagreement with that for the a_{1u} orbital.

4.2 ^{17}O ENDOR of $H_2^{17}O_2$-Oxidized Coupound I

Horseradish peroxidase Compound I was prepared with $H_2^{17}O_2$ (18). The ENDOR spectra taken at g = 1.5 show (Fig. 14), in addition to the ^{1}H and ^{14}N resonances, a strong new feature at 11 MHz. This may be interpreted as an unresolved set of 2I = 5 quadrupole lines from ^{17}O, with their center at $\nu_+ = A_2^0 + \nu_0$. Assignment to ν_+, rather than $\nu_- = A_2^0 - \nu_0$, is based on measurements at two microwave frequencies. In addition, the unobserved partner resonance centered at ν_-, is expected to be less intense. This interpretation requires that quadrupole coupling be small, and is supported by molecular orbital calculations (62) from which we predicted the spread of these lines to be less than the ENDOR linewidth (~1.5 MHz).

Upon changing B_0 to the low field edge of the EPR signal, g ~ 2.4, the ENDOR spectrum of HRPI(^{17}O) Fig. 14C is obtained. In addition to the proton and nitrogen signals, an intense ^{17}O resonance, absent in a HRPI(^{16}O) spectrum, is again seen. It is also assigned as the ν_+ transition, by the means described.

To fully characterize the ^{17}O hyperfine tensor, ENDOR spectra were collected at fields across the entire envelope of the EPR spectrum. They were analyzed in terms of an S = 1 oxy-ferryl moiety spin-coupled to the S = 1/2 porphyrin radical gave an ^{17}O tensor of axial symmetry (18; Table 3). These results are in agreement with extended Hückel (62) and X-α (63) calculations when the latter are corrected by a trivial factor of two that was lost in the definition of spin density (63).

4.3 Discussion

Observation of ^{17}O ENDOR in HRP Compound I (18), along with the ^{1}H and ^{14}N ENDOR results (17), unambiguously established the composition of this center. The spin-triplet Fe^{IV} portion of Compound I retains an oxygen atom from the oxidant, and is indeed an oxy-ferryl (Fe^{IV}-O) moiety, with the absence of a large proton coupling in HRPI arguing against protonated (Fe^{IV}-O-H); the spin-coupled doublet species is the porphyrin π-cation radical.

Analysis of the ^{17}O and ^{57}Fe hyperfine constants allows further insights into the properties of this center. The axial symmetry of the observed ^{17}O and also the ^{57}Fe (as determined by the Mössbauer studies) hyperfine tensors naturally suggests that we interpret the data in terms of a triplet oxy-ferryl center whose axis lies normal to the porphyrin cation plane. Taking the two odd electrons of (Fe^{IV}-O) to be in antibonding π-molecular orbitals,

$$\psi_x = (1-c^2)^{1/2} \; d_{xz}^{Fe} - cp_x^O$$

$$\psi_y = (1-c^2)^{1/2} \; d_{yz}^{Fe} - cp_y^O$$

[24]

the measured coupling constants led to the estimate $c^2 \sim 0.25$, corresponding to an oxy-ferryl center whose unpaired odd-electrons are substantially delocalized between the two atoms through d-p bonding, in excellent accord with theoretical expectations (63).

5.   SUBSTRATE INTERACTIONS: ACONITASE (H. Beinert, M.-C. Kennedy, M. Emptage)

The enzyme aconitase (citrate (isocitrate) hydro-lyase) catalyzes the stereospecific interconversion of citrate and isocitrate via the dehydrated intermediate cis-aconitate, where the α- and β-carbons are derived from oxalo-acetate and the γ carbon from acetyl CoA:

Citrate   cis-Aconitate   Isocitrate

The active site of aconitase contains a $[4Fe-4S]^{2+}$ cluster that performs a catalytic function, rather than its far more common use in electron transport (64). The cluster loses one specific iron ion, designated Fe_a, during routine purification to produce an inactive $[3Fe-4S]^+$ structure. However, iron is readily reincorporated under reducing conditions and activity is regained. The diamagnetic $[4Fe-4S]^{2+}$ cluster can be reduced to give the paramagnetic $[4Fe-4S]^+$ (S = 1/2) EPR-active state ($g_{1,2,3}$ = 2.06, 1.93, 1.86) that binds substrate strongly, with 30% retention of activity (64,65).

Mössbauer spectroscopy provided evidence that the cluster in both the 2+ and 1+ oxidation states binds substrate at Fe_a. Substrate binding to the 1+ form was shown to produce pronounced shifts in the EPR g-values. Hyperfine broadening of the EPR signal in the presence of ^{17}O labeled substrate and solvent water indicated that solvent H_2O (or OH^-) and/or the -OH group of substrate can bind to the $[4Fe-4S]^+$ cluster (66), but indicated that carboxyl groups of substrate do not bind.

As part of the effort to elucidate the mechanism by which a [4Fe-4S] cluster acts as a Lewis acid/base, catalyzing dehydration and rehydration reactions, we initiated ^{17}O, ^{13}C, and ^1H ENDOR study of substrate, inhibitor, and solvent binding to the reduced cluster (22,23). The compounds employed that act as substrate and inhibitors are illustrated in Scheme I. Unless stated otherwise, the term substrate will stand for either citrate, isocitrate, cis-aconitate or their mixtures; H_xO will signify H_2O or OH^- when they cannot be distinguished. These studies directly demonstrated that H_xO and the hydroxyl of substrate can bind to the cluster and further disclosed stereo-specific binding of a single carboxyl group. The consequence of this investigation has been a reformulation of the aconitase mechanism of function.

5.1 Binding of H_xO and Substrate Hydroxyl

Fig. 15 shows the ^{17}O ENDOR spectrum of $[4Fe-4S]^+$ in presence of citrate and $H_2^{17}O$, obtained by monitoring the g_2 EPR signal (22). A single-crystal ($\Delta m = \pm 1$) ENDOR spectrum for an ^{17}O ($I = 5/2$), will occur at the frequencies defined by Eqns. 11 or 18. When $A^O/2 > \nu_0 > 3P(2I-1)/2$, the ENDOR pattern consists of two groups of $2I = 5$ lines each, centered at $A^O/2$ and separated by $2\nu_0$. Within each group, the lines are separated by $3P^O$. The features shown in Fig. 1 clearly correspond to the pattern expected for an ^{17}O ENDOR signal as predicted by Eqn. 1, with two groups of $2I = 5$ lines each. The higher frequency group is more intense, as is often the case (2) but elements of the lower frequency (ν_-) quintet are readily observed and the separation between the two groups at 365 mT is exactly $2\nu_0^O = 4.215$ MHz. The pattern is centered at $A^O/2$ which corresponds to $A^O = 8.65$ MHz. From the field dependence, the hyperfine coupling is essentially isotropic. From the separation of the lines within each group, 3P, an effective value of $P = 0.28$ MHz is obtained. Similar signals from $H_2^{17}O$ coordinated to the $[4Fe-4S]^+$ cluster are also seen in the presence of analogues trans-aconitate and nitroisocitrate (22).

No signal was seen that could be assigned to the exchangeable -OH of the substrate citrate, which suggested that substrate is largely bound to the

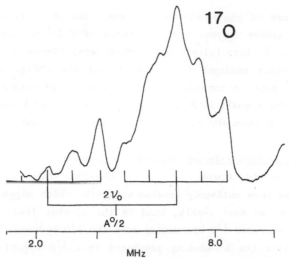

Fig. 15. ^{17}O endor spectrum at g = 1.88 for $[4Fe-4S]^{1+}$ aconitase in the presence of citrate in $H_2^{17}O$ solution. The pattern expected for ^{17}O endor (eq 18) is indicated; the smaller splittings correspond to $3P^O$. Experimental Conditions: T = 2K; B_O = 0.3650 T, modulation amplitude, ~ 0.3 mT; modulation frequency, 100 kHz; microwave frequency, 9.53 GHz; microwave power, 1 μW; time constant, 0.032 s; endor scan rate, 3.0 MHz/s; 3432 scans. The background has been corrected by subtraction of a straight line (Ref. 22).

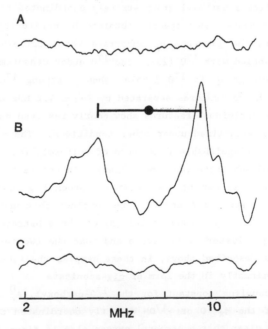

Fig. 16. ^{17}O endor spectra at g = 1.85 of reduced aconitase in the presence of substrate whose -COO$^-$ groups have been individually labeled with ^{17}O. (A) label at α-carboxyl; (B) at β-carboxyl; (C) at γ-carboxyl. (•) indicates $A^O/2$ and (|—|) indicates $2\nu_O$ Conditions: T = 2K; B_O = 0.3815 T; modulation amplitude, ~ 0.4 mT; modulation frequency, 100 KHz; microwave frequency, 9.92 GHz; microwave power, 10 mW; rf power, 10 W; scan rate, 3 MHz/s; time constant, 0.032 s; ~ 1600 scans (Ref. 23).

enzyme in the form of cis-aconitate. However, the observation of ^{17}O endor resonances from active enzyme in the presence of $C-1-^{17}OH$ labeled nitroiso-citrate (1-hydroxy-2-nitro-1,3-propanedicarboxylate) (Scheme I), an inhibitory reaction-intermediate analogue (67), showed that the -OH group of substrate can indeed coordinate to the cluster. This complex of enzyme and inhibitor also gave ^{17}O endor signals in $H_2^{17}O$ enriched solvent, which suggests that the cluster might simultaneously coordinate the -OH of the substrate and H_xO of the solvent (22).

5.2 Binding of Substrate Carboxyl to the Cluster

^{17}O hyperfine line broadening was not observed when the three carboxyl groups of citrate were uniformly labeled with ^{17}O, which suggested that these groups do not, or at most weakly, bind to the cluster (66). However, com-parisons of the measured ^{17}O broadening from the ^{17}O hydroxyl-labeled inhibi-tor and $H_2^{17}O$ with the broadening predicted from ^{17}O hyperfine interaction parameters obtained by endor spectroscopy revealed inherent limitations in the broadening measurements (22). This indicated to us that the issue of carboxyl coordination could be resolved only by endor examination of the ^{17}O carboxyl-labeled samples. We found that despite the negative EPR results, the addition of carboxyl-labeled citrate to the reduced active enzyme does indeed give ^{17}O endor signals from a carboxyl group strongly coordinated to the cluster.

Fig. 16 presents endor spectra obtained by monitoring the g_2 EPR signal of $[4Fe-4S]^+$ in the presence of substrate whose carboxyl groups have been individually labeled with ^{17}O (23). The ^{17}O endor measurements with substrate whose β carboxyl group is ^{17}O labeled show a strong ^{17}O pattern (Fig. 16, spectrum B) with two features separated by $2\nu_0^0 = 4.2$ MHz and centered at $A_2^0/2$ = 7.5 MHz. The individual features show poorly resolved quadrupole splittings that are better visualized under other conditions. The ^{17}O spectrum in B is essentially indistinguishable from those of uniformly (α,β,γ) and doubly (α,β) carboxyl-labeled substrate (data not shown). In contrast, no ^{17}O endor signal was observed when either of the terminal carboxyl groups $(\alpha$ or $\gamma)$ was ^{17}O labeled (Fig. 2, spectra A and C). We conclude that under our experimental conditions, the carboxyl group at C-2 (β) of the substrate interacts with Fe_a of the $[4Fe-4S]^+$ cluster of aconitase and that the two terminal groups $(\alpha$ and $\gamma)$ do not. As described above, in these samples the substrate is thought to be bound predominantly in the form of cis-aconitate.

The ^{17}O coupling constant for the ^{17}O-carboxyl, $A_2^0 = 15$ MHz, is almost double that of the $H_x^{17}O$ or $-^{17}OH$ directly coordinated to the cluster (22), and thus it is clear that a carboxyl oxygen also is directly coordinated. The ^{17}O endor spectrum taken at g_3 (not shown) has the appearance of a simple Larmor-split doublet, which suggests that only one oxygen of the carboxyl group is bound. In general, the two carboxyl oxygens would be magnetically inequivalent even if they were symmetrically coordinated to Fe, and one would expect to see two such doublets at g_3.

To confirm and extend the above measurements, we examined reduced aconitase in the presence of citrate enriched (99%) with ^{13}C in the β-carboxyl (23). An ENDOR spectrum of this sample (Fig. 17) shows a signal from the ^{13}C label, a doublet centered near the Larmor frequency of ^{13}C (3.9 MHz) and split by a small hyperfine coupling constant, $A_2^C = 1$ MHz. In contrast, a sample enriched with ^{13}C at all positions except C-2 and its adjacent carboxyl shows no ^{13}C signal. These observations provide further evidence that only the β-carboxyl of the substrate binds to the cluster in these samples.

^2H and ^3H isotope-exchange experiments (68,69) have shown that the proton extracted from the α-carbon of citrate is stereospecifically returned to the β- carbon upon the formation of isocitrate. This requires that the cis-aconitate intermediate be able to adopt two alternative geometries for binding to the enzyme, one that leads to isocitrate upon hydration and another that leads (back) to citrate. This process involves a switch between coordination of α- and β-carboxyl groups.

The results reported above provide evidence that, at equilibrium, cis-aconitate is bound in the citrate mode (i.e., through the β-carboxyl). This could mean either that the α-carboxyl does not bind at all or that it does so only in an undetectable minority of molecules. To determine whether the active site can accommodate a substrate bound to the cluster through the α-carboxyl (i.e., in the isocitrate mode), we resorted to the use of the nitro-analogue of isocitrate, which lacks the β-carboxyl and therefore cannot be bound in the citrate mode (70). Thus, carboxyl binding to Fe_a might then occur via the α-carboxyl. At pH 8.5 this analogue is bound as the carbanion, and we had shown in previous work (22) that Fe_a binds the hydroxyl at C-1 as well as H_xO from the solvent. Nitroisocitrate was labeled with ^{17}O in the C-1 hydroxyl or in both the C-1 hydroxyl and the C-1 carboxyl, which corresponds to the α carboxyl of substrate.

Fig. 18 shows ENDOR spectra of the enzyme in the presence of nitroisocitrate doubly (carboxyl and hydroxyl) or singly (hydroxyl) labeled with ^{17}O (23). The spectra have been normalized for differences in the absolute EPR intensities. The sample with singly labeled inhibitor (Fig. 4, spectrum B) has the ill-defined pattern reported for ^{17}O-hydroxyl at this field position and under these conditions (22). The doubly labeled sample shows numerous additional well-defined features with peaks separated by $2\nu_O^O$ as expected for an additional ^{17}O (Fig. 4, spectrum A). This must be associated with the ^{17}O-carboxyl group at C-1, which thus is shown to bind to the cluster if that at C-2 is not available. The ^{17}O hyperfine coupling for this group is $A_2^O \approx 13$ MHz, similar to that for the β carboxyl of substrate.

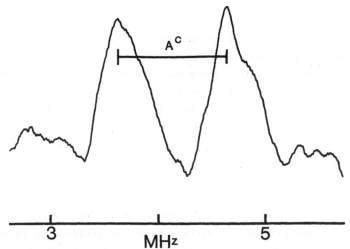

Fig. 17. ^{13}C endor spectrum at g - 1.85 of reduced aconitase in the presence of substrate labeled with ^{13}C in the β carboxyl. The ^{13}C hyperfine coupling, A^C is indicated by ($|-|$). A sweep artifact shifts the center of the pattern slightly to high frequency from ν_C - 3.93 MHz. Conditions: same as in Fig. 16 except $B_0 = 0.3675$ T; modulation amplitude, 0.1 mT; microwave frequency, 9.53 GHz, microwave power, 1 mW; 1360 scans (Ref. 23).

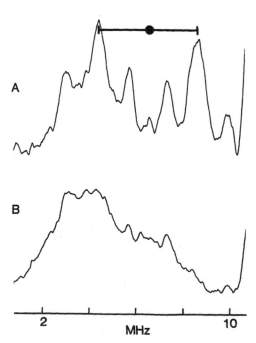

Fig. 18. ^{17}O endor spectrum at g - g_y - 1.87(2) of reduced aconitase plus ^{17}O labeled nitroisocitrate. (A) doubly labeled: $-^{17}OH$ and $-^{17}O_2^-$ at C-1 (α position); (B) singly labeled: $-^{17}OH$; ($\cdot$) and $|-|$ as in Fig. 16. Conditions: as in Fig. 16 except field 0.3675 T; modulation amplitude, 0.1 mT; microwave frequency, 9.53 GHz, microwave power, 1 mW; 1360 scans (Ref. 23).

5.3 Discussion

Our ENDOR data decisively shows that at equilibrium only the β-carboxyl (Fig. 15) of substrate is coordinated to the paramagnetic [4Fe-4S]$^+$ cluster along with H_xO of the solvent. As there is no indication for binding of -OH from substrate, we interpret this to mean the cis-aconitate is the predominant species bound and that it is bound in the citrate mode (i.e., through the β-carboxyl). In contrast, the nitro-analogue of isocitrate, lacking the carboxyl at C-2 (corresponding to β), is coordinated to the cluster through the C-1(α)-carboxyl (i.e., in the isocitrate mode) along with -OH at C-1 and H_xO from the solvent. Thus, it is obvious that the enzyme can accommodate substrates in these two alternative modes, as expected from previous mechanistic arguments. These observations fit well with the model for aconitase action proposed by Gahan et al. (71) that involves carboxyl coordination as a key element.

The details obtained from the endor work, in addition to past results, suggest the following mechanism for aconitase (72). In this scheme the R group on the substrate is CH_2COO^-, -B: represents the amino acid side-chain that stereospecifically transfers a proton between citrate and isocitrate, and X is either water or a protein ligand. In the presence of cis-aconitate, bound water can freely exchange with solvent water. The [4Fe-4S] cluster acts both as a Lewis acid to facilitate the dehydration of citrate and isocitrate, and as an activator of the carbon adjacent to the bound carboxyl group of cis-aconitate for attack by a bound hydroxyl. Together, the stereochemistry of the reaction, the specific transfer of the proton from citrate to isocitrate, and the coordination of a single carboxyl and adjacent hydroxyl group of substrate require that cis-aconitate disengage from the active site and rotate 180° before completing the catalytic cycle.

6.   RIESKE [2Fe-2S] CENTER OF PHTHALATE DIOXYGENASE (J.A. Fee, D.P. Ballou)

The Rieske protein was first identified by J.S. Rieske and co-workers in 1964 who recorded its EPR and optical spectra and recognized them to be quite different from those of other iron-sulfur proteins (73,74). Subsequent studies during the next quarter of a century established the presence of a [2Fe-2S] cluster (75), but failed to identify the structural differences distinguishing it from the ferredoxin-type [2Fe-2S] clusters that are coordinated to proteins by four cysteine ligands. In this context, the distinguishing properties of the Rieske center are an EPR spectrum having $g_{av} \sim 1.91$ ($g_1 = 2.01$; $g_2 = 1.92$; $g_3 = 1.76$) compared with $g_{av} \sim 1.96$ for ferredoxins (See 76,77 for review).

A battery of physical observations including visible, circular dichroism, EPR, endor, Mössbauer, and Raman spectroscopies indicates that the Rieske-type for the Rieske cluster structure in which terminal S and N atoms are provided by cysteine and histidine imidazole.

With this background we performed 9 and 35 GHz endor studies on globally and selectively ^{15}N-enriched phthalate dioxygenase (PDO) prepared from Pseudomonas (P.) cepacia (21). The results show unequivocally that two imidazole nitrogens from histidine coordinate to the cluster. Through analysis of the polycrystalline endor spectra (6,8,9), we established that there are two histidines coordinated to Fe^{+2}, described the coordination geometry about this ion, and determined the bonding parameters of the two nitrogen ligands.

6.1 Ligand Identification

Four different samples of P. cepacia phthalate dioxygenase were prepared:

Sample 1: ^{14}N (natural abundance)

Sample 2: ^{15}N uniformly labelled ($\geq$ 99%)

Sample 3: ^{15}N-Histidine ($\geq$ 95%) in ^{14}N background

Sample 4: ^{14}N-Histidine in ^{15}N background ($\geq$ 99%)

Each has a specific isotopic labelling pattern, with histidine-enriched samples obtained from an auxotrophe, strain DB0110.

The frozen-solution EPR spectrum of PDO is characteristic of a g-tensor with rhombic symmetry; the g_i were given above. Endor measurements were performed at X-Band ($\nu(M) \sim$ 9.4 GHz) across the EPR spectrum of PDO from P. cepacia grown either on natural abundance (^{14}N) or ^{15}N-enriched media. As previously reported (19), the endor spectrum at g_2 of natural abundance PDO (Sample 1) shows a broad pattern of ^{14}N resonances centered at $\sim$ 3.5 MHz (Fig. 19). No attempts at detailed analysis of this signal were made in the original report; this will be done below. Spectra of ^{15}N-enriched PDO (Sample 2) taken at g_2 exhibit two sharp doublets that are centered at 2.55 and 3.35 MHz; they can be assigned to ^{15}N because each shows a splitting of $2\nu(^{15}N) =$ 3.1 MHz. An endor pattern taken at g_2 corresponds to a distribution of orientations and hyperfine anisotropy and might, in principle, cause a single ^{15}N to give more than a single pair of peaks. However, an endor spectrum (see below) taken at g_1 corresponds to a unique molecular orientation and also exhibits two doublets; this proves that the ^{15}N endor signals represent two magnetically distinct nitrogenous ligands coordinated to the [2Fe-2S] cluster.

These two nitrogenous ligands are identified as histidines by endor measurements on PDO specifically labeled with ^{15}N-histidine. Fig. 20A is a Q-band endor spectrum showing the ^{15}N resonances of ^{15}N-histidine-labeled PDO (Sample 3). The Q-band peaks at $\nu_+ =$ 8.2 and 9.2 MHz correspond (Eqn. 2) to the $\nu_+ =$ 4.1 and 4.8 MHz lines at X-Band (Fig. 19). The spectrum for uniformly ^{15}N-labeled PDO (Sample 2) is identical (Curve B), demonstrating that the two nitrogenous ligands are histidines.

Comparison of Curve C (Sample 4) and Curve D (Sample 1) provides additional proof that both nitrogenous ligands are histidine residues. Curve C is the spectrum of PDO labeled with $(^{15}NH_4)_2SO_4$ and ^{14}N-histidine (Sample 4); Curve D is the spectrum of ^{14}N-PDO (natural isotopic abundance; Sample 1). Sharp features from ^{14}N at 7.4, 6.5, and 5.6 MHz are visible both in curve D, the spectrum of ^{14}N PDO, and in curve C, that from PDO with His(^{14}N) in a ^{15}N background. The absence of additional ^{15}N signals in curve C, where non-histidyl nitrogens have been labelled with ^{15}N, indicates that the only nitrogen ligands to the [2Fe-2S] center are from histidine.

The ^{14}N resonances of Sample 4 (Curve C) are more resolved than those of Sample 1 (Curve D); this appears to result from unidentified relaxation processes associated with the bulk ^{14}N nuclei of the protein.

6.2 Determination of ^{15}N Hyperfine Tensors

Determination of the ^{15}N hyperfine tensors by simulation of endor X-Band spectra measured across the EPR envelope allows us to characterize the Fe-N bonding and to describe the orientation of the Fe-N bonds with respect to the

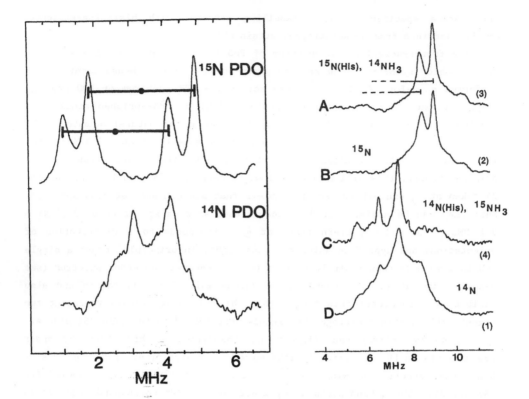

Fig. 19. (Left) Endor spectra of ^{15}N-enriched (Sample 2) and ^{14}N-natural abundance (Sample 1) PDO from P. cepacia taken at $g_2 = 1.92$. The assignments to A/2 (•) and $2\nu_N$ (|—|) of each doublet are indicated. Conditions: $B_o = 0.3596$ (^{14}N), 0.3588 T (^{15}N); T = 2K; microwave power, 10 μW; 100-kHz field modulation, 0.1 mT; rf power, 20 W; rf scan rate, 2.5 MHz/s, 1000 scans (Ref. 21).

Fig. 20. (Right) Endor spectra at $g_2 = 1.92$ and Q-Band frequency of the PDO protein extracted from (A) auxotroph of P. cepacia grown on a medium of ^{15}N histidine and $(NH_4)_2SO_4$ containing natural abundance nitrogen (Sample 3); (B) normal P. cepacia grown on ^{15}N labelled $(NH_4)_2SO_4$ (Sample 2); (C) auxotroph of P. cepacia grown on a medium of histidine containing natural-abundance nitrogen and ^{15}N-labelled $(NH_4)_2SO_4$ (Sample 4); (D) normal P. cepacia grown on $(NH_4)_2SO_4$ containing natural- abundance nitrogen (Sample 1). For ease of reference the traces are labelled with the sample number in parentheses. Conditions: T = 2K; microwave power, 50 μW; microwave frequency, 35.3 GHz; 100-kHz field modulation, 0.32 mT; rf power, 20 W; scan rate, 2.5 MHz/s, 300 scans (Ref. 21).

g-tensors. Fig. 21 shows the ν_+ feature of the two ^{15}N doublets recorded at various positions along the EPR envelope; the ν_- partners (not shown) give an identical pattern at a frequency that is lower by $2\nu(^{15}N) = 3.1$ MHz.

The principal values and relative orientations of the hyperfine coupling tensors for the two ^{15}N ligands have been determined from simulations of the full set of endor spectra following the procedures described above. Spectra at all observing fields could be simulated using the two ^{15}N hyperfine tensors given in Table 4. As seen in the selected data presented in Fig. 21, the simulations fully reproduce the positions of the two sharp resonances, as well as the breadth of the pattern to higher frequency and other resolved features.

TABLE 4

Hyperfine and Quadrupole Tensor Principal Values (MHz) and Orientations (degrees) Relative to g-tensor Principal Axes for the Histidyl Ligands to the [2Fe-2S] Cluster of Phthalate Dioxygenase from P. cepacia.[a] and to the Heme of Met-Mb[b]

Principal values		Site 1	Site 2	Met-Mb
$\underset{\sim}{A}(^{15}N)$:	A_1	4.6(2)	6.4(1)	−11.31(6)
	A_2	5.4(1)	7.0(1)	−11.66(6)
	A_3	8.1(1)	9.8(2)	−16.17(15)
$\underset{\sim}{P}(^{14}N)$:	P_1	0.85(8)	0.80(8)	0.81(6)
	P_2	0.45(8)	0.35(8)	0.31(6)
	P_3	−1.3	−1.15	−1.12(6)
Euler angles[c]				
	α	0(30)	0(10)	
	β	35(5)	50(5)	
	γ	0(10)	0(40)	

a) The absolute signs of the A_i are not determined; for an individual site all have the same sign. The relative signs of the P_i are determined by simulations that use $\underset{\sim}{A}(^{14}N) = (g(^{14}N)/g(^{15}N))\,\underset{\sim}{A}(^{15}N)$; the absolute signs come from comparison with those of met-Mb. P_3 is calculated from $P_1 + P_2 + P_3 = 0$; its relative error is the sum of those for P_1 and P_2. Data from Ref. 21.
b) Scholes et al. (80). The tabulated $A_i(^{15}N)$ were calculated from the published $A_i(^{14}N)$ using the nuclear g-factors and the labelling of the tensor components has been made to conform with ours.
c) Euler angles relate the $\underset{\sim}{A}$-tensor to the reference g-frame (Fig. 2).

For neither site were the simulations particularly sensitive to the angle α that defines the orientation of the Fe-N bond with respect to the g_1-g_3 plane; relative orientations of $0 \leq |\alpha| \leq 30°$ for both sites could be accommodated without any visible changes in the simulated ^{15}N endor pattern, provided corresponding minor modifications were made in the A_i. However, a value of α in the vicinity of 90° is not permitted in either case. Moreover,

582

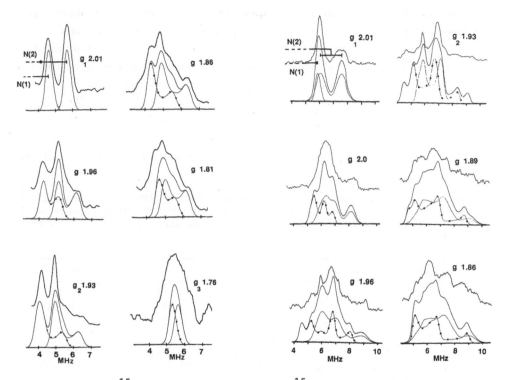

Fig. 21. (Left) ^{15}N endor spectra of the two ^{15}N sites of phthalate dioxgenase (Sample 1) showing the ν_+ resonances at selected magnetic fields. Computer simulations, including those for the individual ^{15}N and their sum, employ the hyperfine interaction parameters of Table 4 and eq 22 are shown below the corresponding spectra; the bold line is the sum of simulations for N(1) and N(2). The component linewidths were taken to be equal for the two sites (0.2 MHz) and independent of angle; relative intensities for the individual sites are normalized. Differential broadening of the simulated endor peaks were seen upon inclusion of an epr linewidth of more than 2.0 mT; all spectra were simulated with epr linewidths of 2.0 mT. Experimental conditions are as described in Fig. 19, with a microwave frequency of 9.6 GHz and g—values as indicated. Each spectrum is the average of 1000 scans (Ref. 21).

Fig. 22. (Right) ^{14}N endor spectra of PDO (Sample 4) showing the ν_+ resonances at selected magnetic fields. Computer simulations that employ the parameters listed in Table 4 are shown below the corresponding spectra. Simulations for the individual ^{14}N plus their sum are shown. The endor linewidths were taken as 0.2 and 0.5 MHz for N(1) and N(2), respectively, independent of orientation except at g_2, where a linewidth of 0.2 MHz was used for both; relative amplitudes were adjusted for best fit. See Legend to Fig. 21 (Ref. 21).

[14]N analyses presented below require $0 \leq |\alpha(2)| \leq 10°$. Crystal field analysis of the g-tensors of [2Fe-2S] clusters has raised the question as to the orientation of the g-tensor axes with respect to the plane containing the four protein ligands to the cluster (76). The present data clearly indicate that the ligand plane most closely corresponds with the g_1-g_3 plane.

The two ^{15}N tensors have nearly axial symmetry with the unique axis assigned to the Fe-N bond. Thus, simulations should not be influenced by rotations of the tensor about this bond, as described by the angle γ, and this is the case.

6.3 Analysis of Fe-[15]N Bonding

Analysis of the isotropic and the anisotropic components of the hyperfine coupling confirms that both nitrogens are coordinated to iron and can be used to characterize the bonding. The hyperfine tensors for N(1) and N(2) both are effectively of axial symmetry (Table 4); the considerable isotropic component ($A_{iso} = (1/3)\Sigma A_i$; $A_{iso}(1) = 6$ MHz; $A_{iso}(2) = 7.7$ MHz) arises from s-orbital density on the two nitrogens, which proves that they are covalently bonded to Fe. The fact that the couplings for the two nitrogens are similar in magnitude proves that they are associated with two different histidines directly coordinated to iron, and not with the coordinated and remote nitrogen of a single such ligand: the magnitude of the hyperfine coupling to the remote nitrogen of metal-coordinated histidyl imidazole is reduced by more than twenty-fold from that of the coordinated nitrogen.

The $\underline{A}(i)$-tensors, $i = 1,2$ of Table 4 are defined in terms of the nuclear hyperfine interaction with the total spin, $S = 1/2$ of the spin-coupled [Fe^{+3}, Fe^{+2}] pair. The spin-coupling model of Gibson et al. (78) can be used to relate these tensors to the fundamental hyperfine tensors, $\underline{a}(i)$, that describe the interaction of nitrogen with the spin of the isolated Fe^{+2} ion:

$$\underline{a} = -3\underline{A}/4. \qquad [25]$$

Justification for taking both ligands to be coordinated to Fe^{+2} is given below (for coordination to Fe^{+3} the relation is $\underline{a} = 3\underline{A}/7$). The isotropic component of the hyperfine coupling, $a_{iso} = -3A_{iso}/4$, is related to f_s, the fraction of unpaired electron spin that resides in the 2s nitrogen orbital; the anisotropic part of the ^{15}N hyperfine coupling has a local contribution, a_{2p}, that arises from unpaired density in the 2p orbital on nitrogen (f_{2p}), and from direct dipolar coupling to the ferrous ion, a_{3d}.

Our approximate analysis of the ^{15}N hyperfine constants gave $f_{2s} \sim f_{2P} \sim$ 2% for both nitrogen ligands; this is nicely self-consistent in that $f_{2p}/f_{2s} \sim$ 1, which is reasonable for the sp^n hybrid of an imidazole nitrogen (79,80).

6.4 Analysis of ^{14}N endor Spectra and ^{14}N Quadrupole Tensors

The knowledge of the two ^{15}N hyperfine tensors gives the corresponding ^{14}N tensors through Eqn. 3. This knowledge along with the enhanced resolution of the ^{14}N endor patterns from PDO with (^{14}N)-histidine in a ^{15}N background (Sample 4; Fig. 20C) give us the further opportunity to analyze the ^{14}N spectra and to determine the ^{14}N quadrupole coupling tensors. This determination permits us to define the orientation of the histidine rings with respect to the NFeN plane and gives the quadrupole tensor components, which are additional bonding parameters.

^{14}N endor spectra of Sample 4 were recorded at various positions along the EPR envelope at Q-band frequency. The ν_+ features (Eqn. 1) are presented in Fig. 22. The starting point for simulating these spectra was as follows. 1) The principal values of $\underline{A}(^{14}N)$ were calculated from those of the $\underline{A}(^{15}N)$ using Eqn. 3. 2) The $\underline{A}$ and $\underline{P}$ tensors were taken as coaxial; because $\underline{A}$ is essentially axial this corresponds to assigning a principal axis of $\underline{P}$ to lie along A_3. This assumption is valid for the histidine coordinated to Fe^{+3} in metmyoglobin (80). 3) The Euler angles α and β given in Table 4 were used. 4) The principal values of the quadrupole tensor were taken to be the same for N(1) and N(2). 5) Finally, the quadrupole tensor is traceless, $P_1 + P_2 + P_3 = 0$ (2). With these restrictions the ^{14}N simulations depend on four unknown parameters: two principal values of $\underline{P}$, namely P_1 and P_2, and $\gamma(1)$ and $\gamma(2)$, the rotations of the two tensors about their Fe-N bond.

The process was initiated by assigning the single-crystal-like $\nu_{+,m}(^{14}N)$ lines at g_1 (Fig. 22) from the corresponding $\nu_+(^{15}N)$ lines (Fig. 21). Use of the $\nu_+(^{15}N)$ transition frequencies at X-band (Fig. 21; 5.6, 4.55 MHz) and the ratio of nuclear g-factors predicts that the $\nu_+(^{14}N)$ transitions at Q-band should be pairs of lines centered at 6.9 and 6.2 MHz and split by the quadrupole interaction (Eqn. 18). Examination of Fig. 22 for $g = g_1$ yields quadrupole splittings of $3P_{obs} \sim 1.5$ MHz for N(2) and ≤ 0.1 MHz for N(1). Initially taking $\gamma(1) = \gamma(2) = 0$, a set of $[P_1,P_2]$ values that reproduced P_{obs} for each ligand at g_1 was determined. For a given value of the splitting at g_1, this set is a straight line in a plot of P_1 vs. P_2, Fig. 23. This is implicit in Eqn. 17a, but is seen best by rewriting the more approximate

$$P_2 = \frac{C_1}{C_2} P_1 + \frac{g^2A^2}{C_2}P_{obs} \qquad [26]$$

Eqn. 17b to give Eqn. 26 (corrected from Ref. 21), where

$$C_1 = [(\underline{\ell}\cdot\tilde{g}\tilde{\underline{A}}\cdot U_3\cdot{}^g\underline{A}\cdot\underline{\ell}) - (\underline{\ell}\cdot g^g\tilde{\underline{A}}\cdot U_1\cdot{}^g\underline{A}\cdot\underline{\ell})]$$

$$C_2 = [(\underline{\ell}\cdot g^g\tilde{\underline{A}}\cdot U_2\cdot{}^g\underline{A}\cdot\underline{\ell}) - (\ell\underline{\ell}\cdot g^g\tilde{\underline{A}}\cdot U_3\cdot{}^g\underline{A}\cdot\underline{\ell})] \qquad [27]$$

$$U_1 = \tilde{\underline{M}}(^1{}_0{}_0)\underline{M} \qquad \tilde{U}_2 = \underline{M}(^0{}_1{}_0)\underline{M} \qquad U_3 = \tilde{\underline{M}}(^0{}_0{}_1)\underline{M}$$

All the other symbols are as described in Section 2. Taking into account uncertainties in peak positions for each ^{14}N, the possible $[P_1,P_2]$ pairs are encompassed by a narrow band of points flanking the lines plotted in Fig. 23. On the assumption that the principal values of $\underline{P}$ are the same for the two ^{14}N, the region near the intersection of the lines for N(1) and N(2), of extent ~ ± 0.1 MHz in P_1 and P_2, gave a set of trial values for $[P_1,P_2]$ (Fig. 23). Parameter pairs within this region were next tested to see if they could reproduce the overall spread of the pattern observed at g = 1.92 (Fig. 22). Further selection of the best parameter sets involved simulations of endor spectra at multiple fields. This procedure was then repeated over a grid of values for $[\gamma(1),\gamma(2)]$, thereby determining the ranges of the parameters $[P_1,P_2,\gamma(1),\gamma(2)]$ that yielded the best fits. Finally, the requirement that the two quadrupole tensors be identical was relaxed and further simulations were done; these simulations in addition explored a grid of $[\alpha(1),\alpha(2)]$ values, $0 \leq |\alpha(i)| \leq 30$. The parameters best reproducing the experiments are listed in Table 4 and the simulations are shown in Fig. 22.

The simulations reproduce the observations in a quite reasonable manner, particularly for $g \geq g_2$, where the experimental signal/noise is best. It is noteworthy that the tensor values <u>independently</u> arrived at are almost identical to those for the axial histidine bound to the heme of aquomet-Mb (Table 4). In particular, the components along the Fe-N bonds, P_3, correspond, as they should. Because of this agreement we assume that we may use the Mb results to deduce the physical orientation of the imidazole planes in PDO relative to the quadrupole axes. In Mb, the tensor component perpendicular to this plane is $P_2 = 0.31$ (80); we assign the same for PDO. Thus, for $\alpha = 0$ the parameter γ corresponds to the dihedral angle between the imidazole and the g_1-g_3 planes, with the latter corresponding to the N(1)FeN(2) plane.

6.5 <u>Site of Histidine Ligation</u>

The data from Mössbauer spectroscopy (81) suggested that the two non-cysteine ligands to the [2Fe-2S] cluster of <u>Thermus</u> Rieske protein are coordinated to the ferrous site, and such unsymmetrical coordination is consistent with the Raman measurements (82). However, the Mössbauer data could not rule out ligation to different iron sites if the two non-cysteine ligands were very dissimilar. The determination by endor spectroscopy that two ligands coordinated to the cluster are nitrogens from histidine and that these have similar hyperfine parameters indicates that both are bonded to Fe^{2+}.

6.6 <u>Structure of the Cluster</u>

The analysis of the ^{15}N endor data indicates that the NFeN and g_1-g_3 planes correspond (Table 4; $\alpha \approx 0$). This leads to a geometric model (Fig. 24) in which the four protein-donated ligands and two iron ions lie in the g_1-g_3 plane with either g_3 or g_1 lying along the Fe-Fe vector. With the former

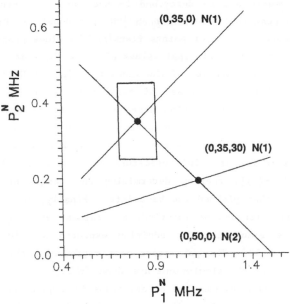

Fig. 23. Plot of P_1 vs. P_2 such that the pair of values yields the quadrupole splitting of 0 MHz and 1.5 MHz for N(1) and N(2) respectively at g_1 (Fig. 22). The lines are calculated using eq 26 and ^{14}N hyperfine tensors calculated from the ^{15}N tensors in Table 4. The numbers in the brackets are the Euler angles α, β and γ. The intersection points represent values of P_1 and P_2 such that quadrupole coupling tensors of the two nitrogens have identical principal values. The box represents the range of uncertainty in determining the intersection point due to the uncertainty of $\pm$ 0.1 MHz in measuring the quadrupole splitting for N(1) and N(2) at g_1. Note the difference in scale for the abscissa and ordinate.

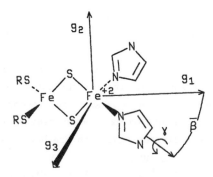

Fig. 24. Structure of the Rieske-type [2Fe-2S] cluster of PDO as determined by endor spectroscopy. The [2Fe-2S] core and the (g_1-g_2) plane lie in the paper. The angles γ and $\bar{\beta}$, with $\beta = \pi/2-\beta$, are given in Table 4.

assignment the angle between Fe-N and Fe-Fe vectors is β, and the bite angle subtended by the two Fe-N vectors is $\beta(1) + \beta(2) \sim 85(10)°$. This assignment is favored by a crystal-field analysis of the g-values for [2Fe-2S] clusters with $g_{av} \sim 1.91$ (76). In contrast, if the Fe-Fe vector corresponds to g_1, as appears to be the case for ordinary [2Fe-2S] ferredoxins with $g_{av} \sim 1.96$, then the angle between Fe-N and Fe-Fe vectors is $\bar{\beta} = (\pi/2)-\beta$ and the bite angle is $\bar{\beta}(1) + \bar{\beta}(2) \sim 95(10)°$. Because the slightly larger bite angle seems more attractive, the latter assignment has been used in Fig. 24; however, ^{57}Fe endor studies or single-crystal measurements will be required for a final determination. In either case, the torsion angle twisting an imidazole about the Fe-N bond and out of the NFeN plane is γ, which appears to be relatively small for both of the histidines (Table 4). These results have been complemented through the determination of Fe-S and Fe-N bondlengths by EXAFS measurements (83).

The resulting geometry at Fe^{+2}, if idealized, would have tetrahedral coordination with the two histidyl imidazoles lying within the NFeN plane. Although this idealized geometry would have unphysically short contacts between the two imidazoles, the range of acceptable values for the $\alpha(i)$, $\beta(i)$, and especially $\gamma(i)$, $i = 1,2$ easily is sufficient to permit a reasonable geometry. Indeed, the simple observation that the quadrupole splittings observed at g_1 differ sharply for N(1) and N(2) (Fig. 22) although the principal values do not, by itself demonstrates that the geometry at Fe^{+2} is not ideally symmetric; in our fits this is indicated by the different values of β and is further allowed by the appreciable uncertainties in the α and γ angles.

Interestingly, although the same endor linewidths can be used to describe the spectra for ^{15}N(1) and ^{15}N(2) (Fig. 21), this is not so for ^{14}N(1) and ^{14}N(2); this can be seen immediately in the spectrum at g_1, where the linewidth for ^{14}N(2) is greater than that for ^{14}N(1) (Fig. 22). We interpret this to mean that histidine (2) has a greater conformational mobility than histidine (1) when the protein is in fluid solution, and that this leads to a distribution in values of $\gamma(2)$ in the frozen solutions under study; this distribution causes a spread in quadrupole splittings and an apparent increase in linewidth. The absence of any effects of the $\gamma(2)$ distribution on the ^{15}N spectra is a consequence of the axial symmetry of the hyperfine tensors and the invariance of the spectra to rotations about the Fe-N bond (angle γ).

6.7 Summary

1) The use of 9 and 35 GHz endor spectroscopy on selectively ^{15}N-enriched phthalate dioxygenase protein samples prepared from wild-type P. cepacia and histidine auxotrophs has unequivocally demonstrated that two imidazole nitrogens from histidine residues are coordinated to Fe of the Rieske [2Fe-2S] cluster. This is in contrast to the classical ferredoxin-type [2Fe-2S]

588

centers in which all ligation is by sulfur of cysteine residues. No other nitrogen couplings to the iron cluster were revealed by endor.

2) The combination of data from endor, Mössbauer, and Raman spectroscopies indicates that both histidines are coordinated to the Fe^{+2} site of the reduced cluster.

3) The ^{15}N hyperfine tensor for each coordinated nitrogen has been determined (Table 4). Analysis of these tensors shows that each can be reasonably interpreted in terms of an sp^n hybrid orbital on histidyl nitrogen coordinated to Fe^{+2}. Analysis of the ^{14}N endor spectra shows the quadrupole tensors for the two ^{14}N ligands to be nearly identical to those of the histidine ^{14}N coordinated to the heme-iron of met-Mb.

4) The orientations of the $\underline{A}$ and $\underline{P}$ tensors relative to the g-tensor is consistent with roughly tetrahedral coordination at Fe^{+2}. The ligand plane as defined by the two Fe and the coordinating N-atoms provided by the protein is found to correspond to the g_1-g_3 plane. The analyses suggest that the dihedral twist of the two imidazole rings relative to the NFeN plane, which causes the ligands to look like propeller blades, is relatively small for both.

5) Evidence is presented that the two histidines have different conformational mobilities.

6) Because the ^{14}N and ^{1}H endor patterns for the Thermus and mitochondrial Rieske proteins are virtually indistinguishable from those of PDO (19,20) we infer that the structure presented here is generally applicable to Rieske-type centers.

ACKNOWLEDGMENTS

Some of the senior investigators whose talents have made possible the development of ENDOR as an incisive tool for the study of metalloenzymes are listed in the section headings; others are to be found in the references listed in Table 1. Likewise, the junior investigators at Northwestern University and other institutions, whose decisive contributions cannot be overpraised, are so listed. This work has been made possible by the generous support of the National Institutes of Health, the National Science Foundation, The United States Department of Agriculture, and the Petroleum Research Foundation.

REFERENCES

1   G. Feher, Phys. Rev., 114 (1959) 1219-1244.
2   A. Abragram and B. Bleaney in Electron Paramagnetic Resonance of Transition Ions, Clarendon Press, Oxford, 1970.
3   N.M. Atherton in Electron Spin Resonance, Halstead, New York, 1973.
4   B.M. Hoffman, R.A. Venters, J.E. Roberts, M.J. Nelson and W.H. Orme-Johnson, J. Am. Chem. Soc., 104 (1982) 4711-4712.

5   R.A. Venters, M.J. Nelson, P.A. McLean, A.E. True, M.A. Levy,
    B.M. Hoffman and W.H. Orme-Johnson, J. Am. Chem. Soc., 108 (1986)
    3487-3498.
6   A.E. True, M.J. Nelson, R.A. Venters, W.H. Orme-Johnson and B.M. Hoffman,
    J. Am. Chem. Soc., 110 (1988) 1935-1943.
7   G.H. Rist and J.S. Hyde, J. Chem. Phys., 52 (1970) 4633-4643.
8   B.M. Hoffman, J. Martinsen and R.A. Venters, J. Mag. Res., 59 (1984)
    110-123.
9   B.M. Hoffman, R.A. Venters and J. Martinsen, J. Mag. Res., 62 (1985)
    537-542.
10 B.M. Hoffman and R.J. Gurbiel, J. Magn. Reson., in Press.
11 R.G. Rist, J.S. Hyde and T. Vanngard, Proc. Nat. Acad. Sci. USA, 67
    (1970) 79-86.
12 J.E. Roberts, T.G. Brown, B.M. Hoffman, and J. Peisach, J. Am. Chem.
    Soc., 102 (1980) 825-829.
13 J.E. Roberts, J.F. Cline, V. Lum, H.B. Gray, H. Freeman, J. Peisach,
    B. Reinhammer and B.M. Hoffman, J. Am. Chem. Soc., 106 (1984) 5324.
14 B.M. Hoffman, J.E. Roberts and W.H. Orme-Johnson, J. Am. Chem. Soc., 104
    (1982) 860-862.
15 P.A. McLean, A.E. True, M.J. Nelson, S. Chapman, M.R. Godfrey, B.-K. Teo,
    W.H. Orme-Johnson and B.M. Hoffman, J. Am. Chem. Soc., 109 (1987)
    943-945.
16 C.P. Scholes, in Multiple Electron Resonance Spectroscopy, M.M. Dorio and
    J.H. Freed, Eds., Plenum Press, New York (1979) Chapter 8, 297-329.
17 J.E. Roberts, B.M. Hoffman, R. Rutter and L.P. Hager, J. Biol. Chem., 256
    (1981) 2118-2121.
18 J.E. Roberts, B.M. Hoffman, R. Rutter and L.P. Hager, J. Am. Chem. Soc.,
    103 (1981) 7654-7656.
19 J.F. Cline, B.M. Hoffman, W.B. Mims, E. LaHaie, D.P. Ballou and J.A. Fee,
    J. Biol. Chem., 260 (1985) 3251-3254.
20 J. Telser, B.M. Hoffman, R. LoBrutto, T. Ohnishi, A-L. Tsai, D. Simpkin,
    and G. Palmer, FEBS Letters, 214 (1987) 117-121.
21 R.J. Gurbiel, C.J. Batie, M. Sivaraja, A.E. True, J.A. Fee, B.M. Hoffman,
    and D.P. Ballou, Biochem., 28 (1989) 4861-4871.
22 J. Telser, M.H. Emptage, H. Merkle, M.C. Kennedy, H. Beinert, and B.M.
    Hoffman, J. Biol. Chem., 261 (1986) 4840-4846.
23 M.C. Kennedy, M. Werst, J. Telser, M.H. Emptage, H. Beinert, and B.M.
    Hoffman, Proc. Nat. Acad. Sci. USA, 84 (1987) 8854-8858.
24 A. Schweiger, Struct. and Bonding, 51 (1982) 1-123.
25 M.M. Dorio and J.H. Freed, Eds., Multiple Electron Spin Resonance
    Spectroscopy, Plenum Press, New York, 1979.
26 L. Kevan and L.D. Kispert, Electron Spin Double Resonance Spectroscopy,
    John Wiley & Sons, New York, 1976.
27 K. Mobius and W. Lubitz, in: Biological Magnetic Resonance, L.J. Berliner
    and J. Reuben, Eds., Plenum Press, New York, 1987, Vol 7, Chapter 3,
    129-247.
28 J. Telser, M.J. Benecky, M.W.W. Adams, L.E. Mortenson, and B.M. Hoffman,
    J. Biol. Chem., 262 (1987) 6589-6594.
29 J. Telser, M.J. Benecky, M.W.W. Adams, L.E. Mortenson, and B.M. Hoffman,
    J. Biol. Chem., 261 (1986) 13536-13541.
30 G. Wang, M.J. Benecky, B.H. Huynh, J.F. Cline, M.W.W. Adams, L.E.
    Mortenson, B.M. Hoffman, and E. Munck, J. Biol. Chem., 259 (1984) 14328.
31 J.F. Cline, P.A. Janick, L.M. Siegel, and Brian M. Hoffman, Biochem., 24
    (1985) 7942-7947.
32 J. Cline, B. Reinhammer, P. Jensen, R. Venters, and B.M. Hoffman, J.
    Biol. Chem., 258 (1983) 5124-5128.
33 J.F. Cline, P.A. Janick, L.M. Siegel, and B.M. Hoffman, Biochem., 25
    (1986) 4647-4654.
34 J. Cline, B. Reinhammer, P. Jensen, R. Venters, and B.M. Hoffman, J.
    Biol. Chem., 258 (1983) 5124-5128.

35   B.M. Hoffman, in: "Biological Chemistry of Iron", H.B. Dunford, et al. editors, D. Reidel & Co., 391-403 (1982).

36   B.M. Hoffman, J.E. Roberts, C.H. Kang and E. Margoliash, J. Biol. Chem., 256 (1981) 6556-6594.

37   J. Peisach, W.E. Blumberg, E.T. Lode and M.J. Coon, J. Biol. Chem., 246 (1971) 5877-5881.

38   We choose to work in a right-handed coordinate system in which $g_1' > g_2' > g_3'$. This is identical to that of the fine-structure Hamiltonian (Eqn. 1) for $\lambda < 0$ with the axes corresponding as $(x,y,z) \longleftrightarrow (1,2,3)$. If the convention $\lambda > 0$ is adopted, then our working coordinate system is related to that of Eqn. 1 by a 90° rotation about the z-axis.

39   Our treatment of coordination transformations follows, J. Mathews and R.L. Walker, Mathematical Methods of Physics, W.A. Benjamin, Inc., New York, 1965.

40   K.-A. Thuomas and A. Lund, J. Magn. Reson., 18 (1975) 12-21.

41   M. Iwasaki, J. Magn. Reson., 16 (1974) 417-423.

42   B.M. Hoffman and R.J. Gurbiel, J. Magn. Reson., (1989), in press.

43   F.K. Kneubuhl, J. Chem. Phys., 33 (1960) 1074-1078.

44   W.H. Orme-Johnson, Ann. Rev. Biophys. Chem., 14 (1985) 419-459.

45   M.J. Nelson, P.A. Lindahl and W.H. Orme-Johnson, in: G.L. Eichhorn and L.G. Marzilli, Eds., Advances in Inorganic Bioch., New York Elsevier, (1982) 1-40.

46   L.E. Mortenson and R.N.F. Thorneley, Ann. Rev. Biochem., 48 (1979) 387-418. 47B.H. Huynh, E. Münck and W.H. Orme-Johnson, Biochim. Biophys. Acta, 576 (1979) 192-203.

48   M.J. Nelson, M.A. Levy and W.H. Orme-Johnson, Proc. Natl. Acad. Sci., U.S.A., 80 (1983) 147-150.

49   M.K. Eidsness, A.M. Flank, B.E. Smith, A.C. Flood, C.D. Garner, and S.P. Cramer, J. Am. Chem. Soc., 108 (1986) 2746-2747.

50   See Refs. 15 and 51 for further background references.

51   A.E. True, P. McLean, M.J. Nelson, W.H. Orme-Johnson, and B.M. Hoffman, J. Am. Chem. Soc., in press.

52   T.R. Hoover, A.D. Robertson, R.L. Cerny, R.N. Hayes, J. Imperial, V.K. Shah, and P.W. Ludden, Nature, 329 (1987) 855-857.

53   W.D. Hewson and L.P.Hager (1979) in: The Porphyrins, Dolphin, D., ed., Academic Press, New York, Vol. 7, pp. 295-332.

54   J.H. Dawson, Science, 22 (1988) 433-439.

55   C.W. Schulz, P.W. Devaney, H. Winkler, P.G. DeBrunner, N. Doan, R. Chiang, R. Rutter and L.P. Hager, FEBS Lett., 103 (1979) 102-105.

56   J. Fajer, D.C. Borg, A. Forman, D. Dolphin and R.H. Felton, R.H. J. Am. Chem. Soc., 92 (1970) 3451-3459.

57   J. Fajer and M.S. Davis, in: The Porphyrins, D. Dolphin, Ed., Academic Press, New York, 1979, Vol. 4, 197-256.

58   I. Morishima and S. Ogawa, S., Biochem., 17 (1978) 4384-4388.

59   I. Morishima and S. Ogawa, J. Am. Chem. Soc., 100 (1978) 7125-7128.

60   G.N. La Mar and J.S. de Ropp, J. Am. Chem. Soc., 102 (1980) 395-397.

61   W. Gordy, W., Theory and Applications of Electron Spin Resonance, Wiley, New York, 1980.

62   L.K. Hanson, C.K. Chang, M.S. Davis and J. Fajer, J. Am. Chem. Soc., 103 (1981) 663-670.

63   D.A. Case, J. Am. Chem. Soc., 107 (1985) 4013-4015.

64   T.A. Kent, M.H. Emptage, H. Merkle, M.C. Kennedy, H. Beinert and E. Münck, J. Biol. Chem., 260 (1985) 6871-6881.

65   M.H. Emptage, J.-L. Dreyer, M.C. Kennedy and H. Beinert J. Biol. Chem., 258 (1983) 11106-11111.

66   M.H. Emptage, T.A. Kent, M.C. Kennedy, H. Beinert and E. Münck, Proc. Natl. Acad. Sci. USA, 80 (1983) 4674-4678.

67   J.V. Schloss, D.J.T. Porter, H.J. Bright and W.W. Cleland, Biochem., 19 (1980) 2358-2362.

68   I.A. Rose and E.L. O'Connell, J. Biol. Chem., 242 (1967) 1870-1879.

69   J.F. Speyer and S.R. Dickman, J. Biol. Chem., 220 (1956) 293-308.

70   R.R. Ramsay, J.L. Dreyer, J.V. Schloss, R.H. Jackson, C.J. Coles,
      H. Beinert, W.W. Cleland and T.P. Singer, Biochem., 20 (1981) 7476-7482.
71   L.R. Gahan, J.M. Harrowfield, A.J. Herlt, L.F. Lindoy, P.O. Whimp and
      A.M. Sargeson, J. Am. Chem. Soc., 107 (1985) 6231-6242.
72   M.H. Emptage, in: Metal Clusters in Proteins, L. Que, Ed., ACS Symposium,
      Series, No. 372, 1987, 343-371.
73   J.S. Rieske, D.H. Machennan and R. Coleman, Biochem. Biophys. Res.
      Commun., 15 (1964) 338-344.
74   J.S. Rieske, W.S. Zaugg and R.E. Hansen, J. Biol. Chem., 239 (1964b)
      3023-3030.
75   S.P.J. Albracht and J. Subramanian, Biochim. Biophys. Acta, 462 (1977)
      36-48.
76   P. Bertrand, B. Guigliarelli, J.-P. Gayda, P. Beardwood and J.F. Gibson,
      Biochim. Biophys. Acta, 831 (1985) 261-266.
77   J.A. Fee, D. Kuila, M.W. Mather and T. Yoshida, Biochimica et Biophysica
      Acta, 853 (1986) 153-185.
78   J.F. Gibson, D.O. Hall, J.H.M. Thornley and F.R. Whatley, Proc. Nat.
      Acad. Sci. USA, 56 (1966) 987-990.
79   T.G. Brown and B.M. Hoffman, Mol. Phys., 39 (1980) 1073-1109.
80   C.P. Scholes, A. Lapidot, R. Mascarenhas, T. Inubushi, R.A. Isaacson and
      G. Feher, J. Am. Chem. Soc., 104 (1982) 2724-2735.
81   J.A. Fee, K.L. Findling, T. Yoshida, R. Hille, G.E. Tarr, D.O. Hearshen,
      W.R. Dunham, E.P. Day, T.A. Kent and E. Münck, E. J. Biol. Chem., 259
      (1984) 124-133.
82   D. Kuila, J.A. Fee, J.R. Schoonover and W.H. Woodruff, J. Am. Chem. Soc.,
      109 (1987) 1559-1561.
83   H.-T. Tsang, C.J. Batie, D.P. Ballou and J.E. Penner-Hahn, Biochem.,
      (1989), submitted.

CHAPTER 16

ENDOR AMPLITUDE SPECTROSCOPY

Marina BRUSTOLON and Ulderico SEGRE

1.   INTRODUCTION

The amplitude of a spectroscopic signal may depend upon a variety of factors. In a linear spectroscopic technique the signal intensity at a given frequency ω is proportional to the number of systems with a pair of energy levels separated by $\Delta E = \hbar\omega$, to the population difference Δn between these levels and to the strength of the oscillating electromagnetic field inducing the transition (1). Conventional EPR is a linear spectroscopic method, provided that the paramagnetic probes are dilute and the microwave (mw) field has low intensity (2). If these conditions are not fulfilled, the amplitude will depend upon additional factors. The intensity of a saturated EPR line is determined by the competition between the radiation induced transition rates and the relaxation rates, which try to maintain the equilibrium population difference.

ENDOR spectroscopy is intrinsically a non-linear technique, and its amplitude results from the balance of many competing mechanisms (3). The relaxation rates, among other factors, strongly affect the signal intensity, from which therefore their value can be obtained.

The aim of this contribution is to shed light on the kind of information that can be extracted from the analysis of the amplitude of the ENDOR signals of radicals in the solid state. It will be shown how it is possible: i) to study internal motions which are too fast or too slow to affect the EPR lineshapes; ii) to obtain the principal directions of hyperfine tensors from the spectra in disordered matrices.

The content of the chapter is the following.

ENDOR enhancement: The factors affecting the intensity of the signal are considered. The definition of ENDOR enhancement E is given and the explicit expressions for it are derived for some simple cases.

ENDOR in the solid state: The applicability of the general theory of the previous section to ENDOR in the solid state is discussed. The spin relaxation mechanisms in this state are briefly reviewed.

Studies of internal motion: The exchange between inequivalent protons causes some non-secular terms of the spin Hamiltonian to fluctuate, activating nuclear, electron or cross nuclear-electron relaxation paths. It is shown, by referring to some specific cases, that the enhancement is greatly affected by these mechanisms, for both very slow or very fast conformational motions.

Studies of radicals in disordered matrices: The peak intensities in the spectra of these radicals depend on the value of the magnetic field B_0. One can select different molecular orientations by varying B_0, and so different ENDOR lineshapes are obtained. Computer simulation of the spectra can then be used to obtain the relative principal directions of the hyperfine tensors.

2. ENDOR ENHANCEMENT

The NMR of paramagnetic species can be detected only when the electron spin relaxation rate is very fast, so that the nuclear spin sees only a time averaged local field due to the interaction with the electron spin *(1)*. ENDOR, on the contrary, is feasible when the electron spin relaxation is quite slow and it is possible to detect the nuclear spin transitions by monitoring the saturation level of the EPR transition when a radio frequency (rf) field is swept across the nuclear resonance. When the rf field is at resonance, the EPR transition is desaturated since a new relaxation path has been opened, see Fig. 1. As a consequence, the level of the EPR absorption increases by an amount which is proportional to the variation of the electron spin relaxation rate *(3)*. In fact, an ENDOR spectrometer records the increment of the EPR signal as a function of the rf frequency.

The ENDOR fractional enhancement E is defined as the ratio:

$$E = \frac{I(on) - I(off)}{I(off)} \qquad [1]$$

where $I(off)$ and $I(on)$ are the peak intensities of the homogeneously broadened EPR signal with the radiofrequency driving a nuclear transition respectively off and on. In solution ENDOR experiments the measurement of the enhancement value would require the determination of the peak amplitudes of the EPR absorption signals with and without the radiofrequency on resonance *(4)*. However, such a determination is normally prevented by a too low intensity of the unmodulated EPR signal. The E value could nonetheless be obtained from the peak-to-peak intensities of the Zeeman-modulated EPR signals, by exploiting the relationships between the absorption and derivative Lorentzian lineshapes. On the other hand, in a solid state ENDOR experiment the determination of E values as defined above is not feasible, since for each setting of the magnetic field on the inhomogeneously broadened EPR signal, different spin packets are brought into resonance. In this case one is still able to extract from the ENDOR and EPR spectra a relative and average ENDOR enhancement. The experimental determination of the average enhancement is accomplished by obtaining the ratio of the peak intensity of the absorption ENDOR signal over the peak-to-peak intensity of the Zeeman-modulated EPR signal *(5)*.

There are some considerations worth noting on the modulation scheme of the ENDOR signal. The widely used rf frequency modulation gives rise to a derivative lineshape. In this case the peak-to-peak intensity of the signal in a derivative form would depend also on the intrinsic ENDOR linewidth. Therefore, in the case of a single crystal the ENDOR intensity measurements cannot be done on frequency-modulated signals, but they must be done on amplitude-modulated ones. On the other hand, in powder ENDOR spectra dominated by an axially symmetric interaction, the first derivative of the step shoulder, corresponding to the parallel value of the tensor, has the shape of a single crystal absorption line *(6)*. Therefore in this case a frequency-modulated ENDOR spectrum is suitable for intensity determinations.

The amplitude modulation of the radiofrequency corresponds to a periodic switching on and off of the irradiating field. Therefore, the modulation frequency must be lower than the slowest path of electron spin relaxation in order to avoid transient effects, which can strongly affect the ENDOR intensities.

2.1 Density matrix equations

It should be clear from the previous discussion that, to account for the ENDOR intensity and lineshape, it is essential to carefully consider the effect of the various relaxation mechanisms affecting the spin populations. The rigorous treatment is accomplished by using the density matrix method *(7,8)*. The reader is referred to the original papers for a complete description of this method. In this section we will only report the most significant results. In particular the expressions for the enhancement will be given for some commonly found limiting conditions.

According to the quantum theory *(1)*, the statistical expectation value of any observable Q can be computed from the knowledge of the density matrix operator ρ by the equation:

$$< Q >= \text{Tr}\{\rho Q\} \tag{2}$$

and the time evolution of ρ obeys the differential equation:

$$\dot{\rho}(t) = -i[H(t), \rho(t)] + \mathcal{R}(\rho - \rho_0) \tag{3}$$

where H is the spin Hamiltonian including the interaction with the oscillating fields:

$$H(t) = H_0 + \epsilon(t) \tag{4}$$

and $\mathcal{R}$ is the relaxation operator *(9)*. $\mathcal{R}$ must have negative eigenvalues to ensure that, at long times, the density matrix eventually reaches its equilibrium value ρ_0 when $\epsilon(t) = 0$. The various elements of the density matrix have different physical meanings. The value of a diagonal element ρ_{aa} gives the population of the state a (note that $\sum_a \rho_{aa} = 1$), while an off diagonal element ρ_{ab} is related to the intensity of the transition between the states a and b. We are interested in the power absorption due to the transition between a pair of levels connected by the mw field, and the equation for the relevant density matrix element is:

$$\dot{\rho}_{ab} = -i[H, \rho]_{ab} + \sum_{cd} \mathcal{R}_{ab,cd}(\rho - \rho_0)_{cd}. \tag{5}$$

Because of the off-diagonal terms of the Hamiltonian, this equation is coupled to other elements of ρ, and a linear system of differential equations must be solved. The elements of the relaxation matrix are expressed in terms of the transition linewidths and the transition probabilities between the spin levels, i.e. $\mathcal{R}_{ab,ab} = -(1/T_2)_{ab} = -\Gamma_{ab}$ and $\mathcal{R}_{aa,bb} = W_{ab}$ (note that $\mathcal{R}_{aa,aa} = -\sum_c W_{ac}$). For a two-level system corresponding to one $S = \frac{1}{2}$ spin coupled to the magnetic field B_0, two equations for $\rho_{\frac{1}{2}\frac{1}{2}}$ and $\rho_{\frac{1}{2}-\frac{1}{2}}$ must be solved and the relaxation matrix elements are $\mathcal{R}_{\frac{1}{2}-\frac{1}{2},\frac{1}{2}-\frac{1}{2}} = -1/T_2 = -\Gamma$ and $\mathcal{R}_{\frac{1}{2}\frac{1}{2},\frac{1}{2}\frac{1}{2}} = -\mathcal{R}_{\frac{1}{2}\frac{1}{2},-\frac{1}{2}-\frac{1}{2}} = -1/2T_1 = -W$.

2.2 Four-level system

Let us consider now the simple four-level scheme corresponding to one electron spin $(S = \frac{1}{2})$ and one nuclear spin $(I = \frac{1}{2})$, interacting with the magnetic field B_0 and with each other through a contact coupling. The Hamiltonian is:

$$H_0 = \omega_e S_z - \omega_n I_z + a\mathbf{I} \cdot \mathbf{S} \qquad [6]$$

where $\omega_e = \gamma_e B_0$ and $\omega_n = \gamma_n B_0$. In the high field approximation the eigenvectors are:

$$|1> = |\alpha_e \alpha_n>, \quad |2> = |\alpha_e \beta_n>, \quad |3> = |\beta_e \alpha_n>, \quad |4> = |\beta_e \beta_n> . \qquad [7]$$

The interaction with the mw and rf fields is given by the time dependent Hamiltonian:

$$\epsilon(t) = (\gamma_e S_x - \gamma_n I_x)(B_m \cos \omega_m t + B_r \cos \omega_r t) \qquad [8]$$

where B_m and B_r are the amplitudes of the mw and rf oscillating fields. The previous relation can be substituted by an effective interaction:

$$\epsilon(t) = \gamma_e B_m S_x \cos \omega_m t + \gamma_n B_r^e I_x \cos \omega_r t \qquad [9]$$

where the rf intensity B_r^e has been modified by the 'hyperfine enhancement effect', which is due to the interaction between the rf field and the electron spin. This effect can dramatically modify the intensity of the rf field acting on the nucleus (3).

For this system the 4×4 transition probability matrix $\mathbf{W}$ has up to six different elements: two electron rates W_e', W_e'' $(\Delta M_I = 0)$; two nuclear rates W_n', W_n'' $(\Delta M_S = 0)$; and two cross rates W_x', W_x'', which are often referred as the flip-flop rate $(\Delta M_I = -\Delta M_S)$ and the flop-flop rate $(\Delta M_I = \Delta M_S)$. These rates originate from the fluctuations of the environment of the paramagnetic probe or from the dynamics of the probe itself. A particular fluctuation mechanism can affect some terms of the spin Hamiltonian, so that the corresponding relaxation rate is activated. If, for instance, the isotropic contact coupling:

$$H_c = a\mathbf{I} \cdot \mathbf{S} = a[I_z S_z + \frac{1}{2}(I_+ S_- + I_- S_+)] \qquad [10]$$

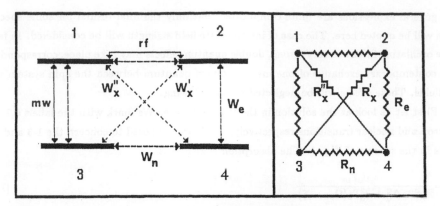

Figure 1. Left: the energy levels for an electron spin interacting with a nuclear $I = \frac{1}{2}$ spin. The broken arrows indicate the relaxation rates. The transitions driven by the mw and rf field are also indicated. In the high field approximation the spin states are given by Eqn. [7]. Right: an electric circuit analogue to the energy level scheme $(R_i = 1/W_i)$.

fluctuates between two or more different values, the non-secular terms $I_\pm S_\mp$ will drive a flip-flop relaxation rate. If instead the Electron Nuclear Dipolar (END) interaction is modulated, all the relaxation paths of the previous scheme will be affected (with different weights), since the END interaction contains the operators $I_z S_\pm, I_\pm S_z, I_\pm S_\mp, I_\pm S_\pm$ (8). A relaxation mechanism inducing transitions between a pair of levels is particularly efficient when the frequency of the motion is of the order of magnitude of the energy difference between the levels.

The goal of this section is to show how it is possible to obtain the expression for the ENDOR intensity in terms of the elements of the transition rate matrix $\mathbf{W}$. Freed (7,8) succeeded in deriving a general expression by looking for a stationary solution of Eqn. [5], whose time dependence is given by:

$$\rho_{ab}(t) = Z(B_m, B_r^e, \omega_r, \omega_m)e^{i\omega_m t} \qquad [11]$$

where the real and imaginary part of $Z = Z' + iZ''$ give the EPR dispersion and absorption signals respectively. The quantity Z is a function of the experimental settings, i.e. the mw and rf intensities and frequencies. The intensity of the EPR absorption can be expressed as a function of the mw frequency with B_m as a parameter:

$$I_{epr}(\omega_m) = Z''(B_m, B_r = 0; \omega_m) \qquad [12]$$

while the ENDOR intensity is a function of the rf frequency, but it depends also upon all the other parameters:

$$I_{endor}(\omega_r) = Z''(B_m, B_r^e, \omega_m; \omega_r) - I_{epr}(\omega_m). \qquad [13]$$

The general expressions are quite complicated, but only the final results for some specific cases will be quoted here. The case of intermediate field strength will be considered. In fact, if the oscillating fields are very strong, double quantum effects may take place, corresponding to a contemporary exchange of one mw and one rf quantum between the spin system and the fields. These effects will be neglected in the following.

First let us look at the solution in the four-level case. We mark with the labels 0,1 the electron and nuclear transitions respectively, which are supposed to concern the 1-3 and 1-2 levels in the scheme of Fig. 1. The absorption is given by:

$$Z'' = \frac{\omega_e d_0 \Gamma_0}{\Gamma_0^2 + \Delta_0^2 + d_0^2 \Gamma_0 (\Omega_{00} - \xi)} \tag{14a}$$

$$\xi = \frac{d_1^2 \Omega_{01}^2 \Gamma_1}{\Gamma_1^2 + \Delta_1^2 + d_1^2 \Gamma_1 \Omega_{11}} \tag{14b}$$

where $\Delta_0 = \omega_m - \omega_e - \frac{1}{2}a$, $\Delta_1 = \omega_r + \omega_n - \frac{1}{2}a$ are the frequency offsets and $d_0 = \frac{1}{2}\gamma_e B_m$, $d_1 = \frac{1}{2}\gamma_n B_r^e$ are the transition moments.

The quantities $\Omega_{\mu\nu}$ in Eqn. [14] play a role similar to the spin-lattice relaxation time in a two-level resonance experiment. They may be computed from the transition probability matrix $\mathbf{W}$, whose elements are the transition rates, by taking its cofactor C and double cofactors $C_{\mu\nu}$:

$$\Omega_{\mu\nu} = 2C_{\mu\nu}/C. \tag{15}$$

The cofactor C is the determinant obtained by deleting a row and a column of $\mathbf{W}$ (all the cofactors are equal because of the properties of $\mathbf{W}$); the double cofactor $C_{\mu\nu}$ is obtained by deleting the pairs of rows and columns concerned in the μ-th and ν-th transitions. The fractional enhancement E has been defined in Eqn. [1] and can be computed from Eqns. [12-14] as the ratio between the ENDOR and EPR signal at exact resonance:

$$E = I_{endor}(\Delta_1 = 0)/I_{epr}(\Delta_0 = 0). \tag{16}$$

In the limiting case of moderate electron and nuclear saturation, $d_0^2 \Omega_{00} > \Gamma_0$ and $d_1^2 \Omega_{11} > \Gamma_1$, it follows that:

$$E = \frac{\xi/\Omega_{00}}{1 - \xi/\Omega_{00}} \tag{17a}$$

$$\xi = \Omega_{01}^2/\Omega_{11}. \tag{17b}$$

This is a result of paramount importance, since it shows that *the limiting ENDOR enhancement is a function of the spin-lattice transition probabilities W_{ij} only.*

The analytical expressions of the enhancement in terms of the different transition probabilities are given elsewhere (8). Two specific cases are worth considering here, both corresponding to the mw and rf fields set on the 1-3 and 1-2 transition, respectively.

i) The cross rates are null, $W_x' = W_x'' = 0$, and the nuclear and electron rates are symmetric, $W_n' = W_n''$ and $W_e' = W_e''$. Then Eqn. [17] reads:

$$E = \frac{W_e W_n}{2(W_e + W_n)^2}.$$
[18]

The enhancement vanishes in the extreme limits $W_n \ll W_e$ and $W_n \gg W_e$, and its maximum value is $E - 0.125$ at $W_n = W_e$.

ii) Only the electron rates $W_e' = W_e''$ and the flip-flop rate W_x' are activated. Then:

$$E = \frac{W_x'}{W_e}$$
[19]

and the enhancement varies proportionally to W_x'. We note that in this case the relaxation paths are not symmetric. As a consequence, if the 3-4 nuclear transition is driven instead of the 1-2, the enhancement vanishes. Furthermore, the intensity of the ENDOR lines for this hypothetical system would be reversed by setting the magnetic field on the low field EPR line or on the high field one. In the first case, only the low frequency ENDOR line will be detected, while only the high frequency line will be present in the other case.

The effect of the different rates W_{ij} in determining the ENDOR enhancement is more easily understood by making use of the analogy between the relaxation network and an electric circuit (3). In fact, it can be shown that computing the limiting enhancement is just the same as computing the increase of the electric current in the circuit shown in Fig. 1 when the 1-2 branch is shorted.

2.3 Many-level systems

The results of the previous paragraphs can be easily extended to more complex level schemes. First consider the case of an electronic spin interacting with N non-equivalent nuclei with $I = \frac{1}{2}$, so that the nuclear transitions are well separated. The value of the fractional enhancement is still given by Eqn. [17], the only difference being that the transition matrix now has dimensions $2^{N+1} \times 2^{N+1}$.

On the other hand, in the case of one nucleus with $I > \frac{1}{2}$ or of many equivalent $I = \frac{1}{2}$ nuclei, the nuclear transitions are degenerate to first order. The enhancement is given by Eqn. [17a], but the quantity ξ now has a different expression. For an m-fold degenerate nuclear transition one must compute the saturation matrix S, whose elements are:

$$S_{\mu\nu} = d_\mu d_\nu \Omega_{\mu\nu} \qquad (\mu, \nu = 0, m)$$
[20]

and the related $\mathbf{F}$ matrix:

$$F_{ij} = S_{ij} + \delta_{ij}\Gamma_i \qquad (i, j = 1, m).$$
[21]

Then ξ is given by:

$$d_0^2 \xi = \sum_{i,j=1}^{m} \frac{S_{0i} F^{ij} S_{j0}}{|F|} \qquad [22]$$

where F^{ij} is the i,j cofactor of $\mathbf{F}$ and $|F| = \sum_{ij} F_{ij} F^{ij}$ is the value of the determinant of $\mathbf{F}$. Note that, in the case of high nuclear saturation, the nuclear linewidths in Eqn. [21] may be neglected, and both ξ and the enhancement are functions of the transition probabilities only.

3. ENDOR IN THE SOLID STATE

The theory for the ENDOR enhancement, described in the previous section, was developed originally for ENDOR in solution. However, it can also be applied to solid state ENDOR because no assumptions about the nature of the relaxation mechanisms have been made in deriving the previous expressions. The only conditions to be fulfilled relate to the magnitude of the elements of the relaxation matrix $\mathcal{R}$, the linewidths Γ and the transition probabilities W:

$$\Gamma_i < |a| \qquad [23a]$$

$$\Omega_{ii} d_i^2 > \Gamma_i. \qquad [23b]$$

These equations imply that the hyperfine lines are well separated and that there is an appreciable degree of saturation. In this case the theory for the enhancement of an ENDOR line is formally the same in solution or in the solid state, and we can safely use the expressions for the limiting ENDOR enhancement that have been derived previously. The origin of the spin relaxation is different, however, and in this section the various mechanisms will be briefly reviewed.

3.1 Spin relaxation

An EPR line in a solid system is given by the contribution of a large number of spin packets, each of them due to an ensemble of spins characterized by the same precessional frequency. Each spin packet can exchange energy with the lattice, or with other spin packets. The two phenomena are characterized by two different rates, which are referred to as the *spin relaxation* rate and the *spin diffusion* rate.

The *electron spin-lattice* relaxation rate W_e of radicals in solids is due to the interaction with the phonons of the matrix and to the intermolecular and intramolecular motions (10). The spin system can exchange energy with the lattice by emitting a phonon at the resonance frequency (direct process) or by interacting simultaneously with two lattice modes (Raman process). In the latter process, the exchange of energy between the lattice and the spin system involves the modulation of the spin Hamiltonian by the inelastic scattering of a

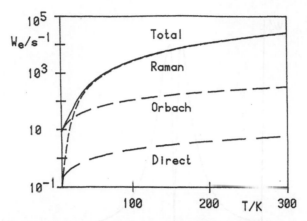

Figure 2. The temperature dependence of the three processes contributing to the spin-lattice relaxation rate for an organic radical in the solid state.

phonon or by the beats between two phonons at their frequency difference. A third process can take place when an excited doublet state of the radical lying within the energy spectrum of the phonon band is present (Orbach process). This is again a two phonon process, one phonon being absorbed by the spin system, and a second of slightly different frequency being emitted.

These different processes have different temperature dependences. The electron spin-lattice relaxation in a series of free radicals trapped in molecular crystals was studied quite extensively in the past (11), and it was found that, at low temperatures, the direct process and the Orbach process (when present) dominate the W_e rate. At high temperatures ($T > 100$ K) W_e is mainly due to the Raman process. When T is greater then the Debye temperature of the lattice Θ_D, the Raman contribution varies as T^2 (the so called phonon bottleneck). The relevant spin relaxation mechanism is through the modulation of the spin-orbit coupling constant. Modulation of the hyperfine coupling constants is in general much less effective, with the exception of fluorinated radicals.

The temperature variation of the different spin-lattice relaxation processes for an organic radical in the solid state is displayed in Fig. 2. The curves are calculated by the semiempirical expression obtained by Dalton *et al.* for the cyclobutyl-1-carboxylic acid radical (11).

The *nuclear spin-lattice* relaxation rate W_n in paramagnetic systems is completely dominated by the interaction with the electron spin. In solution the modulation of the END interaction due to tumbling determines the magnitude of W_n (8). For a nuclear spin in a radical trapped in a solid matrix, residual motions of the radical in the matrix can strongly affect the nuclear spin-lattice relaxation times of the matrix nuclei coupled to the radical by a purely dipolar superhyperfine interaction (12). Intramolecular motions involving nuclear

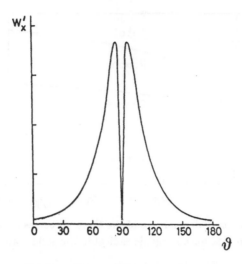

Figure 3. The angular dependence of the cross relaxation rates W_x for an α proton, with the magnetic field in the plane perpendicular to the $2p$ orbital. From *(13)*.

exchange (like methyl and methylene rotations, or conformational transitions in alicyclic compounds) introduce an exchange rate between pairs of spin states, which acts in the high field limit as a binuclear flip-flop spin operator with $\Delta M_{I_1} = -\Delta M_{I_2}$.

The *electron-nuclear cross* relaxation rates can arise from an intramolecular motion modulating the isotropic hyperfine coupling spin Hamiltonian (flip-flop rate W'_x), or the dipolar hyperfine coupling (both flip-flop and flop-flop rate W''_x).

It is worth noting that the distinction between the pure electronic and the cross relaxation rate is meaningful as long as the spin states are given by the high field expressions of Eqn. [7]. However, if the electron and nuclear spin functions are mixed, the relaxation rates too are combined. As an example, Fig. 3 shows the calculated angular dependence of the cross rates for an α-proton as the direction of magnetic field varies in the principal plane xy, z being the direction of the $2p$ orbital and x that of the C–H bond *(13)*. When the magnetic field lies along one of the principal directions of the coupling tensor, the off-diagonal terms of the hyperfine Hamiltonian vanish and the spin states are given by the high-field expressions. As the magnetic field moves from the principal directions, the $|M_s, \alpha_n>$ and $|M_s, \beta_n>$ functions mix and the 'forbidden' EPR transitions become partially 'allowed'. As a consequence, the rates W'_x and W''_x grow, and the intensities of the two ENDOR lines vary in proportion to these rates.

3.2 Spin diffusion

The spin diffusion process exchanges energy between different spin packets. For radicals trapped in a single crystal in a particular site the spin packets are distinguished only by the different spin eigenfunctions of the nuclear spins coupled to the electron spin. On the other hand, for radicals in a disordered matrix there is a continuous distribution of spin packets. In fact, different orientations of the radicals with respect to the magnetic field correspond to different resonance conditions because of the anisotropies of the magnetic interactions.

It should be noted that the direct interaction of the radicals throughout the host matrix can give rise to spin diffusion in both cases. However the spin packets can be connected by additional relaxation processes. When the spin packets correspond to different nuclear spin eigenfunctions, spin diffusion can take place inside the same radical through internal relaxation paths. On the other hand, the diffusion between spin packets due to different molecular orientations can be originated by the radical motion inside the matrix: this process, for instance, is exploited by Saturation Transfer EPR *(14)*.

In the case of radicals in organic molecular crystals at low temperature, the spin-lattice relaxation rate is slower than the spin diffusion process *(15)*. Therefore, the saturation is transferred to all the spin packets in a time short with respect to W_e^{-1}. On the other hand, at $T > 100$ K, the spin diffusion is slower than the spin-lattice relaxation, and therefore it is possible to selectively saturate the spin packets due to a particular orientation of the radicals in a disordered matrix *(16)*. The ENDOR spectrum is then dominated by the contribution of the probes whose orientation has been selected by the resonance condition.

4. STUDIES OF INTERNAL MOTIONS

It is well known that the conformational dynamics of a paramagnetic molecule may affect the EPR lineshape when the motion modulates a spectral feature, for instance a coupling constant *(3)*. The lineshape is sensitive to the motion when the correlation time τ is intermediate between the two limits:

$$\tau < 1/\Gamma' \tag{24a}$$

$$\tau > \Gamma''/(\delta\omega)^2 \tag{24b}$$

where $\delta\omega$ is the separation between the spectral lines affected by the motion, and Γ' and Γ'' are the intrinsic linewidths characteristic of the EPR spectrum in the slow and fast limits. If the motion is slower than the limit of Eqn. [24a] or faster than the limit of Eqn. [24b], then the EPR spectrum is not affected by these dynamical effects.

The ENDOR lineshape also has been analysed to obtain information on the internal motions *(17)*. The range of correlation times which are effective in modifying the lineshape is given again by Eqns. [24], the only difference being that the ENDOR intrinsic linewidths are usually smaller than the EPR ones. Therefore, shorter or longer correlation times should be accessible to ENDOR lineshape studies.

The modulation of the magnetic interactions can affect also the transition probability rates and, therefore, the ENDOR enhancement. Under certain circumstances, this effect

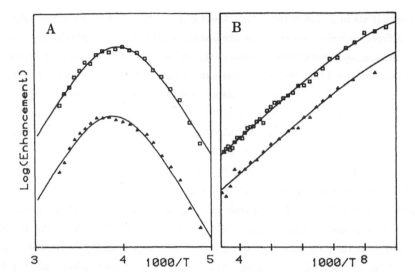

Figure 4. ENDOR enhancement for radical **I** (A) and **II** (B) as a function of the temperature. Squares, single crystal data; triangles, powder sample. From *(5)*.

is very conspicuous, so that a very fast or a very slow dynamics may be studied beyond the limits given by Eqns. [24]. As a typical example of a very fast motion that has been studied by this method, we shall consider the hindered rotation of methyl groups in radicals obtained by irradiation in molecular solids. When the methyl group is bound to a carbon atom bearing spin density, the proton coupling constant is given by the expression *(2)*:

$$a(\beta) = a_0 + a_2 \cos^2 \beta \qquad\qquad [25]$$

where β is the dihedral angle between the axis of the *2p* orbital of the carbon bearing the spin density and the methyl C–H bond direction. The coupling constant $a(\beta)$ is modulated by the methyl rotation, and therefore the contact spin Hamiltonian [10] contains the time dependent terms inducing flip-flop transitions. The a_0 and a_2 constants depend on the spin density on the α-carbon atom, and therefore the efficiency of this spin relaxation mechanism is proportional to the same quantity. When the motion is fast and the three protons are equivalent, the W'_x relaxation rate induced by this mechanism depends on the methyl rotation rate τ^{-1} according to the relation *(5)*:

$$W'_x = \frac{a_2^2}{16} \frac{\tau}{1 + \omega_e^2 \tau^2} \qquad\qquad [26]$$

and it is maximum when the rotation rate is equal to the electron Zeeman frequency ω_e. The dipolar hyperfine interaction for β protons is much less important, being only 10% of the contact coupling.

It can be shown that in the case of three equivalent protons, when the flip-flop relaxation dominates the desaturation rates, the ENDOR spectrum has unbalanced intensities, as in the case of a four-level system discussed in Section 2.2. When the low field EPR lines are saturated, the low frequency ENDOR line is more intense. This property allows unambiguous identification of rotating methyl groups in complex compounds of biological interest *(18)*.

Moreover the ENDOR enhancement is strongly temperature dependent because of the Arrhenius relationship for the methyl rotation rate:

$$\tau^{-1} = \tau_\infty^{-1} e^{-V/kT} \tag{27}$$

where V is the height of the potential barrier.

It has been possible to study by this approach the methyl rotation dynamics in a series of radicals bearing a methyl group (**I–IV**) in the fast motion regime. The temperature dependence of the enhancement for radicals **I** and **II** is displayed in Fig. 4. The data have been fitted to the theoretical expressions to give the potential barrier values, with the assumption that the temperature dependence of the spin-lattice relaxation W_e can be neglected *(5,19)*.

The presence of the oxygen atoms in radicals **III** and **IV** gives rise to a faster relaxation of the nuclear spins, making it impossible to saturate the nuclear transitions in the temperature range used for the experiments *(19)*. Therefore, the condition given by Eqn. [23b] is no longer satisfied and the enhancements depend not only on the relaxation rates, but also on the transition moments d_i's. In this case a convenient quantity to be fitted to obtain the dynamical parameters of the motion is the temperature dependence of the ratio of the enhancements for the low and high frequency ENDOR lines obtained on irradiating the same EPR line. In Table I the values of the barrier heights for the methyl rotation obtained by the present method are reported.

606

TABLE 1
Height of the potential barrier V for methyl rotation in radicals I–IV (kJmol^{-1}).

I	17.3	II	3.2	III	3.0	IV	1.3

A different application of this type of investigation has been performed in the opposite limit of very slow internal motion (20). The radical $\dot{C}H_2COO^-$ was produced by high energy irradiation of Zn acetate dihydrate. At low temperature ($T < 100$ K) the motion of the methylene group is frozen and two inequivalent α protons are observed. Since the anisotropic hyperfine coupling is large, the spin states are a linear combination of the high field states with the same value of M_s. As the temperature increases, the methylene hindered rotation is activated, but initially it remains too slow to affect the EPR or ENDOR lineshapes. The ENDOR enhancement, however, strongly grows. This behaviour can be explained by adding the exchange term to the equation of motion for the density matrix, Eqn. [3]. This term induces transitions between all the nuclear levels and it mixes the electronic and cross relaxation rates. Since the relaxation rates in solids at these temperatures are quite slow, a small exchange term is able to produce a dramatic effect on the ENDOR enhancement.

5. ENDOR IN DISORDERED MATRICES

The ENDOR spectra in disordered matrices may be recorded with quite different aims: If the paramagnetic molecule is used to probe its environment, then one looks at the resonance lines near the free proton frequency ω_H, which are due to the nuclei distant from the unpaired electron (distant or matrix ENDOR); otherwise, when the probe itself is the object of the investigation, the crystalline powder or the frozen solution are often used because of the difficulty in obtaining a single crystal sample, or because of the low sensitivity of liquid solution ENDOR.

It is well known that it is possible, in principle, to obtain the principal values of the dipolar coupling tensors from EPR spectra in disordered matrices. However, this information is often overwhelmed by unresolved hyperfine interactions that broaden the spectra.

The greater spectral resolution of ENDOR offers the possibility of obtaining more information. In fact, in the limit of a first order treatment of the hyperfine interactions, the positions of the features of the lineshape in disordered matrices give the principal values of the interaction tensors. Moreover, in some cases ENDOR allows one to obtain information on the relative orientations of the magnetic tensors by studying the relative amplitudes of the spectral features.

Let us consider a radical containing one proton with axially symmetric hyperfine coupling to the electron spin. If the coupling is small, to first order the frequencies of the ENDOR

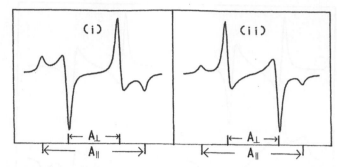

Figure 5. ENDOR powder lineshape for a proton with axially symmetric hyperfine interaction; (i) $A_\| = 2A_\perp$; (ii) $A_\| = -2A_\perp$.

lines are given by:

$$\omega_\pm(\theta) = \omega_H \pm \frac{1}{2}[A_\| \cos^2(\theta) + A_\perp \sin^2(\theta)] \tag{28}$$

where θ is the angle between the magnetic field and the symmetry axis. In an isotropically disordered sample all orientations are equally probable. If moreover the saturation of the electron spin levels is the same irrespectively of the molecular orientation, then all the radicals contribute in an equal way to the ENDOR absorption profile. The lineshape will then appear as shown in Fig. 5. It is worth noting that, if the contact coupling vanishes, the lineshape is just the same as that of the EPR spectrum of a disordered axially symmetric triplet system in the high field approximation. We note that the first derivative of the step shoulder corresponding to the parallel orientation has the shape of an absorption line, as it was anticipated in Section 2.

These characteristic shapes give immediately the principal values of the hyperfine tensor since they arise from the angular dependence of the resonance conditions leading to a rapidly growing spectrum intensity in proximity of the turning points of the magnetic interaction. This kind of spectrum is called *powder-like* ENDOR and it is due to a partial saturation of the electron spin transitions of all the spin packets present in the powder EPR spectrum. Such a saturation is possible only in the case of very small magnetic interactions giving rise to the piling up of all the spin packets in quite narrow EPR lines, or in the case of a rapid spin diffusion propagating the electron spin polarization to all the spin packets in a time short with respect to the spin relaxation times.

On the other hand, if the EPR powder spectrum is characterized by a strongly anisotropic magnetic interaction, in principle it is possible to select different molecular orientations in the rigid matrix by setting the magnetic field at different positions in the EPR spectrum. The ENDOR absorption in this case is due only to the molecules oriented in such a way to be at resonance with the mw field, provided that the spin diffusion is sufficiently slow to be neglected. When these conditions are fulfilled, one obtains *crystal-like* spectra.

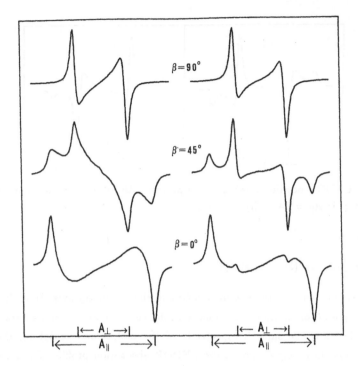

Figure 6. ENDOR lineshapes for a proton with a purely dipolar axial **A** tensor in a radical with axial **g** tensor. β is the angle between the symmetry axes of the two tensors. Left, $\sigma = 20$ MHz; right, $\sigma = 40$ MHz.

In many cases the ENDOR spectra in disordered systems are intermediate between powder and crystal-like shapes.

5.1 Simulation of the disordered ENDOR spectra

Let us recall first some results about the EPR spectra in disordered matrices. The absorption intensity at a given frequency ω is the sum of all the contributions coming from the radicals at different orientations; the larger is the frequency offset from the resonance frequency ω_e for a particular orientation, the lesser will be its contribution. The powder EPR lineshape therefore is given by the expression:

$$I_{epr}(\omega) = \int g[\omega - \omega_e(\theta, \phi)] \sin\theta \, d\theta \, d\phi \qquad [29]$$

where $g(x)$ is a lineshape function, usually gaussian because of unresolved hyperfine splittings. When the EPR transition is split by the hyperfine coupling, a sum over all the hyperfine components must be included into Eqn. [29], which therefore reads:

$$I_{epr}(\omega) = \sum_M \int g[\omega - \omega_e^{(M)}(\theta, \phi)] \sin\theta \, d\theta \, d\phi \qquad [30]$$

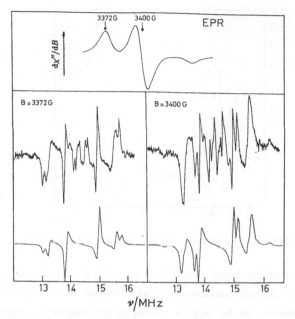

Figure 7. EPR and proton ENDOR spectra for Tempone in a polycrystalline matrix. Upper traces, experimental; lower traces, simulated. From *(16)*.

The ENDOR lineshape will be more complicated, since, as has been shown in Section 2, its absorption intensity depends upon the fulfillment of the two resonance conditions that govern the electron and nuclear transitions. We consider the lineshape for a proton, which is given by the superposition of the $M_s = \pm\frac{1}{2}$ transitions. Therefore we can write the following expression:

$$I_{endor}(\omega_r) = \sum_{\pm 1/2} \int P(\theta,\phi) h[\omega_r - \omega_n^\pm(\theta,\phi)] \sin\theta \, d\theta \, d\phi \qquad [31]$$

where $h(x)$ is the ENDOR lineshape function, and $P(\theta,\phi)$ is an orientation dependent weight factor. In general $P(\theta,\phi)$ should be affected by the orientation dependence of the hyperfine enhancements, of the relaxation mechanisms and of the resonance conditions. We consider here isotropic relaxation mechanisms and a *limiting* ENDOR enhancement, which is independent of the transition moments (see Eqns. [17]). Therefore the weight factor depends only on the offset between the mw field and the resonance frequencies for the electron spin:

$$P(\theta,\phi) = \sum_M g[(\omega_m - \omega_e^{(M)}(\theta,\phi)]. \qquad [32]$$

We assume for the function $g(x)$ a gaussian form:

$$g(x) = (1/2\pi\sigma)^{\frac{1}{2}} e^{-x^2/2\sigma^2}. \qquad [33]$$

610

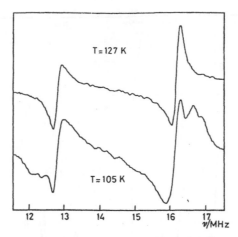

Figure 8. ENDOR spectra of Tempone in a toluene d-8 glassy matrix. From *(16)*.

The shape of the spectrum depends sensitively upon the actual value of the width σ. When σ goes to zero, only the spin packets exactly at resonance with the mw frequency give rise to the ENDOR absorption; for large σ values, on the contrary, a powder-like lineshape is obtained, since all the probes contribute equally to the spectrum, irrespective of their orientation. In Fig. 6 some simulations are reported. They have been computed according to Eqns. [31-33] for a model system that has a quite large g tensor anisotropy ($g_\parallel - g_\perp = .03$ corresponding to a spread of about 5 mT in the X-band EPR spectrum). The magnetic field B_0 is set on the feature of the EPR spectrum corresponding to the $g_\parallel$ value.

5.2 Applications of the angular selection method

Many different systems have been studied so far by the angular selection method in disordered matrices: transition metals complexes *(21,22)* and metalloproteins *(23)*, in which the g tensor anisotropy is exploited; organic biradicals *(24)*, in which the source of magnetic anisotropy is the electron-electron interaction fine tensor; nitroxide radicals *(16,25)*, which present well separated spin packets in the EPR spectrum thanks to the strongly anisotropic nitrogen hyperfine coupling ($A_{N\parallel} - A_{N\perp} \approx 90$ MHz).

Crystal-like proton ENDOR spectra of nitroxides can be easily detected. In Fig. 7 the spectra of a powder sample of Tempone (**V**) dissolved in a diamagnetic matrix of 2,2,4,4-Tetramethylcyclobutanedione (TMCD) are shown *(16)*, together with the simulated spectra obtained by the method previously described (lower traces). The lack of any residual motion in the host matrix gives rise to a crystal-like spectrum with features at well defined frequencies, whose amplitudes change on saturating different spin packets in the EPR spectrum. The principal values of the proton hyperfine tensors are in agreement with those obtained in a single crystal analysis of the same doped compound (Tempone in TMCD) *(26)*.

When the same nitroxide is dissolved in a glassy matrix of toluene, the presence of

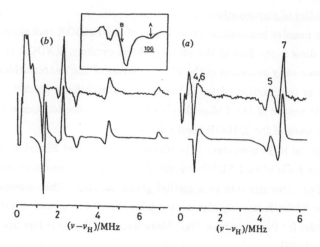

Figure 9. EPR (insert) and proton ENDOR spectra for radical **VI** in toluene d-8 at $T = 105$ K. Upper traces, experimental; lower traces, simulated. From *(25)*.

intra- and inter-molecular motions gives rise to fast spectral diffusion. In fact the spectra reported in Fig. 8 are powder-like since they do not depend on the feature of the EPR spectrum being saturated.

V   **VI**

An interesting comparison can be made with the ENDOR spectra in the same glassy matrix of other more bulky nitroxide radicals derived from indolinone, benzimidazole and quinoline *(25)*. In these radicals there is a substantial delocalization of the spin density from the N-O group to the conjugated phenyl ring. As an example we report in Fig. 9 the spectra of the indolinone nitroxide radical (**VI**). In this case the spectra are crystal-like since the more bulky guest cannot undergo fast reorientation motions. The strong dependence of the ENDOR amplitudes on the feature of the EPR spectrum being saturated allows the determination of the relative orientations of the nitrogen and protons hyperfine tensors, and therefore it is possible to attribute the measured dipolar tensors to the different protons.

The nuclear–electron distances in spin-labeled compounds have been measured by this method, showing that ENDOR spectroscopy provides an accurate noncrystallographic technique for structure determination of immobilized molecules in frozen solution *(27)*.

5.3 ENDOR studies of photosynthetic systems

The most impressive biochemical applications of the ENDOR techniques in the recent years have been done in the field of the photosynthetic systems. ENDOR spectra in frozen matrices have given early evidences of the dimeric nature of the Bacteriochlorophyll species which is present in the reaction center in bacterial photosynthesis *(28,29)*. The more controversial dimeric nature of the Chlorophyll in the P700+ reaction centers has been also discussed on the basis of the ENDOR results *(30,31)*.

The EPR signal in Photosystem II in plants is often called Signal II. It was initially attributed, on the basis of an ENDOR investigation, to a Plastosemiquinone radical anion with the hyperfine structure due to a methyl group *(32,33)*. This attribution has been challenged by an ENDOR investigation on the trimethylsemiquinone radical, which can be considered a model for Plastoquinone *(34)*. More Recently Signal II has been attributed to a Tyrosine radical *(35)*.

A large amount of work on the photosynthetic systems has been done also by ENDOR and TRIPLE in solution *(36)*. ENDOR in solution is considered advantageous with respect to ENDOR in frozen matrix because of its better resolution allowing to measure very precisely the isotropic hyperfine couplings. From the latter ones, it is possible to deduce the spin density distribution in the positions directly bonded to a proton or a methyl group *(37)*.

On the other hand, ENDOR in frozen matrix could in principle give more information, since the anisotropic tensors depend on the total spin distribution. However, the mere measurement of the frequencies of the main features it is not sufficient for extracting all the information. One should attempt also an analysis of the shape and amplitude of the spectrum. Our feeling is that, although the complexity of biochemical objects makes this task difficult, it might be worth trying to exploit all the information contained in the frozen matrix ENDOR spectra.

ACKNOWLEDGMENT

This work was supported in part by the C.N.R. through the Centro Studi Stati Molecolari Radicalici ed Eccitati and in part by the Ministero della Pubblica Istruzione.

REFERENCES

1 A. Abragam, *The Principles of Nuclear Magnetism*, Oxford Univ. Press, London, 1961.

2 C.P. Poole, *Electron Spin Resonance*, J. Wiley and Sons, New York, 1983.

3 N.M. Atherton, *Electron Spin Resonance*, Ellis Horwood, Chichester, 1973.

4 S.D. Leniart, in: M.M. Dorio and J.H. Freed (Eds), *Multiple Electron Resonance Spectroscopy*, Plenum, New York, 1979, pp. 5–72.

5 M. Brustolon, T. Cassol, L. Micheletti and U. Segre, *Molec. Phys.*, **57** (1986) 1005–1014.

6 S. Lee, *Phys. Rev. B*, **23** (1981) 6151–6153.

7 J.H. Freed, *J. Chem. Phys.*, **43** (1965) 2312–2332.

8 J.H. Freed, in: M.M. Dorio and J.H. Freed (Eds), *Multiple Electron Resonance Spectroscopy*, Plenum, New York, 1979, pp. 73–142.

9 A.G. Redfield, *Adv. Magn. Reson.*, **1** (1966) 1–32.

10 M.K. Bowman and L. Kevan, in: L. Kevan and R.N. Schwartz (Eds), *Time Domain Electron Spin Resonance*, J. Wiley and Sons, New York, 1979, pp. 67–105.

11 L.R. Dalton, A.L. Kwiram and J.A. Cowen, *Chem. Phys. Lett.*, **17** (1972) 495–499.

12 L. Kevan and L.D. Kispert, *Electron Spin Double Resonance Spectroscopy*, J. Wiley and Sons, New York, 1976, pp. 165–253.

13 M. Brustolon and T. Cassol, *J. Magn. Reson.*, **60** (1984) 257–267.

14 L.R. Dalton, B.H. Robinson, L.A. Dalton and P. Coffey, *Adv. Magn. Reson.*, **8** (1976) 149–259.

15 L.R. Dalton and A.L. Kwiram, *J. Chem. Phys.*, **57** (1972) 1132–1145.

16 M. Brustolon, A.L. Maniero and U. Segre, *Molec. Phys.*, **55** (1985) 713–721.

17 H. Kurreck, B. Kirste and W. Lubitz, *Electron Nuclear Double Resonance Spectroscopy of Radicals in Solution*, VCH, New York, 1988, pp. 209–227.

18 H. Kurreck, M. Bock, N. Bretz, M. Elsner, H. Kraus, W. Lubitz, F. Müller, J. Geissler and P.M.H. Kroneck, *J. Am. Chem. Soc.*, **106** (1984) 737–746.

19 M. Brustolon, T. Cassol, L. Micheletti and U. Segre, *Molec. Phys.*, **61** (1987) 249–255.

20 M. Brustolon, A.L. Maniero and U. Segre, *Molec. Phys.*, **65** (1988), 447–453.

21 G.H. Rist and J.S. Hyde, *J. Chem. Phys.*, **52** (1970) 4633–4643.

22 G.C. Hurst, T.A. Henderson and R.W. Kreilick, *J. Am. Chem. Soc.*, **107** (1985) 7294–7299.

23 B.M. Hoffman, this book, Chapter 15.

24 H. van Willigen and C.F. Mulks, *J. Chem. Phys.*, **75** (1981) 2135–2140.

25 M. Brustolon, A.L. Maniero, U. Segre and L. Greci, *J. Chem. Soc., Faraday Trans. I*, **83** (1987) 69–75.

26 M. Brustolon, A.L. Maniero and C. Corvaja, *Molec. Phys.*, **51** (1984) 1269–1281.

27 G.B. Wells and M.W. Makinen, *J. Am. Chem. Soc.*, **110** (1988) 6343–6351.

28 G. Feher, A.J. Hoff, R.A. Isaacson and C.C. Ackerson, *Ann. N.Y. Acad. Sci.*, **244** (1975) 239–259.

29 J.R. Norris, H. Scheer, M.E. Druyan and J.J. Katz, *Proc. Natl. Acad. Sci.*, **71** (1974) 4897–4900.

30 A.J. Hoff, *Phys. Rep.*, **54** (1979) 75–200.

31 P.J. O'Malley and G.T. Babcock, *Proc. Natl. Acad. Sci. USA*, **81** (1984) 1098–1101.

32 P.J. O'Malley and G.T. Babcock, *Biochim. Biophys. Acta*, **765** (1984) 370–379.

614

33 P.J. O'Malley, G.T. Babcock and R.C. Prince, *Biochim. Biophys. Acta*, **766** (1984) 283–288.

34 M. Brustolon, D. Carbonera, M.T. Cassol and G. Giacometti, *Gazz. Chim. Ital.*, **117** (1987) 149–153.

35 B.A. Barry and G.T. Babcock, *Proc. Natl. Acad. Sci. USA*, **84** (1987) 7099–7103.

36 K. Möbius, W. Lubitz and M. Plato, this book, Chapter 13.

37 M. Plato, E. Tränkle, W. Lubitz, F. Lendzian and K. Möbius, *Chem. Phys.*, **107** (1986) 185–196.

CHAPTER 17

PULSED ENDOR

KLAUS-PETER DINSE

1. INTRODUCTION

Electron-Nuclear Double Resonance (ENDOR) was introduced as early as 1956 (1) to extract hyperfine splittings from inhomogeneously-broadened EPR spectra, and it has since then proven to be an extremely versatile method. In a simple picture, ENDOR signals can be expected whenever the frequency- (or field-) dependence of an EPR transition on a particular nuclear spin quantum number m_I is larger than the homogeneous width of the EPR transition, measured for instance by a Hahn echo sequence.

In the classic cw-type ENDOR experiment (including its generalizations to multiple resonance like TRIPLE) the population difference of a particular pair of spin sublevels ($\Delta m_S = 1$, $\Delta m_I = 0$) is probed by monitoring the corresponding EPR absorption amplitude. In the limit of small deviations from the equilibrium density matrix, ENDOR is probing the dependence of the diagonal density matrix elements on the various rf fields. In this context, "small deviation " means that EPR as well as NMR transitions are driven only up to a saturation factor close to one.

In a cw-type experiment, the system's relaxation rates in the multi-level spin system obviously determine the appropriate amplitudes of the coherent fields. In addition, the magnitude of the ENDOR effect, i.e. the relative change of the EPR absorption, is a complicated function of the various relaxation rates, which in general cannot be optimized independently. Although a satisfactory description of cw-ENDOR has been obtained in the meantime. allowing for a quantitative prediction of the ENDOR effect, even a careful adjustment of the experimental parameters could not lead to experimental ENDOR intensities significantly above the few percent level. This sensitivity level is clearly sufficient for the study of a large class of organic radicals in solution and also of non-transient radicals in solid matrices, but obviously the method is not well adapted to the study of transient radicals or to situations, where the relaxation parameters cannot be adjusted by changing for instance the temperature or the nature of the matrix.

A new version of the ENDOR experiment was proposed as early as 1965, when Mims realized (2) that a time-resolved ENDOR scheme originating from a 3-pulse (stimulated) echo sequence is essentially free from the restrictions to balance relaxation and induced transition rates. Details of the experiment are described below. Here it is sufficient to note that again only populations within the multi-level spin system are resonantly transfered. Relaxation bottle necks are avoided by repeating the pulse sequence with a rate small compared with the slowest relaxation rate. Depending on populations only, the time scale for the irradiation of the nuclear spin transitions is given approximately by T_{1e}, thus allowing for low rf amplitudes. In addition, the inherently time-limited irradiation of the NMR transitions with its resulting Fourier broadening is least important if the irradiation interval is only limited by the longest kinetic parameter of the system.

A further generalization of ENDOR was proposed by Brown et al. (3), who realized that rf-driven nuclear spin transitions cannot only be used to change populations but can also be utilized to create rf-dependent phase changes of the observed EPR transition. The most simple realization of this concept was to replace the continous microwave field probing the EPR transition by a two-pulse excitation. The resulting Hahn echo is a sensitive indicator for any changes in the average EPR resonance condition, i.e. the accumulated phase of the magnetization vector in the two time intervals of the experiment is compared. The same basic idea was later realized in a much more general fashion by Mehring et al. (4,5). In this kind of experiment the size of the ENDOR effect, i.e. the relative change of the echo intensity, can approach 100% values, if the pulse distance (being limited by the transverse relaxation time T_{2e} of the EPR transition) is long enough to accumulate a rf-driven phase change of π. A realistic upper limit for T_{2e} is 10 μs, thus limiting the minimum observable hf splitting to about 100 kHz. actually the same limit as obtained for cw-ENDOR. Unfortunately, this 2-pulse ENDOR scheme requires the resonant change of a nuclear spin state on this time scale. For non-hf-enhanced nuclear transition moments (vide infra), the μs time scale enforces the use of rf amplitudes in the 10 to 50 G range (rotating frame) even for the most favorable proton case.

An important further variant of the Mims ENDOR scheme was proposed by Davies (6), who replaced the two first microwave pulses in the stimulated echo sequence by a single pulse. As is shown below, a single pulse of a finite length also produces a periodic magnetization pattern in an inhomogeneous absorption line, although its period under normal experimental conditions is much longer than the 2-pulse generated pattern. Clearly, this pattern also can be changed by resonant rf fields, and the effect can be monitored again with an echo sequence.

Finally, the most general scheme for pulsed ENDOR is the one realized by performing a selective coherence transfer from the electron spin subsystem to the nuclear spin reservoir with subsequent time evolution under a controlled spin Hamiltonian, as was shown by Mehring and coworkers

(7). The essence of the ENDOR idea is retained by observing the nuclear spin coherence indirectly by transfering it back to the electron spin sub-system with a final mixing pulse. In this picture, pulsed ENDOR bears great similarity to multi-dimensional NMR.

Obviously, all pulsed ENDOR schemes so far mentioned enable the detection of transient species. Pulsed ENDOR is generally a non-equilibrium method, free of the requirement for detailed balancing of resonantly-induced rates and relaxation rates. In addition, pulsed ENDOR allows for synchronization with, for instance, optical excitation schemes, clearly an attractive property for studies in photophysics and photobiology.

2. THEORY OF PULSED ENDOR

In the following section a brief outline of the various ENDOR schemes is given to provide the reader with an intuitive model of the experiment. In most cases predictions from such a simple picture are reliable. For a quantitative analysis computer simulations are advisable, especially if - as is often the case - the spectral widths of the microwave pulses are compar-able to the inhomogeneous absorption profile.

2.1 Mims ENDOR

Since its introduction in 1965, the ENDOR scheme based on the 3-pulse (stimulated) echo sequence was investigated in detail (8-11) not only in magnetic reasonance but also in the optical regime (8). Its relative

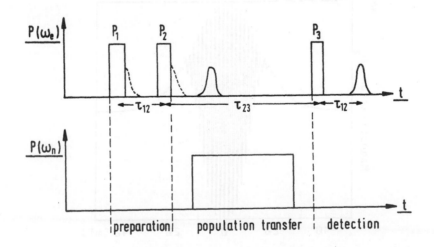

Fig. 1 Mims-ENDOR sequence. In the preparation phase, a periodic pat-tern is created in the absorption line shape function by P_1 and P_2. This pattern is read by the final pulse P_3 in form of the echo amplitude as a function of ω_n.

ease of implementation on a pulsed EPR apparatus is apparent from the timing sequence depicted in Fig. 1. Three different time ranges can be defined, which are termed in analogy to multi-dimensional NMR as *preparation, population transfer,* and *observation* phases, respectively.

In the preparation phase, two microwave pulses of length $t_p(i)$ and distance τ_{12} are used to create a non-Boltzmann population pattern in the multi-level spin system.

In the limiting cases of excitation pulses with a Rabi frequency $\omega_1 = \gamma B_1$ either small or large compared to the line width $\Delta\omega_{1/2}$ of the EPR transition, analytical expressions for the resulting magnetization pattern have been derived (12,13). Here we quote the result for the more frequent case $\Delta\omega_{1/2} \gg \omega_1$:

$$M_Z/M_0 = \cos\varphi \sin^2\psi \cos^2\varphi (1-\cos\Theta)^2 - \cos\varphi \sin^2\psi \sin^2\Theta$$
$$+ 2\sin\varphi \sin^2\psi \cos\psi \sin\Theta (1-\cos\Theta) \qquad [1]$$
$$+ (\cos^2\psi + \sin^2\psi \cos\Theta)^2 .$$

Here we defined $\psi = \arctan(\omega_1/\Delta\omega)$, $\varphi = \tau_{12}\cdot\Delta\omega$, and $\Theta = (\omega_1^2 + \Delta\omega^2)\cdot t_p^{1/2}$, with $\Delta\omega$ denoting the resonance frequency offset.

In the general case, often met with organic radicals and high-power pulse apparatus leading to $\omega_1 \sim \Delta\omega_{1/2}$, numerical solutions of the Bloch equations have to be obtained, and Fig. 2 shows the resultant pattern for

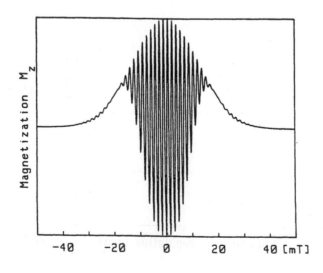

Fig. 2. Field dependence of M_Z after the preparation phase. The microwave field B_1 was chosen to give a rotation angle for the resonant spin packet of $\omega_1 t_p = \pi/2$. Actual values used were B_1 = 0.5 mT; $\Delta\omega_{1/2}$(FWHM) = 3 mT (Gaussian) ; $t_p(i)$ = 17.8 ns; τ_{12} = 0.2 μs.

a representative choice of experimental parameters.

It is important to note that the periodic absorption pattern is determined by the pulse distance τ_{12} and can be controled in the experiment within the limits set by T_{2e} of the driven EPR transition.

The relatively simple frequency dependence of M_z cannot be correlated with a correspondingly simple population pattern for the nuclear spin states of a particular m_s-manifold, because of a missing one-to-one correspondence of *energy differences* and *absolute energies* in the spin system. For this reason it is not possible to discuss the ENDOR effect simply by correlating ENDOR frequencies with the absorption period of M_z.

In general, the pattern ("absorption grating") in M_z can be detected with a third microwave pulse. As indicated in Fig. 1, the magnetization pattern leads to an echo that is delayed by τ_{12} with respect to the third pulse. Its intensity is directly proportional to the Fourier component of the magnetization at frequency $\omega = 1/\tau_{12}$, and is therefore susceptible to non-resonant thermalization of the absorption pattern and/or to a modification of M_z by rf-induced transitions in the m_I-sublevels. This is the basis of the Mims-ENDOR experiment.

Although there is no simple way to correlate the M_z-pattern with populations in the m_s-manifold that are directly affected by the rf-transition, it is nevertheless possible to relate the hyperfine interaction (hfi) parameter with the magnetization period $1/\tau_{12}$. We approximate the spin system by a truncated Hamiltonian:

$$H = \omega_e S_z + a S_z I_z - \omega_n I_z \qquad [2]$$

which is appropriate for a dominating electron Zeeman interaction and a nearly isotropic hfi tensor.

As is indicated in Fig. 3, it is possible to construct a closed "transition frequency loop", correlating ENDOR frequencies with EPR resonance frequencies, monitored by M_z. Obviously we have:

$$\omega_1 + (\omega_n - a/2) - \omega_2 - (\omega_n + a/2) = 0 \qquad [3]$$

leading to

$$(\omega_1 - \omega_2) = a. \qquad [4]$$

From [3] we conclude that either one of the possible ENDOR transitions in the m_s-manifold leads to a transfer of magnetization in EPR space by a. As a result, there will be (to first order) no observable change in the magnetization pattern, if

$$a\tau_{12} = 2\pi n \quad (n = 1,2,..) \qquad \text{(a in radians/s).} \qquad [5]$$

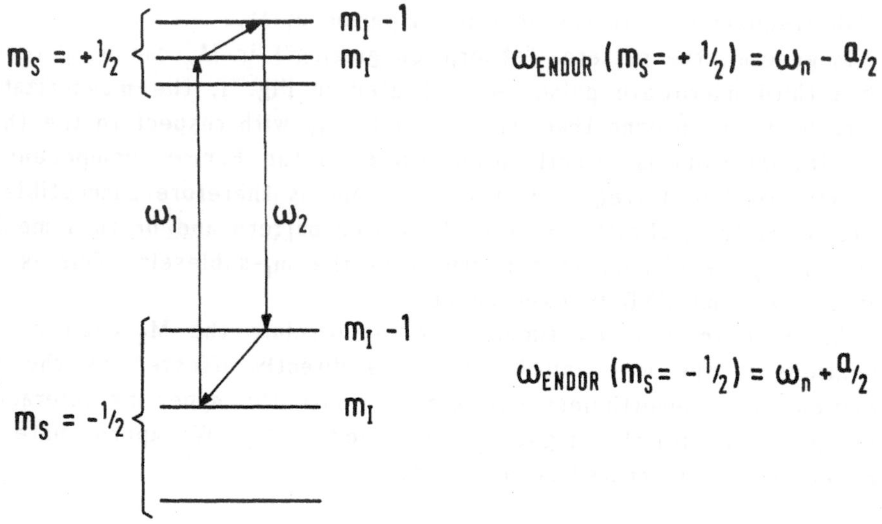

$$\omega_{ENDOR}\left(m_S = +\tfrac{1}{2}\right) = \omega_n - \tfrac{a}{2}$$

$$\omega_{ENDOR}\left(m_S = -\tfrac{1}{2}\right) = \omega_n + \tfrac{a}{2}$$

Fig. 3. Energy level diagram according to [2]. A sequence of allowed EPR and ENDOR transitions is indicated.

Liao and Hartmann (8) derived an expression for the ENDOR intensity in the limit of $\Delta\omega_{1/2} \ll \omega_1$:

$$I(ENDOR) \sim \{1 - \cos(a\tau_{12})\} \qquad [6]$$

giving an analytical expression for the well-documented "blind spot" behaviour of Mims-ENDOR. (See also the contribution of Schweiger in this volume.) As was shown by Höfer (14), in the opposite limit $\Delta\omega_{1/2} \gg \omega_1$, Eqn. [6] is modified to

$$I(ENDOR) \sim \{1 - \cos(a\tau_{12})\}(1 - \langle S_z \rangle) \qquad [7]$$

where $\langle S_z \rangle$ describes the deviation from the thermal magnetization induced by a single pulse of width t_p. Due to this additional factor the periodic pattern is restricted to roughly the spectral range affected by ω_1.

2.2 Davies ENDOR

As is well known, a periodic magnetization pattern within an inhomo-geneously-broadened absorption line can also be produced with a single pulse of lenght t_p and Rabi frequency ω_1. In complete analogy to the Mims ENDOR sequence, this periodic pattern can be detected in principle during a time interval of duration T_{1e}, because the information is stored as $M_z(\omega)$. The single-pulse read out has to be replaced by an echo sequence, however, because the transient signal, induced by a single pulse, has the same time structure as the FID after the preparation pulse and would therefore be lost in the dead time of the spectrometer. (The decay time of the FID for $\Delta\omega_{1/2} \gg \omega_1$ is approximately given by t_p.) Zero deviation in the modulated absorption pattern from the thermalized values are determined by $\omega_{eff} \cdot t_p = 2\pi n$, leading to an approximate period of $\delta\omega = 2\pi/t_p$.

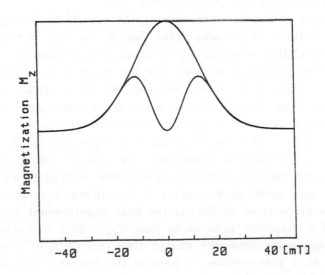

Fig. 4. Field dependence of M_z after preparation with a single pulse of width $t_p = 17.8$ ns and $\omega_1 t_p = \pi/2$. The Gaussian line has a width of $\Delta\omega_{1/2} = 3$ mT (FWHM).

Using the same set of experimental parameters as in Fig. 2, a completely different pattern is obtained, as shown in Fig. 4. The anticipated modulation structure is barely visible, and the effect of the preparation pulse is best described by a "saturation" hole in the absorption line. It should be stressed that in this experiment the monitoring echo is basically insensitive to the modulation pattern, but is rather determined by the central hole in $M_Z(\omega)$.

Whereas [6,7] give an analytical expression for the expected Mims-ENDOR amplitude as a function of the preparation pulse distance, for the Davies sequence only an approximate relation between a and t_p can be derived. Eqns. [6,7] originate from the nearly undamped periodic magnetization pattern (if we assume $\omega_1 \gg \tau_{12}^{-1}$; $\Delta\omega_{1/2} \gg \omega_1$), and this pattern is directly converted into the intensity of the stimulated echo. Missing this modulation period for the single pulse preparation, only limiting cases can be treated easily. For $a \ll t_p^{-1}$, the ENDOR-induced change in frequency of a specific EPR absorption packet $\Delta\omega$ will be small compared to the width of the "saturation hole" defined in Fig. 4. Approximating the deviation from the thermal equilibrium value of M_Z by a triangular function from 0 to $2 \cdot \sqrt{3} \cdot \omega_1$, a linear dependence I(ENDOR) $\sim$ a is predicted. This linear relation holds as long as $a \ll \omega_1$. The ENDOR signal will reach its maximum value for $a \approx \omega_1$ and then stay constant. In the same simple picture the absolute ENDOR signal will increase for a fixed value of a linearly with ω_1 up to $a \approx \omega_1$, originating from the linear dependence of affected spin packets on ω_1 in the (infinitely) broadened absorption line. Beyond this, again a nearly constant absolute ENDOR signal is predicted.

Comparing this general behaviour with results for the Mims-ENDOR scheme, no differences are seen in the limit of $\tau_{12} = 0$. In reality, however, instrument dead time restricts τ_{12} to values greater than 100 ns, still leaving an order-of-magnitude difference in the modulation period of M_Z. Even neglecting the sensitivity modulation, also the total signal intensity can be smaller in the Mims-ENDOR scheme, because the stimulated echo decay time τ_s is strongly dependent on τ_{12} (15). In most experimental cases it might be difficult to reach the limit $\tau_s \approx T_{1e}$ with $\tau_{12} \gg 100$ ns. As a result, the time interval τ_{23} might be restricted more than for the Davies sequence. In that case the spectral resolution is no longer limited by the life time of the nuclear spin sublevels but rather by τ_s. (Here it is assumed that the single pulse preparation of ΔM_Z is the best experimental approximation to the goal of a detection echo decay time T_{1e}.) This limitation in spectral resolution originates from the limited time the ENDOR transition can be irradiated. As a consequence, in general the Davies preparation sequence should be preferred.

2.3 Two-pulse ENDOR

As was mentioned in the introduction, it is also possible to monitor the rf-induced variation of the electron-spin coherence directly with an ordinary Hahn echo sequence (3). In contrast to the other schemes discussed so far, which utilize manipulations of the diagonal elements of the density matrix, here non-diagonal elements are probed directly. As a consequence the observation time is limited by T_{2e}, and therefore the spectral resolution for transitions within the nuclear spin manifold is necessarily restricted to the homogeneous EPR line width. In general $T_{2e} \ll T_{2n}$, so that this imposes a serious drawback for the use of the method. Although Mehring and co-workers have demonstrated in a series of beautiful experiments (4,5) the feasibility of this ENDOR scheme and have shown in particular that relative ENDOR intensities of 200% can be realized for favourable cases, this general limitation cannot be surpassed.

In addition in this phase-sensitive ENDOR experiment the highest demand for the rf amplitude has to be met, because ENDOR transitions have to be excited using Rabi frequencies ω_2 with $\omega_2 \cdot T_{2e} \approx \pi$. Using an average $T_{2e} = 2\mu s$, $\omega_2 \approx 1.5 \cdot 10^6 s^{-1}$ is obtained. The corresponding rf amplitude amounts to $B_2 = 5.5$ mT (rotating frame, $\gamma_p = 2\pi \cdot 10^7 s^{-1}T^{-1}$), a value used in a high-power pulsed NMR spectrometer.

Both reasons seem to exclude the use of a two-pulse echo sequence for the detection of rf transitions in general experiments.

2.4 Coherence transfer ENDOR (CT-ENDOR)

Although not used in our discussion of the various ENDOR schemes, the 2-pulse preparation step also leads to coherences between the nuclear spin sublevels within a specific m_S-manifold. These non-diagonal terms in the density matrix, which are detectable following the second microwave pulse, decay with their individual $T_{2n}(i)$, where i denotes the various transitions. A general limit for T_{2n} is given by the life time of the electron sublevel to which the nuclear levels are connected.

The nuclear spin coherence can be read out *indirectly* by monitoring the stimulated echo amplitude. This technique known as ESEEM is reviewed in chapters I to V. The method has found widespread use because of its ease of implementation and its capability to extract hfi information without the use of rf-fields. Its frequency resolution is limited by τ_s, the decay time of the stimulated echo, which only under somewhat extreme experimental conditions approaches T_{1e}.

The drawback of the ESEEM approach is given by the interdependence of the creation of *nuclear spin coherence* and of the *population modulation*, essential for the subsequent detection of the stimulated echo. The time interval for which nuclear coherence can be observed, is therefore limited by spectral diffusion within the EPR line, a process that does not necessarily contribute to T_{2n}.

Mehring and coworkers (7) developed an ENDOR scheme that decouples these steps, i.e. the creation and detection of nuclear spin sublevel coherence is independent from sublevel populations. In its essence CT-ENDOR is the "pure" realization of ENDOR, i.e. the EPR transition only monitors the developement in time of the nuclear spin coherence, and because both coherent fields are not present simultaneously, the resulting rf-induced spectrum is limited in its spectral resolution only by the inherent $T_{2n}(i)$.

Fig. 5 shows the principle of the experiment. The first microwave pulse, which is assumed to be frequency-selective (i.e. resonant for instance with $|1\rangle$-$|3\rangle$), creates a population difference in the nuclear sublevels of the order of the electronic Boltzmann factor. This population difference is utilized with a prompt rf pulse at the ENDOR frequency $|3\rangle$-$|4\rangle$ to start nuclear spin coherence, which can be visualized as transverse magnetization rotating at frequency ω_{34}. If the rf pulse is chosen as $\pi/2$-pulse, equal populations of levels $|3\rangle$, $|4\rangle$ are obtained in combination with a maximum transverse nuclear magnetization. It should be noted that the nuclear spin precession cannot be seen directly just by applying a further microwave pulse, because this would only be sensitive to population oscillations.

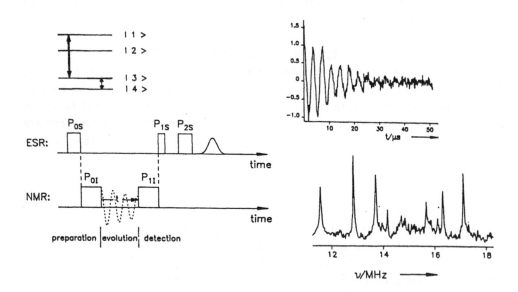

Fig. 5. Pulse sequence of the CT-ENDOR scheme in a simple 4-level system. The sequentially recorded FID is measured for a specific value of the rf frequency, the ENDOR spectrum shown is synthesized from several spectra covering approximately 2 MHz each. This spectrum overlaying was necessary because of the limited B_2. From Ref. 7.

Instead. the nuclear spin FID has to be recorded indirectly, i.e. with a further resonant rf-pulse (preferably of lenght $\pi/2$) the transverse magnetization is converted into a population difference in $|3\rangle, |4\rangle$, which in turn is detected by the final microwave pulse. (It should be noted, however, that in principle the nuclear spin precession could be monitored also with the NMR coil.) Due to spectrometer dead time, the EPR population difference is detected via a 2-pulse echo with fixed delay τ, as in the Davies ENDOR scheme. As apparent from the indirect detection scheme, which is also frequently used in ODMR experiments (16), the nuclear spin FID is recorded point by point. Subsequent Fourier transformation results in a spectrum of nuclear spin sublevel transitions, i.e. the ENDOR spectrum.

Although very elegant in principle, the method is characterized by several subtleties that render the routine use of it difficult. Firstly, the nuclear spin FID should be started (and its final value converted) by a rf-pulse with spectral width as large as possible . Assuming a $\pi/2$-pulse length of 1 μs, still only 1 MHz of the ENDOR spectrum is covered, necessitating not only a point by point recording in time but also an overlap of excitation intervals in frequency. Nevertheless, even to reach this $\pi/2$-pulse length. a rf amplitude of 6 mT is required, considering the transition moment of protons. Secondly, in order to avoid frequency ambiguity in the Fourier spectrum of a single FID, quadrature read-out is essential. This can be realized by recording two FID´s with $\pi/2$-phase- shifted rf read-out pulses. being equivalent to "quadrature" information in 2D-NMR. Finally, recording of the FID necessitates scanning of the microwave pulse distance. This in turn leads to a "trivial" decay of the EPR echo, seen in the Fourier spectrum as a convolution of the ENDOR line width with this decay time. "Trivial" in this context means that this decay is not depending on rf-induced transitions but rather on the vanishing of the population hole due to spin diffusion or T_{1e} processes.

2.5 Further experimental schemes

The extended time sequence of a pulsed ENDOR experiment obviously opens the possibility to vary different parameters additionaly. In particular not only one rf frequency but also two can be applied during the evolution period. This analog of the cw TRIPLE experiment (17,18) not only opens the possibility to identify transitions that have at least one energy level in common (i.e., to select for transitions of one particular paramagnetic center) but furthermore enables the determination of the relative sign of coupling constants (14). The method has been demonstrated not only for radicals in irradiated malonic acid with rather narrow ENDOR transitions but also for paramagnetic centers in amorphous trans-$(CH)_x$ with ENDOR lines extending over several MHz (14). I therefore believe that multi-frequency ENDOR bears great potential for the investigation of powdered samples.

A variant of CT-ENDOR is given by a scheme proposed by Schweiger and coworkers (19), in which the damped modulation of the nuclear spin sublevel population difference, originating from resonant excitation of an ENDOR transition, is monitored *continously* with a weak microwave probing field. In such an experiment, which is described in some detail in chapter VI of this book, the non-diagonal element denoting the electron spin coherence is coupled via the $\Delta m_I = \pm 1$ ENDOR transition to another nuclear hyperfine level, and a characteristic transient nutation driven by the rf field can be observed. The use of the method is predominantly given by the possibility to detect the number of equivalent nuclei via the dependence of the transition moment on the total spin in a coupled representation. It should be noted that, as in conventional cw ENDOR, the number of equivalent nuclei is not apparent from the ENDOR line intensity. The superposition of different transition moments leads to a characteristic structure in the otherwise simply periodic transient nutation pattern, as was demonstrated by Schweiger et al..

As was also shown by Schweiger's group, a change of the microwave frequency/ or external Zeeman field can be utilized to introduce a further dimension for the display of the linear ENDOR spectra. This variant of pulsed ENDOR is described in chapter VI .

3. EXPERIMENTAL PRAXIS OF PULSED ENDOR

The combination of rf- and microwave fields at a sample is a non-trivial task as has been learned from cw-ENDOR experience. In one aspect, the conditions for pulsed ENDOR are relaxed, however. Originating from the demand of a short dead time in the microwave channel, the use of low-Q cavities is mandatory (22). The incorporation of a rf-coil for B_2 generation therefore is much easier because of the low Q for the basic microwave structure. As was shown by Mehring and coworkers (4,5,7), even direct wrapping of a mini rf-coil on the crystal can be tolerated.

Changing the cw-excitation to pulsed schemes with rise times in the 100 ns range poses no problems, in particular because the rf field amplitude B_2 is comparable in both experiments. Considering for the moment Davies- or Mims-Endor schemes, typical time windows for the excitation of ENDOR transitions are of the the order of 10 μs. This can be converted to a maximum necessary Rabi frequency of 50 kHz, i.e. a 1 mT amplitude of B_2 in the rotating frame would be sufficient for protons with their gyromagnetic ratio $\gamma_p/2\pi$ = 43 MHz/T. This value of B_2 seems to be a reasonable design parameter, although a multi-purpose apparatus also has to consider other nuclei of biological interest like ^{14}N, ^{2}D, ^{63,65}Cu, ^{55}Mn etc. Their generally smaller γ_I is in some cases compensated by a hyperfine enhancement factor η, originating from electron-nuclear hfi (23). The enhancement factor η can be approximated by comparing the actual ENDOR frequency ν_{ENDOR} with the Zeeman frequency ν_N of the bare nucleus as

$$\eta \approx \nu_{ENDOR} / \nu_N \qquad\qquad\qquad [8]$$

indicating that in general ENDOR transitions of a large normalized frequency can easily be excited even by weak rf-fields. This direct scaling of the Rabi frequency is observed, however, only for an approximate isotropic hfi tensor (7,23,24).

3.1 Probehead design

Rather than refering to previous designs (4,5,7,11), a system centered about a loop-gap cavity, optimized for easy sample access, laser illumination and temperature control, is described. The design idea is similar to the set-up suggested by Biehl (25) and introduced by Pfenninger et al. (26). As seen in Fig. 6, the rf-field is generated by a coil, being wrapped around a standard Helium finger dewar (Oxford Instruments, ESR 10). This dewar consists of a double-wall evacuated quartz finger and a sample holder as an inner heat shield. This sample holder was modified to be the microwave resonator, whithout changing its standard dimensions. The structure of the X-band resonator is formed by a thin (~1 μm) silver coating on the inside of the sample holder of 15 mm length. This conducting cylinder is interrupted by two gaps parallel to the cylinder axis of 1 mm width, forming a structure with C_2 symmetry. The gaps are bridged on the outside with two strips of varying width (~ 2 mm), which are used for fine tuning of the resonance frequency. This frequency is roughly determined by the diameter of the inner cylinder in combination with the metal shield dimensions (28). The cavity Q obtained was in the range of 150, allowing the use of 15 ns micro-wave pulses. The power-to-field conversion factor of our set-up was 0.07 $mT/W^{1/2}$, somewhat smaller than the values given for the smaller resonators (26). The silver coating of approximately 1 μm thickness is practically transparent to rf-fields up to 100 MHz and therefore allows the use of the outside coil. The requirement to scan the frequency of B_2 over a wide range (for example from 0 to 50 MHz) necessitates the use of a low-Q broad-band rf circuit similar to that used in cw-ENDOR. In the simplest possible design, the frequency-dependent impedance of the coil (8 turns, 11 mm diameter, 15 mm length) is connected in series with 50 Ω. A field B_2 = 1 mT is generated at 15 MHz with an rf power of 200 W. Although the frequency response of this set-up is not perfect ($|\omega L| \approx$ 100 Ω at 15 MHz), this simple scheme works reliably and even partially compensates for the frequency-dependent enhancement factor considering the Zeeman frequency of protons in a field of 0.35 T. Impedance matching is improved by using the B_2-coil as part of a resonant π transformer (Collins transformer). In principle the generator impedance of 50 Ω is transformed with a discrete $C_1/L/C_2$ combination into a terminating 50 Ω resistance of appropriate power rating.

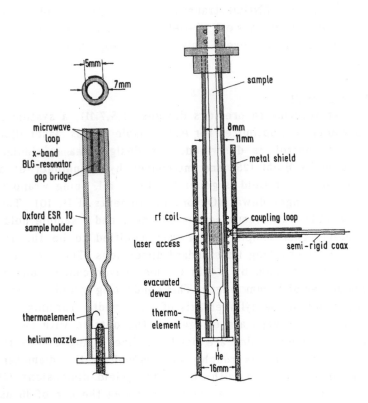

Fig. 6. Integrated ENDOR cavity using a commercial temperature control unit. The sample holder is shown as insert to indicate the dimensions of the microwave cavity.

3.2 Experimental results

An impressive application of pulsed ENDOR on a metastable para-magnetic state of biological interest was described already in 1983 (11). In an exemplary way it highlights the possible merits of pulsed ENDOR. In their study, the authors investigated the hfi of the inner-ring protons of free-base porphin in the photo-excited triplet state, which has a life time of 25 ms to 1 s, depending on the orientation-dependent mixture of zero-field states. This rather long life time imposes no restrictions on the theoretically possible spectral resolution in the ENDOR spectrum. From experience in cw-ENDOR, it is expected that spectral resolution is limited by the crystal defect-induced orientation distribution of the guest molecules. Typical values for proton ENDOR $\Delta\omega_{1/2}$ are 10 kHz (FWHM) in molecular crystals (27). Consequently, in the Mims-ENDOR sequence utilized, the microwave pulse separation τ_{12} was chosen as 160 μs, the rf irradiation time was 150 μs.

Fig. 7 reproduces the experimental spectrum, and although the ENDOR effect amounted to 50% of the echo intensity, extensive averaging (40 scans) was mandatory to obtain a decent signal-to-noise ratio. Originating from the fact that the paramagnetic state is of triplet origin, hfi is observed in

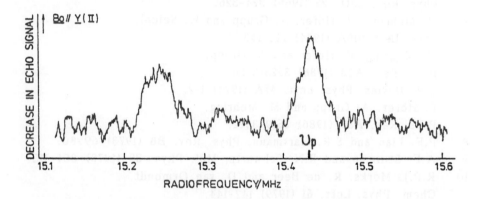

Fig. 7. Pulsed Proton ENDOR spectrum of the inner protons of free-base porphin in n-octane. The lines correspond to transitions in the $m_S=+1$ and $m_S=0$ manifold of electron spin states. From Ref. 11.

two distinct subsets of the electron spin manifold. The $m_S=0$ sublevel contributes to first order no hyperfine field at the nuclei and therefore provides a convenient reference with repect to line width and transition moment, i.e. one can approximately use $\eta = 0$, if the electron Zeeman energy is large compared to the zero-field splitting.

In the future most applications of pulsed ENDOR will focus on disordered systems. From pulsed EPR experiments we know that T_{1e} and T_{2e} are not shortened significantly by disorder effects so that pulsed ENDOR schemes stay operative also in these systems. As was pointed out for instance by Thomann (29), pulsed ENDOR should in general be feasible also at room temperature. The problem of spectral analysis of randomly oriented molecules, however, has to be solved for every specific problem. Most progress can be expected for nuclei with quadrupole couplings large compared to their nuclear Zeeman energy, allowing the identification of different sites in complex molecules.

ACKNOWLEDGEMENT

I wish to thank my collegues who generously informed me about their progress in this rapidly developing field. My coworker Dr. T. Prisner contributed the calculations for the field-dependent magnetization pattern under various conditions of microwave excitation. A. Zahl and M. Plüschau designed the loop-gap resonator for pulsed ENDOR.

REFERENCES

1   G. Feher, Phys. Rev. **103** (1956) 834-835.
2   W.B. Mims, Proc. Roy. Soc. **A283** (1965) 452-457.
3   I.M. Brown, D.J. Sloop and D.P. Ames,
     Phys. Rev. Lett. **22** (1969) 324-326.
4   M. Mehring, P. Höfer, A. Grupp and H. Seidel,
     Phys. Lett **106A** (1984) 146-148.
5   M. Mehring, P. Höfer and A. Grupp,
     Phys. Rev. **A33** (1986) 3523-3526.
6   E.R. Davies, Phys. Lett. **47A** (1974) 1-2.
7   P. Höfer, A. Grupp and M. Mehring,
     Phys. Rev. **A33** (1986) 3519-3522.
8   P.F. Liao and S.R. Hartmann, Phys. Rev. **B8** (1973) 69-80.
9   A.E. Stillman and R.N. Schwartz, Mol. Phys. **35** (1978) 301-313.
10   R.P.J. Merks, R. de Beer and D. van Ormondt,
     Chem. Phys. Lett. **61** (1979) 142-144.
11   W.A.J.A. van der Poel, D.J. Singel, J. Schmidt and J.H. van der Waals,
     Mol. Phys. **49** (1983) 1017-1028.
12   W.B. Mims, K. Nassau and J.D. McGee,
     Phys. Rev. **123** (1961) 2059-2069.
13   W.B. Mims in: S. Geschwind (Ed.), Electron Paramagnetic Resonance:
     Electron Spin Echoes, Plenum Press, New York 1972, pp. 263-351
14   P. Höfer, Thesis, University Stuttgart 1988
15   S.M. Janes and H.C. Brenner, Chem. Phys. Lett. **95** (1983) 23-29.
16   H.C. Brenner, in: R.H. Clarke (Ed.), Triplet State ODMR Spectroscopy:
     Energy transfer and coherence effects in ODMR, Wiley, New York, 1982,
     pp. 185-256.
17   K.P. Dinse, R. Biehl and K Möbius,
     J. Chem. Phys. **61** (1974) 4335-4341.
18   R. Biehl, M. Plato and K. Möbius,
     J. Chem. Phys. **63** (1975) 3515-3522.
19   C. Gemperle, A. Schweiger and R.R. Ernst,
     Chem. Phys. Lett. **145** (1988) 1-8.

20   R.M. Shelby, C.S. Yannoni and R.M. Macfarlane,
     Phys. Rev. Lett. **41** (1978) 1739-1742.

21   H. Brunner and K.P. Dinse, Bull. Magn. Reson. **2** (1981) 110-111.

22   A.D. Trifunac, R.G. Lawler, D.M. Bartels and M.C. Thurnauer,
     Prog. Reaction Kinetics, **14** (1986) 43-156.

23   S. Geschwind, in: A.J. Freeman and R.B. Frankel (Eds), Hyperfine
     Interactions: Special Topics in Hyperfine Structure in EPR, Academic
     Press, New York, 1967 , pp. 225-286.

24   L.R. Dalton and A.L. Kwiram, J. Chem. Phys. **57** (1972) 1132-1145.

25   R. Biehl, private communication

26   S. Pfenninger, J. Forrer, A. Schweiger and T. Weiland,
     Rev. Sci. Instrum. **59** (1988) 752-760.

27   C. von Borczyskowski, M. Plato, K.P. Dinse and K. Möbius,
     Chem. Phys. **35** (1978) 355-366.

28   W. Froncisz and J.S. Hyde, J. Magn. Reson. **47** (1982) 515-521.

29   H. Thomann, 30th Rocky Mountain Conference, Denver 1988

CHAPTER 18

OPTICALLY-DETECTED MAGNETIC RESONANCE OF TRIPLET STATES

ARNOLD J. HOFF

1. TRIPLET STATES

Triplet states are with few exceptions (notably dioxygen) (photo)excited metastable states of usually aromatic, organic molecules. This type of molecules abounds in all biological material and the triplet state is an important probe of the molecular structure. Triplets have certain physical properties that make them eminently useful to identify reactants in photoreactions, to probe structural environment, binding etc., and to serve as gentle perturbers. Studying the system's response then gives important clues to its structure and function.

In this chapter I will first introduce the physical characteristics of the triplet state. I will then briefly discuss a variant of the technique of electron paramagnetic resonance (EPR): optically-detected magnetic resonance (ODMR) in zero magnetic field, which is one of the most important spectroscopic tools to study the triplet state. Most attention will be devoted to absorbance-detected magnetic resonance, ADMR, which technique has taken a real flight in the past few years. By ADMR it is possible to monitor low-temperature triplet absorbance difference spectra with unparalleled accuracy and sensitivity. For coupled pigment systems such as encountered in photo-synthetic membranes this has opened up a whole new field for the study of pigment interaction, interpretation of optical spectra, etc.

The remainder of this chapter is devoted to the applications of ODMR in biological research. I will only briefly discuss applications in biochemistry, since these have been reviewed recently (1,2), and focus on the applications of the various forms of ODMR in photosynthesis.

1.1 Physics of the Triplet State

Optical transitions of aromatic molecules in the visible spectral region are usually $\pi - \pi^*$ absorptions of the π-electrons or $n - \pi^*$ transitions of the lone 2p-pair of heteroatoms (oxygen, nitrogen, etc.). Depending on the spin pairing of the two unpaired electrons, the excited

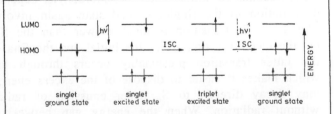

Fig. 1. Distribution of electrons over the highest occupied (HOMO) and lowest unoccupied (LUMO) molecular orbital of singlet and triplet states. ISC, intersystem crossing. From (132).

state is either a singlet state (spins antiparallel) or a triplet state (spins parallel). This is illustrated in Fig. 1. The unexcited molecule is in the singlet ground state; i.e., all electronic orbitals are occupied by a pair of electrons with opposite spin. On excitation by a photon of sufficient energy, one of the electrons of the highest occupied molecular orbital (HOMO) of a molecule may jump to the next higher orbital, the lowest unoccupied molecular orbital (LUMO). During the excitation process the spin state of the excited electron is preserved, because of the law of conservation of angular momentum, so that the excited molecule is still in a $\underline{S} = 0$ singlet (S_1) state and would remain in the singlet manifold if there were no coupling between spin and orbital angular momentum. However, the spin moment and the magnetic moment generated by the orbital motion interact magnetically, and through this spin-orbit interaction orbital angular momentum of the electrons may be converted into spin angular momentum without violating the conservation law. This means that there is a certain probability that the spin vector in the LUMO is inverted (this is allowed by the Pauli principle as the unpaired electrons are in different orbitals). The sum of all individual electron spin vectors now adds up to unity, $\underline{S} = 1$, and the multiplicity $2\underline{S} + 1$ is three, i.e. we have a triplet state, the ground state T_0 of the triplet manifold. (Sometimes the lowest triplet state is denoted T_1, to indicate that it is an excited state). The probability of the spin inversion or intersystem crossing, ISC, depends on the strength of the spin-orbit interaction, which for a single atom is proportional to the nuclear charge Z. The pigments that will concern us in the section on applications contain only relatively light atoms, so that spin-orbit coupling is weak and ISC in the LUMO is a relatively slow process, comparable with that of deexcitation by fluorescence; for example the yield of ^{3}Chl states in vitro is about 65 %.

In some important cases, triplet states are formed by radical recombination reactions. The radicals may be generated from a singlet excited state that is created by illumination, or by ionizing radiation. In photosynthesis under conditions that normal, forward electron transport is blocked, such a recombinational triplet state is formed with high efficiency.

Once the molecule is in a triplet state it may remain there for a long time (microseconds to milliseconds for chlorophyls, depending on the temperature) as deexcitation to the singlet ground state again involves a 'forbidden' spin flip. The $T_0 \rightarrow S_0$ transition is much slower than the $S_1 \rightarrow T_0$ transition, because (1) the electronic orbitals of the initial and the final state are different, and (2) the latter transition presumably occurs through higher-energy triplet states, which makes it easier to dispose of the excess energy as heat. The triplet state may decay directly to S_0 under emission of radiation (phosphorescence) or without radiation. When the energy gap between the S_1 and T_0 states is comparable to $k_B T$ (k_B is Boltzmann's constant and T is the temperature), the triplet may decay to S_0 via S_1 with emission of delayed fluorescence (Fig. 2).

The energy of the T_0 state is usually appreciably lower than that of the S_1 state. This is because the two unpaired electrons have the same spin quantum number and, according to the Pauli principle, cannot move in the same electronic orbital. On average the two electrons are farther apart in the triplet state than in the singlet state, hence the energy of Coulombic repulsion is less

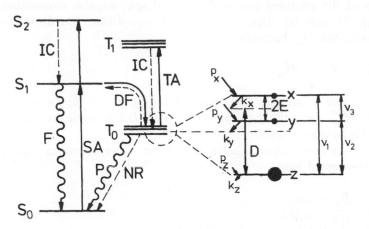

Fig. 2. Energy level diagram of the singlet and triplet manifold. S_0, singlet ground state; S_1 and S_2, singlet excited states; T_0 and T_1, first and second excited triplet states; SA and TA, singlet and triplet absorption, respectively; F and DF, fluorescence and delayed fluorescence, respectively; P, phosphorescence; NR, nonradiative transition; IC, internal conversion. Enlarged T_0 levels: X, Y and Z, eigenenergies of the dipole-dipole interaction; D and E, zero-field splitting parameters. Downward arrows to the right, populating probabilities; to the left, decay rates; filled circles, equilibrium populations. ν_1, ν_2 and ν_3 are the frequencies corresponding to the $(|D| \pm |E|)/h$ and $2|E|/h$ transitions, respectively.

and the state energy is lower. The decrease in repulsion energy is accounted for by introducing the electrostatic exchange energy, $-J$. For two unpaired spins on one molecule the exchange energy is usually negative $(J > 0)$ and the triplet state lies lower than the excited singlet state. This means that the wavelength of phosphorescence emission is longer than that of fluorescence emission.

1.2 The Triplet Spin Hamiltonian in Zero Magnetic Field

The triplet spin hamiltonian without external magnetic field comprises interactions involving the magnetic moment of the electrons. These are two-fold: spin-spin coupling and spin-orbit coupling. The main contribution to the spin-spin coupling operator, $\hat{H}_{SS}$, is the classical magnetic dipole-dipole interaction between two electrons:

$$\hat{H}_{SS} = \frac{3}{4} \cdot \frac{g^2\beta^2\mu_o}{4\pi} \left(\frac{\underline{s}_1 \cdot \underline{s}_2}{r^3} - \frac{(\underline{s}_1 \cdot \underline{r})(\underline{s}_2 \cdot \underline{r})}{r^5} \right) \qquad [1]$$

with g the electronic g-value, β the electronic Bohr magneton, $\underline{s}_1$, $\underline{s}_2$ the magnetic moments of the two electrons, $\underline{r}$ their distance vector and μ_o the permeability of vacuum. Eqn. 1 can be rearranged to

$$\hat{H}_{SS} = \underline{\hat{S}} \cdot \underline{D} \cdot \underline{\hat{S}} \qquad [2]$$

where $\underline{D}$ is a tensor operator whose elements consist of integrals over the coordinates of the electrons and $\hat{\underline{S}}$ is the total spin angular momentum operator $\hat{\underline{S}} = \hat{\underline{S}}_1 + \hat{\underline{S}}_2$. $\underline{D}$ can be diagonalized by a coordinate transformation to its principal axes, and $\hat{H}_{SS}$ becomes

$$\hat{H}_{SS} = -X\hat{S}_x^2 - Y\hat{S}_y^2 - Z\hat{S}_z^2 \qquad [3]$$

where X, Y and Z are the principal values of $\underline{D}$ and $\hat{S}_u$ (u = x, y, z) the components of $\hat{\underline{S}}$ along the principal axes of $\underline{D}$. Often, these axes coincide with the molecular symmetry axes.

In a two-electron approximation the triplet wave functions can be written in symmetry-adapted form

$$
\begin{aligned}
|T_x\rangle &= 2^{-\frac{1}{2}}(\beta_1\beta_2 - \alpha_1\alpha_2) \\
|T_y\rangle &= 2^{-\frac{1}{2}}i(\beta_1\beta_2 + \alpha_1\alpha_2) \\
|T_z\rangle &= 2^{-\frac{1}{2}}(\alpha_1\beta_2 + \beta_1\alpha_2)
\end{aligned}
\qquad [4]
$$

where α and β are the eigenfunctions of the component of the spin operator $\hat{\underline{s}}$ along the z direction, $\hat{s}_z$.

The functions $|T_u\rangle$ (u = x,y,z) belong to different irreducible representations of the point group C_{2v}, which is assumed to be a subgroup of the symmetry point group of the molecule; they are eigenfunctions of $\hat{S}^2$ with eigenvalue 2 and are to a first approximation degenerate if we neglect the spin-spin interaction. The hamiltonian [2] lifts the degeneracy, and it turns out that $|T_u\rangle$ are eigenfunctions of $\hat{H}_{SS}$ with eigenvalues X, Y and Z. The $|T_u\rangle$'s have the property

$$\hat{S}_u|T_u\rangle = 0, \qquad \hat{S}_x|T_y\rangle = -\hat{S}_y|T_x\rangle = i|T_z\rangle. \qquad [5]$$

The second relation holds for cyclic permutation of the subscripts x,y,z. Thus, the triplet component $|T_u\rangle$ is an eigenfunction of the operator $\hat{S}_u$ with eigenvalue zero, i.e. $|T_u\rangle$ corresponds to a situation where the spin angular momentum vector lies in the coordinate plane u = 0. From Eqn. 5 it further follows that there is no net magnetic dipole moment associated with any of the triplet substates in zero magnetic field

$$\gamma\hbar \langle T_u|\hat{\underline{S}}|T_u\rangle = 0 \qquad [6a]$$

but that there is a transition dipole moment present between any two of the triplet substates

$$\gamma\hbar \langle T_x|\hat{S}_y|T_z\rangle = i\gamma\hbar \quad \text{(cyclic)}, \qquad [6b]$$

where γ is the gyromagnetic ratio of the electron. This is an important result as it shows that in zero magnetic field, population can be transferred from one triplet sublevel to another by applying a resonant electromagnetic field. From Eqn. 6b it follows that the transition probability is proportional to

$$|\gamma\hbar\langle T_z|\underline{B}_1\cdot\hat{\underline{S}}|T_y\rangle|^2 = (\gamma\hbar B_{1x})^2 \quad \text{(cyclic)} \tag{7}$$

where $\underline{B}_1$ is the amplitude of the magnetic component of the driving field, $\underline{B} = \underline{B}_1\cos\omega t$. From [7] it follows that the microwave transition $T_u \approx T_s$ (u,s = x,y,z) is polarized with transition moment along $\underline{w} = \underline{u} \times \underline{s}$. This allows to perform microwave-selection spectroscopy, to which we will turn later in this chapter.

The resonance frequencies follow from the eigenenergies of $\hat{H}_{SS}$, viz, X, Y and Z. Because $X + Y + Z = 0$ (the trace of $\underline{D}$ is zero), it is customary to express the energies in two independent parameters D and E, the zero-field splitting (ZFS) or fine structure parameters:

$$D = -\frac{3}{2}Z, \quad E = -\frac{1}{2}(X - Y) \tag{8}$$

with by convention $|D| \geq 3|E|$. The physical meaning of the zero-field splitting parameters is that they represent averages over the spatial coordinates x', y', z' of the distance vector $\underline{r}$ of the two unpaired electrons:

$$D = \frac{3}{4}\cdot\frac{g^2\beta^2\mu_o}{4\pi}\left\langle\frac{r^2 - 3z'^2}{r^5}\right\rangle \quad \text{and} \quad E = \frac{3}{4}\cdot\frac{g^2\beta^2\mu_o}{4\pi}\left\langle\frac{x'^2 - y'^2}{r^5}\right\rangle \cdot \tag{9}$$

Thus, E is a measure of the deviation from axial symmetry about the z axis. The relative order of the energy levels depends on the sign of D and E. For a flat molecule such as chlorophyl, one would expect D to be positive. (The z axis is the axial symmetry axis and is perpendicular to the plane of the molecule, so that the z' component of $\underline{r}$ is on the average much smaller than $|\underline{r}|$.) For a rod-like molecule, such as a bi-radical, D will be negative.

2. OPTICAL DETECTION OF MAGNETIC RESONANCE, ODMR

Continuous illumination will generate an equilibrium population of the triplet sublevels given for not too high light levels by

$$N_u = p_u K N/k_u, \quad \Sigma_u N_u = N_T, \quad u = x,y,z \tag{10}$$

where in a commonly adopted notation (1) p_u is the probability to transit from the singlet excited state to the u-th triplet sublevel with $\Sigma_u p_u = 1$, k_u the decay rate that governs de-excitation from the u-th sublevel back to the singlet ground state, K the overall rate of population of the triplet state, N_T its total population and N the number of photoexcitable molecules. Both p_u and k_u are determined by molecular symmetry. The N_u obtained for about equal p_u and $k_x, k_y \gg k_z$ are depicted in Fig. 2. Let us now assume that we have almost totally inhibited spin-lattice relaxation by working at very low temperature. If we then switch on a microwave field of a frequency corresponding to a transition between the y and the z level, the field will transfer population from the heavily-populated, slowly decaying z level to the much less populated, fast decaying y level. Obviously, this transferred population will not remain there but quickly decay to the singlet ground state. This will lead to enhanced phosphorescence if the triplet decays radiatively. Furthermore, a new equilibrium will be established that for a strong enough microwave field is given by $N_y' = N_z' = p'KN/k'$ with $p' = \frac{1}{2}(p_y + p_z)$ and $k' = \frac{1}{2}(k_y + k_z)$. If we take $p_y \approx p_z$ it is then immediately seen that $(N_y' + N_z') < (N_y + N_z)$, because for $k_y > k_z$, $2/(k_y + k_z) < 1/k_y + 1/k_z$. Although the x level will sense the new equilibrium via the photogeneration cycle of the triplet state, this is a second order effect, and we have $\Sigma N_u' = N_T' < N_T$. Because the concentration of the singlet excited state will be negligibly low when conventional light sources are used, this means that the concentration of the singlet ground state is enhanced. This will lead to enhanced fluorescence, and enhanced singlet ground state absorbance as both these phenomena are proportional to the ground state population, while the absorption of the triplet T_o ground state will be decreased.

In the above we have the essence of ODMR. A sample is continuously illuminated at liquid helium temperatures, preferably below 2.1 K, and simultaneously irradiated by microwaves of a frequency ν not far from that corresponding to one of the triplet sublevel spacings: $\nu_{1,2} = (|D| \pm |E|)/h$ and $\nu_3 = 2E/h$ (Fig. 2). The frequency ν is slowly scanned across one of the frequencies $\nu_{1,2,3}$ while either the phosphorescence, delayed fluorescence or absorbance of the sample is monitored. When ν is close to or precisely equal to $\nu = \nu_{1,2,3}$ the ensemble of triplet state is in resonance with the microwave field and the fluorescence (FDMR), phosphorescence (PDMR) or absorbance (ADMR) will be enhanced, or diminished, depending on the relative values of p_u and k_u.

An important advantage of ODMR in zero magnetic field compared to conventional EPR in high field, where the absorption of microwaves is monitored, is that the optical probing occurs with quanta of much higher energy than the microwave quantum. (E(orange light, 600 nm = 6×10^{-5} cm)/E(microwaves, 3 cm) = 5×10^4.) This enhances detector sensitivity enormously. Secondly, since the modes of detection and excitation are decoupled, one is insensitive to noise sources due to the microwaves (as e.g. amplitude fluctuation), especially when the transition is saturated. (Note that the optical signal does not disappear upon saturation as does the microwave absorption in microwave detection; this is another advantage of optical detection.) Thirdly, compared to high-field EPR of triplet states one has in zero-field resonance much narrower lines and a concomitant increase in sensitivity, since the anisotropy in resonance condition associated with an applied magnetic field is absent. Finally, the possibility to probe the resonance at various wavelengths gives especially for ADMR much new information (section 2.3).

In the next section the theory of ODMR will be summarized. To promote continuity with the literature I have adopted with minor modification the notation of Maki (1) and globally followed his lucid overview of slow-passage and transient ODMR. For in-depth discussions of the various aspects of quantitative ODMR and references to the original literature the reader is referred to the treatise edited by Clarke (2).

2.1 Quantitative Description of ODMR

2.1.1 Time dependence. For continuous illumination the three triplet sublevels, and the singlet ground and excited states form a coupled five-level system, which is described by a set of coupled, first-order linear differential equations:

$$\dot{N}_o(t) = - k_o(N_o(t) - N_1(t)) + k_1 N_1(t) + \Sigma_u k_u N_u(t) \qquad [11a]$$

$$\dot{N}_1(t) = \quad k_o(N_o(t) - N_1(t)) - (k_1 + k_2)N_1(t) \qquad [11b]$$

$$\dot{N}_u(t) = \quad k_2 p_u N_1(t) - (k_u + \sum_{v \neq u} (W_{uv} + P_{uv}))N_u(t)$$

$$\quad + \sum_{v \neq u} (W_{vu} + P_{vu})N_v(t), \quad u,v = x,y,z \qquad [11c]$$

$$N_o(t) + N_1(t) + \Sigma_u N_u(t) = N. \qquad [12]$$

Here, N_u, $N_T = \Sigma_u N_u$ and N are defined before, N_o and N_1 are the population of the singlet ground state and the singlet excited state, respectively, p_u the populating probabilities, the k's are defined in Fig. 2, W_{vu} and $W_{uv} = W_{vu}$ exp$[(E_u - E_v)/k_B T]$ are the spin-lattice relaxation rates to and from the u-th triplet sublevel from and to the v-th sublevel, and $P_{uv} = P_{vu}$ the rate of microwave transition between sublevels u and v. The set of Eqns. 11,12 can be analytically solved with only one simplifying assumption: The population of the singlet excited state is under all but the most extreme cases of continuous illumination much smaller than the total number of molecules: $N_1 \ll N$, so that Eqn. 12 reduces to

$$N_o + \Sigma_u N_u \approx N_o + N_T \approx N \qquad [13]$$

and we may write with $K = k_2 k_o/(k_o + k_1 + k_2)$

$$- k_o(N_o(t) - N_1(t)) + k_1 N_1(t) \approx - K N_o(t) \qquad [14a]$$

$$k_2 P_u N_1(t) \approx K P_u N_o(t). \qquad [14b]$$

Substituting [14a] and [14b] in [11a] and [11c], respectively, we get with [13] a set of three independent differential equations. Their steady-state solution gives the ODMR under slow-passage conditions from which the ODMR frequency and line shape are determined; their analytic solution yields expressions for transient ODMR from which the molecular decay rates are extracted.

2.1.2 Slow-passage ODMR. The solution of Eqns. 11a-c,12 for steady-state conditions is readily obtained for conditions that the W's are negligible:

a. Absence of microwaves

$$N_1^0 = k_0 N_0/(k_0 + k_1 + k_2) = K(N - N_T)/k_2 \tag{15a}$$

$$N_u^0 = NK(p_u/k_u)/(1 + K\Sigma_u p_u/k_u) \tag{15b}$$

$$N_T^0 = \Sigma_u N_u = NK\Sigma_u(p_u/k_u)/(1 + K\Sigma_u p_u/k_u). \tag{15c}$$

When K is much smaller than the k_u's we may expand [15b,c] into a series, giving

$$N_u^0 = NK(p_u/k_u)(1 - K\Sigma_u p_u/k_u + ...) \tag{16a}$$
$$N_T = NK\Sigma_u(p_u/k_u)(1 - K\Sigma_u p_u/k_u + ...). \tag{16b}$$

For small K Eqn. 16a reduces to Eqn. 10. Note that the fraction of molecules in the triplet state, N_T/N, depends on K but that the relative steady-state sublevel population, N_u^0/N_T^0, does not.

b. Saturating microwaves. The steady-state equations for resonant microwaves saturating the u ≈ v (u,v = x,y,z) transitions are readily obtained from Eqns. 15 and 16 by substituting for p_u and p_v, ½($p_u + p_v$) and k_u and k_v, ½($k_u + k_v$), while the third p and k remain unaltered. When two transitions, e.g. u ≈ v and u ≈ w (u,v,w = x,y,z) are saturated, we have $k_u = k_v = k_w = k = 1/3 \Sigma_u k_u$ and $p_u = 1/3$ for all u. The changes in triplet population for microwaves saturating the u ≈ v transition are then to first order in K given by

$$\Delta N_u^{uv} = N_u^{uv} - N_u^0 \approx -(k_u/k_v)\Delta N_v^{uv} \approx NK(k_u p_v - k_v p_u)/k_u(k_u + k_v) \tag{17a}$$
$$\Delta N_T^{uv} \approx \Delta N_u^{uv} + \Delta N_v^{uv} \approx NK(p_u/k_u - p_v/k_v)(k_u - k_v)/(k_u + k_v) \tag{17b}$$

whereas the change in population of the third level is zero to first order in K.

From Eqns. 16,17 we see that the slow-passage ODMR signal for fluorescence or absorbance detection, which is proportional to $\Delta N_0 = -\Delta N_T$, is given by

$$S^{uv}(FDMR, ADMR) \; \alpha \; (N_u^0 - N_v^0)(k_u - k_v)/(k_u + k_v). \tag{18}$$

Thus, to obtain an FDMR or ADMR signal both the equilibrium populations and the molecular decay rates of the two triplet sublevels connected by microwaves must be unequal. The sign of the FDMR or ADMR response depends on the relative magnitudes of N_u, N_v and k_u, k_v. Note that the microwave-induced change in singlet absorbance (sometimes called SADMR) is opposite in sign to that of the triplet absorbance (TADMR). Note also that generally triplet formation leads to complete bleaching of the $S_1 \leftarrow S_0$ transition. Since the $T_n \leftarrow T_0$ transitions almost always have a much lower molar extinction coefficient than the $S_1 \leftarrow S_0$ transition, there will be generally little interest in using TADMR instead of SADMR for the determination of the ZFS parameters and the decay rates.

The proportionality constant for S^{uv}(FDMR) contains a factor Kk_2^{-1} since the microwave-induced change in fluorescence is proportional to the microwave-induced change in S_1 population, ΔN_1^{uv}, which from Eqns. 15,17b is given by

$$\Delta N_1^{uv} = Kk_2^{-1}\Delta N_o^{uv} \approx - Kk_2^{-1}\Delta N_T^{uv} \approx -K^2 k_2^{-1}N \text{ funct}(p_{u,v};\ k_{u,v}). \qquad [19]$$

Thus, for low light conditions such that Eqns. 17a,b are valid, S(FDMR) is proportional to the square of the light flux, in contrast to S(ADMR), which depends linearly on the light flux (provided $K \ll k_u$ for all u holds). This makes it advantageous to use high illumination for FDMR experiments.

The phosphorescence P is given by $P = c\Sigma_s k_s^r N_s$, where c is an instrumental constant and k_s^r the radiative decay rate. For saturating microwaves, $k_u^{r,uv} = k_v^{r,uv} = \frac{1}{2}(k_u^r + k_v^r)$. With Eqn. 17a we then obtain for the phosphorescence-detected ODMR signal

$$S^{uv}(\text{PDMR}) \quad \alpha \quad \Sigma_s k_s^r \Delta N_s^{uv}, \quad s = x,y,z \qquad [20]$$

which on substitution of Eqn. 17a becomes

$$S^{uv}(\text{PDMR}) \quad \alpha \quad (N_v^o - N_u^o)(k_u^r/k_u - k_v^r/k_v)k_u k_v/(k_u + k_v). \qquad [21]$$

Thus, a PDMR response requires that the equilibrium populations and the radiative quantum yields k_u^r/k_u, k_v^r/k_v, rather than the decay rates, are unequal.

2.1.3 <u>Transient ODMR</u>. The time dependence of the ODMR signal in response to a change in the condition of microwave irradiation (switching them on or off, or applying pulses) is given by the solution of Eqns. 11,12, simplified by the neglect of the S_1 population. The general analytic solution is given in the Appendix; for the present discussion it suffices to note that the response of the system on switching on or off resonant microwaves connecting $x \approx z$ or $y \approx z$ or both is given by the sum of three exponentials

$$N_i(t) = \Sigma_j \xi_{i,j} \exp(\lambda_j t) + \eta_i \quad j = 1,2,3; \quad i = o,x,y,z \qquad [22]$$

where ξ_j, λ_j and η_i are rather cumbrous functions of the rates of decay, spin-lattice relaxation and microwave-induced transitions and the rate K (proportional to the light flux) of populating the triplet state. For $K \to 0$ and $W \approx 0$ and in the absence of microwaves, the triplet sublevels are uncoupled and each sublevel decays after some perturbation according to

$$N_u(t) = N_u(\infty) - [N_u(\infty) - N_u(0)]\exp(-k_u(t)). \qquad [23]$$

<u>Pulsed microwaves</u>. If one perturbs the system described by Eqns. 11, 13 and 14 slightly by a pulse of microwaves resonant between two triplet sublevels, u and v, the return to equilibrium is governed by an equation similar to Eqn. 22 (see the Appendix for explicit relations). In the absence of spin-

lattice relaxation and for low light fluxes and a pulse duration that is much shorter than the fastest triplet sublevel decay time, Eqn. 22 reduces to a good approximation to two exponentials with characteristic rates given by

$$\lambda_1 (K) = - k_u - p_u K$$ [24a]
$$\lambda_2 (K) = - k_v - p_v K$$ [24b]

$$K \to 0; \ u \neq v = x, \ y \ or \ z$$

and with amplitudes of opposite sign. Provided K is low enough, the third sublevel is not perturbed (Eqn. 17). A typical response curve is shown in Fig. 3. Note that the extrapolation for $K \to 0$ is non-trivial: Relations 24a,b are valid only for really low light intensities, for which the signal-to-noise ratio is poor, especially for fluorescence detection, Eqn. 19. They are therefore not suitable to determine $p_{u,v}$; these probabilities can be better determined by measuring the relative amplitude of the ODMR signal under various different conditions of microwave saturation (section 2.1.4).

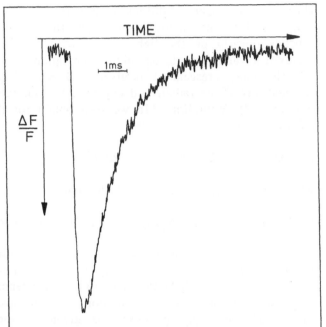

Fig. 3. Transient absorbance-detected ODMR signal following a 20 µs microwave pulse at 467 MHz (|D| − |E| transition). Sample: reaction centers of *Rb. sphaeroides* R-26 with reduced quinone acceptor. The response is essentially bi-phasic, with two exponentials of opposite amplitude that reflect the decay times of the y (fast) and z (slow) triplet sublevels.

The amplitude of the pulse response to first order in K follows from Eqn. 23 for the appropriate boundary condition: $N_u(\infty) = NK p_u/k_u$ and $N_u(0) - N_u(\infty) = f(p_v/k_v - p_u/k_u)$, where f is a parameter describing the effect of the pulse on the population of sublevels u and v: $f = \frac{1}{2}$ for saturation (high power and/or long pulse), $f = 1$ for inversion of N_u and N_v (the maximal effect). Normally, $f << \frac{1}{2}$. With $N_T(t) = \Sigma_u N_u(t)$, the change $\delta N_T(t)$, and thus S(F,ADMR), is to first order in K given by

$$S(F,ADMR) \ \alpha \ \delta N_T(t) = N_T(t) - \Sigma_u N_u(\infty) \approx f(N_v^o - N_u^o)(e^{-k_u t} - e^{-k_v t})$$ [24c]

while for the phosphorescence response the exponentials are weighted by the radiative decay rates:

$$S(PDMR) \propto cf(N_v^o - N_u^o)(k_u^r e^{-k_u t} - k_v^r e^{-k_v t}). \tag{24d}$$

Determination of the average decay rate. When saturating microwaves are applied simultaneously to two of the three ODMR transitions, the triplet collapses to one level with average decay rate $k = 1/3 \, \Sigma_u k_u$. The set of the simplified Eqns. 11,12 reduces to an equilibrium reaction, which immediately yields

$$N_T(t) = NK(K + k)^{-1}[1 - \exp(-(K + k)t)] \tag{25}$$

for the triplet built up after the onset of illumination at $t = 0$. Eqn. 25 permits the evaluation of k by extrapolating to $K \to 0$. Obviously, the same equation applies when the triplet levels are coupled not by microwaves but by spin-lattice relaxation at higher temperatures (3). In fact, the latter method to determine k is to be preferred, since especially for randomly-oriented samples the condition of microwave saturation is difficult to obtain (see below).

Onset of saturating microwaves. When saturating microwaves are applied to the $u \approx v$ transition at $t = 0$, the characteristic rates of attaining the new equilibrium are for $K \to 0$, $\frac{1}{2}(k_u + k_v)$ and k_w $(u,v,w = x,y,z)$. With Eqns. 17,23 we obtain for the time dependence of the connected sublevels, keeping in mind that $N_u^{uv}(0) = N_v^{uv}(0) = \frac{1}{2} NK(p_u/k_u + p_v/k_v)$

$$N_u^{uv}(t) \approx N_u^{uv}(\infty) - (N_u^{uv}(\infty) - N_u^{uv}(0))\exp(-\tfrac{1}{2}(k_u + k_v)t) \tag{26a}$$

$$= NK[(p_u + p_v)/(k_u + k_v) - \tfrac{1}{2}(p_u/k_u - p_v/k_v)(k_u - k_v)(k_u + k_v)^{-1}$$
$$\exp\left(-\tfrac{1}{2}(k_u + k_v)t\right)] \tag{26b}$$

and similarly for $N_v^{uv}(t)$, whereas as before the population of the third sublevel remains constant. Note that the coefficients and the rate constants are to first and zero-th order in K, respectively. (Eqn. 24 would give a rate constant to first order in K.) The time dependence of the total triplet population is $N_T^{uv}(t) = N_u^{uv}(t) + N_v^{uv}(t)$. From Eqns. 17b,26 it follows that the difference $\delta N_T^{uv}(t)$ between $N_T^{uv}(t)$ and N_T before the application of microwaves (the latter equal to $N_T^{uv}(0)$), to which the FDMR and ADMR signals are proportional, is given by

$$S^{uv}(F,ADMR)(t) \propto \delta N_T^{uv}(t) = \Delta N_T^{uv}[(1 - \exp\left(-\tfrac{1}{2}(k_u + k_v)t\right)] \tag{27}$$

with ΔN_T^{uv} given by Eqn. 17b.

The time dependence of the PDMR signal on the onset of saturating microwaves between u and v follows from Eqn. 20

$$S^{uv}(PDMR)(t) - S^{uv}(PDMR)(\infty) = c \; \Sigma_s k_s^r \; \Delta N_s^{uv}(t)$$

$$= c \sum_{s=u,v} k_s^r \; [N_s^{uv}(\infty) - \tfrac{1}{2}(N_u^o + N_v^o)]exp(-\tfrac{1}{2}(k_u + k_v)t)$$

$$= \tfrac{1}{2}c(k_u^r + k_v^r)(N_u^o - N_v^o)(k_v - k_u)(k_u + k_v)^{-1}exp(-\tfrac{1}{2}(k_u + k_v)t) \qquad [28]$$

with $S^{uv}(PDMR)(\infty)$ given by Eqn. 21. It follows that immediately after switching on the microwaves the PDMR signal is not zero, as in FDMR and ADMR, but is equal to $S(PDMR)(0) = \tfrac{1}{2}c(N_u^o + N_v^o)(k_v^r - k_u^r)$. This is a consequence of the immediate response of the phosphorescence intensity to the change in the sublevel population brought about by the microwaves, in contrast to the change in S_o population that needs to be built up by the photocycle.

Recovery from saturating microwaves. Substituting in Eqn. 26a $N_s^{uv}(\infty)$ for $N_s^{uv}(0)$ as the new initial condition at $t = 0$, and $N_s^o = NKp_s/k_s$ for $N_s^{uv}(\infty)$, one obtains the recovery curve after switching off the saturating $u \approx v$ microwaves:

$$N_s(t) \approx NK[p_s/k_s+((p_u+p_v)/(k_u+k_v) - p_s/k_s)exp(-k_s t)]; \; s = u,v. \qquad [29]$$

Again the population of the third sublevel remains unaltered. For the change $\delta N_T(t)$ in total triplet population we obtain, using [29]

$$\delta N_T(t) = N_T(t) - N_T(\infty) \approx NK \sum_{s=u,v} [(p_u+p_v)/(k_u+k_v) - p_s/k_s]exp(-k_s t)$$

$$= (N_v^o - N_u^o)(k_u + k_v)^{-1}(k_v exp(-k_u t) - k_u exp(-k_v t)). \qquad [30]$$

Thus, the decay curve $S(F,ADMR)$ is the sum of two exponentials with opposite amplitudes and rates equal to the molecular decay rates (all assuming $K \to 0$). For $S(PDMR)$ we have to weight the exponentials by the radiative decay rates. Multiplying with $k_u k_v/k_u k_v$ the preexponential factors k_v and k_u then become proportional to the radiative quantum yields k_u^r/k_u and k_v^r/k_v, respectively. Compared with Eqn. 30, this may be of advantage when one of the decay rates is much smaller than the other, since the quantum yields, being ratios, are bound to be comparable (1). Note that this applies also when comparing with the response to a microwave pulse (Eqn. 24d).

No optical excitation. As mentioned above, Eqn. 23 is strictly true only for very low values of K, since only in the limit of zero optical excitation are the triplet sublevels truly decoupled (provided there is no spin-lattice relaxation). Thus, an ideal situation would be measuring microwave-induced transients for $K = 0$. This can obviously only be done for phosphorescence and delayed-fluorescence detection; the technique is called microwave-induced delayed phosphorescence (MIDP) (4) or luminescence (MIDL) (5). Essentially, after photogeneration of the triplet state, the light is switched off at time $t = 0$ and at time $t = t'$ a short pulse of microwaves resonant between sublevels u and v is applied. Population is transferred, and the resultant difference in phosphorescence recorded. The pulse-induced difference in population $\Delta N_u^P(t')$ is given by $\Delta N_u^P(t') = f(N_v(t') - N_u(t'))$ and the associated change in phosphorescence by $\Delta P_u = ck_u^r \Delta N_u^P(t')$, and vice versa for ΔP_v, so that the total pulse-induced change in phosphorescence is

$$\Delta P_{pulse} = \Delta P_u + \Delta P_v = cf(k_u^r - k_v^r)(N_v^o exp(-k_v t') - N_u^o exp(-k_u t')) \qquad [31a]$$

which for $k_u \gg k_v$ and $k_u t' \gg 1$ reduces to

$$\Delta P_{pulse} = cf(k_u^r - k_v^r)N_v^o exp(-k_v t'). \qquad [31b]$$

Thus, by measuring the amplitude of the pulse-induced change in phosphorescence as a function of t', k_v is easily evaluated. For $k_u \gg k_v$ the transient phosphorescence ΔP_{pulse} decays with the fastest rate constant, k_u, so that the MIDP experiment yields accurate values of both decay rates. When $k_u \approx k_v$ a bi-exponental fit has to be carried out. Note that the microwave pulse needs not to be saturating, but must be short compared to the fastest decay time for the above simple analysis to apply. The method works best for rather disparate $k_{u,v}$; the light switching has to be done much quicker than either of the decay times. The initial populations $N_{u,v}^v$ need not be the equilibrium ones, so in principle a laser flash may be used.

Other schemes. A great many variants of the above schemes to measure the decay rates are possible, in some of which the spin-lattice rates are explicitly introduced (see e.g. Refs. 1,2). (Note that W may critically depend on the solvent (6,7).) Often, the treatment then gets more complex and one has to take recourse to numerical simulation along the lines discussed in the Appendix to evaluate the parameters of interest.

2.1.4 Populating probabilities. In the preceding section expressions are given to evaluate radiative and total decay rates from experimental kinetic traces. Here, a few experiments are discussed that allow to determine the populating probabilities. We will limit ourselves to fluorescence, c.q. absorbance detection.

Response to the onset of illumination. The light is switched on at time $t = 0$ with saturating microwaves connecting two triplet levels in different ways, say $u \approx v$ and the double resonance $x \approx y \approx z$ combination, while the fluorescence F is monitored. With Eqns. 13 and 15c we then have

$$\frac{F^{uv}(0)/F^{uv}(\infty) - 1}{F^{xyz}(0)/F^{xyz}(\infty) - 1} = \frac{N_o^{uv}(0)/N_o^{uv}(\infty) - 1}{N_o^{xyz}(0)/N_o^{xyz}(\infty) - 1} = \frac{N/N_o^{uv}(\infty) - 1}{N/N_o^{xyz}(\infty) - 1}$$

$$= (k/NK)\, NK\, \Sigma_s^{uv} p_s/k_s = k\, \{2(p_u + p_v)/(k_u + k_v) + p_w/k_w\} \qquad [32]$$

and similarly for F^{uw} or F^{vw} and F (no microwaves). The triplet excitation rate K is now eliminated and from a set of three equations the p's can be solved; as a check $\Sigma_u p_{u,exp}$ should be close to unity. $F^{xyz}(\infty)$ and $F^{uv}(\infty)$ are best measured by continuously irradiating the sample with $u \approx v$ microwaves and switching the u ≈ w (or v ≈ w) microwaves on and off; the difference signal can be signal-averaged, yielding accurate relative values of F(∞) (8).

Microwave switching under continuous illumination. The above method only works when the light can be switched on faster than the fastest sublevel decay rate. When this is difficult, e.g. for bacteriochlorophyl triplets, a variant may be used in which the relative fluorescence intensity under various conditions of saturating microwave irradiation is measured. For example, for low K

$$(F^{uv}(\infty) - F^{xyz}(\infty)) / (F^{uw}(\infty) - F^{xyz}(\infty)) = (1 - k\Sigma_s^{uv}p_s/k_s) / (1 - k\Sigma_s^{uw}p_s/k_s) \quad [33]$$

with $p_u^{uv} = p_v^{uv} = \frac{1}{2}(p_u + p_v)$, $k_u^{uv} = k_v^{uv} = \frac{1}{2}(k_u + k_v)$ etc. and similarly for F^{vw}. With $\Sigma_u p_u = 1$ we again have three independent linear equations from which the p_u's are solved. Note that we now need all three ODMR resonances.

Triplet states in a pigment bed. In biological material it often happens that pigments are densely packed and that triplet states are stable on traps that are populated by energy transfer. The fluorescence intensity will be low when the traps are open (the fluorescence is quenched by triplet formation on the trap) and high when the traps are filled. Manipulating the concentration of closed traps by resonant microwaves then results in a change of fluorescence intensity that is opposite in sign compared to the FDMR signal of 'free' pigments. The fluorescence quenching can be described by a Stern-Volmer type relation:

$$1/\Phi = A + B \text{ [fraction of open traps]} = A + B (N - N_T)/N \quad [34]$$

where Φ is the fluorescence quantum yield and A and B are experimental constants. We can then write (9)

$$N_T^\dagger(\infty)/N_0^\dagger(\infty) = (1/\Phi_o - 1/\Phi)/(1/\Phi - 1/\Phi_m) \quad [35]$$

where the dagger sign denotes a particular way of connecting the triplet sublevels with microwaves, e.g. u ≈ v,w , and Φ_o and Φ_m are the fluorescence quantum yields for all traps open and all traps closed: $1/\Phi_o = A + B$, $1/\Phi_m = A$. With $F = I\Phi$ (I is the light flux) and $\gamma = F_m/F_o$ we have

$$N_T^\dagger(\infty)/N_0^\dagger(\infty) = \gamma(F^\dagger - F_o)/(\gamma F_o - F^\dagger). \quad [36]$$

As before, the left hand side of Eqn. 36 can be expressed in the molecular decay rates and populating probabilities, so that with known decay rates, the p_u's can be evaluated (9).

Phosphorescence detection. Dividing the phosphorescence response to a pulse of microwaves under cw illumination for t = 0 (Eqn. 24d) by the MIDP response to an identical pulse for t = t' and $k_u \gg k_v$ (Eqn. 31b) we obtain

$$\Delta P^{uv}(0)/\Delta P^{uv}(t') \approx (N_v^0 - N_u^0)/N_v^0 \exp(-k_v t') \quad [37]$$

and similarly for ΔP^{uw} (provided $k_u \gg k_v$). The prefactors are evaluated from a semi-log plot vs. t', and from them the relative populating probabilities p_u/p_v and p_u/p_w can be calculated for known sublevel decay rates (10).

2.1.5 <u>A note of caution</u>. A good many of the methods discussed in the previous sections rely on saturation, i.e. equalization of the populations of one, or two transitions. This requires a very low temperature to inhibit spin-lattice relaxation and fairly large microwave powers to make the microwave-induced transition rate much larger than the fastest of the molecular decay rates. The latter requirement is often not met when the output of microwave sweepers (20 – 40 mW) is used without amplification. As a result large errors of more than a factor of four can be made in the evaluation of the molecular decay rates (11,12). It is absolutely necessary to verify saturation by evaluating the decay rates as a function of applied microwave power. Preferably one should use pulse methods (4,8) to determine the k_u's.

When randomly-oriented samples are used, such as normally is the case for biological material, 100 % saturation can never be achieved. This is a consequence of the polarization of the microwave transitions: The transition probability is proportional to $\cos^2\beta$, where β is the angle between the microwave field $\underline{B}_1$, and the transition moment. Hence, molecules whose transition moment is close to perpendicular to $\underline{B}_1$ (this is a sizeable fraction in view of the $\sin\beta$ distribution), have a low probability to be microwave-excited and their sublevel population is not, or only very slowly affected by the microwaves. This effect is quite noticeable, even at microwave powers exceeding 1 W at 1.2 K (12). For such samples the pulse methods seem to be the only reliable way to measure the decay rates.

Another pitfall in the determination of the k_u's is the dependence of the apparent decay rates on K, the rate of triplet formation, which is proportional to the light flux. This applies equally to equilibrium and pulse methods, except of course for the MIDP technique. As mentioned, extrapolation to $K \to 0$ is not trivial because of the poor signal-to-noise ratio attendant with the requirement that K is much smaller than the slowest molecular decay rate (which can be less than 1 s^{-1}). Most workers prefer to fit a curve of k_u vs. K, using relations such as outlined in the Appendix, but even then the fit in the $K \approx 0$ regime is often ambiguous.

2.2 <u>Line Shape, Hole-Burning and Double Resonance</u>

The ZFS parameters of a triplet state are sensitive to its environment. The larger the interaction with the environment is, the more spread will one find in the values of |D| and |E|. This translates in inhomogeneous broadening of the ODMR lines, which are often close to or even a perfect Gaussian. In molecular crystals ODMR line widths can be as narrow as 1 MHz, but in glassy matrices they often exceed 100 MHz. Such inhomogeneous broadening is demonstrated by so-called hole-burning (13) in which a microwave transition, say the |D| – |E|, is irradiated with constant power at a precisely defined, fixed frequency within the ODMR line, whereas the transition is simultaneously swept with a second modulated source of variable frequency. At the first frequency the ODMR will show a dip, because then the sublevel populations

648

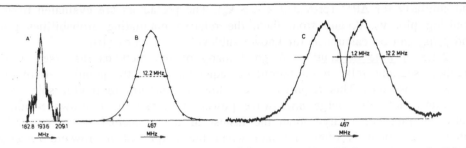

Fig. 4. (a) Double resonance (EEDOR) of the triplet state of *Rb.
sphaeroides* employing fluorescence detection. The first microwave field
was set at 467 MHz, resonant with the |D| − |E| transition, while the
second microwave field was scanned from 183 to 210 MHz. (b) The
|D| − |E| resonance of same. Crosses: computed Gaussian normalized to the
experimental curve. The slight deviation to lower frequency is due to the
earth's magnetic field. (c) Hole-burning experiment on the 467 MHz
resonance of (b). One microwave field was set at 467 MHz while the
frequency of a second field was slowly swept through the resonance. From
(87).

are already more or less equalized and additional power has comparatively little
effect. An example is shown in Fig. 4. The width of the 'hole' is twice the
homogeneous line width or equal to the frequency interval corresponding to the
field intensity of the 'burning' microwaves $|\underline{B}_1|$, whichever is the largest. (The
latter situation is undesirable and should be avoided.)

When a hole is found in the burned (e.g. |D| − |E|) transition, the |D| + |E|
transition does not show a hole, but is somewhat decreased in intensity. This is
because the particular combination of |D| and |E| values that correspond to the
precisely defined frequency of the hole in the |D| − |E| line sum to values that
spread across the whole |D| + |E| line.

The ODMR line width is not very sensitive to the band width of optical
excitation. Usually a broad optical band corresponds to a broad ODMR line
because similar environmental interactions are at work (14-16). However,
selecting a narrow band width of optical excitation (e.g. by using a laser) does
not produce significant narrowing of the ODMR line (17), because generally
there is little correlation (the effect of a similar perturbation in electronic
molecule-solvent interactions on the triplet wave function is different from that
on the singlet wave function). In contrast to this lack of correlation, there are
slight correlated shifts of the ZFS values when the wavelength of detection is
scanned across the phosphorescence band (18-20). When this correlation shows
a discontinuity, it is indicative of the presence of more than one triplet site
(e.g. tryptophanes in a protein).

In addition to the hole-burning double resonance experiment performed on
one ODMR transition, one may carry out a double resonance experiment at two
different ODMR frequencies: electron-electron double resonance, EEDOR.
Saturating one, say the u ≈ v, transition equalizes the N_u^{uv} and N_v^{uv} populations.

The population difference $N_u^{uv} - N_w^o$ is then increased or decreased compared to $N_u^o - N_w^o$ (or vice versa for N_v), with a concomitant change in the intensity of the $u \approx w$ or $v \approx w$ transition. This is often useful to enhance the ODMR line corresponding to two sublevels whose equilibrium populations in the absence of microwaves are nearly equal. An example is shown in Fig. 4. In addition EEDOR allows to discriminate between ODMR resonances belonging to the same triplet state (the same site) when in a single resonance experiment more than three ODMR lines are recorded. The latter double resonance experiment is usually carried out by irradiating one transition with amplitude-modulated microwaves at fixed frequency, and measuring with cw microwaves the other transitions while applying lock-in detection at the modulated frequency. Only those transitions belonging to the same triplet as the first transition will then show up.

2.3 Optical-Microwave Double Resonance

Once the ODMR lines of a triplet have been determined, the resonance frequencies are known precisely, and one can investigate the dependence of the intensity of a particular resonance line on the probing wavelength. Thus, one irradiates the sample with (amplitude-modulated) (21) resonant microwaves of sufficient, preferably saturating intensity, and monitores the (lock-in detected) photodetector output as a function of the probe beam wavelength. The resulting spectra may be called microwave-induced phosphorescence, fluorescence or absorbance spectra, abbreviated by MIP, MIF and MIA spectra, respectively. For one particular triplet state, the shape of these spectra does not depend on which resonance line, $v_{1,2}$ or v_3, is selected. Obviously, if more than one triplet state is present, the MI spectra provide another means to sort out which resonances belong to the same triplet state. Conversely, MI spectroscopy allows the unraveling of complex optical spectra.

In biological systems, the phosphorescence is often weak, and only MIF and MIA spectra can be recorded. Besides their use for discriminating between various triplet states, the MIF spectra are useful to identify the triplet-carrying molecule, as examplified by the elegant studies of Beck et al. (22,23) on photo-induced triplet states in bacterial photosynthetic membranes.

MIA spectra are a case apart, since they provide much more information than the MIP or MIF spectra. As will be discussed in the next section, they represent the difference of the singlet ground state, 'normal', absorbance spectrum and the spectrum for the system when a triplet state is present. They are therefore known as triplet-minus-singlet absorbance difference (T − S) spectra, rather than being labeled by the MIA acronym.

2.3.1 Triplet-minus-singlet (T − S) absorbance difference spectra. When a triplet state is present, the absorbance spectrum contains the following contributions:

1. The <u>unperturbed</u> singlet ground state absorbance spectrum ($S_n \leftarrow S_o$ transitions) of all molecules that are <u>not</u> in the triplet state and that do not interact with the molecule that <u>is</u> in the triplet state.
2. The <u>perturbed</u> singlet ground state spectrum of those molecules in a molecular aggregate (comprising proteins) that are not in the triplet state

but do interact with the triplet-carrying molecule. Generally this interaction will be different when this particular molecule is in the triplet state from that when the molecule is in the singlet ground state.

3. The absorbance spectrum of the triplet state itself, consisting of $T_n \leftarrow T_o$ transitions.

With square-wave, on-off amplitude-modulated microwaves, the MIA spectrum represents the difference in absorbance of the sample for microwaves on and microwaves off (Fig. 5). It can be shown (24) that this difference is proportional to the difference in absorbance with and without the triplet state present. In other words, the MIA spectrum represents the difference of the absorbance of the sample with all molecules in the singlet ground state and that when all molecules of one particular type are excited into the triplet state whose ODMR resonance is being monitored.

It is important to note that other triplet states with different values of |D| and |E| and consequently different ODMR resonance frequencies may be present without showing up in the MIA $\equiv$ T − S spectrum. Their absorbance is not changed by the microwaves and therefore their

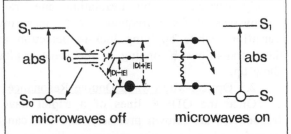

Fig. 5. Principle of absorbance-detected magnetic resonance. Filled circles denote relative equilibrium population of the triplet sublevels, open circles that of the ground state. A saturating microwave field connecting two triplet sublevels (corrugated arrow) leads to a new (here higher) equilibrium value of the singlet ground state population, hence to a change in the absorbance. The same principle holds for fluorescence detection, while phosphorescence is also enhanced by the microwave field.

contribution to the absorbance cancels in the ADMR-monitored T − S difference spectrum. On the other hand, recording T − S spectra for different ADMR resonance frequencies provides a means to discriminate the resonances belonging to one and the same triplet states, since in general contributions 2 and 3 will be different for different triplet states.

The ADMR-monitored T − S spectrum has several-fold interest. For non-interacting triplet states it provides a very accurate triplet absorbance spectrum, since that is given by adding the 'normal' singlet ground state spectrum to the T − S spectrum. For interacting triplet states, for example present in a photosynthetic pigment-protein complex, it records these interactions very sensitively and thus provides a unique means to study pigment configuration. This will be illustrated in section 3.2, where we will also discuss applications of linear dichroic (LD) T − S spectroscopy, the principle of which is explained in the next section.

2.3.2 <u>Linear dichroic T – S spectroscopy</u>. The microwave transitions between the u and v triplet sublevels are polarized along $\underline{w} = \underline{u} \times \underline{v}$. This is analogous to an optical transition, whose transition dipole moment usually has a well-defined direction in the molecular frame. Often, the direction of the triplet magnetic resonance transition moments are not as well-known. In chlorophyls, for example, one may be reasonably certain that the z transition moment is perpendicular to the molecule and that the x and y transition moments lie in the plane of the macrocycle, but the precise direction in the plane of the latter was until recently not known. As will be shown below, LD-(T – S) spectroscopy provides a means to ascertain the directions of the magnetic transition moments. With this knowledge one may then derive from the LD-(T – S) spectra precise structural information on molecular aggregates.

In optical spectroscopy, the transition probability for a transition with transition dipole $\underline{p}$ is proportional to $|\underline{E}|^2|\underline{p}|^2 \cos^2\beta$, where β is the angle between $\underline{p}$ and the $\underline{E}$ electric vector of the (polarized) incident light. A similar relation holds for the magnetic microwave transitions between the triplet sublevels. Thus, for a microwave transition moment $\underline{\mu}_{mw}$ and an angle β between $\underline{\mu}_{mw}$ and the $\underline{B}_1$ magnetic vector of the (polarized) microwave field, we have a transition probability $|\underline{B}_1|^2|\underline{\mu}_{mw}|^2\cos^2\beta$. It follows that molecules oriented with $\underline{\mu}_{mw}$ more or less parallel to $\underline{B}_1$ have a much higher transition probability than those oriented about perpendicular to $\underline{B}_1$. (Of course, for $\beta = 90°$, the transition probability is exactly zero.) Hence, for random excitation to the triplet state, molecules oriented in an angular interval $d\beta$ close to $\beta = 0$ will experience a much higher change in their relative triplet concentration upon the application of (polarized) resonant microwaves than molecules in an interval $d\beta$ close to $\beta = 90°$. Consequently, the distribution of triplet states, which was isotropic before the application of the microwaves, becomes <u>axially anisotropic</u> with the axis parallel to $\underline{B}_1$ when resonant microwaves are switched on.

The microwave-induced anisotropy in the triplet state distribution can be interrogated with a beam of polarized light. For example, let us assume that the optical transition moment $\underline{p}$ is parallel to $\underline{\mu}_{mw}$, that the microwaves decrease the triplet concentration, and that we interrogate at a wavelength where the singlet ground state has an absorption band and the triplet state does not absorb. Then, for light polarized parallel to $\underline{B}_1$ we will measure a lower transmittance than for light polarized perpendicular to $\underline{B}_1$. (Along $\underline{B}_1$ there are fewer triplets, hence more singlet ground states, than perpendicular to $\underline{B}_1$). Obviously, the difference in transmittance (which for small changes can be taken equal to the difference in absorbance ΔA (24)) will depend on the angle α between $\underline{p}$ and $\underline{\mu}_{mw}$. In the above example, the sign of $\Delta A = A_{//} - A_{\perp}$ would be reversed if not $\alpha = 0$ as assumed, but $\alpha = 90°$. Going from $\alpha = 0$ to $\alpha = 90°$, at a given angle ΔA must become zero. This is the magic angle $\alpha = 54.7°$ (for which $3\cos^2\alpha - 1 = 0$, see Eqn. 38). Thus, from the magnitude of ΔA relative to the magnitude of A we should be able to directly derive α (Fig. 6).

It will be recognized that the above description of the microwave-induced selection in the triplet state distribution is very similar to that of <u>photoselection</u>. We can therefore partake of the formalism derived for that technique

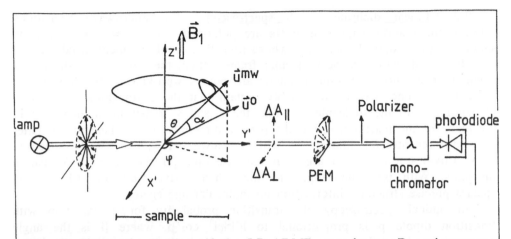

Fig. 6. Schematic drawing of the LD-ADMR experiment. $\underline{B}_1$, microwave field vector; $\underline{\mu}_{mw}$, microwave transition moment; $\underline{\mu}_o$, optical transition moment; x',y',z', laboratory frame; PEM, photoelastic modulator. Unpolarized light becomes elliptically polarized because of the anisotropic transmittance of the sample induced by the microwave field resonant with an ODMR transition. The ellipticity is analyzed by the PEM and the polarizer. Adapted from (133).

(see e.g. (25)) to calculate the functional relationship between ΔA and α. In doing so, we must of course average over all positions of the molecules with respect to $\underline{B}_1$, assuming a random initial distribution. To simplify this averaging we further assume that the triplets are isotropically excited. This is not strictly true, as the light beam does not excite molecules oriented such that their optical transition moment for triplet excitation is parallel to the direction of propagation of the light. Nevertheless, because of energy transfer among differently-oriented pigments before the excitation is trapped onto the triplet state, and because of scattering at sample cell walls and at impurities and cracks inside the sample, isotropic excitation proves to be a good approximation. Finally, we have employed unpolarized probe light and interrogate the difference $A_{//} - A_{\perp}$ <u>after</u> the light has passed the sample, using a photoelastic modulator and a polarizer. Taking all this into account it can be shown that the ratio R in intensity of the LD-(T − S) spectrum (which represents the difference $A_{//} - A_{\perp} = (T - S)_{//} - (T - S)_{\perp}$) and the T − S spectrum is given by (26,27)

$$R = \frac{LD\text{-}(T - S)}{T - S} = \frac{3 \cos^2\alpha - 1}{\cos^2\alpha + 3}.$$ [38]

Eqn. 38 is plotted in Fig. 7, together with plots of the LD-(T − S) and T − S intensities versus α. It is seen that R is quite sensitive to α, so that with proper calibration of the T − S and LD-(T − S) spectra, α can be determined quite accurately. Of course, all this applies rigourously only for single absorbance bands. If bands with different directions of $\underline{p}$ vis-à-vis $\underline{\mu}_{mw}$ overlap, R will have some intermediate value and one will have to simulate the complete LD-(T − S) and T − S spectra to obtain values for the various α's.

In photoselection, one customarily extrapolates to zero intensity of the exciting, selecting light beam. This is because the transition, and therefore photoconversion, probability is proportional to $|\underline{E}|^2\cos^2\beta$. Even for β close to 90°, photoconversion will be appreciable if the field strength $|\underline{E}|$ is high enough. In that case, almost all molecules will be photoconverted, regardless of their orientation, and selection is lost. In microwave

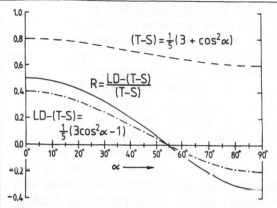

Fig. 7. The dependence of the amplitude of the ADMR-monitored T − S and LD-(T − S) spectra and their ratio R as defined by Eqn. 38 on the angle a between the optical and the microwave transition moment. α = 54.7 is the magic angle for which the microwave-induced linear dichroism is zero (R = 0). From (133).

selection, the relevant field strength is that of the $\underline{B}_1$ field. Thus, to ensure proper selection one has to measure R as a function of $|\underline{B}_1|$ and extrapolate for $|\underline{B}| \rightarrow 0$. This is best done by taking the slope of a graph of the LD-(T − S) versus the T − S intensity as a function of $|\underline{B}_1|$. The shape of the LD-(T − S) spectrum is not dependent on the intensity of $\underline{B}_1$, so that for the study of the orientation dependence of the influence of the triplet state on neighboring pigments, expressed as band shifts, bleachings and appearing bands in the LD-(T − S) spectrum, one works best at comparatively high $\underline{B}_1$ amplitudes.

The angle α in Eqn. 38 refers to one particular $\underline{\mu}_{mw}$, say that corresponding to the $\underline{x}$ polarized y $\approx$ z transition at frequency (|D| − |E|)/h. Tuning the microwaves to the (|D| + |E|)/h frequency, that is the x $\approx$ z or $\underline{y}$ polarized transition, allows the recording of LD-(T − S) spectra and the determination of R for a different α, viz. the angle α_y between $\underline{p}$ and $\underline{y}$. When $\underline{p}$, $\underline{x}$ and $\underline{y}$ lie in one plane, α_y = 90° − α_x, since the triplet spin axes $\underline{x}$, $\underline{y}$ and $\underline{z}$ span a cartesian coordinate frame. If $\underline{p}$, $\underline{x}$ and $\underline{y}$ are not coplanar, the two measurements uniquely define the orientation of $\underline{p}$ in the $\underline{x},\underline{y},\underline{z}$ coordinate frame. This is a great advantage over the ordinary photoselection experiment, where one determines just one angle between two transition moments, which leaves one with a conical ambiguity. Note that it suffices to record LD-(T − S) spectra for just two of the three possible ADMR transition frequencies. Since the orientations of all $\underline{p}$'s are determined in one and the same coordinate frame, their mutual angular dependence immediately follows.

2.4 <u>Electron Spin Echoes in Zero Field</u>

Electron spin echo spectroscopy in zero field has a fairly long history (see for a review Ref. 28). Detection has been done monitoring the echo via a microwave receiver (29), or optically monitoring the phosphorescence (30), fluorescence (31) or absorbance (32) (generally labeled ODESE). The technique is a little different from high-field ESE: In zero field there is no magnetization along a preferred, magnetic axis but rather a spin alignment, i.e. a population difference between the triplet sublevels. For zero-field ESE one adopts the description of a two-level system of Feynman, Vernon and Hellwarth (33), in which the population difference is an element of the 'FVH-vector' that takes the place of the magnetization vector $\underline{M}$ and whose time development can be described by an equation isomorphous with the Bloch equation. Thus, its time evolution can be visualized in the rotating frame just as the motion of $\underline{M}$ in high-field ESE with one important exception: Its refocused 'y-component', the echo, which is analogous to the refocused M_y component, cannot be detected directly as in high-field ESE, but has to be converted to the 'z-axis' (the population difference component of the FVH-vector) by a third 90° pulse.

The above principle is explained in Fig. 8. The first $\pi/2$ pulse of microwaves resonant with the $y \approx z$ transition creates a coherent superposition of the T_y and T_z sublevels and simultaneously equalizes their population. This results in a decrease in transmittance of the sample because the triplet now decays with the average of the fast T_y and the slow T_z decay rates, which leads to a temporary decrease in the triplet population, hence to an increase in singlet ground states. The time evolution of the transmittance response signal is governed by the individual triplet sublevel decay rates and the light flux; the change in transmittance is monitored by a boxcar whose 300 μs wide gate covers a large part of the transient. The π pulse refocuses the electron spins, and after a delay τ the Hahn echo is produced, which is converted into a population difference by the $\pi/2$ probe pulse. Without irreversible loss of phase, and for infinitely narrow pulses of sufficient intensity to cover the whole ODMR line width, the π pulse would have no effect on the transmittance and the transmittance at the time of the echo should then be the same as if no pulses were applied. In practice, the pulses have a finite width, the cavity does not admit the pulsed microwave field immediately, and loss of phase occurs with a characteristic time T_2. Therefore, the measured echo intensity corresponds to a decrease in transmission of at most 25 % compared to the level for $t \gg 2\tau$ where coherence is lost and the response of the transmittance corresponds to that for incoherent pulsed excitation of the $y \approx z$ transition.

ODESE has up to now found little application in biological research, in spite of its potential to accurately measure relaxation times T_1 and T_2, and to observe ODMR of fast-decaying triplet states and of triplet states at higher temperature where spin-lattice relaxation is important. In section 3.2.5 we will briefly discuss fluorescence- and absorbance-detected ODESE of triplet states in the bacterial photosystem.

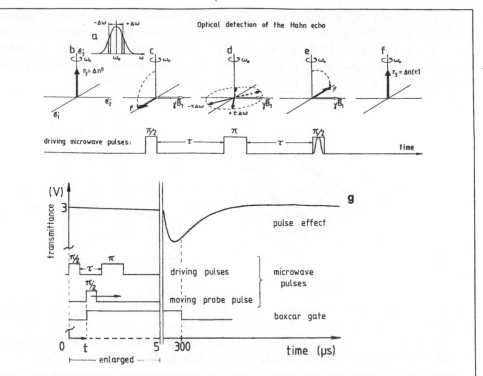

Fig. 8. The principle of ODESE. (a) The inhomogeneous ODMR transition; ω_o, the resonant frequency. (b) $\underline{e}'_1$, $\underline{e}'_2$, $\underline{e}'_3$, the FVH rotating frame; r_3, the component of the FVH-vector representing the population difference of the two triplet sublevels. (c) A microwave field $\underline{B}_1$ applied for $\pi/2\gamma\underline{B}_1$ s rotates r_3 around the $\underline{e}_1$ axis. (d) For τ s the ensemble of r_3 vectors is fanning out because of the $\Delta\omega$ off-set in resonance frequency resulting from the in-homogeneous broadening. After reversal of the fanning-out motion by a π-pulse, the r_3 vectors refocus τ s later, at which time a third, $\pi/2$ pulse is applied (e), which rotates the r_3 vector from the $-\underline{e}_1$ axis back to the $\underline{e}_3$ axis, i.e. the loss of coherence of the spin system is rendered observable as a change in population difference, Δn (f). (g) Response of the transmission to two fixed $\pi/2 - \pi$ driving pulses and a $\pi/2$ probe pulse. The transient change in transmission is monitored with a boxcar whose gate is set at the 'peak' of the response curve. (a) – (f), from (133); (g) from (32).

2.5 Instrumentation

The minimal requirements for an ODMR experiment are a source of light, one of microwaves, a cryostat and a detector. Un up-to-date discussion of these parts can be found in Ref. 1. Here we will limit ourselves to a few notes and a discussion of special arrangements for ADMR and LD-ADMR.

The best excitation light source in the visible or infrared region is an incandescent lamp (e.g. a tungsten-iodine projection lamp) powered by a current-stabilized DC power supply. Mercury or xenon lamps are more intense, certainly in the UV region, but suffer from instabilities in the intensity I, often to a level of $\Delta I/I = 10^{-3}$ or worse. The same holds for laser excitation. Adequate filtering should be provided (34). When the excitation beam is broad-banded, it may also serve as probe beam at various wavelengths.

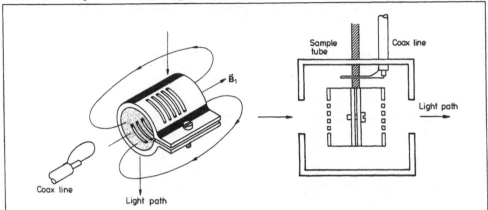

Fig. 9. Loop-gap resonator used for LD-ADMR and pulsed-ADMR experiments. The cavity resonance frequency is tuned by the screw governing the width of the gap. The $\underline{B}_1$ field is along the cavity's cylinder axis. Dimensions: inner diameter, 20 mm; height, 60 mm. From (27).

Microwaves are supplied preferably by a sweep unit (now available with a range of 0.1 - 27 GHz) with provisions to scan the frequency over the desired range and to amplitude- or frequency-modulate the output. Usually, the output (20 - 40 mW) is enough for regular ODMR at liquid helium temperatures without amplification. For measuring the ODMR lines the microwaves are fed into a broad-band resonator, e.g. a helix, that can admit the frequency range of interest. When very broad ODMR lines need to be scanned, attention should be paid to frequency-dependent reflections of the microwaves (due primarily to mismatch between helix and conductor), which may considerably affect the $\underline{B}_1$ field intensity inside the helix, and thus the ODMR intensity. Also, without leveling provisions, the sweeper output may considerably depend on frequency, even in a relatively narrow range. When in doubt that certain features are artificial, another helix of different physical characteristics should be used for comparison. If one is interested in the (probe) wavelength dependence in MI spectroscopy, the microwave frequency is set exactly at resonance and a narrow-band cavity may be used. We have used a loop-gap resonator (see the chapter by Hyde and Froncisz) with slots for optical access (Fig. 9).

The cryostat is preferably a double-walled helium bath cryostat equipped with at least two windows. Helium boils at 4.2 K; scattering by the bubbles then prohibits optical detection. By lowering the pressure above the helium bath the temperature of the helium can be lowered to below 2.1 K, the so-called lambda point, below which helium is superfluid and bubbles disappear completely. Thus, adequate pumping facilities should be provided. For FDMR at 4.2 K a simple He-vessel may be used in which a light pipe with sample compartment at the end is lowered (35). Such a light pipe may also be used in a bath cryostat (36).

The optical detector depends on the optical mode. For phosphorescence and fluorescence a photomultiplier is used. For ADMR, one may employ a strong probe beam (in fact the excitation beam may serve as such), providing a sufficiently high number of transmitted quanta that a photodiode may be used (up to 1100 nm, silicium; 1100 nm − 2 μm, germanium). To observe the resonance by fluorescence or phosphorescence, the wavelength of detection should be separated from the wavelength of excitation by adequate filtering. For ADMR, either a filter or, when the probe wavelength dependence is monitored, a monochromator is used.

The sensitivity of the ODMR spectrometer for slow-passage experiments can be considerably enhanced by modulating the amplitude of the microwaves and applying frequency-selective amplification of the photodetector signal combined with lock-in detection. Noise is then reduced to that corresponding with the passband of the amplifier-lock-in detector combination, leading to an increase in the signal-to-noise ratio of several orders of magnitude. For example, in our ADMR spectrometer a $\Delta A/A$ ratio of better than 5×10^{-7} is routinely achieved. Obviously, the modulating frequency has to be less than the slowest sublevel decay rate. For slowly decaying triplets ($k < 10$ s^{-1}), one may be better off by scanning the line with unmodulated microwaves and signal averaging. For kinetic measurements broad-banded amplification of the detector signal and signal averaging are used.

Finally, as in all modern spectrometers the instrument is interfaced to a small, dedicated computer that handles monochromator setting, data collection and storage, and carries out simple operations as taking the $\Delta I/I$ ratio (important when recording the probe wavelength dependence).

A schematic diagram of our present ADMR set-up is shown in Fig. 10. With small modifications it can be used for PDMR and FDMR as well.

2.5.1 Instrumentation for LD-(T − S) spectroscopy. The instrumentation for LD-(T − S) spectroscopy is quite similar to that of isotropic T − S spectroscopy and we can refer to Fig. 10 for a description. First of all we need polarized microwaves, therefore a simple helix is not suitable. We use a split-ring or loop-gap cavity as described by Hardy and Whitehead (37). This design has the advantage that for the rather long wavelength of the microwaves (between 50 and 100 cm) one still has a cavity of manageable dimensions, which fits easily into a four-window liquid helium cryostat. The $\underline{B}_1$ field is polarized along the vertical, therefore is perpendicular to the horizontal light beam. The probe light is unpolarized. The $\underline{B}_1$ field induces an ellipticity ($\equiv T_{//} - T_{\perp}$) in the transmitted light T, which is detected via a photoelastic modulator (PEM) after the sample followed by a polarizer. The PEM rotates the ellips spanned by the unequally-transmitted light vectors $\underline{E}_{//}$ and $E_{\perp}$ (with respect to $\underline{B}_1$) by 180° at a frequency of 50 kHz. The analyzer converts this polarization modulation into an amplitude modulation at 100 kHz that is

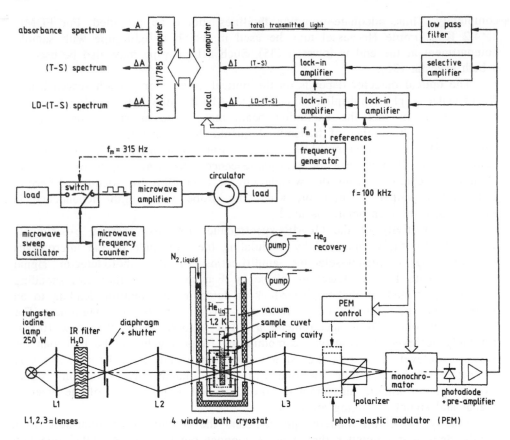

Fig. 10. Block scheme of the ADMR and LD-ADMR set-up. From (133).

proportional to $T_{//} - T_{\perp} \approx A_{//} - A_{\perp}$ (T, transmission; A, absorbance) for small differences. The microwaves are as in $T - S$ spectroscopy modulated at low frequency (say 315 Hz), so that the light intensity falling on the photodiode is doubly modulated at 100 kHz and 315 Hz. Demodulation at 315 Hz combined with suitable electronic filtering gives the normal $T - S$ signal. Double demodulation at 100 kHz and 315 Hz gives the $(T - S)_{//} - (T - S)_{\perp} \equiv$ LD-$(T - S)$ difference signal. This procedure is illustrated in Fig. 11. The signal-to-noise ratio is enhanced by inserting selective amplifiers in both modulation channels. Scanning the monochromator yields simultaneously the $T - S$ and the LD-$(T - S)$ spectra. As before, the signals are divided by the intensity I to correct for changes in lamp output, monochromator sensitivity, etc. as a function of wavelength. To correctly evaluate the ratio R (Eqn. 38) we must mutually calibrate the $T - S$ and LD-$(T - S)$ signals. This calibration includes amplification factors, lock-in sensitivities, etc. It is best done by simulating the transmitted modulated light by a modulated light-emitting diode (LED). The amplitude of the 100 kHz modulation at specific instrument settings

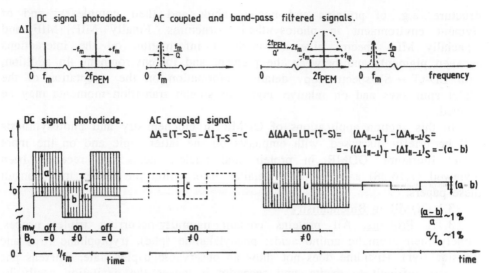

Fig. 11. Double modulation scheme used in LD-ADMR. Upper part: frequency spectrum of the signal and its side bands before and after AC coupling and filtering. Lower part: The photodiode signal is modulated at the frequency of the photoelastic modulator, f_{PEM} = 100 kHz, and that of the microwave field modulation, $f_m \approx 350$ Hz; Q and Q', Q-factors of the electronic filters c.q. selective amplifiers; ΔA, microwave-induced change in absorbance; $\Delta A \approx \Delta I$, the change in transmittance (here negative). Filtering at f_m and low-frequency demodulation yields the T − S spectrum. Filtering at $2f_{PEM} \pm f_m$, demodulation at $2f_{PEM}$ and subsequently at f_m yields the LD-(T − S) spectrum. From (133).

---

can then be accurately compared with that of the 315 Hz modulation. Care has to be taken to avoid as much as possible ellipticities induced by extraneous sources, such as the lamp, cryostat windows, sample cell, etc. In our set-up these extraneous ellipticities amounted to less than 1 % of the LD-(T − S) signal.

3. APPLICATIONS OF ODMR IN BIOLOGY

The principal results of an ODMR experiment are: 1) proof of the photo-production of one or more triplet states, 2) the determination of highly accurate values of the ZFS parameters, 3) the determination of the individual sublevel molecular decay rates, 4) the determination of the sublevel populating probabilities, 5) information on environmental interactions through line shape analysis and hole-burning, 6) spectral information via optical-microwave double resonance experiments.

The zero-field splitting parameters and the sublevel populating probabilities and decay rates are fingerprints of molecular structure. They are also sensitive to environmental effects, which in addition to line shape analysis and hole-burning allows one to use the triplet state as a probe of secondary and tertiary structure, e.g. of proteins and nucleic acids and their complexes, and of pigment environment in photosynthetic structures. Finally MIP, MIF and especially MIA spectra give a wealth of information on the interactions between pigments and between the pigment and its environment. In addition, with LD-(T − S) spectroscopy detailed information on the orientation of the triplet spin axes and on relative angles of singlet transition moments may be derived.

In this section applications of ODMR in biochemistry and photosynthesis will be briefly discussed, with emphasis on the latter topic and on the more recent literature. ODMR in protein and nucleic acids has recently been reviewed (1,16,38) and for more detailed information the reader may consult these papers.

3.1 ODMR in Biochemistry

3.1.1 Proteins. All proteins contain naturally-occurring triplet probes, namely their aromatic amino acids: phenylalanine (phe), tryptophane (trp) and tyrosine (tyr). Histidine does not show an observable triplet state, whereas phe is firstly difficult to excite, and secondly it looses the excitation easily by energy transfer, so that in practically all applications of ODMR in protein research, mostly from the groups of Kwiram and Maki (Seattle and Davis, USA, respectively), attention was focused on the study of the triplet states of tryptophanes and tyrosines in vitro and in various proteins using phosphorescence for the detection of the magnetic resonance. Often, in proteins that contain several trp and/or tyr, energy transfer occurs to the aromatic residue with lowest energy, so that only one trp or tyr is observed. In some cases, however, the individual aromatic residues could be distinguished, for example trp and tyr in azurin (39), trp's in bovine serum albumin (40) and in lysozyme (41) and tyr's in triply point-mutated lysozyme of the bacteriophage T_4, which contains no trp (42).

Aromatic residues close to active sites of enzymes are expected to be sensitive to substrate or dye binding. Triplet-singlet energy transfer may occur, e.g. from trp to proflavin bound to chymotrypsin (43), which allowed detection of ODMR resonances of trp via the delayed fluorescence of proflavin, or triplet-triplet transfer between substrate or dye and aromatic residue. Also signals from coenzymes, such as NAD^+ and NADH, are readily observed by ODMR and give information on the conformation of the cofactor in the enzyme (44).

In addition to the k_u's and the values of D and E, the ODMR line width is an important probe of protein structure. Generally, the lines are broad when the triplet is on a solvent-exposed residue, as various conformers are then frozen in upon cooling (45,46). The ODMR line width then reflects the range of conformational fluctuations at physiological temperatures but now not averaged, as in experiments at ambient temperatures: It is as it were a 'freeze frame' snapshot rather than an uninterpretable blur (16).

The MIP spectrum can be probed as a function of microwave frequency within the site-broadened ODMR line; this provides some insight in the energies of the conformers.

3.1.2 <u>Nucleic acids</u>. Since the nucleic acid bases contain aromatic rings they are at low temperature (< 80 K) readily excited into the triplet state, which can be investigated by ODMR monitoring the phosphorescence. A large number of papers, mostly from the groups of Maki and Clarke (Boston) have been devoted to the study of the isolated bases and of base stacking in DNA, RNA, various polynucleotides, transfer RNA's, etc. (1,38).

It has been established from the measurement of D and E, for example, that in poly(dA) : poly(dT) and in poly(dG) : poly(dC) the phosphorescence stems from the T and G moiety, respectively, the excitation energy being transferred to the lowest triplet energy. In DNA the phosphorescence originates from the neutral T base. In yeast tRNA, the phosphorescence below 80 K is mainly due to the modified base wybutosine (yW), which lies adjacent to the 3'-end of the anticodon and is a useful probe of codon-anticodon binding. In *E. coli* tRNA thiouridine emits short-lived phosphorescence to the red of the normal purine, pyrimidine bases, and again serves as a specific internal label for ODMR studies of binding, etc. In proflavin-DNA complexes the proflavin emits delayed fluorescence when DNA is excited; this delayed fluorescence is modulated when the triplet sublevels of DNA are brought into resonance with a microwave field and serves as a probe of energy transfer along the DNA molecule. The binding of certain covalent adducts, such as aflatoxin B_1 (47-49) and epoxybenzopyrenes (50-52), to DNA has been investigated with ODMR. These complexes form a promising system to study the mechanistics of chemical carcinogenesis. It has been established, for example, that the lowering of the ZFS parameters and the marked red shift of the (−) enantiomer of the <u>anti</u>-benzo(α)pyrenediol expoxide adduct compared to the (+) enantiomer correlates with their respective biological activity (52).

3.1.3 <u>Heavy-atom effects</u>. The triplet yield can be appreciably enhanced by complexing a closed-shell heavy metal component such as Ag^+ or methyl mercury (II) (CH_3Hg^+) to aromatic residues or nucleic acid bases (1,16,46,53). Because of the large nuclear mass (Z number) of e.g. mercury, spin-orbit interaction and therefore the intersystem crossing efficiency is dramatically enhanced. Orbital overlap between the aromatic residue or base and the metal compound provides the required interaction, which is spin-sublevel selective (54,55). Subtle changes in the binding of the heavy-metal component, for example induced by base stacking or by protein conformational changes, may give rise to large changes in the populating probabilities and the decay rates of the individual triplet sublevels. These changes often interact in parallel, making slow-passage ODMR less sensitive to the effect than rapid-passage ODMR, where the frequency is swept fast across the resonance. For the nucelic acid bases it is of advantage to use heavy-atom-substituted derivatives, such as bromouracil, etc. (56). Similar heavy-metal effects were observed for heavy counterions in naphtalene-containing micelles (57,58).

3.1.4 <u>Protein-polynucleotide interactions</u>. The sensitivity of the triplet properties (D, E, p_u, k_u, line width) to the environment makes ODMR a very useful tool to study protein-DNA (RNA) interactions. Especially the ability to discriminate individual aromatic residues makes this method more powerful than studies of the fluorescence or the phosphorescence, which are spectrally insensitive to site and are also difficult to separate kinetically in different components. More recent work is now focusing on the binding of proteins or amino acids to DNA, RNA, tRNA, etc. (1,59-71).

In a series of papers Maki and coworkers have mapped the interactions of *Escherichia coli* gene- or plasmid-encoded single-stranded DNA-binding protein with single-stranded polynucleotides (61-67), in some of the work making use of the elegant site-directed mutagenesis technique to replace the three trp's 40, 58 and 38 one-by-one with phenylalanine (61,66) or phe 60 with alanine (65). Strong indications were obtained for stacking interactions of phe 60 and trp 40 and 54 but not 88 with the nucleotide bases. A similar stacking interaction was found for trp 35 of the p10 single-stranded nucleic acid binding protein of murine leukemia viruses (68), and for the trp's of gene 32-protein of the bacteriophage T4 (69,70). In the complexes of Eco RI endonuclease with decanucleotides, however, such trp stacking interactions were not found (71).

3.1.5 <u>Protein-lipid interactions</u>. A few papers have recently appeared in which the interaction of proteins with lipids is explored (72-75). It was found that the single trp 3 of porcine, bovine and equine pancreatic phospholipase A_2 (PA_2) is involved in binding of PA_2 with n-hexadecylphosphocholine micelles. ZFS- parameters titrations indicated that porcine and equine PA_2 have similar binding affinity, while bovine PA_2 binds much more weakly to the lipid-water interface.

Binding of free oleic acid to bovine serum albumin involves trp 134 but not trp 212 (75). The effects of binding on the phosphorescence 0-0 band and the ZFS parameters of trp 134 was attributed to Stark effects caused by a protein conformational change near trp 134 in the albumin-oleate complex. The trp 134 spectroscopic properties reflect the specificity of binding of various fatty acids with regard to differences in chain length or saturation.

3.1.6 <u>Energy transfer</u>. Energy transfer processes can be grouped into two classes: dipole-dipole-induced transitions (Förster mechanism) and exchange-mediated processes. The former are long-range ($1/R^3$ with R the donor-acceptor distance) and depend on spectral overlap of emission and absorbance spectra, whereas the latter are short-range ($\alpha \exp(-\beta R)$, with $\beta \approx 1.7$ Å^{-1}) and depend on orbital overlap of the singlet and triplet wave functions. Singlet-triplet ($S_1 \rightarrow S_o$ (D), $T_o \leftarrow S_o$ (A)) and triplet-triplet ($T_o \rightarrow S_o$ (D), $T_o \leftarrow S_o$ (A)) processes are spin-forbidden. Consequently, the molar extinction coefficient of the acceptor absorbance spectrum is low and the Förster spectral overlap integral very small. Thus, these processes occur mainly at short distances where the exchange (Dexter) mechanism predominates, which is independent of spectral overlap. Triplet-singlet transfer is equally spin-forbidden, but if the quantum efficiency of phosphorescence is high enough, spectral overlap is sufficient to make the time-integrated probability of transfer comparable to spin-allowed singlet-singlet transfer. Triplet-singlet $T_o \rightarrow S_o$, $S_1 \leftarrow S_o$ processes may give rise to delayed fluorescence (luminescence) of the acceptor. Since the lifetime of the donor triplet state can be manipulated by resonant microwaves, ODMR transitions within the donor can than be measured by monitoring the acceptor luminescence intensity. An application of this method on horse liver alcohol dehydrogenase has recently been published (5).

Triplet-triplet transfer has been studied in a benzophenone (D), naphtalene (A) system in dodecyl sulphate micelles (76-78). It occurs between different trp's (126, 138 and 158) in wild-type bacteriophage T_4 lysozyme and can be monitored making use of the slight differences in phosphorescence peak wavelength and ZFS parameters of the three trp's that were found in work on mutant T4 lysozyme in which trp residues were displaced by tyr (79). The results were compared with the crystal structure of lysozyme.

3.2 ODMR in photosynthesis.

The use of ODMR in photosynthesis research has proved to be a particularly fruitful field of its application in biology. By far the most important triplet state in photosynthetic membranes is that generated on the primary electron donor by radical recombination under conditions that forward electron transport is blocked by the prereduction of one of the acceptors (or by its physical deletion):

$$DA_1A_2^- \xrightarrow{h\nu} D^*A_1A_2^- \xrightarrow{<3\ ps} D^+A_1^-A_2^- \longrightarrow {}^3DA_1A_2^- \cdot \tag{1}$$

Reaction (1) takes place in the so-called reaction center (RC), a specialized pigment-protein complex. The structure of two bacterial RC has been elucidated in atomic detail by X-ray crystallography (80-83). D is a bacteriochlorophyl (BChl) dimer, A_1 is a bacteriopheophytin (Φ_A), A_2 is a quinone, Q_A. There are another Φ molecule, Φ_B, and two so-called accessory BChls (B_B and B_A) whose function is as yet unclear. The pigments subscripted with A and those subscripted with B are arranged in two chains that show C_2 symmetry with D on the symmetry axis.

In reaction (1) a photon first generates the excited singlet state of the primary donor[†] D, D*, either directly or by energy transfer from a so-called antenna pigment-protein complex. In less than 3 ps (established in photosynthetic bacteria, probable in the plant photosystems), an electron is transferred from D* to the first acceptor A_1, creating a radical pair consisting of the cation D^+ and the anion A_1^-. From A_1^- the electron cannot travel further down the acceptor chain, because A_2 is prereduced (it cannot normally accept two electrons) or deleted from the RC. The radical pair $D^+A_1^-$ is not stable. It lives about 20 to 50 ns, and then recombines to either the singlet excited or ground state of D, or to its triplet state, ^3D. The latter reaction is almost 100 % effective at cryogenic temperatures. Mind that the charge separation reaction proceeds from a singlet state, ^1D, which has zero spin angular

---

[†] The name of the primary donor varies with organism and with photosystem. We will use the generic label D, but for specific organisms it is convenient to use the label P with a number that designates the peak of the red-most absorbance band of D. Thus, P860 in certain photosynthetic bacteria, P700 in photosystem I of plants, P680 in photosystem II, etc.

momentum. The law of conservation of angular momentum does not allow conversion to a triplet state (which has a spin of $\underline{S} = 1$) without some exchange of angular momentum from the electronic orbital motion. This cannot take place in the short time (3 ps) available. The radical pair, however, lives long enough to allow the conversion $^1[D^+A_1^-] \rightarrow {}^3[D^+A_1^-]$, mostly because these states are almost isoenergetic (the exchange energy in the radical pair is very small) and the conversion may be driven by hyperfine interactions between electron and nuclear spins. The triplet radical pair then recombines to ^3D (again because of conservation of angular momentum). The details of the radical pair singlet-triplet interconversion will not concern us here, for an extensive review see e.g. (84).

3.2.1 <u>FDMR spectroscopy</u>. The primary donor triplet state was discovered by Dutton et al. (85) in RC of a photosynthetic bacterium by high-field EPR. The first paper employing ODMR was by Clarke et al. (86) who presented the results of an FDMR experiment on whole cells of the photosynthetic bacterium *Rhodobacter sphaeroides* R-26. Shortly thereafter, a number of papers from the groups of Clarke, Schaafsma (Wageningen, The Netherlands), Wolf (Stuttgart, FRG) and from Leiden (The Netherlands) showed that with FDMR accurate values of |D| and |E| of the reaction center triplet and of triplets on plant antenna pigments and (bacterio)chlorophyl precursors could be obtained. Small, but well-observable differences in the ZFS parameters of the reaction center triplet, for example, showed that the primary electron donor was subtly different in closely related species (87).

The values of the k_u of the reaction center triplet that were obtained by FDMR were at first somewhat contentious, but the discrepancies were later proved to be due to the use of different methods, the saturation method being in this case much less accurate than the pulse method (11,12) (see section 2.1).

The earlier work up to 1984 on plant photosynthetic material and on bacterial photosystems has been reviewed in (1,11,88) and will not be further discussed here. Later applications of FDMR included FDMR of triplet states in antenna complexes of several photosynthetic bacteria (89-93) and of plant photosystem I (94,95), and the first experiments on isolated bacterial reaction centers (96-98).

FDMR measurements on light-harvesting pigments of antenna complexes of *Rb. capsulatus* A1a$^+$pho$^-$ (90), *Rb. capsulatus* A1a$^+$ and *Rb. sphaeroides* R26.1 (91) yielded triplet states with distinctly different ZFS parameters. In (90), three different triplets were found with the EEDOR technique; in (91) Triton X-100 detergent treatment reduced the RC triplet signal sufficiently to observe two different triplets without the double resonance technique. It was proposed that the two triplet belonged to the two BChl molecules present per subunit antenna protein.

FDMR measurements of light-harvesting complexes from barley yielded several ODMR signals of positive and negative sign associated with chlorophyls emitting at different wavelengths (94,95). The ZFS parameters were compared with model Chl *a* complexes; it was suggested that the positive FDMR signals were associated with (Chl *a*.H$_2$O)$_2$ dimers and hexacoordinate Chl *a* of the

Chl a-2pyridine type, whereas the negative FDMR signal resembled that of Chl a.$2H_2O$ complexes.

Fluorescence of isolated bacterial reaction centers is weak, and FDMR experiments difficult to carry out. Using the MIF technique, Den Blanken et al. (96) were nevertheless able to discern two MIF bands, one at 905 and one at 935 nm. The $|D|$ and $|E|$ values corresponding to the 905 nm band are those of the reaction center triplet, ^3D. The sign of the 905 nm band, however, is negative, which is unexpected if the fluorescence originates from D. Reversal of sign occurs when the fluorescing pigment is not D itself, but an associated pigment from which singlet excitation may travel to D by singlet energy transfer (section 2.1.4). It is unlikely that this F-905 pigment is one of the reaction center accessory BChl pigments or one of the P-860 monomers whose absorbance appears to shift to 803 nm when a ^3D state is formed that is localized on one of the dimer constituents, because in that case the Stokes shift would amount to 100 nm, a highly unusual value for the chlorophyls. The 905 nm fluorescence may stem from a weak concentration of antenna bacterio-chlorophyls (less than a few percent of P-860) that are strongly bound to the reaction center protein. Conversion of P-860 to the triplet state closes the reaction center and removes it as a strong quencher, which may strongly enhance the fluorescence of such an antenna pigment. Modulation of the concentration of ^3D by resonant microwaves may then very well result in a detectable MIF signal, even when the concentration of F-905 is well below that of P-860.

The sign of the 935 nm fluorescence indicates that it originates from P-860 proper, but the $|D|$ and $|E|$ values are not those of the bulk P-860 (as measured with ADMR or high-field EPR). It was concluded that this fluorescence stems from a minority of reaction centers with a slightly different configuration, perhaps caused by the isolation procedure. Charge separation from these altered P-860 complexes may be somewhat less fast than in the intact, bulk P-860, giving rise to appreciable fluorescence that is modulated by the presence of ^3D. It was noted that the maximum of the positive MIF signal, when corrected for the presence of F-905, was close to the main reaction center fluorescence at 920 nm. (Note that this is blue-shifted by about 10 nm compared to the fluorescence of whole cells, measured at 4.2 K.) Thus, it appears to be possible that the main fluorescence intensity of reaction centers is due to P-860 with a somewhat different structure than P-860 in the intact cell. Hence, caution should be exerted in analyzing experiments in which the fluorescence of isolated reaction centers is monitored as a probe of the primary photoprocesses.

Recent fluorescence-detected ODMR experiments on isolated reaction centers of *Rhodopseudomonas (Rps.) viridis* bore out similar sign reversal of the ODMR transition as a function of monitoring wavelength as observed for *Rb. sphaeroides* R-26 (97). However, the sign of the fluorescence-detected ODMR signal of whole cells was positive, in agreement with (96,98). This poses a problem, as the authors' interpretation of the sign reversal predicts a negative sign of the ADMR signal of reaction centers, contrary to observation (99). Note in this respect that the sign of ADMR-monitored T – S spectra (see below)

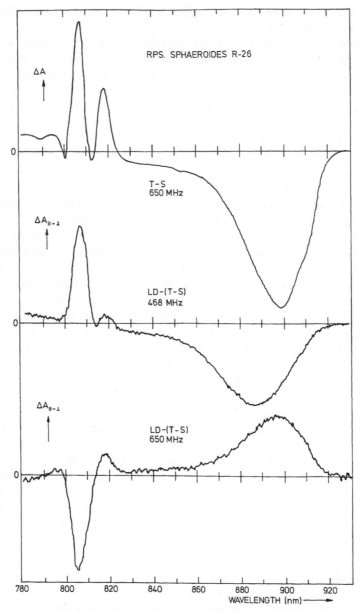

Fig. 12. T − S spectrum and LD-(T − S) spectra at 1.2 K for the two ADMR transitions of *Rb. sphaeroides* R-26. From (121).

does not reflect the sign of the ADMR signal perse (24,27,99). The discrepancy between the signs of the antenna fluorescence- and absorbance-monitored ODMR signals in *Rps. viridis* was addressed by Den Blanken et al. (99) and tentatively attributed to energy transfer from a vibrationally excited S_1-state of the antenna pigments to the reaction center.

3.2.2 <u>ADMR spectroscopy of reaction centers</u>. In this section we will focus on the application of absorbance-detected magnetic resonance (ADMR), a variant of ODMR that after a first rather unpromising experiment on a pigment solution (100), has been developed in Leiden to a high degree of sophistication (reviewed recently in (26,101)).

The great advantage of the ADMR variant above fluorescence or phosphorescence detection is that it can always be applied, regardless of quantum yields of emission, provided the lifetime of the triplet state is not too short (this holds for all cw ODMR techniques) and a sufficient optical density can be attained (OD $\approx$ 0.7 gives maximal signal). Both conditions hold for the photosynthetic triplets, and in the first application of ADMR to isolated bacterial reaction centers (102) it was shown that the sensitivity of ADMR was several orders higher than that of FDMR on the same material.

The high sensitivity of the ADMR method opened the way to studies of numerous isolated reaction centers, pigment solutions, etc., both of bacterial and plant origin. Accurate values of $|D|$, $|E|$ and the k_u were determined (Tables 1 and 2). From the tables it is readily seen that the values of $|D|$ for the primary donor triplet in bacterial reaction centers are lower by 20 - 30 % compared to that of the isolated bacteriochlorophyl pigment. At first, this was attributed to delocalization of the triplet state over the two BChls making up the primary donor and attempts were made to derive the structure of the BChl dimer from the in vivo and in vitro values of $|D|$, $|E|$ and the k_u (103,104), in which an angle of 48° between the two BChl planes was derived (103). The more accurate values of the k_u for BChl obtained by ADMR (105), however, showed that the dimer structure was compatible with a more or less parallel dimer. The recently resolved X-ray structure of crystals of RC of *Rps. viridis* and *Rb. sphaeroides* R-26 (80-83) has fully sustained this result. From recent EPR data on the reaction center triplet state in single crystals (106) and from ADMR spectroscopy (107) it was concluded that, at least in *Rps. viridis*, the triplet state is largely localized on one of the dimer BChl, viz. D_A. The deviation of the $|D|$ value between the in vivo and in vitro BChl triplet was ascribed to admixture of charge transfer (CT) states of the form $^3[BChl^+. BChl^-]$ to the monomeric ^3BChl state. Less than a few percent admixture is sufficient to lower $|D|$ by the required amount (106). A postulated CT state $^3[D^+B_A^-]$, with B_A one of the accessory BChls, is rendered unlikely by the recent observation (108,109) that the values of $|D|$ and $|E|$ are virtually independent of temperature in the range 4.2 - 75 K.

In plant RC the values of $|D|$ and $|E|$ are practically the same as those of monomeric chlorophyl in solution (Table 1). This can be explained by either one of three possibilities: i) the primary donor of both plant photosystems I and II is a monomeric Chl *a* molecule, ii) the primary donor is a plane-parallel sandwich (Chl $a)_2$ dimer with a fully delocalized triplet state (strong exciton coupling), iii) the primary donor is a dimer on which the triplet state is fully localized on one of the monomeric constituents and does not have CT admixture. In the latter case the term dimer in the sense of two *interacting* molecules obviously applies only to the singlet and possibly oxidized states of

TABLE 1

Representative zero field splitting parameters ($cm^{-1} \times 10^4$) of triplet states of (bacterio)chlorophyls in vivo and in vitro

Pigment and organism	|D|	|E|	Reference
Bacteriochlorophyl			
Rhodobacter (Rb.) sphaeroides			
Strain R-26, cells	187.2 ± 0.2	31.2 ± 0.2	87
Reaction centers	188.0 ± 0.4	32.0 ± 0.4	102
Strain 2.4.1	185.9 ± 0.6	32.4 ± 0.3	9
Rhodospirillum (R.) rubrum			
Strain S1, cells	187.8 ± 0.6	34.3 ± 0.3	87
Strain FRI, cells	187.9 ± 0.6	34.3 ± 0.3	9
Rhodobacter (Rb.) capsulatus			
Strain ATC 23872, cells	184.2 ± 0.6	30.3 ± 0.3	9
Chromatium (C.) vinosum	177.4 ± 0.6	33.7 ± 0.3	9
Strain D, cells			
Chloroflexus (Cfl.) aurantiacus			
Reaction centers	197.7 ± 0.7	47.3 ± 0.7	115
Prosthecochloris (P.) aestuarii			
Reaction centers	208.3 ± 0.7	36.7 ± 0.7	128
Rhodopseudomonas (Rps.) viridis[†]			
Cells	156.2 ± 0.7	37.8 ± 0.7	135
Reaction centers	160.3 ± 0.7	39.7 ± 0.7	135
BChl *a* in methyltetrahydrofuran	230.2 ± 2.0	58.0 ± 2.0	105
BChl *b* in methyltetrahydrofuran	221.0 ± 2.0	57.0 ± 2.0	105
Chlorophyl			
Photosystem I particles	281.7 ± 0.7	38.3 ± 0.7	119
Photosystem II particles	285.5 ± 0.7	38.8 ± 0.7	120
Chl *a* in methyltetrahydrofuran	281.0 ± 6.0	39.0 ± 3.0	136
Chl *b* in n-octane	320	40	134

[†] Contains BChl *b* instead of BChl *a*.

the primary donor. At present it is difficult to choose between the three possible explanations. It has recently become clear that the reaction center of photosystem II is quite homologous to that of the purple bacteria, both with respect to protein structure and the prosthetic groups. Since in the reaction center of these bacteria the primary donor is certainly dimeric, one is inclined to take the donor P680 in PS II also dimeric. Arguments in favor of this derive from the observation of a narrow EPR line of P680$^+$ (line width about $1/\sqrt{2}$ times that of monomeric Chl a^+ (110,111)) and the interpretation of the optical singlet absorbance difference spectrum P680$^+$ − P680 (112). Remarkably, the orientation of ^{3}P680 in the photosynthetic membrane is different from that found for bacteria (113). Thus, the P680 dimer might consist of two non-parallel Chl a's, with the triplet state localized on a Chl a that is oriented more or less perpendicular to the postulated C_2 axis of the reaction center (case iii). A way to check this would be the study of triplet orientation by LD-ADMR. Preliminary results show indeed LD-(T − S) spectra that could be interpreted to result from a triplet localized on a non-parallel dimer (E.J. Lous and K. Satoh, unpublished results).

TABLE 2
Triplet sublevel decay rates in s^{-1} of triplet states of (bacterio)chlorophyls in vivo and in vitro

Pigment and organism	k_x	k_y	k_z	Reference
Rb. sphaeroides Strain R-26, cells	9,000 ± 1,000	8,000 ± 1,000	1,400 ± 200	87
R. rubrum Strain S1, cells	8,000 ± 700	7,200 ± 700	1,350 ± 150	87
Cfl. aurantiacus Reaction centers	12,660 ± 750	14,290 ± 800	1,690 ± 50	115
P. aestuarii Reaction centers	6,790 ± 500	3,920 ± 300	1,275 ± 100	128
Rps. viridis Cells	≤16,000	≤16,000	≤2,600	135
Reaction centers	13,700 ± 900	16,100 ± 1300	2,420 ± 90	135
BChl *a* in MTHF	11,950 ± 700	15,900 ± 1300	1,635 ± 50	105
BChl *b* in MTHF	12,400 ± 900	14,900 ± 1300	1,300 ± 30	105
PS I particles	990 ± 100	1,010 ± 100	93 ± 5	119
PS II particles	930 ± 40	1,088 ± 50	110 ± 5	120
Chl *a* in n-octane	661 ± 89	1,255 ± 91	241 ± 15	134
Chl *b* in n-octane	268 ± 34	570 ± 54	34 ± 4	134

The values of the k_u of the two plant photosystems obtained by ADMR are practically identical to those of monomeric Chl a (Table 2). Again, this points to a triplet state localized on a non- or weakly-interacting dimer or monomer. The line widths of the $|D| \pm |E|$ transitions are quite similar to those of the bacterial RC; the lines are approximately Gaussian and are presumably inhomogeneously broadened as a result of slight, frozen-in conformational differences.

3.2.3 T – S spectroscopy of reaction centers. ADMR was first used to record T – S spectra in 1982 by Den Blanken et al. (24). It was immediately clear that this technique permitted the recording of low temperature T – S spectra far more accurately than was possible with conventional flash techniques. For the first time the complicated structure in the 800 nm region

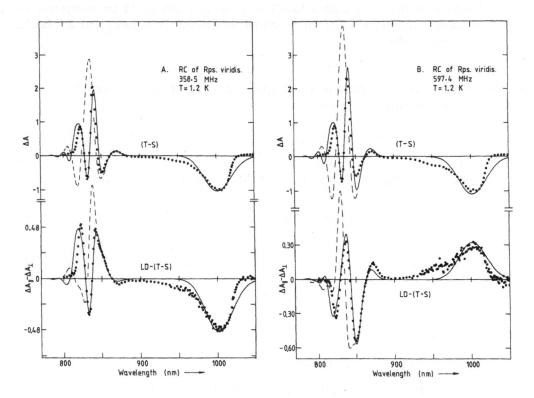

Fig. 13. T – S spectra and LD-(T – S) spectra at 1.2 K of *Rps. viridis* for the two ADMR transitions at $|D| - |E|$ (A) and $|D| + |E|$ (B). Dots: measured spectra. Drawn line: simulation assuming the triplet state of D localized on the D_A monomer. Dashed line: simulation assuming ^3D to be localized on D_B. From (107).

for *Rb. sphaeroides* and the 830 nm region for *Rps. viridis* could be accurately measured (Figs. 12,13). In these regions, the accessory BChls absorb. The absorption of the dimer BChl are shifted to the red, partly as a result of the strong excitonic coupling between the two BChls. If a (partly) localized triplet state is formed on D, then the excitonic coupling is greatly diminished. For 100 % localization one BChl is in the triplet state and absorbs only little in the red and near-infrared region, and one is in the singlet state, which is more or less unperturbed compared to the in vitro state because of the absence of strong excitonic coupling. The result is that ^{3}D will then mostly absorb in the 800 or 830 nm region (for *Rb. sphaeroides* and *Rps. viridis*, respectively), which absorption shows up as a strong positive band in the T – S spectrum (24). The two smaller features at the long- and short-wavelength side of the positive band were attributed to band shifts of the two accessory BChls induced by the change D → ^{3}D (24).

The bleaching at 890 nm (*Rb. sphaeroides*) or 990 nm (*Rps. viridis*) is due to the disappearance of the red-shifted long-wavelength band of D that is shifted to the red because of excitonic coupling between the two BChls of D. As stated above, exciton coupling is much weaker in the triplet state of ^{3}D, irrespective whether the triplet is localized or not. Therefore, in a localized ^{3}D state, part of the BChl absorption is bleached and part shifts back to the 'normal' wavelength of BChl absorption in a protein matrix.

With minor differences the above qualitative picture holds for all purple bacteria investigated (114). The T – S spectra of green photosynthetic bacteria are more complex and have only been tentatively interpreted (115,116), with the exception of the green gliding bacterium *Chloroflexus (Cfl.) aurantiacus*, whose RC is very similar to that of purple bacteria (117,118).

The T – S spectra of reaction centers of the plant photosystems are shown in Fig. 14 (119,120). Both are characterized by a strong bleaching of donor bands at 703 and 682 nm for photosystem I and II, respectively. At the blue side of this bleaching a small positive band appears, at a wavelength close to that of the absorption of Chl *a* in vitro. This band was attributed to the appearing absorption of a monomeric Chl *a* molecule belonging to a primary donor dimer on which the triplet state was localized on the second Chl *a*, analogous to the assignment of the strong appearing band in the T – S spectra of the purple bacteria. Thus, the plant T – S spectra support the notion that in both plant photosystems the primary donor is a dimeric Chl *a* complex.

3.2.4 <u>Linear dichroic T – S spectroscopy</u>. As explained in section 2.4 LD-(T – S) spectroscopy is very similar to that of photoselection, in which an oriented distribution of e.g. photo-oxidized primary donors is produced by exciting the immobilized unoriented sample with a beam of linearly-polarized light of a wavelength corresponding to a particular absorption band, e.g. that of D. It has the advantage that, in contrast to photoselection, the dichroic spectrum can be recorded with respect to two axes of reference that are perpendicular to each other. This reduces considerably the ambiguity of unraveling the orientation of the various transition moments (tm) in the T – S spectrum. Complex T – S spectra such as those of the photosynthetic RC, where many overlapping features are present such as appearing bands, bleachings, band shifts to the red and to the blue, can only be interpreted with certainty if the corresponding LD-(T − S) spectra are available. Although a consistent interpretation is sometimes possible without a full spectral simulation (121), the full power of T − S and LD-(T − S) spectroscopy is realized when the experimental spectra can be compared with spectra simulated with exciton theory, which describes the interaction between pigment molecules brought about by the electrostatic dipole-dipole coupling between their electronic transition moments. For two identical molecules this interaction causes a shift and a splitting of the excited state, which is now composed of two levels with (apart from the shift) energies given by the original energy plus or minus the interaction energy. The corresponding wave functions are the symmetric (+) and anti-

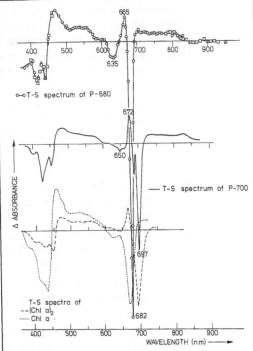

Fig. 14. T − S spectra at 1.2 K of particles of photosystem I (P700), photosystem II (P680), and of Chl *a* monomer and dimer (at −78°C). From (102).

symmetric (−) combination of the original wave functions. The intensity of each component of the split absorption band depends on the geometry of the dimer: $|M(\pm)|^2 = |M|^2(1 \pm \cos \alpha)$ or $|M(+)|^2/|M(-)|^2 = \cot^2(\alpha/2)$, where M and $M(\pm)$ are the original and the perturbed tm's, respectively, and α is the angle between the original tm's. The theory is readily extended to n pigments, each pigment interacting differently with the other n − 1 pigments (122-124). For a known pigment configuration the absorption spectrum of the excitonically-coupled spectrum is then readily obtained. Generally the resulting bands are mixtures of the original uncoupled bands, with tm's that are vectorial combinations of the original tm's.

Bacterial RC and also the RC of plant photosystem II contain six pigments (2 (B)Chl of the primary donor, 2 accessory (B)Chl and 2 (bacterio) pheophytins, Φ). Exciton calculations based on the crystal structure of the RC of *Rps. viridis*, which has recently become available (80), yielded satisfactory simulations of the ADMR-monitored T − S and LD-(T − S) spectra (107,125,126) of these RC (Fig. 13). The spectral fits allowed to draw quite an array of conclusions, the most important of which are

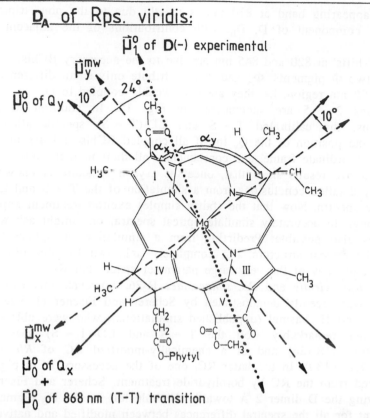

Fig. 15. Results of ADMR-monitored T – S and LD-(T – S) spectroscopy of *Rps. viridis*. The orientation of the microwave transition moments, the Q_y transition moments of the monomer and of the primary donor dimer (D), and the T – T transition moment of ^3D have been determined in the molecular frame of the BChl of D that is bound to the L subunit of the RC (D_A). For details, see Ref. 107. From (133).

1) The Q_y transition moment of monomeric BC in vitro is oriented very close to the N_I-N_{III} axis.

2) The triplet is localized on one of the BChl components of D, D_A.

3) The y spin axis of the localized triplet state is very close to the N_I-N_{III} axis at an angle of 10°, the x and y spin axes lie approximately in the plane of the BChl macrocycle. (Sublevel ordering y, x, z for D,E > 0.)

4) The tm of the long-wavelength band of D is oriented close to the y spin axis of the triplet state making an angle of 24° with the N_I-N_{III} axis, and lies close to the plane of the macrocycle.

5) ^3D has a triplet-triplet absorption at about 872 nm, which is oriented along the N_{II}-N_{IV} axis.

6) The appearing band at 830 nm is mostly due to the non- triplet-carrying BChl component of D, D_B, with contributions of the adjacent accessory BChl.

7) The 'shifts' at 820 and 845 nm are due to the accessory BChls.

8) The two Φ pigments, Φ_A and Φ_B, contribute only small difference bands in the 790 nm region. i.e. they are only weakly coupled to D.

Conclusions 2 - 5 are summarized in Fig. 15. In addition to the above conclusions, the calibrated T – S and LD-(T – S) spectra allowed us to calculate the position of the Q_y tm of all coupled BChls and Φ's in the x, y, z spin axes coordinate frame, and consequently all their mutual angles.

The above results show that, once the crystal structure is known, one can draw very detailed conclusions from a simulation of the T – S and LD-(T – S) difference spectra. Now that the (fairly simple) exciton treatment appears to be good enough to accurately simulate optical spectra, one might ask whether the reverse is also possible: Predicting from a simulation of optical (difference) spectra the crystal structure. For completely unknown RC this obviously is a tall order, in view of the very large parameter space. For RC closely related to those of *Rps. viridis* and *Rb. sphaeroides* R-26, however, this can indeed be done, as was recently demonstrated by Scherer and Fischer (126,127), Vasmel (118,128) and H. Vasmel (unpublished simulations), who were able to simulate and predict remarkably well the T – S and LD-(T – S) spectra of *Cfl. aurantiacus* (118,126) and of a chemically-modified RC of *Rb. sphaeroides* R-26 (127,129,130). In the latter RC, one of the accessory BChl pigments, B_B, is removed from the RC by borohydride treatment. Scherer and Fischer showed that moving the D dimer 2 Å towards the hole left by the B_B pigment sufficed to account for all the spectral differences between modified and native RC.

3.2.5 <u>ODESE of reaction centers</u>. An optically-detected electron spin echo experiment is the technique of choice to measure spin relaxation times, especially the longitudinal or spin-spin relaxation time T_2. Spin-spin dephasing is influenced by the presence of close-lying triplet states, as e.g. excitonically-split states in a coupled dimer (mini-exciton (131)). From the temperature dependence of T_2 one may, under favourable conditions, derive a value of the excitonic coupling, which in turn provides information on the dimer structure. This, of course, is of great interest for bacterial RC, where the primary donor is a BChl dimer.

In (31) FDESE was applied to measure T_2 at 1.5 K of free base porphin and of RC of *Rb. sphaeroides* R-26 in whole cells. For the porphin, an accurate value of T_2 could be measured, but for the RC the signal-to-noise ratio was rather poor and only an estimate of T_2 could be made.

The development of ADMR, with the dramatic increase in signal-to-noise compared to FDMR, prompted us to combine the three-pulse ODESE technique of Nishi et al. (31) with absorbance detection (32). The experimental echo trace is depicted in Fig. 16. The good signal-to-noise ratio allowed an accurate determination of T_2, yielding a value of 1.16 μs, independent of temperature over the range 1.2 - 2.1 K. This means that the excitonic coupling is in excess of 1.5 cm^{-1}, perhaps not surprising in light of the considerable optical spectral

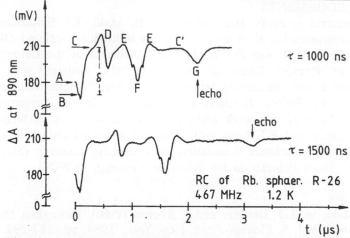

Fig. 16. Absorbance-detected ODESE of the triplet state of reaction centers of *Rb. sphaeroides* R-26, monitored at the |D| − |E|, 467 MHz resonance. τ is the delay betweeen the $\pi/2$ and π driving pulses, the probe $\pi/2$ pulse is scanned from $t = 0$ onward (Fig. 8). The features A − G are changes in absorbance resulting from coherent (D,G) or incoherent effects of the driving pulses. G is the echo (Fig. 8g). D is the so-called 'midway' echo for which the probe $\pi/2$ pulse acts as the second driving pulse, and the π pulse as the probe pulse (see Refs. 32,137). From (32).

shifts engendered by the exciton coupling of the primary donor dimer (107,125,126). We expect that future developments combining pulsed photo-excitation with ADESE will sufficiently extend the temperature range to allow the determination of the temperature dependence of T_2.

4. CONCLUSIONS AND PROSPECTS

1) The triplet state is a versatile probe of structure and function in biology on a molecular level.

2) ODMR of triplet states is a powerful tool to study the structure of proteins, including active sites, pigment environment, etc., of nucleic acids and of protein-nucleic acid complexes.

3) ADMR permits to record isotropic and linear dichroic triplet-minus-singlet absorbance difference spectra of pigment-protein complexes with unparallelled accuracy and sensitivity. Spectral simulation provides a wealth of information on the structure of, and coupling between, chromophores.

4) Applications of ODMR in biochemistry will focus on protein- nucleic acid interactions and protein-lipid interactions, those in photosynthesis on the study of the structure of reaction centers and of antenna complexes, on energy transfer, and on relating the optical properties of reaction centers and antenna complexes to their structure and function via ADMR-monitored triplet absorbance difference spectroscopy.

ACKNOWLEDGEMENTS

I am grateful to Profs. H.C. Brenner, A.H. Maki, T.J. Schaafsma and H.C. Wolf for communicating their recent work in biologically-related ODMR. I am indebted to Profs. J.H. van der Waals and J. Schmidt of the Leiden Centre for the Study of Excited States of Molecules, who assisted with equipment, laboratory space and stimulating interest and to Mrs. Tineke Veldhuyzen, Linda Gorsen and Mr. Freddy Bischoff for their help in the preparation of the manuscript. Much of the work carried out in Leiden was performed by Drs. H.J. den Blanken and E.J. Lous with skill and unstinted enthusiasm under the auspices of the Netherlands Foundation for Chemical Research (SON), financed by the Netherlands Foundation for Scientific Research (NWO).

REFERENCES

1 A.H. Maki, in: L.J. Berliner and J. Reuben (Eds.), Biological Magnetic Resonance. Vol. 6, Plenum Press, New York, 1984, pp. 187-294.
2 R.H. Clarke (Ed.), Triplet State ODMR Spectroscopy, Wiley- Interscience, New York, 1982.
3 M. Nakamizo and T. Matsueda, J. Molec. Spectr., 27 (1968) 450-460.
4 J. Schmidt, W.S. Veeman and J.H. van der Waals, Chem. Phys. Lett., 4 (1969) 341-346.
5 J.G. Weers and A.H. Maki, Biochemistry, 25 (1986) 2897-2904.
6 H.C. Brenner and V. Kolubayev, J. Lumin., 39 (1988) 251-2577.
7 S. Gosh, M. Petrin and A. Maki, Biophys. J., 49 (1986) 753-760.
8 W.G. van Dorp, W.H. Schoemaker, M. Soma and J.H. van der Waals, Mol. Phys., 30 (1975) 1701-1721.
9 A.J. Hoff and H. Gorter de Vries, Biochim. Biophys. Acta, 503 (1978) 94-106.
10 I.Y. Chan and B.N. Nelson, J. Chem. Phys., 62 (1975) 4080-4088.
11 A.J. Hoff, in: R.H. Clarke (Ed.), Triplet State ODMR Spectroscopy, Wiley-Interscience, New York, 1982, pp. 367-425.
12 A.J. Hoff and B. Cornelissen, Mol. Phys., 45 (1982) 413-425.
13 M. Leung and M.A. El-Sayed, Chem. Phys. Lett., 16 (1972) 54-459.
14 J.P. Lemaistre and A.H. Zewail, Chem. Phys. Lett., 68 (1979) 296-301, 302-308.
15 J. van Egmond, B.E. Kohler and I.Y. Chan, Chem. Phys. Lett. 34 (1975) 423-426.
16 A.L. Kwiram, in: R.H. Clarke (Ed.), Triplet State ODMR Spectroscopy, Wiley-Interscience, New York, 1982, pp. 427-478.
17 R.L. Williamson and A.L. Kwiram, J. Phys. Chem., 83 (1979) 3393-3397.
18 J.U. von Schutz, J. Zuclich and A.H. Maki, J. Am. Chem. Soc., 96 (1974) 714-718.
19 A.L. Kwiram, J.B.A. Ross and D.A. Deranleau, Chem. Phys. Lett., 54 (1978) 506-509.
20 R.L. Williamson and A.L. Kwiram, J. Chem. Phys., 88 (1988) 6092-6106.
21 M.A. El-Sayed, D.V. Owens and D.S. Tinti, Chem. Phys. Lett., 6 (1970) 395-399.
22 J. Beck, G.H. Kaiser, J.U. von Schutz and H.C. Wolf, Biochim. Biophys. Acta, 634 (1981) 165-173.

23 J. Beck, J.U. von Schutz and H.C. Wolf, Z. Naturforsch., 38c (1983) 220-229.

24 H.J. den Blanken and A.J. Hoff, Biochim. Biophys. Acta, 681 (1982) 365-374.

25 A. Vermeglio, J. Breton, G. Paillotin and R. Cogdell, Biochim. Biophys. Acta, 501 (1978) 514-530.

26 A.J. Hoff, in: M.E. Michel-Beyerle (Ed.), Antennas and Reaction Centers of Photosynthetic Bacteria. Structure, Interaction and Dynamics, Springer-Verlag, Berlin, 1985, pp. 150-163.

27 H.J. den Blanken, R.F. Meiburg and A.J. Hoff, Chem. Phys. Lett., 105 (1984) 336-342.

28 J. Schmidt and J.H. van der Waals, in: L. Kevan and R.N. Schwarts (Eds.), Time Domain Electron Spin Resonance, Wiley-Interscience, New York, 1979, pp. 343-398.

29 J. Schmidt, Chem. Phys. Lett., 14 (1972) 411-414.

30 W.G. Breiland, C.B. Harris and A. Pines, Phys. Rev. Lett., 30 (1973) 158-161.

31 N. Nishi, J. Schmidt, A.J. Hoff and J.H. van der Waals, Chem. Phys. Lett., 56 (1978) 205-207.

32 E.J. Lous and A.J. Hoff, Chem. Phys. Lett., 140 (1987) 620-625.

33 R.P. Feynman, F.L. Vernon and R.W. Hellwarth, J. Appl. Phys., 28 (1957) 49-52.

34 M. Kasha, J. Opt. Soc. Am., 38 (1948) 929-934.

35 S.J. van der Bent, P.A. de Jager and T.J. Schaafsma, Rev. Sci. Instrum., 47 (1976) 117-212.

36 I.Y. Chan, in: R.H. Clarke (Ed.), Triplet State ODMR Spectroscopy, Wiley-Interscience, New York, 1982, pp. 1-24.

37 W.H. Hardy and L.A. Whitehead, Rev. Sci. Instrum., 57 (1981) 213-216.

38 A.H. Maki, in: R.H. Clarke (Ed.), Triplet State ODMR Spectroscopy, Wiley-Interscience, New York, 1982, pp. 479-557.

39 K. Ugurbil, A.H. Maki and R. Bersohn, Biochemistry, 16 (1977) 901-907.

40 S.Y. Mao and A.H. Maki, Biochemistry, 26 (1987) 3106-3114.

41 R.L. Williamson and A.L. Kwiram, Biochem. Biophys. Res. Commun., 125 (1984) 974-979.

42 S. Gosh, L.-H. Zang and A.H. Maki, Biochemistry, in the press.

43 A.H. Maki and T.-T. Co, Biochemistry, 15 (1976) 1229-1235.

44 J.B.A. Ross, K.W. Rousslang, A.G. Motten and A.L. Kwiram, Biochemistry, 18 (1979) 1808-1813.

45 M.V. Hershberger, A.H. Maki and W.C. Galley, Biochemistry, 19 (1980) 2204-2209.

46 K.L. Bell and H.C. Brenner, Biochemistry, 21 (1982) 799-804.

47 A.L. Kwiram, Y.C. Liu, M.N. Farquhar and E.S. Smuckler, Biochem. Biophys. Res. Commun. 83 (1978) 1354-1359.

48 S.M. Lefkowitz, H.C. Brenner, D.G. Astorian and R.H. Clarke, FEBS Lett., 105 (1979) 77-80.

49 A.L. Kwiram, Bull. Magn. Reson., 2 (1981) 35-38.

50 S.M. Lefkowitz and H.C. Brenner, J. Am. Chem. Soc., 103 (1981) 5257-5259.

51 S.M. Lefkowitz and H.C. Brenner, Biochemistry, 21 (1982) 3735-3741.

52 V. Kolubayev, H.C. Brenner and N.E. Geacintov, Biochemistry, 26 (1987) 2638-2641.
53 L.-H. Zang, S. Gosh and A.H. Maki, Biochemistry, in the press.
54 S. Gosh, M. Petrin, A.H. Maki and L.R. Sousa, J. Chem. Phys., 87 (1987) 4315-4323.
55 S. Gosh, M. Petrin, A.H. Maki and L.R. Sousa, J. Chem. Phys., 88 (1988) 2913-2918.
56 T.A. Chai and A.H. Maki, Biochim. Biophys. Acta, 799 (1984) 171-180.
57 M. Petrin, A.H. Maki and S. Gosh, Chem. Phys. Lett., 128 (1986) 425-431.
58 S. Gosh, M. Petrin and A.H. Maki, J. Phys. Chem., 90 (1986) 5206-5215.
59 T.A. Chai and A.H. Maki, J. Biol. Chem., 259 (1984) 1105-1109.
60 M.R. Taherian, K.F.S. Luk and A.H. Maki, Biochemistry, 23 (1984) 6614-6618.
61 M.I. Khamis, J.R. Casas-Finet, A.H. Maki, J.B. Murphy and J.W. Chase, FEBS Lett., 211 (1987) 155-159.
62 M.I. Khamis, J.R. Casas-Finet and A.H. Maki, Biochemistry, 26 (1987) 3347-3354.
63 M.I. Khamis, J.R. Casas-Finet and A.H. Maki, J. Biol. Chem., 262 (1987) 1725-1733.
64 J.R. Casas-Finet, M.I. Khamis, A.H. Maki, P.P. Ruvolo and J.W. Chase, J. Biol. Chem., 262 (1987) 8574-8583.
65 J.R. Casas-Finet, M.I. Khamis and A.H. Maki, FEBS Lett., 220 (1987) 347-352.
66 M.I. Khamis, J.R. Casas-Finet, A.H. Maki, J.B. Murphy and J.W. Chase, J. Biol. Chem., 262 (1987) 10938-10945.
67 L.-H. Zang, A.H. Maki, J.B. Murphy and J.W. Chase, Biophys. J., 52 (1987) 867-872.
68 J.R. Casas-Finet, J.R., N.-I. Jhon and A.H. Maki, Biochemistry, 27 (1988) 1172-1178.
69 M.I. Khamis and A.H. Maki, Biochemistry, 25 (1986) 5865-5872.
70 J.R. Casas-Finet, J.-J. Toulme, R. Santus and A.H. Maki, Eur. J. Biochem., 172 (1988) 641-646.
71 N.-I. Jhon, J.R. Casas-Finet, A.H. Maki and P. Modrich, Biochim. Biophys. Acta, 949 (1988) 189-194.
72 S.Y. Mao, A.H. Maki and G.H. de Haas, FEBS Lett., 185 (1985) 71-75.
73 S.Y. Mao, A.H. Maki and G.H. de Haas, Biochemistry, 25 (1986) 2781-2786.
74 S.Y. Mao, A.H. Maki and G.H. de Haas, FEBS Lett., 211 (1987) 83-88.
75 S.Y. Mao and A.H. Maki, Biochemistry, 26 (1987) 3576-3582.
76 S. Gosh, M. Petrin and A.H. Maki, J. Phys. Chem., 90 (1986) 1643-1647.
77 S. Gosh, M. Petrin and A.H. Maki, J. Phys. Chem., 90 (1986) 5206-5215.
78 M. Petrin, A.H. Maki and S. Gosh, Chem. Phys. Lett., 128 (1986) 425-431.
79 S. Gosh, L.-H. Zang and A.H. Maki, J. Chem. Phys., 88 (1988) 2769-2775.
80 J. Deisenhofer, O. Epp, K. Miki, R. Huber and H. Michel, Nature (London), 318 (1985) 618-624.
81 C.-H. Chang, D. Tiede, J. Tang, U. Smith, J. Norris and M. Schiffer, FEBS Lett., 205 (1986) 82-86.
82 J.P. Allen, G. Feher, T.O. Yeates, H. Komiya and D.C. Rees, Proc. Natl. Acad. Sci. USA, 84 (1987) 5730-5734, 6162-6166.

83 T.O. Yeates, H. Komiya, A. Chirino, D.C. Rees, J.P. Allen and G. Feher, Proc. Natl. Acad. Sci. USA, 85 (1988) 7993-7997.
84 A.J. Hoff, Quart. Rev. Biophys., 14 (1981) 599-665.
85 P.L. Dutton, J.S. Leigh and M. Seibert, Biochem. Biophys. Res. Commun., 46 (1972) 406-413.
86 R.H. Clarke, R.E. Connors, J.R. Norris and M.C. Thurnauer, J. Am. Chem. Soc., 97 (1975) 7178-7179.
87 A.J. Hoff and J.H. van der Waals, Biochim. Biophys. Acta, 423 (1976) 615-620.
88 T.J. Schaafsma, in: R.H. Clarke (Ed.), Triplet State ODMR Spectroscopy, Wiley-Interscience, New York, 1982, pp. 292-365.
89 J. Beck, J.U. von Schutz and H.C. Wolf, Chem. Phys. Lett., 94 (1983) 141-146.
90 J. Beck, J.U. von Schutz and H.C. Wolf, Chem. Phys. Lett., 94 (1983) 147-151.
91 A. Angerhofer, J.U. von Schutz and H.C. Wolf, Z. Naturforsch., 40c (1985) 379-387.
92 A. Angerhofer, J.U. van Schutz and H.C. Wolf, in: M.E. Michel-Beyerle (Ed.), Antennas and Reaction Centers of Photosynthetic Bacteria. Structure, Interaction and Dynamics, Springer-Verlag, Berlin, pp. 78-80.
93 A. Angerhofer, J.U. von Schutz and H.C. Wolf, in: J. Biggins (Ed.), Progress in Photosynthesis Research. Vol. 1. Martinus Nijhoff, Dordrecht, pp. 427-430.
94 G.F.W. Searle, R.B.M. Koehorst, T.J. Schaafsma, B.L. Moller and D. von Wettstein, Carlsberg Commun., 46 (1981) 183-194.
95 T.J. Schaafsma, G.F.W. Searle and R.B.M. Koehorst, J. Mol. Struct., 79 (1982) 461-464.
96 H.J. den Blanken, G.P. van der Zwet and A.J. Hoff, Biochim. Biophys. Acta 681 (1982) 375-382.
97 A. Angerhofer, J.U. von Schutz and H.C. Wolf (1984), Z. Naturforsch., 39c 91984) 1085-1090.
98 F.G. van Wijk and T.J. Schaafsma, in: C. Sybesma (Ed.), Advances in Photosynthesis Research. Vol. II. Martinus Nijhoff/Dr. W. Junk, The Hague, pp. 173-176.
99 H.J. den Blanken, A.P.J.M. Jongenelis and A.J. Hoff, Biochim. Biophys. Acta, 725 (1983) 472-482.
100 R.H. Clarke and R.E. Connors, Chem. Phys. Lett., 33 (1975) 365-368.
101 A.J. Hoff, in: L.A. Staehelin and C.J. Arntzen (Eds.), Photosynthesis III. Photosynthetic Membranes and Light Harvesting Systems. Vol. 19 Encyclopedia of Plant Physiology, New Series, Springer-Verlag, Berlin, 1986, pp. 400-421.
102 H.J. den Blanken, G.P. van der Zwet and A.J. Hoff, Chem. Phys. Lett., 85 (1982) 335-338.
103 R.H. Clarke, R.E. Connors, H.A. Frank and J.C. Hoch, Chem. Phys. Lett., 45 (1977) 523-528.
104 W. Hagele, D. Schmid and H.C. Wolf, Z. Naturforsch., 33a (1978) 94-97.
105 H.J. den Blanken and A.J. Hoff, Chem. Phys. Lett., 96 (1983) 343-347.
106 J.R. Norris, C.P. Lin and D.E. Budil, J. Chem. Soc. Faraday Trans. 1, 83 (1987) 13-27.
107 E.J. Lous and A.J. Hoff, Proc. Natl. Acad. Sci. USA, 84 (1987) 6147-6151.

108 A.J. Hoff and I.I. Proskuryakov, Chem. Phys. Lett., 115 (1985) 303-310.

109 J. Ulrich, A. Angerhofer, J.U. von Schutz and H.C. Wolf, Chem. Phys. Lett., 140 (1987) 416-420.

110 D.F. Ghanotakis and G.T. Babcock, FEBS Lett., 153 (1983) 231-234.

111 A.J. Hoff, Phys. Rep., 54 (1979) 75-200.

112 H.J. van Gorkom, Doctoral Dissertation, University of Leiden, 1976.

113 A.W. Rutherford, Biochim. Biophys. Acta, 807 (1985) 189-201. 114J.A. Dijkman, H.J. den Blanken and A.J. Hoff, Isr. J. Chem., in the press.

115 H.J. den Blanken, H. Vasmel, A.P.J.M. Jongenelis, A.J. Hoff and J. Amesz, FEBS Lett., 161 (1983) 185-189.

116 A.J. Hoff, H. Vasmel, E.J. Lous and J. Amesz, in: J.M. Olson, J.G. Ormerod, J. Amesz, E. Stackebrandt and H.G. Truper (Eds.), Green Photosynthetic Bacteria, Plenum Press, New York, pp. 119-126.

117 H. Vasmel, R.F. Meiburg, J. Amesz and A.J. Hoff, in: J. Biggins (Eds.), Progress in Photosynthesis Research. Vol. 1, Martinus Nijhoff, Dordrecht, pp. 403-406.

118 H. Vasmel, Doctoral Dissertation, University of Leiden, 1986.

119 H.J. den Blanken and A.J. Hoff, Biochim. Biophys. Acta, 724 (1983) 52-61.

120 H.J. den Blanken, A.J. Hoff, A.J.P.M. Jongenelis and B.A. Diner, FEBS Lett., 157 (1983) 21-27.

121 A.J. Hoff, H.J. den Blanken, H. Vasmel and R.F. Meiburg, Biochim. Biophys. Acta, 806 (1985) 389-397.

122 A.S. Davidov, Theory of Molecular Excitons, Plenum Press, New York, 1981.

123 M. Kasha, H.S. Rawls and M.A. El-Bayoumi, Pure Appl. Chem., 11 (1965) 371-392.

124 R.M. Pearlstein, in: Govindjee (Ed.), Photosynthesis. Vol. I: Energy Conversion in Plants and Bacteria, Academic Press, New York, pp. 293-329

125 E.W. Knapp, P.O.J. Scherer and S.F. Fischer, Biochim. Biophys. Acta, 852 (1986) 295-305.

126 P.O.J. Scherer and S.F. Fischer, Biochim. Biophys. Acta, 891 (1987) 157-164.

127 P.O.J. Scherer and S.F. Fischer, Chem. Phys. Lett., 137 (1987) 32-36.

128 H. Vasmel, H.J. den Blanken, J.T. Dijkman, A.J. Hoff and J. Amesz, Biochim. Biophys. Acta, 767 (1984) 200-208.

129 D. Beese, R. Steiner, H. Scheer, B. Robert, M. Lutz and A. Angerhofer, Photochem. Photobiol., 47 (1988) 293-304.

130 A. Angerhofer, D. Beese, A.J. Hoff, E.J. Lous and H. Scheer, in: G. Singhal (Ed.), Applications of Molecular Biology in Bioenergetics of Photosynthesis, Narosa Publishing House, New Delhi, 1989, pp. 197-203.

131 B.J. Botter, C.J. Nonhof, J. Schmidt and J.H. van der Waals, Chem. Phys. Lett., 43 (1976) 210-216.

132 A.J. Hoff, in: G. Blauer and H. Sund (Eds.), Optical Properties and Structure of Tetrapyrroles, Walter de Gruyter & Co., Berlin, 1985, pp. 453-473.

133 E.J. Lous, Doctoral dissertation, Leiden, 1988.

134 R.H. Clarke and R.H. Hofeldt, J. Chem. Phys., 61 (1974) 4582-4587.

135 H.J. den Blanken and A.J. Hoff, Chem. Phys. Lett., 98 (1983) 255-262.

136 A.J. Hoff, Govindjee and J.C. Romijn, FEBS Lett., 73 (1977) 191-196.
137 R. Vreeker, E.J. Lous and A.J. Hoff, Chem. Phys. Lett., in the press.

APPENDIX

General solution of the ODMR equations

In this appendix an analytical solution of the time dependence of the singlet and triplet population in an ODMR experiment is given with no limiting assumptions, except that the population of the singlet excited state is assumed to be negligibly small. The solution, although analytic, is not very elegant but lends itself well to implementation on a personal computer.

On condition that the population of the singlet excited state is very low, $N_1 \ll N$, the general ODMR equations are

$$\frac{dN_o}{dt} = -K \, N_o + \Sigma_u k_u N_u \tag{A1}$$

$$\frac{dN_u}{dt} = p_u K \, N_o - k_u N_u + \sum_{s \neq u} [N_s(P_{su} + W_{su}) - N_u(P_{us} + W_{us})] \tag{A2}$$

$$N_o + \Sigma_u N_u = N_o + N_T = N; \qquad u,s = x,y,z \tag{A3}$$

with K, N, and the (time dependent) N_o, N_u, N_T as defined in section 2.1, and $P_{su} = P_{us}$ and $W_{su} = W_{us} \exp[(E_s - E_u)/k_B T]$ the rate of microwave-induced transitions and of spin-lattice relaxation between the s and the u triplet sublevel with energy E_s and E_u, respectively. Here, k_B is Boltzmann's constant and T the temperature.

Since there is no particular advantage in irradiating all three possible microwave transitions simultaneously, we may assume without loss of generality that $P_{xy} = 0$. To simplify we set $P_{xz} = P_{zx} = r$ and $P_{yz} = P_{zy} = q$, and denote the fractional populations N_o/N, N_x/N, N_y/N and N_z/N by S, X, Y and Z, respectively. With Eqn. A3 we then obtain the matrix equation

$$\dot{\underline{x}}(t) = \underline{\underline{A}} \, \underline{x}(t) + \underline{f} \tag{A4}$$

with

$$\underline{\underline{A}} = \begin{pmatrix} -(K + k_z) & k_x - k_z \\ p_x K - r - W_{zx} & -(k_x + 2r + W_{xy} + W_{xz} + W_{zx}) \\ p_y K - q - W_{zy} & -(q - W_{xy} + W_{zy}) \end{pmatrix}$$

$$\begin{pmatrix} k_y - k_z \\ -(r - W_{yx} + W_{zx}) \\ -(k_y + 2q + W_{yx} + W_{yz} + W_{zy}) \end{pmatrix}, \; \underline{f} = \begin{pmatrix} k_z \\ r + W_{zx} \\ q + W_{zy} \end{pmatrix} \; \text{and} \; \underline{x} = \begin{pmatrix} S \\ X \\ Y \end{pmatrix}.$$

We have eliminated Z, because for FDMR and ADMR it is handy to have N_o/N directly. For PDMR one may set up $\underline{A}$ by eliminating S.

The standard solution of Eqn. A4 is

$$\underline{x}(t) = \exp[\underline{A}(t - t_o)]\underline{c} + \exp(\underline{A}t)\int_{t_o}^{t} \exp(-\underline{A}s)\underline{f}(s)ds \qquad [A5]$$

with $\underline{c} = \underline{x}(t_o)$. We may take $t_o = 0$, and $\underline{f}(s) = \underline{f}$ so that Eqn. A5 reduces to

$$\underline{x}(t) = \exp(\underline{A}t)(\underline{c} + \underline{A}^{-1}\underline{f}) - \underline{A}^{-1}\underline{f} . \qquad [A6]$$

Exp$(\underline{A}t)$ is calculated with the aid of

$$\exp(\underline{A}t) = \alpha_2\underline{A}^2t^2 + \alpha_1\underline{A}t + \alpha_o\underline{I}, \qquad [A7]$$

with $\underline{I}$ the unit matrix.

The coefficients α_j ($j = 0,1,2$) are determined from the set of equations

$$\alpha_2\lambda_i^2t^2 + \alpha_1\lambda_it + \alpha_o = \exp(\lambda_it), \quad i = 1,2,3 \qquad [A8]$$

where λ_i are the eigenvalues of $\underline{A}$. Eqn. A8 can be written in matrix form

$$\underline{\Lambda} = \underline{A}.\underline{\alpha} \qquad [A9]$$

with

$$\underline{A} = \begin{pmatrix} \lambda_1^2t^2 & \lambda_1t & 1 \\ \lambda_2^2t^2 & \lambda_2t & 1 \\ \lambda_3^2t^2 & \lambda_3t & 1 \end{pmatrix} , \quad \underline{\alpha} = \begin{pmatrix} \alpha_2 \\ \alpha_1 \\ \alpha_o \end{pmatrix} \quad \text{and} \quad \underline{\Lambda} = \begin{pmatrix} \exp(\lambda_1t) \\ \exp(\lambda_2t) \\ \exp(\lambda_3t) \end{pmatrix} .$$

Substituting the α_j solved from [A9] in [A7] we obtain for the elements of the matrix $\exp(\underline{A}t)$:

$$\begin{aligned}
(\exp(\underline{A}))_{nm} = \left(\frac{\det\underline{A}}{t^3}\right)^{-1} & [\{(\underline{A}^2)_{nm}(\lambda_2 - \lambda_3) + A_{nm}(\lambda_3^2 - \lambda_2^2) \\
& + \delta_{nm}(\lambda_2^2\lambda_3 - \lambda_3^2\lambda_2)\}\exp(\lambda_1t) + \{(\underline{A}^2)_{nm}(\lambda_3 - \lambda_1) \\
& + A_{nm}(\lambda_1^2 - \lambda_3^2) + \delta_{nm}(\lambda_3^2\lambda_1 - \lambda_3\lambda_1^2)\}\exp(\lambda_2t) \\
& + \{(\underline{A}^2)_{nm}(\lambda_1 - \lambda_2) + A_{nm}(\lambda_2^2 - \lambda_1^2) \\
& + \delta_{nm}(\lambda_1^2\lambda_2 - \lambda_2^2\lambda_1)\}\exp(\lambda_3t)],
\end{aligned} \qquad [A10]$$

where δ_{nm} is the Kronecker delta and A_{nm} is a matrix element of $\underset{\sim}{A}$. The eigenvalues of $\underset{\sim}{A}$ are obtained in the usual way from

$$\det(\underset{\sim}{A} - \lambda \underset{\sim}{I}) = -(\lambda^3 + B\lambda^2 + C\lambda + D) = 0 \qquad [A11]$$

with

$$B = -\text{trace } \underset{\sim}{A}, \quad C = \sum_{i<j}(A_{ii}A_{jj} - A_{ij}A_{ji}), \quad D = \sum_i(-1)^i A_{i1}\det(\underset{\sim}{M}_{i1}),$$

where $\underset{\sim}{M}_{i1}$ is the i1-minor of matrix $\underset{\sim}{A}$ and i,j = 1,2,3. Eqn. A11 is solved by substitution of $\lambda = \mu - 1/3\ B$, yielding

$$\mu^3 + m\mu + n = 0 \qquad [A12]$$

with

$$m = -\frac{1}{3} B^2 + C, \quad n = \frac{2}{27} B^3 - \frac{1}{3} BC + D.$$

The roots of Eqn. A12 are

$$\mu_i = \rho \cos \theta_i; \quad i = 1,2,3 \qquad [A13]$$

with

$$\rho = (-\frac{4}{3}\ m)^{\frac{1}{2}}, \quad \theta_i = \frac{1}{3} \arccos(-\frac{4n}{\rho^3}) + (i - 1)\frac{2\pi}{3} \quad \text{and} \quad \lambda_i = \mu_i - \frac{1}{3} B.$$

The solution of [A6] has the general form

$$x_i(t) = \sum_{j=1}^{3} \xi_{i,j} \exp(\lambda_i t) + \eta_i \qquad [A14]$$

with $\eta_i = (\underset{\sim}{A}^{-1} \cdot \underset{\sim}{f})_i$.

Thus, the response of the four-level system on switching on or off one (or a combination of) microwave frequency(ies) is, in general, described by three exponentials. Only on certain limiting conditions is the response described by one exponential (microwaves on for q and/or r $\gg$ k_u) or two exponentials (microwaves off for K $\gg$ k_u). Note that in most cases the relaxation rates W_{us}, W_{su} will be almost equal and can be set to one rate W. For work at very low temperatures, close to 1.5 K, W can be set to zero. The explicit solution of Eqn. A11 for that case is given in (12).

The time response of the system after perturbation by application of a pulse of resonant microwaves is given by Eqns. A6, A10-A14 with r = q = 0. Eqn. A14 then reduces to the form given in (8), now however with the decay rates λ_j and the coefficients $\xi_{i,j}$ and η_i explicitly given. For K $\rightarrow$ 0 and W = 0 the roots of Eqn. A11 reduce for $q = r = 0$ to $\lambda_i = -k_u$, $u = x,y,z$, as is immediately seen from $\underset{\sim}{A}$ in Eqn. A4. $\xi_{i,j}$ and η_i then approach $N_i(\infty)/N$ and $(N_i(0) - N_i(\infty))/N$, respectively. For small K the λ_i's approach the values $-k_u$ along the line

$$\lambda_i(K) = -k_u - p_u K, \quad K \rightarrow 0. \qquad [A15]$$

This means that in the low-light regime, the measured decay rates are proportional to the light flux, and that one must extrapolate to $K = 0$ to obtain the true molecular decay rates.

CHAPTER 19

RYDMR - THEORY AND APPLICATIONS

WILHELM LERSCH and MARIA E. MICHEL-BEYERLE

1. INTRODUCTION

Many photochemical and photophysical processes involve the formation of very short-lived para-magnetic states. This is especially true for photoinduced electron transfer reactions in biological and biomimetic compounds, which produce radical pair intermediates living only a few nanoseconds or less. Electron paramagnetic resonance (EPR) is a primary source of information on the kinetics of such transient states and the magnetic interactions of the unpaired electron spins involved. Conventional c.w. EPR spectrometers permit the observation of paramagnetic intermediates down to the microsecond time scale. A much higher time resolution of several tens of nanoseconds is achieved by using pulsed EPR and spin echo techniques. However, this time resolution is still not sufficient to detect species with lifetimes of ten nanoseconds or less, characteristic e.g. for radical ion pair states involved in the early steps of photosynthesis (1).

In order to trace the EPR spectrum of spin-correlated pairs of paramagnetic species on a nano-second time scale a special technique can be applied known as Reaction Yield Detected Magnetic Resonance (RYDMR) (2-4). In the RYDMR experiment - as in time-resolved EPR - resonant microwaves are irradiated onto the sample during the lifetime of the paramagnetic state of interest. The effect of the microwaves, however, is not detected as a change of the magnetization of the sample due to transitions between the spin substates of the pairs of paramagnetic species investigated, but instead as a change of the product yield of some spin selective reaction of the pairs. This detection technique has two major advantages:

1) The product that is monitored in the RYDMR experiment is usually much longer lived than the paramagnetic precursor state.

2) The observable can be any measurable property of the product, e.g. its optical extinction, such that the spectroscopic method used for detection of the RYDMR signal is typically not of magnetic resonance type.

Thus, by fundamentally decoupling the stages of inducing microwave transitions and detecting the effect of the microwaves, the RYDMR method circumvents the instrumental limitations imposed on time domain EPR experiments yielding enhanced time resolution. At the same time, the sensitivity of RYDMR can be much higher than that obtained in EPR measurements. By detecting the lumines-cence of electronically excited reaction products via the single-photon counting technique, for example, radical pairs at a concentration of only a few tens per sample can be investigated (5).

In this chapter we will outline the principles of the RYDMR technique and review some of its applications. In 1981 RYDMR was used for the first time to elucidate the spin dynamics of a transient radical ion pair state involved in the light-induced electron transfer cascade in bacterial pho-tosynthesis (6). This application will be reviewed here in some detail since it yields a salient example of how RYDMR can contribute to the understanding of electron transfer mechanisms.

2. PRINCIPLES OF THE METHOD

In the RYDMR experiment the EPR spectrum of a pair of correlated paramagnetic species is traced by observing changes in the amount of a reaction product of the pair under the influence of resonating microwaves. (A pair of paramagnetic species will be termed "correlated" if for some reason the expectation value of the projection of its two magnetic moments onto each other does not vanish.) To illustrate this principle of the RYDMR technique and to point out requirements for the technique to be applicable, let us consider a radical pair initially created in the overall spin singlet state. For simplicity, we will assume that the components of the pair do not move relative to each other such that interactions between the unpaired electron spins do not depend on time. This will be termed a "localized" radical pair.

Suppose that the radical pair finally undergoes a reaction (e.g. recombination) and that the type of product formed depends on the spin multiplicity of the pair, i.e. is different for singlet- and triplet-phased pairs, respectively (Fig. 1). For the radical pair to acquire triplet character starting from the initial singlet state there must be some intersystem crossing (ISC) mechanism. This can be either the so-called "radical pair mechanism" known from Chemically Induced Dynamic Electron and Nuclear Polarization (CIDEP and CIDNP) (5,7) or some paramagnetic relaxation mechanism (8,9). Singlet-triplet transitions, however, can occur only if the energy separation between the singlet and the triplet states is small, i.e. if it does not exceed by much the strength of the ISC matrix element. This restriction sets limits on the strength of the exchange interaction between the unpaired electrons, which separates the singlet from the triplet manifold (see section 4.1). A further requirement is that the lifetime of the radical pair must be sufficiently long for the ISC mechanism to become effective before the radical pair is consumed in a reaction.

Consider now the situation depicted in the left part of Fig. 1. Without an external magnetic field (and for moderate splittings of the triplet states due to the magnetic spin dipole interaction in the radical pair) all three triplet sublevels are populated from the initial singlet state via ISC. As soon as one turns on an external magnetic field two of the triplet states start to separate from the singlet state and the remaining triplet state. At field strengths considerably exceeding the magnitude of the ISC matrix element the two outer triplet sublevels ($|T_+>$ and $|T_->$) are no longer populated to any significant extent due to the large $|S> \longleftrightarrow |T_\pm>$ energy gap (middle part of Fig. 1). In this situation triplet products are formed only via the $|T_0>$ substate of the triplet manifold, which remains in approximate degeneracy with the singlet state. As a consequence, the triplet product yield decreases and the singlet yield concomitantly increases upon application of an external magnetic field. This sensitivity of a radical pair reaction to the magnitude of an external magnetic field (5), which has been termed MARY (MAgnetic field dependence of the Reaction Yield) (10), is a prerequisite for the RYDMR technique to be applicable as was pointed out by Frankevich (4).

Let us proceed to the case where microwaves are irradiated onto the sample to induce EPR-transitions between the spin sublevels of the radical pair at high magnetic field strengths (right part of Fig. 1). The only transitions that are spin-allowed are those between $|T_0>$ and $|T_+>$ or $|T_->$. Since $|T_0>$ interacts with the singlet state via ISC, microwave induced transitions from $|T_0>$ to $|T_+>$ or $|T_->$

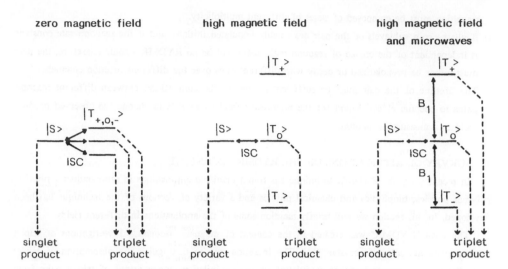

Fig. 1. Singlet and triplet energy levels of a localized radical pair (for vanishing exchange and magnetic spin dipole interaction between the unpaired electrons) at zero and high static external magnetic field, respectively. At zero (high) magnetic field all three triplet sublevels (only the $|T_0\rangle$ sublevel) exchange(s) population with the singlet state via radical pair intersystem crossing. Dashed arrows indicate spin selective recombination pathways of the radical pair. Transitions induced by a resonant microwave field $\underline{B_1}$ (right part of the figure) can change the relative amounts of singlet and triplet recombination products as explained in the text.

mix some singlet character into the split-off triplet states $|T_+\rangle$ and $|T_-\rangle$ thus effectively reopening a pathway for $|S\rangle \rightarrow |T_+\rangle$ transitions. This enhances the population of the triplet states and concomitantly leads to an increase (decrease) of the triplet (singlet) product yield. (This explanation applies to situations where the strength of the microwave field, the $|S\rangle \longleftrightarrow |T_0\rangle$ energy gap and the lifetime broadening of the spin states are small compared to the value of the ISC matrix element. Other limiting cases are discussed in sections 4.1 and 4.3.) Thus, by monitoring e.g. the triplet yield and varying the magnetic field across resonance, the EPR transitions of the radical pair can be observed. One must bear in mind, however, that the observable of a RYDMR experiment is very different from the one of an EPR experiment. Therefore, although the same transitions are detected in both types of experiments, they may show up differently in the EPR and the RYDMR spectrum of the radical pair.

Obviously, not every short-lived radical pair and more generally not every pair of correlated paramagnetic species will be accessible to the RYDMR technique. The main requirements for the technique to be applicable can be summarized as follows (cf.(4)):

1) The pair of correlated paramagnetic species investigated must be subject to at least two competing reactions leading to different reaction products. The choice of reaction path depends on the spin multiplicity of the pair.

2) The idea of the RYDMR experiment is to change the reaction path of a pair by affecting its spin multiplicity in a microwave field. Since microwave radiation by itself cannot change the multiplicity of the pair and since transitions within a given multiplicity as such do not alter the yield of the reaction corresponding to that multiplicity, at least one of the states involved in a RYDMR

transition must be composed of states of different multiplicity.

3) If all the spin sublevels of the pair are equally populated initially and if the reaction rate constant is independent of the choice of reaction path, there will be no RYDMR signal. Therefore, the pair must either be prepolarized or decay with different rates over the different reaction channels.

4) The lifetime of the pair must be sufficiently long for the competition between different reaction paths to become effective and for the microwave field to be able to change the observed product yield to a measurable amount.

3. SURVEY OF APPLICATIONS OF THE RYDMR TECHNIQUE

In recent years the RYDMR technique has found manifold applications in semiconductor physics, polymer physics, biophysics and chemical physics and a variety of versions of the technique has been developed. In this section we will briefly mention some of the applications in different fields.

The term "RYDMR" was created in the context of magnetic resonance investigations of triplet exciton pairs in molecular crystals (11). The intention was to distinguish the observation of triplet EPR spectra via magnetoresonant modulation of the annihilation rate constant of triplet pairs from their detection in an ODMR (Optically Detected Magnetic Resonance) experiment where resonant microwaves cause a redistribution of spin sublevel populations of an isolated triplet molecule (4). In solid state physics such a distinction is not made, although ODMR studies of spin-dependent recombination processes in semiconductors (12-13) are clearly of a RYDMR type. Also in chemical physics the nomenclature concerning RYDMR experiments is not unequivocal. Depending on the system studied and on the detection technique several names have been used to designate experimental methods that - from a more general point of view - belong to the RYDMR type. Examples for this are: ODMR (5,14), ODESR (Optically Detected ESR) (15,16), ODMR RP (ODMR of Radical Pairs) (17), CIDNP detected ESR (18-20), SNP (Stimulated Nuclear Polarization) (21-23) and PYESR (Product Yield Detected ESR) (24-26).

The application of RYDMR in semiconductor physics dates back to 1970 where it was observed that the recombination rate of excess carriers can be influenced by changing their spin orientation (or that of paramagnetic recombination centers) in a resonant microwave field (12,13). In later years a variety of semiconducting materials has been investigated with this technique (14).

In chemical physics RYDMR has been applied to observe magnetic resonance spectra of triplet excitons, charge transfer states and paramagnetic impurities in molecular crystals and to study the spin-dependent recombination dynamics of geminate radical pairs in a liquid. Reviews are available in (4,5,15). Continuing interest in these applications has led to the development of new detection techniques such as CIDNP detected ESR (19-21) and SNP (22-24). In the latter versions of the RYDMR experiment the EPR spectrum of an intermediate radical pair is traced by monitoring changes of the nuclear spin polarization of diamagnetic reaction products in the presence of microwaves resonating with the unpaired electron spins. As a new application of the RYDMR technique, radical pairs in micelles have been investigated in recent years (25-28). Valuable information has been obtained about magnetic interactions within correlated pairs of radicals in a situation where - in contrast to pairs in a liquid - diffusion of the partners is limited by the geometry of a micelle (29).

In polymer physics RYDMR was applied to explore the conductivity mechanism of weakly doped polyacetylene (30). In biophysics, finally, it is an important tool for characterizing the spin-dependent recombination dynamics of the primary radical ion pair state in photosynthesis. This latter application will be reviewed in detail in the next section since it provides useful insight into how RYDMR contributes to the understanding of electron transfer processes by probing the spin dynamics of transient radical pair intermediates.

4. INVESTIGATION OF THE FIRST LIGHT-INDUCED RADICAL ION PAIR STATE IN BACTERIAL PHOTOSYNTHESIS

Photosynthesis is the mechanism by which plants and certain bacteria convert light energy into chemical energy. Following the absorption of light in antenna pigments the excitation energy is transferred to specialized, membrane bound pigment-protein complexes called reaction centers. There, in a sequence of electron transfer steps, the excitation energy is converted into a transmembrane electrochemical potential gradient. In a complex scheme of slower reactions this gradient drives the formation of chemical energy carriers, like ATP and glucose (31).

After the successful isolation of reaction centers from photosynthetic bacteria (32), another "quantum step" in the history of photosynthesis were the crystallization of reaction centers from two bacterial species (*Rhodopseudomonas viridis* and *Rhodobacter sphaeroides*) in recent years (33-35) and the subsequent x-ray structure analysis to a resolution of below 3 Å (36-38) yielding invaluable infor-

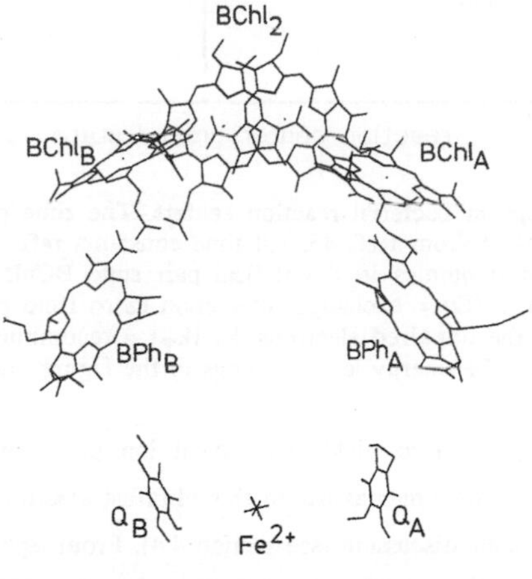

Fig. 2. Arrangement of cofactors in the reaction center from the purple bacterium *Rhodobacter sphaeroides*, strain R-26. Being held in a fixed geometry by the protein environment, the pigments mediate the light-driven charge separation across the photosynthetic membrane. The direction of the electron flow is from top to bottom in the figure. (This figure was kindly provided by Dr.J.Allen; see also Ref. 37.)

mation on the structural basis of the primary charge separation in bacterial photosynthesis.

Fig. 2 shows the arrangement of cofactors in the reaction center from the purple bacterium *Rhodobacter sphaeroides* (37), which - from previous spectroscopic work - is the best characterized reaction center at present (31). For simplicity, the protein surrounding the cofactors and determining their relative orientations and distances is omitted in Fig. 2. Two nearly symmetrical branches of pigments (A and B) can be discerned taking off from a "special pair" of bacteriochlorophyll molecules ($BChl_2$), which has already been postulated from earlier spectroscopic evidence (39,40,41). Next to the special pair are two "accessory" bacteriochlorophyll (BChl) monomers followed in each branch by a bacteriopheophytin (BPh) and a quinone (Q) molecule. A non-heme bound Fe^{2+}-ion located opposite of the dimer is bridging the quinones, Q_A and Q_B.

Upon optical excitation of the dimer charge separation occurs in a sequence of fast electron transfer steps from top to bottom in Fig. 2. Only the A-branch of pigments seems to be involved in the charge separation process, however (42). The excited singlet state $^1BChl_2^*$ of the dimer decays

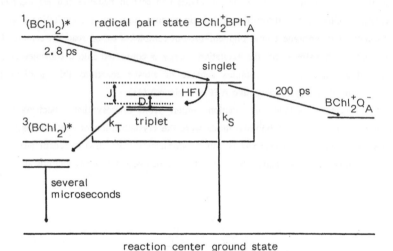

Fig. 3. Electron transfer steps in bacterial reaction centers. The time constant of 2.8 ps for the first electron transfer step was taken from Ref. 45. All time constants refer to room temperature. The parameters governing the spin dynamics in the radical pair state $BChl_2^+ BPh_A^-$ are defined as follows: HFI = hyperfine interaction; J (D) = exchange interaction (zero field parameter of the magnetic spin dipole interaction) between the unpaired electrons; k_S (k_T) = recombination rate constant for singlet-(triplet-)phased radical pairs. The energy level spacings in the figure are not true to scale.

within a few picoseconds (43-46) to yield the radical ion pair state $BChl_2^+ BPh_A^-$ (Fig. 3). Most probably, the bridging $BChl_A$ monomer assists in this ultrafast electron transfer process, although the specific mechanism is still under discussion (see section 4.4). From Bph_A^- the electron moves on to the quinone Q_A within $\simeq$ 200 ps (1) and from there to Q_B in another 100 μs (1) (not shown in Fig. 3). The latter electron transfer completes the light-induced charge separation in the reaction center.

A series of intermediate radical ion pair stages is involved in the photoinduced cascade of electron transfer steps described above. To characterize the longer lived ones of these, e.g. $BChl_2^+ Q_A^-$, conventional EPR techniques are adequate. The first radical ion pair state $BChl_2^+ BPh_A^-$, however, escapes observation with any magnetic resonance technique owing to its short lifetime of only 200 ps under native conditions. By either prereducing or extracting Q_A the electron transfer from BPh_A to Q_A can be blocked. In this case the radical ion pair state $BChl_2^+ BPh_A^-$, termed P^F, is stabilized for 10-20 nanoseconds (1) decaying via spin selective recombination of the separated charges (Fig. 3) to form either the singlet ground state or the lowest triplet state of the system (i.e. the dimer triplet

^{3}BChl$_2^*$). This longer lifetime is still not sufficient to permit observation of P^F in time-resolved EPR or electron spin echo measurements (6). However, in 1981 a clear RYDMR signal was obtained from P^F (6) demonstrating the applicability of the RYDMR technique. In fact, from the results of earlier MARY and ESP (Electron Spin Polarization) measurements (48,49) the success of the RYDMR experiment could have been predicted (cf. section 2; for a review of ESP in photosynthesis see the chapter by P.J.Hore in this book).

The application of RYDMR to the state P^F deserves special interest for two reasons:

1) In contrast to most other pairs of correlated paramagnetic species none of the components of the pair BChl$_2^+$ BPh$_A^-$ is mobile as the result of the fixation of the pigments in the protein matrix. Such a localized radical pair gives the unique opportunity to study details of the spin-spin interactions between the unpaired electrons that otherwise are obscured by some kind of random relative motion of the spins.

2) Determination of the temperature dependence of the exchange interaction J between the unpaired electrons in the state P^F is of key importance for the understanding of the mechanism of the first electron transfer step ^{1}BChl$_2^* \rightarrow$ BChl$_2^+$ BPh$_A^-$ (see section 4.4). Being particularly sensitive to the strength of the exchange interaction, the RYDMR experiment is superior to other methods (e.g. MARY (50)) in providing a reliable value for J.

In the following subsections we will review theoretical and experimental aspects of the application of RYDMR to the radical ion pair state P^F and discuss recent results for reaction centers from *Rhodobacter sphaeroides*, which is the only photosynthetic organism that has been investigated with the RYDMR technique so far.

4.1 Theory of RYDMR for localized radical pairs

In this subsection we will briefly develop the theoretical background required for the interpretation of RYDMR signals from the radical pair state P^F. For localized radical pairs spatial motion of the radicals need not be considered such that we are left with the task to analyze the spin dynamics of the pair. Our first goal will be therefore to specify the spin Hamiltonian that properly characterizes the magnetic interactions of the unpaired spins under the conditions of a RYDMR experiment.

Originating in the first excited singlet state of the dimer the radical pair BChl$_2^+$ BPh$_A^-$ is created in an overall spin singlet state. In section 2 we argued that for the RYDMR technique to be applicable there must be some mechanism mixing triplet character into the initial singlet state. Since spin relaxation processes are of no importance during the 10-20 ns lifetime of the state P^F, ISC must be caused by the so-called radical pair mechanism (5,7,46). In this mechanism the unpaired spins are assumed to experience different local magnetic fields due to differences in a) the hyperfine interaction with the magnetic nuclei of their respective neighborhoods and b) the Zeeman interaction with an external magnetic field as a result of unequal g-factors of the two radicals. By precessing at different frequencies about their local fields the spins dephase relative to each other. This gives rise to transitions from singlet to triplet and vice versa.

Besides the hyperfine interaction and the interaction with the static external magnetic field $\underline{B}_0$ applied in the RYDMR experiment we have to consider interactions between the unpaired spins. These are the exchange interaction J (J=singlet-triplet splitting, singlet below triplet for J>0) and the magnetic spin dipole interaction (zero field parameters D and E), both of which are stationary for a localized radical pair. Including further the interaction of the spins with the microwave field $\underline{B}_1(t)$ we can write down the following spin Hamiltonian

$$\hat{H}=g_1\beta(\underline{B}_0+\underline{B}_1(t))\hat{\underline{S}}_1 + g_2\beta(\underline{B}_0+\underline{B}_1(t))\hat{\underline{S}}_2 + \sum_n^{(1)} A_n\hat{\underline{I}}_n\hat{\underline{S}}_1 + \sum_n^{(2)} A_n\hat{\underline{I}}_n\hat{\underline{S}}_2 +$$

$$+J\left(\hat{\underline{S}}_1 \hat{\underline{S}}_2 - \frac{1}{4}\right) + \left(\hat{\underline{S}}_1+\hat{\underline{S}}_2\right) \mathbf{D} \left(\hat{\underline{S}}_1+\hat{\underline{S}}_2\right). \qquad [1]$$

$\hat{\underline{S}}_i$ (i=1,2) is the spin of the i-th radical with components $\hat{S}_{ix}$, $\hat{S}_{iy}$ and $\hat{S}_{iz}$. g_i is the g-factor of the i-th radical taken to be isotropic. β is Bohrs electronic magneton. The sum $\sum_n^{(i)}$ runs over all nuclei on the i-th radical. A_n ($\hat{\underline{I}}_n$) is the isotropic hyperfine coupling constant (the spin) of the n-th nucleus on the i-th radical. Anisotropic contributions to the hyperfine interaction are disregarded for simplicity. The only anisotropic contribution to the spin Hamiltonian considered here is the last term in [1] involving the spin dipole tensor $\mathbf{D}$. The neglect of the nuclear Zeeman interaction will be justified below.

Let us introduce the total spin $\hat{\underline{S}}=\hat{\underline{S}}_1+\hat{\underline{S}}_2$ and the operators $\hat{P}^S=(1/4)-\hat{\underline{S}}_1\hat{\underline{S}}_2$ and $\hat{P}^T=1-\hat{P}^S$, which project onto the singlet and the triplet manifold of spin states, respectively. Let us further take $\underline{B}_0$ to point in the z-direction $\underline{B}_0=B_0\underline{z}$ and $\underline{B}_1(t)=2B_1\underline{x}$ cosωt to lie perpendicular to it along the x-axis (ω is the angular frequency of the microwaves). Let us finally assume that the static magnetic field B_0 is so strong that - apart from terms involving the microwave field - only terms commuting with $\hat{S}_z$ need to be considered in [1]. In this high field case the nuclear spin configuration is a constant of motion. Consequently, the nuclear Zeeman interaction has no relevance to the spin dynamics and can be disregarded. (Note, however, that this reasoning depends on the neglect of anisotropic parts of the hyperfine interaction.) For convenience, we represent the spin Hamiltonian in a frame rotating around $\underline{B}_0$ at the frequency of the microwave field (48). We thus obtain

$$\hat{H}_{rot}^{(k)}=g\beta B_z^{(k)}\hat{S}_z + g\beta B_1\hat{S}_x + a^{(k)}(\hat{S}_{1z}-\hat{S}_{2z}) - J\hat{P}^S + D_{zz}\left(\frac{3}{2}\hat{S}_z^2-\hat{P}^T\right), \qquad [2]$$

where (k) denotes a certain configuration of the nuclear spins and

$$g\beta B_z^{(k)}=g\beta B_0 - \hbar\omega + \frac{1}{2}\left[\sum_n^{(1)} A_n m_n^{(k)}+\sum_n^{(2)} A_n m_n^{(k)}\right], \qquad [3]$$

$$a^{(k)}=\frac{1}{2}\left[\Delta g\beta B_0+\sum_n^{(1)} A_n m_n^{(k)} - \sum_n^{(2)} A_n m_n^{(k)}\right]. \qquad [4]$$

$m_n^{(k)}$ is the z-projection of the n-th nuclear spin in the k-th nuclear spin configuration. $g=(1/2)(g_1+g_2)$

is the average g-factor of the radicals; $\Delta g = g_1 - g_2$ is the difference of the g-factors. For the radical pair state P^F, Δg is on the order of 10^{-3}. With respect to the microwave field the difference of the g-factors is therefore neglected. The component $D_{zz} = \underline{z}\, D\underline{z}$ of the spin dipole tensor can be expressed by the zero field parameters D and E and the Euler angles θ and ψ that determine the orientation of the principal axes of the spin dipole tensor with respect to the direction of $\underline{B}_0$:

$$D_{zz} = \left(\cos^2\theta - \frac{1}{3}\right) D - \sin^2\theta \, \cos2\psi \, E. \tag{5}$$

To properly describe the spin dynamics of a localized radical pair we also have to account for spin selective recombination processes. In fact, recombination from P^F leads either to the singlet ground state of the reaction center (rate constant k_S in Fig. 3) or to the first excited triplet state of the dimer (rate constant k_T), which can be distinguished from the ground state by e.g. its optical transitions. A convenient way to determine the time evolution of the spin state of a radical pair under the influence of spin selective recombination processes is provided by the stochastic Liouville equation for the spin density operator $\hat{\rho}$ (46,49),

$$\frac{d\hat{\rho}}{dt} = -\frac{i}{\hbar}\,[\hat{H},\hat{\rho}] - \frac{1}{2}\,k_S(\hat{P}^S\hat{\rho}+\hat{\rho}\hat{P}^S) - \frac{1}{2}\,k_T(\hat{P}^T\hat{\rho}+\hat{\rho}\hat{P}^T), \tag{6}$$

where $[\hat{A},\hat{B}]$ denotes the commutator of $\hat{A}$ and $\hat{B}$. $\hat{\rho}$ and $\hat{H}$ are related either to the laboratory frame or to the rotating frame. By introducing the non-Hermitian Hamiltonian (50,51)

$$\hat{H}' = \hat{H} - i\hbar(k_S/2)\hat{P}^S - i\hbar(k_T/2)\hat{P}^T, \tag{7}$$

[6] can be rewritten to yield

$$(d\hat{\rho}/dt) = -(i/\hbar)(\hat{H}'\hat{\rho}-\hat{\rho}\hat{H}'^\dagger). \tag{8}$$

($\hat{H}'^\dagger$ is the Hermitian conjugate of $\hat{H}'$.) In the rotating frame, where $\hat{H}_{rot}^{(k)}{}'$ ([2] and [7]) does not depend on time, [8] can be readily solved:

$$\hat{\rho}_{rot}^{(k)}(t) = \exp\{-(i/\hbar)\hat{H}_{rot}^{(k)}{}'t\}\,\hat{\rho}_{rot}(0)\,\exp\{(i/\hbar)(\hat{H}_{rot}^{(k)}{}')^\dagger t\}. \tag{9}$$

For a radical pair born in the singlet state $\hat{\rho}_{rot}(0)$ is given by

$$\hat{\rho}_{rot}(0) = \hat{P}^S/Tr\hat{P}^S, \tag{10}$$

where the trace Tr... extends over all electronic and nuclear spin states.

Let us suppose that the RYDMR spectrum is recorded at a time t after creation of the radical pairs by monitoring the yield Φ_T of triplet recombination products as a function of the strength B_0 of the static magnetic field in the presence of microwaves of field strength B_1. For a single nuclear spin configuration k and a definite orientation of the spin dipole tensor **D** with respect to $\underline{B}_0$ (angles θ and ψ in Eqn. 5) the triplet yield can be calculated from [9] as follows (46,50):

$$\Phi_T^{(k)}(B_0,B_1,t;\theta,\psi) = k_T \int_0^t \text{Tr}\{\hat{P}^T \hat{\rho}_{rot}^{(k)}(B_0,B_1,t';\theta,\psi)\} \, dt'. \tag{11}$$

The dependences of Φ_T and $\hat{\rho}_{rot}^{(k)}$ on the parameters B_0,B_1,t,θ,ψ and on the label k are explicitly displayed in [11]. In general, the triplet yield has to be averaged over suitable distributions of the nuclear spin configuration (k) and of the orientations θ and ψ yielding the averaged triplet yield (50)

$$\Phi_T(B_0,B_1,t) = \langle \Phi_T^{(k)}(B_0,B_1,t;\theta,\psi) \rangle_{k,\theta,\psi}, \tag{12}$$

where the brackets $\langle...\rangle_{k,\theta,\psi}$ symbolize the averaging procedure. Experimentally, the absolute triplet yield [12] is difficult to determine. A more convenient quantity is the "relative triplet yield" defined as the ratio of the triplet yield with and without microwaves applied to the sample:

$$\text{Relative Triplet Yield} = \Phi_T(B_0,B_1,t)/\Phi_T(B_0,0,t). \tag{13}$$

To actually calculate the triplet yield, we first have to choose a set of spin states that forms the basis for a matrix representation of the spin operators. Since the nuclear spin configuration is a constant of motion we only have to find a basis for a system of two coupled electron spins. A suitable choice of basis set will be $\{|S\rangle, |T_+\rangle, |T_0\rangle, |T_-\rangle\}$ where the triplet states are quantized along $\underline{B}_0$. In this basis the spin Hamiltonian $\hat{H}_{rot}^{(k)\prime}$ is represented by the 4x4 matrix (50)

	$|T_+\rangle$	$\|S\rangle$	$|T_0\rangle$	$|T_-\rangle$
$|T_+\rangle$	$g\beta B_z^{(k)} + D_{zz}/2 - i\hbar k_T/2$		$g\beta B_1$	
$\|S\rangle$		$-J - i\hbar k_S/2$	$a^{(k)}$	
$|T_0\rangle$	$g\beta B_1$	$a^{(k)}$	$-D_{zz} - i\hbar k_T/2$	$g\beta B_1$
$|T_-\rangle$			$g\beta B_1$	$-g\beta B_z^{(k)} + D_{zz}/2 - i\hbar k_T/2$

$$. \tag{14}$$

In fact, the whole task of calculating the triplet yield can be reduced to the solution of a 4x4 problem. Marginally, we note that the coupling scheme expressed in [14] forms the basis for the illustration of the RYDMR effect in Fig. 1.

In spite of the low dimensionality of the problem the calculations leading to the triplet yield are very extensive. In most cases they cannot be done without a computer. Suppose, however, that ISC induced by the matrix element $a^{(k)}$ in [14] can be treated as a small perturbation of the spin dynamics of the radical pair. In that case perturbation theory can provide useful analytical expressions for the triplet yield (50). Making the simplifying assumption that the spin dipole interaction has negligible influence on the spin dynamics ($D_{zz} \rightarrow 0$) the relative triplet yield to lowest order in $a^{(k)}$ is given by (50,52,53):

$$\frac{\Phi_T(B_0, B_1)}{\Phi_T(B_0, 0)} = 1 + \frac{1}{2} \frac{B_1^2}{B_{eff}^2} \left(\frac{J^2 + \kappa^2}{(J + g\beta B_{eff})^2 + \kappa^2} + \frac{J^2 + \kappa^2}{(J - g\beta B_{eff})^2 + \kappa^2} - 2 \right). \qquad [15]$$

B_{eff} is the resultant magnetic field in the rotating frame,

$$B_{eff} = [B_1^2 + (B_0 - \hbar\omega/(g\beta))^2]^{1/2}, \qquad [16]$$

whereas $\kappa/\hbar$ is the arithmetic mean of the recombination rates,

$$\kappa = \frac{\hbar}{2} (k_S + k_T). \qquad [17]$$

The relative triplet yield [15] is the ratio of the accumulated triplet yields for $t \rightarrow \infty$.

The perturbation theory approach has turned out to be highly useful for qualitative studies of the RYDMR effect (50,53). In addition, a quantitative analysis of RYDMR signals from the radical ion pair state P^F in reaction centers from *Rhodobacter sphaeroides* was successfully based on Eqn. 15 (52) suggesting that the conditions of small $a^{(k)}$ and small D_{zz} required for the application of [15] are approximately fulfilled in this case.

4.2 Experimental techniques

RYDMR spectra of the radical ion pair state P^F in reaction centers from *Rhodobacter sphaeroides* have been detected optically by monitoring the effect of resonant microwaves on either the triplet recombination yield (6,57-59) or the lifetime of P^F (60,61). In both cases detection of the RYDMR signal involves observation of the resonant variation of light-induced absorbance changes of the sample at some suitable wavelength. Time-resolved (6,59-61) and quasi-stationary (50,58) versions of the experiment have been realized. To illustrate the RYDMR technique, let us look in some detail at one particular experimental set-up (Fig. 4). This set-up was used to measure the time-resolved RYDMR spectra of quinone-depleted reaction centers from *Rhodobacter sphaeroides*, strain R-26, shown in Fig. 5.

Except for the microwave part the technique to be described is a typical pump-probe sequence well known from time-resolved optical spectroscopy. In the set-up depicted in Fig. 4 radical pairs are created by exciting the dimer of the reaction center in the Q_x-band (at 600 nm (31)) with a 15 ns laser pulse from an excimer-pumped dye laser (dye laser 1, pulse energy $\simeq$ 30 μJ). The output of a second, synchronously pumped dye laser (dye laser 2) is optically delayed and passes through the sample about 100 ns after the first pulse. The second pulse, the energy of which is orders of magnitude smaller than that of the first, serves to probe the amount of triplet recombination products by monitoring the bleaching of the Q_y-band of the dimer (at 865 nm (31)) at a time when all the radical pairs created by the first pulse have recombined but before the triplet state of the dimer has

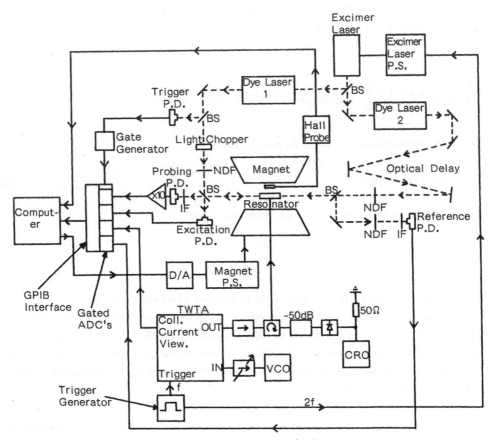

Fig. 4. Block diagram of an experimental set-up used for time-resolved RYDMR measurements. Abbreviations: P.S.=Power Supply; P.D.=Photo-Detector; NDF=Neutral Density Filter; IF=Interference Filter; BS=Beamsplitter; f=repetition frequency of the microwave pulses. Dashed lines represent light paths, full lines show electrical connections. For details of the experimental procedure, see text.

decayed to the ground state (cf. Fig. 3). During the 10–20 ns lifetime of the radical pairs microwaves supplied by a solid state device and amplified to an output level of up to 1 kW by a pulsed TWT amplifier (pulse width $\simeq 2\mu s$) are directed onto the sample. The microwave resonator used in the set-up is of a loop-gap type resonating around 3 GHz (59) (for a review of applications of loop-gap resonators see the chapter by J. Hyde and W. Froncisz in this book). To obtain the RYDMR spectrum, the signal from the probing photo-detector is recorded with gated ADC's while the static magnetic field is varied stepwise across the resonance.

In fact, the actual experimental procedure is a bit more complicated. To eliminate drifts, a double modulation technique is used. By means of a light chopper the excitation light is periodically turned on and off. The corresponding signals from the probing photo-detector are subtracted from each other in the computer to yield the "difference extinction" of the sample, i.e. the change of the absorbance of the sample at 865 nm upon optical excitation. In addition to modulating the excitation light, the microwaves are turned on and off on alternate laser shots such that the difference extinction is measured alternately with and without microwaves applied to the sample. (The state of the microwaves is monitored via one of the gated ADC's as shown in Fig. 4.) The ratio of the respective difference extinctions of the sample with and without microwaves is a direct measure of the relative triplet yield (Eqn. 13). To compensate for energy fluctuations of the laser pulses the output energies of the two dye lasers are monitored by the excitation and the reference photo detector, respectively. In the computer the signal from the probing photo-detector is suitably normalized to these energies. Data acquisition, signal averaging and control of the magnetic field is performed by the computer. The repetition rate of the experiment ($f \simeq 10$ Hz) is determined by the free-running trigger generator depicted in the lower left part of Fig. 4.

Experimental set-ups comparable to the one described have been used by other groups to record time-resolved RYDMR spectra of either the state P^F in bacterial reaction centers (6,61) or radical pairs in micellar solution (28).

4.3 Discussion of experimental results for reaction centers
from the purple bacterium *Rhodobacter sphaeroides*

Up to now, applications of the RYDMR technique to the radical ion pair state P^F have been confined to reaction centers from the purple bacterium *Rhodobacter sphaeroides* (wild type and strain R-26, respectively). In this subsection we will review some of the results and discuss them in the light of the RYDMR theory developed above.

RYDMR spectra of quinone-depleted reaction centers from *Rhodobacter sphaeroides* were first measured by Norris and collaborators (6,57). It was shown that application of a resonant microwave field results in an enhancement of the triplet yield of up to 8% at low microwave powers (57). At high microwave powers, however, the signal was observed to invert corresponding to a decrease of the triplet yield in the presence of microwaves (57). In fact, the shape of the RYDMR spectrum exhibits a complex dependence on the microwave field strength as was predicted theoretically (53,54) and verified only recently by the sequence of high resolution spectra shown in Fig. 5 (59).

In section 2 we gave a reasoning for the enhancement of the triplet yield at low microwave powers. It is not applicable here, however, since it requires the $|S\rangle \longleftrightarrow |T_0\rangle$ splitting and the lifetime broadening of the spin states to be small compared to the ISC matrix element, both of which conditions are not fulfilled in reaction centers from *Rhodobacter sphaeroides*. On the contrary, the $|S\rangle \longleftrightarrow |T_0\rangle$ splitting determined by the exchange and the magnetic spin dipole interaction and the lifetime broadening of the spin states determined by k_S and k_T turn out to be comparable to or even larger than the ISC matrix element governed by the hyperfine interaction and the Δg-effect. In this case the appropriate explanation for the resonant increase of the triplet yield at low microwave powers is the following (53). In strong static magnetic fields $|T_0\rangle$ is the only triplet sublevel interacting with the singlet precursor state of the radical pair. The triplet yield therefore depends strongly on the size of the $|S\rangle \longleftrightarrow |T_0\rangle$ energy gap, which acts as a barrier for $|S\rangle \rightarrow |T_0\rangle$ transitions. By coupling $|T_0\rangle$ with $|T_+\rangle$ and $|T_-\rangle$ the microwave field effects a distribution of $|T_0\rangle$-character over all three triplet states. As a consequence for some portion of $|T_0\rangle$ the energy barrier to the singlet state may be lowered to such an extent that the triplet yield as a whole increases beyond its value in the absence of microwaves. This, for instance, is the case at field strengths B_0 where $|T_+\rangle$ (or $|T_-\rangle$) becomes degenerate with the singlet state (in the rotating frame) such that transitions from $|S\rangle$ to the portion of $|T_0\rangle$ that is mixed into $|T_+\rangle$ (or $|T_-\rangle$) become isoenergetic. (These arguments apply as long as neither the microwave field strength nor the lifetime broadening of the spin states does exceed the value of the $|S\rangle\longleftrightarrow|T_0\rangle$ energy gap determined at $B_1=0$.)

To understand the decrease of the triplet yield at high microwave powers, we again have to look closely at the energetics of the spin states in the rotating frame. Let us take the microwave field $\underline{B}_1$ to point along the x-axis in the rotating frame and let us quantize the triplet spin states along $\underline{B}_1$ instead of along $\underline{B}_0$. This is a natural thing to do at high microwave powers where at resonance the Zeeman interaction of the spins with the microwave field becomes the dominant interaction in the rotating frame ($g\beta B_1 \gg |a^{(k)}|, |J|, |D_{zz}|$). By this change of quantization axis, the triplet sublevel $|T_0\rangle_{\underline{B}_0}$ quantized along $\underline{B}_0$ is decomposed into the states $|T_+\rangle_{\underline{B}_1}$ and $|T_-\rangle_{\underline{B}_1}$ quantized along $\underline{B}_1$. This means that to go over from the initial singlet state of the radical pair to the triplet sublevel $|T_0\rangle_{\underline{B}_0}$ (which at high static magnetic field strengths is the only triplet sublevel populated from the singlet state) an energy gap proportional to the Zeeman energy $g\beta B_1$ of the states $|T_\pm\rangle_{\underline{B}_1}$ in the microwave field has to be overcome by the singlet-triplet mixing mechanism. This leads to a decrease of the triplet yield with increasing B_1 completely analogous to the decrease of the triplet yield with increasing B_0 at low static magnetic field strengths discussed in section 2.

Details of the structure of the RYDMR spectrum as a function of the microwave field strength can be well understood in terms of the theory outlined above. By simulating RYDMR spectra values can be assigned to the parameters entering the theory. Fig. 5 shows results of a thorough study of the RYDMR lineshape as a function of the microwave field strength for quinone-depleted reaction centers from *Rhodobacter sphaeroides*, strain R-26. Due to the large number of nuclei contributing to the hyperfine interaction in the radical pair state P^F the spectrum does not exhibit any resolved hyperfine structure. Splittings due to the exchange or the magnetic spin dipole interaction between the unpaired electrons are also absent from the spectrum, which thus consists of a single broad line with variable shape depending on the microwave field strength. The full curves represent theoretical fits to the spectra carried out with the formalism developed in Eqns. 2-14. Let us look at some details of the simulation procedure.

In Eqn. 12 the hyperfine fields $2(\sum_n^{(i)} A_n m_n^{(k)})$ of the two radicals (i=1,2) were averaged over Gaussian distributions with second moments of 0.95 mT and 1.3 mT for $BChl_2^+$ and BPh_A^-, respectively, corresponding to the EPR linewidths of these radicals (62,63). The g-values $g_{BChl_2^+}=2.0026$ and $g_{BPh_A^-}=2.0036$ were taken from the literature (62,63). The zero field parameters D and E of the magnetic spin dipole interaction [5] were evaluated from crystal structure data (64) and corresponding calculated spin densities (65) for the reaction center from *Rhodopseudomonas viridis* (which is homologous to the one from *Rhodobacter sphaeroides* (37)) resulting in D=-0.52 mT and |E|=0.005 mT. According to previous experiences in simulating RYDMR spectra for the state P^F (50,55) such a small

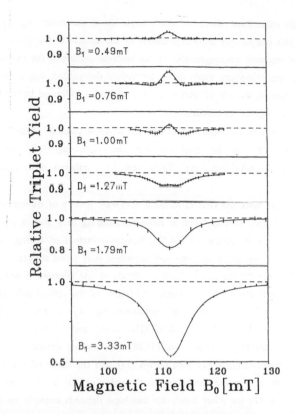

Fig. 5. Experimental RYDMR spectra of quinone-depleted reaction centers from *Rhodobacter sphaeroides*, strain R-26, for different microwave field strengths B_1 at a sample temperature of 276±1 K. Reaction centers suspended in Tris buffer (pH 8.0, 0.1% NP40) were adjusted to an optical density of 1.1 at 865 nm. The spectra were taken with the RYDMR spectrometer of Fig. 4 in the way described in the text (microwave frequency: 3.1 GHz). Each data point constitutes an average of 1000 measurements (error bars=95% confidence limits). The full curves are theoretical fits to the data. Each spectrum was adjusted separately by varying the fit parameters B_1, k_S, k_T and J. The following sets of parameters were finally used to simulate the spectra (from top to bottom; for comparison, the experimentally determined values of B_1 are also listed):

B_1=0.49 mT (B_1^{exp}=0.57±0.17 mT), k_S=0.050 ns^{-1}, k_T=0.40 ns^{-1}, J=1.26 mT;
B_1=0.76 mT (B_1^{exp}=0.93±0.28 mT), k_S=0.050 ns^{-1}, k_T=0.40 ns^{-1}, J=1.21 mT;
B_1=1.00 mT (B_1^{exp}=1.13±0.34 mT), k_S=0.067 ns^{-1}, k_T=0.43 ns^{-1}, J=1.03 mT;
B_1=1.27 mT (B_1^{exp}=1.28±0.38 mT), k_S=0.055 ns^{-1}, k_T=0.55 ns^{-1}, J=0.90 mT;
B_1=1.79 mT (B_1^{exp}=1.30±0.50 mT), k_S=0.056 ns^{-1}, k_T=0.49 ns^{-1}, J=0.89 mT;
B_1=3.33 mT (B_1^{exp}=3.32±1.00 mT), k_S=0.072 ns^{-1}, k_T=0.49 ns^{-1}, J=1.25 mT.

dipole interaction has negligible influence on the amplitude and the shape of spectra from disordered samples. This interaction was therefore disregarded in the simulations.

Each of the spectra in Fig. 5 was adjusted separately by optimizing the parameters k_S, k_T, J and B_1 in a least squares fit procedure (the microwave field strength B_1 was varied only within the uncertainty limits of its experimental determination). The fits obtained were not unique, however. For instance, the spectrum at the lowest microwave power could be simulated equally well with the set of parameters B_1=0.29 mT, k_S=0.076 ns^{-1}, k_T=0.21 ns^{-1}, $|J|$=0.90 mT and the one indicated in the caption to Fig. 5. On the other hand, there was not a single set of parameters k_S, k_T and J fitting all the spectra at a time. This can be due to a) degradation of the sample in the course of the measurements, b) the neglect of intrinsic inhomogeneities of the reaction centers as for example a distribution of the value of J or c) approximations made in the theoretical description, e.g. neglect of anisotropic interactions. As a compromise, the sets of parameters k_S, k_T and J indicated in the caption to Fig. 5 were chosen to lie as close together as possible.

According to the numerical fits described above, the smallest possible ranges to which the free parameters k_S, k_T and J can be confined are the following: 0.05 ns^{-1}≤k_S≤0.07 ns^{-1}, 0.4 ns^{-1}≤k_T≤0.6 ns^{-1} and 0.9 mT≤$|J|$≤1.3 mT. These can be compared with results from earlier RYDMR measurements.

Norris and collaborators were the first to measure RYDMR spectra of quinone-depleted reaction centers (6,57). To evaluate their spectra taken at room temperature they used a theoretical treatment analogous to the one outlined in section 4.1 (54,57). However, instead of fitting the RYDMR lineshapes at various microwave powers in the way just described they fitted selected properties of these spectra (e.g. the linewidth and the maximum signal amplitude at low microwave powers) together with a) data from MARY experiments and b) the measured lifetime of the radical pair state P^F (57). They arrived at the following limits for the free parameters k_S, k_T, J and D of their fit (57): 0.05 ns^{-1}≤k_S≤0.09 ns^{-1}, 0.5 ns^{-1}≤k_T≤0.6 ns^{-1}, 1.2 mT≤$|J|$≤2.0 mT and 4.0 mT≤$|D|$≤6.0 mT. The latter range of $|D|$ has been questioned in later work (53,58) even before detailed crystal structure data have become available. Specifically, it has been shown (53) that the shape of the RYDMR spectrum averaged over all orientations is rather insensitive to the value of $|D|$ if the other parameters lie in the ranges indicated above. On the other hand, the lineshape depends strongly on the orientation of the principal axes of the spin dipole tensor with respect to the external magnetic field (53). Therefore, even in the absence of crystal structure data, accurate information on $|D|$ should be obtainable from linear dichroic measurements of the RYDMR signal. For reaction centers, however, such measurements have not been carried out so far.

Hoff and coworkers measured quasi-stationary RYDMR spectra of quinone-depleted reaction centers at various temperatures (58) using low microwave powers and low static magnetic field strengths. They found that the shape of the spectrum was practically constant over the temperature range 160K-295K. To evaluate their spectra they used the perturbation theory approach outlined in section 4.1 (Eqns. 15-17). By fitting the spectrum taken at 211K (fit parameters: B_1, k_S+k_T and J; $|D|$ was estimated to be small, $|D|$≤2.0 mT) they obtained the following limits (55): 0.25 ns^{-1}≤k_S+k_T≤0.27 ns^{-1} and 0.96 mT≤$|J|$≤1.06 mT. Here the sum k_S+k_T of the recombination rates comes out a factor of two smaller than in the sets of parameters cited above. Among other things this can be due to a) different measuring conditions (quasi-stationary versus time-resolved measurements (50); application of low static magnetic fields (B_0<20 mT) versus application of high static magnetic fields (B_0>100 mT)), b) different formalisms used to fit the spectra (perturbation theory approach versus full theoretical treatment), c) ambiguity of the resulting parameter values (see above) or d) different characteristics of the samples as a result of e.g. different aging conditions (67).

In spite of small discrepancies concerning the recombination rates and the spin dipole interaction all three RYDMR experiments discussed above yielded remarkably similar results for the (absolute value of) the exchange interaction J in quinone-depleted reaction centers. This indicates the special sensitivity of the RYDMR method to this parameter (50,54,57). Another way to obtain information on the value of $|J|$ in a RYDMR experiment is to measure the relative triplet yield as a function of B_1 for fixed B_0 ("B_1-spectrum") instead of as a function of B_0 for fixed B_1 ("RYDMR spectrum") (53,54,56,57). If, as in Fig. 5, the triplet yield in the center of resonance ($B_0=\hbar\omega/g\beta$) increases at low microwave powers and decreases at high powers the corresponding B_1-spectrum exhibits a maximum. This indicates that with increasing B_1 one of the triplet states becomes degenerate with the singlet state in the rotating frame resulting in a maximum enhancement of the triplet yield. The position of the maximum is determined by the singlet-triplet splitting in the rotating frame at $B_1=0$ mT, which - in the center of resonance and for vanishing spin dipole interaction   is given by J, provided that contributions from the hyperfine interaction can be neglected to lowest order as seems to be the case for reaction centers from *Rhodobacter sphaeroides*. Thus, by measuring the value of B_1 corresponding to the maximum enhancement of the triplet yield an estimate (more precisely: a lower bound) for the value of $|J|$ can be obtained (53,54,56). For quinone-depleted reaction centers from *Rhodobacter sphaeroides*, strain R-26, the maximum of the B_1-spectrum was found to lie at $B_1=1.3\pm0.2$ mT (68) indicating that $|J|\geq1.3\pm0.2$ mT.

For substantial strengths of the magnetic spin dipole interaction also the sign of J can be determined in a RYDMR experiment (relative to the sign of D). In the case of reaction centers from *Rhodobacter sphaeroides*, however, the smallness of $|D|$ defeats its determination. Information on the recombination rates k_S and k_T has also been extracted from MARY experiments and on the parameters J and D from MARY and ESP experiments. For a review of results of these experiments and a comparison with RYDMR results see Ref. 66.

For the interpretation of the parameters k_S, k_T and J in the context of electron transfer theory (section 4.4) it is of special interest to know their values as a function of temperature. The temperature dependence of the RYDMR signal of quinone-depleted reaction centers from *Rhodobacter sphaeroides*, strain R-26, has been studied over the range 160K - 295K by Hoff and coworkers (58) and over the range 220K - 290K by Norris and coworkers (69). The signal intensity was observed to vary inversely with temperature, whereas, as mentioned above, the lineshape at low microwave powers turned out to be practically temperature independent (58). It was concluded that the exchange interaction does not vary with temperature and that the change of the signal intensity can be related to an increase of the absolute triplet yield at lower temperatures (58), which in turn can be traced back to an activated behaviour of the singlet recombination rate k_S (70-72).

A completely different temperature dependence of the RYDMR signal was found for prereduced reaction centers from *Rhodobacter sphaeroides*, strain R-26, (55,58). Between 236K and 293K the RYDMR signal inverts as a function of temperature as shown in Fig. 6. To interpret this sequence of spectra it was necessary to assume a) that the absolute value of the exchange interaction drops continuously from about 0.93 mT at 236K to about 0.33 mT at 293K (55) and b) that the sum k_S+k_T of the recombination rates increases from about 0.23 ns^{-1} at 236K to about 0.38 ns^{-1} at 279K and decreases for higher temperatures to about 0.30 ns^{-1} at 293K (55).

So far, we have shown that the RYDMR experiment can provide insight into spin selective recombination pathways and magnetic interactions in the radical pair state P^F. The question is now:

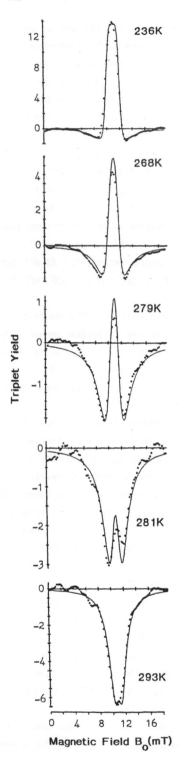

Triplet Yield (y-axis)

Magnetic Field B_0 (mT)

Fig. 6. Temperature dependence of the RYDMR signal of prereduced reaction centers from *Rhodobacter sphaeroides*, strain R-26. The spectra were taken at low microwave powers (≤ 40 W) under conditions of continuous photolysis with the RYDMR spectrometer described in Ref. 58. Lines: Theoretical RYDMR spectra calculated from [15] in the way detailed in Ref. 55. From (55).

What kind of information can be extracted from the parameters k_S, k_T and J obtained in a RYDMR experiment? In the following subsection we shall mention some conclusions that have been based upon the knowledge of (the temperature dependences of) these parameters and relate to a) mechanisms and energetics of the electron transfer steps populating and depleting the state P^F and b) conformational changes of the reaction center protein as a function of temperature.

4.4 What can we learn from the parameters J, k_S and k_T?

One of the most challenging problems in bacterial photosynthesis is the mechanism of the primary electron transfer from the excited dimer $^1BChl_2^*$ to the first spectroscopically resolved acceptor molecule BPh_A. The time scale of this electron transfer over a center-to-center distance of 17 Å is limited to several picoseconds in order to compete favourably with the back transfer of the excitation energy to the antenna and its loss due to other radiative and radiationless decay routes. All the mechanisms proposed so far attribute a central role to the accessory bacteriochlorophyll $BChl_A$ that - according to the x-ray structure (Fig. 2) - is located between the dimer and BPh_A. Among several mechanisms advanced for the primary electron transfer are the two-step sequential electron transfer (47,73,74)

$$^1BChl_2^* \, BChl_A \, BPh_A \xrightarrow{k_1} BChl_2^+ \, BChl_A^- \, BPh_A \xrightarrow{k_2}$$

$$\xrightarrow{k_2} BChl_2^+ \, BChl_A \, BPh_A^-, \tag{18}$$

where $BChl_2^+ \, BChl_A^- \, BPh_A$ acts as a transient ionic state and $k_2 \gg k_1$ since no intermediate ion pair $BChl_2^+ \, BChl_A^-$ is detectable (75), and a one-step superexchange mechanism (42,76)

$$^1BChl_2^* \, BPh_A \xrightarrow[BChl_A]{k} BChl_2^+ \, BPh_A^-, \tag{19}$$

where $BChl_A$ assists in the electron transfer by enhancing the electronic coupling between $^1BChl_2^*$ and $BChl_2^+$ BPh_A^-.

The information contained in J and k_T can be used to distinguish between different models. For example, the temperature dependence of J provides information on the potential prevalence of the two-step mechanism [18] by yielding limits on the energy level of the intermediate ion pair state $BChl_2^+$ $BChl_A^-$ BPh_A (77,78). Applied to the reaction center of *Rhodobacter sphaeroides*, the temperature independence of J (55,70) disfavours a sequential two-step mechanism (77,78).

The different absolute values and temperature dependences of k_S and k_T can be rationalized in terms of the different energetics involved in the recombination from the singlet- and the triplet-phased radical pair state P^F, respectively. Specifically, the singlet recombination process leading to the reaction center ground state proceeds either directly in a highly exothermic reaction or via the excited singlet state $^1BChl_2^*$ of the dimer, the latter constituting a highly activated recombination pathway. In contrast, triplet recombination to the triplet state $^3BChl_2^*$ of the dimer is an almost activationless process involving a much smaller energy gap than the singlet recombination process (0.17 eV versus 1.14 eV at room temperature) (72,79-81).

The decrease of $|J|$ with increasing temperature observed for prereduced reaction centers from *Rhodobacter sphaeroides* (section 4.3) has been interpreted in terms of temperature dependent structural changes (55). These can modify the value of $|J|$ by changing the electronic coupling between $^1BChl_2^*$ ($^3BChl_2^*$) and $^1P^F$ ($^3P^F$) and/or the energy differences between the relevant states (78,82).

In conclusion, these short remarks show that the localized radical pair state P^F in bacterial reaction centers provides a salient example of how RYDMR can contribute to the understanding of electron transfer mechanisms and of structural rearrangements induced by the variation of external parameters, as e.g. temperature. The encouraging results obtained so far may stimulate comparative RYDMR investigations of reaction centers from other species and also of rigid model compounds specifically designed to mimic the photosynthetic charge separation.

ACKNOWLEDGEMENT

We are very grateful to Dipl. Phys. E. Lang for his assistance in measuring and interpreting the RYDMR spectra of Fig. 5. We are also highly indebted to Prof. K. Möbius and Dr. F. Lendzian for helping us with the design and the development of the microwave part of our RYDMR spectrometer (Fig. 4) and to Prof. M. Bixon for stimulating discussions. Financial support by the Deutsche Forschungsgemeinschaft (SFB 143) is gratefully acknowledged.

REFERENCES

1 W.W. Parson and B. Ke, in: Govindjee (Ed.), Photosynthesis. Volume 1: Energy Conversion by Plants and Bacteria, Acad. Press, New York, 1982, pp. 331-385.
2 S.I. Kubarev and E.A. Pshenichnov, Chem. Phys. Lett., 28 (1974) 66-67.
3 E.L. Frankevich and A.I. Pristupa, Pisma Zh. Eksp. Teor. Fiz., 24 (1976) 397-400.
4 E.L. Frankevich and S.I. Kubarev, in: R.H. Clarke (Ed.), Triplet State ODMR Spectroscopy, John Wiley, New York, 1982, pp. 137-183.
5 Yu.N. Molin (Ed.), Spin Polarisation and Magnetic Effects in Radical Reactions, Elsevier, Amsterdam, 1984.

6 M.K. Bowman, D.E. Budil, G.L. Closs, A.G. Kostka, C.A. Wraight and J.R. Norris, Proc. Natl. Acad. Sci. USA, 78 (1981) 3305-3307.

7 L.T. Muus, P.W. Atkins, K.A. McLauchlan and J.B. Pedersen (Eds.), Chemically Induced Magnetic Polarization, Reidel, Dordrecht, 1977.

8 B. Brocklehurst, Nature, 221 (1969) 921-923.

9 H. Hayashi and S. Nagakura, Bull. Chem. Soc. Jpn., 57 (1984) 322-328.

10 W. Lersch, A. Ogrodnik and M.E. Michel-Beyerle, Z. Naturforsch., 37a (1982) 1454-1456.

11 E.L. Frankevich, A.I. Pristupa and V.I. Lesin, Chem. Phys. Lett., 47 (1977) 304-308.

12 D. Lepine and J.J. Prejean, Proc. 10th Int. Conf. Phys. Semicond., Boston, 1970.

13 D. Lepine, Phys. Rev. B, 6 (1972) 436-441.

14 B.C. Cavenett, Adv. Phys., 30 (1981) 475-538.

15 A.D. Trifunac and R.G. Lawler, Magn. Res. Rev., 7 (1982) 147-174.

16 S.N. Smirnov, V.A. Rogov, A.S. Shustov, S.V. Sheberstov, N.V. Panfilovitch, O.A. Anisimov and Yu.N. Molin, Chem. Phys., 92 (1985) 381-387.

17 V.O. Saik, O.A. Anisimov and Yu.N. Molin, Chem. Phys. Lett., 116 (1985) 138-141.

18 A.B. Doktorov, O.A. Anisimov, A.I. Burshtein and Yu.N. Molin, Chem. Phys., 71 (1982) 1-8.

19 E.G. Bagryanskaya, Yu.A. Grishin and R.Z. Sagdeev, Chem. Phys. Lett., 113 (1985) 234-237.

20 E.G. Bagryanskaya, Yu.A. Grishin, R.Z. Sagdeev and Yu.N. Molin, Chem. Phys. Lett., 114 (1985) 138-142.

21 E.G. Bagryanskaya, Yu.A. Grishin, R.Z. Sagdeev, T.V. Leshina, N.E. Polyakov and Yu.N. Molin, Chem. Phys. Lett., 117 (1985) 220-223.

22 E.G. Bagryanskaya, Yu.A. Grishin, R.Z. Sagdeev and Yu.N. Molin, Chem. Phys. Lett., 128 (1986) 417-419.

23 E.G. Bagryanskaya, Yu.A. Grishin, N.I. Avdievitch, R.Z. Sagdeev and Yu.N. Molin, Chem. Phys. Lett., 128 (1986) 162-167.

24 S.A. Mikhailov, K.M. Salikhov and M. Plato, Chem. Phys., 117 (1987) 197-217.

25 M. Okazaki and T. Shiga, Nature, 323 (1986) 240-243.

26 M. Okazaki, S. Sakata, R. Konaka and T. Shiga, J. Chem. Phys., 86 (1987) 6792-6800.

27 M. Okazaki, T. Shiga, S. Sakata, R. Konaka and K. Toriyama, J. Phys. Chem., 92 (1988) 1402-1404.

28 A.I. Grant, K.A. McLauchlan and S.R. Nattrass, Mol. Phys., 55 (1985) 557-569.

29 N.J. Turro, Proc. Natl. Acad. Sci. USA, 80 (1983) 609-621.

30 E.L. Frankevich, D.I. Kadyrov, I.A. Sokolik, A.I. Pristupa, V.M. Kobryanskii and N.Y. Zurabyan, Phys. Stat. Sol. (B), 132 (1985) 283-294.

31 M. Okamura, G. Feher and N. Nelson, in: Govindjee (Ed.), Photosynthesis. Volume 1: Energy Conversion by Plants and Bacteria, Acad. Press, New York, 1982, pp. 195-272.

32 D.W. Reed and R.K. Clayton, Biochim. Biophys. Res. Commun., 30 (1968) 471-475.

33 H. Michel, J. Mol. Biol., 158 (1982) 567-572.

34 J.P. Allen and G. Feher, Proc. Natl. Acad. Sci. USA, 81 (1984) 4795-4799.

35 P. Gast and J.R. Norris, FEBS Lett., 177 (1984) 277-280.

36 J. Deisenhofer, O. Epp, K. Miki, R. Huber and H. Michel, Nature, 318 (1985) 618-624.

37 J.P. Allen, G. Feher, T.O. Yeates, H. Komiya and D.C. Rees, Proc. Natl. Acad. Sci. USA, 84 (1987) 5730-5734.

38 J.P. Allen, G. Feher, T.O. Yeates, H. Komiya and D.C. Rees, Proc. Natl. Acad. Sci. USA, 84 (1987) 6162-6166.

39 J.R. Norris, R.A. Uphaus, H.L. Crespi and J.J. Katz, Proc. Natl. Acad. Sci. USA, 68 (1971) 625-628.

40 G. Feher, A.J. Hoff, R.A. Isaacson and L.C. Ackerson, Ann. N.Y. Acad. Sci., 244 (1975) 239-259.

41 W.W. Parson, Ann. Rev. Biophys. Bioeng., 11 (1982) 57-80.

42 M.E. Michel-Beyerle, M. Plato, J. Deisenhofer, H. Michel, M. Bixon and J. Jortner, Biochim. Biophys. Acta, 932 (1988) 52-70.

43 M.G. Rockley, M.W. Windsor, R.J. Cogdell and W.W. Parson, Proc. Natl. Acad. Sci. USA, 72 (1975) 2251-2255.

44 K.J. Kaufmann, P.L. Dutton. T.L. Netzel, J.S. Leigh and P.M. Rentzepis, Science, 188 (1975) 1301-1304.

45 J.-L. Martin, J. Breton, A.J. Hoff, A. Migus and A. Antonetti, Proc. Natl. Acad. Sci. USA, 83 (1986) 957-961.

46 J.-L. Martin, J. Breton, J.C. Lambry and G. Fleming, in: J. Breton and A. Vermeglio (Eds.), The Photosynthetic Bacterial Reaction Center - Structure and Dynamics, NATO ASI Series, Series A: Life Sciences, Vol. 149, Plenum Press, New York, 1988, pp. 195-203.

47   R. Haberkorn, M.E. Michel-Beyerle and R.A. Marcus, Proc. Natl. Acad. Sci. USA, 76 (1979) 4185-4188.
48   A.J. Hoff, Quart. Rev. Biophys., 14 (1981) 599-665.
49   A.J. Hoff, Quart. Rev. Biophys., 17 (1984) 153-282.
50   W. Lersch, Thesis, TU München, 1987.
51   G.E. Pake and T.L. Estle, The Physical Principles of Electron Paramagnetic Resonance, W.A. Benjamin, Reading/Mass., 1973.
52   R. Kubo, Adv. Chem. Phys., 15 (1969) 101-127.
53   W. Lersch and M.E. Michel-Beyerle, Chem. Phys., 78 (1983) 115-126.
54   J. Tang and J.R. Norris, Chem. Phys. Lett., 94 (1983) 77-80.
55   D.A. Hunter, A.J. Hoff and P.J. Hore, Chem. Phys. Lett., 134 (1987) 6-11.
56   W. Lersch and M.E. Michel-Beyerle, Chem. Phys. Lett., 136 (1987) 346-351.
57   J.R. Norris, M.K. Bowman, D.E. Budil, J. Tang, C.A. Wraight and G.L. Closs, Proc. Natl. Acad. Sci. USA, 79 (1982) 5532-5536.
58   K.W. Moehl, E.J. Lous and A.J. Hoff, Chem. Phys. Lett., 121 (1985) 22-27.
59   W. Lersch, F. Lendzian, E. Lang, R. Feick, K. Möbius and M.E. Michel-Beyerle, (J. Magn. Res.), in press.
60   M.R. Wasielewski, C.H. Bock, M.K. Bowman and J.R. Norris, J. Am. Chem. Soc., 105 (1983) 2903-2904.
61   M.R. Wasielewski, J.R. Norris and M.K. Bowman, Faraday Discuss. Chem. Soc., 78 (1984) 279-288.
62   J.D. McElroy, G. Feher and D.C. Mauzerall, Biochim. Biophys. Acta, 267 (1972) 363-374.
63   M.Y. Okamura, R.A. Isaacson and G. Feher, Biochim. Biophys. Acta, 546 (1979) 394-417.
64   J. Deisenhofer, private communication.
65   M. Plato, private communication.
66   A.J. Hoff, Photochem. Photobiol., 43 (1986) 727-745.
67   W. Lersch, A. Ogrodnik and M.E. Michel-Beyerle, in: A.H. Zewail (Ed.), Photochemistry and Photobiology. Volume 2, Harwood Acad. Publ., Chur, 1983, pp.987-993.
68   J.R. Norris, C.P. Lin and D.E. Budil, J. Chem. Soc., Faraday Trans. 1, 83 (1987) 13-27.
69   D.E. Budil, J.H. Tang and J.R. Norris, Biophys. J. (Abstr.), 49 (1986) 585a.
70   A. Ogrodnik, N. Remy-Richter, M.E. Michel-Beyerle and R. Feick, Chem. Phys. Lett., 135 (1987) 576-581.
71   D.E. Budil, S.V. Kolaczkowski and J.R. Norris, in: J. Biggens (Ed.), Progress in Photosynthesis Research, Volume 1. Proceedings of the 7th International Congress on Photosynthesis, 1986, Martinus Nijhoff Publ., Dordrecht, 1987, pp. 25-28.
72   A. Ogrodnik, M. Volk, R. Letterer, R. Feick and M.E. Michel-Beyerle, (Biochim. Biophys. Acta), in press.
73   R.A. Marcus, Chem. Phys. Lett., 133 (1987) 471-477.
74   S.V. Chekalin, Y.A. Matveetz, A.Y. Shkuropatov, V.A. Shuvalov and A.P. Yartzev, FEBS Lett., 216 (1987) 245-248.
75   G.R. Fleming, J.-L. Martin and J. Breton, Nature, 333 (1988) 190-192.
76   M. Bixon, J. Jortner, M. Plato and M.E. Michel-Beyerle, in: J. Breton and A. Vermeglio (Eds.), The Photosynthetic Bacterial Reaction Center - Structure and Dynamics, NATO ASI Series, Series A: Life Sciences, Vol. 149, Plenum Press, New York, 1988, pp.399-420.
77   M. Bixon, J. Jortner, M.E. Michel-Beyerle, A. Ogrodnik and W. Lersch, Chem. Phys. Lett., 140 (1987) 626-630.
78   M. Bixon, M.E. Michel-Beyerle and J. Jortner, (Israel J. Chem.), in press.
79   A. Ogrodnik, M. Volk and M.E. Michel-Beyerle, in: J. Breton and A. Vermeglio (Eds.), The Photosynthetic Bacterial Reaction Center - Structure and Dynamics, NATO ASI Series, Series A: Life Sciences, Vol. 149, Plenum Press, New York, 1988, pp. 177-184.
80   S.G. Boxer, R.A. Goldstein, D.J. Lockhart, T.R. Middendorf and L. Takiff, in: J. Breton and A. Vermeglio (Eds.), The Photosynthetic Bacterial Reaction Center - Structure and Dynamics, NATO ASI Series, Series A: Life Sciences, Vol. 149, Plenum Press, New York, 1988, pp. 165-176.
81   R.A. Goldstein, L. Takiff and S.G. Boxer, Biochim. Biophys. Acta, 934 (1988) 253-263.
82   M.E. Michel-Beyerle, M. Bixon and J. Jortner, Chem. Phys. Lett., 151 (1988) 188-194.

SPIN LABEL STUDIES OF THE STRUCTURE AND DYNAMICS OF LIPIDS AND PROTEINS IN MEMBRANES

DEREK MARSH and LÁSZLÓ I. HORVÁTH

1. MOTIONAL EFFECTS IN SPIN LABEL EPR SPECTRA

One of the principal uses of spin label EPR is in studying the mobility of nitroxide–labelled molecules. The sensitivity to rotational motion arises from modulation of the angular anisotropy of the hyperfine interaction of the unpaired electron spin ($S = 1/2$) with the nitrogen nuclear spin ($I = 1$), and of the Zeeman interaction of the unpaired electron spin with the static magnetic field, B, of the spectrometer. Both the hyperfine interaction, A, and the electron g–factor, g, are tensor quantities. If the rotational motion is of the appropriate time–scale, specified by the rotational correlation time, the modulation of the tensor anisotropies will be detected in the EPR spectrum.

1.1 Time–scale of nitroxide EPR spectroscopy

The positions of the EPR absorption lines of nitroxide free radicals depend on the orientation of the radical with respect to the spectrometer magnetic field, in a manner specified by the spin Hamiltonian tensors. To a reasonable approximation at a microwave frequency of 9 GHz, the angular dependence of the spin Hamiltonian for a nitroxide radical can be written as:

$$\mathcal{H} = g_0\beta B_z S_z + a_0 I_z S_z + \frac{1}{3}(3\cos^2\theta - 1)[\Delta g\beta B_z S_z + \Delta A I_z S_z] \quad [1]$$

where S and I denote the spin and nuclear angular momenta, respectively; $B = (0, 0, B_z)$ is the magnetic field vector, β is the Bohr magneton, and θ is the angle between the magnetic field direction and the nitroxide z–axis. Pseudosecular terms are neglected in Eqn. 1 and the hyperfine and g–tensors have been separated into an isotropic (a_0 and a_0) and anisotropic (Δg and ΔA) part. The spin Hamiltonian tensor values, $g = (g_{xx}, g_{yy}, g_{zz})$, $A = (A_{xx}, A_{yy}, A_{zz})$ and their isotropic and anisotropic parts, are given in Table 1 for different doxyl nitroxides in various host matrices. The values are all rather similar. However, the absolute values of the tensor elements (indicated by a_0 and g_0) vary somewhat from host to host, since these are dependent on the polarity of the environment in which the nitroxide is located.

In a rigid and randomly ordered environment the spectra from radicals with different orientations, θ, relative to the magnetic field direction are superimposed and a so–called powder spectrum is obtained. This is illustrated in Figs. 1a and 1c, for spectra of a semi–rigid system recorded at two different microwave frequencies. Maxima and minima are obtained at the turning points in the angular dependence of the spectra, corresponding to the principal values of the spin Hamiltonian tensors. At X–band microwave frequency ($B \approx 0.33T$), the anisotropy in the Zeeman interaction is relatively small and the spectrum is dominated by the hyperfine anisotropy alone (Fig. 1a). At Q–band microwave frequency ($B \approx 1.28T$), however, the anisotropy of the Zeeman interaction is comparable to that of the hyperfine interaction, and the turning points from the principal elements of the g–tensor are also resolved (Fig. 1c). Measurements at higher microwave frequencies, such as Q–band, allow determination of the full orientational properties of the spin label radical, since the g–tensor displays the full non–axial anisotropy, whereas the hyperfine tensor has only axial anisotropy.

TABLE 1

Spin Hamiltonian tensors for doxyl nitroxide free radicals. I:doxyl propane in tetra-methylcyclobutanedione (1). II: doxyl stearic acid in soap bilayer (2). III: doxyl stearic acid bound to bovine serum albumin (3). IV: doxyl cholestane in cholesteryl chloride crystal (3).

Tensor element	I	II	III	IV
A_{xx} (mT)	0.59	0.695	0.626	0.629
A_{yy} (mT)	0.54	0.535	0.580	0.584
A_{zz} (mT)	3.29	3.30	3.35	3.190
a_0 (mT)	1.47	1.51	1.52	1.47
ΔA (mT)	2.73	2.69	2.75	2.58
δA (mT)	0.025	0.080	0.023	0.023
g_{xx}	2.0088	2.00872	2.0088	2.0090
g_{yy}	2.0058	2.00616	2.0061	2.0060
g_{zz}	2.0021	2.00270	2.0027	2.0024
g_0	2.0056	2.00586	2.0059	2.0058
Δg	−0.0052	−0.00474	−0.0048	−0.0051
δg	0.0015	0.00128	0.0013	0.0015

The principal axes of the nitroxide radical have x directed along the N-O bond and z perpendicular to the C-N(O)-C plane (i.e. parallel to the axis of the p_π-orbital).

$$a_0 = \tfrac{1}{3}(A_{xx} + A_{yy} + A_{zz}) \qquad g_0 = \tfrac{1}{3}(g_{xx} + g_{yy} + g_{zz})$$
$$\Delta A = A_{zz} - \tfrac{1}{2}(A_{xx} + A_{yy}) \qquad \Delta g = g_{zz} - \tfrac{1}{2}(g_{xx} + g_{yy})$$
$$\delta A = \tfrac{1}{2}(A_{xx} - A_{yy}) \qquad \delta g = \tfrac{1}{2}(g_{xx} - g_{yy})$$

In liquids and liquid crystals, the nitroxide radicals undergo random reorientations and consequently the last two terms in Eqn. 1 become time–dependent. In this instance, one can write formally:

$$\mathcal{H}(t) = \langle \mathcal{H}_0 \rangle + \{\mathcal{H}(t) - \langle \mathcal{H}_0 \rangle\} \qquad [2]$$

where the $\langle \mathcal{H}_0 \rangle$ term consists of the time–averaged electronic Zeeman and hyperfine interactions, and the second term describes the time–dependent fluctuations of these interactions about this average. This motional averaging is seen from comparison of the immobilized spin label spectra in Figs. 1a and 1c, with those of the same labels in a fluid lipid environment in Figs. 1b and 1d, respectively.

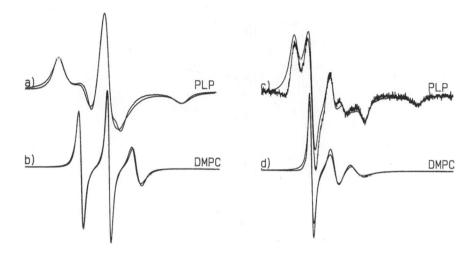

Fig. 1. Experimental (full line) and simulated (dashed line) EPR spectra of spin-labelled phosphatidic acid in environments of different mobility, recorded at X-band (9.1 GHz) and at Q-band (34.2 GHz) microwave frequencies. a) spectrum of phosphatidic acid spin label bound to the myelin proteolipid protein (PLP), recorded at 9.1 GHz. b) spectrum of phosphatidic acid spin label in fluid dimyristoyl phosphatidylcholine (DMPC) bilayers, recorded at 9.1 GHz. c) spectrum of phosphatidic acid spin label bound to PLP, recorded at 34.2 GHz. d) spectrum of phosphatidic acid spin label in fluid DMPC bilayers, recorded at 34.2 GHz. Each spectrum was recorded at 30°C, with a scan width of 10 mT at 9.1 GHz and of 12 mT at 34.2 GHz.

The application of spin label EPR spectroscopy to the description of molecular dynamics in biological systems relies principally on the motional modulation of the anisotropy of the electronic Zeeman and hyperfine interactions. In principle, a minimum of two cases must be considered. In the first, the dominant features are described by the $\langle \mathcal{H}_0 \rangle$ term in Eqn. 2, and the time–dependent terms may be treated as perturbations. Accordingly, a narrow three–line spectrum, corresponding to the $m_I = +1, 0, -1$ manifolds of the ^{14}N nucleus, is observed in such cases. In the second case, if the molecular motions are slow, only partially averaged spectra are observed. A series of EPR spectra, representing the above cases, which

710

are characterized by the rotational correlation time, τ_R, is shown in Fig. 2. In the correlation time range of 10^{-11} to 10^{-7} s, the lineshape is sensitive to the molecular motion, and the correlation time τ_R can be determined from the spectral shapes and linewidths, as will be discussed later.

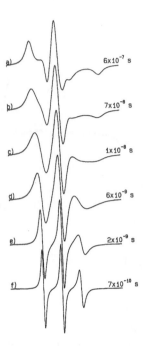

Fig. 2. Series of simulated 9 GHz EPR spectra of nitroxide free radicals with increasing (top to bottom) rotational correlation times, τ_R. In the fast $(\tau_R \approx 10^{-10}$ s $< \tau_{lim})$ and slow $(\tau_R \approx 10^{-7}$ s $> \tau_{lim})$ motional regimes, motionally averaged three-line spectra and powder pattern lineshapes, respectively, are observed. The rotational correlation time, τ_R, is determined from the spectral lineshapes and linewidths. Total plot width = 10 mT.

It is possible to define a limiting time–scale, τ_{lim}, which represents the slowest molecular motions to which the conventional EPR spectra are sensitive. This time–scale is determined by the spectral anisotropy, $\Delta B \left[\approx (A_{zz} - A_{xx}), (g_{xx} - g_{zz})\beta B \right]$, of the nitroxide group and therefore is characteristic of the spin label method:

$$\tau_{lim} = \frac{\hbar}{\Delta B(\theta)} \approx 6.8 \times 10^{-9}\text{s} .$$ [3]

Since the g–tensor and hyperfine tensor anisotropies have opposite signs, different limiting time–scales can be assigned to each nuclear manifold (cf. Eqn. 1), and consequently the corresponding regions in the spectrum display different motional sensitivities. This differential sensitivity is determined by the relative strengths of the Zeeman and hyperfine interactions, and therefore is dependent on the microwave frequency. It is particularly enhanced in the case of Q–band (35 GHz) spectra. This can be seen from the much stronger differential line broadening of the Q–band spectrum in Fig. 1d, compared with the spectrum of the same system at X–band in Fig. 1b.

The sensitivity of the conventional spin label EPR spectra to molecular motion is determined by transverse relaxation processes (T_2 processes) and is limited by the spectral anisotropies to motions faster than $\approx 10^{-8} - 10^{-7}$ s. This time–scale for conventional EPR is optimal for studying the rotational motions of spin–labelled lipids in fluid biological membranes. The region around the lower limit for conventional EPR is referred to as the slow motion regime. The sensitivity of the spin label method can be extended to correlation times beyond the conventional slow motion regime by studying the saturation properties of the EPR spectra, which are determined by the spin–lattice relaxation time, T_1. This is done in the saturation transfer EPR spin label method, and in this case the optimum motional sensitivity is on the time–scale: $\tau_{opt}(STEPR) \approx T_1 \approx 10^{-6} - 10^{-5}$ s. Practically, the lower limit of sensitivity is found for correlation times $\approx 10^{-3}$ s, and the upper limit borders almost on the lower limit for conventional spin label EPR. The saturation transfer EPR time–scale is optimal for studying the rotational motion of spin–labelled integral proteins in membranes.

1.2 Bloch equations and relaxation

A theoretical description of spin label EPR spectra can be given by a classical phenomenological approach using the Bloch equations of motion for the macroscopic magnetization of the spins. This approach can be shown to be formally equivalent to the quantum mechanical description using the spin density matrix, which will be referred to later.

The macroscopic electron spin magnetization, $\mathbf{M}$, precesses in an external magnetic field, $\mathbf{B}$, and is assumed to relax exponentially to its equilibrium value, $\mathbf{M_0}$. The time dependence of the individual components of the magnetization is given by (4):

$$\frac{dM_x}{dt} = \gamma(\mathbf{M} \times \mathbf{B})_x - \frac{M_x}{T_2} \tag{4a}$$

$$\frac{dM_y}{dt} = \gamma(\mathbf{M} \times \mathbf{B})_y - \frac{M_y}{T_2} \tag{4b}$$

$$\frac{dM_z}{dt} = \gamma(\mathbf{M} \times \mathbf{B})_z - \frac{M_z - M_0}{T_1} \tag{4c}$$

where the first terms on the right–hand side of each equation represent the torque experienced by the magnetic moment in the magnetic field and $\gamma = g\beta/\hbar$ is the gyromagnetic ratio of the electron. The second terms represent the rate at which the magnetization relaxes to its equilibrium value, $(0, 0, M_0)$, where it is assumed that the static component of the magnetic field lies along the z−axis. Two relaxation times are introduced: T_1 is the longitudinal or spin–lattice relaxation time, and T_2 is the transverse or spin–spin relaxation time. EPR resonance absorption can be induced by a small circularly polarized magnetic field, $\mathbf{B_1} = B_1(\cos\omega t, -\sin\omega t, 0)$, applied perpendicular to the static field and rotating in the x, y−plane at the angular resonance frequency, $\omega = \omega_0 = \gamma B_0$, of the precessing magnetization. In the so–called rotating frame, which rotates around the z−axis at the angular frequency, ω, the magnetization precesses around the B_1 field. The Bloch equations in the rotating frame are:

$$\frac{du}{dt} = (\omega_0 - \omega)v - \frac{u}{T_2} \tag{5a}$$

$$\frac{dv}{dt} = -(\omega_0 - \omega)u + \gamma B_1 M_z - \frac{v}{T_2} \tag{5b}$$

$$\frac{dM_z}{dt} = -\gamma B_1 v - \frac{M_z - M_0}{T_1} \tag{5c}$$

where u and v are the x and y components of the magnetization in the rotating frame: $M_x = u\cos\omega t + v\sin\omega t$. The steady state (slow passage) solutions of Eqns. 5a–5c for the magnetization describe the lineshape and the saturation properties of the absorption (v) and dispersion (u) EPR signals. These are given by:

$$u = M_0 \frac{\gamma B_1 T_2^2(\omega_0 - \omega)}{1 + T_2^2(\omega_0 - \omega)^2 + \gamma^2 B_1^2 T_1 T_2} \tag{6a}$$

$$v = M_0 \frac{\gamma B_1 T_2}{1 + T_2^2(\omega_0 - \omega)^2 + \gamma^2 B_1^2 T_1 T_2} \; . \tag{6b}$$

At low microwave powers ($\gamma^2 B_1^2 \ll T_1 T_2$) one obtains the well–known Lorentzian lineshape for the absorption spectrum, and at high microwave powers the effects of saturation are described by the saturation parameter, $\gamma^2 B_1^2 T_1 T_2$, in the denominator. The Lorentzian lineshape is centered about the angular frequency ω_0, and has a half–width at half–height given by: $\Delta\omega_{1/2} = 1/T_2$. The peak–to–peak width of the first derivative EPR spectrum in a magnetic field display is given by:

$$\Delta H_{pp} = \frac{2\hbar}{\sqrt{3}g\beta} T_2^{-1} \tag{7}$$

where $\hbar$ is Planck's constant and other symbols have their usual meaning.

1.3 Lineshape for exchanging spins

For studying motional effects in nitroxide spectra, the exchange between different environments or orientations has to be considered. This is done by adding the rate equations for the exchange of the spin magnetization to the Bloch equations (5). Using a two–site model, the rate equations for the spin magnetizations, M_A and M_B, associated with the two environments are given by:

$$\frac{dM_A}{dt} = -\tau_A^{-1} M_A + \tau_B^{-1} M_B \qquad [8a]$$

$$\frac{dM_B}{dt} = \tau_A^{-1} M_A - \tau_B^{-1} M_B \qquad [8b]$$

where τ_A^{-1} and τ_B^{-1} are the probabilities per unit time for transfer of magnetization from state A to state B and vice–versa. At exchange equilibrium, the ratio of the transfer rates is related to the steady state populations of the two components:

$$\frac{\tau_A^{-1}}{\tau_B^{-1}} = \frac{1-f}{f} \qquad [9]$$

where f is the fraction of component A. The steady–state Bloch equations for the complex transverse magnetizations: $\hat{M}_A = u_A + iv_A$ and $\hat{M}_B = u_B + iv_B$, (cf. Eqns. 5a and 5b) can then be written as:

$$[(\omega - \hat{\omega}_A) + i\tau_A^{-1}]\hat{M}_A - i\tau_B^{-1}\hat{M}_B = -\gamma B_1 M_0 f \qquad [10a]$$

$$[(\omega - \hat{\omega}_B) + i\tau_B^{-1}]\hat{M}_B - i\tau_A^{-1}\hat{M}_A = -\gamma B_1 M_0 (1-f) \qquad [10b]$$

where $\hat{\omega}_A$ and $\hat{\omega}_B$ are the complex angular resonance frequencies of components A and B, respectively: $\hat{\omega}_A = \omega_{0,A} - iT_{2,A}^{-1}$ and $\hat{\omega}_B = \omega_{0,B} - iT_{2,B}^{-1}$. The total complex magnetization, $\hat{M} = \hat{M}_A + \hat{M}_B$, can then be calculated, and the intensity of the EPR signal is proportional to the imaginary (v) part of $\hat{M}$ (see e.g. ref. 6).

A series of two component spectral simulations is shown in Fig. 3, to illustrate the effects of exchange (or angular rotation) on the lineshape. For very slow exchange the spectra consist of the independent components, centered at their individual resonance frequencies, $\omega_{0,A}$ and $\omega_{0,B}$, and having their intrinsic linewidths, $T_{2,A}^{-1}$ and $T_{2,B}^{-1}$. For slow rates of exchange $(\tau_A^{-1}, \tau_B^{-1} \ll \omega_{0,A} - \omega_{0,B})$ the linewidths begin to increase by an amount given essentially by the exchange lifetime:

$$T_{2,eff}^{-1}(A) = T_{2,A}^{-1} + \tau_A^{-1} \qquad [11a]$$

$$T_{2,eff}^{-1}(B) = T_{2,B}^{-1} + \tau_B^{-1} . \qquad [11b]$$

For slightly higher rates of exchange $(\tau_A^{-1}, \tau_B^{-1} < \omega_{0,A} - \omega_{0,B})$ the lines begin to move together. In this regime the amount by which the spectral splitting, $\omega_{0,A} - \omega_{0,B}$, is decreased is given by (7):

$$(\omega_A - \omega_B)_{ex} = (\omega_{0,A} - \omega_{0,B})[1 - 8\tau^{-2}(\omega_{0,A} - \omega_{0,B})^{-2}]^{\frac{1}{2}} \qquad [12]$$

714

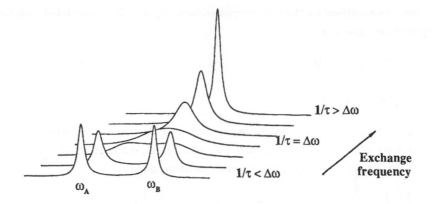

Fig. 3. Effect of two-site exchange on the absorption spectral lineshapes. The two resonances A and B are assigned to molecules in two different environments (or, in general to two different paramagnetic species). The observed linebroadening and subsequent exchange narrowing depends on the relative magnitudes of the exchange frequency, $1/\tau_{ex}$, and the frequency separation, $\Delta\omega = \omega_A - \omega_B$, of these two states.

where it is assumed that the two components have equal populations. As the exchange rate increases further, the lines move closer together and then collapse to a single, broad line. In this condition of fast exchange $(\tau_A^{-1}, \tau_B^{-1} > \omega_{0,A} - \omega_{0,B})$ the collapsed line is centered about the mean position:

$$\langle\omega_0\rangle = f\omega_{0,A} + (1-f)\omega_{0,B} \tag{13}$$

and the linewidth is the mean of the two linewidths, but with an additional contribution from lifetime broadening which is given by:

$$T_{2,ex}^{-1} = f^2(1-f)^2(\omega_{0,A} - \omega_{0,B})^2(\tau_A + \tau_B) . \tag{14}$$

Finally at extremely fast exchange rates $(\tau_A^{-1}, \tau_B^{-1} \gg \omega_{0,A} - \omega_{0,B})$, the spectrum is no longer sensitive to the exchange rate and consists of a single Lorentzian line centered at $\langle\omega_0\rangle$ and with a linewidth that is the weighted mean of the two components:

$$\langle T_2^{-1}\rangle = fT_{2,A}^{-1} + (1-f)T_{2,B}^{-1} . \tag{15}$$

These spectral changes with exchange rate summarize all the qualitative features which are observed in the dependence of nitroxide EPR spectra on the rotational mobility of spin–labelled molecules. They also describe the spectral effects of exchange of spin–labelled molecules between different environments, and qualitatively the effects of Heisenberg spin exchange between spin–labelled molecules.

2. ANISOTROPIC AND SLOW ROTATIONAL DIFFUSION

If the rotational diffusion of a spin labelled molecule is sufficiently rapid, i.e. the rotational correlation time is sufficiently short, $[\tau_R \leq \hbar/(A_{zz} - A_{xx}), \hbar/(g_{xx} - g_{zz})\beta B \approx 6.8 \times 10^{-9}$ s], the angular anisotropies in the conventional EPR spectra will be wholly or partially averaged. As seen in the previous section, the extent of averaging can depend on the rate of the angular rotational motion, and in addition the spectral lines are subject to a broadening that is dependent on the rate of motion. In anisotropic media such as membranes, the extent of averaging will also depend on the angular amplitude of the motion. The principal parameters which may be determined from the spin label EPR spectrum are therefore the order parameters (or orienting potentials) and the correlation times (or diffusion coefficients) associated with the rotational motion.

2.1 Rotational diffusion coefficients and correlation times

The rotational motions of spin–labelled molecules can be analyzed by hydrodynamic methods (see e.g. refs. 8, 9). The rotational diffusion coefficient is given by the Einstein equation:

$$D_R = \frac{kT}{f_R} \qquad [16]$$

where f_R is the rotational frictional coefficient which depends on the particle shape and size and on the viscosity of the medium, and other symbols have their usual meaning. The rotational correlation times measured by spin label EPR are defined by the correlation functions associated with the modulation of the spectral anisotropies. For isotropic rotation, the correlation functions involve the second order spherical harmonics, $Y_2^M(\Omega)$, and the correlation times, $\tau_R \equiv \tau_{20}$, are related to the diffusion coefficient by (see e.g. ref. 9):

$$\tau_R = \frac{1}{6D_R}. \qquad [17]$$

For a spherical particle in an isotropic medium of viscosity η, the frictional coefficient is given by the familiar expression:

$$f_{R,0} = 6\eta V \qquad [18]$$

where V is the volume of the sphere. This expression can be used to determine the effective microviscosity in the cytoplasm of cells from measurements of the rotational correlation time of small, near spherical, water–soluble spin labels, e.g. TEMPONE [2,2,6,6–tetramethyl–4–piperidine–N–oxyl] (10,11). For TEMPONE, $V = 1.131 \times 10^{-22}$ cm^3, corresponding to a spherical radius of 3 Å. As will be seen later, Eqn. 18 is also routinely used in determining correlation time calibrations for saturation transfer EPR spectra.

For non–spherical particles, the frictional coefficient is a tensorial quantity. If the particle has axial symmetry, the frictional coefficient has two principal values, $f_{R\parallel}$ and $f_{R\perp}$, which correspond to rotation around and perpendicular to the principal axis, respectively. Expressions for prolate and oblate ellipsoids in an isotropic medium have been derived by Perrin (12). The frictional coefficients for rotation about the two semiaxes can be related to that of a sphere, $f_{R,0}(=6\eta V)$, of equivalent volume, V, by the shape factors, F_R. For rotation about the a and b semiaxes:

$$f_{R,a} = f_{R,0} F_{R,a} \qquad\qquad [19a]$$

$$f_{R,b} = f_{R,0} F_{R,b} . \qquad\qquad [19b]$$

For prolate ellipsoids, the shape factor corresponding to rotation about the principal axis is: $F_{R\parallel} \equiv F_{R,a}$. This approaches the value of 2/3 as the a–axis becomes very long. For oblate ellipsoids: $F_{R\parallel} \equiv F_{R,b}$, and in the limit of large a/b : $F_{R,a} = F_{R,b} = 3a/4\pi b$, which also implies that $F_{R\perp} = F_{R\parallel}$. The complete equations and numerical values for the shape factors of ellipsoids of revolution, in the general case, can be found in ref. (13).

For axial anisotropic rotation, the correlation functions required for analysis of the spin label EPR spectra are the generalized spherical harmonics, i.e. the normalized Wigner rotation matrices, $\mathcal{D}_{K,M}^{L}(\Omega)$, where the usual spherical harmonics are given by $K = 0$. The corresponding correlation times, τ_{LK} are then given by (see e.g. refs. 9,14):

$$\frac{1}{\tau_{LK}} = D_{R\perp} L(L+1) + (D_{R\parallel} - D_{R\perp})K^2 \qquad\qquad [20]$$

where $D_{R\parallel}$ and $D_{R\perp}$ are the diffusion coefficients for rotation about and perpendicular to the principal axis, respectively. The diffusion coefficients are the fundamental parameters characterizing the rotational motion, but for convenience they may also be defined in terms of empirical correlation times, $\tau_{R\parallel} \equiv 1/6D_{R\parallel}$ and $\tau_{R\perp} \equiv 1/6D_{R\perp}$, in analogy with Eqn. 17 for isotropic rotation. The latter are then related to the true correlation times by:

$$\tau_{R\parallel} = \frac{2\tau_{20}\tau_{22}}{3\tau_{20} - \tau_{22}} \qquad\qquad [21a]$$

and

$$\tau_{R\perp} = \tau_{20} \qquad\qquad [21b]$$

where the τ_{LK} are the fundamental quantities characterizing the spin relaxation behaviour.

2.2 Orientational potentials and order parameters

Membranes are anisotropic structures and the molecules within them are constrained by an orienting potential. The director for this orienting torque is usually the membrane normal. In general, the orientational pseudopotential can be expanded in terms of spherical harmonics. For axially symmetric systems, the terms up to second order, have the following angular dependence (see e.g. ref. 14):

$$U(\phi, \theta) = \varepsilon_1 \cos^2 \theta + \varepsilon_2 \sin^2 \theta \cos 2\phi \qquad [22]$$

where $(\phi, \theta) \equiv \Omega$ are the Euler angles relating the principal diffusion axis to the director axis. The first term in Eqn. 22 is the leading term and corresponds to the usual Maier–Saupe (15) pseudopotential.

The extent of ordering induced by the orienting potential can be specified by order parameters that are time averages of the direction cosines of the principal diffusion axis with respect to the director (16). In Cartesian coordinates, the elements of the ordering tensor are defined by:

$$S_{ij} = \frac{1}{2}(3\langle \cos \theta_i \cos \theta_j \rangle - \delta_{ij}) \qquad [23]$$

where $i, j = x, y, z$, the latter being the Cartesian axes in the diffusion tensor frame, and the angular brackets indicate an average over the orientational distribution. The orthogonality relation between the direction cosines requires that the trace of the order tensor shall be zero:

$$\sum_i S_{ii} = 0 \qquad [24]$$

and the definition in Eqn. 23 also implies that the order tensor is symmetric: $S_{ij} = S_{ji}$. The averages over the orientational distribution may by performed using the orientational potential, $U(\Omega)$, in Eqn. 22. For instance, the S_{zz} component is the weighted average of the Legendre polynomial, $P_2(\cos \theta)$:

$$S_{zz} = \langle P_2(\cos \theta) \rangle = \frac{\int P_2(\cos \theta) e^{-U(\Omega)/kT} d\Omega}{\int e^{-U(\Omega)/kT} d\Omega} . \qquad [25]$$

For the simple model of random diffusion in a cone of fixed angle, θ_c, Eqn. 25 yields: $S_{zz} = \frac{1}{2} \cos \theta_c (1 + \cos \theta_c)$, and from Eqn. 24 for axial symmetry: $S_{xx} = S_{yy} = -\frac{1}{2} S_{zz}$.

The usefulness of the order parameter formalism arises because, in the fast motional regime, the positions of the spin label EPR lines are determined by the values of the spin Hamiltonian tensors, averaged over the angular motion of the spin–labelled group (cf. Eqn. 13 for two–site exchange). From the angular transformation properties of the tensors (16), it is found that the principal components of the motionally averaged g–tensor are:

$$g_{\parallel} = g_0 + \frac{2}{3} \Delta g \, S_{zz} + \frac{2}{3} \delta g (S_{xx} - S_{yy}) \qquad [26a]$$

$$g_{\perp} = g_0 - \frac{1}{3} \Delta g \, S_{zz} - \frac{1}{3} \delta g (S_{xx} - S_{yy}) \qquad [26b]$$

where the principal axis ($\parallel$) is the director axis of the orienting potential. The definition of the g-tensor elements Δg and δg is given in Table I. Similar equations hold for the motionally averaged hyperfine tensor, but since the static tensor is almost axial ($\delta A \approx 0$), simpler expressions can be used:

$$A_\parallel \approx a_0 + \frac{2}{3}\Delta A \, S_{zz} \qquad\qquad [27a]$$

$$A_\perp \approx a_0 - \frac{1}{3}\Delta A \, S_{zz}. \qquad\qquad [27b]$$

Thus, the principal order parameter S_{zz} can be obtained simply from the measured hyperfine splittings $A_\parallel$ and $A_\perp$. If the g-values are also measured, the S_{xx} element and the $S_{yy} = -(S_{xx} + S_{zz})$ element may also be determined. The latter give some idea of the asymmetry $\eta = (S_{xx} - S_{yy})/S_{zz}$ of the ordering tensor.

In the slow motion regime, the spectra are determined by the full orientational distribution and the order parameters have no special significance. They can be determined from Eqn. 25 if the strength of the orientational potential is known, but it is the latter, not the order parameters, which is required for the simulation of the slow motion spectra.

2.3 Slow and fast motional regimes

The fast and slow motional regimes of conventional spin label EPR spectroscopy are specified by the correlation time ranges: 10^{-11} s $< \tau_R < 10^{-9}$ s and 10^{-9} s $< \tau_R < 10^{-7}$ s, respectively. They are differentiated by the methods required to simulate the spectra and by the information which can be determined directly from the spectrum.

In the fast motional regime, the line positions are not dependent on the rate of motion, but only on the amplitude of motion, i.e. on the order parameters. This regime corresponds to the Eqns. 13 and 14 of the two-site exchange problem, i.e. $\tau_R < \tau_{lim}$. The spectra may be simulated using time dependent perturbation theory, based on the formulation of the spin Hamiltonian given in Eqn. 2. This is commonly known as motional narrowing theory, and has been applied to the simulation of lipid spectra in highly fluid bilayers by Schindler and Seelig (2, 17). The perturbation calculations for the motional narrowing limit indicate that the modulation of the hyperfine and g-value anisotropies leads to a differential broadening of the three ^{14}N hyperfine lines, by an amount which is dependent on the rate of motion. The dependence of the linewidth on the nuclear spin quantum number, $m_I = 0, \pm 1$, is given in general by (18):

$$\Delta H(m_I) = A + B m_I + C m_I^2. \qquad\qquad [28]$$

This is a generalization of the Eqn. 14 for two-site exchange, where the term $(\omega_{0,A} - \omega_{0,B})^2$ is now replaced by all quadratic terms involving the g-value and hyperfine anisotropies. This implies that the terms A and B, but not C, will be dependent on the microwave frequency. Such considerations are particularly important in spin label experiments at 35 GHz. From Eqn. 14 it is clear that all three of the linewidth coefficients depend linearly on the rotational correlation time, τ_R. However, only the coefficients B and C are useful in determining τ_R, because the A term also contains the intrinsic linewidth, which is not known. For isotropic rotation, the B and C terms give independent measurements of the correlation time, τ_R. In the isotropic case, the ratio $\mid C/B \mid$ is ≈ 1 for a microwave frequency of 9 GHz; departures of this ratio from unity definitely indicate anisotropic rotation. In the case of anisotropic rotation in an isotropic medium, the ratio $\mid C/B \mid$ may take the following values at X–band: $0.5 \leq \mid C/B \mid \leq 1.5$ for preferential rotation about the nitroxide x–axis; $0.8 \leq \mid C/D \mid \leq 8.8$ for preferential rotation about the y–axis; and $\mid C/B \mid \leq 1$ for preferential rotation about the z–axis. Analytical (and numerical) expressions for the correlation times τ_{20} and τ_{22}, in terms of the B and C coefficients, can be found in ref. (19). In the case of anisotropic rotation in an orienting medium, the coefficients will depend not only on the rate of motion, but also on the amplitude of motion. Not just the order parameter derived from the second order Legendre polynomial, $\langle P_2(\cos\theta)\rangle$, (cf. Eqn. 23) is involved, but also that from the fourth order Legendre polynomial, $\langle P_4(\cos\theta)\rangle$, (see e.g. ref. 2). In the oriented case the linewidths will also be anisotropic with respect to the magnetic field orientation.

In the slow motional regime the line positions are not strongly dependent on the amplitude of motion, but depend on the rate of motion (cf. Eqn. 12). The linewidths increase with increasing rate of motion (see Eqn. 11), in contrast to the fast motion regime where the linewidths decrease with increasing rates of motion (see Eqn. 14). Simulations for isotropic rotational diffusion in the slow motional regime indicate that the line broadening and the decrease in outer hyperfine splitting with decreasing rotational correlation time, correspond reasonably well with Eqns. 12 and 13, respectively. Rotational correlation time calibrations deduced from spectral simulations for slow isotropic motion have been presented based on these equations (14):

$$\tau_R = a_{m_I}\left[\frac{\Delta H(m_I)}{\Delta H_R(m_I)} - 1\right]^{b_{m_I}} \tag{29a}$$

$$\tau_R = a\left[1 - \frac{A'_{zz}}{A^R_{zz}}\right]^b \tag{29b}$$

where $\Delta H(m_I)$ are the linewidths at half-height of the outer hyperfine extrema, $2A'_{zz}$ is the outer hyperfine splitting, and both are normalized to the rigid limit values: $\Delta H_R(m_I)$ and $2A^R_{zz}$, respectively. The calibration constants are established by simulation, but for the two site exchange model take the values: $b_{m_I} = -1$, $a_{m_I} = 2.19 \times 10^{-8}$ s, and $b = -1/2$, $a = 1.78 \times 10^{-9}$ s. Empirical calibrations for $\tau_{R\parallel}$, based on spectral simulations for slow axially anisotropic rotation, have been given in ref. (20). For slow motion, the calibration constants depend on the motional model used in the simulations, i.e. strong jump, Brownian diffusion or free diffusion. This is not the case in the fast motional limit, where the spectra are insensitive to the particular motional mode which gives rise to the motional averaging.

The spectral simulations in the slow motion regime must be performed with the full quantum mechanical stochastic Liouville equation (14), or with the diffusion–coupled Bloch equations (21) which are the classical equivalent of the density matrix formalism. These general methods are valid in the fast motional regime, as well as in the slow motional regime. However, perturbation theory is computationally far less demanding and therefore is preferred for the fast motional regime.

2.4 Chain rotational isomerism

The overall rotational motion of a spin–labelled lipid molecule can be described by the rotational diffusion coefficients, $D_{R\parallel}$ and $D_{R\perp}$, and by the orientational potential which orders the long axis of the molecule. For spin labels attached to the lipid chains, the segmental motion arising from $trans - gauche$ isomerism must also be taken into account. $Gauche^{\pm}$ conformations are generated from the planar all–$trans$ configuration by rotations of $\pm 120°$ about a C–C single bond. In the nomenclature of ref. (22), the $trans$ conformation is defined as anti–periplanar and the $gauche^{\pm}$ conformations are defined as $\pm$anti–clinal. The $trans - gauche$ isomerism can be modelled by a jump process (equivalent to the two–site exchange model), which is characterized by the transition rate, τ_J^{-1}, between the different conformers.

For sp^3 bond symmetry, the possible orientations that a spin–labelled segment may take up, are specified by the edges of a tetrahedron. These six independent orientations account for all possible combinations of $trans$ and $gauche$ conformations of the various chain segments (23). The chain segment configurations can be specified by the conformation of the segment, $trans(t)$ or $gauche^{\pm}(g^{\pm})$, relative to the preceding segment, and the orientation of the segment, θ, relative to the bilayer normal (24). In this notation the allowed orientations are: $(t, 0°)$, $(g^{\pm}, 60°)$, two $(t, 60°)$, and $(g^{+}, 90°)$; and six conjugate, indistinguishable orientations which are: $(t, 180°)$, $(g^{\pm}, 120°)$, two $(t, 120°)$, and $(g^{-}, 90°)$, respectively. The first three distinct orientations require minimally a rotation about only one C–C bond, whereas the other three orientations require minimally a rotation about two adjacent C–C bonds. Therefore, in addition to the jump rate, τ_J^{-1}, the populations, $n_1 - n_6$, of the six independent segmental orientations are required to describe the effects of rotational isomerism on the EPR spectra of spin–labelled lipid chains.

For the simple situation in which only the first three orientations, $(t, 0°)$ and $(g^{\pm}, 60°)$, are appreciably populated, the resulting segmental order parameter is defined by:

$$S_{t-g} = \frac{9\,n_t - 1}{8} \tag{30}$$

where $n_t \equiv n_1$ is the population of the $(t, 0°)$ orientation, and the population of the $(g^{\pm}, 60°)$ orientations is simply: $n_g \equiv n_{2,3} = \frac{1}{2}(1 - n_t)$. The total molecular order parameter for a chain segment, including long axis motion, is then:

$$S_{mol} = S_{t-g} \, S_{zz} \tag{31}$$

where S_{zz} is the order parameter of the chain long axis, given by Eqn. 25. For situations involving all possible $trans-gauche$ populations, $n_1 - n_6$, S_{t-g} takes a more complicated form than in Eqn. 30, but Eqn. 31 still holds. In this case, the tensor S_{t-g} can be diagonalized to yield the segmental order parameter, $S_{\pi'\pi'}$, which also may be used to characterize the $trans - gauche$ isomerization (23).

2.5 Spectral simulation

In the fast, or motional narrowing, regime spectra may be simulated using perturbation theory, based on Eqn. 2. The time-dependent modulation of the magnetic anisotropies is expressed in terms of the autocorrelation function of the perturbation Hamiltonian, $\mathcal{H}_1$:

$$G(\tau) = \langle \mathcal{H}_1^*(t) \, \mathcal{H}_1(t + \tau) \rangle \, . \tag{32}$$

The frequency dependence of the electron spin relaxation times is then given directly in terms of the spectral densities, $J(\omega)$, which are Fourier transforms of the autocorrelation functions:

$$J(\omega) = \int_{-\infty}^{+\infty} G(\tau) \, e^{i\omega\tau} \, d\tau \tag{33}$$

where ω is the angular frequency. According to the rotational diffusion equation, the required functions are simply autocorrelation functions of the Wigner rotation matrices, and are characterized by exponential decays specified by the correlation times given in Eqn. 20. In ordered systems, the evaluation of the autocorrelation functions also requires knowledge of the orientational pseudopotential, $U(\Omega)$, given for example by Eqn. 22. Such a perturbation treatment has been presented by Schindler and Seelig (2, 17), who showed that it could be applied successfully to the simulation of the spectra of spin-labelled lipid molecules in soap bilayers.

The motion of spin-labelled phospholipid molecules in membranes is found to contain important contributions in the slow motion regime of conventional nitroxide EPR spectroscopy (23, 25). In this latter case, a more comprehensive treatment than perturbation theory is required. This is done by combining the exact equation of motion for the magnetic spins with the rotational diffusion equation, yielding the stochastic Liouville equation (see e.g. ref. 14). The equation of motion for the electron spins is expressed in terms of the spin density matrix, ϱ (see e.g. ref. 26):

$$\frac{\partial \varrho}{\partial t} = -i[\mathcal{H}(t), \varrho]$$ [34]

where $\mathcal{H}(t)$ is the total time-dependent spin Hamiltonian, including the time-varying magnetic field that induces the EPR transitions, and the square brackets denote the commutator. Combination with the rotational diffusion equation leads to the stochastic Liouville equation, in which the time dependence is then included solely in the elements of the density matrix (27):

$$\frac{\partial \varrho(\Omega, t)}{\partial t} = -i[\mathcal{H}(\Omega), \varrho(\Omega, t)] - \Gamma_\Omega\, \varrho(\Omega, t)$$ [35]

where the spin Hamiltonian, $\mathcal{H}(\Omega)$, depends on the orientation and conformation of the spin-labelled molecule, specified by the Euler angles, Ω. Here Γ_Ω is the Markov operator which appears in the rotational diffusion equation:

$$\Gamma_\Omega = \mathbf{L} \cdot \mathbf{D_R} \cdot \mathbf{L} + \mathbf{L} \cdot \mathbf{D_R} \cdot \mathbf{L}\, \frac{U(\Omega)}{kT}$$ [36]

where $\mathbf{L}$ is the operator which generates an infinitesimal elementary rotation (formally equivalent to the dimensionless angular momentum operator), and $\mathbf{D_R}$ is the rotational diffusion tensor. Solutions are sought for the departures of the density matrix from its equilibrium value, ϱ_0, which have a time dependence of the form $e^{i\omega t}$, where ω is the angular frequency of the microwave magnetic field that induces the transitions. This leads to a series of coupled linear equations that must be solved numerically, most conveniently using the Lanczos algorithm (28). Further details of the method can be found in refs. (14) and (29). A detailed application to the simulation of the EPR spectra of lipid spin labels in membranes is given in refs. (23) and (25). The application of similar methods to the simulation of saturation transfer EPR spectra is given in ref. (30).

3. SATURATION TRANSFER EPR

Saturation transfer EPR is used for studying the slow rotational diffusion of spin-labelled macromolecules or of spin labelled phospholipids in gel–phase bilayers. The saturation transfer EPR method extends the range of sensitivity of conventional spin label EPR to the rotational correlation time range 10^{-7} to 10^{-3} s. The method requires detection (usually second harmonic) in phase quadrature with the field modulation, and partially saturating microwave powers must be used. Experimental calibrations using model systems under standardized conditions are needed to determine the correlation times. This calibration method requires modification in the case of anisotropic diffusion. The analysis of multi–component spectra is best undertaken with the integral method.

3.1 <u>Instrumental calibration</u>

Saturation transfer EPR spectra are much more sensitive to the instrumental settings than are conventional EPR spectra, because they depend critically on the degree of saturation, passage conditions, and quadrature phase settings. Correspondingly, they are also strongly dependent on the homogeneity of the microwave and modulation fields, which in conventional EPR are set at levels that have negligible influence on the spectral lineshapes. A standardized measurement protocol has been given in ref. (31). This is summarized in Table 2. Attention to such instrumental factors is crucial since rotational correlation times are routinely determined relative to standard spectra. The calibration methods and instrumental settings are reviewed in an itemized way below.

TABLE 2

Recommended standard conditions for the measurement of STEPR spectra (V_2' display) of samples contained in 1 mm ID capillaries (31).

[Spin Label] (μM)	Sample length (mm)	$\langle B_1^2 \rangle^{\frac{1}{2}}$ (mT)	$B_m(0)$ (mT)
$\geq$30	5	0.025	0.5
$\leq$30	>20	0.025	1.0

The microwave frequency is 9.1 GHz, and the microwave detection is set to absorption. The modulation frequency is 50 kHz, the detection frequency is 100 kHz (second harmonic), and the detection phase is set in quadrature to the modulation.

Field Inhomogeneities: Since the saturation transfer lineshapes and intensities depend on the strength of the microwave and modulation fields, it is important that these should be uniform over the sample. The inhomogeneities of the microwave and modulation fields have been studied in detail for the standard TE$_{102}$ rectangular cavity by Fajer and Marsh (32). Both fields have their maximum value at the centre of the cavity and drop off steeply towards the top and bottom of the cavity. The microwave field profile is determined by the resonant mode of the cavity and the modulation field profile by the dimensions of the modulation coils. The solution to this problem is to use small samples centred accurately in the cavity. Samples 5 mm in length in 1 mm I.D. capillaries are recommended for the TE$_{102}$ cavity at 9 GHz (31,32). Centring the sample is achieved most accurately by maximizing the height of the conventional spectrum on the oscilloscope. Line samples in capillaries or planar samples in flat cells will not yield STEPR spectra free from distortion by field inhomogeneities. The cylindrical TM$_{110}$ microwave cavity can accommodate larger samples and has larger modulation coils than the rectangular TE$_{102}$ cavity. Therefore this cavity affords a larger region of homogeneous microwave and modulation field than does the TE$_{102}$ cavity. The loop gap resonator also has a more uniform field over a greater proportion of its active length than does the TE_{102} cavity. In addition, its smaller size affords a more homogeneous modulation field over the active sample length for a given size of modulation coil (see chapter by Hyde in this book).

Microwave Field Intensity: The square of the microwave field intensity at the sample, B_1^2, is directly proportional to the incident microwave power, P. Both the power output from the klystron and the settings of the attenuator in the microwave bridge can be calibrated using a power meter and a standard attenuator (32). The microwave field at the centre of the cavity can be measured using the saturation broadening of the conventional EPR spectrum from an aqueous point sample (2 mm long) of peroxylamine disulphonate (PADS, 0.9 mM PADS in 10 mM nitrogen–saturated K_2CO_3). The peak–to–peak linewidth is given by (33):

$$\Delta H_{pp}^2 = \Delta H_{pp}^{\circ}{}^2 + \frac{4}{3} B_1^2 \frac{T_1}{T_2} \qquad [37]$$

where ΔH_{pp}° is the unsaturated peak–to–peak linewidth. Since for aqueous solutions of PADS at room temperature $T_1 \approx T_2$, B_1 can be obtained as a function of power from a plot of ΔH_{pp}^2 against P. Alternatively, a small crystal of N–methylphenazinium tetracyanoquinodimethane (NMP–TCNQ) may be used (34). Other methods of measuring the microwave power intensity are discussed in the book edited by Dalton (30). Measuring the saturation curve of PADS is less appropriate for calibration than the saturation broadening, since the saturation curve is only linear over a very small range of microwave power.

Modulation amplitude: The magnetic field modulation amplitude, B_m, can be measured from the modulation broadening of the conventional EPR spectrum from an aqueous point sample (2 mm long) of peroxylamine disulphonate (PADS, 0.9 mM PADS in 10 mM nitrogen–saturated K_2CO_3). The peak–to–peak modulation amplitude is given by the increase in peak–to–peak linewidth relative to the spectrum recorded at low modulation amplitude (33). Alternatively, a small crystal of N–methylphenazinium tetracyanoquinodimethane (NMP–TCNQ), which has a very small natural linewidth, may be used.

Cavity Q: The cavity Q must be measured to correct for changes in the microwave field intensity at the sample due to differences in dielectric loss in the sample (32). The value of B_1^2 is proportional to the product $P.Q$ and therefore the incident power is adjusted for changes in cavity Q by keeping this product constant. The Q of the cavity can be measured from the frequency separation, $\Delta\nu$, of the points of half–power absorption by observing the cavity dip in the klystron mode, under conditions of critical coupling. The Q is then given by (35):

$$Q = \frac{\nu_0}{\Delta\nu} \qquad [38]$$

where ν_0 is the resonant frequency of the cavity. Microwave frequencies, ν, are measured with a digital frequency counter.

Quadrature Phase Setting: Because the phase of the modulation field over the sample is inhomogeneous due to eddy currents, the phase setting depends on the composition, shape, size and position of the sample. The receiver phase is therefore best set by the "self–null" method (36). For this reason, small samples, centred in the cavity are again most desirable; these give rise to the best nulls. In the self–null method, the out–of–phase spectrum of the sample of interest is recorded at low power (< 1 mW) and the detector phase adjusted to give the minimum signal. Typical out–of–phase nulls are less than 1 % of the intensity of the out–of–phase spectrum recorded at the power at which the STEPR spectrum is recorded (corresponding to $B_1 = 0.025$ mT). The phase setting procedure is illustrated in ref. (37). Alternatively, the null phase may be determined by extrapolation of the signal height on either side of the null position; different methods are described in refs. (38, 39). Saturation transfer EPR requires a high level of phase setting accuracy and phase stability of the receiver. The increase in amplitude of the out–of–phase signal for a phase error of $\Delta\phi$ is equal to $V_2 \sin \Delta\phi$, where V_2 is the intensity of the in–phase signal. Since the latter is typically 10 times greater than that of the out–of–phase signal, a phase error of 1° will already introduce an approximately 20 % in–phase admixture to the STEPR spectrum. Changes in the receiver gain and modulation amplitude may also cause slight shifts in the receiver phase, and for this reason it is desirable to record the STEPR spectrum with the same receiver settings as used to set the phase.

3.2 Rotational diffusion coefficients of membrane proteins

The rotational motions of spin–labelled macromolecules can be analysed by hydro-dynamic methods, as indicated previously in the discussion of conventional spin label EPR spectra. For instance, Eqns. 16–18 applying to the rotation of spherical particles in isotropic media, are used routinely for determining correlation time calibrations of saturation transfer EPR spectra. Spin–labelled haemoglobin in glycerol–water solutions of known viscosity is normally used as the calibration system (36).

Similar considerations to those applied above for the anisotropic rotation of non-spherical particles in isotropic media hold also for the uniaxial rotation of integral proteins in biological membranes. In this case the frictional coefficient, $f_{R\|}$, for rotation about the membrane normal is required. For uniaxial rotation about the axis of a right circular cylinder, embedded in a membrane of effective viscosity η_0, the frictional coefficient is given by (40):

$$f_{R\|,0} = 4\pi\eta_0 a^2 h \qquad [39]$$

where a is the radius of the cylinder and h is the cylinder length which is embedded in the membrane. For integral proteins of non–circular cross–section a shape factor, $F_{R\|}$, may be defined:

$$f_{R\|} = f_{R\|,0}/F_{R\|} \qquad [40]$$

where $f_{R\parallel,0}(=4\eta_0 V)$ is the frictional coefficient of a right circular cylinder of equivalent intramembranous volume, V. For a cylinder of elliptical cross–section of semi–axes a and b, the shape factor is (41):

$$F_{R\parallel} = \frac{2(b/a)}{1+(b/a)^2}. \qquad [41]$$

The dependence of the shape factor on axial ratio, b/a, for proteins of ellipsoidal intramembranous cross–section is given in Fig. 4. For an axial ratio $a/b = 2$ the shape factor is $F_{R\parallel} = 0.8$, and $F_{R\parallel} = 0.5$ for an axial ratio $a/b = 3.7$.

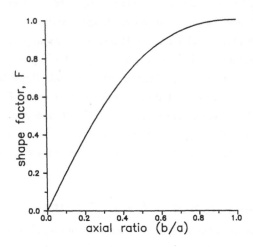

Fig. 4 Dependence of the shape factor, $F_{R\parallel}$, for the uniaxial rotational diffusion of a cylindrical protein on the axial ratio, b/a, of its elliptical cross-section. Calculated according to Eqn. 41.

3.3 Rotational correlation time calibrations

Correlation time calibrations of the saturation transfer spectrum are routinely given in terms of lineheight ratios in the low–field (L''/L), central (C'/C) and high–field (H''/H) diagnostic regions of the spectrum (36), and in terms of the first integral of the STEPR spectrum (42). The calibrations are obtained from isotropically rotating spin–labelled haemoglobin in glycerol–water solutions of known viscosity. Eqns. 16–18 are used, with a value for the volume of the haemoglobin molecule of $V = 1.0216 \times 10^{-19}$ cm^3.

The lineheights L and H in the low-field and high–field regions of the spectrum are measured at the z turning points, defined by the magnetic field orientation relative to the nitroxide axes of $\theta = 0°$. The lineheights L'' and H'' in the low–field and high–field diagnostic regions are measured at points 1/3 of the way from the z turning point to the average $x - y$ turning point in the $m_I = +1$ and -1 manifolds, respectively. These points correspond to the magnetic field orientation $\theta = 35°$ [i.e. $\frac{1}{2}(3\cos^2\theta - 1) = \frac{1}{2}$] and azimuthal angle $\phi = 45°$. The lineheight C is measured at the x turning point of the $m_I = 0$ manifold (i.e. $\theta = 90°$ and $\phi = 0°$), and the lineheight C' at the position half–way between the x and y turning points (i.e. $\theta = 90°$ and $\phi = 45°$). These definitions were given by Robinson and Dalton (43), in order to have an exact comparison with simulated spectra.

The normalized integral of the saturation transfer spectrum is defined as the first integral of the second harmonic, absorption out–of–phase spectrum (V_2' display), divided by the second integral of the conventional in–phase spectrum (V_1 display):

$$I_{ST} = \frac{\int V_2'(B)dB}{\iint V_1(B)\, d^2B} \; .$$ [42]

In this way, the integrals are independent of the spectrometer gain and modulation amplitude settings (see ref. 42).

The correlation time calibrations can be expressed in terms of the inverse functions relating the correlation times to the measured spectral parameters. Analytical best fit functions have been given in terms of polynomials involving either the spectral parameter itself or its logarithm (44). The correlation time is then expressed in terms of the logarithm:

$$\log_{10} \tau_R^{eff} = \sum_n a_n \left(\frac{P'}{P}\right)^n$$ [43]

and

$$\log_{10} \tau_R^{eff} = \sum_n a_n \left(\log_{10} \frac{P'}{P}\right)^n$$ [44]

where a_n are the polynomial coefficients of the best fit, which are listed in Table 3 for the linear and logarithmic forms of the experimental STEPR parameter, P'/P, which is either the lineheight ratio or integral. The linear and logarithmic parameter calibrations yield slightly different results because of the different fits. For example, for a lineheight ratio $L''/L = 1.0$, the linear calibration yields a correlation time of $\tau_R^{eff} = 5.6 \times 10^{-5}$ s, and for the logarithmic calibration yields a value of $\tau_R^{eff} = 6.2 \times 10^{-5}$ s.

3.4 Anisotropic rotational diffusion

Rotational correlation times are routinely deduced from experimental STEPR spectra by comparing the diagnostic lineheight ratios in the low–field, central and high–field regions of the spectrum with those obtained from isotropically rotating spin–labelled haemoglobin in

TABLE 3

Polynomial coefficients of Eqns. 43 and 44 for fitting the STEPR calibration curves (44). Calibrations were performed for 5 mm samples in 1 mm ID capillaries, according to the protocol of refs. (31, 32).

STEPR Parameter	Fitted Range	a_0	a_1	a_2	a_3	a_4
L''/L	0.1;2.0	-7.529	10.94	-13.99	7.684	-1.356
C'/C	-1.;1.	-5.333	1.032	-1.269	1.387	1.570
H''/H	0.1;2.0	-8.406	15.68	-20.26	10.99	-2.012
$I_{ST} \times 10^2$	0.1;2.0	-8.265	13.04	-13.37	6.003	-0.928
$\log(L''/L)$	-1.;0.3	-4.207	4.282	6.624	4.779	-
$\log(H''/H)$	-0.9;0.3	-3.929	3.443	4.204	2.994	-
$\log(I_{ST} \times 10^2)$	-0.9;-0.2	-3.005	6.764	-3.058	-29.21	-26.63
$\log(I_{ST} \times 10^2)$	-0.4;0.3	-3.403	5.803	7.464	6.335	-

solutions of known viscosity (36). The outer lineheight ratios, L''/L and H''/H, are sensitive to rotation of the nitroxide z–axis, via modulation of the hyperfine interaction, and the central lineheight ratio, C'/C, is sensitive to rotation of all three nitroxide axes, via modulation of the g–value anisotropy. For anisotropic rotation, the different lineheight ratios will therefore have differential sensitivities, as illustrated in Table 4 for the rotational diffusion of a spin–labelled phospholipid in gel phase lipid bilayers (45). The nitroxide z–axis is oriented along the lipid long molecular axis (cf. Table 1). At low temperatures the effective correlation times, $\tau_R^{eff}(m_I)$, deduced from the different lineheight ratios using the isotropic model system for calibration, are all very similar. Above the pretransition at 25° C, the effective correlation time deduced from the outer lineheight ratios is relatively unchanged, whilst that deduced from the central region decreases abruptly, indicating the onset of rapid anisotropic rotational diffusion about the long axis of the lipid molecule. A differential response of the different lineheight ratios may therefore be used to diagnose anisotropic rotational diffusion (46). Comparison of the integral of the high–field region of the saturation transfer spectrum with that of the total STEPR spectrum may also be similarly used (42).

TABLE 4

Effective rotational correlation times of a spin-labelled phospholipid in gel phase bilayers of dipalmitoyl phosphatidylcholine, deduced from the STEPR lineheight ratios using isotropic reference spectra (45).

Parameter	$\tau_R^{eff}(m_I)$ [s]			
	12°C	25°C	30°C	45°C
$L''/L\,(m_I = +1)$	6×10^{-4}	1×10^{-4}	0.8×10^{-4}	$\approx 10^{-9}$
$H''/H\,(m_I = -1)$	4×10^{-4}	2×10^{-4}	1×10^{-4}	$\approx 10^{-9}$
$C'/C\,(m_I = 0)$	0.8×10^{-4}	5×10^{-6}	0.9×10^{-6}	$\approx 10^{-9}$

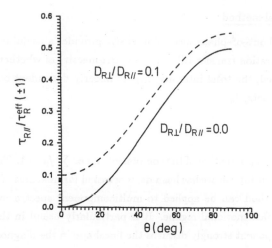

Fig. 5 Dependence of the effective rotational correlation time, $\tau_R^{eff}(\pm 1)$, obtained from the low-field (L''/L) and high-field (H''/H) STEPR lineheight ratios using isotropic calibrations, on the orientation, θ, of the nitroxide z–axis with respect to the principal rotational diffusion axis for axial rotation. $\tau_{R\|} = 1/6D_{R\|}$ is the rotational correlation time for rotation about the principal axis. Data are shown for two values of the rotational diffusion coefficient appropriate to the minor axis: $D_{R\perp} = 0.0$ (full line) and $D_{R\perp} = 0.1\,D_{R\|}$ (dashed line). Curves are calculated according to Eqn. 45.

Simulations of first harmonic phase–quadrature dispersion STEPR spectra by Robinson and Dalton (43) give some guide to the quantitative interpretation of the effective correlation times in terms of the true rotational diffusion parameters. It was found that the effective correlation times deduced from the low–field and high–field regions of the spectrum were very similar, independent of the degree of anisotropy of the motion (cf. Table 4). If the anisotropy of the rotation is great enough and the rotational rates are slow, the dependence of the effective correlation times on the orientation, θ, of the nitroxide z–axis with respect to the rotational diffusion axis, are given by:

$$\tau_R^{eff}(\pm 1) = \frac{1}{3[D_{R\|}\sin^2\theta + D_{R\perp}(1 + \cos^2\theta)]} \qquad [45]$$

where $D_{R\|}$ and $D_{R\perp}$ are the principal elements of the rotational diffusion tensor. The dependence on the nitroxide orientation is given in Fig. 5 for two values of the anisotropy in the rotation diffusion tensor, $D_{R\perp}/D_{R\|}$. Clearly precise measurements require knowledge of the value of the orientation, θ. However, some estimate of whether θ is close to 0° or close to 90° may be obtained from the relative sizes of $\tau_R^{eff}(\pm 1)$ and $\tau_R^{eff}(0)$, or by comparing the high–field and total spectral integrals. In addition, it may be possible to discriminate between protein monomer and oligomer formation on the basis of Eqns. 39–41 without an accurate knowledge of θ.

3.5 STEPR integral method

The integrated out–of–phase spectral intensity provides a useful means for analysing multi–component saturation transfer EPR spectra. Irrespective of whether the spectral components can be resolved, the total intensity, I_{ST}, is linearly dependent on the intensities of the individual components, I_n:

$$I_{ST}^{tot} = \sum_n f_n I_n \qquad [46]$$

where f_n is the fractional population of the component n, i.e. $\sum f_n = 1$. The linear additivity of the saturation transfer intensities has been demonstrated for mixtures of spin–labelled lipid vesicles (42). The method can be applied to multicomponent spectra in which one of the components is in the fast motional regime. It is particularly useful in these cases because, although the fast component strongly disturbs the lineshape in the diagnostic regions for the lineheight ratios, it contributes negligibly to the saturation transfer integral (42,47). The total integral intensity is then given simply by:

$$I_{ST}^{tot} = (1 - f)I_b \qquad [47]$$

where I_b is the desired saturation transfer integral of the slow motion component, and f is the fraction of the fast component which can be determined readily from spectral subtractions with the conventional in–phase spectra.

If all components lie in the slow motional regime, they will not be resolved in the spectral lineshape, but will give separate contributions to the total integral. The calibrated saturation transfer intensity for each component is approximately linearly related to the logarithm of the rotational correlation time (42, 48), i.e.

$$I_{ST} = A \ln(c\tau_R^{eff}) \qquad [48]$$

where A and c are constants from the calibration. Thus, substituting the calibration for the individual saturation transfer intensities, I_n, in Eqn. 46 gives:

$$I_{ST}^{tot} = A \sum_n f_n \ln(c\tau_{R,n}) \qquad [49]$$

where $\tau_{R,n}$ are the effective rotational correlation times of the individual components. This equation is useful if the rotational correlation times of the different components can be related to one another, as in the polymerization, cross–linking or aggregation reactions of macromolecules. In such a case, the effective rotational correlation times of each component will be related, via the rotational diffusion coefficient, to the molecular weight, $M_n = nM_1$, of the rotating species (cf. Eqn. 39):

$$\tau_{R,n} = B \frac{M_n}{F_n} \qquad [50]$$

where F_n is a shape factor (cf. Eqn. 41 and Fig. 4), and B is a constant involving the temperature, membrane viscosity and the partial specific volume of the protein (cf. Eqns. 16, 39). Here n is the degree of polymerization of the monomer of molecular weight M_1. Combining Eqns. 49 and 50, and substituting the calibration for the total integral, the measured effective rotational correlation time of the multi–component system becomes:

$$\ln \tau_R^{eff} = \sum_n f_n \ln \frac{n}{F_n} + \ln(cBM_1).$$ [51]

Alternatively, the correlation time of the total system may be expressed in terms of the effective rotational correlation time of the monomer, $\tau_{R,1}$, which yields a power dependence on the fractional populations of the different multimers:

$$\frac{\tau_R^{eff}}{\tau_{R,1}} = F_1 \prod_n \left(\frac{n}{F_n} \right)^{f_n}$$ [52]

where F_1 is the shape factor for the monomer. The relative importance of the shape factors can be deduced from Fig. 4. For a monomer of circular cross–section, the maximal axial ratio of the equivalent elliptical cross–section for the $n-$mer is $a/b = n$.

4. EXCHANGE PHENOMENA IN SPIN LABEL SPECTRA

Spin labels in different environments display different EPR spectra, reflecting the different molecular mobility in the various environments. Typical examples are spin label molecules free in solution and bound to proteins or membranes, or spin–labelled lipids in fluid lipid bilayers and associated at the intramembranous surface of integral proteins. The spin labels in the different environments will have distinct EPR spectra if the rate of exchange between the two environments is not greater than the intrinsic difference in the spectral splittings (expressed in frequency units) arising from the labels in the different environments. The degree of smearing of the resolution of the two components can be used to measure the exchange rate, using the methods outlined in section 1.3. In the following sections this is applied particularly to the analysis of exchange rates and relative populations in lipid interactions with integral membrane proteins.

4.1 Linewidth analyses

For slow exchange, the spectral components experience an additional Lorentzian broadening that is determined by the exchange lifetime according to Eqns. 11a, b. These equations can be applied directly to the binding of spin–labelled ligands to macromolecules or membranes. The free aqueous spin label has sharp, symmetrical lines and therefore the effects of exchange can be determined most sensitively from this spectral component. The on–rate

732

is then determined from the increase, δH_{pp}^L, in the peak–to–peak Lorentzian linebroadening of the free spin label, in the presence of the binding species:

$$\tau_{on}^{-1} = \frac{\sqrt{3}g\beta}{2\hbar} \delta H_{pp}^L .$$ [53]

From the law of mass action, Eqn. 9, the off–rate is given by:

$$\tau_{off}^{-1} = \tau_{on}^{-1} \frac{1-f}{f}$$ [54]

where f is the fraction of spin label bound. For a diffusion–controlled binding, one finds typically second–order rate constants of $k_2^{diff} \equiv \tau_{on}^{-1}/c_b \approx 7 \times 10^9$ M^{-1} s^{-1} (or greater), where c_b is the molar concentration of the species bearing the binding sites. An increase in peak–to–peak Lorentzian linewidth of 0.01 mT can be readily detected for sharp aqueous spin label spectra. Thus exchange rates can be measured at concentrations of the binding species down to $c_b \approx 0.2$ mM (6).

To obtain the Lorentzian linewidth, correction must be made for the inhomogeneous broadening of the aqueous spectra, due to unresolved proton hyperfine structure. If the inhomogeneous broadening is represented by a Gaussian distribution of Lorentzian lines, the peak–to–peak experimental linewidth is given by (49):

$$\Delta H_{pp} = \frac{\Delta H_{pp}^L}{2} + \left[(\Delta H_{pp}^G)^2 + \frac{(\Delta H_{pp}^L)^2}{4} \right]^{1/2}$$ [55]

where ΔH_{pp}^L and ΔH_{pp}^G are the peak–to–peak first–derivative Lorentzian and Gaussian linewidths, respectively. Therefore, if the Gaussian linewidth is known, the Lorentzian linewidth can be obtained from the measured peak–to–peak linewidths, ΔH_{pp}. For doxyl spin labels $\Delta H_{pp}^G = 0.076$ mT (50) and values for other spin labels can be found in refs. (51, 74). Using these methods, the on–rates and activation energies for the incorporation of spin–labelled lipid monomers into micelles have been measured (50,52). These rates were found to be diffusion–controlled.

4.2 Two–component exchange simulations

Lineshape simulations for two exchanging spectral components can be obtained from solution of the exchange–coupled Bloch equations (Eqns. 10a,b). The out–of–phase component of the total complex magnetization ($\hat{M} = \hat{M}_A + \hat{M}_B$), which corresponds to the resonance absorption, is then given by:

$$v(\omega) = Im \left\{ \frac{f\hat{L}_A + (1-f)\hat{L}_B + \hat{L}_A\hat{L}_B(f\tau_A^{-1} + (1-f)\tau_B^{-1})}{1 - \hat{L}_A\hat{L}_B\tau_A^{-1}\tau_B^{-1}} \right\}$$ [56]

where

$$\hat{L}_A = [(\omega - \omega_A) + i(T_{2,A}^{-1} + \tau_A^{-1})]^{-1} \qquad [57]$$

$$\hat{L}_B = [(\omega - \omega_B) + i(T_{2,B}^{-1} + \tau_B^{-1})]^{-1}. \qquad [58]$$

The simulations can be performed for systems, e.g. lipid–protein complexes, which have individual spectral lineshapes that are more complicated than a simple, single Lorentzian line. This will be considered in greater detail in the following section, 4.3. Clearly, in the limit of no exchange ($\tau_A^{-1} = \tau_B^{-1} = 0$) the lineshape is a weighted sum of the component Lorentzians (cf. Eqn. 56), with the relative proportions reflecting the fractional populations, f. With the onset of exchange, the increase in the observed Lorentzian linewidths is given by Eqns. 11a,b. For more complex single component lineshapes, it is not necessarily expected that Eqn. 53 can be applied directly. However, for spin–labelled lipids in lipid–protein systems, the residual hyperfine anisotropy of the fluid lipid component can be very low and the individual hyperfine lines approximate quite closely to single, first–derivative EPR lineshapes (see e.g. ref. 53). Therefore, in principle, the exchange lifetime might be determined from the broadening of the fluid component in the lipid–protein complex relative to the spectra of the fluid lipids alone, using methods similar to those of section 4.1.

Making the approximation that the linebroadening in Eqn. 55 is dominated by the Gaussian component, ΔH_{pp}^G, the on–rate at the lipid–protein interface, τ_f^{-1}, can be determined from the observed peak–to–peak broadening according to (54):

$$\Delta H_{pp} = \frac{\hbar}{\sqrt{3}g\beta}\tau_f^{-1} + \Delta H_{pp}^G + \Delta H_{pp}^{L\,\circ} \qquad [59]$$

where the last term on the right is the *intrinsic* Lorentzian linewidth. Here, the Gaussian component contains contributions, not only from the unresolved proton hyperfine structure, but also from the residual ^{14}N hyperfine anisotropy. The predicted linear dependence of the linebroadening on the on–rate was confirmed from simulations by analyzing the broadening of the low–field and high–field nitroxide lines of the fluid component as a function of τ_f^{-1} (cf. Fig. 6). Over the range of $\tau_f^{-1} = 0$ to 4×10^7 s^{-1}, the apparent broadening of the low–field nitroxide line displayed a linear dependence with a gradient of 2.9×10^3 mT.s, which is reasonably close to the value of 3.3×10^3 mT.s, calculated from Eqn. 59. The high–field line displayed a lower sensitivity to exchange due to greater residual anisotropy (54).

4.3 Two–component lipid–protein systems

On reconstituting membrane proteins, two–component EPR spectra are obtained from spin–labelled lipids: one component (f) corresponds to the fluid lipid bilayer environment, and the other component (b) corresponds to lipids whose motion is restricted by direct interaction with the intramembraneous surface of the protein (see ref. 53). The reconstituted

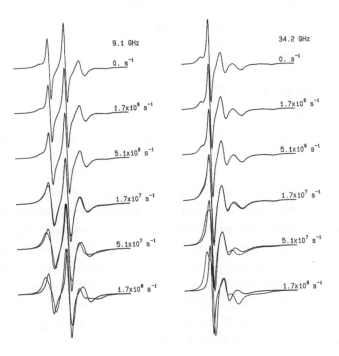

Fig. 6. Effect of exchange on the spectral line shapes of two–component 9 and 35 GHz EPR spectra. Simulated spectra corresponding to the uncorrelated jump model (full line) are shown together with spectra for the correlated jump model (broken line) at various exchange rates. The different asymptotic behaviour in the fast exchange regime is discussed in the text. The spectral intensity of the two components is in the ratio 1:1 $(i.e.\ \tau_b^{-1} = \tau_f^{-1})$. The spin Hamiltonian tensors and single–component linewidths are from ref. (54).

lipid vesicles that are obtained result in samples with an isotropic distribution of director orientations with respect to the static spectrometer magnetic field. The EPR spectrum is thus a weighted sum of spectra $v(\theta)$ corresponding to different orientations, the weighting factor being $p(\theta)\,d\theta = \sin\theta\,d\theta$. To simulate the entire lineshape, a summation must be made over all angular orientations, θ, of the spin–label director axes relative to the magnetic field direction (55). Since the nuclear spin orientation is preserved during the exchange, the line shapes corresponding to the $m_I = -1, 0$, and $+1$ manifolds are calculated separately. Depending on the nature of the reorientation of the molecular axes during the exchange process, two cases can be distinguished.

If the exchange event has no effect on the instantaneous molecular orientation, the angular orientation of the magnetic tensors remains correlated for the two spectral components. In this instance, only transitions corresponding to the same orientations are connected by exchange, i.e. only the fluid, $\hat{L}_f(\theta_i)$ and the motionally restricted, $\hat{L}_b(\theta_i)$, component lineshape positions, corresponding to the same orientation, θ_i, are in exchange. The overall spectrum is then a weighted sum of these mixed lineshapes, with an isotropic distribution of the orientation θ_i. This corresponds to model I described by Davoust and Devaux (55).

An alternative model for the exchange is to assume that the angular orientation of the spin label is uncorrelated before and after the exchange, i.e. there is a random reorientation of the lipid chains on collision with the invaginated protein surface. In this instance, the exchange process takes place between transitions in the fluid component, with the resonant frequency $\omega_f(\theta_i)$, and a distribution of transitions in the motionally restricted component, with resonant frequencies $\sum \omega_b(\theta_j)$. Thus, each Lorentzian lineshape, $\hat{L}_f(\theta_i)$, in the fluid component has to be exchanged with a sum of weighted Lorentzians, $\sum p(\theta_j)\hat{L}_b(\theta_j)$, in the motionally restricted component. Again, the summation must be made over an isotropic distribution of θ_i, as in the correlated exchange model. This corresponds to model II described by Davoust and Devaux (55).

The simulated EPR lineshapes corresponding to the uncorrelated and correlated jump model are compared in Fig. 6. The uncorrelated exchange model yields very similar spectra to those for the correlated exchange model in the slow–exchange regime. This is expected theoretically, since for slow exchange the spectra are not dependent on the resonance position to which the exchange takes place, provided that it is sufficiently far removed from the original resonance to ensure slow–exchange conditions (cf. Eqns. 11a,b). In the fast–exchange limit, the results from the two models differ considerably. For the uncorrelated model, quasi-isotropic spectra are obtained, whereas only partially motionally averaged, anisotropic spectra are obtained for the correlated model. This result is also that expected theoretically for fast exchange. In the uncorrelated model, the resonance positions are in fast exchange with an isotropic distribution and, in the correlated model, fast exchange results in motionally narrowed lines which are located at the mean resonance positions of the two components, for each fixed angular orientation (cf. Eqn. 13). Thus the decreasing spectral anisotropy predicted from the uncorrelated exchange model is entirely different from the motional averaging behaviour of the correlated exchange model. The asymptotic linewidths, however, are the same for both models, in accordance with Eqn. 15.

The spectral parameters needed for the exchange simulations can be adjusted in two stages. First, pure lipid and protein–alone reference spectra are compared with simulated single–component spectra in order to set the principal values of the $g-$ and $A-$tensors and to find the best matching linewidth parameters $(T_{2,m}^{-1})$. Motionally restricted endpoints obtained from spectral subtractions are useful for adjusting the motionally restricted component, whenever delipidated protein–alone spectra are not available. The single–component lipid spectra used to establish these first–level simulation parameters are those recorded at exactly the same temperature as that for the two–component spectrum of the lipid–protein

736

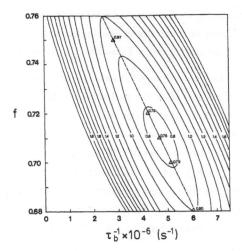

Fig. 7. Contour plot of the r.m.s. fitting error as a function of exchange rate (τ_b^{-1}) and the fraction of motionally restricted component (f). Spectral parameters for best fitting line shapes were selected by a computer search routine (steepest descent) minimizing the fitting error. Data are from ref. (54).

system in question. [This is in contrast to the strategy employed for the analysis of two–component spectra by spectral subtraction, where fluid and motionally restricted components are taken at somewhat lower and higher temperatures, respectively, in order to emulate the effects of exchange (cf. ref. 53)]. In the second stage of simulation, only the fraction of motionally restricted component and the exchange frequency are varied until the amplitudes of the simulated two–component spectrum (Y_{sim}) match the lineshape of the experimental spectrum (Y_{exptl}). Because this matching is never perfect over the entire spectrum, an error function is introduced:

$$\Delta Y^2 = \frac{\sum_i (Y_{sim,i} - Y_{exptl,i})^2}{\sum_i Y_{exptl,i}^2} \tag{60}$$

which has to be minimized by some optimization algorithm (56). Typical contour plots for the least–squares minimization are given in Fig. 7. In principle, the exchange rates are estimated primarily from exchange broadening based on the rather good fits in the low–field and central part of the spectra, which are expected to be most sensitive to exchange. The fits are less satisfactory in the high–field region of the fluid component, possibly due to long–range perturbations by the protein causing an increase in the spectral anisotropy.

4.4 Specificity of lipid–protein interactions and exchange rates

Spectral simulations for different phospholipid and stearic acid spin labels in myelin proteolipid apoprotein–, viral M13 coat protein–, and rhodopsin–lipid recombinants of varying lipid/protein ratio have led to two general conclusions. (i) The off–rate, τ_b^{-1}, for the protein–interacting lipid population is approximately constant, independent of the total lipid/protein ratio. Increases in exchange rate are occasionally observed at high lipid/protein ratios. (ii) The off–rate constants for the different lipid spin probes are inversely proportional to their relative association constants, K_r^{av}, with the protein. This indicates a significantly longer lifetime at the interfacial sites for those lipids which display a selectivity for the protein.

The two independent conjugate variables which can be obtained from the exchange–coupled, two–component simulations are τ_b^{-1} and the fraction, f, of motionally restricted lipid component (cf. Fig. 8). The off–rate, τ_b^{-1}, is the intrinsic parameter governing the lipid exchange, and the on–rate, τ_f^{-1}, is simply related to this by the requirement for material balance (Eqn. 9):

$$\tau_f^{-1} = \tau_b^{-1} \frac{f}{1-f} \; . \tag{61}$$

Values for the intrinsic off–rate for three different lipid–protein systems studied are found to be in the range $\tau_b^{-1} \approx 0.5-1.5 \times 10^7$ s^{-1} at 30° C for lipids which show no selectivity (54,56–58). These values are of the same order of magnitude, but significantly slower, than the rates of translational diffusion in protein–free lipid bilayers. The latter typically have values of $\tau_{diff}^{-1} \approx 4 D_T / \langle r^2 \rangle \approx 7.5 \times 10^7$ s^{-1} for the same lipids at the same temperature, when determined by the methods described in section 5.3. The relatively rapid off–rates explain why the spectra of deuterium–labelled lipids are in fast exchange on the ^{2}H NMR timescale in lipid–protein systems (59,60).

The stoichiometry and selectivity of interaction for the different spin–labelled lipids can be expressed in terms of the equation for equilibrium association with the protein (53,61):

$$\frac{n_f^*}{n_b^*} = \left(\frac{n_t}{N_b} - 1 \right) \frac{1}{K_r^{av}} \tag{62}$$

where $n_f^*/n_b^* = (1-f)/f$ is the ratio of the number of labelled lipids in fluid and interfacial sites, n_t is the total lipid–to–protein ratio, and N_b is the number of interfacial sites on the protein. The dependence of the lipid off–rate and the ratio of fluid to motionally restricted lipid components on lipid/protein ratio in M13 coat protein–dimyristoyl phosphatidylcholine recombinants is given in Fig. 9. These parameters were obtained from the values of τ_b^{-1} and f, respectively, which were derived from simulation of the EPR spectra of spin–labelled lipids. One of the spin–labelled lipids (PA^-) displays a selectivity relative to the host background lipid for interaction with the protein, and the other (PC) displays no selectivity. Whereas the ratio of fluid to motionally restricted lipids depends strongly on the lipid/protein ratio, the off–rates are seen to be independent of lipid/protein ratio. The lipid/protein titration shown in the lower part of the figure conforms to Eqn. 62 and indicates that the two lipids

738

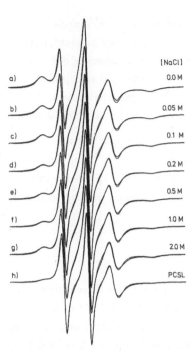

Fig. 8. Experimental (full line) and simulated (broken line) EPR spectra of spin–labelled stearic acid, as a function of NaCl concentration, in myelin proteolipid apoprotein–dimyristoyl phosphatidylcholine recombinants of lipid/protein ratio 37:1 mol/mol, at pH 9.0. (a–g) stearic acid spin–label at the indicated NaCl concentrations; (h) phosphatidylcholine spin–label, PCSL, at 0.1 M NaCl at pH 7.4. T=30° C, scan width=10 mT. R.m.s. deviations between the experimental and simulated spectra are in the range 0.6–1.0 %. The values of τ_b^{-1} and f required for the various simulations are inversely related, whereas the value of τ_f^{-1} remains constant. From ref. (56).

compete for the same $N_b \approx 5$ sites on the protein, but with different relative association constants: $K_r^{av} \approx 1$ for PC and $K_r^{av} \approx 2$ for PA^-. This difference in selectivity of the lipid–protein interaction is directly reflected in the off–rates for the two lipids: that for PA^- is approximately half that for PC, corresponding to a two–fold longer residence time at the lipid–protein interface.

The connection between the lipid selectivity and the exchange rates can be seen by substituting the condition for material balance (Eqn. 61) into the equation for lipid–protein association (Eqn. 62):

$$\frac{\tau_b^{-1}}{\tau_f^{-1}} = \left(\frac{n_t}{N_b} - 1\right)\frac{1}{K_r^{av}} \; . \tag{63}$$

It can be assumed that the on–rate, τ_f^{-1}, is constant for a given lipid/protein ratio, because it is diffusion–controlled. This assumption is justified by the simulations in Fig. 8, where

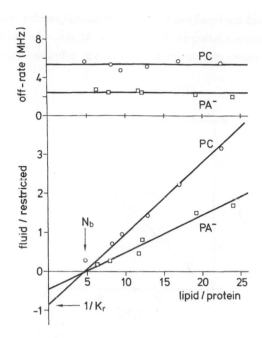

Fig. 9. *Upper*: off–rates, and *lower*: ratio of fluid to motionally restricted components, of spin–labelled phosphatidylcholine (PC) and phosphatidic acid (PA^-) as a function of lipid/protein mole ratio in recombinants of the M13 viral coat protein with dimyristoyl phosphatidylcholine, at 30° C. The values of K_r^{av} and N_b (Eqn. 62) are determined as indicated, from the intercepts on the ordinate and abcissa, respectively, of the lower part of the figure. Data from ref. (58).

the lipid/protein selectivity is dependent on ionic strength, and whereas τ_b^{-1} and f vary, τ_f^{-1} remains constant (56). In addition, measurements of the collision rates between diffuseable ^{14}N–labelled lipids and ^{15}N–labelled chains covalently linked to the protein, have demonstrated directly that the lipid on–rate is diffusion–controlled (62). Hence for different spin–labelled lipids, say L and PC, at fixed lipid/protein ratio and fixed temperature, the ratio of off–rates from Eqn. 63 is:

$$\frac{\tau_b^{-1}(PC)}{\tau_b^{-1}(L)} = \frac{K_r^{av}(L)}{K_r^{av}(PC)} \, . \tag{64}$$

This is exactly the inverse relation observed for the selectivity between PA^- and PC in Fig. 9, and also accounts very well for the dependence of the selectivities and off–rates on ionic strength in Fig. 8 (56). The results of simulations for a range of other lipids are in similar agreement, as summarized in the correlation plot of K_r^{av} $vs.$ τ_b^{-1} that is shown in Fig. 10. These simulations for two different proteins were all performed with the same fluid component, but different motionally restricted components (spectra recorded at 30° C). The data conform to the inverse relation predicted by Eqn. 63, since the product $[(n_t/N_b) - 1].\tau_f^{-1}$ is expected to remain constant if the intrinsic on–rate is diffusion–controlled.

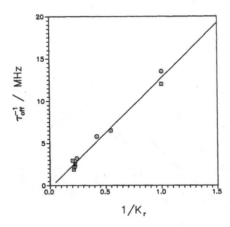

Fig. 10. Correlation diagram between the off–rate constants, τ_b^{-1}, and the relative association constants, K_r^{av}, for two different lipid–protein systems. (o) Myelin proteolipid apoprotein reconstituted in dimyristoyl phosphatidylcholine; (⊓) M13 viral coat protein reconstituted in a mixture of 80 mol% dimyristoyl phosphatidylcholine and 20 mol% dimyristoyl phosphatidylglycerol. Data are from refs. (54,58,63).

5. SPIN–SPIN INTERACTIONS: TRANSLATIONAL DIFFUSION

At high local spin label concentrations magnetic interactions occur between the radical spins. These interactions are either of a magnetic dipole–dipole origin or arise from Heisenberg spin exchange. The exchange interaction is of extremely short range and therefore can be used to measure bimolecular collision frequencies and hence translational diffusion coefficients. The magnetic dipolar interaction depends both on intermolecular separation and the orientation of the inter–spin vector, and can be used for investigating both proximity relationships and translational diffusion.

5.1 Spin exchange interactions

The Heisenberg exchange interaction makes a contribution to the spin Hamiltonian of the form $J S_1.S_2$, and therefore couples the two spin states, $S_1 = \frac{1}{2}$ and $S_2 = \frac{1}{2}$, to a singlet $(S = 0)$ and a triplet $(S = 1)$ state. As a result of this coupling, radicals with oppositely directed spins exchange their spin orientations within the two–spin states. The frequency of this spin exchange depends on the exchange constant, $J/\hbar$. The probability that spin exchange takes place within the collision lifetime, τ_P, of a radical pair is given by (see e.g. ref. 6):

$$p_{ex} = \frac{\frac{1}{2}(J/\hbar)^2\tau_P^2}{1 + (J/\hbar)^2\tau_P^2} \; . \qquad [65]$$

For nitroxide radicals, $(J/\hbar) \approx 6 \times 10^{10}$ s^{-1} (64). Therefore $p_{ex} = \frac{1}{2}$ for collision pair lifetimes of $\tau_P \gg 2 \times 10^{-11}$ s. This latter condition implies off–rates of less than 5×10^{10} s^{-1}, which is likely to be the case for nearly all situations of biological relevance, including the diffusion of small molecules in water.

The effects of spin exchange on the EPR spectrum of interacting spin labels can be calculated using the exchange–coupled Bloch equations introduced in section 1.3. Spin exchange influences the EPR spectrum only if the colliding radicals have different nuclear spin orientations, since only then does the exchange give rise to a shift in resonance frequency. As regards the EPR spectrum, the effective rate of exchange is therefore:

$$\tau_{ex}^{-1} = \frac{2I}{2I + 1} p_{ex} \, \tau_{coll}^{-1} \qquad [66]$$

where τ_{coll}^{-1} is the frequency of collisions. For a nitroxide spin label, with $I(^{14}N) = 1$, only two–thirds of the spin exchanges have an effect on the EPR spectrum.

The version of the two–site exchange–coupled Bloch equations (cf. Eqns. 10a,b) relevant to the different hyperfine manifolds of the spin label are (65, 66):

$$(T_{2,m}^{-1} + 3\nu_{ex})u_m - \nu_{ex} \sum_{m'=-1}^{+1} u_{m'} + (\omega_m - \omega)v_m = 0 \qquad [67a]$$

$$(T_{2,m}^{-1} + 3\nu_{ex})v_m - \nu_{ex} \sum_{m'=-1}^{+1} v_{m'} - (\omega_m - \omega)u_m = -\gamma B_1 M_m^0 \qquad [67b]$$

where the subscripts m and m' refer to the nuclear spin magnetic quantum number ($m_I = 0, \pm 1$). Note that eqns. 67a,b are formulated in such a way that the net spin transfer rate in Eqn. 66 is given by $\tau_{ex}^{-1} = 2\nu_{ex}$ (66).

Simulations based on Eqns. 67a,b reproduce the spectral effects summarized by Eqns. 11a,b and 12–14 of section 1.3, depending on the exchange frequency (67). For slow exchange ($\nu_{ex} \leq 10^{-7}$ s $\ll | \omega_0 - \omega_{\pm 1} |$), corresponding to low spin label concentrations, the hyperfine lines are unshifted, but each is subject to a Lorentzian broadening by an amount τ_{ex}^{-1}. For intermediate exchange frequencies ($\nu_{ex} \leq 5 \times 10^{-7}$ s $< | \omega_0 - \omega_{\pm 1} |$), the outer hyperfine lines move in towards the central line by an amount which is given by an effective reduction in the nitrogen hyperfine splitting constant, A_N, by a factor of $[1 - 8\tau_{ex}^{-2}(\omega_0 - \omega_{\pm 1})^{-2}]^{1/2}$. In the fast exchange limit ($\nu_{ex} \geq 10^{-9}$ s $\gg | \omega_0 - \omega_{\pm 1} |$), the hyperfine lines collapse to a single Lorentzian line, at the position of the $m_I = 0$ hyperfine line and with an additional Lorentzian broadening by an amount $\frac{1}{8}(\omega_0 - \omega_{\pm 1})^2 \tau_{ex}$.

This collapse of the hyperfine structure is a unique feature of the exchange interaction and can be used to distinguish it from the magnetic dipole–dipole interaction (68, 69). The latter produces a monotonously increasing linewidth with increasing strength of spin–spin interaction, which is either Lorentzian in the dynamic case, or approximately Gaussian in the static case. Even incipient exchange narrowing can be detected unambiguously, as is illustrated in Fig. 11. The EPR spectra recorded at three different temperatures reflect increasing rates of translational diffusion and correspond to situations in which: a) both dipole–dipole and exchange interactions contribute a Lorentzian broadening, b) there is incipient exchange narrowing, and c) exchange narrowing dominates. In the first case it is not possible to distinguish between exchange and dipolar interactions by simulation alone. However, already for the intermediate exchange regime (Fig. 11b), the exchange and dipolar interactions can be distinguished by simulation. A pure Lorentzian broadening (dipole–dipole interaction) cannot account satisfactorily for the observed spectrum in this case.

The ^{14}N hyperfine manifolds in the spin label spectra are inhomogeneously broadened by unresolved proton hyperfine structure. The inhomogeneous broadening will be removed by exchange narrowing when the exchange frequency is greater than the proton hyperfine splittings. This is found to occur for exchange frequencies $\tau_{ex}^{-1} \geq 1.2$ MHz for doxyl nitroxides (69). Thus for higher exchange frequencies than this value, inhomogeneous broadening may be neglected in the exchange simulations. For lower spin label concentrations, corresponding to exchange frequencies below this limit, the inhomogeneous broadening must be considered explicitly. The ^{14}N hyperfine structure is in slow exchange in this regime and exchange therefore simply gives a Lorentzian broadening. The total peak–to–peak experimental linewidth can then be decomposed into its Gaussian and Lorentzian contributions using the method described by Bales (51).

The collision frequency, and hence the exchange frequency (Eqn. 66), is linearly related to the spin label concentration. This is now illustrated for one biologically important case: the lateral diffusion of lipid molecules in membranes (see e.g. ref. 9). The mean-square displacement, $\langle r^2 \rangle$, of a lipid molecule in time t' is given by the Einstein relation for two–dimensional diffusion:

$$\langle r^2 \rangle = 4D_T t'. \tag{68}$$

If a lattice model is used for the diffusion process, the diffusion coefficient is then given by:

$$D_T = \frac{1}{4}\nu_j \lambda^2 \tag{69}$$

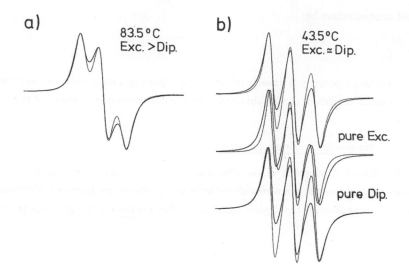

Fig. 11. Simulated (dashed line) and experimental (full line) spin-spin broad-ened EPR spectra of spin-labelled phosphatidylcholine (9.4 mole%) in fluid phosphatidyl-choline bilayer membranes. a) Experimental spectrum recorded at 28.5° C. Spectrum simulated with $\tau_{ex}^{-1} = 2.77 \times 10^7$ s^{-1}, $\Delta H_{pp,L}^{(dd)} = 0.379$ mT. b) experimental spectrum recorded at 43.5° C. Upper spectrum simulated with $\tau_{ex}^{-1} = 3.78 \times 10^7$ s^{-1}, $\Delta H_{pp,L}^{(dd)} = 0.278$ mT. Middle spectrum simulated with $\tau_{ex}^{-1} = 8.02 \times 10^7$ s^{-1}, $\Delta H_{pp,L}^{(dd)} = 0.0$ mT. Lower spectrum simulated with $\tau_{ex}^{-1} = 0.0$ s^{-1}, $\Delta H_{pp,L}^{(dd)} = 0.536$ mT. c) Experi-mental spectrum recorded at 83.5° C. Spectrum simulated with $\tau_{ex}^{-1} = 1.18 \times 10^8$ s^{-1}, $\Delta H_{pp,L}^{(dd)} = 0.048$ mT. From ref. (69).

where $\lambda(\approx 8$ Å) is the lattice spacing (or mean free path) and ν_j is the hopping frequency between adjacent lattice sites. For bimolecular diffusive encounters, the origin of coordinates ($r = 0$) is not a fixed position but is defined by the position of another identical molecule. Then D_T must be replaced by $2D_T$ in Eqns. 68, 69. If a given lipid exchanges with one of its six nearest neighbours, it encounters three new nearest neighbours and the collision frequency is given by (66):

$$\tau_{coll}^{-1} = 3\nu_j c \qquad [70]$$

where c is the mole fraction of spin–labelled lipid. Combining Eqns. 69 and 70, and including the substitution of $2D_T$ for bimolecular collisions, the collision frequency is related to the spin label concentration by:

$$\tau_{coll}^{-1} = \frac{24D_T}{\lambda^2} c \cdot$$ [71]

Thus the collision frequency is linearly related to the spin label concentration, and the translational diffusion coefficient can be obtained from the concentration gradient of the exchange frequency.

5.2 Magnetic dipole–dipole interactions

The effects of magnetic dipole–dipole interactions on the spin label EPR spectrum depends on whether or not the nitroxide radicals are in relative motion on a timescale faster than that characteristic for the dipolar interaction. The characteristic timescale is defined by (70):

$$\tau_D = \frac{2r_c^2}{D_T}$$ [72]

where D_T is the translational diffusion coefficient of the spin labels, and $2r_c$ is the distance of closest approach of the two spin dipoles. The *static limit* is defined by: $\tau_D/T_{2,m} \gg 1$, where $T_{2,m}$ is the transverse electron spin relaxation time. In this case, the spin systems are coupled and give rise to an inhomogeneous linebroadening caused by the local fields of the neighbouring dipoles. The *dynamic limit* is defined by: $\tau_D/T_{2,m} \ll 1$. In this case, the dipolar interaction is modulated by the mutual motions of the spin dipoles and contributes to the transverse and longitudinal relaxation of the electron spins. The transverse dipolar relaxation results in a homogeneous broadening of the spectral lines, uniformly increasing the Lorentzian width of all hyperfine manifolds.

In the case of *static dipolar broadening*, the EPR lineshape depends on the spin concentration range. For an isolated spin label pair, triplet state spectra are obtained. The nuclear spins, $I_1, I_2 = 1$, are coupled to a total spin of $I = 2$, giving rise to 5 hyperfine lines, which may or may not be resolved. The electron spins are coupled to a triplet state, giving rise to two resonances which are split by the dipolar interaction by an amount (71):

$$\Delta B_{dd} = \frac{3}{2} \frac{g\beta}{r^3} (1 - 3\cos^2\theta)$$ [73]

where θ is the angle between the magnetic field direction and the inter-spin vector, $\mathbf{r}$. Thus for an isolated pair of radicals, the separation, r, between the spin label groups can be measured directly from the dipolar splittings of the EPR lines (72). Isotropic Heisenberg spin exchange does not affect the splitting in the EPR spectrum, because it contributes only to the splitting of the singlet and triplet levels, and not to the splitting of the triplet levels between which the EPR transitions take place.

For several interacting spin labels at higher concentrations, the EPR lineshape takes a complicated form which may or may not be approximated by a Gaussian lineshape. An analytical expression, due to Van Vleck, can be given for the second moment of the lineshape which is defined by:

$$M_2 = \langle B - B_0 \rangle^2 \qquad [74]$$

and is given by the mean square deviations of the field positions, B, relative to the centre resonance field, B_0, averaged over the entire lineshape. For a Gaussian lineshape:

$$L_G(B) = \frac{1}{\Delta H_{pp}^G} \sqrt{\frac{2}{\pi}} e^{-2(B-B_0)^2/\Delta H_{pp}^{G\,2}} \qquad [75]$$

the second moment is given simply by (73):

$$M_2 = \left(\frac{\Delta H_{pp}^G}{2}\right)^2 \qquad [76]$$

where ΔH_{pp}^G is the peak-to-peak linewidth in the first derivative spectrum, and is related to the full width at half-height of the absorption spectrum by: $\Delta H_{1/2} = \sqrt{2\ln 2}\, \Delta H_{pp}^G$. Thus for a Gaussian lineshape, the second moment can be obtained directly from the peak-to-peak linewidth. One of the criteria that the lineshape shall be Gaussian is that the fourth and second moments are related by: $M_4/M_2^2 = 3$ (73).

The Van Vleck second moment for dipolar broadening of like spins (e.g. corresponding to the same hyperfine manifold) is given by (73):

$$M_2 = \frac{3}{5} g^2 \beta^2 S(S+1) \sum_j \left(\frac{1}{r_{jk}}\right)^6 \qquad [77]$$

where $S = \frac{1}{2}$ is the electron spin quantum number, and the summation excludes the term $j = k$. This equation applies to a powder sample with random orientation of the inter–spin vectors relative to the magnetic field direction. For unlike spins, such as applies for the different ^{14}N hyperfine manifolds in the spin label spectrum when the dipolar broadening is less than the hyperfine splitting, the second moment is reduced by a factor of $\frac{4}{9}$. Thus, in the case of weak dipolar broadening, the second moment of an individual spin label hyperfine manifold is given by the average of that for like and unlike spins, weighted in the ratio 1:2. The net result is:

$$M_2(SL) = \frac{17}{60} g^2 \beta^2 \sum_j \left(\frac{1}{r_{jk}}\right)^6. \qquad [78]$$

For very strong dipolar broadening, in which the hyperfine structure is lost, the multiplicative factor in Eqn. 78 is $\frac{9}{20}$. For a uniform distribution of spins, the summation in Eqns. 77, 78 can be replaced by an integral, yielding a value of $\frac{4}{3}\pi n(1/b)^3$, where b is the distance of closest approach of the spin dipoles and n is their number density per unit volume. If necessary, the first shell of dipoles can be treated by direct summation, and then b becomes the radius of the boundary of the first shell. As noted above, the experimental second moment can be obtained from the linewidth using Eqn. 76. Since for a uniform distribution, the second moment is directly proportional to the density of spin dipoles, n, this equation predicts that the dipolar broadening will be proportional to the square root of the spin label concentration.

For *static dipolar broadening* at low concentration, statistical theories for high dilutions yield a Lorentzian broadening with a linear dependence on spin concentration (73):

$$\Delta H_{pp}^{(dd)} = \frac{4\pi^2}{9} g\beta\, n \tag{79}$$

where $\Delta H_{pp}^{(dd)}$ is the peak-to-peak linewidth of the first derivative spectrum, and $n(\approx 1/r^3)$ is the number density per unit volume of spin dipoles. Such a concentration dependence is clearly different from that predicted by the moments calculations applicable to higher spin label concentrations.

For the *dynamic limit*, the transverse relaxation induced by magnetic dipole-dipole interactions can be evaluated using motional narrowing theory (70, 73). For two like spins, S (i.e. associated with the same ^{14}N nuclear quantum number, $m_I = m$), the transverse relaxation time is given by:

$$\frac{1}{T_{2,m}^{(dd)}} = \frac{g^4\beta^4}{\hbar^2} S(S+1) [\frac{3}{8} J^{(0)}(0) + \frac{15}{4} J^{(1)}(\omega_m) + \frac{3}{8} J^{(2)}(2\omega_m)] \tag{80}$$

where $J^{(i)}$ are the spectral densities appropriate to the mutual diffusive motion of the two dipoles. For two unlike spins, S (i.e. associated with different nuclear magnetic quantum numbers, m and m'), the transverse relaxation time of the electron spin associated with nuclear quantum number m is given by:

$$\frac{1}{T_{2,m}^{\prime(dd)}} = \frac{g^4\beta^4}{\hbar^2} S(S+1) [\frac{1}{6} J^{(0)}(0) + \frac{1}{24} J^{(0)}(\omega_m - \omega_{m'}) + \frac{3}{4} J^{(1)}(\omega_m) +$$

$$+ \frac{3}{2} J^{(1)}(\omega_{m'}) + \frac{3}{8} J^{(2)}(\omega_m + \omega_{m'})] \ . \tag{81}$$

The probability of interaction between spins of like nuclear magnetic quantum number is $(2I+1)^{-1}$ and that between unlike spins is $2I(2I+1)^{-1}$, where I is the nuclear spin quantum number. Therefore, the net transverse relaxation induced by dipolar interaction is given by:

$$\frac{1}{T_2^{(dd)}} = \frac{(2I+1)^{-1}}{T_{2,m}^{(dd)}} + \frac{2I(2I+1)^{-1}}{T_{2,m}^{\prime(dd)}} \tag{82}$$

where the two terms on the right are specified by Eqns. 80 and 81, respectively.

If it is assumed that $\omega_m^2 \tau_D^2 \gg 1$, which is valid for $D_T \ll 10^{-4}$ cm^2 s^{-1}, all spectral density terms at frequencies ω_m and $(\omega_m + \omega_{m'})$ can be neglected. If it is further assumed that $A^2 \tau_D \ll 1$, where $A(\approx \omega_m - \omega_{m'})$ is the hyperfine splitting constant, the spectral densities at frequencies $(\omega_m - \omega_{m'})$ can be set equal to those at zero frequency. This latter condition requires $D_T \gg 7 \times 10^{-7}$ cm^2 s^{-1}, which is not always attained in practice, but the approximation is retained for the sake of mathematical simplicity. Thus only the spectral density at zero frequency is required. This can be obtained from solution of the three–dimensional translational diffusion equation, and is found to be (73):

$$J^{(n)}(0) = \frac{48\pi}{15^2} \frac{n}{r_c D_T} \tag{83}$$

where n is the number density per unit volume of spin dipoles. The final expression for the dipolar contribution to the transverse relaxation of nitroxide spin labels then becomes:

$$\frac{1}{T_2^{(dd)}} = \frac{19\pi}{450} \frac{g^4 \beta^4}{\hbar^2} \frac{1}{r_c D_T} n \tag{84}$$

which predicts a Lorentzian broadening that is linear in concentration and inversely proportional to the translational diffusion coefficient.

5.3 <u>Line broadening at low concentration</u>

The spectral effects of both dipolar interaction and spin exchange are sensitive to translational diffusion. For high concentrations, i.e. high bimolecular collision rates, the two interactions can be distinguished, since the dipolar interaction leads to a broadening and the exchange interaction to a narrowing of the spectral lines (cf. section 5.1). In general, it is desirable to use low spin label concentrations. However, at low concentrations, both interactions give rise to a Lorentzian broadening of the spectral lines which is linearly dependent on the concentration. In this case they can be distinguished by their opposite dependences on the rate of translational diffusion. For rapid diffusion rates, as for example with small molecules in water, the dipole–dipole interaction is averaged to zero (see e.g. ref. 52). The spin exchange frequency, τ_{ex}^{-1}, can then be obtained directly from the dependence of the peak–to–peak Lorentzian linewidth, ΔH_{pp}^L, on concentration, c:

$$\frac{\tau_{ex}^{-1}}{c} = \frac{\sqrt{3}g\beta}{2\hbar} \frac{d(\Delta H_{pp}^L)}{dc} . \tag{85}$$

The Lorentzian linewidth can be obtained from the experimental linewidth (which is inhomogeneously broadened by unresolved proton hyperfine structure) using the deconvolution procedure described by Bales (51). The second–order collision rate constant is simply related to the collision frequency (Eqn. 66) by: $k_{coll} = \tau_{coll}^{-1}/c$, and the translational diffusion coefficient, D_T, is then obtained from the Smoluchowski equation (52):

$$k_{coll} = 16\pi r_{coll} D_T \tag{86}$$

where r_{coll} is the molecular collision radius and the concentration, c, is expressed in spins per unit volume.

For slower rates of diffusion, as is the case for lipid molecules in membranes, the concentration–dependent Lorentzian broadening is contributed by both the dipolar and exchange interactions. In this case, the contributions from the two interactions can be distinguished by their opposing temperature dependences. Since the diffusion coefficient is given by the Einstein relation: $D_T = kT/f_T$, where the friction coefficient, f_T, is proportional to the effective membrane viscosity, its temperature dependence is expected to be of the form:

$$D_T = D_{T,0}\, T e^{-E_a/kT} \qquad [87]$$

where E_a is the activation energy associated with the effective membrane viscosity. The concentration–dependent gradient of the total Lorentzian broadening [again determined from the total linewidth using the Bales (51) deconvolution procedure] is the sum of the exchange and dipolar contributions:

$$\frac{d[\Delta H_{pp}^L]}{dc} = \frac{d[\Delta H_{pp}^L(ex)]}{dc} + \frac{d[\Delta H_{pp}^L(dd)]}{dc} \qquad [88]$$

where, in this case, the spin–labelled lipid concentration, c, is conventionally taken to be in mole fraction units. The temperature dependence of Eqn. 88 therefore will have the following form (cf. Eqns. 71, 84, 87):

$$\frac{d[\Delta H_{pp}^L]}{dc} = A\, T\, e^{-E_a/kT} + B\, T^{-1}\, e^{E_a/kT} \qquad [89]$$

where A and B are temperature–independent constants representing the exchange and dipolar interactions, respectively. These have the following values for nitroxide spin labels (69):

$$A = \frac{32\hbar}{\sqrt{3}g\beta\lambda^2}\, D_{T,0} \qquad [90]$$

where λ is the elementary translational step for two–dimensional diffusion, and

$$B = \frac{19g^3\beta^3\pi N_{Av}}{225\sqrt{3}\hbar r_c}(M_L\, \bar{v}_L)^{-1}\, D_{T,0}^{-1} \qquad [91]$$

where an effective three–dimensional diffusion is assumed for the spin–labelled lipid chain segment. M_L and $\bar{v}_L$ are the lipid molecular weight and partial specific volume, respectively, which are needed to convert from molar concentration to mole fraction units, and N_{Av} is Avogadro's number. The constants A, B and E_a can be determined by fitting the biphasic temperature dependence of the concentration dependence of the Lorentzian broadening, and hence yield two independent estimates of the translational diffusion coefficient, from the exchange (A) and dipolar (B) terms, respectively. Agreement to within a factor of two is found for the two methods (69), which is satisfactory in view of the quite different models used for the diffusion process. Since the exchange interaction gives a direct measure of the collision rate, it is likely to be the preferred method.

6. CONCLUSION

EPR spectroscopy can be used to study the conformations and dynamics of spin–labelled biomolecules. The methods described in this chapter detail the techniques used to analyze rotational and translational diffusion, the dynamics of molecular exchange, the degree of orientational ordering in membranes, and the dynamic configuration of chain molecules. The application to lipid–protein interactions in membranes yields structural (stoichiometry) and thermodynamic (selectivity) data, as well as dynamic information (exchange rates). Saturation transfer EPR can be used to study the slower rotational diffusion of supramolecular aggregates or proteins in membranes. In addition to the use in determining translational diffusion rates, spin–spin interactions can also yield distance information (intermolecular separations).

REFERENCES

1 P.C. Jost, L.J. Libertini, V.C. Hebert and O.H. Griffith, J. Mol. Biol., 59 (1971) 77-98.

2 H. Schindler and J. Seelig, J. Chem. Phys., 59 (1973) 1841-1850.

3 B.J. Gaffney and H.M. McConnell, J. Magn. Reson., 16 (1974) 1-28.

4 F. Bloch, Phys. Rev., 70 (1946) 460-474.

5 H.M. McConnell, J. Chem. Phys., 28 (1958) 430-431.

6 D. Marsh, in: G. Pifat-Mrzljak (Ed), Supramolecular Structure and Function, Springer-Verlag, Berlin-Heidelberg, 1986, pp. 48-62.

7 H.S. Gutowsky and C.H. Holm, J. Chem. Phys., 25 (1956) 1228-1234.

8 C.R. Cantor and P.R. Schimmel, Biophysical Chemistry. Part II, W. H. Freeman, San Francisco, 1980.

9 D. Marsh, in: C. Hidalgo (Ed), Physical Properties of Biological Membranes and Their Functional Implications, Plenum, New York, 1988, pp. 123-145.

10 A.D. Keith and W. Snipes, Science, 183 (1974) 666-668.

11 B. Schobert and D. Marsh, Biochim. Biophys. Acta, 720 (1982) 87-95.

12 F. Perrin, J. Phys. Radium, 5 (1934) 497-511.

13 S.H. Koenig, Biopolymers, 14 (1975) 2421-2423.

14 J.H. Freed, in: L.J. Berliner (Ed), Spin Labeling, Theory and Applications., Vol. I, Academic Press, New York, 1976, pp. 53-132.

15 W. Maier and A. Saupe, Z. Naturforsch. A, 14 (1959) 882-889.

750

16 A. Saupe, Z. Naturforsch., 19a (1964) 161-171.

17 H. Schindler and J. Seelig, J. Chem. Phys., 61 (1974) 2946-2949.

18 S.A. Goldman, G.V. Bruno, C.F. Polnaszek and J.H. Freed, J. Chem. Phys., 56 (1972) 716-735.

19 D. Marsh, in: P.M. Bayley and R.E. Dale (Eds), Spectroscopy and the Dynamics of Molecular Biological Systems, Academic Press, London, 1985, pp. 209-238.

20 C.F. Polnaszek, D. Marsh and I.C.P. Smith, J. Magn. Reson., 43 (1981) 54-64.

21 R.C. McCalley, E.J. Shimshick and H.M. McConnell, Chem. Phys. Lett., 13 (1972) 115-119.

22 H. Hauser, I. Pascher, R.H. Pearson and S. Sundell, Biochim. Biophys. Acta 650 (1981) 21-51.

23 M. Moser, D. Marsh, P. Meier, K.-H. Wassmer and G. Kothe, Biophys. J., (1989) in press.

24 J.P. Meraldi and J. Schlitter, Biochim. Biophys. Acta, 645 (1981) 183-210.

25 A. Lange, D. Marsh, K.-H. Wassmer, P. Meier and G. Kothe, Biochemistry, 24 (1985) 4383-4392.

26 C.P. Slichter, Principles of Magnetic Resonance, 2nd. Edition, Springer-Verlag, Berlin-Heidelberg, 1978.

27 R. Kubo, in: K.E. Shuler (Ed) Stochastic Processes in Chemical Physics, Advances in Chemical Physics, Wiley, New York, 1969, pp. 101-127.

28 G. Moro and J.H. Freed, J. Chem. Phys., 74 (1981) 3757-3773.

29 D.J. Schneider and J.H. Freed, in (L.J. Berliner, Ed), Spin Labeling. A Third Compendium, Plenum Press, New York, 1989.

30 Dalton, L. R. (Ed), EPR and Advanced EPR Studies of Biological Systems, CRC Press, Boca Raton, 1985.

31 M.A. Hemminga, P.A. De Jager, D. Marsh and P. Fajer, J. Magn. Reson., 59 (1984) 160-163.

32 P. Fajer and D. Marsh, J. Magn. Reson., 49 (1982) 212-224.

33 R.G. Kooser, W.V. Volland and J.H. Freed, J. Chem. Phys., 50 (1969) 5243-5257.

34 A.I. Vistnes and L.R. Dalton, J. Magn. Reson., 54 (1983) 78-88.

35 Poole, C. P., Jr., Electron Spin Resonance. A Comprehensive Treatise on Experimental Techniques, Wiley-Interscience, New York, 1967.

36 D.D. Thomas, L.R. Dalton and J.S. Hyde, J. Chem. Phys., 65 (1976) 3006-3024.

37 D. Marsh, in: E. Grell (Ed), Membrane Spectroscopy, Springer-Verlag, Berlin-Heidelberg-New York, 1981, pp. 51-142.

38 P.A. De Jager and M.A. Hemminga, J. Magn. Reson., 31 (1978) 491-496.

39 J.S. Hyde and D.D. Thomas, Ann. Rev. Phys. Chem., 31 (1980) 293-317.

40 P.G. Saffman, J. Fluid Mech., 73 (1976) 593-602.

41 F. Jähnig, Eur. J. Biophys., 14 (1986) 63-64.

42 L.I. Horváth and D. Marsh, J. Magn. Reson., 54 (1983) 363-373.

43 B.H. Robinson and L.R. Dalton, J. Chem. Phys., 72 (1980) 1312-1324.

44 L.I. Horváth and D. Marsh, J. Magn. Reson., 80 (1988) 314-317.

45 D. Marsh, Biochemistry, 19 (1980) 1632-1637.

46 P. Fajer and D. Marsh, J. Magn. Reson., 51 (1983) 446-459.

47 C.A. Evans, J. Magn. Reson., 44 (1981) 109-116.

48 T.C. Squier and D.D. Thomas, Biophys. J., 49 (1986) 921-935.

49 S.N. Dobryakov and Ya.S. Lebedev, Sov. Phys. Dokl., 13 (1969) 873-875.

50 M.D. King and D. Marsh, Biochemistry, 26 (1987) 1224-1231.

51 B.L. Bales, J. Magn. Reson., 48 (1982) 418-430.

52 M.D. King and D. Marsh, Biochim. Biophys. Acta, 863 (1986) 341-344.

53 D. Marsh, in: A. Watts and J.J.H.H.M. de Pont (Eds), Progress in Protein-Lipid Interactions, Vol. 1, Elsevier, Amsterdam, 1985, pp. 143-172.

54 L.I. Horváth, P.J. Brophy and D. Marsh, Biochemistry, 27 (1988) 46-52.

55 J. Davoust and P.F. Devaux, J. Magn. Reson., 48 (1982) 475-494.

56 L.I. Horváth, P.J. Brophy and D. Marsh, Biochemistry, 27 (1988) 5296-5304.

57 N.J.P. Ryba, L.I. Horváth, A. Watts and D. Marsh, Biochemistry, 26 (1987) 3234-3240.

58 C.J.A.M. Wolfs, L.I. Horváth, D.Marsh, A. Watts and M. A. Hemminga, (1989) to be published.

59 P. Meier, J.-H. Sachse, P.J. Brophy, D. Marsh and G. Kothe, Proc. Natl. Acad. Sci. USA, 84 (1987) 3704-3708.

60 J. Seelig, A. Seelig and L. Tamm, in: P.C. Jost and O.H. Griffith (Eds), Lipid-Protein Interactions, Vol. 2, Wiley-Interscience, New York, 1982, pp. 127-148.

61 J.R. Brotherus, O.H. Griffith, M.O. Brotherus, P.C. Jost, J.R. Silvius and L.E. Hokin, Biochemistry 20 (1981) 5261-5267.

62 J. Davoust, M. Seigneuret, P. Hervé and P.F. Devaux, Biochemistry, 22 (1983) 3146-3151.

63 K.P. Datema, C.J.A.M. Wolfs, D. Marsh, A. Watts and M.A. Hemminga, Biochemistry, 26 (1987) 7571-7574.

64 W. Plachy and D. Kivelson, J. Chem. Phys., 47 (1967) 3312-3318.

65 E. Sackmann and H. Träuble, J. Am. Chem. Soc., 94 (1972) 4492-4498.

66 P.F. Devaux, C.J. Scandella and H.M. McConnell, J. Magn. Reson., 9 (1973) 474-485.

67 J.H. Sachse and D. Marsh, J. Magn. Reson., 68 (1986) 540-543.

68 H.-J. Galla and E. Sackmann, Biochim. Biophys. Acta, 401 (1975) 509-529.

69 J.H. Sachse, M.D. King and D. Marsh, J. Magn. Reson., 71 (1987) 385-404.

752

70 B. Berner and D. Kivelson, J. Phys. Chem., 83, (1979) 1406-1412.

71 D. Marsh and I.C.P. Smith, Biochim. Biophys. Acta, 298 (1973) 133-144.

72 D. Marsh, Biochim. Biophys. Acta, 363 (1974) 373-386.

73 A. Abragam, The Principles of Nuclear Magnetism, Oxford University Press, London, 1961.

74 D. Marsh, in (L.J. Berliner, Ed), Spin Labeling. A Third Compendium, Plenum Press, New York, 1989.

CHAPTER 21

MEASUREMENTS OF THE CONCENTRATION OF OXYGEN IN BIOLOGICAL SYSTEMS
USING EPR TECHNIQUES

HAROLD M. SWARTZ AND JAMES F. GLOCKNER

1. INTRODUCTION

  1.1 Aim and organization of this chapter

  The aim of this chapter is to provide an overview of the
principles of a number of related methodologies that use electron
paramagnetic resonance (EPR) to measure the concentration of
oxygen in biological systems; to indicate the current status of
the field including some representative applications; and to
provide an indication of the future potential of these
techniques. Most of the examples in the chapter have been drawn
from the results of the EPR laboratory at the University of
Illinois because of the convenience of doing so, but the reader
should understand that the development of this important field is
taking place at several leading laboratories throughout the
world. A relatively complete bibliography on EPR oximetry is
included in this chapter (1-66).

  1.2 Why measure oxygen concentrations in biological systems?

  Initially, it might be worthwhile to review briefly why one
would want to make measurements of oxygen concentrations in
biological systems, and what kinds of measurements would be
desirable. Oxygen has a central role in energy metabolism in
biological systems: Although mammalian cells can derive energy
from food sources in the absence of oxygen, the total amount of
energy obtainable is only 1/17 of that available in the presence
of oxygen. The energy demands of most organ systems are too high
to be met for very long by anaerobic pathways--the brain is
especially sensitive in this respect. As a consequence, mammals
have evolved complex and elaborate mechanisms for ensuring that
adequate oxygen concentrations, to the greatest extent possible,
go to the organ systems that have the most critical needs. In
order to understand this function, of course, one needs to be

able to measure oxygen at both the cellular and tissue level over a wide range of oxygen concentrations.

Oxygen, however, is a two-edged sword in biological systems. Because of the strong capability of oxygen to oxidize organic substrates, including virtually all constituents of cells, there also are a number of pathophysiological circumstances in which the concentration of oxygen is a critical parameter. This is especially true for those phenomena that often are termed "free radical pathology" or "oxygen radical pathology" (which might better be termed oxidative pathophysiology because the role of radicals is not always central). In order to understand both the physiological and pathophysiological phenomena, therefore, accurate and facile measurements of oxygen concentrations are desirable. It is becoming increasingly recognized that there is a need to make these measurements of oxygen in as specific and accurate a manner as possible, because both the physiological and pathophysiological phenomena tend to occur at specific sites, usually inside cells.

1.3 Types of oxygen concentrations that can be measured

The measurement of oxygen can occur at several different levels of complexity and specificity. Perhaps the easiest but also the least specific is oxygen utilization. That is, one measures the total amount of oxygen that is consumed per unit time by the animal, cell suspension, etc. In intact mammals, especially in patients, considerable efforts have been made to obtain more pertinent measurements of oxygen concentrations by monitoring the levels of oxygen in the central arteries and veins (67-69). More sophisticated measurements of these vascular concentrations of oxygen related to specific organ systems are desirable but more difficult to obtain (70-76). What is really desired is the oxygen concentration within specific tissues or better yet, within the cells in those tissues. Eventually we need to be able to measure the concentration of oxygen at specific sites within the cell. Methodology to make such measurements has heretofore been unavailable, but there are several recent promising developments in this area.

1.4 The relative value of measuring oxygen concentrations by EPR

There are a number of situations in which it appears that the measurement of oxygen by EPR may have significant advantages over other techniques that are available or potentially available. The specific approaches will be discussed in detail in the balance of this chapter, but for purposes of orientation, it might be useful to delineate here some of the particularly useful aspects of EPR oximetry.

For measurements of oxygen utilization, there are a number of circumstances where EPR oximetry may be advantageous. These are based on the small sample sizes that can be used for such measurements; the sensitivity of the method over a broad range of oxygen concentrations, including oxygen concentrations above ambient; and the nonperturbing nature of these measurements. The latter is particularly important for small samples for which the competing methodology is the Clark electrode: In small samples the Clark electrode potentially is perturbing because of the consumption of oxygen by the electrode, the need for stirring the sample, and the difficulty of obtaining a small enough yet still accurate and reliable probe. It should be noted, however, that Clark electrodes have been used in samples as small as single cells (77–78).

A second area of potentially advantageous application is for the continuous, nonperturbing, and sensitive measurement of oxygen concentrations within complex biological systems such as spheroids or even whole animals. These capabilities include spatial resolution on the order of the dimensions of cells (2). In some of these systems, the currently available alternative technique is the technically complex and potentially perturbing method of inserting Clark electrodes.

The third area where EPR methodology appears to have some particularly favorable characteristics is for the measurement of intracellular oxygen concentrations. This is the concentration of oxygen that one really would like to measure in most situations because the physiological and pathophysiological roles of oxygen occur primarily within cells. The usual measurements of oxygen are made outside of cells; this, of course, is due to technical limitations rather than conceptual blindness. The measurement of intracellular oxygen concentrations by EPR is discussed in some detail in this chapter.

2. PRINCIPLES OF THE MEASUREMENT OF OXYGEN CONCENTRATIONS BY EPR

2.1 Basic phenomenon of interactions of molecular oxygen with free radicals

Although there are several different approaches to using EPR for measuring oxygen concentrations, they are all based on the same physical principles: The paramagnetism of molecular oxygen and the effects of paramagnetic species on the relaxation rates of other paramagnetic species. The ground state of molecular oxygen has two unpaired electrons and a very rapid relaxation rate and therefore can be an effective relaxer of other paramagnetic species. These effects scale, usually linearly, with the concentration of oxygen and thus in a properly calibrated system can provide a relatively direct measurement of the concentration of oxygen. The effects can be observed using EPR spectral parameters that are sensitive to T_1, T_2, or both T_1 and T_2. The latter situation, of course, occurs in virtually all cases, but it is conventional to consider some of the methodological approaches in terms of being primarily T_1 (e.g., microwave power saturation) or T_2 (e.g., effects on line shape) phenomena.

Heisenberg exchange between free radicals and oxygen results in a broadening of EPR spectral lines. Line width (W) is related to T_2 by

$$W = \gamma_e \left(\frac{1}{T_2{}^*} \right) \qquad [1]$$

where

$$\frac{1}{T_2{}^*} = \frac{1}{T_2} + \frac{1}{2T_1} \; . \qquad [2]$$

Most EPR oximetric measurements based on line broadening effects utilize nitroxide spin labels as probes, for a variety of reasons discussed below. Typical nitroxides in aqueous solution have $T_1 \sim 10^{-7}$ sec. and $T_2 \sim 10^{-9}$ sec. , so that T_2 represents the dominant contribution to line width.

The extent of the line broadening due to oxygen is defined by

$$\Delta W_{O_2} = \frac{\omega}{\gamma_e} \left(\frac{2}{\sqrt{3}} \right) = \left(\frac{2}{\sqrt{3}} \right) \frac{4\pi Rp}{\gamma_e} [O_2] \{ D_{NO} + D_{O_2} \} \qquad [3]$$

where

$$\omega = 4\pi Rp \, [O_2] \, \{D_{NO} + D_{O_2}\} \tag{4}$$

is the bimolecular collision rate between oxygen and nitroxide molecules as defined by the Smoluchowski equation. Here R is the interaction radius between the spin label and oxygen, D_{NO} and D_{O_2} are the diffusion coefficients of the nitroxide and oxygen, and p is the probability that relaxation will occur as a result of a collision (p is approximately one in most physiological solutions, but will decrease when the viscosity becomes very high). Thus the bimolecular collision frequency is directly proportional to both the oxygen concentration and the nitroxide line width, allowing convenient calibration for oximetric measurements.

Heisenberg exchange between a fast relaxer (oxygen) and a slowly relaxing species also alters T_1 by providing thermal contact of the slow relaxer to the lattice (24,49). This effect can be used to measure oxygen by noting changes in the saturation properties of free radicals in the presence of oxygen. Microwave power saturation occurs when the rate of energy absorption is greater than the rate at which spins can relax. This will cause a decrease in signal intensity and an increase in line width. In the absence of power saturation the EPR signal (Y') will be proportional to the square root of power. We define the parameter S as the ratio of the EPR signal with saturation to that expected with no saturation. S can be related to the saturation factor Z as follows:

$$\frac{Y'/\sqrt{P}}{Y_o/\sqrt{P}} = S = Z^n = \frac{1}{(1 + B_1{}^2\gamma_e{}^2 T_1 T_2/4)} \tag{5}$$

where n = 3/2 for Lorentzian and 1/2 for Gaussian line shapes. For intermediate line shapes n will vary between these extremes and will depend on the power. $P_{1/2}$, defined as the power at which S = 1/2, can be determined from a plot of S vs. P, and is related to T_1 and T_2 by

$$P_{1/2} = \frac{2^{1/n} - 1}{1/4\gamma_e{}^2 KQ} \left(\frac{1}{T_1 T_2}\right) \tag{6}$$

where K is a constant depending on experimental factors, Q is the filling factor of the cavity, and $KQP = B_1^2$. $P_{1/2}$ is inversely proportional to the product of T_1 and T_2, and thus changes in oxygen concentration will alter both T_1 and T_2 and lead to changes in $P_{1/2}$.

Power saturation is generally a less sensitive technique than line broadening, but it is useful for oximetric measurements involving melanin--melanin is an inhomogeneously-broadened polymer and has a line width of several gauss, so that T_2-dependent line width changes due to oxygen are virtually undetectable.

2.2 Types of paramagnetic probes

2.2.1 Nitroxides. Although in principle any paramagnetic molecule might be used for biological oximetry (since oxygen will affect the EPR spectra of all paramagnetic species) most of the studies have made use of synthetic stable free radicals, the nitroxides (see Fig. 1). The nitroxides have been used because of a combination of properties including: availability in a wide variety of chemical forms; an extensive literature on their use as spin labels in physical and biological systems; a fairly extensive knowledge of their behavior and interactions with biological systems; and relaxation times that are well suited for oximetric measurements (i.e., their relaxation rates are of the same order as the collision frequency of oxygen with nitroxide molecules in solution--this fortuitous agreement is responsible for the strong effect of oxygen on EPR spectra of nitroxides). These properties led to the use of nitroxides in the early EPR oximetry experiments, and it seems likely that nitroxides will continue to be the predominant probe as the method is applied to more complex biological systems including intact animals. This is because the use of EPR oximetry in such complex systems poses a number of requirements as to solubility, permeability, and stability that can best be met by using a versatile set of probes such as the nitroxides. For example, nitroxides of different solubility might be exploited to obtain oxygen concentrations in lipophilic or hydrophilic regions of cells or tissues, while nitroxides that are highly ionic could be used as probes of the intravascular or extracellular environment. As described in greater detail below, more sophisticated techniques may allow selective measurement of the intracellular concentration of

Structure	Simple Name	Chemical Name
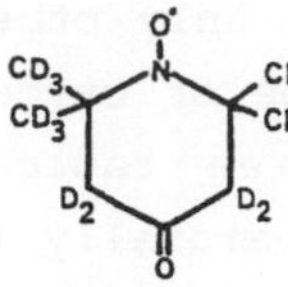	PDT	4-oxo-2,2,6,6-tetramethyl-piperidine-d$_{16}$-1-oxyl
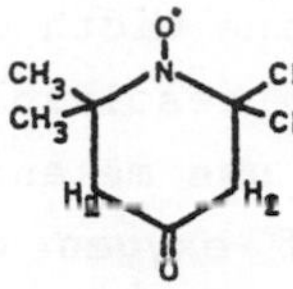	Tempone	4-oxo-2,2,6,6-tetramethyl-piperidine-1-oxyl
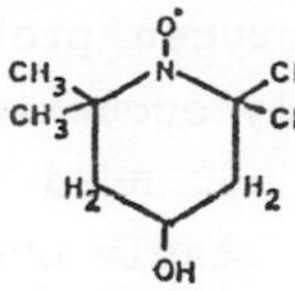	Tempol	2,2,6,6-tetramethylpiperi-dine-N-oxyl-4-ol
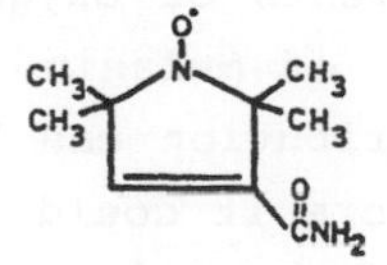	CTPO	3-carbamoyl-2,2,5,5-tetra-methyl-3-pyrroline-1-yloxyl
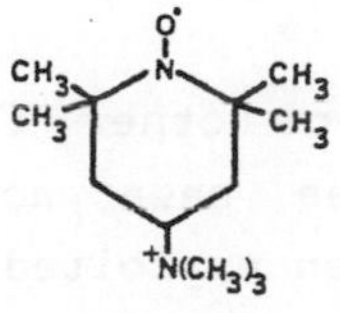	Cat$_1$	4-trimethylammonium-2,2,6,6-tetramethylpiperidine-1-oxyl
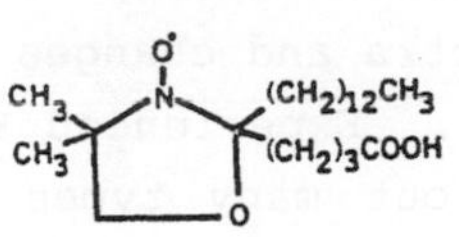	5-Doxyl stearic acid	2-(3-carboxypropyl)-2-tri-decyl-4,4-dimethyl-3-oxazo-pidinyloxy

Fig. 1. Representative nitroxides used for EPR oximetric measurements.

oxygen or even oxygen concentrations in particular regions within the cell.

2.2.2 <u>Other types of paramagnetic probes</u>. The only other free radical that has been used extensively as a probe of oxygen concentration is the naturally occurring stable free radical melanin. This is a paramagnetic polymer that occurs naturally in a wide variety of biological systems including mammalian skin and the brains of primates and some other species. The line width of melanin is relatively broad (0.5-1.0 mT) and has no resolvable superhyperfine structure, and therefore studies that use melanin to measure oxygen are usually based on effects of oxygen on microwave power saturation. As noted below, this is often a less sensitive parameter, but the use of melanin as an oxygen probe does have some potential advantages. It is a naturally occurring substance and therefore, unlike the nitroxides, may not need to be introduced into the system. Also, it is extremely stable over the time periods that would be used for measurements of oxygen concentrations. Another favorable characteristic of melanin is that because it is a pigmented polymer, its distribution can be determined by microscopic examination, and therefore it could be used to report on oxygen concentrations at specific locations in the sample.

The effect of oxygen on the EPR spectra of other free radicals and even paramagnetic metal ions has been noted frequently in the literature and sometimes has been exploited as an indicator of the presence of oxygen. The most commonly noted effects are broadening of the lines of the spectra and changes in the microwave power saturation (5,17,35,36,59). Experienced EPR spectroscopists know that in order to carry out many types of measurements for spectroscopic purposes, oxygen needs to be removed from the sample to eliminate these effects.

Inquiries are often made as to whether or not the concentration of oxygen might be determined by measuring the products of reactions of oxygen such as "oxygen radicals." The answer is generally negative, both because of the low concentration of such species and the lack of a linear relationship between their observable concentration and the concentration of oxygen. It should be noted, however, that studies of such reactions logically should be carried out under circumstances in which the actual concentration of oxygen is known accurately, and achievement of this is difficult without direct measurements of the oxygen concentration by techniques such as EPR oximetry.

2.3 Types of experimental parameters used to measure the concentration of oxygen

2.3.1 Physical parameters. Although, as noted above, all of the physical parameters are based on the same physical principle, i.e., the effects of the paramagnetic molecular oxygen molecule on relaxation times of the paramagnetic probe molecule, several distinct approaches, using different experimental parameters may be distinguished.

Broadening of the principal hyperfine lines. As a consequence of the faster relaxation that occurs in the presence of paramagnetic oxygen, the line width must increase proportionately. For most of the commonly used nitroxides, when the oxygen concentration in solution increases from 0 to 210 µmoles (the concentration of oxygen in a physiological solution in equilibrium with air) the line width increases by approximately 10 µTesla. Empirically, the relationship between the line width and the concentration of oxygen over this range approximates a straight line (Fig. 2) although it is always prudent and usually essential that a calibration curve be constructed carefully, under conditions that reflect the experimental conditions. In practice, the most sensitive and hence usual way to make use of this oxygen-broadening effect is to use perdeuterated nitroxides in order to obtain narrow intrinsic line widths--this ensures that the percentage change in line width due to oxygen broadening is maximal. Such nitroxides in aqueous solutions have a line width of approximately 15 µTesla in the absence of oxygen and 25 µTesla when the solution is equilibrated with air (Fig. 3). Since the relationship between oxygen tension and nitroxide line width is linear over quite an extensive range of oxygen concentrations, this method can, in principle, be used for experimental situations in which the oxygen concentration exceeds that of air, including hyperbaric oxygen concentrations.

Effects on resolution of "superhyperfine" structure. Some of the earliest EPR oximetry experiments used the effect of oxygen on the superhyperfine structure of nitroxides. These are the additional splittings of the principal hyperfine lines that occur as a consequence of the localization of the unpaired

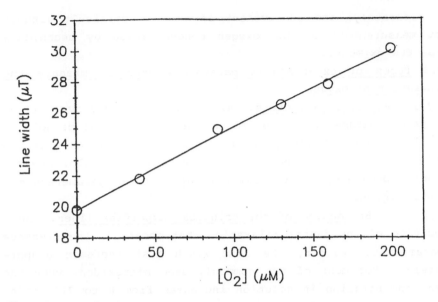

Fig. 2. Variation of line width with oxygen concentration for a solution of 0.1 mM ^{15}N PDT in PBS at 37°C.

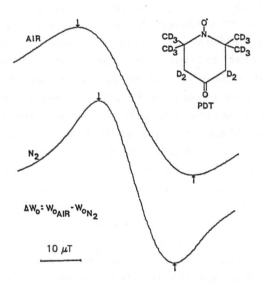

Fig. 3. Demonstration of the sensitivity of the midfield line width of PDT to oxygen-induced line broadening. An aqueous solution of 0.1 mM ^{15}N PDT was drawn into a Teflon sample tube and equilibrated first in air and then in nitrogen. The peak-to-peak line widths are measured between the arrows as in the figure, while the broadening because of oxygen, ΔW, is calculated as shown. From (62).

electron on the protons attached to the ring on which the
nitroxide resides, including the protons of the methyl groups
attached to the carbon atoms adjacent to the nitrogen of the
nitroxide groups. For these measurements an empirical parameter
is typically defined, based on the ratio of the superhyperfine
lines to the principal hyperfine lines (Fig. 4). This method is
usually not applicable to oxygen concentrations above 200–250 μM
because at concentrations higher than these, the superhyperfine
structure is not resolvable. At lower concentrations quite
accurate measurements can be made; again, they should be based on
a suitable calibration curve constructed under appropriate
experimental conditions.

Effects on signal amplitude. Since the nitroxide
concentration is roughly proportional to $I \cdot W^2$ where I is the
signal amplitude and W is the line width, changes in oxygen can
be monitored either by noting changes in line width or in the
signal amplitude. Several studies have used this technique
rather than the measurement of line widths or superhyperfine
parameters (25,47,48). The major advantage of this method is the
ability to follow rapid kinetics by sitting at the top of the
peak and continuously monitoring the oxygen concentration rather
than taking several separate scans--the time response is better,
and subtle changes in oxygen consumption rates are less likely to
be missed. The major difficulty is that any processes affecting
the nitroxide concentration or signal intensity (reduction,
oxidation, temperature, binding of labels to macromolecules,
etc.) must be very carefully accounted for.

A similar technique has recently been reported (6) in which
the second derivative spectrum was used in the same manner--that
is, the height of the second harmonic peak was followed as an
indication of the oxygen concentration. The authors reported the
same advantages and problems discussed above, with the additional
caveat that the height of the second harmonic was a highly nonlinear function of the oxygen concentration.

Microwave power saturation. This is principally a T_1
effect of oxygen, in contrast to the line broadening methods
described above that are based on changes in T_2. In some
circumstances, especially in complex biological systems,
microwave power saturation may be the most effective parameter
for measuring oxygen concentrations. Most organic free radicals,

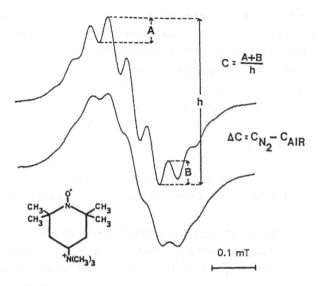

Fig. 4. Demonstration of the oxygen-induced change in the empirical parameter C defined as in the figure. An aqueous solution of 0.15 mM Cat_1 was equilibrated first in air (bottom spectrum) and then in nitrogen (top spectrum). From (62).

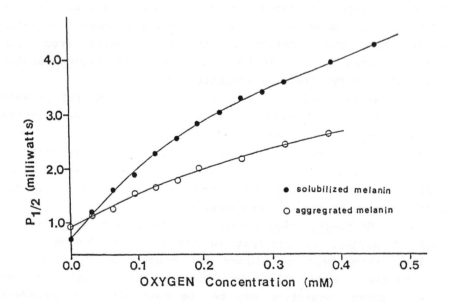

Fig. 5. Plot of $P_{1/2}$ vs. oxygen concentration for solubilized and aggregated DOPA melanin in McCoy's media at room temperature.

under the conditions and temperature ranges likely to be encountered in biological systems, show saturation at readily achievable incident microwave power levels. The presence of paramagnetic molecular oxygen decreases the microwave power saturation, and this effect is proportional to the concentration of oxygen (Fig. 5). The most precise data are obtained by using a full microwave power saturation curve, thereby deriving a saturation parameter that is based on a large number of points. The usual approach is to measure the peak height of the first derivative EPR spectrum as a function of the incident microwave power over a broad range of power levels. It is often possible, however, to get quite useful data by measuring microwave power saturation effects at a few or even two incident microwave powers. This approach may be particularly useful in very complex biological systems, such as intact animals, where relative oxygen concentrations, and especially changes in oxygen concentrations, can be the critically needed data and these can be inferred from changes in the ratios of the intensities at two appropriate microwave powers. This method is applicable to essentially any saturable paramagnetic species and has the advantage that it does not require that the species have narrow lines.

Rapid-passage T_1-sensitive display. Froncisz et al. (18) have reported a novel technique in which the EPR rapid-passage signal is observed when tuned to the dispersion using a loop-gap resonator. The signal shape is then sensitive to changes in T_1. This method can detect oxygen concentrations as low as 0.1 μM, and the sensitivity of the loop gap resonator allows the use of samples as small as 1 μl.

2.3.2 Biological methods. It recently has been shown that the concentration of oxygen in biological systems can be measured by EPR techniques by taking advantage of the oxygen-dependent rates of metabolism of nitroxides and the corresponding hydroxylamines (2,11-13,60,61). Many nitroxides in biological systems will undergo reversible reduction to the hydroxylamines (Fig. 6).

The rate of reduction of some of these nitroxides is a sensitive function of the oxygen concentration. In particular, some nitroxides show a dramatically increased rate of reduction when the oxygen concentration in the cells reaches very low levels (Fig. 7).

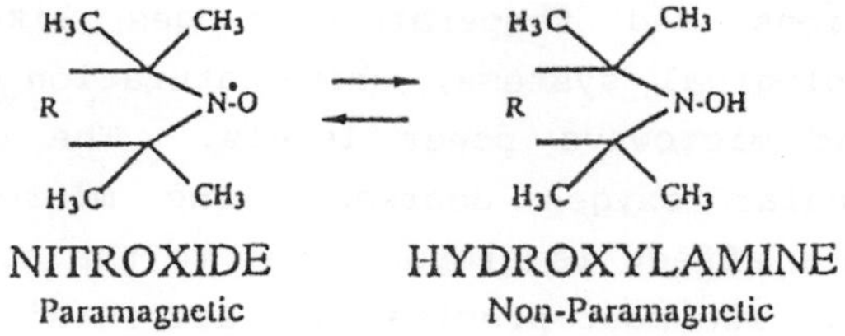

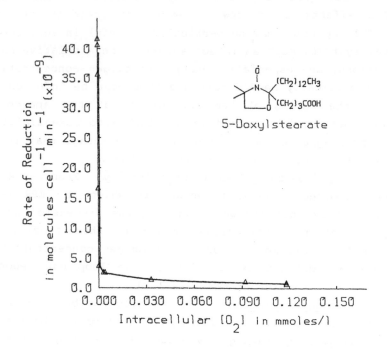

Fig. 6. Reduction of the nitroxide to the non-paramagnetic hydroxylamine, and the reverse oxidation reaction.

Fig. 7. Rate of reduction of 5-Doxyl stearate vs. the intracellular oxygen concentration. Reduction rates are the initial rates, assuming zero order kinetics. Samples consisted of 10^7 TB cells suspended in 100 µl media, to which 2 nmol of 5-DS were added. Spectra were obtained at 37°C. From (61).

Table 1. Reduction of nitroxides and oxidation of corresponding hydroxylamines in TB cells. The units of the rates are 10^6 molecules cell^{-1} min^{-1}. The hydroxylamines were produced by reduction of nitroxides by cells. From (13).

Nitroxides	Oxidation rate	Reduction rate
Water-soluble		
Piperidines		
Tempone	0.0	44
Tempol	0.0	53
Cat$_1$	0.0	2.2
Tempo sulfate	0.0	71
Pyrrolidines		
PCA	1.3	8.0
5-Tempamine	0.3	9.6
3-(Isothiocyanatomethyl)-2,2,5,5-tetramethylpyrrolidine-1-oxyl	1.5	3.3
Oxazolidine		
2N4	0.0	48
Lipid-soluble		
Piperidine		
Tempo stearate	1.7	38
Pyrrolidine		
2-Carboxymethyl-2-tridecyl-5,5-dimethylpyrrolidine-1-oxyl	1.7	23
Oxazolidines		
5-Doxyl stearate	6.0	43
10-Doxyl stearate	5.8	9.6
2N14	15.2	103
7N14	9.9	8.6

768

This has particular practical importance for studies of radiation effects because, fortuitously, the oxygen concentrations of most interest to the radiation biologist (i.e., the concentrations at which there are large changes in radiation sensitivity) are very close to the concentrations where the maximum changes in the rate of reduction of nitroxides occurs. Higher oxygen concentrations appear to have relatively little effect on the rate of reduction of nitroxides, even those nitroxides whose reduction rates show the largest sensitivity to the critical concentration of oxygen.

Another approach is to use the reverse reaction, the oxidation of hydroxylamines back to the nitroxides. This reaction can also occur at a fairly high rate in cells--for some nitroxides the rate of oxidation is comparable to their rate of reduction by cells, and therefore net oxidation can be observed if the hydroxylamines are administered (Table 1). The rate of this reaction is proportional to the concentration of molecular oxygen, so that this technique can be applied over a wide range of oxygen concentrations (Fig. 8) and hence may be a method of choice to measure moderate to high levels of oxygen.

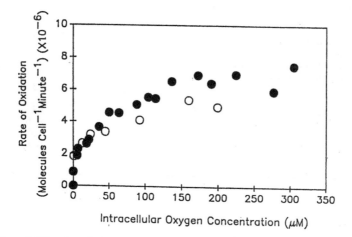

Fig. 8. Effect of intracellular oxygen concentration on the rate of oxidation of the hydroxylamines of 5- and 10-Doxyl stearates. The filled circles represent 5-DS and the open circles 10-DS. The hydroxylamines were produced by anaerobic cellular reduction of the corresponding doxyl stearates. The bulk concentration of the hydroxylamines is 0.05 mM. Oxidation was measured after reintroducing oxygen into the chamber. Samples consisted of 10^7 TB cells suspended in 100 μl of media. From (12).

Both metabolic approaches to the measurement of oxygen provide an indirect but sometimes effective means to measure oxygen concentrations in complex biological systems. The experimental parameter that is measured is the concentration of nitroxides. This can be followed over time by measuring peak heights, if the shape of the EPR spectra does not change over the time of the measurement. If such changes do occur, then more sophisticated measurements such as double integration are required. The occurrence of other factors that can affect the concentration of the nitroxides is another important experimental consideration in these measurements. The problem is conceptually simple but sometimes experimentally complex because these factors include processes that can affect the metabolism of nitroxides, (e.g., alterations in the functional states of enzymes involved in the metabolism of nitroxides and hydroxylamines) and processes that change the distribution of nitroxides (e.g., changes in cell permeability and excretion).

2.4 Experimental considerations in the use of nitroxides to measure concentrations of oxygen

The successful and accurate use of all of these approaches to the measurements of oxygen concentrations requires an awareness of the factors other than oxygen that can affect the parameter (e.g., line shape or the amplitude of the observed EPR signal) used to make the measurement. Among the factors that need to be considered especially are: loss of nitroxides due to metabolism to nonparamagnetic derivatives or from leaving the sensitive volume of the experimental system; changes in the distribution of the nitroxides, especially due to changes in the permeability of cells; line shape changes due to changes in motion such as increased or decreased microscopic viscosity and/or attachment to macromolecules; or changes in relaxation rates due to association with other paramagnetic species such as paramagnetic metal ions. Temperature is an important variable as well, since, according to Eqn. 3, the line broadening due to oxygen is directly proportional to the diffusion coefficient of oxygen, which in turn is a linear function of temperature. In practice, the line widths and signal amplitudes of many of the perdeuterated nitroxides are very sensitive to fairly small variations in temperature. Consideration also needs to be made of the effects of the nitroxide itself, especially in regard to line broadening, because at fairly low concentrations some nitroxides will show significant line broadening whose magnitude is a function of the nitroxide concentration, and hence may change over the course of the experiment. In principle, nitroxides could alter the rate of their metabolism by affecting the functional state of enzymes, but direct experimental studies indicate that this is unlikely to be a significant effect at the concentrations used in typical oximetric experiments.

All of the above factors usually can be taken into account satisfactorily by appropriate control experiments, and in practice the nitroxides have proven to be useful and sensitive probes to measure the concentration of oxygen.

3. APPLICATIONS OF THE EPR TECHNIQUES TO MEASURE THE CONCENTRATION OF OXYGEN IN VARIOUS TYPES OF EXPERIMENTAL SYSTEMS

3.1 Simple solutions

A number of biologically pertinent measurements have been made using nitroxides in relatively simple solutions or suspensions, such as suspensions of melanin (40). For such studies, any of the physical measurement techniques can be used. The only special experimental precaution is consideration of physical or chemical interactions of the probes with the components of the system. Binding of the probes will affect the experimental parameters that are observed, and there is also the potential of reduction or oxidation of the probes, thereby altering their concentrations.

3.2 Measurements of oxygen consumption by functioning biological systems

Such studies, in suspensions of cells or subcellular components, were among the earliest uses of EPR oximetry and they remain a very useful application of the technique. The EPR methods are especially valuable in small samples because the usual method, the Clark electrode, is subject to experimental artifacts due to physical perturbation of the system by the electrode, consumption of oxygen at the electrode, or damage due to the stirring that usually is required with this technique. The typical experimental approach using EPR oximetry is to place an appropriate nitroxide in the system under investigation, seal the system containing the desired initial concentration of oxygen and then measure the utilization of oxygen by observation of the rate of change of the EPR spectrum of the probe, usually using one of the line broadening methods.

3.3 Measurement of oxygen concentrations in extracellular compartments

Unlike the system described previously, these studies usually involve open systems, and the experimental parameter of interest is the oxygen concentration outside of the cells. Most typically this is done in suspensions of cells, but a clever adaptation to the extracellular compartment of whole animals has been described (29). Important experimental concerns in such systems include keeping the probe in the extracellular compartment, keeping the concentration of the probe at an appropriate level, and avoiding nonspecific interactions of the probe that lead to modifications of its motion or chemical interactions. Typically a charged nitroxide is used in order to prevent the probe from crossing the cell membrane, thereby avoiding both the rapid metabolism of the probe by the cell and the entrance of the probe into the intracellular compartment where the oxygen concentration may differ from that of the extracellular compartment. Under some circumstances, alternative methods to keep the probe out of the cellular system have been used, such as attaching the probe to a macromolecule that will not readily enter the cell or placing the probe in a liposome (10). The latter approach may also have value when there are reducing substances, such as ascorbate, in the extracellular medium. Another method to deal with reduction of the probe in the extracellular medium is to add a mild oxidizing agent such as ferricyanide. Ferricyanide in concentrations of 1-5 mM will oxidize hydroxylamines back to nitroxides, but it may also affect the EPR spectra slightly. If ferricyanide is used as the oxidizing agent, it is therefore important to determine its effect on the EPR spectra with appropriate control experiments, because some broadening may occur, especially with positively charged nitroxides.

3.4 Measurements of the concentration of oxygen inside cells (or isolated organelles)

In most situations, the biological phenomena of interest are dependent on the concentration on oxygen inside the cells or organelles. This usually is not measured because readily usable methods to make such measurements have not been available. It has been shown recently, however, that EPR oximetry can be adapted to permit the facile and accurate measurement of average intracellular oxygen concentrations (30,32,62). The usual approach is to use a small, neutral nitroxide such as Tempone that readily crosses cell membranes along with a charged, membrane-impermeable paramagnetic ion which is added to the suspension in order to broaden away the EPR spectra of nitroxides that are outside the cells or membrane bound organelles. As a consequence, only the spectra of the nitroxides residing within the outer membrane of the cell or organelle are readily observable, and therefore the effects of oxygen on the intracellular nitroxides can be observed and the concentration of intracellular oxygen determined. Using these techniques a significant difference between intracellular and extracellular concentrations of oxygen has been demonstrated (19); this finding has several important biological implications and therefore it is likely that the measurement of intracellular oxygen by EPR will be used more frequently.

Recently a method has been described that should allow the measurement of intracellular oxygen without the use of extracellular broadening agents (21). It is based on an approach analogous to that used to measure intracellular calcium concentrations, and involves a nitroxide with several esterified carboxyl groups that is initially neutral and therefore passes into cells. Intracellular esterases then convert it into a polyanion that cannot cross membranes and thus is trapped inside the cell.

With the demonstration of the value and feasibility of measuring the average concentration of oxygen inside cells, attention has turned to the possibility of measuring the concentration of oxygen at specific sites within cells. The concentration of oxygen in membranes can be studied using lipo-philic nitroxides. The measurement of the concentration of oxygen in the cytoplasmic compartment could be accomplished by the use of melanin or nitroxides that become converted to highly charged species after they enter cells (21). These would be of great value if even more specifically located probes could be developed, especially for the study of the concentrations of oxygen in the nucleus and in mitochondria.

3.5 <u>Measurement of concentrations of oxygen in extended biological systems such as tissues or spheroids</u>

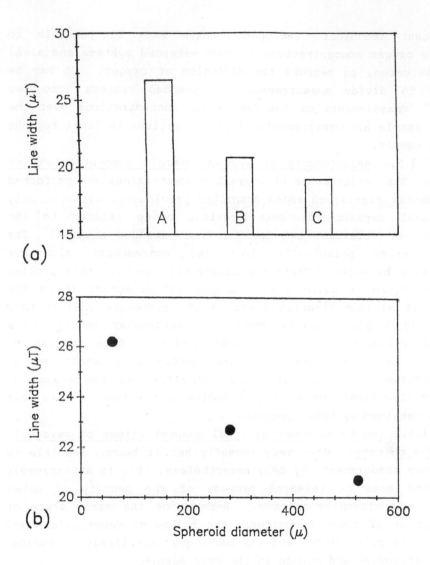

Fig. 9. (a) The line width of an overall EPR spectrum of ^{15}N PDT derived from the live cells inside a spheroid with a hypoxic center.
    A - non-respiring (NaCN present) spheroid in aerated sample.
    B - respiring (no NaCN) spheroid in aerated sample.
    C - respiring (no NaCN) spheroid in sample equilibrated with nitrogen.
(b) Line width of an overall EPR spectrum of ^{15}N PDT derived from the live cells inside spheroids of different sizes. The spheroid samples were equilibrated with air.

Recent technical developments have made it possible to measure oxygen concentrations in such extended systems and also, in some cases, to measure the diffusion of oxygen. It may be useful to divide measurements in extended systems into two types: measurements of average oxygen concentrations over the entire sample and measurements of concentrations in local regions of the sample.

3.5.1 <u>Measurements of average overall concentrations of oxygen</u>. The measurements of overall concentrations are performed in a manner similar to those described previously, with the only additional experimental considerations being related to the problems of obtaining EPR spectra from extended samples. The latter arise principally from the nonresonant microwave absorption by water. Where the sample will permit, this problem can be solved by using a thin sample and orienting it in the region of minimum electric field in the microwave cavity; this region is a plane in the center of rectangular cavity and a central cylinder in a cylindrical cavity. Some samples of interest, such as spheroids, are sufficiently small to be accommodated in a capillary tube and placed in the region of minimum electrical field. Fig. 9 indicates the type of data that can be obtained by this approach.

3.5.2 <u>The measurement of local concentrations of oxygen in extended objects</u>. Only very recently has it become feasible to make such measurements by EPR; nevertheless, this is an extremely promising area of research because of the paucity of other available noninvasive methods. Because of the early state of development of these techniques, the following descriptions are necessarily quite crude, and the techniques are likely to undergo rapid refinement and change in the near future.

One approach is to use imaging techniques under conditions in which the intensity of the image is affected by the concentration of oxygen. This can be done in several ways, utilizing principles that have already been described, such as the effects of oxygen in broadening EPR spectra (2). If the intensity of the image is proportional to the peak height of the local concentration of nitroxide, as is the case in many EPR imaging techniques, then in those areas with higher oxygen concentrations, the image will appear less intense (Fig. 10). An analogous technique has been used to measure the diffusion of

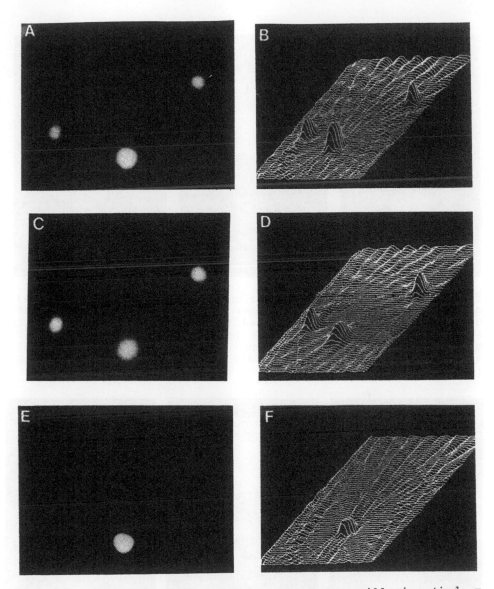

Fig. 10. Cross-sectional images of two glass capillaries (i.d. = 0.63 mm, center to center separation = 5.4 mm) and a Teflon tube (i.d. = 0.81 mm), all containing 0.5 mM ^{15}N PDT in air saturated aqueous solutions, packed in a triangular array. The images were reconstructed using 18 projections (10° intervals), gradient strength = 0.3 T/m, scan range = 10 mT. A, C, and E are spin-density plots and B, D, and F are contour level plots (for picture clarity only every fourth line was used). (A, B) With nitrogen flowing. (C, D) With oxygen flowing. (E, F) Subtraction of image C from image A. From (2).

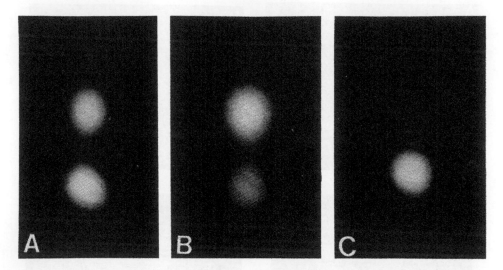

Fig. 11. Cross-sectional images of two capillaries (i.d. = 0.63 mm, center to center = 0.97 mm) containing 1 mM solution of Cat_1. Images were reconstructed from 18 projections, gradient strength = 1 T/m, and microwave power was (A) 5 mW or (B) 150 mW. (C) is the difference image obtained after normalizing (A) and (B) so that the glass capillary had the same intensity. From (2).

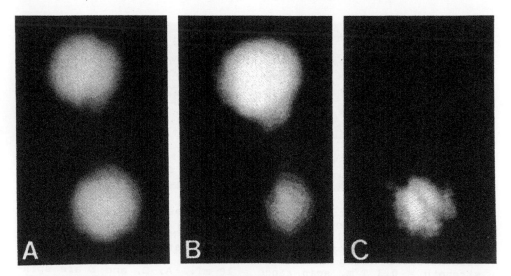

Fig. 12. Cross-sectional images of two tubes (i.d. = 0.81 mm, distance between centers = 6 mm) containing TB cells labelled with 5-DS at 37°C. Images were reconstructed from 12 projections (15° intervals) with gradient strength = 1 T/m.
(A) Both tubes in air. (B) One tube exposed to nitrogen from 7 minutes. (C) Image obtained by subtracting (B) from (A). From (2).

777

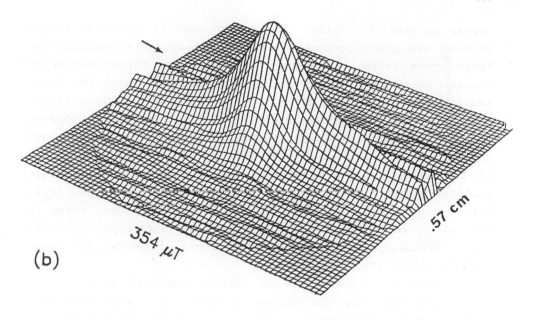

(b)

354 μT

.57 cm

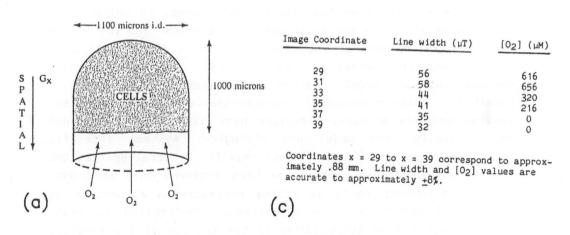

←—1100 microns i.d.—→

S | Gx
P
A
T   CELLS   1000 microns
I
A
L ↓

(a)   O₂   O₂   O₂

Image Coordinate	Line width (μT)	$[O_2]$ (μM)
29	56	616
31	58	656
33	44	320
35	41	216
37	35	0
39	32	0

Coordinates x = 29 to x = 39 correspond to approximately .88 mm. Line width and $[O_2]$ values are accurate to approximately ±8%.

(c)

Fig. 13. (a) Schematic of the capillary of packed cells used for the oximetry experiment. (b) Image of a packed cell sample exhibiting an intracellular ^{15}N PDT signal and exposed to an oxygen gradient along the spatial axis. The location along the spatial axis of the open end of the capillary is indicated by an arrow. (c) Values of line width, oxygen concentration (obtained from a separate calibration), and corresponding image spatial coordinate.

oxygen in tissues (14,15). Another approach is to obtain images at high and low microwave powers and subsequently construct an image whose intensity is proportional to the concentration of oxygen, based on the differences in intensity between the first two images, i.e., those areas with higher oxygen concentrations will have relatively more intense images at high power compared to those areas with lower oxygen tensions, while the images obtained at low microwave power will be essentially independent of the concentration of oxygen but will provide the necessary scaling factor for the local concentration of nitroxides (Fig. 11). Another strategy is to use the oxygen-dependent rates of metabolism of nitroxides to create local alterations in the concentrations of nitroxides based on local variations in oxygen tensions (Fig. 12).

A different approach is to obtain localized spectra with sufficient resolution to observe directly the oxygen-dependent features of the EPR spectra. In principle this can be done most completely by obtaining four-dimensional images, in which three of the dimensions are the usual spatial dimensions and the fourth is a spectral dimension; the result of such a study would be a three dimensional array of EPR spectra. We have recently demonstrated in model systems that such an approach is feasible. Three dimensional spectral-spatial images, with two spatial and one spectral dimension have already been obtained experimentally from model and biological systems (Fig 13). Because of sensitivity limitations, thus far spectral-spatial EPR imaging of biological objects has been achieved only at 9 GHz. At this frequency there is strong nonresonance absorption by water, and therefore these experiments are limited to small objects that can be accommodated in the regions of low electric fields in the EPR resonators. It turns out, however, that even with these limitations it appears feasible to carry out spectral-spatial imaging of oxygen concentrations in a very important biological model for the study of tumors, multicellular spheroids. Because they can model some aspects of diffusion processes that occur in tumors, spheroids are used extensively to study the effects of gradients of oxygen concentrations and other nutrients on the response of cytotoxic agents such as radiation and anticancer drugs; the efficacy of such research has been limited in part because of the lack of a good method to measure oxygen concentrations at different depths in the spheroid. EPR spectral-spatial imaging may soon make such measurements possible, with resolution sufficient to differentiate between the outer layers of cells that are on or near the surface and therefore are well oxygenated, and the underlying hypoxic cells.

3.6 <u>Measurements of the concentration of oxygen in intact animals</u>

In order to carry out EPR oximetry in larger objects such as intact animals, it appears necessary to use lower frequencies. A limited number of studies have been reported that are aimed at measuring concentrations of nitroxides in intact animals using low-frequency spectrometers, with the implicit or explicit indication that these approaches could be used to measure the oxygen tensions (7,28,29,39). It appears feasible to measure both global and local concentrations of oxygen in animals and, if the method to develop nitroxides that selectively localize inside cells is successful, to measure intracellular oxygen concentrations as well.

One of the approaches is to go to "L-band", 1-2 GHz. At this frequency, with appropriate configurations of EPR resonators, whole animals the size of mice can be studied by insertion into the resonator and larger animals can be studied by the use of surface probes based on loop-gap resonators. The measurements in animals inserted into the resonator most frequently have been used to obtain EPR spectra reflecting the average concentration in the whole animal. In principle, however, these spectra could be localized to specific regions either by using a resonator with a limited width of sensitivity such as a wire loop or by means of a magnetic field gradient.

The other approach, which has been aimed specifically at the measurement of oxygen concentrations, is the use of a surface probe which provides localization for the area that is immediately under the probe. Additional localization can be achieved by the use of magnetic field gradients. Recently we have obtained oxygen-dependent changes in EPR spectra in intact mice using surface probes and perdeuterated nitroxides.

At even lower frequencies, it should be feasible to obtain EPR spectra from larger animals at greater depths than seem feasible at 1 GHz (20). Whether or not the sensitivity at such frequencies would be sufficient to obtain measurements other than those reflecting an average value for the entire animal, remains to be seen. A 300 MHz EPR spectrometer has been built and oxygen independent spectra of a specially synthesized nitroxide have been obtained (Fig. 14).

ACKNOWLEDGEMENT   The experimental data used in this chapter is the product of the efforts of a large number of colleagues, and was supported by NIH research grants RR-01811, GM-34250, GM-35534, and NIH training grant CA-09067.

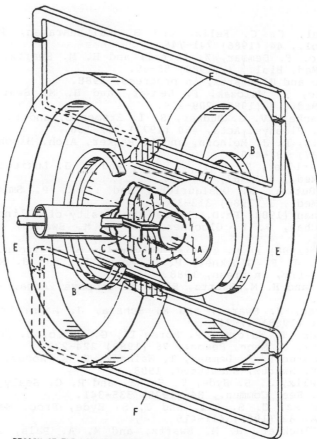

DESIGN OF THE LOW FREQUENCY SPECTROMETER

A. STRIPLINE SAMPLE HOLDER

B. MODULATION COILS

C. CAPACITIVE COUPLING

D. RADIOFREQUENCY SHIELD

E. MAIN MAGNET

F. HORIZONTAL PLANE
   DEFINING COILS

Fig. 14. Schematic diagram of a low frequency EPR spectrometer that may prove useful for oximetric measurements involving tissues and whole animals (20,39). The spectrometer operates at frequencies of ~ 300 MHz with a magnetic field of 9 mT.

REFERENCES

1   E. Ankel, C. C. Felix, and B. Kalyanaraman, Photochem. Photobiol., 44 (1986) 741-746.
2   G. Bacic, F. Demsar, Z. Zolnai, and H. M. Swartz, J. Magn. Reson. Med. Biol., 1 (1988) 55-65.
3   G. Bacic and H. Swartz, in progress, 1988.
4   G. Bacic, T. Walczak, F. Demsar, and H. M. Swartz, Magn. Reson. Med., 8 (1988) 209-219.
5   J. M. Backer, V. G. Budker, S. I. Eremenko, and Y. N. Molin, Biochim. Biophys. Acta, 460 (1977) 152-156.
6   S. Belkin, R. J. Melhorn, and L. Packer, Arch. Bioch. Bioph., 252 (1987) 487-495.
7   L. J. Berliner, H. Fujii, X. Wan, and S. J. Lukiewicz, Magn. Reson. Med., 4 (1987) 380-384.
8   K. W. Butler, R. Deslauriers, and I. C. P. Smith, Magn. Reson. Med., 3 (1986) 312-316.
9   H. C. Chan (1988) Ph.D. thesis, University of Illinois.
10 H. C. Chan, J. F. Glockner, and H. M. Swartz, submitted, 1988.
11 K. Chen (1988) Ph.D. thesis, University of Illinois.
12 K. Chen, J. F. Glockner, P. D. Morse II, and H. M. Swartz, Biochemistry, in press, 1988.
13 K. Chen and H. M. Swartz, Biochim. Biophys. Acta, 970 (1988) 270-277.
14 F. Demsar, H. Swartz, and M. Schara, J. Magn. Reson. Med. Biol., 1 (1988) 17-24.
15 F. Demsar, T. Walczak, P. Morse II, G. Bacic, Z. Zolnai, and H. Swartz, J. Magn. Reson., 76 (1988) 224-231.
16 J. W. Dobrucki, F. Demsar, T. Walczak, R. K. Woods, G. Bacic, and H. M. Swartz, submitted, 1988.
17 C. C. Felix, J. S. Hyde, T. Sarna, and R. C. Sealy, Biochem. Biophys. Res. Commun., 84 (1978) 335-341.
18 W. Froncisz, C. S. Lai, and J. S. Hyde, Proc. Natl. Acad. Sci. USA, 82 (1985) 411-415.
19 J. F. Glockner, H. M. Swartz, and M. A. Pals, submitted, 1988.
20 H. Halpern, D. P. Spencer, J. vanPolen, M. K. Bowman, R. J. Massoth, B. A. Teicher, E. M. Downy, and A. C. Nelson, Int. J. Radiat. Oncol. Biol. Phys., 12 sup. 1 (1986) 117-118.
21 H. Hu, G. Sosnovsky, S. Wenli, N. U. M. Rao, P. D. Morse II, and H. M. Swartz, submitted, 1988.
22 J. S. Hyde, and W. K. Subczynski, J. Magn. Reson., 56 (1984) 125-130.
23 J. S. Hyde and W. Subczynski, in L. J. Berliner (Ed.), Spin Labeling IV. Theory and Applications, Academic Press, Inc., New York, 1988.
24 J. S. Hyde, W. K. Subczynski, W. Froncisz, and C. S. Lai, Bull. Magn. Reson., 4 (1982) 180-182.
25 W. Korytowski, T. Sarna, B. Kalyanaraman, and R. C. Sealy, Biochim. Biophys. Acta, 924 (1987) 383-392.
26 A. Kusumi, W. K. Subczynski, and J. S. Hyde, Proc. Natl. Acad. Sci. USA, 79 (1982) 1854-1858.
27 C. S. Lai, L. E. Hopwood, J. S. Hyde, and S. Lukiewicz, Proc. Natl. Acad. Sci. USA, 79 (1982) 1166-1170.
28 S. J. Lukiewicz, Radio and Microwave Spectroscopy, no. 54 (1985) 37-54.
29 S. J. Lukiewicz and S. G. Lukiewicz, Magn. Reson. Med., 1 (1984) 297-298.

30 P. D. Morse II and H. M. Swartz, Magn. Reson. Med., 2 (1985) 114-127.

31 S. Pajak, L. E. Hopwood, J. S. Hyde, C. C. Felix, R. C. Sealy, V. M. Kushnaryov, and M. C. Hatchell, Exptl. Cell Res., 149 (1983) 513-526.

32 M. A. Pals (1988) Ph.D. thesis, University of Illinois.

33 M. A. Pals and H. M. Swartz, Invest. Radiol., 22 (1987) 497-501.

34 C. A. Popp and J. S. Hyde, J. Magn. Reson., 56 (1984) 125-130.

35 M. J. Povich, J. Chem. Phys., 79 (1975) 1106-1109.

36 M. J. Povich, Analytical Chem., 47 (1975) 346-347.

37 G. M. Rajala, C. S. Lai, G. L. Kolesari, and R. H. Cameron, Life Sciences, 36 (1985) 291-297.

38 K. Reszka and R. C. Sealy, Photochem. Photobiol., 39 (1984) 293-299.

39 G. M. Rosen, H. J. Halpern, L. A. Brunsting, D. P. Spencer, K. E. Strauss, M. K. Bowman, and A. S. Wechsler, Proc. Natl. Acad. Sci. USA, 85 (1988) 7772-7776.

40 T. Sarna, A. Duleba, W. Korytowski, and H. Swartz, Arch. Biochem. Biophys., 200 (1980) 140-148.

41 T. Sarna, W. Korytowski, and R. C. Sealy, Arch. Biochem. Biophys., 239 (1985) 226-233.

42 T. Sarna, I. A. Menon, and R. C. Sealy, Photochem. Photobiol., 39 (1984) 805-809.

43 T. Sarna, A. Menon, and R. C. Sealy, Photochem. Photobiol., 42 (1985) 529-532.

44 T. Sarna and R. C. Sealy, Photochem. Photobiol., 39 (1984) 69-74.

45 R. C. Sealy, T. Sarna, E. J. Wanner, and K. Reszka, Photochem. Photobiol., 40 (1984) 453-459.

46 G. Sosnovsky, N. U. M. Rao, S. W. Li, and H. M. Swartz, submitted, 1988.

47 K. Strzalka, T. Sarna, and J. S. Hyde, Photochem. Photobiophys., 12 (1986) 67-71.

48 K. Strzalka, T. Walczak, T. Sarna, and H. M. Swartz, submitted, 1988.

49 W. K. Subczynski and J. S. Hyde, Biochim. Biophys. Acta, 643 (1981) 283-291.

50 W. K. Subczynski and J. S. Hyde, Biophys. J., 41 (1983) 283-286.

51 W. K. Subczynski and J. S. Hyde, Biophys. J., 45 (1984) 743-748.

52 W. K. Subczynski, J. S. Hyde, and A. Kusumi, submitted, 1987.

53 V. K. Subchinskii, A. A. Konstantinov, and E. K. Ruuge, Biofizika, 23 (1978) 57-60.

54 W. K. Subczynski and A. Kusumi, Biochim. Biophys. Acta, 821 (1985) 259-263.

55 W. K. Subczynski, S. Lukiewicz, and J. S. Hyde, Magn. Reson. Med., 2 (1985) 114-127.

56 H. M. Swartz, Bull. Magn. Reson., 8 (1986) 172-175.

57 H. M. Swartz, in P. S. Allen, D. P. J. Boisvert, and B. C. Lentle (Eds) Magnetic Resonance in Cancer, 169-177, 1986.

58 H. M. Swartz, J. Chem. Soc., Faraday Trans., 1, 83 (1987) 191-202.

59 H. M. Swartz, Acta Biochim. Biophys. Hung., 22 (1987) 277-293.

60 H. M. Swartz, in E. Feig (Ed.) Advances in Magnetic Resonance Imaging, Ablex Publishing Company, Norwood, 1988.

784

61 H. M. Swartz, K. Chen, M. Pals, M. Sentjurc, and P. D. Morse II, Magn. Reson. Med., 3 (1986) 169-174.

62 H. M. Swartz and M. A. Pals, in J. Miquel, H. Weber, and A. Quintanilha (Eds), Handbook of Biomedicine of Free Radicals and Antioxidants, CRC Press, Boca Raton, 1988.

63 H. M. Swartz, M. Sentjurc, and P. D. Morse II, Biochim. Biophys. Acta, 888 (1986) 82-90.

64 D. A. Windrem and W. Z. Plachy, Biochim. Biophys. Acta, 600 (1980) 655-665.

65 R. K. Woods, J. W. Dobrucki, J. F. Glockner, P. D. Morse II, and H. M. Swartz, submitted, 1988.

66 J. J. Yin and J. S. Hyde, Zeitschrift fur Physikalische Chemie, Neue Folge, 153 (1987) 57-65.

67 K. K. Tremper and S. J. Barker, Int. Anesth. Clinics, 25 (1987).

68 W. I. Dorson and B. A. Bogue, Adv. Exptl. Med. Biol., 37A (1973) 251-259.

69 K. Reinhart, T. Kersting, U. Fohring, and M. Schafer, Adv. Exptl. Med. Biol., 200 (1986) 67-73.

70 D. W. Lubbers, Adv. Exptl. Med. Biol., 37A (1973) 45-53.

71 M. Tomura, N. Oshino, B. Chance, and I. Silver, Arch. Biochem. Biophys., 191 (1978) 8-22.

72 R. Araki, M. Tamura, and I. Yamazaki, Cir. Res., 53 (1983) 448-455.

73 F. F. Jobis-Vander Vliet, Adv. Exptl. Med. Biol., 191 (1985) 833-843.

74 E. Fox, F. F. Jobis-Vander Vliet, and H. M. Mitnick, Adv. Exptl. Med. Biol., 191 (1985) 844-855.

75 R. A. Gatenby, L. A. Coia, M. P. Richter, H. Katz, P. J. Moldofsky, P. Engstrom, D. Q. Brown, R. Brookland, and G. J. Broder, Radiology, 156 (1985) 211-214.

76 R. A. Gatenby, H. B. Kessler, J. S. Rosenblum, l. R. Coia, P. J. Moldofsky, W. H. Hartz, and G. J. Broder, Int. J. Radiat. Oncol. Biol. Phys. 14 (1988) 831-838.

CHAPTER 22

g-STRAIN: INHOMOGENEOUS BROADENING IN METALLOPROTEIN EPR

WILFRED R. HAGEN

1.   INTRODUCTION

Biological EPR data are frequently analyzed by means of spectral synthesis. The computation of powder patterns requires knowledge of the angular-dependent expressions for the resonance condition, for the transition probability, and for the inhomogeneous line broadening. The former two follow from a choice of the spin Hamiltonian. The latter one has thus far been guessed at from ad-hoc arguments. As the contribution of the anisotropic line width to the powder shape is important, traditional spectral simulation is quantitatively unreliable. In this chapter the specific effects of inhomogeneous broadening on the powder shape are described first. Subsequently, the physical nature of the broadening is discussed. Then a formal description of the angular-dependency of the line width is developed, and its application is illustrated on several examples. Finally, some suggestions are made for directed research in this still largely unexplored area of biological EPR.

2.   THE PHENOMENON OF g-STRAIN

The study of g-strain is a generic name for two decades of attempts to describe the fine details of continuous-wave EPR spectra from metalloproteins (1-13). The nature of the problem is readily introduced with a simple example: assume a paramagnet to carry a spin of $S = 1/2$ and not to be subject to hyperfine interactions to any significant measure. The continuous-wave EPR spectrum of this substance should be characterizable in terms of a single, three-component g-tensor. Real examples of this situation would be the low-temperature X-band powder patterns resulting from dilute, frozen solutions of, e.g., single-center iron-sulfur proteins or low-spin hemoproteins. When one tries to reproduce these spectra on the basis of a simple Zeeman term one is faced with the problem how to define the spectral line width. Taking this line width to be a scalar, or even when allowing for angular variation in concert with the anisotropy of the g-tensor, always

results in spectral simulations that only approximately fit the
experimental data. An illustration is given in Fig. 1.

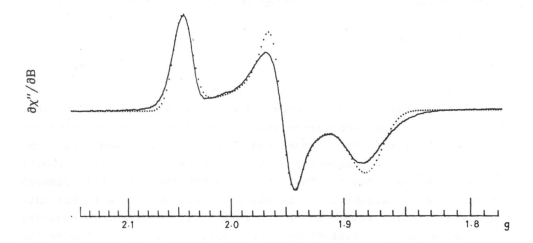

Fig. 1. Traditional simulation (dots) of the X-band spectrum (solid
line) of reduced [2Fe-2S] in spinach ferredoxin, assuming a linewidth anisotropy collinear with the g anisotropy (from Ref. 11).

The residual misfit may be conveniently identified with the g-
strain problem. g-Strain is a broadening mechanism. A g-strain
analysis means to obtain a quantitative description of this
broadening in terms of its angular-dependent magnitude. If one's
scope were to just classify a single substance by means of its
approximate g-values, then the g-strain problem could be ignored.
Frequently, the interpretation of EPR from complex (i.e.
multicenter) metalloproteins goes beyond this level of
sophistication right into the fine details of spectral shapes to
obtain information on, e.g., stoichiometries and mutual
interactions, thereof also distances, between different
paramagnetic centers. As g-strain is a main contributor to these
fine details, a knowledge of its properties is a conditio sine qua
non. The present chapter is about the state of the art in
establishing this condition.

At ambient temperature the EPR spectrum of a metalloprotein
is broadened, if at all detectable, with a Lorentzian line width
that can be associated with a homogeneous spin-lattice relaxation
time T_1. Lowering the sample temperature results in a sharpening
of the spectrum. Below a specific temperature, usually of the

order of 10 to 100 K, the spectral shape becomes independent of the temperature provided due care is taken to avoid passage and/or saturation effects. It is this low-temperature, maximally sharpened-up spectrum that is used for quantitative analysis in search for biologically relevant information. The now inhomogeneously broadened spectrum has a line width that is not commonly related to a spin-spin relaxation time T_2. Namely, the spectral shape is independent of the concentration of the protein. Also, the observed line width is orders-of-magnitude larger than the line width of the corresponding ENDOR spectra. Finally, the line width is not constant with the microwave frequency, or with the static magnetic field, but rather is proportional to these quantities. This latter observation signals that the line width is associated with a Zeeman interaction; it reflects a distribution in g-values hence the name g-strain.

3. DETECTION AND DECONVOLUTION OF g-STRAIN

The length and torsional angle of molecular bonds are distributed about mean values. These distributions result in a distribution of the ligand-field strength at any paramagnetic site. Suppose the width of this distribution to be of sufficient magnitude to cause a measurable broadening effect on an observed g-value. One might intuitively expect this broadening to look rather more like a Gaussian than like the high-temperature Lorentzian distribution. It is indeed the observation of a near-Gaussian EPR line shape (1) with a width approximately proportional to the microwave frequency at and above X-band (2,3) that led Sands and collaborators to coin the name g-strain and relate it to an (as yet unspecified) microdistribution in protein conformation for [2Fe-2S] ferredoxin. This first instance of g-strain detection is also a good example of the experimental difficulties one can expect to be faced with in the characterization of g-strain. Conventional EPR of metalloprotiens is usually detected at a single X-band frequency. Some properties of single-frequency data may indicate the polycrystal spectrum to be g-strain broadened. These properties are (4,5):

i) the spectral shape is independent of the temperature (no T_1 broadening); ii) the line shape is close to Gaussian, far from Lorentzian (no T_1 broadening); iii) the spectral shape is independent of dilution (no T_2 broadening); iv) the line width is smaller in glass than in polycrystal samples (g-strain attenuation); v) the line width increases with externally applied stress (g-strain amplification); vi) the line width tends to be proportional to g^{-2} (rigorously true for isotropic g-strain); vii) the outermost lines are skewed towards zero field and infinite field, respectively (non-collinear g-strain); viii) the derivative-shaped intermediate spectral feature (the "g_y-line") has an asymmetric shape.

However, unresolved hyperfine structure of a sufficiently large number of ligand nuclei will usually give the same result (except for property vii and viii). A definite identification of g-strain requires the collection of data at more than one microwave frequency. In fact, the line width of the [2Fe-2S] ferredoxin was found to significantly deviate from linearity in ν below X-band frequencies. This points to a convolution of g-strain broadening with unresolved superhyperfine splittings. A quantitative analysis now requires a set of low-noise, multifrequency data to be simulated with a broadening independent of ν convoluted with the g-strain broadening linear in ν.

An illustration of two likely situations is given in Fig. 2.

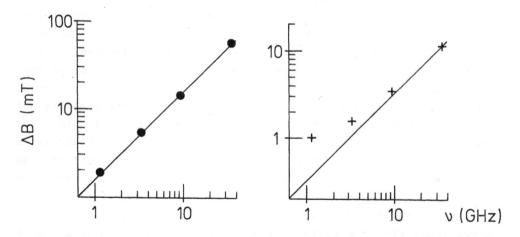

Fig. 2. The line width at the low-field g-value of low-spin heme Fe III in horse heart cytochrome $\underline{c}$ (●) and of [2Fe-2S] in spinach leaf ferredoxin (+) as a function of the microwave frequency (data taken from Refs. 4,5 and W.R. Hagen and S.P.J. Albracht, unpublished).

For two proteins the full width at half height of the low-field peak is plotted as a function of microwave frequency. In the case of the low-spin ferric hemoprotein horse-heart cytochrome $\underline{c}$, the line width is linear in ν all the way from 1 to 35 GHz. The broadening is only caused by g-strain. As a data set for further analysis any one of the four spectra will do. Therefore, one chooses the one with the best signal-to-noise ratio, usually the X-band spectrum. This simple situation occurs because the g-strain broadening for cytochrome c is very pronounced and, therefore, the superhyperfine broadening undetectably small. It is not known whether this also holds for low-spin cytochromes with much sharper spectra such as the cytochrome P-450.

The second trace is from an iron-sulfur protein, [2Fe-2S]$^{1+}$ spinach ferredoxin. Here the linearity with ν breaks down at lower frequencies. In fact, below some 2 GHz the frequency-independent contribution to the line width becomes dominant. Evidence that this contribution is from unresolved proton splitting is summarized in Ref. 5. Reduction of this contribution is thus only possible by a complete deuteration of the whole protein, an undertaking that has indeed been achieved once (12). Even then a finite contribution from superhyperfine broadening from other nuclei remains at lower frequencies. One can make either of two choices: go the hard way and try to generate a unique set of simulations consistent with the complete, multifrequency, low-noise data set. Alternatively, given the much more common situation in which only reliable X-band data are available, one may at least prove/disprove consistency of spectral features with a g-strain interpretation. In case the fine details of the spectrum are not reproducible under a g-strain model, then these are likely to be caused by - and therefore contain information on - dipolar interactions and/or center multiplicity.

4. INTERFERENCE OF g-STRAIN WITH BIOLOGICAL APPLICATIONS OF EPR

The application of EPR techniques to complex metalloproteins usually presupposes a certain amount of analytical information on the system at hand. A sound knowledge of the metal content (type and stoichiometry) of the protein is a good starting point. Unfortunately, this data is frequently much less accurately known then the spectroscopist would want it to be. Especially the number of ions of a given metal per protein molecule is a derived quantity afflicted with a propagative error that reflects the combined difficulties of metal determination in a protein matrix, protein concentration colorimetry, and protein molecular mass determination. Additional complications may arise from the presence of apoprotein. Also, a lack of knowledge on protein subunit composition and stoichiometry can make it difficult to define the minimum active unit of a metalloenzyme.

In the above outlined situation, which is the rule rather than the exception, the EPR experiment becomes the primary analytical step. Intra-molecular stoichiometries of paramagnetic centers are obtained from EPR data independent of the values for protein concentration and molecular mass. These numbers may be used for a (re-)definition of what is the minimal enzymic unit. Also, comparing the spin concentration of individual centers to the protein chemical data may provide corrections to the molecular mass and/or to the assumption of 100% holoprotein.

As the interpretation of the EPR data, from which these biologically relevant conclusions are drawn, is not checked against independently obtained information, the pivotal question becomes to what level of sophistication we understand g-strain, because changes in the powder-spectral shape caused by g-strain broadening mimic center multiplicity and also interfere with spin quantification of individual systems.

Suppose that the principal values of the g-tensors of two centers within a protein differ by an amount in field units less than the average magnitude of the EPR line width. The merging of their spectral components will in general result in asymmetric features in the powder sum spectrum, i.e. features very similar indeed to those found in a single-component, g-strain-broadened spectrum. The two situations may usually not be discriminated except by detailed spectral simulation, if possible in combination with selective oxidation/reduction of individual centers. Note especially that the two phenomena of g-strain broadening and broadening by unresolved g-tensor multiplicity behave identically in frequency-dependent measurements.

Spin quantification of individual spectral components in complex spectra is usually done by single numerical integration of the bell-shaped, or absorption-line shaped, features at the extreme of the powder pattern (14,15) or even by integration of one half of such a feature (16). This quantity can be related to the total second integral of the spectral component (and subsequently to the second integral of the spectrum from an external standard of known concentration) via a proportionality factor that is a simple function of the principal g-values (14,15). The validity of this method is based on two assumptions: i) the line width associated with the principal-axis orientation under consideration is less than the g-anisotropy; ii) the angular dependences of the g-value, the transition probability, and the line width all three exhibit an extremum along that principal axis. A line width from g-strain broadening in general has an angular dependence dramatically different from that of the g-value (vide infra). One of the obvious effects is the skewing of the absorption line shaped features. One or both of the assumptions underlying the single-peak integration method is/are likely to be violated.

The extent of the detrimental effect of g-strain broadening on the single-peak quantification is at present poorly documented. Comparison with total second integration for some low-spin heme (4) and ferredoxin centers (6) indicate the error for quantification of skewed peaks to be typically some 30%. When the second integral is not obtainable no method is available to determine and correct for this error except by a detailed fitting of the g-strained powder pattern.

5. PROTEIN MICROHETEROGENEITY AND MATRIX STRESS

In the 1971 paper of Fritz et al. (1), in which they for the first time described the phenomenon of g-strain, they also provided an intuitive model for its physical nature: protein microheterogeneity. A protein is a molecular structure with a certain degree of flexibility. The structure is distributed about an average conformation. With every possible conformation is associated a certain ligand field at the paramagnetic center. Consequently, the g-value of the paramagnet is distributed, and this is observed as an inhomogeneous line broadening. Note that this model does not say anything about the cause of the conformational distribution and, therefore, about the ultimate cause of the line broadening. It is also not of much help when we want to predict the shape and magnitude of the g-value distribution.

The very name "g-strain" was later an inspiration to me to extend the microheterogeneity model to include the medium of the protein, the frozen solution (4-6). Suppose the matrix, in which the protein is embedded, exerts a significant stress on the protein. This is simply equivalent to the statement that a protein molecule in vacuo is not identical to the same molecule in a (frozen) solution. More specifically, we can as a starting point assume this stress to take the form of a hydrostatic pressure. The stress is communicated via the protein as a strain to the paramagnetic center. This adds a term to the spin Hamiltonian observable as a shift in the effective g-value. When either the pressure itself or the protein structure as a transducer of strain is a distributed quantity then we end up with a line width from a distributed g-value. The first moment of this distribution is not necessarily equal to the g-value in the absence of strain. Therefore, attenuation/amplification of the stress will not only result in a change in line width but also in a shift of the peak position.

Some experimental support for this concept is available. When a frozen aqueous solution (polycrystalline) of ferredoxin is subjected to a mechanical stress, small changes are observed in line width and peak positions (5). Similar effects of opposite sign are observed when the average size of the ice crystals of the polycrystallite is reduced by the addition of organic solvents to the aqueous solution before freezing (cf. Refs. quoted in Ref. 5).

There is also some evidence that the magnitude of g-strain decreases with increasing size of the protein. This indicates that the protein is actually an attenuator – be it with complicated tensorial properties – of the pressure exerted on the paramagnetic center by the enclosing matrix. The effect is illustrated in Table 1 on the width of the high-field line from a set of azido-derivatives of cytochromes as a function of protein size. Since the line width reflects a distribution in the g-value its magnitude must be expressed in units on a g-value scale.

TABLE 1

The high-field EPR line width of azido-cytochromes as a function of the molecular mass

species	molecular mass (kDa)	globule radius[a]	line width[b] (g-value units)
azido-cytochrome c	12.4	2.3	0.103
azido-myoglobin	17	2.6	0.072
azido-hemoglobin	64	4.0	0.067
azido-cytochrome a3	204	5.9	0.050

a) the cube root of the molecular mass, i.e. in units of $[(3/2\pi)kDa]^{1/3}$
b) data taken from Ref. 4

The globule radius is the radius of the protein if it were a perfect sphere. The product of this number and the entry in the last column of Table 1 is approximately a constant. In other words, the g-strain is approximately inversely proportional to the radius of the protein as a globule. This correlation is no less than a small miracle as a protein is of course not a globule of homogeneous density and elasticity and the paramagnet is usually not in the center of the protein. It is probably not too difficult to find examples of metalloproteins in which this correlation is less obvious. However, the trend of decreasing line width (in g-value units) with increasing size of the protein is a frequent observation within a single class of proteins containing similar paramagnetic centers, and provided these centers give rise to a single EPR spectrum.

A limiting case of this phenomenon is observed with model compounds (i.e. hemes, iron-sulfur clusters) that generally exhibit broader EPR lines than the protein-enclosed equivalent structures, even when the former are in "low-strain", glass-forming non-aqueous solvents. For example, a model compound for the class of proteins of Table 1 is chloroprotoporphyrin IX iron III plus excess imidazole in frozen dimethylformamide. The high-field line width is 0.113 g-value units (4) which is larger than for any of the proteins in Table 1.

6. NON-COLLINEARITY OF g-STRAIN AND THE g-TENSOR

Any protein structure inherently lacks symmetry, if only because natural polypeptides are made up of L-amino acids. Any stress exerted on a protein will result in a strain with angular properties that are incompatible with whatever (pseudo-)symmetry was originally present locally at the paramagnetic center. The tensorial properties of the g-shift from the local strain will usually not be identical to the tensorial properties of the g-value. This obvious statement of tensor non-collinearity has several far-reaching corollaries.

Firstly, the algorithm describing the line width, even for the simplest possible system with S = 1/2, I = 0, will be a complicated one. A general mathematical approach to the solution of this problem is described in detail later in this chapter. Clearly, the naked human eye will usually not suffice as a instrument to analyse g-strained spectra in detail.

Secondly, the shape of the powder or polycrystal spectrum will be characterized by the skewings and asymmetries listed above as properties vii and viii of a g-strain broadened spectrum. These features are easily recognizable in experimental spectra (cf. Fig. 1) and appear readily in simulations based on non-collinear g-strain.

Thirdly, the statement of tensor non-collinearity is equivalent to the statement that the cause of g-strain is not to be found in the g-tensor itself. No small support for these statements is found in the literature (17-29) on two decades of unsuccesful attempts to prove the contrary. These references apply to low-spin hemoproteins (17-20), high-spin hemoproteins (21-26), copper proteins (27), and iron-sulfur proteins (28-29). In all these modelings of inhomogeneous broadening, an effective g-value was made distributed by assuming a symmetric distribution in one or more quantities that directly define the effective g-value itself, i.e. crystal field splittings, spin-orbit interactions, ground-state wavefunction coefficients, or coefficients in the (unstrained) spin-Hamiltonian. All these models approximately simulate the magnitudes of line-width anisotropy, however, these never generate the right skewings and asymmetries that are the trade mark of the phenomenon of g-strain.

Note that the quoted list of references (17-29) does not include the much larger number (hundreds?) of described attempts to simulate metalloprotein EPR, not on the basis of an explicit physical model for the line-width anisotropy, but by making the ad-hoc (i.e. unfounded) assumption that this anisotropy is simply identical to that of the g-tensor (30) or to that of a (super-) hyperfine tensor collinear with the g-tensor (31).

7. THE STATISTICS OF g-STRAIN

Attempts to describe g-strain with physical models (17-29) have thus far never been succesful quantitatively when the cause of g-strain was modeled after the cause of the g-tensor. A realistic model that would place the cause of g-strain outside the g-tensor is likely to be extremely involved as it may have to encompass the three-dimensional structure of the metalloprotein itself as well as that of the interface between protein and surroundings. In order not to be stuck at this point in time, we have developed an effective description of g-strain based on statistical arguments.

Its validity lies in the fact that it provides at present the only available quantitative description of g-strained EPR spectra.

In the simplest possible situation the powder, or frozen-solution sample consists of a single, randomly-oriented spin system. In the statistical theory of g-strain (10) this single system is replaced by an ensemble of spin systems. For every particular orientation of the magnetic field there is a distribution of systems with slightly different physical properties. Every individual system still obeys the Zeeman Hamiltonian, $\hat{H} = \beta \underline{B} \cdot \underline{\underline{g}} \cdot \hat{\underline{S}}$, however each with its own g-tensor. By consequence, the g-value detected for a specific magnetic-field orientation is a random variable or, more generally, g is a function of random variables, $g = g(x_1, \ldots x_n)$. Via expansion of the function g in a Taylor series it can be shown (cf. Ref. 32) that the variance of g, σ_g^2 (i.e. the square of the g-strain line width) is approximated by

$$\sigma_g^2 = \sum_{i=1}^{n} (\delta g/\delta x_i)^2 \sigma_i^2 + 2 \sum_{j>i}^{n} (\delta g/\delta x_i)(\delta g/\delta x_j) r_{ij} \sigma_i \sigma_j \qquad [1]$$

in which r_{ij} is the correlation coefficient between x_i and x_j. A particular choice for the random variables x_i gives an explicit expression for the line width of g-strained spectra.

As an example we can take the random variables x_i to be the principal g-values themselves. When l_i are the three direction cosines between the diagonal g-tensor and the lab frame, we have

$$g^2 = \sum_{i=1}^{3} l_i g_i^2 \qquad [2]$$

and the variance of g is (10,32)

$$\sigma_g^2 = (\sum_{i=1}^{3} l_i^4 g_i^2 \sigma_i^2 + 2 \sum_{j>i}^{3} l_i^2 l_j^2 g_i g_j r_{ij} \sigma_i \sigma_j) / 2g^2 \qquad [3]$$

in which the summation is over the three principal axes. This expression contains, in addition to the three principal g-values, three line width parameters σ_i, and three correlation coefficients r_{ij}. In case the random variables g_i are all fully positive correlated, the r_{ij} reduce to unit value. When the random variables g_i are not correlated at all, the second term in Eqn. 3 is zero and the expression begins to look like the simple ad-hoc (i.e lacking theoretical foundation) expressions (30,31) that have been used extensively over the last two decades in "approximate" simulations of biological EPR data. The general form, Eqn. 3, is the effective statistical analogue of the physical models that hypothesize a common cause for g-strain and the g-tensor.

Since we know that the previous description is still too limited to produce the typical skewed and asymmetrical features of g-strain, we further generalize the statistical description by allowing the principal g-values to be functions of random variables. However, we limit this generalization, both to keep the problem tractable and to comply with our physical intuition, by retaining the assumptions that: i) there are no more than three random variables; and ii) their relationship to the g-tensor is linear. Again, the corroboration of these assumptions is in the quality of the resulting spectral simulations.

The most general linear function of three random variables is

$$\underset{\sim}{g} = \underset{\sim}{g_0} + \underset{\sim}{R} \cdot \underset{\sim}{p} \cdot \underset{\sim}{\tilde{R}} \qquad [4]$$

where p is a tensor whose elements are random variables, g_0 is a tensor whose elements do not fluctuate, and R is the three-dimensional rotation that transforms the p principal axis system to the g_0 principal axis system. To identify g-strain, or rather its local effects on the g-tensor, with a three-dimensional tensor p of random variables is – for the time being – as far as we go in our interpretation of the phenomenon of g-strain. The resulting algorithm for the line width will reproduce the broadening effects on the EPR spectrum for any physical phenomenon that is congruous with this statistical effective description. For example, for the future one may think of trying to show plausibility for a randomly distributed, hydrostatic pressure to result in a three-dimensional tensorial quantity at the site of the paramagnetic center non-collinear with the g-tensor.

The algebraic expression for the variance of g, when the g-tensor obeys Eqn. 4, has been explicitly derived in Ref. 10. It has the form

$$g^2 \sigma_g^2 = \sum_{i=1}^{3} A_i l_i^4 + \sum_{j=i}^{3} B_{ij} l_i^3 l_j + \sum_{j>i}^{3} C_{ij} l_i^2 l_j^2 + \sum_{k=j=i}^{3} D_k l_i l_j l_k^2 \ . \quad [5]$$

For three reasons its explicit form is not reproduced here. Firstly, it is a dreadfully long and complicated equation. It involves four different types of summation. Each coefficient, A–D, is a complicated function of the three g-values, the three line-width parameters σ_i, the correlation coefficients r_{ij}, the elements of the rotation matrix $\underset{\sim}{R}$ (each element in itself being a trigonometric expression), and the first moments of the elements of the p-tensor. Secondly, the full equation has been coded in FORTRAN and extensively tested (Refs. 8, 10–12). From this work a drawback has emerged in the use of Eqn. 5, even in the limiting case of full correlation, i.e. when $|r_{ij}| = 1$. Variation of the rotation elements R_{ij}, while keeping the σ_{ij}'s constant, rapidly changes the apparent width of the three main spectral features in the powder spectrum. This effect rather seriously interfers with minimization procedures. Thirdly, it is possible to write Eqns 3 and 5 out in matrix form. This does not only result in a much more compact notation but it also leads to the derivation of an algorithm that is practical in numerical analysis.

8. A MATRIX FORMULATION OF g-STRAIN

Eqn. 3, the expression for the variance of g when the principal g-values are random variables, can be written out in matrix notation as

$$
g^2 \sigma_g{}^2 = (l_1{}^2 g_1, l_2{}^2 g_2, l_3{}^2 g_3)
\begin{bmatrix}
\sigma_1{}^2 & r_{12}\sigma_1\sigma_2 & r_{13}\sigma_1\sigma_3 \\
r_{12}\sigma_1\sigma_2 & \sigma_2{}^2 & r_{23}\sigma_2\sigma_3 \\
r_{13}\sigma_1\sigma_3 & r_{23}\sigma_2\sigma_3 & \sigma_3{}^2
\end{bmatrix}
\begin{bmatrix}
l_1{}^2 g_1 \\
l_2{}^2 g_2 \\
l_3{}^2 g_3
\end{bmatrix}
\qquad [6]
$$

or in shorthand

$$
g^2 \sigma_g{}^2 = \underset{\sim}{\tilde{\Lambda}} \cdot \underset{\sim}{P} \cdot \underset{\sim}{\Lambda} . \qquad [7]
$$

The matrix $\underset{\sim}{P}$ is the covariance matrix (33) of the function of random variables, g.

Taking this as a starting point, it can be shown (10) that Eqn. 5, the expression for the variance of g when the principal g-values are functions of random variables, is written out in matrix notation as

$$g^2 \sigma_g{}^2 = {}_2\tilde{\underline{\Lambda}} \cdot {}_2\underline{R} \cdot {}_2\underline{P} \cdot {}_2\tilde{\underline{R}} \cdot {}_2\underline{\Lambda} \; . \tag{8}$$

The rotation matrix ${}_2\underline{R}$ is not the rotation matrix $\underline{R}$ used in Eqn. 4. In group-theoretical language $\underline{R}$, or better ${}_1\underline{R}$, is a 3 x 3 matrix in the irreducible representation $D^{(1)}$ of the rotation group (cf. 10,34). The pre-subscript "2" in Eqn. 8 is to indicate that ${}_2\underline{R}$ is a 5 x 5 matrix in the irreducible representation $D^{(2)}$ of the rotation group. Similarly, ${}_2\tilde{\underline{\Lambda}}$ is a weighted form of the unit length basis vector for $D^{(2)}$

$$_2\tilde{\underline{L}} = \{ \; (3l_3{}^2 - 1)/2, \; \sqrt{3}(l_1{}^2 - l_2{}^2)/2, \; \sqrt{3}l_1 l_2, \; \sqrt{3}l_1 l_3, \; \sqrt{3}l_2 l_3 \; \} \tag{9}$$

where $\tilde{\underline{\Lambda}}$ is obtained from $\tilde{\underline{L}}$ by redefining the elements with

$$\lambda_i = l_i \sqrt{g_i} \; . \tag{10}$$

Using the deconvolution of Euler rotations

$$_2\underline{R}(\alpha,\beta,\gamma) = {}_2\underline{R}(\gamma) \cdot {}_2\underline{R}(\beta) \cdot {}_2\underline{R}(\alpha) \tag{11}$$

the matrices of Eqn. 8 can be written out explicitly (10) as

$$_2\underline{P} = \begin{bmatrix} \sigma_1{}^2 & (\sigma_2{}^2 - \sigma_1{}^2)/\sqrt{3} & 0 & 0 & 0 \\[2mm] (\sigma_2{}^2 - \sigma_1{}^2)/\sqrt{3} & (2\sigma_1{}^2 + 2\sigma_2{}^2 - \sigma_3{}^2)/3 & 0 & 0 & 0 \\[2mm] 0 & 0 & \rho_{123} & 0 & 0 \\[2mm] 0 & 0 & 0 & \rho_{132} & 0 \\[2mm] 0 & 0 & 0 & 0 & \rho_{231} \end{bmatrix} \tag{12}$$

in which

$$\rho_{ijk} = (\sigma_i{}^2 + 2r_{ij}\sigma_i \sigma_j + \sigma_j{}^2 - \sigma_k{}^2) \; / \; 3 \tag{12a}$$

$$_2\underset{\sim}{R}(\beta) =$$

$$
\begin{bmatrix}
1-3(\sin^2\beta)/2 & \sqrt{3}(\sin^2\beta)/2 & 0 & -\sqrt{3}\cos\beta\sin\beta & 0 \\
\sqrt{3}(\sin^2\beta)/2 & 1-(\sin^2\beta)/2 & 0 & \cos\beta\sin\beta & 0 \\
0 & 0 & \cos\beta & 0 & \sin\beta \\
\sqrt{3}\cos\beta\sin\beta & -\cos\beta\sin\beta & 0 & 1-2\sin^2\beta & 0 \\
0 & 0 & -\sin\beta & 0 & \cos\beta
\end{bmatrix}
\quad [13]
$$

$$
_2\underset{\sim}{R}(\alpha) =
\begin{bmatrix}
1 & 0 & 0 & 0 & 0 \\
0 & 1-2\sin^2\alpha & -2\sin\alpha\cos\alpha & 0 & 0 \\
0 & 2\sin\alpha\cos\alpha & 1-2\sin^2\alpha & 0 & 0 \\
0 & 0 & 0 & \cos\alpha & -\sin\alpha \\
0 & 0 & 0 & \sin\alpha & \cos\alpha
\end{bmatrix}
\quad [14]
$$

with an equivalent expression for $_2\underset{\sim}{R}(\gamma)$ by replacing γ for α in Eqn. 14.

Although the matrix equation Eqn. 8 may be a bit more transparant than its algebraic counterpart, Eqn. 5, it is still a very complicated, bulky expression. We have only little experience (cf. 11) with a FORTRAN-coded Eqn. 8 in the analysis of real data. There is at present no saying whether its application will be practical. Its actual importance lies in the following combined facts: i) for the special case of full correlation, Eqn. 8 reduces to a much less complicated form that contains only 3 x 3 matrices; ii) the application of this reduced form in spectral simulation turns out to be practical; iii) the g-strain of all spectra of metalloproteins thus far analyzed is fit by this reduced form.

Thus, for $|r_{ij}| = 1$, Eqn. 8 reduces to

$$g^2 \sigma_g^2 = (_1\underset{\sim}{\tilde{\Lambda}} \cdot {}_1\underset{\sim}{R} \cdot {}_1\underset{\sim}{P} \cdot {}_1\underset{\sim}{\tilde{R}} \cdot {}_1\underset{\sim}{\Lambda})^2 \qquad [15]$$

in which $_1\tilde{\Lambda}$ is the g-weighted form of the unit length basis vector for $D^{(1)}$

$$_1\tilde{\underline{L}} = (l_1, l_2, l_3) \tag{16}$$

and $_1\underline{R}$ is the familiar 3 x 3 rotation matrix in the irreducible representation $D^{(1)}$. The definition of the $_1\underline{P}$ matrix is

$$_1\underline{P} = \begin{bmatrix} r_{23}\sigma_1 & 0 & 0 \\ 0 & r_{13}\sigma_2 & 0 \\ 0 & 0 & r_{12}\sigma_3 \end{bmatrix} . \tag{17}$$

That Eqn. 15 is practical becomes obvious when we write the matrix product $_1\underline{R} \cdot _1\underline{P} \cdot _1\tilde{\underline{R}}$ as a single 3 x 3 real, symmetric matrix of g-strain parameters:

$$_1\underline{R} \cdot _1\underline{P} \cdot _1\tilde{\underline{R}} = \begin{bmatrix} \Delta_{11} & \Delta_{12} & \Delta_{13} \\ \Delta_{12} & \Delta_{22} & \Delta_{23} \\ \Delta_{13} & \Delta_{23} & \Delta_{33} \end{bmatrix} \tag{18}$$

which finally gives us then a simple expression for the g-strain line width

$$\sigma_g = | \; l_1{}^2 g_1 \Delta_{11} + l_2{}^2 g_2 \Delta_{22} + l_3{}^2 g_3 \Delta_{33} + $$
$$+ 2 l_1 l_2 \sqrt{g_1 g_2} \Delta_{12} + 2 l_1 l_3 \sqrt{g_1 g_3} \Delta_{13} + 2 l_2 l_3 \sqrt{g_2 g_3} \Delta_{23} \; | / g . \tag{19}$$

The key advantage of Eqn. 19 is that the absolute values of the parameters Δ_{11} to a good approximation are equal to the apparent line widths on a g-value scale of the three features of the powder spectrum. Therefore, their approximate value (apart from a sign) can be read directly from the experimental spectrum. This then only leaves the magnitude of the off-diagonal elements Δ_{ij} to be determined by simulation, which can be done efficiently because their variation will not (contrary to what we encountered in using, e.g., Eqn. 5) rapidly shift and/or broaden the powder spectral features. Their main effect on the spectrum is to produce the skewings of absorption-type lines typical for g-strain.

What remains to be determined are the signs of the six Δ's. In principle, there are 64 possible permutations of these six signs. However, by a combination of theoretical arguments, simple experiments, and a few quick, approximate simulations one can usually reduce this to only two possibilities (cf. Ref. 5). For example, half of the permutations are pairwise identical to the other half with respect to the resulting value of σ_g because of the absolute value bars in Eqn. 19. Also, from the periodicity of Eqn. 19 over half the unit sphere it follows (5) that σ_g is invariant under a pairwise change in the signs of Δ_{13} and Δ_{23}. Furthermore, the relative signs of the diagonal elements, Δ_{ii}, can be determined from the shifts in the positions of the three powder-spectrum features upon increasing the g-strain with an applied external stress, or decreasing the g-strain by means of decreasing the size of ice microcrystals with organic solvent (5). Thus, in practice the approximate value for the six variables g_i and $|\Delta_{ii}|$ are read from the spectrum, the number of permutations in the signs of Δ_{ii} and Δ_{ij} is reduced, as indicated above, and the simulation is optimized with the three $|\Delta_{ij}|$ as free parameters. Subsequent diagonalization of the matrix in Eqn. 18 characterizes the g-strain in terms of the principal standard deviations σ_l and the rotation with respect to the g-tensor.

9. SYNTHESIS OF g-STRAINED POWDER SPECTRA

Powder spectra are simulated by summing the individual spectra for many different orientations of the spin system with respect to the externally applied magnetic field. This is done by numerical integration over the surface Ω of a unit sphere (35). The commonly used spectral shape is given by (14)

$$s(B) = \Omega^{-1} \int_\Omega \int_B f'(B)\,dB\,d\Omega \qquad [20]$$

in which $f'(B)$ is a normalized derivative line shape function at the resonance field $B = h\nu/g\beta$ with a line width ΔB and an intensity I_B proportional to the transition probability (cf. Refs 11 and 14 and Refs. quoted therein)

$$I_B = g^{-1} \sum_{i=1}^{3} (g_i{}^2 - g^2 l_i{}^2 g_i{}^4) \; . \qquad [21]$$

Because of the periodicity of Eqns 2 and 21 the integration is carried out over 1/8 of the unit sphere, i.e. in polar angles

$$\Omega^{-1} \int_{\Omega} d\Omega = (2/\pi) \int_{\phi=0}^{\pi/2} \int_{\cos\theta=0}^{\pi/2} d\cos\theta d\phi . \qquad [22]$$

This traditional description of EPR powder patterns relies on three intrinsic ad-hoc assumptions:

A-1) the angular dependence of the line width ΔB is known and is periodic over 1/8 unit sphere. For example, one frequently uses, on the basis of some intuitive arguments (30)

$$\Delta^2 B = \sum_{i=1}^{3} l_i^2 \Delta^2 B_i \qquad [23]$$

or (31)

$$\Delta^2 B = g^{-4} \sum_{i=1}^{3} l_i^2 g_i^4 \Delta^2 B_i . \qquad [24]$$

A-2) the line shape is symmetrical "in B-space", i.e. on a magnetic-field scale;

A-3) the line shape $f(B)$ is a Gaussian; $f'(B)$ is the derivative of a Gaussian.

As a set of corollaries to the statistical theory of g-strain, developed in the preceeding sections, we now replace these ad-hoc assumptions by:

A-i) the angular dependence of the line width σ_g is periodic over 1/2 unit sphere as it is given by Eqn. 19 in the case of full correlation;

A-ii) the line shape from g-strain is symmetrical "in g-space", i.e. on a g-value scale;

A-iii) the line shape $f(g)$ is a Gaussian.

Thus, we calculate the EPR-absorption spectrum in g-space as

$$s(g) = (1/2\pi) \int_{\phi=0}^{2\pi} \int_{\cos\theta=0}^{\pi} \int_{g} I_g f(g) dg d\cos\theta d\phi \qquad [25]$$

in which

$$I_g = gI_B \qquad\qquad\qquad [26]$$

and

$$f(g) = (\sigma\sqrt{2\pi})^{-1}\exp[\ -(1/2)\{(g - \bar{g})/\sigma\}^2\ . \qquad\qquad [27]$$

The simulation of the field-swept, first-derivative spectrum is then obtained by transposing $s(g)$ to B-space and subsequently taking its derivative with respect to B

$$s(B) = \hat{O}_{BB} \cdot (\hat{O}_{gB}(s(g)))\ . \qquad\qquad [28]$$

The transposition from g-space to B-space involves, to a very good approximation (4), simply the multiplication of every point of the digital powder spectrum in g-space with its associated g-value. This operation encompasses the combined effects of Eqn. 26 (i.e. a multiplication by g^{-1}) and a renormalization of the line width (i.e. a multiplication by g^2). The latter factor appears when we write the line width in B-space as a difference of two resonance conditions (4),

$$2\Delta B = (B + \Delta B) - (B - \Delta B) = (h\nu/\beta)[(g - \sigma_g)^{-1} - (g + \sigma_g)^{-1}] =$$

$$= (h\nu/\beta)[2\Delta g/(g^2 - \Delta^2 g)] \simeq constant\ x\ g^{-2} \quad (g^2 \gg \Delta^2 g)\ . \qquad [29]$$

The proposition that the g-strain line shape is a Gaussian, symmetrical in g-space, is at present unproven (cf. Ref. 10 for a discussion of this problem, and Ref. 4 for some possible exceptions). However, a number of very close fits to experimental data have been obtained under this assumption (4-6,11,12,36-38). It is also less consequential than the earlier assumption of a Gaussian in B-space, as in g-strained spectra the individual line shape is somewhat blurred away by the skewing effects caused by the non-colinearity of g-strain and the g-tensor.

10. NUMERICAL ANALYSIS OF g-STRAIN

In addition to producing asymmetries and skewings the g-strain line width algorithm, Eqn. 19, essentially differs from the earlier ad-hoc descriptions, Eqns. 23 and 24, in that it potentially produces nodes. In other words, for particular molecular orientations the g-strain line width can be zero (4). Possible theoretical and/or experimental consequences of this statement have thus far not been explored to any detail. In practice, the phenomenon may be hard to observe because there is always something that will abolish this singularity, e.g., a line width from lifetime broadening or from unresolved (super-) hyperfine structure. For the computer synthesis of g-strained powder spectra the occurrence of nodes has two practical consequences.

Firstly, the powder spectral simulator has to be protected from a zero-divide in Eqn. 27 by means of a residual line width. As the computation of the line shape f(g) for every molecular orientation in Eqn. 25 involves the sampling on a mesh in g-space, no g-strain information is lost when this residual line width is put equal to one mesh unit.

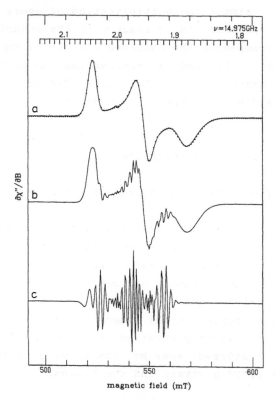

Fig. 3. The P-band spectrum (solid trace a) of ^{2}H, [2^{56}Fe–2S]$^{1+}$ in <u>Synechococcus</u> <u>lividus</u> simulated with g-strain: (a) dotted trace, simulation based on 301 x 299 orientations, (b) simulation based on 41 x 39 orientations, (c) difference spectrum x 4 (from Ref. 11).

Secondly, the residual line width is usually an order-of-magnitude smaller than the smallest of the line widths along the principal g-tensor axes. Compared to a traditional simulation, Eqn. 22, this means that the number of steps to be made in both the polar angles θ, ϕ must be raised by an order of magnitude in order to obtain a true powder pattern, i.e. to overcome mosaic-artifact ripples, as illustrated in Fig. 3.

Both the traditional integration, Eqn. 22, and the g-strain integration, Eqn. 25, are coded as three nested loops. The total computation time of a powder spectral simulator is approximately proportional to the product of the number of steps in each of these loops. Thus, because of the occurrence of zero-strain orientations, a g-strain simulation requires some two orders-of-magnitude more computation time than a traditional simulation. In order to keep the use of a g-strain simulator practical a number of tricks can be implemented to reduce the integration time by one to two orders-of-magnitude without significant loss in accuracy (11):

A judicious choise of the starting point $(\cos\theta_1, \phi_1)$ of the numerical integration over the unit sphere and of the incremental steps $d\cos\theta$ and $d\phi$ can avoid the sampling of possibly equivalent orientations in different octants. The number of steps in $\cos\theta$ and ϕ are m_θ and m_ϕ, and we choose

$$d\cos\theta = - 2/m_\theta$$
$$\cos\theta_1 = 1 - 3/(2m_\theta)$$
$$d\phi = 2\pi/m_\phi$$
$$\phi_1 = (7/8)d\phi \qquad\qquad [30]$$

in which m_θ and m_ϕ must be odd. This accounts for a midpoint integration for a uniform mesh over the complete surface of the unit sphere. However, for the eightfold set of potentially equivalent points in all octants only one occurs in the integration.

A second useful trick is to allow for some mosaic rippling and thus decreasing the required number of molecular orientations to be computed. After the generation of the full powder absorption pattern, the simulation is Fourier transformed, cut-off filtered, differentiated, and Fourier back-transformed. As the mosaic-artifact ripples have higher frequencies in the Fourier space than the EPR powder pattern they are eliminated in the back-transform.

The extent to which filtering is allowed without eroding the EPR information depends somewhat on the shape of the powder pattern, but especially on the signal-to-noise ratio of the experimental data. The fully worked-out example of a [2Fe-2S] ferredoxin spectrum can be found in Ref. 11.

Finally, the computation of the line shape function $f(g)$ is in the very inner loop of the simulator. It involves a repeated call to the time-consuming exponential function. Considerable time can be saved when this is replaced by a repeated call to a previously tabulated shape function (11).

11. EXAMPLES OF g-STRAIN ANALYSIS

With high-quality experimental data the g-strain algorithm, Eqn. 19, can give extremely good fits. This statement is illustrated by comparing the g-strain fit in Fig. 3 with that of a traditional fit in Fig. 1. The quality of the fit in Fig. 3 is further strengthened in Ref. 12 where it is shown to be valid for experimental data taken at microwave frequencies from 3 to 15 GHz. A result of such a simulation procedure is an accurate set of numbers for the principal g-values, which differ significantly from the peak positions and zero crossing of the powder pattern. These numbers may be of use in theoretical interpretation (cf. Ref. 39) but otherwise do not specifically bear relevance to any biological problem. Indeed, the spectrum in Fig. 3 was chosen as a model data set to test the g-strain algorithm because the system was considered to be well characterized (12).

A different situation is illustrated by the spectra in Fig. 4 that are from the two different iron-sulfur clusters in the multicenter ([2Fe-2S], [2Fe-2S], Mo-cofactor, FAD) enzyme xanthine oxidase. After elimination of the contribution of Mo(V), the spectrum can be accurately fit as a 1 : 1 sum of two g-strained, magnetically isolated $S = 1/2$ systems (37). This analysis not only resolved the contribution of each Fe/S center to each point in the spectrum, but also established that – contrary to several previous determinations – both centers were "regular" in having two out of three principal g-values below the free electron value.

Another example in which the determination of the g-values has proven important, is the g-strain simulation of the spectrum from the model prismane cluster [6Fe-6S]. The g-strain simulation indicated a g-value as low as 1.2 where no particular feature was visible in the powder pattern. Using this low g-value in the Aasa-

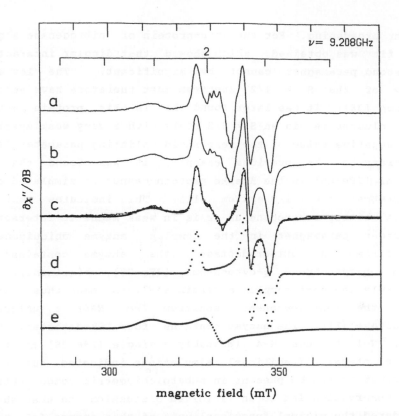

Fig. 4. A one-to-one composite g-strain fit (dotted trace c) to
the spectrum of centers Fe/S I and Fe/S II in xanthine-reduced
bovine milk xanthine oxidase. The molybdenum-free spectrum (solid
trace c) was obtained as a difference of spectra taken at two
microwave powers (traces a and b). Traces d and e are the
individual simulations of the spectra from Fe/S I and Fe/S II,
respectively (from Ref. 37).

Vänngård intensity expression (14) quantification would give an S =
1/2 concentration equal to the weighed-in amount of prismane
(13,40).

A good g-strain fit is also a sign that the spectrum is to all
likelihood not complicated by other interactions. For example, a
perfect fit to the 3Fe spectrum of Thermus thermophilus ferredoxin
([3Fe-4S], [4Fe-4S]) was used as an argument that the protein is
homogeneous as observed from the site of the 3Fe center. This
argument was important in the subsequent analysis on reduced
protein of the spectrum from the 4Fe cluster in weak magnetic
interaction with the 3Fe cluster (36).

In many instances unusual powder shapes and substoichiometric
spin integrations have been "explained" as reflecting the effects

of dipolar broadening. For the iron-protein of nitrogenase a good
g-strain fit was obtained, which showed that dipolar interaction
with a second paramagnet cannot be significant. The low spin
intensity of the S = 1/2 spectrum must therefore have another
explanation (38). It was later found that in this protein part of
the Fe/S cluster is in an S = 3/2 state with a very weak spectrum
due to a negative value of the zero-field splitting parameter (41).

Contrary to the previous example, it was found that the
spectrum of [2Fe-2S] in the Rieske protein cannot be simulated as a
single, g-strained S = 1/2 system (11). This indicates that the
cluster is either inhomogeneous or is in weak magnetic interaction
with another paramagnet in the complex enzyme ubiquinone :
ferricytochrome c oxidoreductase (the enzyme contains: b-
cytochrome, b-cytochrome, c-cytochrome, [2Fe-2S], ubiquinone).

Probably the most complex g-strain analysis made thus far is
that of the four-component spectrum from NADH : ubiquinol
oxidoreductase (11). The enzyme contains four EPR-detectable Fe/S
clusters, N-1 through N-4 (probably a single [2Fe-2S] and three
[4Fe-4S] clusters). Traditional simulations indicated one or two
of the clusters to be present in substoichiometric concentration,
and this observation led to an involved discussion on what should
be considered the minimal functional unit of this enzyme (cf. Refs.
quoted in 11). In Fig. 5 it is shown that a good fit to the

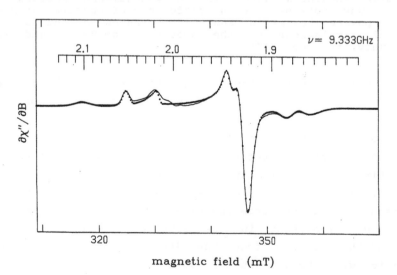

Fig. 5. The spectrum (solid trace) of NADH-reduced bovine heart
NADH:Q oxidoreductase was simulated as a 1:1:1:1 composite (dotted
line) of four g-strained ferredoxin spectra (from Ref. 11).

overall spectrum is obtained with the g-strain algorithm, assuming an exact 1 : 1 : 1 : 1 stoichiometry of all centers. In other words, the EPR spectrum does not provide indication that the functional unit of this enzyme is more complex than one encompassing a single species of each of the four Fe/S centers.

12. POSSIBLE FUTURE DEVELOPMENTS

A first possibility: The examples of g-strain analysis described above were aimed at the accurate identification of component powder patterns. There was no particular concern with the physical nature of what causes the g-strain. It was indeed the very complexity of this problem that led us to develop the statistical theory of g-strain as an effective description of a phenomenon that is at present poorly understood within a physical framework. Specifically, the question whether structural information is extractable from a knowledge of g-strain would seem to call for a long-term research effort.

A second possibility: The g-strain algorithm described in this chapter applies to $S = 1/2$ systems subject to the simple spin Hamiltonian $\hat{H} = \beta \ \underline{B} \cdot \underline{g} \cdot \hat{\underline{S}}$. The EPR of many biological systems is described by more complex spin Hamiltonians, e.g., when hyperfine and/or superhyperfine interactions are present, when dipolar interaction is significant, or when the spin $S > 1/2$. At present no rigorous description of inhomogeneous broadening in these systems is available, therefore, spectral simulation is not quantitative.

It is likely that, in addition to the g-value, also the other coefficients in the spin Hamiltonian, e.g., hyperfine constants, zero-field splitting parameters, can be described as functions of random variables. For example, a distribution in the central hyperfine splitting would result in the individual peaks for different nuclear orientations to have different line widths. This phenomenon is, in fact, frequently observed for the hyperfine lines in spectra from metalloproteins and model compounds containing vanadium, cobalt, copper, or molybdenum (enriched in ^{95}Mo). A few attempts have been made to analyze these patterns in cupric proteins and model compounds (4,27,42), however, only for the parallel part of (semi-) axial spectra. To my knowledge the only available full powder spectral syntheses are attempts by myself to fit multifrequency EPR data from the hydrated cupric ion and from cytochrome $\underline{c}$ oxidase (6). These simulations were based on an early version of the g-strain algorithm (4), extended with a distribution in the first-order hyperfine splitting, plus a possible non-collinearity of the g-tensor and the hyperfine tensor. The bottom line of this work was that, even for the simple model system $^{65}Cu(H_2O)_6^{2+}$, it proved impossible to obtain a single quantitative fit of the multifrequency data set.

A similar situation exists for high-spin Kramers and non-Kramers systems, where the zero-field splitting appears to be a distributed quantity (4,43) and no g-strain type modeling has yet been initiated. In summary, the quantitative analysis by spectral simulation of systems, subject to more than an electronic Zeeman interaction only, is at present a virginal area of research.

A third possibility: The importance of g-strain is probably not limited to EPR. For any spectroscopy in which use is made of the EPR phenomenon, one can ask the question as to the observability of g-strain effects, or, from a different perspective, to what extent a quantitative data analysis is hampered by ignoring the g-strain phenomenon. A relevant area is the analysis, by spectral synthesis, of ENDOR data from polycrystalline frozen metalloprotein solutions. Here the powder pattern is described traditionally, i.e. by a distribution symmetric in B-space (cf. Eqn. 20). Also, the EPR line width is commonly taken to be isotropic. The approach is described in detail in the chapter by B.M. Hoffman in this book. It now appears that g-strain effects, specifically the possible occurrence of zero-strain orientations, can greatly affect the ENDOR spectra taken where the EPR resonance is set anywhere in the middle (i.e not at an extreme) of the EPR spectrum (12). Again, the consequences of g-strain for EPR-related spectroscopies are essentially unexplored to date.

A fourth possibility: Paramagnetic relaxation in magnetically dilute solids proceeds essentially via a modulation of the crystal field by elastic waves that is experienced by the spin system as a time-varying electromagnetic field by virtue of the spin-orbit coupling. As the effective spin-orbit-lattice Hamiltonian, which describes these concomitant transitions of the spins and the phonons, is formally equivalent to the time derivative of the Hamiltonian describing the g-shift from static strains, the suggestion has been made that anisotropy in the spin-lattice relaxation in metalloproteins can be linked to the anisotropy of g-strain (5,6,44). It has indeed been possible, on the basis of some elementary modeling, to deduce a semi-quantitative description of anisotropic T_1-relaxation in a [2Fe-2S] ferredoxin from a knowledge of the g-strain (5). This problem likewise deserves to be the subject of a wider and more thorough analysis in the future.

ACKNOWLEDGEMENTS

A lasting collaboration with W.R. Dunham provided the basis for a major part of this chapter. Over the years many others also helped me with the subject, especially S.P.J. Albracht, D.O. Hearshen, and R.H. Sands.

REFERENCES

1   J. Fritz, R.E. Anderson, J.A. Fee, G. Palmer, R.H. Sands, J.C.M. Tsibris, I.C. Gunsalus, W.H. Orme-Johnson and H. Beinert, Biochim. Biophys. Acta, 253 (1971) 110-133.
2   R.H. Sands and W.R. Dunham, Quart. Rev. Biophys., 7 (1975) 443-504.
3   L.H. Strong, Ph.D. thesis, University of Michigan, 1976.
4   W.R. Hagen, J. Magn. Reson., 44 (1981) 447-469.
5   W.R. Hagen and S.P.J. Albracht, Biochim. Biophys. Acta, 702 (1982) 61-71.
6   W.R. Hagen, Ph.D. thesis, University of Amsterdam, 1982.
7   D.O. Hearshen, W.R. Dunham, R.H. Sands and H.J. Grande, in: C. Ho (Ed.), Electron Transport and Oxygen Utilization Elsevier, New York, 1982, pp. 395-398.
8   D.O. Hearshen, Ph.D. thesis, University of Michigan, 1983.
9   J.R. Pilbrow, J. Magn. Reson., 58 (1984) 186-203.
10 W.R. Hagen, D.O. Hearshen, R.H. Sands and W.R. Dunham, J. Magn. Reson., 61 (1985) 220-232.
11 W.R. Hagen, D.O. Hearshen, L.J. Harding and W.R. Dunham, J. Magn. Reson., 61 (1985) 233-244.
12 D.O. Hearshen, W.R. Hagen R.H. Sands, H.J. Grande, H.L. Crespi, I.C. Gunsalus and W.R. Dunham, J. Magn. Reson. 69 (1986) 440-459.
13 W.R. Hagen, in: S. Papa, B. Chance and L. Ernster (Eds.), Cytochrome Systems. Molecular Biology and Bioenergetics, Plenum, New York 1987, pp. 459-466.
14 R. Aasa and T. Vänngård, J. Magn. Reson., 19 (1975) 308-315.
15 S. de Vries and S.P.J. Albracht, Biochim. Biophys. Acta 546 (1979) 334-340.
16 S.P.J. Albracht, G. Dooijewaard, F.J. Leeuwerik and B. van Swol, Biochim. Biophys. Acta, 459 (177) 300-317.
17 P. Eisenberger and P.S. Pershan, J. Chem. Phys., 47 (1967) 3327-3333.
18 C. Mailer and C.P.S. Taylor, Can. J. Biochem., 50 (1972) 1048-1055.
19 A. Dwivedi, W.A. Toscano and P.G. Debrunner, Biochim. Biophys. Acta, 576 (1979) 502-508.
20 J.C. Salerno, J. Biol. Chem., 259 (1984) 2331-2336.
21 G.A. Hecklé, D.J.E. Ingram and E.F. Slade, Proc. Roy. Soc. B, 169 (1968) 275-288.
22 E.F. Slade and D.J.E. Ingram, Nature, 220 (1968) 785-785.
23 R. Calvo and G. Bemski J. Chem. Phys., 64 (1976) 2264-2265.

24   A.S. Brill, F.G. Fiamingo and D.A. Hampton, in: P.L. Dutton, J.S. Leigh and A. Scarpa (Eds.), Frontiers of Biological Energetics. Vol. 2, Acad. Press, New York, 1978, pp. 1025-1033.
25   D.A. Hampton and A.S. Brill, Biophys. J., 25 (1979) 301-311.
26   D.A. Hampton and A.S. Brill, Biophys. J., 25 (1979) 313-322.
27   A.S. Brill, Transition Metals in Biochemistry, Springer, Berlin, 1977.
28   B. Guigliarelli, J.-P. Gayda, P. Bertrand and C. Moore, Biochim. Biophys. Acta, 871 (1986) 149-155.
29   B. Guigliarelli, C. Moore, P. Bertrand and J.-P. Gayda, J. Chem. Phys., 85 (1986) 2774-2778.
30   T.S. Johnston and H.G. Hecht, J. Mol. Spectrosc., 17 (1965) 98-107.
31   J.H. Venable, in: A. Ehrenberg, B.G. Malmström and T. Vänngård (Eds.), Magnetic Resonance in Biological Systems, Pergamom Press, Oxford, 1967, pp. 373-381.
32   A. Papoulis, Probability Random Variables and Stochastic Processes, McGraw-Hill, New York 1965.
33   R. von Mises, Mathematical Theory of Probability and Statistics, Acad. Press, New York, 1964.
34   E.P. Wigner, Group Theory, Acad. Press, New York, 1959.
35   F.K. Kneubühl, J. Chem. Phys., 33 (1960) 1074-1078.
36   W.R. Hagen, W.R. Dunham, M.K. Johnson and J.A. Fee, Biochim. Biophys. Acta, 828 (1985) 369-374.
37   R. Hille, W.R. Hagen and W.R. Dunham, J. Biol. Chem., 260 (1985) 10569-10575.
38   W.R. Hagen, W.R. Dunham, A. Braaksma and H. Haaker, FEBS Lett., 187 (1985) 146-150.
39   P. Bertrand and J.-P. Gayda, Biochim. Biophys. Acta, 579 (1979) 107-121.
40   M.G. Kanatzidis, W.R. Hagen, W.R. Dunham, R.K. Lester and D. Coucouvanis, J. Am. Chem. Soc., 107 (1985) 953-961.
41   W.R. Hagen, R.R. Eady, W.R. Dunham amd H. Haaker, FEBS Lett., 189 (1985) 250-254.
42   W. Froncisz and J.S. Hyde, J. Chem. Phys., 73 (1980) 3123-3131.
43   W.R. Hagen, W.R. Dunham, R.H. Sands, R.W. Shaw and H. Beinert, Biochim. Biophys. Acta, 765 (1984) 399-402.
44   W.R. Hagen, Biochim. Biophys. Acta, 708 (1982) 82-98.

CHAPTER 23

EPR OF IRON - SULFUR AND MIXED - METAL CLUSTERS IN PROTEINS

ISABEL MOURA, ANJOS MACEDO and JOSÉ J. G. MOURA

1. INTRODUCTION

The iron-sulfur proteins play a diversified role in biological systems, being involved in electron transfer as well as directly interacting with substrates. The study of the structure-function relationships in this class of proteins has been one of the most active research fields involving multi-disciplinary sciences such as Biology, Biochemistry, Chemistry and Physics.

Most of the spectroscopic tools used to reveal the properties of the iron-sulfur centers are related with the study of different oxidation states, which have characteristic magnetic properties. This enables the extraction of important information by the use of different magnetic resonance techniques, namely EPR, MCD, NMR, MB and magnetic susceptibility measurements. Among these techniques EPR has a large potential application, not only being now generally available, but in addition, it can give access to important parameters generally required in biochemical work: type of center, active-site stoichiometry and composition, kinetic and thermodynamic parameters or the EPR signal can be used as a simple intensive property of the system (for example, the determination of mid-point redox potentials following the intensity of an EPR signal as function of the poised solution potential). Also EPR is optimally complemented by/and related to MB and MCD spectroscopies and magnetic susceptibility measurements.

Basic clusters have been described in these proteins: [2Fe-2S], [3Fe-4S] and [4Fe-4S] centers, the simplified structures of which are indicated in Fig. 1. The cores are generally paramagnetic ($S=1/2$) or diamagnetic (except for the reduced [3Fe-4S] core). Rubredoxin-type centers contain one iron atom tetrahedrally coordinated to four cysteinyl residues and no labile sulfur, and are frequently added to the basic iron-sulfur unities. This core involves high-spin ferric and ferrous atoms ($S=5/2$ and $S=2$).

However, the field is now rapidly developing and different stoichiometries in yet unknown structures are under consideration, as well as ground states with different multiplicity, i.e., $S=3/2$, $5/2$ and $7/2$. The general picture is becoming more and more complex and the possibility of variability in cluster coordination, with replacement of the cysteinyl residues by other amino acids (namely the participation of O or N from the amino-acid side chain) must be taken into account.

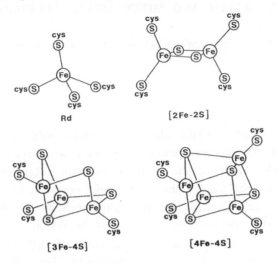

Fig. 1. Structure of the iron-sulfur centers.

In certain cases, the proteins and enzymes under study contain one of the above clusters. More frequently, however, these clusters are associated with other prosthetic groups such as hemes, flavins, molybdenum, vanadium, nickel, selenium, thiamine diphosphate and chlorophyll, or different types of iron-sulfur clusters are found in the same protein.

The discovery of the basic [3Fe-4S] structure is relatively recent (1,2) and the subsequent observation of the possibility of conversion of a [3Fe-4S] cluster into a [4Fe-4S] one (3,4,5) or into a mixed-metal cluster of the general type [M, 3Fe-4S] has opened new research fields (6,7,8).

Early X-ray crystallographic data on a 3Fe containing protein (*Azotobacter vinelandii* Fd) implied that the iron and sulfur atoms in this structure could form an alternative planar structure (9). However, new X-ray data on the same protein (10) and in *Desulfovibrio gigas* FdII (11) demonstrated that the [3Fe-4S] is a unique cubane-like structure missing one iron atom in one of the cube corners as indicated in Fig. 1.

In this chapter we will focus on the EPR properties of the basic iron-sulfur structures and of the newly formed mixed-metal clusters. A particular emphasis is put on the unusual high-spin states recently established in this group of proteins.

Previous review articles and references therein, give detailed background information on the subject as well as indications on the use of EPR for iron-sulfur cluster identification (12,13,14,15).

2. EPR SPECTRA OF BASIC IRON - SULFUR STRUCTURES

Paramagnetic forms of the majority of the well characterized iron-sulfur structures are S=1/2 (except for rubredoxin-type centers and the reduced 3Fe core). Their EPR signals are relatively simple when compared with the EPR features of other biologically relevant transition metals. At X-band frequency most of the hyperfine interactions are unresolved (although ^{57}Fe, ^{77}Se, ^{33}S isotopes may induce measurable broadenings) and most of the data analysis is confined to the detection of a fingerprint of the centers or to quantitative work.

Assignment of the spin is, for half-integer spin systems, usually straighforward since the highest intensity arises commonly from the $\pm 1/2$ doublet centered at g-values given approximately by 2S+1. However, in the case of higher spin multiplicity, the problem may become more difficult and the sign and magnitude of the zero-field splitting parameter, D, must be known. For example, an inverted spin multiplet is obtained if the D value is negative, the Kramers' doublet with $m_S = \pm 1/2$ being the highest energy level. Analysis of the EPR temperature dependence is required in order to populate differentially the higher-energy doublets involved and obtain information on the sign of D.

The magnetic properties of half-integer spin systems (S>1/2) can be described by the following spin Hamiltonian:

$$\hat{H}_e = D[\hat{S}_z^2 - 1/3\ S(S+1) + E/D\ (\hat{S}_x^2 - \hat{S}_y^2)] + g_0 \cdot \hat{\underline{S}} \cdot \underline{B} \qquad [1]$$

where D and E are parameters describing the zero-field splitting of the spin multiplet. It is assumed that the Zeeman interaction is isotropic ($g_0 \sim 2$). The energy levels are calculated under conditions that the zero field-splitting is larger than the energy involved in the EPR transition ($|D| >> g_0\ \beta\ B$), a condition that is valid for the iron-sulfur clusters studied here. Then, each Kramers' doublet can be treated separately, assuming a fictitious spin, $\hat{\underline{S}}' = 1/2$. The properties of each doublet can be described by a term $\hat{\underline{S}}'.\ g'.\underline{B}$, where the principal g'-values of the g'-tensor depend only on the parameter E/D. The g'-values can be computed directly from Eqn. 1 and Fig. 2 shows the g'-values of the Kramers' doublets as function of E/D for the spin systems S=3/2, 5/2 and 7/2. The analysis of these diagrams is useful for the interpretation of the EPR data in systems with high spin multiplicity and they are extensively used below.

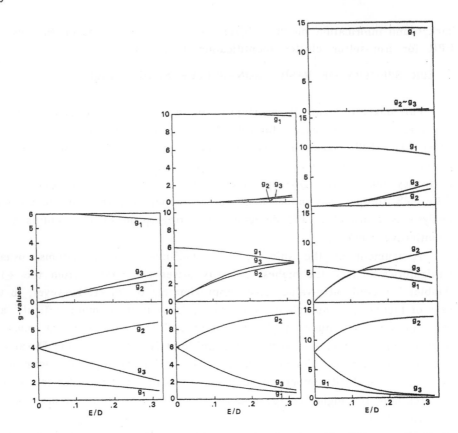

Fig. 2. Plot of the g'-value (g_1', g_2' and g_3' of each Kramer doublet) as function of E/D for three spin systems: S=3/2, S=5/2, S=7/2.

2.1 Rubredoxin-type center

Rubredoxins (Rd) are the simplest proteins that may be included in this class. The active center is characterized by the absence of labile sulfur and the presence of one iron atom linked in a tetrahedral arrangement to the sulfur atoms of four cysteinyl residues, with well-characterized spacing (cys-x-x-cys) (16).

A variation of the basic rubredoxin type core is present in desulforedoxin (Dx) (17). The tetrahedral cysteinyl iron coordination is maintained but two of the cysteinyl ligands are adjacent in the amino acid sequence, imposing different geometrical constraints in the active center (18).

In the oxidized form, rubredoxin contains one high-spin ferric atom (magnetic moment μ=5.73 Bohr magneton) (19). The EPR spectrum is highly conserved for the different isolated rubredoxins (12,15). Resonances at g'=4.3 and g'=9.4 are observed below 20K (Fig. 3). These resonances can be interpreted (20) as arising from EPR transitions within the ground and first excited doublet state (middle Kramers' doublet) of the high-spin ferric center (S=5/2) with a high degree of rhombic distortion (E/D=0.28). As the temperature is lowered the g'=4.3 signal decreases in intensity, as the middle Kramers' doublet becomes thermally depopulated and the g'=9.4 resonance originating from the ground state increases (Fig. 3). The Dx EPR spectrum (Fig. 3) reveals also a high-spin ferric state (19,21) with a nearly

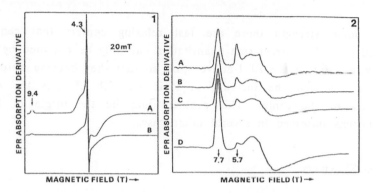

Fig. 3. 1 . Comparison of the EPR spectra of rubredoxin from *D.gigas* (A) and *Clostridium pasteurianum* (B). Experimental conditions: temperature, 12 K; microwave power, 20 mW; frequency, 9.256 Hz; modulation amplitude, 1 mT; gain, 100. 2 . EPR spectra of *D.gigas* desulforedoxin at the indicated temperatures and the following power and gain: (A) 200 μW, 2×10^4 (B) 20 μW, 10^4 (C, D) 20 μW, 6.3×10^3. Adapted from Ref. 16 and 19.

axial symmetry (E/D=0.08). The principal g'-features are observed at 7.7 and 5.7 and broad components at 4.1 and 1.8 (Fig. 3). The observed g'-values calculated for a high-spin ferric ion, associated with the three Kramers' doublets for a E/D=0.08, are shown in Fig. 4. The resonances at 4.1, 7.7 and 1.8 are attributed to the ground state doublet ($\pm 1/2$) and the resonance at g'=5.7 to the middle doublet ($\pm 3/2$). When the temperature is decreased (Fig. 3) the signal at g'=7.7 increases while the signal at g'=5.7 decreases. The zero-field splitting was estimated to be 2cm^{-1} (21).

At low salt concentrations a surprising result is observed in the EPR spectrum of oxidized Dx (17): the EPR signal quantifies only up to $\sim$ 0.5 spins/1Fe. Increasing the ionic strength increases the number of spins quantified by EPR to $\sim$ 1 spin/1Fe. Complementary MB studies revealed that

818

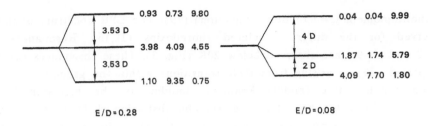

Fig. 4. Schematic representation of the energy levels of an S=5/2 spin multiplet for E/D=0.28 and E/D=0.08 and respective g'-values of each Kramers' doublet.

at low ionic strength there are fast relaxing centers that can not be detected by EPR. Correct spin quantitation can only be obtained by changing the environment of the EPR-active spins so that they become detectable by EPR. In this case the increase in ionic strength changes the relaxation properties of the center. This problem may be the origin of low EPR quantitations observed in several other cases.

2.2 [2Fe-2S] centers

This center contains two iron atoms and two labile sulfurs and each iron atom is bound to the polypeptide chain by two cysteine residues.

EPR played a crucial role on the elucidation of the structure of this type of center (22, 23). It can exist in two stable oxidation states. In the oxidized form the $[2Fe-2S]^{+2}$ center has a diamagnetic ground state (S=0), containing two ferric sites. This state is EPR silent. Upon one electron reduction the cluster has an S=1/2 ground state and is EPR active (the redox potential is generally lower than -300 mV). Rhombic or axial signals can be observed with one of the g values close to 1.94.

Gibson and co-workers (23) developed a model taking into account an antiferromagnetic coupling (through the labile sulfurs) of the two iron atoms (one ferric site and one ferrous site) to explain the observed EPR signature. This model was later confirmed by Dunham et. al. (24) and Munck et al. (25). In this model the spin coupling between the Fe^{3+}-site (S_1=5/2) and the Fe^{2+}-site (S_2=2) in the +1 state of a [2Fe-2S] center is described by the Heisenberg Hamiltonian $\hat{H} = J \, \hat{S}_1 . \hat{S}_2$. The [2Fe-2S] cluster in the +1 state has localized Fe^{2+} and Fe^{3+} valence states and the coupling is antiferromagnetic.

More recently this structure was confirmed by X-ray diffraction studies of the ferredoxin from *Spirulina platensis* (26,27,28).

Some [2Fe-2S] containing proteins (i.e. Rieske center) were reported in the literature to have unusual redox properties (29). Their redox potential was placed between +150 and +330 mV, a value that is 400-600 mV higher than the reported values for the [2Fe-2S] center in plant ferredoxins. A protein isolated from *Thermus thermophilus*, that contains two of these centers was demonstrated to have only four cysteinyl residues, which are insufficient to bind two cores (eight cysteinyl residues are required for the binding). Other ligands are then involved in the coordination of the clusters, and nitrogen atoms are alternative ligands that may be responsible for the unusual cluster properties (29).

2.3 [3Fe-4S] centers

Eight years ago the [3Fe-4S] structure was introduced through the spectroscopic studies of the *A. vinelandii* Fd (2), aconitase (30) and *D.gigas* Fd (FdII) (1). Since then this cluster was shown to be wide-spread in different biological systems. This discovery had a tremendous impact being responsible for the renewed interest on the iron-sulfur protein field.

The spectroscopic features of *D.gigas* Fd II have been described in detail by different methods and it will be used here as a model.

FdII is a tetramer from identical monomers (57 amino acids and 6 cysteines). By combined EPR and Mössbauer studies it was demonstrated that FdII is spectroscopically pure and contains a single [3Fe-4S] cluster per monomeric unit (1).

In contrast to FdII, another form of ferredoxin, FdI, was shown by the conjunction of EPR and MB techniques to contain variable amounts of a [3Fe-4S] center (10-30 % depending on preparation), together with a majority of [4Fe-4S] centers (3).

FdII preparations are EPR active in both oxidized and reduced form. The EPR spectrum of the oxidized form accounts for 1 spin/3Fe and is shown in Fig. 5. The spectrum can be reasonably fitted choosing $g_1=2.02$, $g_2=2.00$ and $g_3=1.97$ (using line-broadening of 1.5, 3.5 and 8 mT at g_1, g_2 and g_3, respectively).

MB spectroscopy shows that all the three iron atoms are associated with the S=1/2 EPR signal. They are all high-spin ferric (S=5/2) participating in an antiferromagnetic coupling. The analysis of the EPR spectrum of oxidized [3Fe-4S] clusters with a $g_{av}>2$ is not enough to distinguish a [3Fe-4S] cluster from a [4Fe-4S] center in the +3 oxidation state, that is, the state

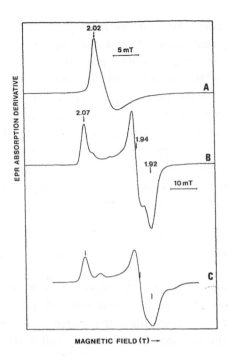

Fig. 5. EPR spectrum of (A) Fd II, oxidized form (B) Fd I, reduced form and (C) ^{57}Fe-reconstituted Fd II in the reduced form Experimental conditions: (A, B) microwave power, 10 μ W; T, 8 K; modulation amplitude, 0.4 mT; (C) microwave power, 2 mW; modulation amplitude, 1 mT and T, 18 K. Adapted from Ref. 3.

of oxidized HiPIP (see below), and it is not surprising that the EPR signals of the [3Fe-4S] center have been early interpreted as "typical" HiPIP signals.

Reduction by one electron, E_m=-130mV (vs. NHE) yields a state with integer cluster spin. Thomson and co-workers (31) have studied reduced FdII with low-temperature magnetic circular dichroism and inferred that the ground manifold has S=2. MB studies of the reduced cluster has shown that one of the Fe sites remains high-spin Fe^{3+} and two of the Fe sites share equally the electron that enters the cluster upon reduction, having an oxidation level $Fe^{2.5+}$ (3). This is an example of a mixed-valence system with one localized site (Fe^{3+}) and a delocalized pair ($Fe^{+2.5}$-$Fe^{+2.5}$).

MB studies also indicated that the two lowest electron spin states are separated by about 0.35 cm^{-1}. Fig. 6 shows an energy level diagram pertinent to the S=2 system of FdII. These observations suggested that an EPR transition at X-band should be detected. Indeed reduced FdII shows extremely broad and weak low field signals stretching virtually to zero magnetic field (32). This signal is absent in oxidized Fd II. A similar signal was also observed for the [3Fe-4S] cluster of a ferredoxin from *T. thermophilus* and attributed to a "$\Delta m_s = 4$" transition. The observation of such a signal down to zero field implies D ~ 0.30 cm^{-1}. The splitting of the two states is given by $\Delta = 2D \, (x^{1/2} - 1)$ with $x = 1 + 3 \, (E/D)^2$.

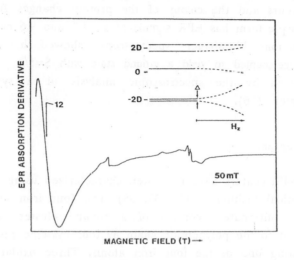

Fig. 6. EPR spectrum of a 22mM sample of reduced Fd II (32). The low-field resonance is that indicated in the insert. The weak features at 150 mT are a minor (<1%) impurity, and the signal around 330 mT belongs to a [4Fe-4S]$^{1+}$ cluster (<1% of total clusters in the sample). EPR conditions: T, 9 K; microwave power, 5 mW; modulation amplitude, 1 mT. Insert - Energy level diagram for a S=2 manifold for D=-2.5cm^{-1} and E/D=0.23. Dashed lines indicate response of the levels to a field along z. The EPR transition occurs between the two lowest levels. Adapted from Ref. 32.

The S=2 state was studied by EPR and MCD for the reduced [3Fe-4S] cluster of *T.thermophilus* Fd and *A.vinelandii* Fd I (33, 34). An extremely broad signal at low-field in the X-band EPR spectrum of reduced *T.thermophilus* Fd was detected but absent for the *A.vinelandii* Fd I. This is interpreted in terms of the $\Delta m_s = 4$ EPR transition not being possible when the splitting of the $m_s = \pm 2$ doublet is greater than the microwave energy ($\Delta \sim 0.3$ cm^{-1} at X-band). In conclusion the observation of the low-field resonance for S=2 is an important fingerprint of the reduced [3Fe-4S] center, but its absence does not preclude the presence of this center.

The highly-resolved MB spectra of the reduced [3Fe-4S] cluster were analysed using S=2 spin Hamiltonian. An important result of the analysis is the observation (28) that the magnetic hyperfine coupling constant of the Fe^{3+} site is positive, showing that its local spin is antiparallelly coupled to the system spin, and its magnitude suggests a spin of 9/2 for the delocalized dimer. The data has been further analysed with a spin-coupling model that takes into account Heisenberg exchange and valence delocalization (32). Another form of the [3Fe-4S] cluster can be produced under alkaline conditions from the [3Fe-4S] of aconitase (35). Under these conditions a linear form occurs and the colour of the protein changes from brown to purple. The purple form has EPR signals at g'=4.3 and 9.6 characteristic of high-spin ferric ions. Mössbauer spectroscopy showed that the three iron sites are exchange-coupled to yield a ground state with S=5/2. The data was further supported by the spectroscopic analysis of a synthetic linear trinuclear cluster (36).

2.4 [4Fe-4S] centers

The [4Fe-4S] centers have been well characterized using a wide variety of physicochemical techniques (15, 37, 38). The four iron and four sulfur atoms occupy alternate corners of a cubane cluster. The center is covalently bound to the polypeptide chain by four cysteine residues, each of them coordinating one of the four iron atoms. Three oxidation states are known for this type of center $[4Fe-4S]^{1+,2+,3+}$. In the oxidation state with total charge +2 (the +2 state, $2Fe^{3+}$ and $2Fe^{2+}$) this center as S=0 and gives no electron paramagnetic (EPR) signal (see Footnote 1).

In ferredoxins this center can be reduced to the +1 state (formal oxidation states $1Fe^{3+}$ and $3Fe^{2+}$) and in this state the cluster has S=1/2 and an EPR signal (axial or rhombic) can be observed. One of the g-values is smaller than 2, typically 1.94 (the "g=1.94" type signal). The redox potential associated with this transition is negative (around -400 mV).

In other type of proteins, the so called high-potential iron proteins (HiPIP), the [4Fe-4S] center can be oxidized into the +3 state ($3Fe^{3+}$, $1Fe^{2+}$). In the +3 state the spin of the center is S=1/2 and an EPR signal is detectable around g=2.02. The redox potential associated with this transition is positive (around +300 mV).

There are no known examples of proteins that in their native conformation can stabilize all the three oxidation states of this cluster.

Footnote 1 - The charge of the cluster is calculated taking into account the charge of each of the iron atoms and considering that each of the labile sulfurs have a -2 charge.

In Fig. 5 the EPR spectrum of reduced Fd I is presented with principal g-features at 2.04, 1.94 and 1.92, typical values of a reduced $[4Fe-4S]^{1+}$ cluster.

More complex EPR spectra can be observed for the [4Fe-4S] centers in the +1 state if there is more than one paramagnetic center per molecule. Examples are known in the literature of such interactions between such centers that perturb the EPR spectra from wich, in some cases, distances between redox centers can be obtained (39). The perturbing effect is better seen comparing the EPR spectra of a center with the other center in a non-magnetic and in a magnetic state. It can be observed, for example, for the 2[4Fe-4S] ferredoxin from *C.pasteurianum* (Fig. 7) by comparison with the EPR spectrum of a single [4Fe-4S] cluster (Fig. 5). The effect of spin-spin interaction is reflected in the splitting of the spectral lines. The interaction was identified as a combination of dipolar and exchange interaction (40).

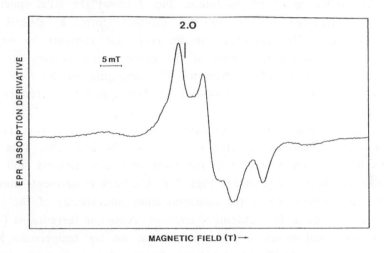

Fig. 7. EPR spectrum of reduced *C.pasteurianum* ferredoxin. Experimental conditions: microwave power, 30 μW; T, 8 K; modulation amplitude, 0.4 mT.

3. INTERCONVERSIONS BETWEEN IRON-SULFUR CENTERS

The geometrical similarities found in the arrangement of the cluster atoms, when [3Fe-4S] and [4Fe-4S] centers are compared, led to considering the transformation of one structure into the other. The problem was already encountered in the reductive activation of aconitase (30). The active form contains a single [4Fe-4S] core, but a [3Fe-4S] center is present in inactive preparations. Also, recent X-ray analysis of *D.gigas* FdII indicates that three cysteinyl residues ligate the iron atoms to the polypeptide chain and a fourth cysteinyl residue is available nearby ready to form a fourth S-Fe bond when required (11).

Most of the interconversion pathways were demonstrated in *D.gigas* FdII. Incubation of dithionite reduced *D.gigas* Fd II (containing only [3Fe-4S] cores) with Fe^{2+} and S^{2-}, under anaerobic conditions and in the presence of dithiothreitol converts these centers into [4Fe-4S] ones. EPR is an ideal tool to follow the time course of the cluster conversion by the increase of the characteristic "g=1.94" signal, indicative of the formation of [4Fe-4S] clusters . When a 5-fold excess of iron and sulfide is added, an inter-conversion of ~70% of the centers takes place after one hour of incubation time.

Another way of converting the [3Fe-4S] center into a [4Fe-4S] center is to reconstitute the active core of apoferredoxin, by the addition of iron and sulfide under controlled conditions. Fig. 5 shows the EPR spectrum of reduced reconstituted Fd II. In the oxidized form a g=2.02 signal characteristic of a [3Fe-4S] center is observed that accounts for only 0.05 spins/molecule, showing that most of the reconstituted centers are of the [4Fe-4S] type as can be deduced from the EPR fingerprint in the reduced form. The principal g-values at 2.07, 1.94 and 1.92 are typical of a reduced $[4Fe-4S]^{1+}$ cluster.

The EPR spectrum of reconstituted FdII (Fd_R) or interconverted FdII contains a minor species detected at g-values 2.04 and 1.86. This species accounts for approximately 20% of the total spin concentration and is not observable in native FdI (compare Figs. 5 B, C). Such components may reveal different micro-environments (or conformational alterations) of the [4Fe-4S] cluster, as observed in *Thermodesulfobacterium commune* ferredoxins (41).

Thomson and coworkers (42) reported that the low temperature MCD of ferricyanide-treated 2x[4Fe-4S] Fd from *C.pasteurianum* is very similar to that of Fd II. Using the [4Fe-4S] cluster of reconstituted *D.gigas* Fd II the ferricyanide oxidation studies were extended to another system. The MB and EPR data showed unambiguously that a [4Fe-4S] to [3Fe-4S] cluster conversion can indeed be achieved by the oxidative procedure (3).

The incubation step of the [3Fe-4S] center with Fe^{2+} enables the introduction of a specific label in the cluster as demonstrated by the conjunction of MB and EPR measurements. Obviously the [57]Fe labelling represents an advantage for spectral assignments and simulations (3,4). Recently Thomson et al. (43) suggested that in the *D.africanus* Fd III the three iron center could be converted into a [4Fe-4S] center. This protein contains in the native state a [3Fe-4S] and a [4Fe-4S] center and only 7 cysteine residues, as detected by amino-acid sequence analysis. The magnetic properties of the newly interconverted [4Fe-4S] cluster are unusual and it was suggested that they may be due to the presence of a non-cysteinyl ligand, probably a carboxylate.

4. MIXED METAL CLUSTERS OF THE TYPE [M, 3Fe-4S] (M=Co, Zn and Cd)

Using the procedures described for the interconversion of [3Fe-4S] into [4Fe-4S] clusters, cubane-like structures containing three iron atoms and one extra metal atom: [M, 3Fe-4S] were described (6, 7).

[Co, 3Fe-4S] - The single [3Fe-4S] cluster (enriched and non-enriched in ^{57}Fe) was used as a starting material. The cobalt ion was successfully introduced into a mixed metal cluster; it assumed a paramagnetic configuration into the fourth site of the cubane structure as demonstrated by the combination of metal analysis, EPR and MB measurements. The newly formed cluster, [Co, 3Fe-4S], can exist in two oxidation states.

In the oxidized state the cluster exhibits an S=1/2 EPR signal with g-values at 1.98 (g_1), 1.94 (g_2) and 1.82 (g_3) (Fig. 8). The g_1-feature shows

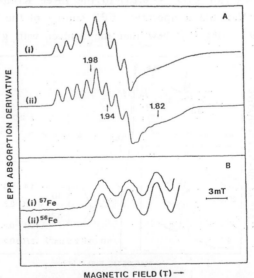

Fig. 8. EPR spectra of the oxidized CoFe cluster. (A) Lower trace: ^{56}Fe, T, 40 K; microwave power, 2 mW; modulation amplitude, 0.5 mT. Upper trace: spectral simulation using the parameters quoted in the text. (B) Expanded view of low field portion of the spectrum using samples containing ^{56}Fe (I=0) and ^{57}Fe (I=1/2). Adapted from Ref. 6.

eight, well-resolved ^{59}Co, hyperfine lines ($^{59}A_1$=4.4 mT) and the ^{59}Co hyperfine structure is broadened by ^{57}Fe (~0.6 mT) indicating that Fe and Co reside in the same complex. Quantitation of the EPR spectra at 40 K yields 1 spin/Co.

Upon one electron reduction (E_m= -220 mV vs.NHE) the cluster has an integer spin (S>0) as indicated by MB studies of the reduced cluster. No EPR spectrum was observed for the reduced form.

The MB results of the oxidized form suggest the presence of a Fe^{3+} site and again a delocalized pair (Fe^{3+} - Fe^{2+}). Upon one electron reduction the third iron is formally Fe^{2+} (6).

[Zn, 3Fe-4S] - The procedure of converting [3Fe-4S] into [Zn, 3Fe-4S] involves the reduction of the native cluster followed by anaerobic incubation with excess zinc (II) (7). The product can be studied directly without further purification or removal of the excess of Zn^{2+}. The EPR and MB spectra of both preparations are identical. The conversion into a mixed metal cluster is complete.

The incubation product is EPR active yielding reproducible spectra as shown in Fig. 9. The observed resonances at g'=9.8, 9.3, 4.8 and ~3.8 belong to a system with S=5/2 (7). The lower panel in Fig. 9 shows the association of the resonances (for E/D ~ 0.25) to the three Kramers' doublets of the S=5/2 system. Variable temperature EPR studies of the low field resonances at g'=9.8 and 9.3 (Fig. 9) show that the doublet with g'= 9.8 is the ground

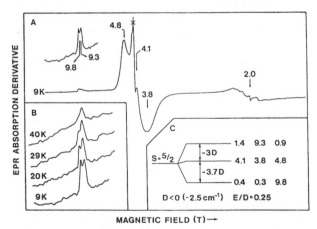

Fig. 9. EPR spectra of the reduced ZnFe cluster. (A) Spectra recorded at 8 K; microwave power, 100μW; modulation amplitude, 1 mT. (B) Expanded region around g=9-10: microwave power, 1 mW; modulation amplitude at different temperature, 1 mT. (C) Energy level diagram for a S=5/2 manifold for D=-2.7 cm^{-1} and E/D=0.25.

state and that the zero-field splitting parameter is D $\sim$-2.7 cm^{-1}. The sharp, derivative-type resonance at g'=4.3 belongs to a second S=5/2 species (this specie accounts for less than 10 % of total spin concentration).

With the Mössbauer information (7) the core oxidation state of the cluster was determined to be 1+. Formally two Fe^{2+} and one Fe^{3+} ion can be described. Since the cluster has S=5/2, the system is in a Kramers' state, and therefore, the Zn must be divalent (as expected). This suggests that the conversion process consists of the incorporation of Zn^{2+} into the core of the reduced [3Fe-4S] cluster followed by a one-electron reduction, the electron being supplied by the excess dithionite.

[Cd, 3Fe-4S] - Another diamagnetic metal can be incorporated in the fourth site. The cluster has also an S=5/2 spin state. Preliminary MB studies performed on this cluster showed that its very similar to the [Zn, 3Fe-4S]$^{1+}$ cluster (8).

The above data describes the three iron sites of Fd II in conditions where the fourth site is empty or occupied by Fe^{2+}, Co^{2+}, Zn^{2+} and Cd^{2+}. EPR spectroscopy proved to be a valuable tool for probing the cluster conversions and for the detection of unusual spin states in the mixed-metal clusters. Putting together this information with MB spectroscopic data, a more precise description of the metal-ion oxidation states was achieved.

The delocalized pair Fe^{2+}-Fe^{3+} present in the reduced [3Fe-4S] center is almost not modified in the series [3Fe-4S]$^{1+}$, [Co,3Fe-4S]$^{1+}$ and [4Fe-4S]$^{2+}$. Meanwhile the ferric site becomes more ferrous in this series. In the [4Fe-4S]$^{2+}$ the ferric site belongs to a second pair Fe^{2+}-Fe^{3+} (distinguishable from the first one in terms of MB spectroscopy). Comparing with [3Fe-4S]$^{1+}$ we can say that the [4Fe-4S]$^{2+}$ core consists of two pairs with spin S$_{12}$=S$_{34}$=9/2 (with dominant double exchange), which are then antiferromagnetically coupled to give the observed S=0 system spin.

In the series [4Fe-4S]$^{1+}$, [Co, 3Fe-4S]^{0} and [Zn, 3Fe-4S]$^{1+}$ the third site is formally Fe^{2+}. In the [4Fe-4S]$^{1+}$ the two Fe of the (Fe^{2+}-Fe^{2+}) pair have less Fe^{2+} characteristics (8). This indicates that in a spin coupling model of [4Fe-4S]$^{1+}$ besides the superexchange terms a double exchange term should be considered to connect the Fe^{2+}-Fe^{3+} pair with the pair of Fe^{2+}-Fe^{2+}.

The knowledge obtained with the new mixed-metal clusters together with the advance made in describing the spin coupling of delocalized mixed-valence dimers with double exchange interaction will help the elucidation of the electronic structure of new clusters relevant in the bioinorganic field.

5. HIGH - SPIN STATES IN IRON - SULFUR CLUSTERS

The reduced tetranuclear iron-sulfur clusters [4Fe-4S] display in most cases an S=1/2 ground spin state, in proteins as well as in synthetic analogues (44, 45). A discovery of considerable significance in the field of synthetic and biological iron-sulfur clusters is the existence of different spin multiplicity for reduced tetranuclear centers. In addition to the conventional S=1/2 ground state, states with S=3/2, 5/2 and 7/2 have been described in protein and enzymes. Holm and co-workers (46) recently demonstrate S=3/2 states in synthetic reduced cubane structures. Three categories of ground spin states behaviour were proposed in terms of the EPR and MB spectroscopic studies:

a) pure S=1/2 or 3/2 spin states;

b) physical mixtures of S=1/2 and 3/2 spin states as indicated by the composite spectra;

c) spin-admixed ground state (S=1/2 + S=3/2).

No admixed systems have been reported in biological materials. It seems that stabilization of a given ground spin state is highly sensitive to the environment, but there is as yet no evidence for the structural features and/or ligand conformations that control directly the cluster spin state.

5.1. S=3/2 and 7/2 spin states in nitrogenase components

Nitrogenase is composed of two redox components: Fe-protein and MoFe-protein. The MoFe-protein contains the MoFe-cofactor plus iron-sulfur centers of yet unknown origin, named P-clusters (see below). The Fe-protein is the active reductant of the MoFe-protein, and the electron transfer process requires the hydrolysis of bound ATP. There is spectroscopic evidence that the Fe-protein contains a single [4Fe-4S] cluster.

Native Fe-protein displays a "g=1.94" type signal (g-values at 2.05, 1.94 and 1.88). Quantitation of this signal yields a spin concentration of 0.33 spin/4Fe (47,48,49,50,51). MB studies revealed that aproximately 50% of the cluster belong to a normal EPR active "g=1.94" type signal and also indicate that all the sample exists in an half-integer spin state. Susceptibility measurements support the presence of S=3/2 species (51). These observations led to a more careful analysis of the EPR spectra in the low field region, where broad resonances were detected around g ~ 5.0. Also solvent effects on the EPR spectra clearly indicate the coexistence of different EPR-active species (Fig. 10).

Addition of ethyleneglycol increases the "g=1.94" features (with decline of the low field resonances) that correspond to aproximately 1 spin/4Fe. In 0.4 M urea the "g=1.94" features are drastically reduced (0.08 spin/4Fe) but well-defined and resolved resonances are detected around g-values of 5.8 and 5.1. The dependence of g'-values on the E/D parameter enables the assignment of this S=3/2 spin system. For a E/D ~ 0.22 the diagram predicts g'-values at 2.65, 5.10 and 1.75 for the excited doublet and 1.36, 1.10 and 5.74 for the ground doublet. The g'-values at 5.8 and 5.1 originate from different doublets. The temperature dependence of the resonance intensities (the feature at g'=5.8 increases in relation to that at g'=5.15 when the temperature is lowered), enables an estimation of the negative D value (~ - 1.5 to -3 cm^{-1}) (53).

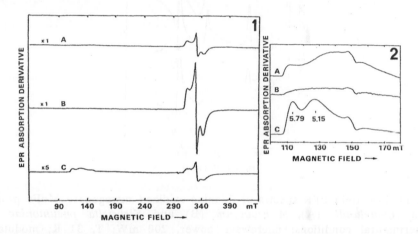

Fig. 10. EPR spectra of reduced Fe-protein from *A.vinelandii* 1. (A) reduced with dithionite; (B) reduced with dithionite and 50% (v/v) ethyleneglycol; (C) reduced with dithionite and 0.4 M urea. Experimental conditions: microwave power, 0.2 mW; T, 8.8 K; modulation amplitude 1 mT. 2. Enlarged low-field region. Adapted from Ref. 52.

The Mg-ATP- and Mg-ADP-bound forms also show a similar coexistence of S=1/2 and 3/2 spin states. A minor S=5/2 species (D ~ 1-3 cm^{-1}, E/D ~ 0.32) was also detected in these forms, and assigned to have its origin in the [4Fe-4S] center, the possibility that it originated from adventitious high-spin iron (III) impurities being discarded (54).

The P-clusters present in the MoFe-protein of nitrogenase are of unknown structure and have unusual magnetic properties (55,56). In thionine-oxidized protein magnetic susceptibility measurements and MCD, NMR and MB studies, indicated previously the presence of an half-integer spin state, probably

S=5/2 (57). This EPR-active state was not revealed for a long time. A recent EPR analysis of samples oxidized by the addition of solid thionine reveal the presence of well-defined signals in the g=10-8 region that were interpreted to belong to an inverted multiplet structure of an S=7/2 spin system (58) (Fig. 11).

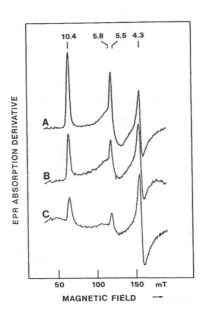

Fig. 11. Low-field EPR spectra of thionine-oxidized nitrogenase MoFe proteins from *A.vinelandii* (A), *A.chroococcum* (B) and *Klebsiella pneumoniae* (C). Experimental conditions: microwave power, 200 mW; T, 31 K; modulation amplitude, 1.2 mT. Adapted from Ref. 58.

Fig. 12 indicates a schematic representation of the energy levels for S=7/2, with D<0 (inverted). For axial symmetry (E=0) only the transition corresponding to the highest Kramers' doublet is expected (g'$_3$=2 and g'$_2$$_1$ =8). For a finite rhombic distortion (E≠0) transitions within the other Kramers' doublets become possible and the lines at g'=10.4 and 5.5 and that at g'=5.8 are interpreted as transitions within the fourth and the third doublet, respectively. These EPR transitions behave as expected since they are temperature dependent and the intensities at g'=10.4 and 5.5 vary in concert and differently from the line at g'=5.8. At low temperature all the three lines should disappear, since they represent excited states. By studies of the temperature dependence values of D=-3.7$\pm$0.7 cm^{-1} and E=0.16$\pm$0.1 cm^{-1} were calculated. At very low temperature broad features around g'=14 are observed and assigned to the lowest Kramers' doublet $(S=\pm 7/2)$.

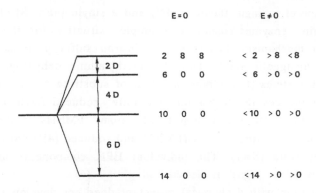

Fig. 12. Schematic representation of the energy levels of an $S=7/2$ spin multiplet for $E=0$ and $E\neq0$ showing the g' values of each Kramers' doublet.

5.2. Selenium derivatives of [4Fe-4S] clusters

Unusual spin-states are observed when selenium replaces labile sulfur in [4Fe-4S] reduced cores prepared from *C.pasterianum* ferredoxin (59,60). EPR signals from $S=3/2$ and $7/2$ spin-states, in addition to those from the usual $S=1/2$ state, are observed. Reduced selenium-substituted Fd displays a complex EPR spectrum with features in a wide range of the magnetic field. Rhombic features in the $g=2$ region (g-values at 2.103, 1.940 and 1.888) were attributed to the $S=1/2$ spin state since they are reminiscent of the EPR signals of reduced native ferredoxin. This signal accounts for ~ 0.8 spin/cluster. Broad lines in the $g=4$ region were assigned to the ground doublet of the $S=3/2$ species (60). The $S=7/2$ state gives rise to an isotropic signal at $g'=5.17$ and two weak peaks at $g'=10.11$ and 12.76. The $g'=5.17$ line was assigned by the analysis of unusual lines in the MB spectrum, to sublevels of a $S=7/2$ spin state. The temperature behaviour of the low-field lines as well as the $g'=5.17$ line suggest that they arise from transitions associated with excited levels.

Selenium replacements in *Bacillus polymyxa* ferredoxin (containing one single [4Fe-4S] center per polypeptide chain) only show a $S=1/2$ EPR signal in the reduced state. Thus, the high-spin multiplicity spin states must arise from specific interactions between the *C.pasteurianum* ferredoxin polypeptide chain and the reduced $[4Fe-4Se]^{1+}$ cluster (60).

5.3. Triplet states in [4Fe-4S] clusters

The presence of a single flavin (FMN) and a single [4Fe-4S] cluster per subunit in certain enzymes constitute a simple situation for the study of electron transfer mechanisms in complex flavin-iron sulfur proteins and may yield unique paramagnetic states. Di-and trimethylamine dehydrogenases are examples of such systems (61). Upon addition of substrate a very rapid two-electron transfer induces the formation of a fully reduced flavin (FMNH$_2$), which is followed by a slow one-electron transfer from FMNH$_2$ to the [4Fe-4S] center. These two interacting centers (FMNH and reduced [4Fe-4S] core) give rise to a triplet state (S=1). The individual EPR components are barely visible and a new EPR signal is detected at g'~4.0 (61).

Partial reduction with a chemical reductant does not develop the triplet state and individual EPR signals originating from FMNH (radical signal) and reduced [4Fe-4S] ("g=1.94" signal) are detected. Further reduction generates FMNH$_2$ and a reduced iron-sulfur core. Thus the binding of the substrate seems to impose an important conformational alteration that favors a strong spin-spin interaction between the redox centers.

An attempt was made to understand, describe and simulate the EPR spectra of the S=1 triplet state originating from the exchange interaction between the two S=1/2 spin systems (62). The fits were greatly improved after introduction of a distribution of E and D values, which may arise from substrate effects ("g-strain").

6. UNKNOWN STRUCTURES

In addition to the new spectroscopic features described in iron-sulfur centers, many experimental observations indicate that unknown centers with yet unrevealed structural aspects may exist.

An example is the analysis of the unique spectroscopic data that have been detected in the iron-only containing hydrogenases (named [Fe] hydrogenases), implicating a novel structure for the H$_2$-binding site.

Characteristic EPR spectra of *D.vulgaris* [Fe] hydrogenase recorded in the g=2 region are collected in Fig. 13 (63). They originate from reaction with substrate (H$_2$) or with chemical reductants. The nearly isotropic signal at g=2.02 is observed in the native preparations of the enzyme. Its low intensity has been taken to indicate that it represents a [3Fe-4S] center arising from conversion (degradation) of [4Fe-4S] centers. Two rhombic signals are observed in intermediate redox states of the enzyme. A rhombic

"2.06" signal (g-values at 2.06, 1.96 and 1.89) developing at a very unusual redox potential for a [4Fe-4S] cluster (around 0 mV), reaches a maximal intensity of 0.7 spin/mole. The signal is observable up to 40 K. Another rhombic signal at g=2.10, 2.04 and 2.00 starts appearing when the "2.06 signal" decreases in intensity, reaches its maximal intensity at around -300 mV (0.4 spin/mole) and disappears below -320 mV. A complex signal is observed in the fully reduced state typical of spin-spin interacting [4Fe-4S] clusters (63,64).

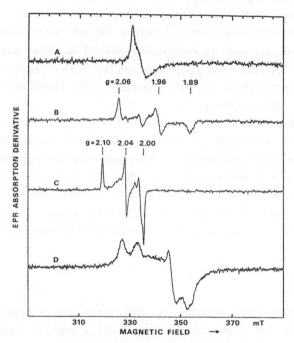

Fig. 13. EPR spectra of *D.vulgaris* hydrogenase at different redox states: (A) the "isotropic" 2.02 signal, (B) the rhombic 2.06 signal, (C) the rhombic 2.10 signal, (D) the complex signal of the reduced hydrogenase. Adapted from Ref. 63.

The ensemble of the EPR data obtained indicates the presence of two [4Fe-4S] cores plus a third one, responsible for the 2.06 and 2.10 rhombic signals and the interaction with the hydrogen molecule. The rhombic 2.10 signal was also observed in *Megasphaera elsdenii* (65) and *C.pasteurianum* (66)

Footnote 2 - Another signal, "axial 2.06" also must be added to the EPR species observable in *D.vulgaris* hydrogenase. The species is observable when the reduced enzyme is reacted with CO or oxidizing dyes (63).
[Fe] hydrogenases representing a common intermediate. MB, EPR and ENDOR

[Fe] hydrogenases representing a common intermediate. MB, EPR and ENDOR studies of *C.pasteurianum* hydrogenase indicate that the center responsible for the rhombic 2.10 signal contains 3-4 iron atoms (67). Recent Raman spectroscopic studies on the oxidized state of this enzyme showed characteristic bands of a [2Fe-2S] cluster (68). An H_2-binding site constituted by a [2Fe-2S] center magnetically coupled to a single iron was then postulated. However, based on chemical analysis Hagen et al. (69) proposed that the atypical center has 6 iron atoms.

The rhombic 2.06 signal associated center has unique properties with emphasis on the mid-point potential that is 300 mV more positive than usual for [4Fe-4S] centers and its observability at 40 K. The only [Fe-S] center that exhibits EPR and redox properties similar to this rhombic 2.06 signal is the [4Fe-4S] center found in APS reductase (70). However, the ligation of this center is still unknown.

Another unique EPR signal is observed in *D.vulgaris* hydrogenase: a sharp signal at g=5.0, observable in reoxidized *D.vulgaris* hydrogenase samples in addition to the rhombic 2.10 signal. The signal was interpreted as a $\Delta m_s = 4$ transition of an S=2 system, only observable with high microwave power (69).

At the present moment, the significance of the above observations are still puzzling and difficult to place in a mechanistic framework, but the ensemble of the "contradictory" data implies the presence of a novel center.

Acknowledgments

During the years working in this field we acknowledge valuable discussions with our close collaborators: B. H. Huynh, J. LeGall, E. Munck, V. Papaefthymiou, K. Surerus and A. V. Xavier and support from the Instituto Nacional de Investigação Científica, Junta Nacional de Investigação Científica e Tecnológica (Portugal) and CEE-BAP n. 0259-P-(TT). We also thank I. Ribeiro and L. Nóbrega for carefully typing this manuscript and I.Pacheco for skillful technical help.

REFERENCES

1 B. H. Huynh, J. J. G. Moura, I. M. Moura, T. A. Kent, J. LeGall, A. V. Xavier and E. Munck, J. Biol. Chem., 255 (1980) 3242-3244.
2 M. K. Emptage, T. A. Kent, B. H. Huynh, J. Rawlings, W. H. Orme-Johnson and E. Munck, J. Biol. Chem., 255 (1980) 1793-1796.
3 J.J.G. Moura, I. Moura, T. A. Kent, J. D. Lipscomb, B. H. Huynh, J. LeGall, A. V. Xavier and E. Munck, J. Biol. Chem., 257 (1982) 6259-6267.

4   T. A. Kent, I. Moura, J. J. G. Moura, J. D. Lipscomb, B. H.   Huynh, J. LeGall, A. V. Xavier   and E. Munck, FEBS   Letters, 138 (1982) 55-58.

5   I. Moura, J.J. G. Moura and A. V.   Xavier,   in: Sund, C. Veeger (Eds), Mobility and   Recognition in cell Biology,   Walter de   Gruyter Co., Berlin, New York 1983, pp.83-101.

6   I. Moura, J. J. G. Moura, E. Munck, V. Papaefthymiou   and   J. LeGall, J. Am. Chem. Soc. 108 (1986) 349-351.

7   K. Surerus, E.   Munck, I.   Moura, J. J. G. Moura   and J. LeGall, J. Am. Chem. Soc., 109 (1987) 3805-3807.

8   E. Munck, V.   Papaefthymiou, K. K. Surerus   and J. J. Girerd, Proceding of the 194th ACS National Meeting, New Orleans, Louisiana, 30 August-4 September, 1987.

9   C. D. Stout, D. Gosh, V. Paltabhi and A. H. Robbins, J. Biol. Chem. 255 (1980) 1797-1800.

10   G. H. Stout, S. Turley, L. Sieker and L. H. Jensen, Proc.   Natl.   Acad. Sci., USA, 85 (1988) 1020-1022.

11   C. R. Kissinger, E. T. Adman, L. C.   Sieker and L. H. Jensen, J. Am. Chem. Soc. (1988) in press.

12   H. Beinert and A. J. Thomson,   Arch. Biochim. Biophys., 222 (1983) 333-361.

13   I. Moura and J.J.G. Moura (1982), The Biological Chemistry of Iron, NATO Advanced Study Institutes Series, D. Reidel Publishing Company, Dordrecht, Holland.

14   W. H. Orme-Johnson and N.R. Orme-Johnson, in: T. G. Spiro, (Ed.) Iron-Sulfur Proteins Vol. 4 in the Metal Ions in Biology Series, John Wiley and Sons, New York, 1982, pp. 67-96.

15   A.V. Xavier, J.J.G. Moura and I. Moura, Struc. Bond., 43 (1981) 187-213.

16   J. LeGall, J. J. G. Moura, H. D.   Peck, Jr. and A. V. Xavier, in: T.G.Spiro (Ed.), Iron-sulfur proteins, volume 4 in the Metal Ions in Biology Series, 1982, pp. 177-248.

17   I. Moura, M. Bruschi, J. LeGall, J.J.G. Moura and A.V. Xavier, Biochem. Res. Commun. 75 (1977) 1037-1044.

18   M. Bruschi, I. Moura, L. Sieker, J. LeGall and A.V. Xavier, Biochem. Biophys. Res. Comun., 90 (1979) 596-605 .

19   I. Moura, A.V. Xavier, R. Cammack, M. Bruschi, and J. LeGall, Biochim. Biophys. Acta, 533 (1978) 156-162.

20   W.E. Blumberg, in: A. Ehrenberg, B.C. Malmstrom and T. Vanngard (Eds.), Magnetic Resonance in Biological Systems   Pergamon, Oxford, 1967, pp.119-133.

21   I. Moura, B. H. Huynh, R. P. Hausinger,J. LeGall, A. V. Xavier and E. Munck, J. Biol. Chem. 255 (1980) 2493-2498.

22   H. Brintzinger, G. Palmer, and R.H. Sands, Proc. Natl. Acad. Sci. USA 55 (1966) 397.

23   J.F. Gibson, D.O. Hall, J.H.M. Thornley and F.R. Whatley, Proc. Natl. Acad. Sci., USSA. 56 (1966) 987.

24   W. R. Dunham, A. J. Beardeu, I.T. Salmeen, G. Palmer, R.H. Sands, W.H. Orme-Johnson and H. Beinert, Biochim. Biophys. Acta, 253 (1971) 134-152.

25   E. Munck, P.G. Debruner, J.C.M. Tsibris and I.C. Gunsalus, Biochemistry,

11 (1972) 855-863.

26 K. Fukuyama, T. Hase, S. Matsumoto, T. Tsukihara, Y. Katsube, N. Tanaka, M. Kakudo, K. Wada and H. Matsubara, Nature, 286 (1980) 522.

27 K. Ogawa, T. Tsukihara, H. Tahara, Y. Katsube, Y. Matsuura, N. Tanaka, M. Kakudo, K. Wada and H. Matsubara, J. Biochem., 81 (1977) 529.

28 T. Tsukihara, K. Futuyama, M. Nakamura, Y. Katsube, N. Tanaka, M. Kakudo, K. Wada, T. Hase and H. Matsubara, J.Biochem. 90 (1981) 1763.

29 J.A. Fee, K.L. Findling, T. Yoshida, R. Hille, G.E. Tarr, D.O. Hearshen, W.R. Durham, E.P. Day, T.A. Kent and E. Munck, J.Biol.Chem., 259 (1984) 124-133.

30 T. A. Kent, J. L. Dreyer, M. C. Kennedy, B. H. Huynh, M.H. Emptage, H. Beinert and E. Munck, Proc. Natl. Acad. Sci., USA, 79 (1982) 1096-1100.

31 A.J.Thomson, A.E.Robinson, M.K.Johnson, J.J.G.Moura, I.Moura, A.V.Xavier and J.LeGall, Biochim. Biophys. Acta, 670 (1981) 93-100.

32 V. Papaefthymiou, J.J. Girerd, I. Moura, J.J.G. Moura and E. Munck, J. Am. Chem. Soc., 109 (1987) 4703-4710.

33 W.R. Hagen, W. R. Dunham, M. K. Johnson and J. A. Fee, Biochim. Biophys. Acta, 828 (1985) 369-374.

34 M. K. Johnson, D. E. Bennett, J. A. Fee and W. V. Sweeney, Biochim. Biophys. Acta, 911 (1987) 81-94.

35 M.C.Kennedy, T.A. Kent, M. Emptage, H. Merkle, H. Beinert and E. Munck, J. Biol. Chem., 259 (1984) 14463-14471 .

36 J.J. Girerd, G.C. Papaefthymiou, A.D. Watson, E. Camp, K.S. Hagen, N. Edelstein, R.B. Frankel and R.H. Holm, J. Am. Chem. Soc., 106 (1984) 5941-5947.

37 T.G. Spiro, Ed., "Iron-Sulfur proteins", John Wiley and Sons, New York, 1982.

38 H. Matsubara, Y. Katsube and K. Wada, Eds. "Iron-sulfur protein research", Japan Scientific Societies Press, Springer-Verlag (1987).

39 R.Cammack, in: H. Matsubara, Y. Katsubara, Y. Katsube and K. Wada (Eds.) Iron-Sulfur Protein Research, 1987, pp.40-55, Japan Scientific Societies Press, Tokyo and Springer-Verlag, N.Y.

40 R. Mathews, S. Charlton, R. H. Sands and G. Palmer, J. Biol. Chem., 249 (1974) 4326.

41 B. Guigliarelli, P. Bertrand, C. More, P. Papavassiliou, E.C. Hatchikian and J. P. Gayda, Biochim. Biophys. Acta, 810 (1985) 319-324.

42 A. J. Thomson, A. E. Robinson, M. K. Johnson, R. Cammack., K. K. Rao and D.O. Hall, Biochem. Biophys. Acta, 637 (1981) 423-432.

43 A. J. Thomson, S. J. George, F. A. Armstrong and C. Hatchikian, Proceedings of the XIII International Conference on Magnetic Resonance in Biological Systems, Madison, Wisconsin, USA, 19 August, 1988.

44 R. Cammack, D. P.E. Dickson and C. E. Johnson in: Lovenberg, W. (Ed.) Iron-Sulfur Proteins, vol. III, 1977, pp. 283--329, Academic Press, New York.

45 R. W. Lane, A. G. Wedd, W. O. Gillum, E. J. Laskowki, R. H. Holm, R. B. Frankel and G. C. Papaefthmiou, J. Am. Chem. Soc., 99 (1977) 2350-2352.

46 M.J. Carney, G.C. Papefthymiou, K. Spartalian, R.B. Frankel and R. Holm, J. Am. Chem. Soc. (1988), in press.

47   W.H. Orme-Johnson, L.C. Davis, M.T. Henzl, B.A. Averill, N.R. Orme-Johnson, E. Munck and R. Zimmermann 1977 in: Newton, W., Postgate, J.R.(Eds.), Recent Developments in Nitrogen Fixation, 1977, pp 133-178, Academic Press, N.Y.

48   G.L. Anderson and J.B. Howard, Biochemistry, 23 (1984) 2118-2122.

49   G. Palmer, J.S. Multani, W.C. Cretney, W.G. Zumft and L.E. Mortenson, Arch. Biochem. Biophys. 153 (1972) 325-332.

50   W.G. Zumft, G. Palmer and L.E. Mortenson, Biochim. Biophys. Acta, 292 (1973) 413-421.

51   H. Haaker, A. Braaksma, J. Cordewener, J. Klugkist, H. Wassink, H. Grande, R.R. Eady and C. Veeger in: C. Veeger, W. E.Newton, (Eds.), Advances in Nitrogen Fixation Research, 1984, pp. 123-131, Nijhoff/Junk, The Hague Netherlands.

52   P. Lindahl, E.P. Day, T.A. Kent, W. H. Orme-Johnson and E. Munck, J. Biol. Chem., 260 (1985) 11160-11173.

53   W.R. Hagen, R.R. Eady, W.R. Dunham and H.Haaker, FEBS Lett., 189 (1985) 250-254.

54   P.A. Lindahl, N. J. Gorelik, E. Munck and W. H. Orme-Johnson, J. Biol. Chem., 262 (1987) 14945-14953.

55   R. Zimmerman, E. Munck, W. J. Brill, V.K. Shah, M.T. Henzl, J. Rawlings and W.H. Orme-Johnson, Biochim. Biophys. Acta, 537 (1978) 185-207.

56   E. Munck, H. Rhodes, W.H. Orme-Johnson, L.C. Davis, W.J. Brill and V.K. Shal, Biochim. Biophys. Acta, 400 (1975) 32-53.

57   B.H. Huynh, M.T. Henzl, J.A. Christner, R. Zimmerman, W.J. Orme-Johnson and E. Munck, Biochim. Biophys. Acta, 623 (1980) 124-128.

58   W.R. Hagen, H. Wassink, R.R. Eady, B.E. Smith and H. Haaker, Eur. J. Biochem., 169 (1987) 457-465.

59   J. Gaillard, J. M. Moulis, P. Auric and J. Meyer, Biochemistry, 25 (1986) 464-468.

60   J. M. Moulis, P. Auric, J. Gaillard and J. Meyer, J. of Biol. Chem., 259 (1984) 11396-11402.

61   R.C. Stevenson, W.R. Dunham, R. H. Sands, T.P. Singer and H. Beinert, Biochim. Biophys. Acta, 869 (1986) 81-88.

62   R.C. Stevenson, J. Magn. Res,. 57 (1984) 24-42.

63   D.S. Patil, J.J.G. Moura, S. H. He, M. Teixeira, B. C. Prickril, D. V. DerVartanian, H. D. Peck, Jr., J. LeGall and B. H. Huynh, J. of Biol. Chemistry, 263 (1988) 18732-18738.

64   H. J. Grande, W. R. Dunham, B. Averill, C. Van Dijk & R. H. Sands, Eur. J. Biochem. 136 (1983) 201-207.

65   C. Van Dijk, H. J. Grande, S. G. Mayhew and C. Veeger, Eur. J. Biochem., 107 (1980) 251-261.

66   D. L. Erbes, R. H. Burris and W. H. Orme-Johnson, Proc. Natl. Acad. Sci. USA, 72 (1975) 4795-4799.

67   C. Wang, M. J. Benecky, B. H. Huynh, J. F. Cline, M. W. W. Adams, L. E. Mortenson, B. M., Hoffman and E. Munck, J. Biol. Chem. 259 (1984) 14328-14331.

68   K. A. Macor, R. S. Czernuszewicz, M.W.W. Adams and T. G. Spiro, J. Biol. Chem. 262 (1987) 9945-9947.

838

69   W. R. Hagen, A. Van   Berkel-Arts, K. M. Kruse-Wolters, W. R. Dunham and
     C. Veeger, FEBS Lett. 201 (1986) 158-162.
70   J. Lampreia, I. Moura, G.   Fauque, A. V. Xavier, J. LeGall, H. D. Peck,
     Jr. and J. J. G. Moura, Rec. Trav. Chim. Pays-Bas, 106 (1987) 234.

CHAPTER 24

EPR SPECTROSCOPY OF MANGANESE ENZYMES

GARY W. BRUDVIG

1. INTRODUCTION

Manganese is one of the most abundant transition metals in the earth's crust and is an essential metal ion for a variety of enzymes [reviewed in (1-3)]. Only the Mn(II) ion is stable in aqueous solution at neutral pH. Consequently, the vast majority of manganese enzymes utilize only the Mn(II) oxidation state. Usually, Mn(II) serves as a Lewis acid and often a variety of divalent metal ions will function nearly equally well. Indeed, in many cases it is unclear whether Mn(II) is actually the physiological metal ion. Most of the Mn(II) enzymes have a single Mn(II) ion per active site. There are also a few examples of proteins in which two Mn(II) ions are in close proximity such as enolase (4), dimanganese S-adenosylmethionine synthase (5) and dimanganese concanavalin A (6).

Mn(II), having a high-spin d^5 configuration, is not readily studied by optical spectroscopy. However, mononuclear Mn(II) is always EPR visible and, therefore, EPR spectroscopy of Mn(II) has been extensively utilized to probe metal ion binding sites in proteins and nucleic acids. [See (3) for an excellent review on EPR of Mn(II).]

Only a small number of enzymes have been identified that utilize oxidation states of manganese higher than +2. These redox-active manganese enzymes include the manganese superoxide dismutase that contains a mononuclear active site (7) and the manganese catalase that contains a binuclear active site (8); x-ray crystal structures are available for both of these enzymes (9,10). Manganese is also utilized in the water-oxidation reaction catalyzed by photosystem II (11-12). Photosystem II contains four manganese ions per active site.

In this chapter, I give an overview of the EPR properties of mono- and multinuclear manganese complexes. Most of the past studies have focused on Mn(II). The EPR properties of mononuclear Mn(II) are now quite well understood. Until recently, however, the higher oxidation states of manganese had not been extensively studied. But due to the recognition that high-valent manganese plays an important role in several enzymes, there has been a flurry of activity in the synthesis and characterization of high-valent manganese complexes, especially multinuclear complexes. The EPR and magnetic properties of many of these inorganic complexes have been determined and these data provide a base of information from which to understand the EPR properties of the redox-active manganese enzymes.

2. MONONUCLEAR MANGANESE

2.1 Spin Hamiltonian

The dominant contributions to the spin Hamiltonian of a mononuclear manganese ion are the electron spin-spin interactions (zero-field splitting, $\hat{H}_{zfs}$), the nuclear hyperfine interaction, $\hat{H}_{hf}$, and the electron Zeeman interaction, $\hat{H}_z$:

$$\hat{H} = \hat{H}_{zfs} + \hat{H}_{hf} + \hat{H}_z = h[D(\hat{S}_z^2 - 1/3\ \hat{S}^2) + E(\hat{S}_x^2 - \hat{S}_y^2)] + h\hat{I}\cdot\underline{A}\cdot\hat{S} + \beta\hat{S}\cdot\underline{g}\cdot\underline{B} \quad [1]$$

where D and E are the axial and rhombic zero-field splitting parameters, $\underline{A}$ is the ^{55}Mn nuclear hyperfine coupling tensor, $\underline{g}$ is the g tensor, $\underline{B}$ is the magnetic field vector, h is Planck's constant, β is the electron Bohr magneton, and S and I refer to the electron and nuclear spin angular momenta, respectively. Contributions to the spin Hamiltonian from the nuclear quadrupole and quartic zero-field splitting terms are usually unnecessary to provide an adequate description of the properties of manganese ions. The nuclear Zeeman interaction is also usually negligible and does not contribute directly to the EPR spectrum, but it is responsible for the appearance of "forbidden" transitions.

In some cases, superhyperfine coupling to the nuclear spins of ligand atoms must be included. However, due to the ionic nature of manganese ions coordinated to proteins, the electron spin is highly localized on the manganese ion. Consequently, the magnitude of the superhyperfine coupling is usually very small and very seldom is resolved in X-band measurements. Indeed, it is usually difficult to even detect line broadening from directly-coordinated paramagnetic nuclei at X-band. Frequently a clear-cut demonstration of line broadening requires Q-band measurements to minimize second-order effects that broaden the lines (see Fig. 1).

^{55}Mn is the only stable isotope of manganese and has a nuclear spin I = 5/2. This leads to the well-known six-line spectrum of Mn(II) in solution (Fig. 1). Manganese is invariably high spin in all of its oxidation states that occur in biological systems giving S = 5/2 for Mn(II), S = 2 for Mn(III), and S = 3/2 for Mn(IV). The fact that manganese ions always have more than one unpaired electron means that the EPR properties of manganese depend greatly on the coordination environment.

2.2 Mn(II)

I will only give a brief overview of Mn(II) EPR; a more comprehensive treatment can be found in (3). The orbital angular momentum of Mn(II) is largely quenched because of its high-spin d^5 configuration. Consequently, Mn(II) has a nearly isotropic g value close to 2.0. In aqueous solution, Mn(II) exists in the hexaaquo form with an octahedral geometry. Because of the high symmetry and rapid tumbling, the zero-field splitting term is zero for hexaaquo Mn(II). Therefore, the 36 states that arise from the possible combinations of $|m_s, m_I\rangle$ are essentially degenerate in zero applied magnetic field (a small internal field is present due to the nuclear spin). There are 30 allowed transitions with $\Delta m_S = \pm 1$ and $\Delta m_I = 0$. Using [1] and assuming an isotropic g value and ^{55}Mn hyperfine coupling constant, the resonant field, B_{res}, in the absence of a zero-field splitting is given in [2] to second order:

$$B_{res} = (2g\beta)^{-1}\{K + [K^2 - 2(hA_0)^2[35/4 - m_I^2 + m_I(2m_S + 1)]^{1/2}\}\tag{2}$$

where $K = h\nu - hA_0m_I$, A_0 is the isotropic nuclear hyperfine coupling constant [negative for Mn(II)] and ν is the microwave frequency. The spectrum consists of a sextet with each of the lines composed of five transitions (Fig. 1). Due to the second-order terms, the six lines are not equally spaced and do not have equal linewidths. The second-order effects are minimized, and the lines are, therefore, narrower, if the EPR signals are measured at a higher frequency (Fig. 1B). Q-band measurements are frequently employed to improve the resolution in Mn(II) EPR spectra. Q-band measurements have proven especially important in studies of line broadening by ^{17}O-labeled ligands to Mn(II) [see (3) and references therein].

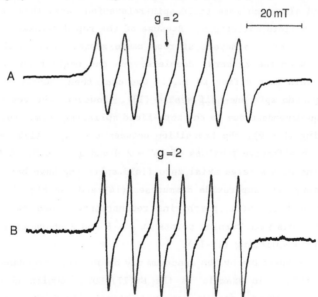

Fig. 1. X-band, 9 GHz, (A) and Q-band, 35 GHz, (B) EPR spectra of Mn(H$_2$O)$_6^{2+}$. The spectra were obtained from 0.1 mM aqueous solutions of MnCl$_2$ at 20°C [from (3)].

When Mn(II) is bound to a protein the zero-field splittings are not zero due to slow tumbling and the reduced symmetry of the metal ion binding site. In this case, the zero-field splitting tensor defines a molecular principal-axes system and the resonant field will depend on the orientation of the molecule in the laboratory frame. The result is that when D is nonzero the fine-structure transitions: $m_S = \pm 5/2 \leftrightarrow m_S = \pm 3/2$, $m_S = \pm 3/2 \leftrightarrow m_S = \pm 1/2$, and $m_S = +1/2 \leftrightarrow m_S = -1/2$, will split apart. Two limits can be considered depending on the magnitude of D: $D \ll h\nu$ and $D \gg h\nu$.

When $D \ll h\nu$ all of the fine-structure transitions are accessible with conventional EPR instrumentation. But the $m_S = \pm 5/2 \leftrightarrow m_S = \pm 3/2$ and $m_S = \pm 3/2 \leftrightarrow m_S = \pm 1/2$ transitions will be very broad and usually are not observed for Mn(II) bound to proteins. However, the $m_S = +1/2 \leftrightarrow m_S = -1/2$ transition will remain fairly sharp, centered near $g = 2.0$, and readily observable, even in liquid samples at physiological temperatures. This is frequently the situation for Mn(II) in proteins. Although the EPR spectrum in this case qualitatively resembles that of hexaaquo Mn(II), only part of the signal is present in the well-resolved lines near $g = 2.0$. The integrated intensity of this sharp signal will be smaller than that for an equal concentration of hexaaquo Mn(II) by roughly a factor of four.

When $D \gg h\nu$ the $m_S = \pm 5/2$, $\pm 3/2$, and $\pm 1/2$ levels are split into three Kramers' doublets separated by 4D and 2D, respectively. This leads to a large anisotropy and also much more rapid spin-relaxation rates than found for Mn(II) with small zero-field splittings. Because of the rapid relaxation rate, the EPR signals are too broad to detect at room temperature. Due to the large energy separation between the Kramers' doublets, one can neglect the matrix elements connecting levels that differ in $|m_S|$. One can, then, consider the $m_S = \pm 1/2$ levels as a pseudo spin one-half system (13). However, the resonant field will be orientation dependent due to the zero-field splitting term. For an axial zero-field splitting (E = 0), the transition between the $m_S = \pm 1/2$ levels will give a spectrum with effective g values of $g_{\parallel}' = 2.0$ and $g_{\perp}' = 6.0$. A few examples of Mn(II)-proteins with a large axial zero-field splitting have been observed, most notably the manganese superoxide dismutase [(14) and see Fig. 2]. Although not naturally occurring, Mn(II)-porphyrins reconstituted into heme proteins also exhibit EPR signals with turning points at $g = 6$ and 2 characteristic of a large axial zero-field splitting (15-16).

In the case where $D \approx h\nu$ the spectrum may be severely broadened with no well-resolved features. An example is the Mn(II)-EDTA complex which, at X-band, exhibits a very broad, poorly resolved signal, even at low temperatures. Interestingly, at Q-band the $m_S = +1/2 \leftrightarrow m_S = -1/2$ transition can be observed at room temperature illustrating the transition from a case where $D \approx h\nu$ to a case where $D < h\nu$ on going from X-band to Q-band (3).

2.3 Mn(III)

The zero-field splitting of Mn(III) often precludes its detection when conventional EPR instrumentation is used (an example is shown in Fig. 2A) or at best leads to very broad EPR signals. This is due to the fact that the zero-

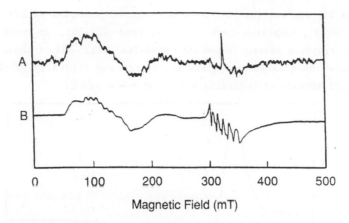

Fig. 2. EPR spectra of manganese superoxide dismutase recorded at 17 K [from (14)]: (A) freshly-isolated enzyme at 6.25 times higher gain than (B), and (B) after addition of dithionite. The freshly-isolated enzyme is mostly in the Mn(III) state, which does not exhibit an EPR signal; a weak signal at g = 6 (≈100 mT) is present in (A) due to a small amount of the Mn(II) form. Upon adding dithionite, the enzyme is reduced to the Mn(II) form, which shows a turning point at g = 6 characteristic of Mn(II) with a large axial zero-field splitting; the six-line signal near g = 2 is due to Mn(II) released from the protein after adding dithionite and this six-line signal obscures the turning point at g = 2 from Mn(II) bound to the protein.

field splitting interaction splits the five levels into two Kramers' doublets with $m_S = \pm 1$ and $m_S = \pm 2$, respectively, and a nondegenerate level with $m_S = 0$. With a sufficiently large zero-field splitting term, none of the allowed $\Delta m_S = \pm 1$ transitions will occur at magnetic fields available with conventional instrumentation. Further, unless the zero-field splitting is small with respect to the Zeeman interaction, the EPR signals will be extremely broad. It is rarely the case that the zero-field splitting is sufficiently small, as this requires a near cubic or higher symmetry, an unlikely situation for the d^4 Mn(III) ion bound to a protein that will undergo a Jahn-Teller distortion away from cubic symmetry. No EPR signals from a mononuclear Mn(III) in a protein have yet been reported.

2.4 Mn(IV)

The zero-field splittings for Mn(IV) will split the four spin levels into two Kramers' doublets with $m_S = \pm 1/2$ and $m_S = \pm 3/2$. Even when the zero-field split-tings are large, as is typically the case for Mn(IV), the $m_S = +1/2 \leftrightarrow m_S = -1/2$ transition will be observed. As for Mn(II) with a large zero-field splitting (discussed above), a Mn(IV) ion with large zero-field splittings can be treated as a pseudo spin one-half species (13). In this case, the effective g values for a system with axial symmetry will be $g_\parallel' = 2.0$ and $g_\perp' = 4.0$ (Fig. 3).

Thus, far no clear-cut case of a mononuclear Mn(IV) species in a naturally-occurring enzyme has been detected by EPR (although a possible case is in

844

photosystem II, see section 5). However, numerous examples of EPR signals from inorganic Mn(IV) complexes have been reported (17-20), in some cases from manganese-porphyrins reconstituted into proteins (20). All of these mononuclear Mn(IV) species have large zero-field splittings and typically exhibit axially-symmetric EPR signals with turning points at g = 4 and 2.

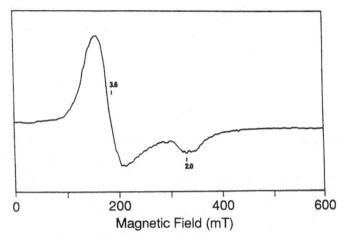

Fig. 3. X-band EPR spectrum of Mn(IV)-tetraphenylporphyrin in CH_2Cl_2 recorded at 195 K. The signal has turning points at g = 4 and 2 characteristic of an S = 3/2 species with a large axial zero-field splitting. Reprinted with permission from (17). Copyright 1988 American Chemical Society.

3. EPR OF BINUCLEAR MANGANESE

  3.1 Dipolar and exchange Hamiltonians

    When two manganese ions are in close proximity, dipolar and/or exchange couplings between the two ions must also be considered. Depending on the magnitude of the couplings, the EPR properties may be dominated by these new terms. The dipolar, $\hat{H}_{dip}$, [3] and Heisenberg exchange, $\hat{H}_{ex}$, [4] Hamiltonians describe the interaction between two manganese ions, A and B, with electron spins of S_A and S_B, respectively:

$$\hat{H}_{dip} = \hat{S}_A \cdot \underline{T} \cdot \hat{S}_B \qquad\qquad [3]$$
$$\hat{H}_{ex} = J \hat{S}_A \cdot \hat{S}_B = 1/2\ J\ (\hat{S}^2 - \hat{S}_A^2 - \hat{S}_B^2) \qquad\qquad [4]$$

where $\underline{T}$ is the dipolar interaction tensor and J is the isotropic exchange coupling constant. (In [4], J is positive for antiferromagnetic coupling and negative for ferromagnetic coupling.) The elements of $\underline{T}$ depend on both the distance and orientation of the two manganese ions. The magnitude of the dipolar interaction is $\mu_0 g^2 \beta^2 / 4\pi R^3$ where R is the interion distance. The magnitude of J depends on the degree of orbital overlap between the two ions, which depends greatly on the ligands between the two manganese ions. In the absence of a bridging ligand, J is expected to decrease exponentially with the distance of separation of the two ions due to the exponential dependence of the wavefunctions on distance (21). However, when the two manganese ions share a common bridging ligand the exchange interaction can be very large and often is the dominant term in the spin Hamiltonian (22).

The EPR spectrum of a binuclear manganese complex depends dramatically on the magnitude of the exchange coupling with respect to A, the nuclear hyperfine coupling constant of ^{55}Mn. The EPR spectrum can be extremely complex when J is on the same order as A (23), due to the large number of spin states that are thermally accessible [1296 eigenstates for two Mn(II)] that will give rise to an enormous number of overlapping transitions. However, when the exchange coupling is large the EPR spectrum can be fairly straightforward to interpret.

When the exchange coupling is large, one must begin by considering the coupled spin states that arise from the exchange interaction. As a result of the exchange coupling, the effective g value and hyperfine-coupling constants are changed in the coupled system from the values of the isolated ions (24).

Consider the case of two manganese ions with a large isotropic exchange interaction. The exchange interaction will generate a ladder of spin levels with energies given in [5] and illustrated in Fig. 4:

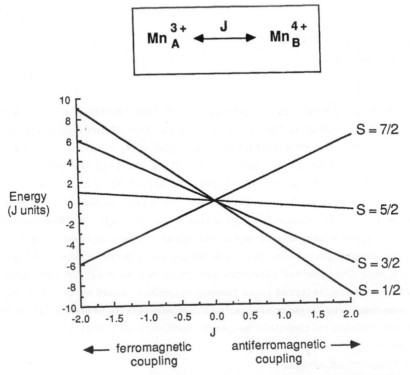

Fig. 4. Energy levels given by [5] for an exchange coupled Mn(III)-Mn(IV) dimer.

$$\text{Energy} = E = 1/2 \ J \ [\ S(S + 1) - S_A(S_A + 1) - S_B(S_B + 1) \ ]. \tag{5}$$

Fig. 4 depicts the energies of the spin states that arise from either antiferromagnetic or ferromagnetic exchange coupling of a Mn(III)-Mn(IV) dimer. For a variety of μ-oxo-bridged dimanganese model compounds, discussed below, the exchange coupling is typically antiferromagnetic. When the dimer is in a mixed-valence state, this leads to a ground state that has S = 1/2. If an antiferromagnetically exchange-coupled dimer is not in a mixed-valence state, then the ground state has S = 0. Consequently, for strongly antiferromagnetically exchange-coupled manganese dimers, EPR signals are normally only observed in the mixed-valence states.

The ^{55}Mn nuclear hyperfine interaction in the coupled system is determined by the projection of the nuclear hyperfine coupling tensors of each individual ion on the total spin (24). If we let g_A, g_B, $\underline{A}_A$, and $\underline{A}_B$ be the single ion g and $\underline{A}$ tensors, respectively, the tensors for the coupled dimer, g^c and $\underline{A}_i{}^c$, can be defined in terms of the constant C,

$$C = [S_A(S_A + 1) - S_B(S_B + 1)]/S(S + 1) \tag{6}$$

so that,

$$g^c = [(1 + C)/2]g_A + [(1 - C)/2]g_B \tag{7}$$
$$\underline{A}_A{}^c = [(1 + C)/2]\underline{A}_A \tag{8}$$
$$\underline{A}_B{}^c = [(1 - C)/2]\underline{A}_B. \tag{9}$$

As a result of the exchange interaction, the ^{55}Mn hyperfine couplings will be changed from the values of the free ions. In the case of two equivalent manganese ions both having the same oxidation state (giving $S_A = S_B$), C = 0 and the ^{55}Mn hyperfine coupling in the coupled system will be one-half the value of the free ion. For a mixed-valence manganese dimer, the hyperfine couplings will depend on the oxidation states of the manganese ions. The hyperfine couplings are tabulated in Table 1 for the ground S = 1/2 state of a strongly antiferromagnetically exchange-coupled mixed-valence manganese dimer. The observation of a distinct hyperfine coupling constant for each manganese ion in a mixed-valence dimer requires that the unpaired electrons are localized on a single manganese ion on the EPR time scale, referred to as trapped valences. Based on data from manganese model complexes with ligands similar to those possible in proteins, it is expected that mixed-valence multinuclear manganese complexes in proteins will have trapped valences.

3.2 Examples of spectra

There are a number of proteins with two Mn(II) ions in close proximity.

TABLE 1

^{55}Mn hyperfine coupling constants for the S = 1/2 state of an exchange-coupled manganese dimer as given in [8] and [9].

	S_A	S_B	C	$\underline{A}_A{}^c$	$\underline{A}_B{}^c$
Mn(II)-Mn(III)	5/2	2	11/3	7/3 $\underline{A}_A$	− 4/3 $\underline{A}_B$
Mn(III)-Mn(IV)	2	3/2	3	2 $\underline{A}_A$	− $\underline{A}_B$
Mn(IV)-Mn(V)	3/2	1	7/3	5/3 $\underline{A}_A$	− 2/3 $\underline{A}_B$

Examples include enolase (4), dimanganese S-adenosylmethionine synthase (5) and dimanganese concanavalin A (6). In all of these cases, the two Mn(II) ions are only weakly coupled. This results in a complex EPR spectrum and, indeed, the observation of a highly structured EPR spectrum from a Mn(II) protein is often taken as an indication of two Mn(II) ions in close association.

Magnetic susceptibility and multi-frequency EPR measurements have been made on the dimanganese concanavalin A (6). These measurements have shown that the two Mn(II) ions are antiferromagnetically exchange coupled with J = 1.8 cm^{-1}. The X- and Q-band EPR spectra of the dimanganese concanavalin A are shown in Fig. 5. For two equivalent Mn(II) ions, one expects an eleven hyperfine line pattern from the ^{55}Mn nuclear hyperfine interaction. An apparent eleven-line pattern is seen in the wings of the EPR spectrum from dimanganese concanavalin A (Fig. 5A), although not all of the hyperfine lines are discerned. Note also that the ^{55}Mn hyperfine splitting is 4.7 mT, which is one-half of the value for a mononuclear Mn(II), as predicted for an exchange-coupled Mn(II)-Mn(II) dimer. However, the EPR spectrum from dimanganese concanavalin A is more complex than a simple eleven-line signal. The complexity may arise because more than one spin state is thermally populated due to the small exchange coupling. The EPR spectra in Fig. 5, therefore, probably arise from superpositions of transitions from more than one spin manifold. The spectra may be further complicated by zero-field splitting interactions or dipolar coupling between the two Mn(II) ions. A detailed analysis of the EPR spectra from dimanganese concanavalin A, and spectra for other di-Mn(II) proteins, will require further work, perhaps utilizing spectral simulations.

Very few proteins have been found that contain a binuclear manganese center in which the manganese ions have valencies higher than +2. And only one enzyme in this group has been extensively studied, that being the manganese catalase. A manganese-containing ribonucleotide reductase (26) has also been reported to contain a high-valent binuclear manganese center. It has been argued that the two manganese ions in the manganese catalase have μ-oxo-di-μ-carboxylato bridging ligands based on similarities of the electronic spectra of the enzyme and model compounds (27). A similar structure is probably also present in the manganese-

848

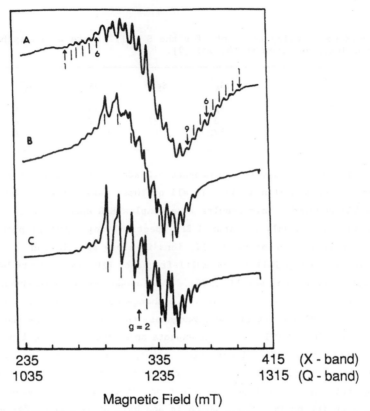

235   335   415   (X - band)
1035   1235   1315   (Q - band)

Magnetic Field (mT)

Fig. 5. X-band (A) and Q-band (B,C) EPR spectra of dimanganese concanavalin A at 120 K. The arrows in (A) indicate some of the lines of the eleven-line pattern expected for a Mn(II) dimer. On the low-field side, six of the lines are discerned. On the high-field side, nine of the lines are indicated. The vertical bars in (B) and (C) indicate interfering signals from mononuclear Mn(II). (C) differs from (B) in that the microwave power was 0.3 mW instead of 40 mW. Reprinted with permission from (6). Copyright 1987 American Chemical Society.

containing ribonucleotide reductase based on analogy with the iron-containing form of this enzyme.

The manganese catalase appears to contain a binuclear manganese active site (8,10) in which the manganese ions are antiferromagnetically exchange coupled (28). Although the manganese catalase is not very well characterized at this time, there are a large number of synthetic binuclear manganese complexes [reviewed in (29,30)], whose magnetic properties are well defined, that can be used as models. I will focus on antiferromagnetically exchange-coupled complexes because these appear to be most relevant to the high-valent binuclear manganese centers in proteins.

For an antiferromagnetically exchange-coupled mixed-valence dimer, the ground state has S = 1/2. An EPR signal from the S = 1/2 state is easily observable in a glass at low temperature; an EPR signal is even observable in a solution sample at room temperature if the exchange coupling is sufficiently large. The ^{55}Mn nuclear hyperfine splittings are well resolved unless the exchange coupling is small [J < 5 cm^{-1}, (31)], giving what is often referred to as a sixteen-line spectrum. There are, actually, 36 transitions from the S = 1/2 state of a mixed-valence manganese dimer, but overlapping transitions give a sixteen-line pattern in many cases (Figs. 6,7). The sixteen-line EPR signal can be understood as a sextet of sextets arising from two inequivalent manganese hyperfine splittings. The inequivalent hyperfine splittings arise from the exchange interaction (Table 1), which causes the splitting from the lower valent ion to be roughly twice that of the higher valent ion. The result is that many of the transitions are superimposed and leads to the overall sixteen-line pattern.

Assuming an isotropic g value and hyperfine-coupling constants, the resonance condition for the transitions of the S = 1/2 state of a strongly antiferromagnetically exchange-coupled Mn(III)-Mn(IV) dimer is given in [10] to second order (25):

$$h\nu = g\beta B + m_{IA}A_A^\circ + m_{IB}A_B^\circ + \{(A_A^\circ)^2[35/4 - m_{IA}^2] + (A_B^\circ)^2[35/4 - m_{IB}^2]\}/2g\beta B \qquad [10]$$

where A_A° and A_B° are given in Table 1.

Fig. 6 shows the EPR spectrum of di-μ-oxo-tetra-bipyridyl Mn(III)-Mn(IV).

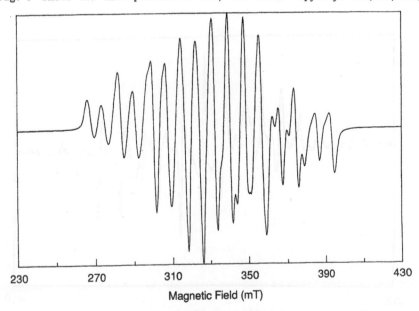

Fig. 6. X-band EPR spectrum of a frozen acetone solution of di-μ-oxo-tetra-bipyridyl Mn(III)-Mn(IV) at 10 K [prepared as in (32)].

This complex has a very large antiferromagnetic exchange coupling [$J = 300$ cm^{-1}, (32)] and only the ground $S = 1/2$ state is populated at 10 K. Based on spectral simulations, it was determined that the Mn(III) and Mn(IV) ions have hyperfine couplings of 16.7 mT and 7.9 mT, respectively, and that the dimer has an axial g tensor with $g_\parallel \approx g_\perp$ and isotropic hyperfine coupling constants (32). Note that the unequal ^{55}Mn hyperfine couplings for the individual ions ($A_A{}^c \neq 2A_B{}^c$) cause some of the superimposed lines to be resolved, as can be seen on the high-field side of the spectrum in Fig. 6.

The projection of each ion's ^{55}Mn hyperfine coupling depends on the oxidation states of the manganese ions (Table 1). As can be seen in Table 1, the ^{55}Mn hyperfine interaction in the coupled system is expected to be largest for a Mn(II)-Mn(III) dimer and to decrease with increasing oxidation state. One might expect to be able to assign the oxidation states of manganese on this basis. However, the hyperfine-coupling constants of the individual ions vary over a significant range depending on the oxidation state, coordination number, geometry, and ligand type (33). Hence, the magnitude of the hyperfine splitting in the spectrum from a mixed-valence dimer is not a reliable indicator of oxidation states in general. However, the hyperfine splittings may be used to obtain the oxidation states if one compares complexes with very similar structure (34).

The EPR spectrum of a mixed-valence state of the manganese catalase from Lactobacillus plantarum (Fig. 7) has recently been reported (28). A sixteen-line

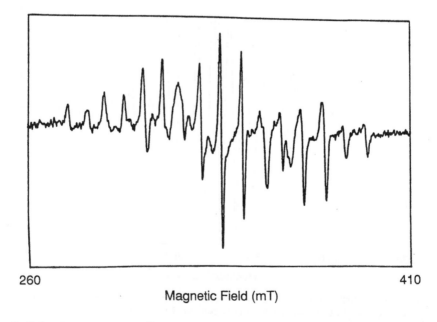

260   410

Magnetic Field (mT)

Fig. 7. X-band EPR spectrum from L. plantarum manganese catalase recorded at 6 K. Reprinted with permission from (28). Copyright 1988 American Chemical Society.

signal is observed which is characteristic of an antiferromagnetically exchange-coupled manganese dimer. From a spectral simulation, the following isotropic constants were obtained: $g = 2.0075$, $A_A{}^c = 0.0144$ cm^{-1} and $A_B{}^c = 0.0076$ cm^{-1}. The spectrum from the manganese catalase appears to arise from a Mn(III)-Mn(IV) dimer based on a comparison of the hyperfine splittings with values reported for model complexes. However, in view of the uncertainty in the coordination number and ligands to manganese in the manganese catalase, the EPR signal shown in Fig. 7 may, alternatively, arise from a Mn(II)-Mn(III) dimer.

The mixed-valence manganese dimers studied to date have, generally, been found to have nearly isotropic g values and hyperfine-coupling constants. However, several very interesting Mn(II)-Mn(III) complexes have been recently reported that exhibit moderate g anisotropy and well-resolved hyperfine splitting for only one of the three orientations (35). The development of g anisotropy and lack of well-resolved hyperfine structure on all of the turning points probably arises because of an asymmetrical structure. It will be important to determine whether the EPR spectra observed for multinuclear manganese centers in proteins also exhibit similar properties.

4. EPR OF TETRANUCLEAR MANGANESE

Photosystem II contains four manganese ions that are probably arranged as a μ-oxo-bridged tetranuclear cluster, although there is some support for a binuclear or a trinuclear manganese cluster (see section 5). Although the manganese site in photosystem II remains poorly characterized, several tetranuclear μ-oxo-bridged manganese complexes have been prepared whose magnetic properties are understood in more detail (36-38).

The number of spin states for a manganese tetramer is enormous (1,679,616 eigenstates for a Mn(II) tetramer; and 331,776 eigenstates for a Mn(IV) tetramer). Unless the exchange couplings are large enough to reduce the number of thermally populated levels, the EPR spectrum of a manganese tetramer will be hopelessly complicated. Fortunately, the exchange couplings of μ-oxo-bridged high-valent manganese complexes can be quite large, and in some cases the EPR and magnetic properties may be reasonably straightforward to explain.

Six isotropic exchange-coupling constants are required to model the magnetic properties of a manganese tetramer. Generally, magnetic data are not sufficient to define six independent coupling constants. In order to use a spin-coupling model to account for the magnetic properties, one must usually rely on symmetry arguments to reduce the number of parameters. For some of the model complexes, the symmetry is imposed by the structure. In this case, vector-coupling rules can be used to generate an analytical expression for the energies of the spin levels (39). In other cases, for example when the structure is unknown, one may be able to account for the magnetic properties with a simplified model and, thereby, obtain evidence for an effective symmetry of the complex.

Model complexes with a Mn_4O_2 "butterfly" core (Fig. 8) have been synthesized with oxidation states of $Mn(II)_2$-$Mn(III)_2$, $Mn(II)$-$Mn(III)_3$, and $Mn(III)_4$. The magnetic properties have been analyzed in detail using variable-temperature magnetic susceptibility measurements for the $Mn(II)_2$-$Mn(III)_2$ and $Mn(III)_4$ complexes (36).

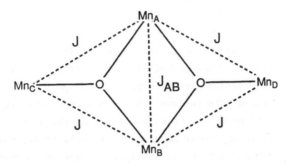

Fig. 8. Structure and exchange-coupling constants of the Mn_4O_2 "butterfly" core.

In order to model the magnetic properties of the Mn_4O_2 complexes, the following exchange-coupling scheme has been used. The exchange coupling between a wing-tip manganese ion and either of the central manganese ions is assumed to be equivalent, $J_{AC} = J_{BC} = J_{AD} = J_{BD} = J$. In this case, the number of exchange-coupling constants is reduced to three. The simplified exchange Hamiltonian is given in [11]:

$$\hat{H}_{ex} = J_{AB} \, \hat{S}_A \cdot \hat{S}_B + J_{CD} \, \hat{S}_C \cdot \hat{S}_D + J(\hat{S}_A \cdot \hat{S}_C + \hat{S}_A \cdot \hat{S}_D + \hat{S}_B \cdot \hat{S}_C + \hat{S}_B \cdot \hat{S}_D). \qquad [11]$$

The energies can be solved analytically by using the Kambe vector-coupling method (39). Making the substitution of $\hat{S}' = \hat{S}_A + \hat{S}_B$ and $\hat{S}^* = \hat{S}_C + \hat{S}_D$ gives the exchange Hamiltonian in [12]:

$$\hat{H}_{ex} = 1/2 \, J_{AB}(\hat{S}'^2 - \hat{S}_A^2 - \hat{S}_B^2) + 1/2 \, J_{CD}(\hat{S}^{*2} - \hat{S}_C^2 - \hat{S}_D^2)$$
$$+ 1/2 \, J(\hat{S}^2 - \hat{S}'^2 - \hat{S}^{*2}). \qquad [12]$$

The energies for this system are given in [13]:

$$E(S,S',S^*) = 1/2 \, J_{AB} \, [S'(S' + 1) - S_A(S_A + 1) - S_B(S_B + 1)]$$
$$+ 1/2 \, J_{CD} \, [S^*(S^* + 1) - S_C(S_C + 1) - S_D(S_D + 1)]$$
$$+ 1/2 \, J \quad [S(S + 1) - S'(S' + 1) - S^*(S^* + 1)]. \qquad [13]$$

For the Mn_4O_2 "butterfly" structure, it can be further assumed that the exchange coupling between the two wing-tip manganese ions is negligible, $J_{CD} = 0$.

From a fit of [13] to the temperature dependence of the molar magnetic susceptibility, values for g, J and J_{AB} were determined for both the $Mn(II)_2$-$Mn(III)_2$ and $Mn(III)_4$ "butterfly" complexes (36). For the $Mn(II)_2$-$Mn(III)_2$ complex, g = 1.70, J = 3.94 cm^{-1} and J_{AB} = 6.24 cm^{-1}. The ground state has S = 2 and there are six other spin states within 15 cm^{-1} of the ground state. For the $Mn(III)_4$ complex, g = 2.0, J = 15.6 cm^{-1} and J_{AB} = 47.0 cm^{-1}. The ground state has S = 3 and there are two excited S = 2 states within about 15 cm^{-1} of the ground state.

Note that the exchange coupling between the central two manganese ions in the Mn_4O_2 "butterfly" complexes, which have a di-μ-oxo-bridged structure, is rather large and antiferromagnetic. The exchange coupling between a wing-tip manganese ion and a central manganese ion, which have a μ-oxo-μ-carboxylato-bridged structure, is also antiferromagnetic, but smaller in magnitude. The exchange coupling between the two wing-tip manganese ions is small. This combination of exchange couplings leads to a high-multiplicity ground state for the Mn_4O_2 "butterfly" complexes.

Unfortunately, neither of the two Mn_4O_2 "butterfly" complexes that have been studied in some detail is an odd-electron species. For the $Mn(III)_4$ complex, no EPR signals are detected, even at liquid helium temperatures. The EPR spectrum for the $Mn(II)_2$-$Mn(III)_2$ complex exhibits many fine-structure transitions that have not yet been assigned. The EPR and magnetic properties of the odd-electron $Mn(II)$-$Mn(III)_3$ "butterfly" complex have not been analyzed in detail because a crystal structure has not yet been obtained for this complex.

Two $Mn(III)_3$-$Mn(IV)$ complexes have recently been reported that have a Mn_4O_3Cl "cubane"-like core (40). One of these complexes has a crystallographically imposed C_3 symmetry (40b). This allows one to derive a vector-coupling model by first coupling together the spin operators for the three $Mn(III)$ ions and then coupling the resultant spin to the $Mn(IV)$ ion. Magnetization data show that the ground state is isolated and has S = 9/2. Together with this information and a fit to temperature-dependent magnetic susceptibility data, the following parameters were obtained: J_1 = +53.6 cm^{-1}, J_2 = −24.2 cm^{-1}, and g = 1.86, where J_1 and J_2 are the exchange couplings between $Mn(III)$-$Mn(IV)$ and $Mn(III)$-$Mn(III)$, respectively. A moderately large antiferromagnetic exchange interaction is mediated between the $Mn(III)$-$Mn(IV)$ pairs by the di-μ-oxo-μ-carboxylato bridging ligands; in this complex, as well as the complexes discussed above, antiferromagnetic exchange coupling is observed for di-μ-oxo-bridged $Mn(III)$-$Mn(IV)$. Interesting, a moderately large ferromagnetic exchange interaction is mediated between the $Mn(III)$-$Mn(III)$ pairs by the μ-oxo-μ-chloro bridging ligands. The overall combination of exchange couplings leads to a high-multiplicity ground state.

The manganese complex in photosystem II exhibits two EPR signals that probably arise from S = 1/2 and S = 3/2 states, respectively (see section 5). Such low-multiplicity ground states appear to be inconsistent with the exchange-coupling scheme expected for a tetrameric manganese complex having a Mn_4O_2 "butterfly" structure or a Mn_4O_3Cl "cubane"-like structure. Therefore, we are led to consider another exchange-coupling scheme to account for the magnetic properties of manganese in photosystem II.

Next, consider a manganese tetramer composed of two dimers in which the interdimer exchange-coupling constants are all equivalent (Fig. 9). Call one pair of manganese ions A and B and the second pair of manganese ions C and D. The exchange Hamiltonian for this case is the same as already given in [11] and the energies of the spin levels are given in [13]. However, in this case, the exchange coupling between Mn_C and Mn_D, J_{CD}, need not be small.

S_A, S_B, S_C and S_D depend on the oxidation states of the manganese ions in the tetranuclear complex. Consider the two, odd-electron, cases: $Mn(III)_3$-$Mn(IV)$ and $Mn(III)$-$Mn(IV)_3$. In each case, one of the pairs of manganese ions will consist of $Mn(III)$ and $Mn(IV)$, with $S_A = 2$ and $S_B = 3/2$, respectively. As discussed in section 5, extended x-ray absorption fine structure (EXAFS) data point to a di-μ-oxo-bridged structure for manganese in photosystem II. Based on data from model complexes (32,41), a di-μ-oxo-bridged $Mn(III)$-$Mn(IV)$ pair is expected to be strongly antiferromagnetically exchange coupled. Therefore, for a manganese tetramer consisting of a dimer of dimers in which one dimer is $Mn(III)$-$Mn(IV)$, the low-lying spin levels are expected to be those with S' = 1/2, where S' denotes the electron spin quantum number of the $Mn(III)$-$Mn(IV)$ pair. Eqn. [14a] gives the energies for a $Mn(III)_3$-$Mn(IV)$ complex,

$$E(S,S',S^*) = 1/2\ J_{AB}\ [S'(S' + 1) - 39/4]\ +\ 1/2\ J_{CD}\ [S^*(S^* + 1) - 12]$$
$$+\ 1/2\ J\ \ [S(S + 1) - S'(S' + 1) - S^*(S^* + 1)] \qquad [14a]$$

which reduces to Eqn. [14b] if S' = 1/2

$$E(S,S'=1/2,S^*) = -\ 9/2\ J_{AB}\ +\ 1/2\ J_{CD}\ [S^*(S^* + 1) - 12]$$
$$+\ 1/2\ J\ [S(S + 1) - S^*(S^* + 1) - 3/4]. \qquad [14b]$$

Eqn. [15a] gives the energies for a $Mn(III)$-$Mn(IV)_3$ complex,

$$E(S,S',S^*) = 1/2\ J_{AB}\ [S'(S' + 1) - 39/4]\ +\ 1/2\ J_{CD}\ [S^*(S^* + 1) - 15/2]$$
$$+\ 1/2\ J\ \ [S(S + 1) - S'(S' + 1) - S^*(S^* + 1)] \qquad [15a]$$

which reduces to Eqn. [15b] if S' = 1/2

$$E(S,S'=1/2,S^*) = -9/2 \ J_{AB} + 1/2 \ J_{CD} \ [S^*(S^* + 1) - 15/2]$$
$$+ 1/2 \ J \ [S(S + 1) - S^*(S^* + 1) - 3/4]. \tag{15b}$$

Fig. 9 gives the energies of the states with $S' = 1/2$ for a Mn(III)-Mn(IV)$_3$ complex in the limit where J_{AB} is large and positive (strong antiferromagnetic coupling between the Mn(III) and Mn(IV) ions). A nearly identical figure is obtained for a Mn(III)$_3$-Mn(IV) complex except that five additional states with $S^* = 4$ must be included (compare [14b] and [15b]). For either a Mn(III)-Mn(IV)$_3$ or a Mn(III)$_3$-Mn(IV) complex, when J_{AB} is large and positive and J/J_{CD} falls in the range from -4 to 1, the lowest energy level has the same quantum numbers (Fig. 9). Consequently, EPR measurements may not allow one to unambiguously determine the oxidation states of manganese in a tetranuclear cluster. As will be discussed in section 5, the exchange-coupling scheme for a manganese tetramer consisting of

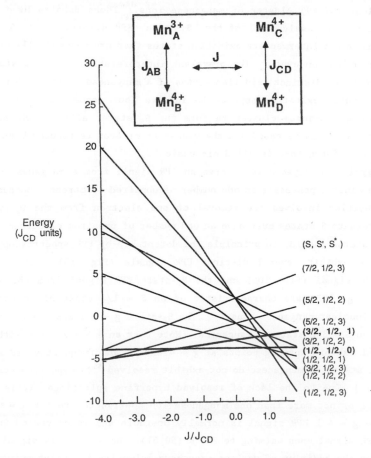

Fig. 9. Lowest energy levels given by [15b] for an exchange-coupled Mn(III)-Mn(IV)$_3$ tetramer in the limit where J_{AB} is large and positive.

a dimer of dimers appears to provide the best description of the magnetic properties of manganese in photosystem II.

5. EPR OF MANGANESE IN PHOTOSYSTEM II

Photosystem II is unique among manganese enzymes in several respects. It is the only enzyme that contains four manganese ions per active site and its biological function, the oxidation of water, absolutely requires manganese. In contrast, many manganoenzymes function with a variety of metal ions and, in other cases, distinct enzymes exist that contain a different metal ion but catalyze the same reaction. For example, alternate naturally-occurring forms of superoxide dismutase and catalase exist that contain iron instead of manganese.

Although four manganese ions per photosystem II are required for O_2-evolution activity (reviewed in 11,12,42,43), there are currently debates over whether the active site of water oxidation consists of a binuclear (44-45), trinuclear (46), or tetranuclear (47-49) cluster of manganese ions. These debates have arisen because of different assignments of the S_2-state EPR signals. The S_i states (i = 0,4) denote the intermediate oxidation states that are sequentially produced in the water oxidation cycle. The prevalent interpretation of the S states is that they represent distinct oxidation states of a manganese cluster. Hence, the mechanism of water oxidation appears to involve four sequential one-electron oxidations of a manganese cluster to form the S_4 state, after which water is rapidly oxidized, O_2 is released and the manganese cluster is reduced back to the S_0 state (see 12 for a more detailed discussion).

In general, one expects to observe an EPR signal from a manganese cluster whenever the cluster possesses an odd number of unpaired electrons. Because each S-state transition involves the removal of one electron from the O_2-evolving complex, alternate S states must have an odd number of unpaired electrons. These alternate S states should, in principle, be detectable by EPR spectroscopy. The S_2 state does exhibit several distinct EPR signals (Fig. 10). The S_2-state multiline EPR signal (Fig. 10A) and the S_2-state EPR signal from the ammonia derivative (Fig. 10B) are characteristic of an S = 1/2 state of an exchange-coupled multinuclear manganese cluster. The S_2-state g = 4.1 EPR signal has been assigned to an S = 3/2 state of manganese. [Note that an S = 3/2 state with axial symmetry will exhibit turning points at g = 4 and 2, and frequently EPR signals from S = 3/2 states of manganese do not exhibit resolved ^{55}Mn hyperfine structure (see Fig. 3).] Due to the lack of resolved hyperfine splittings, it is not as clear whether or not more than one manganese ion contributes to the g = 4.1 EPR signal. The g = 4.1 EPR signal is normally unstable and is converted into the multiline EPR signal upon warming to 200 K (50,51). However, this signal can be stabilized by the addition of various exogenous molecules including amines (52), fluoride (50), and sucrose (53).

Two interpretations of the identity of the species that give rise to the S_2-state multiline and g = 4.1 EPR signals have been proposed (44,47). Both EPR signals are proposed to arise from manganese centers. The difference in interpretations concerns the number of manganese ions involved in each paramagnetic species.

One view is that three distinct manganese centers function in the water oxidation process: two mononuclear manganese centers and a binuclear manganese center (44). It was proposed (44) that the binuclear center gives rise to the multiline EPR signal, whereas, one of the mononuclear manganese ions gives rise to the g = 4.1 EPR signal. In order to explain the stabilization of the g = 4.1 EPR signal by exogenous molecules, Hansson et al. (44) proposed that the g = 4.1 EPR signal arises from a mononuclear Mn(IV) in redox equilibrium with a binuclear

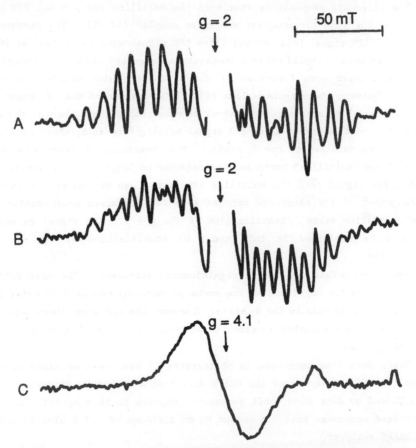

Fig. 10. Light-induced S_2 state EPR spectra from photosystem II membranes: (A) multiline EPR signal from an untreated sample; (B) multiline EPR signal produced in the presence of 100 mM NH_4Cl; (C) g = 4.1 EPR signal from an untreated sample. Experimental conditions are as in (47,54). In (A) and (B), the sample was warmed in the dark for 1 min to 0°C after forming the S_2 state to allow reoxidation of the electron acceptor, Q_A.

manganese species. Hence, in the S_2 state, it was proposed that either the mononuclear manganese is oxidized to Mn(IV) and gives a g = 4.1 EPR signal or the binuclear manganese center is oxidized to Mn(III)-Mn(IV) and gives a multiline EPR signal. The reduction potentials of these two species must be comparable in the cases when the g = 4.1 EPR signal is stabilized by the addition of sucrose, fluoride or amines. In this model, a further oxidation of the system to the S_3 state should leave <u>both</u> the binuclear and mononuclear centers oxidized. However, no EPR signal is observed from the manganese complex in the S_3 state. For this result to be compatible with the redox-equilibrium model, the binuclear and mononuclear species must have either a large magnetic or redox coupling in the S_3 state. Hence, the redox-equilibrium model requires that these two manganese species must be very close together.

The alternate proposal is that both the multiline and g = 4.1 EPR signals arise from the same tetranuclear manganese complex (47,51). The conversion of the g = 4.1 EPR signal into the multiline EPR signal upon incubation at 200 K in the dark can then be explained by a temperature-dependent structural change of the manganese cluster upon formation of the S_2 state, which alters the exchange couplings between the manganese ions (47). Generation of the S_2 state below 200 K may not allow such a rearrangement to occur rapidly and, hence, the g = 4.1 EPR signal could be viewed as an EPR signal arising from an S_1-state conformation that has been oxidized to the S_2 state. The structural difference between the "g = 4.1" and "multiline" conformations need not be large; the conversion of the g = 4.1 EPR signal into the multiline EPR signal can be understood as a small rearrangement of the manganese complex into its preferred conformation in the higher oxidation state. Stabilization of the g = 4.1 EPR signal by exogenous molecules is explained in this model by stabilization of the "g = 4.1" conformation.

These two models may not be significantly different. The main difference between them is the magnitude of the exchange coupling proposed to exist between the four manganese ions in the S_2 state. However, the different manganese centers must be very close together in order to account for the absence of an EPR signal from the S_3 state.

EXAFS data from manganese in photosystem II have been obtained by several groups (55-57). Analyses of the EXAFS data indicate a Mn-Mn distance of 2.7 Å, which, based on data from model compounds, appears to be diagnostic of a di-μ-oxo-bridged manganese pair. A second Mn-Mn distance of 3.3 Å also is indicated from EXAFS analyses.

One could imagine a number of arrangements of the four manganese ions in photosystem II that contain a di-μ-oxo-bridged structure. Moreover, the observation of two different Mn-Mn distances by EXAFS, could only be accounted for by one of the following three arrangements of manganese: a manganese trimer plus one isolated (more than 3.3 Å away) mononuclear manganese center, two isolated inequivalent manganese dimers, or a manganese tetramer.

In order to distinguish between these possibilities, one must consider the EPR properties of the S_2 and S_3 states. Measurements have been made of the temperature dependence of the S_2-state multiline (44,45,47) and g = 4.1 (47) EPR signals in order to determine whether the EPR signals arise from the ground state or from a thermally-populated state. The S_2-state g = 4.1 EPR signal exhibits Curie-law behavior characteristic of an EPR transition from a ground state or from a system in the high-temperature limit (47). It seems probable that the g = 4.1 EPR signal arises from a ground S = 3/2 state of a manganese species. The S_2-state multiline EPR signal has been assigned to either a ground or low-lying S = 1/2 state of a multinuclear mixed-valence manganese species (44,45,47); and the S_3 state does not exhibit an EPR signal. These assignments severely restrict the possible arrangements of manganese in photosystem II and also provide important information on the structure of the manganese complex.

On the basis of a similarity of the ^{55}Mn nuclear hyperfine couplings in the S_2-state multiline EPR signal with those of EPR signals from manganese dimer model complexes (compare Fig. 10A with Fig. 6), it has been suggested that the S_2-state multiline EPR signal arises from the S = 1/2 state of an antiferromagnetically exchange-coupled mixed-valence manganese dimer (44,58). This assignment has led to proposals of a variety of models in which a manganese dimer is the catalytic site for water oxidation. Fig. 4 depicts the energies of the spin states that arise from either antiferromagnetic or ferromagnetic exchange coupling of a Mn(III)-Mn(IV) dimer. For di-μ-oxo-bridged Mn(III)-Mn(IV) model complexes, the exchange coupling is typically antiferromagnetic, in which case the ground state has S = 1/2. Observation of a Curie-law behavior for the temperature dependence of the S_2-state multiline EPR signal is consistent with this EPR signal arising from the ground S = 1/2 state of an antiferromagnetically exchange-coupled manganese dimer (45).

The assignment of the S_2-state multiline EPR signal to a manganese dimer must be considered in light of the assignment of the S_2-state g = 4.1 EPR signal to a ground S = 3/2 state of a manganese species. From Fig. 4, it is apparent that the S_2-state g = 4.1 EPR signal cannot arise from a magnetically-isolated exchange-coupled mixed-valence manganese dimer. Several assignments of the S_2-state g = 4.1 EPR signal have been proposed. The g = 4.1 EPR signal could arise from a mononuclear high-spin Mn(IV) which has S = 3/2 [(44) and see Fig. 3], from an S = 3/2 state of an exchange-coupled trinuclear manganese complex (46), or from an S = 3/2 state of an exchange-coupled tetranuclear manganese complex (47). All of these possibilities are supported by observations of EPR signals near g = 4 from well-characterized manganese model complexes.

If we return to the possible arrangements of manganese in photosystem II that are consistent with the EXAFS data, one can rule out the possibility that the manganese ions are arranged as two isolated, inequivalent manganese dimers because this arrangement cannot account for the S_2-state g = 4.1 EPR signal. This leaves two possible arrangements of the manganese ions: a manganese trimer plus a mononuclear manganese center, or a manganese tetramer. Consequently, one should look for an assignment for the S_2-state multiline EPR signal from either a mixed-valence manganese trimer or tetramer rather than from a mixed-valence manganese dimer.

The ^{55}Mn nuclear hyperfine couplings for the S_2-state multiline EPR signal strongly resemble those for mixed-valence manganese dimers. However, the ^{55}Mn nuclear hyperfine couplings for a manganese trimer or tetramer depend on the projection of each manganese ion's nuclear hyperfine coupling tensor on the total spin of the coupled system (see [6]-[9] and Table 1 for the results for a manganese dimer). For a tetranuclear cluster, the projection of the nuclear hyperfine tensor of each individual ion on the total spin is more complicated than the case of a dimeric cluster. However, if one can use vector-coupling rules to first couple the angular momenta of pairs of manganese ions, it is possible to evaluate the hyperfine splitting in an exchange-coupled tetrameric manganese cluster. In the second stage of this calculation, one can use S' and S^* (as in Eqn. [12]) in place of S_A and S_B in Eqn. [6] to obtain the hyperfine-coupling constants in the coupled tetranuclear cluster. For the dimer-of-dimers model discussed above (Eqns. [14]-[15] and Fig. 9), S^* denotes the coupled spin of Mn_C and Mn_D. From Eqns. [6]-[9], it is apparent that Mn_C and Mn_D will not give any hyperfine splittings when $S^* = 0$. In particular, the dimer-of-dimers model predicts that only the mixed-valence pair will contribute to the ^{55}Mn nuclear hyperfine couplings in the $(S,S',S^*) = (1/2,1/2,0)$ state (Fig. 9) and that, in this state, the tetranuclear cluster will have the same hyperfine couplings as a mixed-valence dimer. Consequently, it is not always possible to determine the number of manganese ions in a multinuclear manganese cluster simply on the basis of ^{55}Mn nuclear hyperfine couplings. Thus, the S_2-state multiline EPR signal (Fig. 10A) can be accounted for by a tetranuclear cluster.

It is not obvious how a manganese trimer could account for the S_2-state multiline EPR signal. Specifically, it is not possible for the nuclear hyperfine couplings of a trimer to be equivalent to those of a mixed-valence dimer, unless the third manganese ion is uncoupled from a mixed-valence pair. For an odd-electron trinuclear cluster, only one manganese ion will contribute significantly to the hyperfine coupling if the two ions with the same valence are strongly antiferromagnetically exchange coupled; whereas, all three manganese ions will contribute substantially to the hyperfine coupling if the exchange-couplings between all three manganese ions are comparable.

Three criteria must be met for a manganese tetramer to account for the properties of the S_2-state EPR signals. For one conformation of the manganese cluster, the ground state should have $S = 3/2$ to account for the $g = 4.1$ EPR signal; for another conformation, the ground state or a low-lying excited state should have $S = 1/2$ to account for the multiline EPR signal; and the [55]Mn nuclear hyperfine interaction in the $S = 1/2$ state giving rise to the multiline EPR signal should be dominated by only two of the manganese ions. As discussed above, this last criterion is satisfied for the $(S,S',S^*) = (1/2,1/2,0)$ state of a manganese tetramer (Fig. 9). It is expected that the exchange couplings in the two conformations that give the S_2-state $g = 4.1$ and multiline EPR signals, respectively, will be similar because these two conformations interconvert at 200 K or below, temperatures at which only minor structural changes are expected to occur. Fig. 9 shows that the $(1/2,1/2,0)$ state, from which the S_2-state multiline EPR signal could arise, is the ground state when J/J_{CD} is between 1 and -2, and that the $(3/2,1/2,1)$ state, from which the S_2-state $g = 4.1$ EPR signal could arise, becomes the ground state when J/J_{CD} is smaller than -2. Consequently, no major changes in the exchange couplings are required in order to explain the conversion of the S_2-state $g = 4.1$ EPR signal into the multiline EPR signal with a manganese tetramer model.

All of the magnetic properties of the S_2 state can be accounted for with a dimer-of-dimers manganese tetramer model. One conclusion that can be drawn is that both antiferromagnetic and ferromagnetic exchange couplings must be present simultaneously in order for this tetramer model to account for the magnetic properties of the manganese complex in the S_2 state. This unusual combination of exchange couplings has been previously observed for "cubane"-like complexes (59-60), and this analogy suggests that the tetrameric manganese complex may have a "cubane"-like structure in the S_2 state (47). An oxo-bridged "cubane"-like manganese tetrameric complex was proposed for the structure of the manganese complex in the S_2 state (47,48). A distorted oxo-bridged "cubane"-like manganese tetramer seems to best account for the arrangement of manganese ions as seen by both EXAFS and EPR in the S_2 state.

ACKNOWLEDGMENTS

I thank Drs. David Hendrickson, James Penner-Hahn and Roger Prince for kindly sending preprints of their work and Dr. Warren Beck for help with the preparation of the figures. This work was supported by the National Institutes of Health (GM32715) and by a Camille and Henry Dreyfus Teacher/Scholar Award.

REFERENCES

1   A.R. McEuen, Inorg. Biochem., 3 (1982) 314-343.
2   C.L. Schram and F.C. Wedler (Eds), Manganese in Metabolism and Enzyme Function, Academic Press, New York, 1986.
3   G.H. Reed and G.D. Markham, in: L.J. Berliner and J. Reuben (Eds),

862

Biological Magnetic Resonance, Vol. 6, Plenum Press, New York, 1984, pp. 73-142.

4   J.C.W. Chien and E.W. Westhead, Biochemistry, 10 (1971) 3198-3203.

5   G.D. Markham, J. Biol. Chem., 256 (1981) 1903-1909.

6   B.C. Anatanaitis, R.D. Brown III, N.D. Chasteen, J.H. Freedman, S.H. Koenig, H.R. Lilienthal, J. Peisach and C.F. Brewer, Biochemistry, 26 (1987) 7932-7937.

7   A.M. Michelson, J.M. McCord and I. Fridovich (Eds), Superoxide and Superoxide Dismutases, Academic Press, New York, 1977.

8   W.F. Beyer Jr. and I. Fridovich, Biochemistry, 24 (1985) 6460-6467.

9   W.C. Stallings, K.A. Pattridge, R.K. Strong and M.L. Ludwig, J. Biol. Chem., 259 (1984) 10695-10699.

10   V.V. Barynin, A.A. Vagin, V.R. Melik-Adamyan, A.I. Grebenko, S.V. Khangulov, A.N. Popov, M.E. Andrianova and B.K. Vainshtein, Sov. Phys. Dokl., 31 (1986) 457-459.

11   G.T. Babcock, in: J. Amesz (Ed), New Comprehensive Biochemistry: Photosynthesis, Elsevier, Amsterdam, 1987, pp. 125-158.

12   G.W. Brudvig, W.F. Beck and J.C. de Paula, Annu. Rev. Biophys. Biophys. Chem., 18 (1989) 25-46.

13   J.E. Wertz and J.R. Bolton, Electron Spin Resonance, McGraw-Hill, New York, 1972, p. 308.

14   J.A. Fee, E.R. Shapiro and T.H. Moss, J. Biol. Chem., 251 (1976) 6157-6159.

15   T. Yonetani, H.R. Drott, J.S. Leigh, Jr., G.H. Reed, M.R. Waterman and T. Asakura, J. Biol. Chem., 245 (1970) 2998-3003.

16   H. Hori, M. Ikeda-Saito, G.H. Reed and T. Yonetani, J. Magn. Res., 58 (1984) 177-185.

17   K.R. Rodgers and H.M. Goff, J. Amer. Chem. Soc., 110 (1988) 7049-7060.

18   D.T. Richens and D.T. Sawyer, J. Amer. Chem. Soc., 101 (1979) 3681-3683.

19   D.P. Kessissoglou, X. Li, W.M. Butler and V.L. Pecoraro, Inorg. Chem., 26 (1987) 2487-2492.

20   H. Hori, M. Ikeda-Saito and T. Yonetani, Biochim. Biophys. Acta, 912 (1987) 74-81.

21   J. Owen and E.A. Harris, in: S. Geschwind (Ed), Electron Paramagnetic Resonance, Plenum, New York, 1972, p. 427.

22   C.J. Cairns and D.H. Busch, Coord. Chem. Rev., 69 (1986) 1-55.

23   R.W. Wilkins and J.W. Culvahouse, Phys. Rev. B, 14 (1976) 1830-1841.

24   R.P. Scaringe, D.J. Hodgson and W.E. Hatfield, Mol. Phys., 35 (1978) 701-713.

25   Ö. Hansson and L.-E. Andréasson, Biochim. Biophys. Acta, 679 (1982) 261-268.

26   A. Willing, H. Follmann and G. Auling, Eur. J. Biochem. 170 (1988) 603-611.

27   J.E. Sheats, R.S. Czernuszewicz, G.C. Dismukes, A.L. Rheingold, V. Petrouleas, J. Stubbe, W.H. Armstrong, R.H. Beer and S.J. Lippard, J. Amer. Chem. Soc., 109 (1987) 1435-1444.

28   R.M. Fronko, J.E. Penner-Hahn and C.J. Bender, J. Amer. Chem. Soc., 110 (1988) 7554-7555.

29   G. Christou and J.B. Vincent, in: L. Que (Ed), Metal Clusters in Proteins, ACS Symposium Series No. 372, American Chemical Society, 1988, pp. 238-255.

30   G.W. Brudvig and R.H. Crabtree, Prog. Inorg. Chem., 37 (1989) in press.

31   H.-R. Chang, S.K. Larsen, P.D.W. Boyd, C.G. Pierpont and D.N. Hendrickson, J. Amer. Chem. Soc., 110 (1988) 4565-4576.

32   S.R. Cooper, G.C. Dismukes, M.P. Klein and M. Calvin, J. Amer. Chem. Soc., 100 (1978) 7248-7252.

33   A. Abragam and B. Bleaney, Electron Paramagnetic Resonance of Transition Ions, Clarendon Press, London, 1970.

34   G.C. Dismukes, J.E. Sheats and J.A. Smegal, J. Amer. Chem. Soc., 109 (1987) 7202-7203.

35   H.-R. Chang, H. Diril, M.J. Nilges, X. Zhang, J.A. Potenza, H.J. Schugar, D.N. Hendrickson and S.S. Isied, J. Amer. Chem. Soc., 110 (1988) 625-627.

36   J.B. Vincent, C. Christmas, H.-R. Chang, Q. Li, P.D.W. Boyd, J.C.

Huffman, D.N. Hendrickson and G. Christou, J. Amer. Chem. Soc., 111 (1989) in press.

37   K. Wieghardt, U. Bossek and W. Gebert, Angew. Chem. Int. Ed. Engl., 22 (1983) 328-329.

38   R.J. Kulaweic, R.H. Crabtree, G.W. Brudvig and G.K. Schulte, Inorg. Chem., 27 (1988) 1309-1311.

39   K. Kambe, J. Phys. Soc., Jpn., 5 (1950) 48-51.

40   a) J.S. Bashkin, H.-R. Chang, W.E. Strieb, J.C. Huffman, D.N. Hendrickson and G. Christou, J. Amer. Chem. Soc., 109 (1987) 6502-6504.
     b) Q. Li, J.B. Vincent, E. Libby, H.-R. Chang, J.C. Huffman, P.D.W. Boyd, G. Christou and D.N. Hendrickson, Angew. Chem. Int. Ed. Engl., 27 (1988) 1731-1733.

41   K. Wieghardt, U. Bossek, B. Nuber, J. Weiss, J. Bonvoisin, M. Corbella, S.E. Vitols and J.J. Girerd, J. Amer. Chem. Soc., 110 (1988) 7398-7411.

42   G.W. Brudvig, in: L. Que (Ed), Metal Clusters in Proteins, ACS Symposium Series No. 372, American Chemical Society, 1988, pp. 221-237.

43   G.M. Choniae, Annu. Rev. Plant Physiol., 21 (1970) 467-498.

44   Ö. Hansson, R. Aasa and T. Vänngård, Biophys. J., 51 (1987) 825-832.

45   R. Aasa, L.-E. Andréasson, G. Lagenfelt and T. Vänngård, FEBS Lett., 221 (1987) 245-248.

46   X. Li, D.P. Kessissoglou, M.L. Kirk, C.J. Bender and V.L. Pecoraro, Inorg. Chem., 27 (1988) 1-3.

47   J.C. de Paula, W.F. Beck and G.W. Brudvig, J. Amer. Chem. Soc., 108 (1986) 4002-4009.

48   G.W. Brudvig and R.H. Crabtree, Proc. Natl. Acad. Sci. USA, 83 (1986) 4586-4588.

49   J.B. Vincent and G. Christou, Inorg. Chim Acta, 136 (1987) L41-L43.

50   J.L. Casey and K. Sauer, Biochim. Biophys. Acta, 767 (1984) 21-28.

51   J.C. de Paula, J.B. Innes and G.W. Brudvig, Biochemistry, 24 (1985) 8114-8120.

52   W.F. Beck and G.W. Brudvig, Biochemistry, 25 (1986) 6479-6486.

53   J.-L. Zimmermann and A.W. Rutherford, Biochemistry, 25 (1986) 4609-4615.

54   W.F. Beck, J.C. de Paula and G.W. Brudvig, J. Amer. Chem. Soc., 108 (1986) 4018-4022.

55   V.K. Yachandra, R.D. Guiles, A.E. McDermott, J.L. Cole, R.D. Britt, S.L. Dexheimer, K. Sauer and M.P. Klein, Biochemistry, 26 (1987) 5974-5981.

56   G.N. George, R.C. Prince and S.P Cramer, Science, 243 (1989) in press.

57   J.E. Penner-Hahn, R.M. Fronko, V.L. Pecoraro, C.F. Yocum and N.R. Bowlby, J. Amer. Chem. Soc. (1989) submitted.

58   G.C. Dismukes and Y. Siderer, Proc. Natl. Acad. Sci. USA, 78 (1981) 274-277.

59   L. Noodleman, J.G. Norman, Jr., J.H. Osborne, A. Aizman and D.A. Case, J.Amer. Chem. Soc., 107 (1985) 3418-3426.

60   W. Haase, L. Walz and F. Nepveu, in: I. Bertini, R.S. Drago and C. Luchinat (Eds), The Coordination Chemistry of Metalloenzymes, D. Reidel, Holland, 1983, pp. 229-234.

CHAPTER 25

EPR SPECTRA OF ACTIVE SITES IN COPPER PROTEINS

EDWARD I. SOLOMON, ANDREW A. GEWIRTH, and T. DAVID WESTMORELAND

1. INTRODUCTION

The field of copper proteins has been strongly influenced by EPR spectroscopy since the early classifications of copper sites based on their EPR spectra [1]. One can generally divide copper active sites into mononuclear and copper cluster centers [2]. Within the mononuclear copper proteins there are two subclasses: the normal or Type 2 copper enzymes, and the blue or Type 1 copper proteins.

The normal copper enzymes contain active sites which exhibit EPR signals that reflect mononuclear, approximately tetragonal copper(II) centers. These proteins, their function, biochemical characterization and EPR parameters of the native copper(II) states are summarized in Table 1.

TABLE 1
Type 2 copper sites.

Enzyme	Reaction	Cu/M.W.(kD)	$g_\parallel$	$g_\perp$	$A_\parallel$ ($\times 10^{-4} cm^{-1}$)	Ref.
Cu Zn SOD	$2O_2^- + 2H^+ \rightarrow$ $O_2 + H_2O_2$	2/31	2.26	2.09 2.03	142	[3]
Diamine Oxidases	$R'CHNR_2 + O_2 + H_2O \rightarrow$ $R'CHO + HNR_2$	2/160-200	2.29	2.06	160	[4]
Dopamine-β-Monooxygenase	dopamine+ascorbate $+O_2 \rightarrow$ noradrenaline + dehyroascorbate + H_2O	4/290	2.270	2.041	190	[5]
Phenylalanine hydroxylase	phenylalanine + $O_2 \rightarrow$ tyrosine + H_2O	1/32	2.32	2.06	155	[6]
Galactose Oxidase	$RCH_2OH + O_2 \rightarrow$ $RCHO + H_2O_2$	1/68	2.227	2.051	182	[7]

The diamine oxidases [4], phenylalanine hydroxylase [6], and galactose oxidase [7] are all believed to contain additional organic cofactors that are involved in catalysis. In the case of diamine oxidases this has been identified as pyrroloquinoline quinone (PQQ), and for phenylalanine hydroxylase this is H_4biopterin. For dopamine-β-monooxygenase there has been some discussion as to whether a copper dimer site may actually be involved in catalysis [8]. However, thus far all reported EPR spectra [9] indicate only isolated mononuclear copper(II) sites.

Of the normal copper enzymes listed in Table I, only Cu Zn SOD is structurally defined [10]. The active site and its absorption and EPR spectra are given in Fig. 1. The copper(II) has a distorted tetragonal geometry

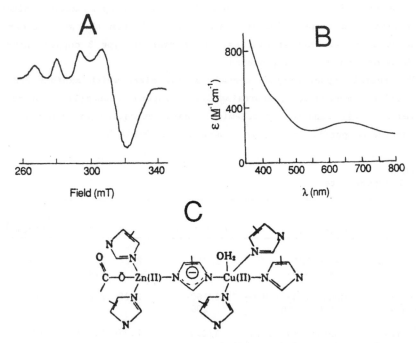

Figure 1. A) X-band EPR, B) optical absorbance, and C) structure of Cu Zn SOD. From [3b].

coordinated by three imidazole ligands and bridged by an imidazolate ion to a tetrahedral zinc(II). As shown in Fig. 1A and 1B, the EPR and optical spectra are consistent with this structure. This ligand field will produce a tetragonal splitting of the d orbitals as sketched in Fig. 2, which would result in one unpaired electron occupying the $d_{x^2-y^2}$ orbital for the d^9 copper(II) ion [11]. For this ground state the EPR spectrum has $g_\parallel > g_{\perp,ave} > 2.00$ and a large $A_\parallel$ splitting of the four Cu hyperfine lines. The absorption spectrum exhibits d→d transitions at approximately 650 nm for a tetragonal field of nitrogen ligands. The low symmetry distortion of the Cu(II) in Fig.

lA is also reflected in the spectra, as a rhombic splitting of the $g_\perp$ region in the EPR (g_y-g_x= 0.06) and intensity enhancement of the Laporte forbidden d→d transitions, which have $\epsilon \sim 300$ M^{-1} cm^{-1} rather than the $\epsilon = 10$-40 M^{-1} cm^{-1} usually observed for tetragonal copper(II) complexes [11].

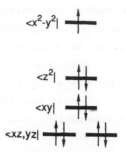

Figure 2. One electron orbital splittings for a tetragonal d^9 complex.

For normal copper(II) proteins the major use of CW EPR spectroscopy thus far has been to identify changes in oxidation state and to probe substrate and small molecule binding to the active site in order to define the role of the copper ion in the catalytic mechanism.

In contrast to the normal mononuclear copper enzymes, blue copper proteins exhibit some very unusual spectral features in comparison to tetragonal copper(II) centers [1,2a,2c]. These are summarized in Fig. 3. For the blue copper sites, the absorption intensity at 600 nm is 3-5,000 M^{-1} cm^{-1} and the A$\|$ value is very small ($\leq 90 \times 10^{-4}$ cm^{-1}). These copper sites are generally involved in mediating electron transfer, and function with quite high reduction potentials [12] and rapid rates of electron transfer [13] relative to tetragonal copper(II) complexes. The original goal of inorganic spectroscopy on the blue copper proteins was to define the origin of these unique spectral features and to use these to generate a "spectroscopically effective" model of this active site [14]. A significant step in this direction came from near-IR circular dichroism spectroscopy, which demonstrated that the d→d transitions were observed at energies down to 5,000 cm^{-1}. Based on simple ligand field arguments this required that the copper site be approximately tetrahedral. In addition all d→d transitions could be accounted for at energies below the 600 nm absorption band in Fig. 3B. Therefore this must be a ligand → copper(II) charge transfer (CT) transition. A variety of chemical and spectroscopic data [15] had indicated that thiolate was a ligand at the copper site and as this is an easily oxidized residue, it could be assigned as the ligand involved in the charge transfer process.

Finally, at this point it also appeared that the geometry was responsible for the small A∥ value of the blue copper site, since D_{2d} distorted tetrahedral $CuCl_4^{2-}$ also exhibits a small parallel hyperfine splitting [16a,b].

In 1978 Professor Hans Freeman published [17] the high resolution crystal structure of the blue copper site in plastocyanin (Fig. 3C). This structure confirmed that the copper is tetrahedral with thiolate sulfur of cysteine 84 bound to the metal ion. The crystal structure further showed this thiolate-copper bond to be rather short and, in addition, identified the remaining three ligands: two imidazole-N of histidine residues 37 and 87, and a thioether S of methionine 92 which has a quite long, 2.9 Å copper-sulfur bond. We further note that the blue copper proteins can be divided into two subgroups based on their EPR spectra. Plastocyanin and azurin are representative of one subgroup that exhibits a close to axial EPR spectrum (g_y-g_x = 0.017) [18], while stellacyanin and the basic blue protein from cucumber belong to a second group exhibiting a large rhombic distortion (g_y-g_x = 0.059, Fig. 4) [19]. The crystal structure of azurin [20] and, at lower

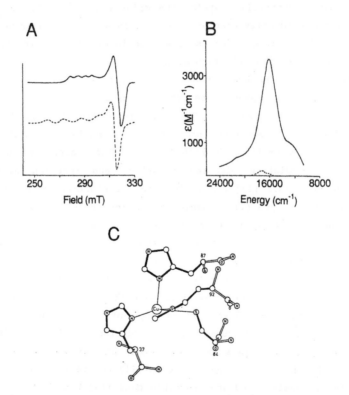

Figure 3. A) X-band EPR and B) optical absorbance of normal Cu(II) (---) vs. blue Cu (———) C) Structure of the plastocyanin active site. Adapted from [17].

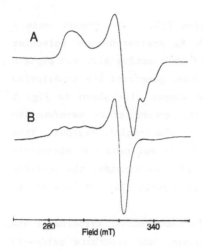

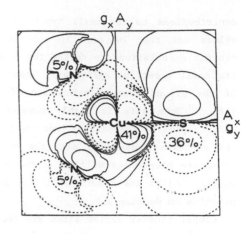

Figure 4. X-band EPR spectra for A) stellacyanin and B) plastocyanin.

Figure 5. Ground state wavefunction for the blue copper (plastocyanin) active site and the orientation of the g^2 and A^2 tensors in the plane containing both nitrogens, the thiolate, and the copper. Adapted from [18].

resolution, the basic blue protein [21] are quite similar to that of plastocyanin given in Fig. 3C. While no crystal structure information is yet available for stellacyanin it is known that there is no thioether in its sequence [22].

The blue copper protein provides an ideal system for defining electronic structure contributions to active site reactivity [18,23]. In particular, EPR spectroscopy probes the half-occupied $d_{x^2-y^2}$ orbital of the copper, its orientation, and covalent delocalization onto the ligands. It is this ground state wavefunction that plays the critical role of taking up the electron in the redox reactivity of this active site. Obtaining an experimental and theoretical description of this ground state wavefunction is the goal of the first part of this review. Section 2 summarizes our basic understanding of the spin Hamiltonian parameters of tetragonal copper(II) sites. In particular, we focus on D_{4h} $CuCl_4^{2-}$ as an electronic structural analogue, and on the factors that contribute to the reduced hyperfine on distortion to a D_{2d} tetrahedral structure [24]. In Section 3 we extend this analysis toward understanding the ground state spectral features of the blue copper active site. First single crystal EPR spectroscopy is used to define the orientation of the ground state in plastocyanin [23a]. Then high energy spectroscopic methods are used to complement ground state studies to probe possible Cu 4p contributions to the small hyperfine interaction [25]. The ground state g values can then be used in combination with Xα scattered wave molecular orbital calculations to define the covalency of this active site and explain its small hyperfine splitting. These studies have generated the description of the ground state wavefunction of the blue copper site shown in Fig. 5 [18]. This site is highly covalent with the ground state wavefunction anisotropically delocalized into the pπ orbital of the thiolate residue. Thus the cysteine appears to play the dominant role in defining the electronic structure properties of this active site. Finally we consider the possible origin [24] of the rhombic splitting in the stellacyanin class of blue copper proteins in Section 3.4.

Copper cluster active sites are found in the hemocyanins, tyrosinases and the multicopper oxidases (laccases, ceruloplasmin, and ascorbate oxidases) [2b,c,d]. The coupled binuclear copper active sites in the hemocyanins and tyrosinases have been shown to be extremely similar [26], and must be considered separately from the cluster sites in the multicopper oxidases (vide infra). Proteins containing the coupled binuclear copper site are summarized in Table 2. For these proteins the deoxy [Cu(I)Cu(I)] site binds dioxygen to form the oxy derivative that has the spectral features presented in Fig. 6. Resonance Raman spectroscopy (which shows that oxygen is bound as peroxide, O_2^{2-}) [29] and X-ray absorption edge spectroscopy [30] demonstrates that the coppers are both in the cupric oxidation state. However, the spectra in Figure 6 are clearly very different from those of tetragonal copper(II). In particular, intense transitions dominate the absorption spectrum (350 nm, ϵ

TABLE 2

Coupled binuclear copper sites.

Protein	Reaction	Sites/M.W.(kD)	Ref.
Arthropod Hemocyanin	deoxy + O_2 $\rightleftharpoons$ oxy	1/70	[27]
Mollusc Hemocyanin	deoxy + O_2 $\rightleftharpoons$ oxy 2 H_2O_2 $\rightleftharpoons$ 2H_2O + O_2	1/50	[27]
Tyrosinase	phenol + 2e$^-$ + O_2 + 2H$^+$ $\rightarrow$ $\underline{o}$-diphenol + H_2O	1/40	[28]
	2$\underline{o}$-diphenol + O_2 $\rightarrow$ 2$\underline{o}$-quinone + 2H_2O		

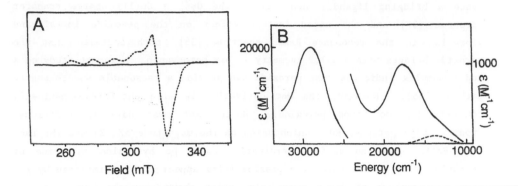

Figure 6. A) X-band EPR and B) optical absorbance of oxy hemocyanin (⸺)
and tetragonal monomeric Cu(II) (-----). Adapted from [2b].

~ 20,000 M^{-1} cm^{-1}; 600 nm, ϵ ~ 1000 M^{-1} cm^{-1}) and the active site has no
detectable EPR signal. Significant insight into the origin of these unique
spectral features became available with the characterization of the met
derivative [31], in which the peroxide of oxy hemocyanin and tyrosinase has
been displaced (i.e. met = [Cu(II)Cu(II)]). The intense absorption bands in
Fig. 6B are not present in the met derivative demonstrating these to be O_2^{2-} →
Cu(II) CT transitions [32]. Met-hemocyanin does, however, exhibit weak
tetragonal Cu(II) d→d transitions at approximately 680 nm (ϵ ~ 225 M^{-1} cm^{-1})
and X-ray absorption edges which confirm the +2 oxidation state for both
coppers [30c]. As with oxy hemocyanin, the met derivative is EPR non-
detectable [31]. A possible explanation for the lack of an EPR signal is an
antiferromagnetic interaction between the two S=1/2 Cu(II) ions, which is
described by the spin Hamiltonian $\hat{H}$ = -2J $\underline{\hat{S}}_1 \cdot \underline{\hat{S}}_2$ and produces the ground state
splitting defined in Fig. 7. Susceptibility studies confirm this possibility
showing the oxy and met sites to be diamagnetic to the sensitivity of SQUID
detection placing a lower limit on -2J of > 600 cm^{-1} for oxy hemocyanin [33]
and > 400 cm^{-1} for met hemocyanin [34]. These large singlet-triplet
splittings require a good superexchange pathway between the copper(II)'s and

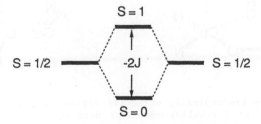

Figure 7. Antiferromagnetic coupling scheme of two S=1/2 centers.

hence a bridging ligand. Analysis of the $O_2^{2-} \rightarrow$ Cu(II) charge transfer spectrum [32] and the mixed isotope effect on the peroxide intraligand vibration in the resonance Raman spectrum [35], both indicate that the peroxide bridges with a μ-1,2 geometry in oxy hemocyanin. A recent study on a model complex indicates that peroxide can provide a reasonable superexchange pathway [36]. However, the met derivative is also antiferromagnetically coupled with no bound peroxide. Hence, met must have an additional superexchange pathway, RO⁻, which based on the magnitude of -2J and the fact that EXAFS shows no sulfur coordination could be hydroxide, alkoxide or phenoxide [32]. The latter two possibilities appear to be eliminated by the crystal structure of deoxy hemocyanin, which shows that conserved residues containing these functional groups are not in the vicinity of the active site [37]. The study of the endogenous bridge, RO⁻, at the coupled binuclear copper site has strongly involved both the EPR non-detectable met form and a second binuclear cupric derivative, the dimer form [31,38], which exhibits an S=1 EPR spectrum. These chemical and EPR spectroscopic studies of [Cu(II)Cu(II)] derivatives are summarized in Section 4.1. These and other spectral studies have strongly supported the spectroscopically effective model for the oxy hemocyanin (and oxy tyrosinase) active site shown in Figure 8A. This site contains two tetragonal Cu(II) ions bridged by peroxide in a cis μ-1,2 binding mode with the second endogenous bridge RO⁻ providing a significant superexchange pathway at the active site. Thus far only the crystal structure of deoxy hemocyanin [Cu(I)Cu(I)] has been reported [37], Fig. 8B, in which each copper is bound to two near (~2.0 Å) and one distant (~2.6 Å) imidazoles in a trigonal antiprism arrangement. It cannot yet be determined whether or not an additional hydroxide or water molecule is present at the deoxy site.

It has proved possible to generate a half-reduced mixed-valence form of the coupled binuclear copper active site in hemocyanin and tyrosinase, the half-met [Cu(II)Cu(I)] derivative, which has an S=1/2 ground state and is thus EPR detectable [39]. On a chemical level this derivative has proved to be

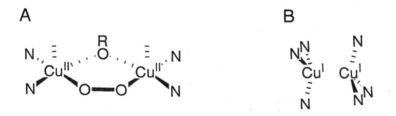

Figure 8. A) Spectroscopically effective active site for oxy hemocyanin and oxy tyrosinase. B) Crystallographically determined deoxy hemocyanin active site.

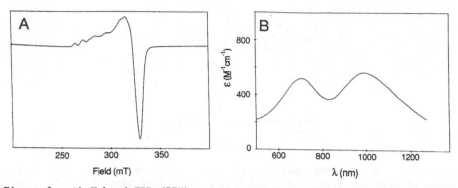

Figure 9. A) X-band EPR (77K) and B) optical absorbance (10K) of half-met-Br⁻ hemocyanin. Adapted from [68].

particularly useful in probing exogenous ligand interactions with the active site, and in demonstrating that the active site in tyrosinase is capable of binding substrate analogues at the copper with a trigonal bipyramidal distortion of the copper(II) geometry [40]. Further, dependent upon the exogenous ligand bound to the half-met site, this derivative exhibits some very unusual spectral features as compared to normal tetragonal Cu(II) complexes [39]. These are summarized for half-met-Br⁻ in Fig. 9, which shows that the parallel hyperfine region is more complex than the normal four line pattern expected for mononuclear Cu(II), and a new fairly intense absorption band appears on the low energy side of the tetragonal ligand field transitions in the optical spectrum. A summary of the analysis of the spectral features of this half-met derivative is presented in Section 4.2 which demonstrates that these spectral features relate to electron delocalization between the coppers in the mixed- valence ground state wavefunction and thus probes bridging ligation at the active site.

The coppers contained in the multicopper oxidases have been traditionally divided into Type 1, Type 2 and Type 3 centers based on EPR features [1,2] (see laccase, Fig. 10). The Type 1 and Type 2 sites have blue and normal tetragonal copper(II) spectral features respectively; the Type 3 is binuclear and EPR non-detectable, and thus has often been associated with the coupled binuclear copper site in hemocyanin and tyrosinase. The most recent biochemical characterization of these enzymes is given in Table 3. All of these enzymes function in the four-electron reduction of dioxygen to water. The goal of research on these enzymes has been to define the interaction between these copper centers and their involvement in this multi-electron irreversible reduction of dioxygen.

There have been significant research advances on the multicopper oxidases over recent years, which have provided important new insight into these active

874

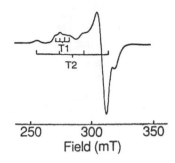

Field (mT)

Figure 10. X-band EPR spectrum of native laccase. Hyperfine features associated with the Type 1 (T1) and Type 2 (T2) sites are indicated. Adapted from [2e].

TABLE 3
Multi-copper oxidases.

	Stoichiometry				
Enzyme	T1	T2	T3	MW (kD)	Ref.
Laccase (tree)	1	1	1	110-140	[41]
Laccase (fungal)	1	1	1	65	[41]
Ascorbate Oxidase	2	2	2	130-140	[42]
Ceruloplasmin	2	1	1-2	125-130	[43]

sites. Spectroscopic studies on deoxy, half-met, met and uncoupled met derivatives of the Type 3 center (in a Type 2 depleted derivative of laccase) demonstrate major differences relative to the coupled binuclear copper site in hemocyanin and tyrosinase [44]. In particular, as briefly summarized in Section 4.2 and in Fig. 11, while there is an endogenous bridge (R'O⁻), exogenous ligands do not appear to bridge the two coppers at the Type 3 site. This correlates with differences in reactivity in that the reduced Type 3 site does not react with dioxygen in the absence of the Type 2 copper [45]. Further, low temperature magnetic circular dichroism (MCD) studies of azide binding to native laccase, which enables one to correlate excited state spectral features with ground state EPR data, have demonstrated that the azide bridges between the Type 3 and Type 2 centers defining a new trinuclear copper cluster [46] site, as pictured in Figure 12. This trinuclear copper cluster site has recently been supported by X-ray structural determination of ascorbate oxidase [47], which shows that the Type 1 center is approximately 13 Å away from the trinuclear copper cluster. This is an exciting new area of research activity in copper Bioinorganic Chemistry. One might anticipate interesting results on the effects of trinuclear interactions on the EPR spectra of the multicopper oxidases to appear in the near future.

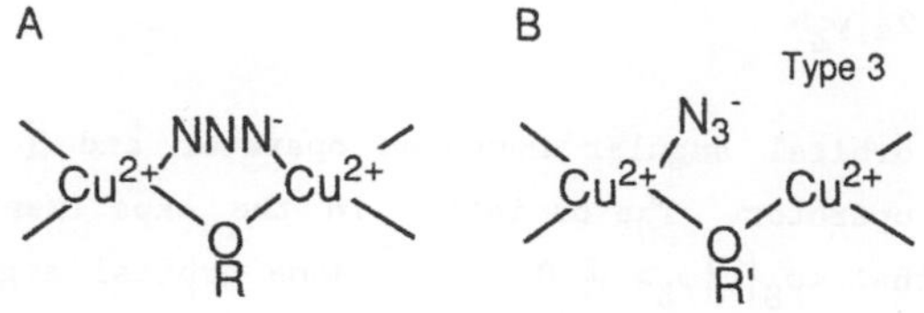

A

B

Type 3

Figure 11. Azide binding at A) the met-hemocyanin site and B) the Type 3 site in Type 2 depleted laccase.

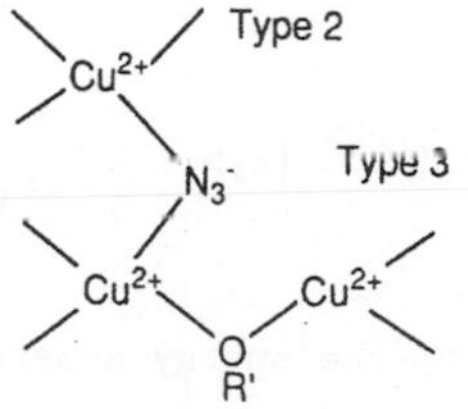

Type 2

Type 3

Figure 12. Trinuclear copper cluster site in native laccase.

2. ELECTRONIC STRUCTURE ORIGIN OF SPIN HAMILTONIAN PARAMETERS IN NORMAL COPPER SITES

The EPR spectrum of a tetragonal copper center can be described by the spin hamiltonian given in Eqn. 1, which contains explicitly the Zeeman and

$$\hat{H} = g\beta\underline{B}\cdot\hat{\underline{S}} + \mathbf{A}^{Cu}\hat{\underline{I}}^{Cu}\cdot\hat{\underline{S}} + \text{higher terms} \tag{1}$$

copper hyperfine couplings terms, respectively. The interpretation of the spin hamiltonian parameters g and $\mathbf{A}^{Cu}$, will be the subject of this section. These parameters are first related to the high- symmetry structurally defined square planar D_{4h} $CuCl_4{}^{2-}$ site to develop the information content available in each spin hamiltonian parameter. This site presents a number of advantages in that its spectroscopy has been studied over a wide energy range and the excited states have been definitively assigned [48]. Once D_{4h} $CuCl_4{}^{2-}$ is understood, this insight can be extended to the distorted tetrahedral D_{2d} $CuCl_4{}^{2-}$ and finally to the blue copper active site in plastocyanin in Section 3.

2.1 g values

The EPR spectrum of an approximately tetragonal copper site is shown in Fig. 1A. The EPR of D_{4h} $CuCl_4{}^{2-}$ is also highly anisotropic with the g values deviating significantly from the free ion value of 2.0023 ($g_\parallel$=2.221, $g_\perp$=2.040) [49]. The experimental g values can be described through ligand field theory. The energy of interaction of an external magnetic field with the ground state wavefunction ψ_g is given by the Zeeman operator in Eqn. 2

$$E_{Zee} = \beta \underline{B} <\psi_g | \hat{\underline{L}} + 2\hat{\underline{S}} | \psi_g> \tag{2}$$

where $\hat{\underline{L}}$ is the orbital angular momentum operator and $\hat{\underline{S}}$ is the total spin angular momentum operator. The deviation in the experimental g values from 2.0023 requires that $<\psi_g | \hat{\underline{L}} | \psi_g> \neq 0$, hence some orbital angular momentum must be mixed into the $d_{x^2-y^2}$ orbital. This modification is brought about by spin-orbit coupling that is given by $\lambda \hat{\underline{L}} \cdot \hat{\underline{S}}$ where λ, the spin-orbit coupling constant, is equal to -830 cm^{-1} for copper. For D_{4h} $CuCl_4^{2-}$ first order perturbation theory gives

$$|\psi_g> = |x^2-y^2> - \lambda \frac{<x^2-y^2|\hat{\underline{L}} \cdot \hat{\underline{S}}|xy>}{E_{xy}} |xy> - \lambda \frac{<x^2-y^2|\hat{\underline{L}} \cdot \hat{\underline{S}}|xz,yz>}{E_{xz,yz}}|xz,yz> \tag{3}$$

where $E_{(xz,yz),xy}$ represents the energy separation between the ground state and the excited $d_{xz,yz}$ and d_{xy} states, respectively. Application of the Zeeman operator in Eqn. 2 to Eqn. 3 gives

$$g_\parallel = g_e - 8\lambda/E_{xy} \quad ; \quad g_\perp = g_e - 2\lambda/E_{xz,yz} \tag{4}$$

as expressions for the g values. Since λ is negative and the denominators are positive these expressions give $g_\parallel > g_\perp > 2.00$. A more general expression for the g values of any $S=1/2$ orbitally nondegenerate ground state $|\psi_g>$ is given in Eqn. 5

$$g_i = g_e - 2\lambda \sum_{e \neq g} \frac{<\psi_g | \hat{L}_i | \psi_e><\psi_e | \hat{L}_i | \psi_g>}{(E_g - E_e)} \tag{5}$$

where i=x,y,z, $|\psi_e>$ are excited state d orbitals and $E_{g,e}$ are the energies of the ground and excited states, respectively. Eqn. 5 indicates that g values can be increased by either increasing the magnitude of the spin-orbit coupling constant or reducing the transition energies. Note however that when λ and the transition energies become comparable in magnitude these perturbation expressions are no longer applicable, and spin-orbit coupling must be added in a complete calculation.

When Eqn. 5 is applied to a pure crystal field wavefunction for D_{4h} $CuCl_4^{2-}$ the g values shown in Table 4 (LFT) are obtained. These ligand field-derived values have substantially larger deviations from g = 2.0023 than do the experimental values, indicating that too much angular momentum has been mixed into the ground state. This arises from the neglect of covalent interactions between the Cu d and the Cl p orbitals that delocalize the half-occupied $d_{x^2-y^2}$ orbital onto centers with less spin-orbit coupling, diminishing the amount of angular momentum mixed into the ground state. In

TABLE 4
Results of g value calculation for D_{4h} $CuCl_4^{2-}$.

	LFT	Xα (Norman radii)	Xα (adjusted)	experiment
$g_\parallel$	2.743	2.144	2.221	2.221
$g_\perp$	2.177	2.034	2.048	2.040
%Cu	100	42	61	

earlier treatments, this neglect of covalency was accounted for through use of the Steven's orbital reduction factors [50], k, where the operator $\hat{L}$ in Eqn. 2 is replaced by an effective operator $k\hat{L}$ with k < 1.0. While agreement with experiment is then obtained, interpretation of these factors in terms of covalent delocalization is somewhat indirect.

Alternatively, a delocalized ground state molecular orbital can be obtained directly utilizing, for example, Self Consistent Field-Xα-Scattered Wave (SCF-Xα-SW) calculations [51]. The partitioned charge decomposition from this calculation can be used in a complete calculation taking into account both ligand field and charge transfer excited states to compute a new spin-orbit corrected ground state. Operation with the Zeeman operator then gives the g values. The Xα-SW calculations used here require a choice of sphere radii around each atomic center. A standard set of criteria are those suggested by Norman [52], where the ratio of atomic sphere radii is determined by the ratio of atomic charges on each center. Results [18] of the g value calculation for D_{4h} $CuCl_4^{2-}$ using the Norman radii calculated partioned charge decomposition are presented in Table 4. As these g values are now too small relative to experiment, it is evident that the Norman radii have overestimated the degree of delocalization and thus reduced the amount of angular momentum mixed into the ground state.

In order to correct for this overestimation of covalent delocalization, the copper to chlorine sphere radius ratio can be adjusted [53,18] from that given by the Norman criteria. By increasing this ratio, electron charge is drawn into the copper sphere, which lowers its effective nuclear charge and thus increases the amount of copper character in the ground state. The Cu/Cl sphere ratio is adjusted until the experimental g values are matched. When good agreement with the g values is obtained for D_{4h} $CuCl_4^{2-}$, the Xα-SW calculation gives 61% Cu $d_{x^2-y^2}$ character in the ground state wave function (Table 4). This amount of delocalization is consistent with other independent experimental estimations of covalency (vide infra).

2.2 Hyperfine coupling

Hyperfine splitting in copper complexes, like the Zeeman splitting, is anisotropic ($A_\parallel \neq A_\perp$ in Fig. 1A) implying that the magnitude of the hyperfine coupling constant is dependent on the orientation of the external magnetic field relative to the molecular axes. For D_{4h} $CuCl_4^{2-}$, $|A_{\parallel}| = 164 \times 10^{-4}$ cm^{-1} and $|A_{\perp}| = 34 \times 10^{-4}$ cm^{-1} [54]. It should be noted that because the hyperfine interaction is a second rank tensor, the magnitude of the coupling but not the sign is available from CW EPR experiments.

The hyperfine coupling determined experimentally can be interpreted in terms of the ground state wavefunction. The magnitude of **A** is related to the electronic structure of the complex via the Abragam and Pryce hamiltonian [55],

$$\hat{A} = K\hat{S} + P_d[\hat{L} + \xi L(L+1)\hat{S} - 3/2\ \xi(\hat{L}(\hat{L}\cdot\hat{S}) + (\hat{L}\cdot\hat{S})\hat{L})] \tag{6}$$

where $\xi=2/21$ for d orbitals, $P_d=g_e g_N \beta_e \beta_N/<r^{-3}>$, K is the Fermi contact term and the operation is over copper terms in the spin-orbit corrected ground-state wavefunction. For the Cu $d_{x^2-y^2}$ ground state, perturbation theory can be used to rewrite Eqn. 6 as

$$A_{\parallel} = P_d[\kappa - (4/7)\alpha^2 + (3/7)\ \Delta g_{\perp} + \Delta g_{\parallel}]$$

$$\tag{7}$$

$$A_{\perp} = P_d[\kappa + (2/7)\alpha^2 + (11/14)\ \Delta g_{\perp}]$$

where α^2 is the metal character in the ground state. The value of P_d can be taken from atomic Hartree-Fock calculations [56] (where a value of 393×10^{-4} cm^{-1} is generally used) or alternatively obtained by operating directly with the r^{-3} operator over the $X\alpha$ wavefunction giving a value of 447×10^{-4} cm^{-1} [24].

There are three contributions to the hyperfine coupling tensor that are generally considered: 1) Fermi contact coupling (which is isotropic), 2) spin dipolar coupling (which is anisotropic and forms a traceless tensor), and 3) orbital dipolar coupling (which is also anisotropic, but not traceless).

2.2.1 <u>Fermi contact coupling. A_F</u>. Fermi contact, the term represented by K or $P_d\kappa$ above, arises as a consequence of the Dirac equation and is given by

$$A_F = (8\pi/3)g_e\beta_e g_N\beta_N \Sigma\ [|\psi_s\!\uparrow(0)|^2 - |\psi_s\!\downarrow(0)|^2] \tag{8}$$

where g_N is the nuclear g value, β_N is the nuclear magneton, and $|\psi_s\!\uparrow\downarrow(0)|^2$ is the unpaired electron density at the nucleus for orbitals with the same ($\uparrow$) and opposite ($\downarrow$) spin as the electron in the half-occupied ground state. Non-zero Fermi contact thus is a consequence of net spin density at the nucleus and there are two mechanisms that can cause this in copper complexes.

Fermi contact will arise if the ground state orbital contains some s character. Tetragonal copper complexes such as D_{4h} $CuCl_4^{2-}$ have nominally a $d_{x^2-y^2}$ ground state, and s mixing into this b_{1g} symmetry orbital is not allowed by group theory. Alternatively, a low symmetry distortion can make 4s mixing allowed. Only a small amount of 4s mixing is necessary to give rise to a substantial contact contribution; an unpaired electron in a ground state made up entirely of 4s copper character would have a contact contribution of 1680×10^{-4} cm^{-1}.

High symmetry complexes that have no 4s character in the half-occupied $d_{x^2-y^2}$ orbital still have appreciable contact contributions due to indirect Fermi contact coupling. The electron in the half-occupied orbital spin polarizes the core s electrons through an exchange interaction. This gives the core s electrons with the same spin as the unpaired electron a different radial probability density than core s electrons with opposite spin, resulting in net spin density at the nucleus. This exchange interaction reciprocally gives rise to a multiplet splitting in the photoemission spectrum of the core s electrons which can be used to experimentally calibrate the degree of indirect polarization among different copper complexes (vide infra). For a Cu 3d electron, the core 1s and 2s shells contribute negative spin density while the 3s shell gives positive spin density. The total indirect contribution is negative as the 2s contribution dominates. A reasonable estimate of the isotropic Fermi contact contribution to the hyperfine for D_{4h} $CuCl_4^{2-}$ is -125×10^{-4} cm^{-1} (vide infra).

2.2.2 <u>Spin dipolar coupling, A_D</u>. The second contribution to hyperfine coupling is the spin dipolar term, which arises from the interaction between the dipoles of the electron and nuclear spin. This interaction is given by

$$\hat{H}_{dipolar} = -g_e\beta_e g_N\beta_N \; [(\hat{\underline{S}}\cdot\hat{\underline{I}})r^{-3} - 3(\hat{\underline{S}}\cdot\underline{r})(\hat{\underline{I}}\cdot\underline{r})r^{-5}] \tag{9}$$

where S and I are the electron and nuclear spins and $\underline{r}$ is the position vector of the electron relative to the nucleus. In the Abragam and Pryce hamiltonian, eqn 6, the operator equivalent expression for this coupling is given by all of the third term and the diagonal elements of the last one. As the hyperfine operator is evaluated only over the copper nucleus, the spin dipolar coupling has a direct dependence on the degree of metal character in the ground state which is denoted by the α^2 factor in Eqn 7. Thus, the magnitude of this term will increase with decreasing covalent delocalization. Operation over the $d_{x^2-y^2}$ ground state orbital with the spin dipolar operator gives the $(-4/7)P_d$ and $(+2/7)P_d$ terms in Eqn. 7. Alternatively, $4p_z$ mixing into the ground state will act in an opposite sense and will tend to diminish the magnitude of the hyperfine coupling. In D_{4h} $CuCl_4^{2-}$ where p_z mixing into the ground state $d_{x^2-y^2}$ orbital is prohibited by group theory, the Xα calculations indicate that a reasonable value for the spin dipolar contribution is -155 X 10^{-4} cm^{-1} for the parallel and +78 X 10^{-4} cm^{-1} for the perpendicular components with 61% $d_{x^2-y^2}$ character in the ground state wavefunction.

2.2.3 <u>Orbital dipolar coupling, A_L</u>. The last term involved in the hyperfine coupling constant is the orbital dipolar term which arises from operation with $P_d\hat{L}$ plus the off-diagonal elements of the last term in Eqn. 6. As with the g value analysis, a nonzero value for this term requires orbital angular momentum mixed into the ground state. In the perturbation expression in Eqn. 7, the orbital dipolar term reduces to the Δg dependent terms and will vary with the amount of spin-orbit coupling and covalent delocalization in the same manner as the g values. However, unlike the Zeeman operator, the hyperfine operator is only evaluated over the copper nucleus. Thus, the perturbation expressions in Eqn. 7 are only correct in the limit of no delocalization. However, the sphere adjustment procedure for the Xα calculations described above, which fits the orbital contribution to the g value, is expected to give a good estimate for this term.

The sum of the three contributions to the hyperfine for D_{4h} CuCl$_4{}^{2-}$ is given at the top of Table 5. An experimental estimate for the Fermi contact is obtained by summing the Xα calculated spin and orbital dipolar contributions and subtracting this from the experimental values. A choice of sign must thus be made for the experimental hyperfine parameters. The presence of substantial quadrupole coupling can give rise to off-axis EPR or ENDOR signals that enable determination of relative sign, and there is substantial evidence that both the parallel and perpendicular components in tetragonal copper complexes have the same sign [57]. With both components negative we obtain -123 x 10^{-4} cm^{-1} as an estimate for the Fermi contact term.

An alternative method of utilizing the experimental hyperfine for D_{4h} CuCl$_4{}^{2-}$ is to combine these with the experimental g values to simultaneously solve the perturbation expressions in Eqn. 7, and thus obtain the degree of delocalization and the Fermi contact. From this approach, one obtains delocalization (α^2 = 67%) and contact ($P_d\kappa$ = -106 x 10^{-4} cm^{-1}) values in reasonable agreement with those given by the Xα analysis.

2.3 <u>Origin of the reduced hyperfine in D_{2d} CuCl$_4{}^{2-}$</u>

While it is apparent that the perturbation expressions work well for the D_{4h} CuCl$_4{}^{2-}$ complex, substantial discussion has been generated using these expressions to explain the small $A_\parallel$ values in distorted tetrahedral copper complexes (D_{2d} CuCl$_4{}^{2-}$, $|A_\parallel|$ = 25 x 10^{-4} cm^{-1}, $|A_\perp|$ = 48 x 10^{-4} cm^{-1}). Early studies by Bates [16a] and Sharnoff [16b], required 12% p_z mixing into the ground state, made allowed by the D_{2d} distortion. Since then, almost every contribution to the hyperfine has been cited as a possible cause of this reduction. This section presents a term-by-term analysis of the reduced parallel hyperfine in D_{2d} $CuCl_4^{2-}$ using, where possible, direct experimental estimates of each term. The results are given in Table 5.

TABLE 5
Calculation of hyperfine parameters for D_{4h} $CuCl_4^{2-}$, D_{2d} $CuCl_4^{2-}$, and plastocyanin.

		g exp	g calc	$\%d_{x^2-y^2}$	A^{Cu} (x 10^{-4}cm^{-1}) exp.	calc.	contribution (x 10^{-4}cm^{-1}) A_F	A_D	A_L
D_{4h}	‖	2.221	2.221	61	164	-164	-123	-155	+144
	⊥	2.040	2.048		35	-28	-123	+78	+17
D_{2d}	‖	2.435	2.435	67	25	-25	-77	-162	+214
	⊥	2.079	2.112		48	+48	-77	+83	+42
Plasto-cyanin	‖	2.226	2.226	42	65	-65	-79	-90	+106
	⊥	2.051	2.067		17	-15	-79	+45	+18

2.3.1 <u>Spin Dipolar Interaction</u>. a) Covalent contributions. The g values have been used to calibrate the Xα calculation and this estimates the degree of covalent delocalization in D_{2d} $CuCl_4^{2-}$. Fitting the Xα calculation to the experimental g‖ gives 67% Cu character in the ground state $d_{x^2-y^2}$ orbital, ~6% more than that found for D_{4h} $CuCl_4^{2-}$. A number of experimental methods are available to check this increase and are listed in Table 6. In all cases, these independent methods indicate that the D_{2d} salt has between 3% and 11% more $d_{x^2-y^2}$ character than D_{4h} $CuCl_4^{2-}$, supporting the g value analysis. This increase in Cu character in the D_{2d} $CuCl_4^{2-}$ indicates that covalent delocalization, acting through the spin dipolar term, should increase, not decrease, the magnitude of the parallel hyperfine coupling constant.

TABLE 6
$CuCl_4^{2-}$ ground-state covalent mixing

	%Cu $d_{x^2-y^2}$ in ground state wavefunction: D_{4h}	D_{2d}
ground state		
g values - Xα adjusted	61	67
copper hyperfine	67	69
chlorine superhyperfine	64	75
photoelectron spectra		
core level-XPS satellite	60	64
valence level-variable energy photoemission	65	68
X-ray absorption	61	71

b) p_z mixing. As mentioned above, substantial p_z mixing acting through the spin dipolar term, has been thought responsible for the reduced hyperfine in D_{2d} $CuCl_4^{2-}$. Xα calculations on this site, however, indicate only ~3% $4p_z$ mixing, far less than is required. An independent experimental probe of the amount of $4p_z$ mixing is obtained from an analysis [24] of near-edge polarized single crystal X-ray absorption spectra for D_{2d} $CuCl_4^{2-}$ shown in Fig. 13. Two bands are observed and the one at 8979 eV is substantially less intense than the transition occurring at 8987 eV. In D_{4h} $CuCl_4^{2-}$, the 8979 eV band has been assigned to a 1s to 3d quadrupole transition, polarized perpendicular to the z axis [58a]. In D_{2d} $CuCl_4^{2-}$, any intensity polarized parallel to the z

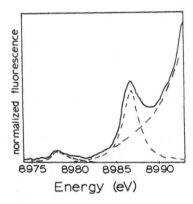

Figure 13. Polarized (E∥z) single-crystal X-ray absorption Cu-K edge spectrum of D_{2d} $CuCl_4^{2-}$. From [24].

axis at this energy will be due to $4p_z$ mixing into the 3d ground state, imparting some electric dipole-allowed character to the transition. To higher energy is the 8987 eV, band which has been assigned to either a 1s to $4p_z$ electric dipole allowed transition or to a 1s to 4p transition simultaneous with a Cl to Cu(II) charge transfer "shake down" transition. Thus the 8987 eV band is representative of either all or part of the available $4p_z$ intensity. The ratio of the intensities of the 8979 and 8987 eV bands therefore gives an upper limit to the amount of $4p_z$ character mixed into the Cu $d_{x^2-y^2}$ ground state. Good fits to the data in Fig. 13 were obtained with the ratio of the 8979 to 8987 eV bands between 5.4 and 5.8%. This experimental finding of less than half the amount of $4p_z$ character required to account for the reduced hyperfine conclusively demonstrates that $4p_z$ mixing is not responsible for the reduction in the parallel hyperfine coupling in D_{2d} $CuCl_4^{2-}$. Table 5 gives the calculated spin dipolar terms, which are greater than those for the D_{4h} salt due to the reduced covalency, which is slightly offset by the very limited $4p_z$ mixing.

2.3.2 <u>Orbital dipolar coupling</u>. Evaluation of the orbital dipolar hyperfine operator over the experimentally calibrated $X\alpha$ ground state wavefunction gives the orbital dipolar terms for D_{4h} and D_{2d} $CuCl_4^{2-}$ listed in Table 5. Since these terms relate to the g value (and in the perturbation formalism reduce to the g value expressions in Eqn. 7) they are expected to be calculated accurately, as the ground state wavefunction is calibrated by the g values. From Table 5, the orbital dipolar contribution for the parallel hyperfine coupling in the D_{2d} salt is ~100 x 10^{-4} cm^{-1} more positive than the corresponding contribution in D_{4h}, a result that reflects the substantially larger g values in the D_{2d} salt. This accounts for about 70% of the reduction in $A_{\parallel}$ on going to D_{2d} [24] and is thus the major cause of the change in hyperfine parameters between D_{4h} and D_{2d} $CuCl_4^{2-}$. This term in the reduction in $A_{\parallel}$ has been cited [16e] by others using different methods of calculation.

The increased orbital dipolar contribution in D_{2d} clearly arises from increased orbital angular momentum in the ground state. The increase is not due to increased ionic character going from D_{4h} to D_{2d}, but rather to the lower energy ligand field transitions in the D_{2d} $CuCl_4^{2-}$ complex.

2.3.3 <u>Fermi contact</u>. The remaining term in the hyperfine hamiltonian is the Fermi contact. If we subtract the sum of the spin and orbital dipolar terms, then the isotropic remainder is the Fermi contact. As with the D_{4h} salt, this subtraction requires a choice for the sign of the hyperfine. Of four possible choices, only one of them, with $A_{\parallel}$ = -25 x 10^{-4} cm^{-1} and $A_{\perp}$ = +48 x 10^{-4} cm^{-1}, gives a remainder on subtraction of the orbital and spin dipolar contributions that is isotropic; all other choices would require at least 9% $4p_z$ mixing, which is inconsistent with experiment. This remainder of -77 x 10^{-4} cm^{-1} can thus be considered [24] the Fermi contact contribution for D_{2d} $CuCl_4^{2-}$ and the 48 x 10^{-4} cm^{-1} change in this term between D_{4h} and D_{2d} $CuCl_4^{2-}$ is the remaining contribution to the reduction in hyperfine between these two complexes (Table 5).

There are several mechanisms that could give rise to this change in Fermi contact. First, there could be a change in the spin polarization of the core electrons between the two salts. An experimental probe of the relative indirect Fermi contact contributions to the hyperfine is obtained from analysis [24] of the Cu 3s XPS satellite splittings for the two complexes, which is presented in Fig. 14. In both these salts, the main line at 123 eV is followed by a satellite at 8 eV higher binding energy. The satellites are split into two components and the magnitude of this splitting, 2.8 eV, is the same for both geometries. The satellite final state corresponds to one hole in the $d_{x^2-y^2}$ orbital and one hole in the Cu 3s core level and the multiplet splitting thus corresponds to the energy difference between the singlet and triplet states. The magnitude of multiplet splitting is related to the value

884

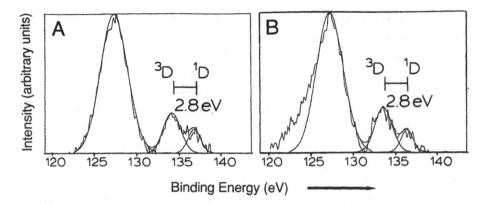

Figure 14. Photoemission spectra from the Cu 3s core level of A) D_{4h} and B) D_{2d} $CuCl_4^{2-}$ with the multiplet splitting indicated. Adapted from [24].

of the exchange integral, G^2, between the 3d and 3s levels through the Van Vleck expression [59]

$$\Delta E = G^2(3s,3d) \; (2S + 1)/(2\ell + 1) \tag{10}$$

where $\ell=2$ and $S=1/2$ in this case. It is this exchange interaction between the 3d and core s electrons that gives rise to the indirect Fermi contact. Thus the multiplet splitting in the copper 3s XPS satellite is directly related to the magnitude of the indirect Fermi contact hyperfine. As no change is observed in the multiplet splitting between the two salts the indirect contribution to the hyperfine from the Cu 3s level must be the same, and is not responsible for the reduction in Fermi contact. Table 7 gives the results [24] of calculations of the core level Fermi contact from a spin polarized $X\alpha$ calculation. The contribution from the 3s level for both salts is the same and differences in core level indirect Fermi contact are not responsible for the reduction in the magnitude of A_F between the two geometries.

The second possible mechanism for the reduction of A_F is increased 4s participation in the ground state giving rise to a direct contact interaction. As 4s mixing is prohibited by symmetry in the D_{2d} geometry used previously, calculations were performed [24] on other possible low symmetry geometries for the distorted tetrahedral site and these results are presented in Table 8. In all cases, no more than 0.15% 4s mixing is calculated, an order of magnitude less than required. As the $X\alpha$ calculations perform well in other cases where 4s mixing is required, this result indicates that direct 4s mixing into the ground state should be eliminated as a possible explanation of the reduction in Fermi contact in D_{2d} $CuCl_4^{2-}$. Alternatively, Tables 7 and 8

TABLE 7
Evaluation of contributions to Fermi contact for D_{4h} and D_{2d} $CuCl_4^{2-}$ ($\times 10^{-4} cm^{-1}$).

level	D_{4h} $CuCl_4^{2-}$ contact (% 4s)	level	D_{2d} $CuCl_4^{2-}$ contact (% 4s)
1s	-8.6	1s	-9.2
2s	-165.4	2s	-170.3
3s	+109.1	3s	+109.3
tot. core	-64.9	tot. core	-70.2
$1a_{1g}$	-0.1 (0.2)	$1a_1$	+1.5 (10)
$2a_{1g}$	-18.8 (14)	$2a_1$	-2.8 (24)
$3a_{1g}$	+15.2 (5)	$3a_1$	+10.8 (0.6)
tot. valence	-3.7	$4a_1$	-3.4 (0.8)
		tot. valence	+6.1
tot.	-68.6 (19)	tot.	-64.0 (35)

TABLE 8
Calculation of Fermi contact for distorted copper complexes.

Complex	(sym)	%s	%p_z	Fermi contact, ($10^{-4} cm^{-1}$) valence	core
$CuCl_4^{2-}$	(D_{2d})		2.76	+6.1	-70.2
$CuCl_4^{2-}$	(C_{2v})	0.01	3.57	+5.5	-69.2
$Zn[Cu](dmi)_2Cl_2$	(C_{2v})	0.09	3.38	+7.7	-71.8
$(enH_2Cl_2)Zn[Cu]Cl_4$	(C_{2v})	0.15	3.08	+8.0	-71.7
$CuCl_4^{2-}$	(D_{4h})			-3.7	-64.9

indicate that the Xα calculations are consistent in predicting a $\sim +10 \times 10^{-4}$ cm^{-1} increase in the indirect contact contribution in D_{2d} $CuCl_4^{2-}$ arising from spin polarization of the filled valence levels due to increased 4s mixing (D_{2d}: 35% 4s; D_{4h}: 19% 4s). When this $\sim 10 \times 10^{-4}$ cm^{-1} increase is scaled by the factor of two that the Xα calculation is empirically found to underestimate the indirect contact, the valence 4s polarization makes a substantial contribution to the change in A_{iso} between D_{4h} and D_{2d} $CuCl_4^{2-}$.

In the above term-by-term analysis of the reduction in the parallel hyperfine coupling between D_{4h} and D_{2d} $CuCl_4^{2-}$, complementary high energy techniques have been especially important in evaluating small contributions to the ground state that could have a large effect on the hyperfine interaction. Approximately 70% of the reduction in hyperfine parameters relates to increased orbital dipolar coupling in the D_{2d} site arising from the presence of greater orbital angular momentum in the ground state due to reduced energies of the ligand field centered states which are primarily involved in spin-orbit coupling. The remaining 30% reduction is most likely due to increased spin polarization of valence 4s character in D_{2d} $CuCl_4^{2-}$.

3. BLUE COPPER GROUND STATE

The analysis of Section 2 has been extended to the ground state spectral features of the blue copper active site. As mentioned in the Introduction, these features (Fig. 3A) are important because the parallel hyperfine coupling is also reduced in blue copper sites relative to normal tetragonal copper complexes and it is important to determine the orientation and delocalization of the ground state wavefunction involved in electron transfer. In this section we develop an experimental and theoretical understanding of the ground state of the blue copper active site and use it to explain the origin of the spin hamiltonian parameters.

3.1 Orientation of the g^2 tensor

The orientation of the g^2 tensor relative to the molecular axes has been addressed through single crystal EPR spectroscopy [23a]. Plastocyanin crystallizes in the orthorhombic space group $P2_12_12_1$ with four molecules per unit cell [17]. Single crystal EPR data were obtained for four rotations ($\underline{B}\perp a$ is given in Fig. 15 left). With the applied field close to the c axis an almost pure $g\|$ spectrum is obtained, while the field along b gives a $g_\perp$ EPR signal. It is important to note that the Cu-S(met) bond lies close to c for all four molecules and thus the electronic z axis must be close to the Cu-S(met) bond. Simulations of these experimental results are included in Fig. 15 (right). A good fit to the experimental data was obtained with A_z parallel to g_z and g_z 5° out of the bc plane and 8° off c.

With four molecules in the unit cell the single crystal EPR spectra alone are not sufficient to specify the orientation of g_z relative to a specific active site. A ligand field calculation using the method of Companion and Komarynsky [60] was then performed [23a] for the blue copper site geometric coordinates (Fig. 3C) to determine the correct orientation. This calculation showed that only one of the four experimental possibilities was consistent with the ligand field. This orientation is reproduced in Fig. 16 where g_z is oriented approximately 5 degrees off the Cu-S(met) bond, which indicates that the half-occupied d_{x2-y2} oribital is within 15° of the plane formed by the remaining three ligands. One feature of the g value calculation described here is the relatively small Steven's orbital reduction factors ($k\|^2 = 0.136$, $k_\perp^2 = 0.241$) required to match the calculated g value to experiment, which indicates a highly covalent ground state (vide infra).

The energy level splitting associated with the crystal field calculation is shown in Fig. 17, where the g^2 coordinate system is used to describe the orbitals. Two axial symmetry subgroups are possible for a distorted

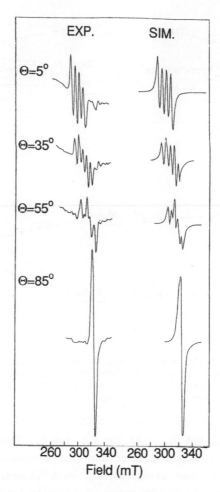

Figure 15. Representative experimental and simulated EPR spectra of a single crystal of plastocyanin. The rotation axis is a ($\underline{B} \perp$ a) and θ is the angle between $\underline{B}$ and c. From [23a].

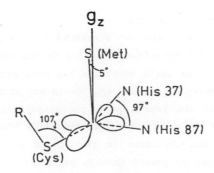

Figure 16. Electronic structural representation of the plastocyanin active site. From [23a].

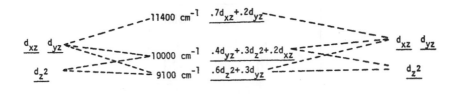

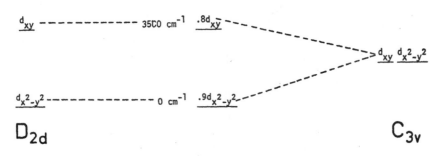

D_{2d} C_{3v}

Figure 17. Ligand field energy diagram for plastocyanin. Center contains energies and electron hole wavefunctions of plastocyanin site. Energy levels obtained by removing rhombic distortions are shown for D_{2d} (left) and C_{3v} (right) limiting geometries. From [23a].

tetrahedral site: C_{3v} and D_{2d}. As the d_{xz} and d_{yz} levels are calculated to be high in energy, an ordering which does not obtain in the D_{2d} limit, the calculation indicates that elongated C_{3v} with rhombic distortions is the correct effective symmetry for the blue copper active site. A recent analysis [23b] of the MCD parameters of plastocyanin confirms the main features of the orbital ordering shown in Fig. 17 and provides further support for the C_{3v} effective geometry.

3.2   Determination of 4p mixing

The effective C_{3v} site symmetry presented a major problem with respect to the origin of the small hyperfine splitting in the EPR spectrum in Fig. 3A. As with D_{2d} $CuCl_4^{2-}$, the small hyperfine has been attributed to $4p_z$ mixing. However, in C_{3v} symmetry $4p_{x,y}$, not $4p_z$, would mix with $d_{x^2-y^2}$. This mixing would contribute with the same sign to the anisotropic spin dipolar term in Eqn. 6 and thus make the magnitude of the hyperfine greater, not smaller.

The amount of $4p_z$ character in the ground state orbital of the blue copper active site can be probed through polarized single crystal absorption spectroscopy [25] at the Cu-K edge, as discussed for D_{2d} $CuCl_4^{2-}$ in Section 2. In Fig. 18, it is clear that all 3d absorption intensity at 8979 eV occurs with the electric vector parallel to the x,y axes (defined by the single crystal EPR g tensor) and no intensity occurs along z. This indicates only x,y electric dipole (and xy quadrupole) moments contribute to the intensity which in turn indicates that no $4p_z$ is mixed in the ground state. Thus, substantial $4p_z$ mixing is eliminated as the source of the reduced parallel hyperfine in the blue copper active site.

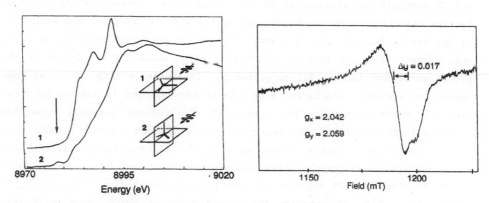

Figure 18. Polarized edge spectra for plastocyanin showing variation in intensity of the 8979 eV transition (arrow) for two different orientations of the active site. From [25a].

Figure 19. Q-band EPR spectrum (ν = 34.282 GHz) of the $g_\perp$ region of plastocyanin. From [18].

3.3 Delocalization

As in the case for $CuCl_4^{2-}$, the g values have been used to calibrate SCF-Xα-SW calculations and obtain a corrected ground state wavefunction for the blue copper site. Xα calculations performed using the Norman radii gave 31% Cu character and g_z = 2.159, substantially lower than the experimental value of 2.226. It is again apparent that the Norman radii overestimate the degree of covalent delocalization. Systematically varying the atomic sphere radii [23b] to give good agreement with the experimental g_z results in a ~10% increase in the amount of copper character in the ground state. The final wavefunction for the blue copper site in plastocyanin is given in Eqn. 11:

$$|\psi_g> = 42\% \; Cu|d_{x^2-y^2}> + 0.8\% \; Cu|4p_{x,y}> + 36\% \; S|p_\pi>$$
$$+ 4\% \; N|p_\sigma> + 0.01\% \; S_{met} \; . \tag{11}$$

A contour diagram of this wavefunction was given in Fig. 5 along with the calculated orientation of the g^2 tensor. In agreement with experiment, g_z is found to lie close to the S(met)-Cu bond with the ground state orbital in the plane perpendicular to this bond.

The ground state wavefunction has a number of significant features. First, as the Steven's orbital reduction factors indicate, the site is highly covalent with the unpaired electron having substantial (36%) density on the p_π orbital of the thiolate sulfur, which is antibonding with respect to $d_{x^2-y^2}$ (42% character) on the copper center. The high degree of delocalization over the cysteine sulfur indicates that the antibonding interaction between it and $d_{x^2-y^2}$ controls the orientation of the ground state. Second, while the site for the calculation described above has C_s symmetry, which precludes either 4s or $4p_z$ mixing into the ground state, calculations performed on sites relaxing this symmetry indicate that very little (1%) 4p character is present in the ground state and this is all $p_{x,y}$, not p_z, in agreement with the X-ray edge absorption studies. Similarly, only 0.1% Cu 4s character is calculated to be present, which cannot be the source of the reduced parallel hyperfine coupling in plastocyanin. The final interesting feature concerns a calculated rhombic splitting of the g values, $\Delta g = 0.017$. The predicted splitting was also indicated in the crystal field calculations described above and motivated us to examine the EPR spectrum of the plastocyanin active site at higher resolution. Fig. 19 presents the Q-band spectrum which shows a rhombic splitting of g_y-$g_x = 0.017$ in agreement with the calculations. This splitting also arises from the anisotropic delocalization over the cysteine sulfur p_π orbital.

The experimentally calibrated ground state wavefunction can now be used to calculate contributions [18] to the hyperfine interaction (Table V). Relative to the D_{4h} $CuCl_4^{2-}$ complex, the blue copper site shows no reduction in the orbital dipolar coupling term, a result that relates to the fact that the two sites exhibit similar g values. Alternatively, the d_{xy} orbital in plastocyanin is much lower in energy that in D_{4h} $CuCl_4^{2-}$ [23b]. Since the g values are similar between the two complexes, this implies that the plastocyanin ground state must have substantially less copper character than D_{4h} $CuCl_4^{2-}$. The results of the Xα calculations in Table V show that the plastocyanin ground state is 20% less localized on the Cu center relative to D_{4h} $CuCl_4^{2-}$ and as a consequence the spin dipolar contribution in the blue copper site is significantly reduced. This accounts for 65% of the difference between the "normal" parallel copper hyperfine in D_{4h} $CuCl_4^{2-}$ and the "anomalous" value for blue copper. A further reduction of 44 x 10^{-4} cm^{-1} in the Fermi contact term is also required by the experimental hyperfine values. The 20% increased delocalization of the blue site would reduce Fermi contact by 41 x 10^{-4} cm^{-1}, close to the required value. Thus, from Table V the reduced parallel hyperfine coupling observed for the blue copper site results from a reduced spin dipolar term arising from substantially increased delocalization in this site relative to D_{4h} $CuCl_4^{2-}$. In contrast, most of the reduction in the parallel hyperfine in D_{2d} $CuCl_4^{2-}$ comes from the orbital dipolar term. Both these sites also require reduced Fermi contact relative to D_{4h} $CuCl_4^{2-}$. In blue copper sites, the reduction of A_F is another consequence of the increased delocalization while in D_{2d} $CuCl_4^{2-}$ the reduced A_F appears to arise from spin polarization of valence 4s electrons.

3.4 Extension to stellacyanin-class EPR spectra

From Fig. 4 the blue copper site in stellacyanin exhibits highly rhombic EPR g values (g_x = 2.018, g_y = 2.077) and A values (A_x = 57 x 10^{-4} cm^{-1}, A_y = 29 x 10^{-4} cm^{-1})[19]. The splitting in g values for this class of proteins is much larger than that found in plastocyanin (Fig. 19) where the rhombic splitting is attributed to covalent delocalization over the thiolate $S(p_\pi)$ orbital. Stellacyanin has three of four ligands in common with plastocyanin, however it does not contain methionine and the fourth ligand is presently unknown.

The highest effective symmetry of a tetrahedron that allows rhombic splitting is C_{2v}. In this symmetry, the perturbation expressions for the g shifts [61] become

$$\Delta g_x = -2\lambda k_x^2 (a - 3^{\frac{1}{2}}b)^2 / E_{xz}$$
$$\Delta g_y = -2\lambda k_y^2 (a + 3^{\frac{1}{2}}b)^2 / E_{yz} \qquad (12)$$
$$\Delta g_z = -8\lambda k_z^2 a^2 / E_{xy}$$

where k_i represents the Stevens orbital reduction factor, a and b are the coefficients of $d_{x^2-y^2}$ and d_{z^2} in the ground state, respectively, and the denominators are the d-d transition energies. From Eqn. 12 there are three mechanisms that could split g_x and g_y: 1) differences in the transition energies between E_{xz} and E_{yz}, 2) differences in the Steven's reduction factors (i.e. excited state delocalization), or 3) d_{z^2} mixing into the ground state wavefunction.

Optical spectra for plastocyanin [62] and stellacyanin are shown in Fig. 20 with the assignments [23b] indicated for plastocyanin. Band 5 is associated with the $d_{x^2-y^2}$ to d_{xz-yz} transition while band 6 is attributed to d_{xz+yz}, which gains intensity due to configuration interaction with the $S_\pi(cys)$ to $d_{x^2-y^2}$ CT transition (Band 4). The splitting between band 5 and band 6 is 1150 cm^{-1} in plastocyanin and 1000 cm^{-1} in stellacyanin. This difference can only account for a rhombic splitting of 0.005 in the g values. Similarly, the roughly equivalent extinction coefficients for these bands suggests that there are no substantial changes in exited state delocalization between the two proteins. Thus the increased rhombic splitting in stellacyanin appears to be due to d_{z^2} mixing into the ground state wavefunction. The degree of rhombic splitting can be expressed as a function

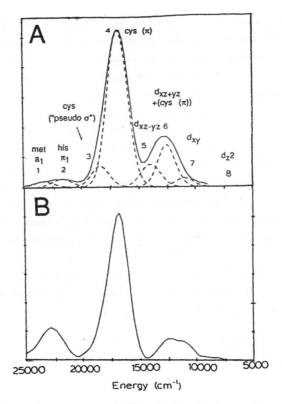

Figure 20. Absorption spectra (25K) of A) plastocyanin and B) stellacyanin films. Gaussian resolution and assignments of plastocyanin bands are indicated. Adapted from [23b].

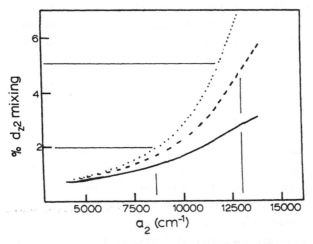

Figure 21. Plot of d_{z^2} mixing into the ground state of stellacyanin as a function of increasing ligand field strength along the Cu-methionine bond direction for three different ratios of ligand field parameters: (——) $\alpha_2 = 4$ α_4; (---) $\alpha_2 = 2$ α_4; (···) $\alpha_2 = 1.5$ α_4. Lines give the range of d_{z^2} mixing considered. From [24].

of d_z2 mixing using Eqn. 12 and for stellacyanin between 1.8% and 3.8% d_z2 mixing is required. A similar analysis utilizing the rhombic split A values yields between 1.1 and 5.2% d_z2 character.

The calculated range of d_z2 mixing can be used to gain insight into the changes between the bonding in stellacyanin and plastocyanin. Increased d_z2 mixing is obtained by increasing the ligand field strength in a direction perpendicular to the g_{xy} plane, i.e. along the Cu-methionine bond direction. Fig. 21 gives a plot of ground state d_z2 character as a function of increasing ligand field strength of the methionine ligand in the plastocyanin site geometry [24]. To reproduce the 2-5% d_z2 mixing, a value of the second power radial integral, α_2, of between 9000 and 12500 cm^{-1} is necessary. In contrast, α_2 is 4000 cm^{-1} for the methionine in plastocyanin.

The increase in ligand field strength has two consequences for the nature of the copper active site in stellacyanin. First, the increase in ligand field strength along the Cu-methionine direction moves g_z ~15° further away from the Cu-X direction, where X is the replacement ligand for methionine in stellacyanin. Second, this analysis requires a stronger ligand field along the Cu-methionine direction. This can be accomplished either by replacing methionine with a stronger field ligand, or by shortening the Cu-methionine bond. This last point is especially important with regard to the recently solved structure of Basic Blue Protein[21], which shares the same ligand set as plastocyanin, but which exhibits a stellacyanin-like EPR spectrum.

4. THE COUPLED BINUCLEAR COPPER SITE

4.1 Triplet EPR signals: met and dimer [Cu(II)Cu(II)] sites

As mentioned in the Introduction, two of the hemocyanin derivatives, met and dimer, have binuclear cupric sites that have proven particularly important in probing the endogenous bridge. Dimer is prepared by NO + O_2 oxidation of deoxy and exhibits (Fig. 22A, top) a triplet EPR signal with an intense broad $\Delta m_s = 1$ transition at $g \simeq 2$ and a weak $\Delta m_s = 2$ transition at $g \simeq 4$. This signal, which corresponds to ~65% of the sites and exhibits negligible exchange coupling ($|2J| < 5$ cm^{-1}), has been simulated and shown to correspond to two dipolar interacting tetragonal copper(II)'s held ~6 Å apart [38]. Addition of excess azide results in the dimer N_3^- form which exhibits a new but similar triplet EPR signal (Fig. 22A, bottom), indicating retention of the same site structure, but with a change in Cu-Cu separation to ~5 Å. Dialysis of either dimer form produces the met derivative [31]. Met-hemocyanin is normally prepared by two-electron oxidation of deoxy [63a] or associative ligand displacement of peroxide from oxy [63b]. In parallel to oxy, more than 90% of the sites in met are EPR nondetectable[31].

894

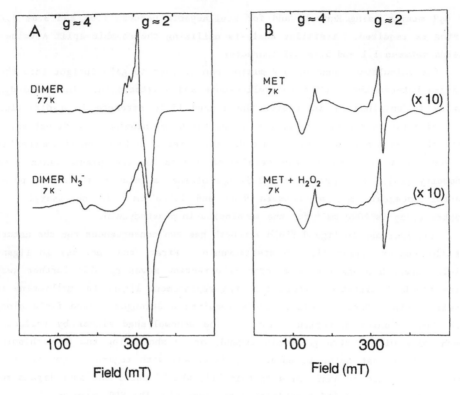

Figure 22. X-band EPR spectra of A) top: dimer hemocyanin (77K), bottom: dimer + excess N$_3^-$ (7K) and B) top: met hemocyanin, bottom: met + ten-fold excess H$_2$O$_2$ (7K). From [65].

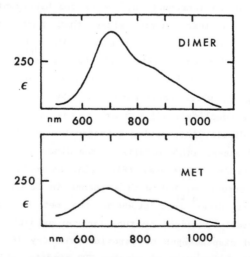

Figure 23. Optical absorption spectra (7K) in the ligand field region of dimer and met hemocyanin. From [2b].

A comparison of the ligand-field optical spectra of dimer and met (Fig. 23) demonstrates that the tetragonal cupric energy levels of met are not significantly perturbed by the interaction between the ions that eliminates the EPR signal [32]. This combined with the magnetic susceptibility data on met and oxy (Section I) demonstrates that a weak interaction between copper orbitals resulting in antiferromagnetic coupling is responsible for the lack of EPR signals. This requires that an endogenous bridging ligand provide an efficient pathway for strong superexchange coupling. The preparation of dimer hemocyanin produces a metastable form of the site where the endogenous bridge is eliminated by certain bridging exogenous anions, which keep the coppers > 5 Å apart. The resulting EPR signals are those from dipolar interacting Cu(II) ions. Removal of these "group 2" exogenous anions allows the bridge to reform; the dimer EPR signals disappear as the sites are converted to met. This model [32] is summarized in Fig. 24.

Figure 24. Representations of the active sites of (left) met and (right) dimer hemocyanin.

Even though met is EPR nondetectable, weak signals at $g \approx 2$ and $g \approx 4$ are associated with this derivative. This spectrum is shown at low temperature and high sensitivity in Fig. 22B (top). While the $g \approx 2$ signal originates from < 3% of the sites, it is difficult to quantify the broad, weak signal most obvious at $g \approx 4$. A key experiment, however, has shown that all the EPR signals associated with met must originate from < 10% of the sites. Mollusc met hemocyanin can be regenerated to oxy with small excesses of peroxide. When Busycon met is regenerated with peroxide to >90% EPR-nondetectable oxy sites, the EPR signal is not affected (Fig. 22B, bottom) [31]. Unfortunately, these EPR signals associated with met have led to confusion about the nature of this derivative and, in particular, a misleading correlation with dimer [64]. Although the $g \approx 4$ regions of both EPR spectra (Fig. 22A,B) have some similarities, the relative magnitude of the $g \approx 2$ region clearly indicates the signals are quite different. Further, quantification of these signals has indicated that most of the sites (>65%) contribute to the dimer signal while the met EPR signals originate from only <10% of the sites.

896

These broad, weak EPR signals associated with met hemocyanin have been shown to originate from a small fraction of the sites that are uncoupled at low pH by competitive displacement of the protonated endogenous bridge by specific exogenous anions [65]. Fig. 25 shows that on lowering the pH these signals appear with the addition of N_3^- or acetate. A quantitative analysis of the pH dependence of this competition allows an estimate of the intrinsic pK_a of the endogenous bridge (RO^-) of > 7.0. This eliminates some possible bridging groups but is consistent with hydroxide, alkoxide, or phenoxide.

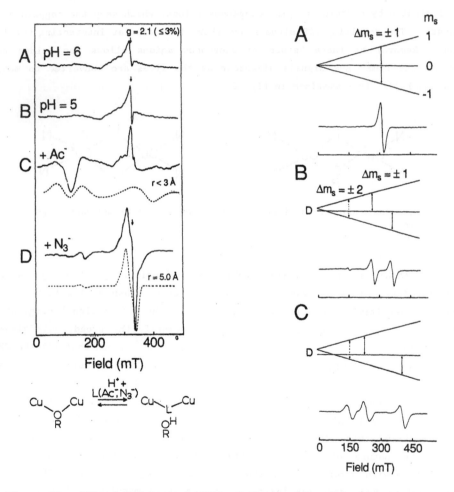

Figure 25. X-band EPR spectra (7K) of A) met, pH = 6.0, B) met, pH = 5.0, C) met + acetate, pH = 5.0, and simulated spectrum (---) and, D) met + azide, pH = 5.0, and simulated spectrum (---).

Figure 26. Splitting of the S=1 state in a magnetic field with A) r > 7Å, B) r ~ 4Å, and C) r ~ 3Å. Transitions and associated EPR spectra are indicated. Adapted from [65].

The broad EPR spectra from several of the chemically perturbed (i.e. varied exogenous ligand) forms of this protonatively uncoupled met site have been simulated in order to probe active site structural features [65]. As the temperature dependence of these signals indicates there is negligible exchange coupling, dipolar interaction between the cupric ions should be the dominant mechanism leading to zero field splitting (D) of the paramagnetic triplet spin state formed by the two S=1/2 ions. This dipolar zero field splitting is given by $D_{dip} = (-3/2)(g^2\beta^2/r^3)(\mu_o/4\pi)$ and the EPR spectrum should therefore reflect the Cu(II)-Cu(II) separation as schematically shown in Fig. 26. However, when g anisotropy, hyperfine interaction, and lower site symmetry are also considered, r cannot be directly determined from the powder EPR signals of dipolar coupled binuclear cupric complexes. Therefore, the EPR spectral simulation calculation developed by J. Pilbrow [66] has been used, which is based on second-order perturbation analysis of a point-dipole model for anisotropic, binuclear S=1/2 systems. When $|2J| < 30$ cm^{-1}, this calculation has been shown to be capable of accurately determining metal-metal separations in the range 3.5 to 4.0 Å and indicating the relative orientation of the magnetic axes on the two coppers [67]; this program has also been used to simulate the EPR spectrum of the dimer hemocyanin derivative[38] (vide supra).

Simulations of the broad dipole-coupled EPR signals associated with met-Ac$^-$ and N$_3^-$ are included as the dashed lines in Figure 24. The EPR parameters obtained from spectral simulations of several protonatively uncoupled met derivatives (Table 9) reflect large differences in Cu(II)-Cu(II) separation: N$_3^-$, r = 5.0 Å; Br$^-$, r = 2.9 Å; Ac$^-$, r = 2.4 Å. The parameters obtained for the acetate signal suggests that both the second-order perturbation analysis and the point-dipole approximation are breaking down due both to comparable magnitudes of the zero field splitting and the microwave energy and to the close approach of the cupric ions. This signal does, however, reflect a

TABLE 9
Best fit simulation parameters for EPR spectra from chemically perturbed met sites.

	Acetate	Acetate + Br$^-$	N$_3^-$
r, Å	2.4 ± 0.1	2.92 ± 0.05	5.0 ± 0.2
ξ, deg	60	90	85
$g_\parallel$	2.05	2.12	2.15
$g_\perp$	2.25	2.01	2.07
$A_\parallel$, x 10^4 cm^{-1}	150	150	120
$A_\perp$, x 10^4 cm^{-1}	20	20	25
$|2J|$, cm^{-1}	1.0 ± 5.0	0.0 ± 5.0	0.0 ± 5.0

shorter Cu-Cu separation than that of the Br⁻ signal. The g and A values used in simulating the Br⁻ and N_3^- spectra are appropriate for two tetragonal cupric ions, and the best fit angle ξ indicates that the principal axis of the g tensor of each ion is oriented approximately perpendicular to the internuclear vector. However, the broad linewidths and lack of hyperfine structure in the experimental spectra do not allow a very accurate estimate of ξ, $g_{x,y,z}$, and $A_{x,y,z}$. Alternatively, simulations of the Br⁻ and N_3^- spectra are quite sensitive to small changes in r leading to a higher degree of confidence in this parameter.

These simulations have indicated that when the endogenous bridge is uncoupled, exogenous anions modify the Cu-Cu distance. Since the relative metal ion separation correlates with expected bridging distances, this trend supports a bridging geometry for exogenous ligand binding at the uncoupled met site. The metal-metal separation determined for acetate is short enough (r < 3.1 Å) to sustain an unidentate or bidentate bridging mode. The r ~ 5 Å determined from the azide signal requires a μ-1,3 bridging geometry for this ligand in the scheme at the bottom of Fig. 25. Though these exogenous anions bridge the copper ions, they do not mediate significant exchange coupling ($|2J| < 7$ cm^{-1}). Poor overlap must therefore exist between the bridging ligand orbitals involved in the superexchange pathway and at least one of the Cu(II) $d_{x^2-y^2}$ magnetic orbitals.

4.2 Mixed-valence EPR signals: half met [Cu(II)Cu(I)] Sites

The half-met derivatives provide significant insight into exogenous ligand binding at the coupled binuclear and Type 3 copper sites [39,44], and contribute to our fundamental understanding of the electronic structure of binuclear metal complexes [68]. Using half-met-Br⁻ as an example (Fig. 9) the EPR signal exhibits more than four hyperfine lines in the g∥ region. The single $g_\perp$ signal in the Q-band spectrum, Fig. 27C, demonstrates that this derivative is homogeneous, and the EPR signal of met-apo-Br⁻, a derivative which contains a single copper(II), [Cu(II)—], active site [39] exhibits a normal pattern in the g∥ region with four hyperfine lines. Thus the additional hyperfine structure of the half-met-Br⁻ site must relate to electron delocalization onto the second copper center. This is quite important as the relative magnitude of the hyperfine coupling constants of the two coppers can directly define the ground state wavefunction in this mixed-valence site and its change with variation in exogenous ligand.

The ground state of a mixed-valence dimer may be described by a wavefunction of the form:

$$|\psi_g\rangle = (1-\alpha^2)^{\frac{1}{2}} \, |\psi[\mathrm{Cu(II)_A Cu(I)_B}]\rangle + \alpha \, |\psi[\mathrm{Cu(I)_A Cu(II)_B}]\rangle \qquad (13)$$

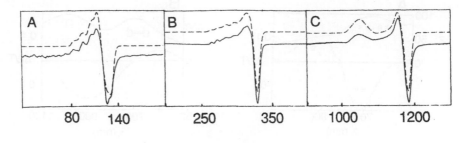

Field (mT)

Figure 27. EPR spectra (–) and similations (--) of half-met-Br⁻ hemocyanin at
A) S-band, B) X-band, and C) Q-band. (Simulation parameters: $g_\parallel$ = 2.312, $g_\perp$ =
2.077, $A_\parallel{}^{Cu}{}_A$ = 92 x 10^{-4}cm⁻¹, $A_\parallel{}^{Cu}{}_B$ = 32 x 10^{-4}cm⁻¹. Adapted from [68].

which expresses the delocalized ground state in terms of completely localized
wavefunctions with the unpaired electron on either Cu_A or Cu_B [69].   The
mixing coefficient α is the delocalization parameter with α^2 representing the
unpaired electron density on Cu_B. Thus far, estimates of α in mixed-valence
complexes have been somewhat indirect and relied on an analysis of the
intervalence transfer (IT) transition,$|\psi_g\rangle \rightarrow |\psi_{IT}\rangle$, where:

$$|\psi_{IT}\rangle = \alpha\ |\psi[Cu(II)_A Cu(I)_B]\rangle - (1-\alpha^2)^{1/2}\ |\psi[Cu(I)_A\ Cu(II)_B]\rangle\ . \tag{14}$$

Hush [69] has shown that the value of α^2 can be obtained from the intensity of
the IT absorption band and is given by Eqn. 15:

$$\alpha^2{}_{opt} = 4.24 \times 10^{-4}\ \epsilon_{max}\ \Delta E_{1/2}/(E_{max}\ R^2)\ . \tag{15}$$

In Eqn. 15 ϵ_{max}, E_{max}, and $\Delta E_{1/2}$ are the maximum molar absorptivity, maxium
energy, and the fullwidth at half maximum, respectively, of the IT band and R
is the Cu-Cu distance in ångstroms.

    A correlation of the low temperature magnetic circular dichroism (LT-MCD)
spectrum with the absorption spectrum allows assignment of the new low energy
band in the absorption spectrum of half-met Br⁻ in Fig. 9 as the IT
transition.   Half-met-NO_2⁻ exhibits absorption and LT-MCD features simply
associated with d→d transitions of a tetragonal Cu(II) site, Fig. 28A.
Comparison to the half-met-Br⁻ spectrum, Fig. 28B, indicates that while the
d→d transitions shift in energy reflecting some change in ligand field at the
Cu(II), there is no significant additional LT-MCD intensity associated with
the new fairly intense low energy absorption band [68].   The LT-MCD spectrum
of a paramagnetic center is dominated by C terms [70]. The general expression

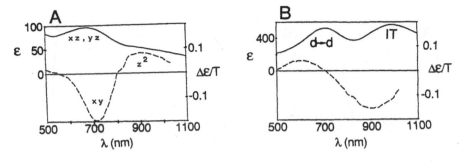

Figure 28. Low temperature (10K) optical absorbance (———) and LT-MCD (- - -) spectra of A) half-met-NO_2^- and B) half-met-Br^-. Adapted from [2d].

for an orientationally averaged C term is given in Eqn. 16 where m_i is the transition dipole in the direction i:

$$I_C \propto g_z\, m_x\, m_y + g_y\, m_x\, m_z + g_x\, m_y\, m_z \ . \tag{16}$$

Since the g_i values are all non-zero, the absence of LT-MCD intensity must result from the lack of two perpendicular nonzero electric dipole transition moments for this new absorption band. As an IT transition involves excitation of an electron from the Cu(I) to the Cu(II), it is polarized along the Cu-Cu vector. Thus, the unidirectional nature of the low energy (~1000 nm) transition in half-met-Br^- allows it to be assigned as the IT transition. Using the observed IT energy, intensity, and halfwidth and a value of R of 3.5 Å for the half-met site (based on EXAFS results on oxy (3.6 Å) and deoxy (3.4Å)) Eqn. 15 gives $\alpha^2_{opt} = 8.0 \times 10^{-3}$.

The S, X and Q-band EPR data in Fig. 27 provide an alternative probe of α^2 from the relative Cu hyperfine coupling constants, $A_\parallel^{Cu_A}$, $A_\parallel^{Cu_B}$. These EPR data were fit using the spin Hamiltonian given in Eqn. 17:

$$\begin{aligned}
\hat{H} = &\ g_z \beta B \hat{S}_z + g_\perp \beta B (\hat{S}_x + \hat{S}_y) \\
&+ A_\parallel^{Cu_A} (\hat{S}_z \hat{I}_z^{Cu_A}) + A_\perp^{Cu_A} (\hat{S}_x \hat{I}_x^{Cu_A} + \hat{S}_y \hat{I}_y^{Cu_A}) \\
&+ A_\parallel^{Cu_B} (\hat{S}_z \hat{I}_z^{Cu_B}) + A_\perp^{Cu_B} (\hat{S}_x \hat{I}_x^{Cu_B} + \hat{S}_y \hat{I}_y^{Cu_B}) \ .
\end{aligned} \tag{17}$$

The Q-band fit gives an accurate estimate of the g values and the S-band data provides the best values for the hyperfine constants. These simulations are included as the dash lines in Fig. 27. The program used to generate the powder patterns (a modified version of SIM14A [71]) requires that the g and A tensors have coincident principal axes. This is in general not the case for the half-met sites and in order to determine the intrinsic hyperfine parameters, the relative orientations of the dimer g tensor and the individual copper **A** tensors must be determined. Analysis of the energy of the IT band has yielded the spectroscopically effective active site geometry in Fig. 29 [68].

In the site, Cu(II)$_A$ is approximately tetragonal while the geometry of Cu(I)$_B$ is distorted tetrahedral. Also indicated are the local coordinate systems at Cu$_A$, Cu$_B$, and the bridging exogenous ligand (vide infra). The molecular g tensor is given by Eqn. 18:

$$g = g_e 1 + 2\Lambda \tag{18a}$$

$$\Lambda_{ij} = - \Sigma_n \langle\psi_g|\hat{L}_i'|\psi_n\rangle \langle\psi_n|\hat{L}_j|\psi_g\rangle (E_n - E_g)^{-1} \tag{18b}$$

In Eqn. 18b the sum is over all n excited states. $\hat{L}_j$ is the jth component of the orbital angular momentum operator and is evaluated with respect to a suitably chosen molecular origin. $\hat{L}_i' = \Sigma_k \lambda_k \hat{L}_{i(k)}$, where the sum is over each atom k, λ_k is the spin-orbit coupling constant for atom k and $\hat{L}_{i(k)}$ is the ith component of the orbital angular momentum operator evaluated with respect to the local coordinate system (Fig. 29) at atom k. The matrix elements of Eqn. 18 are evaluated for the ground state wavefunction given in Eqn. 19:

$$|\psi_g\rangle = a|\psi(Cu_A)\rangle + b|\psi(Cu_B)\rangle + c|\psi(Br^-)\rangle + d|\psi(L)\rangle . \tag{19}$$

In Eqn. 19 the contribution of $|\psi(Br^-)\rangle$ has been treated separately from those of the other ligands ($|\psi(L)\rangle$) since the spin-orbit coupling constant for Br$^-$ (2460 cm^{-1}) is large enough to affect the observed g values. In order to determine the orientation of the molecular g tensor relative to the local

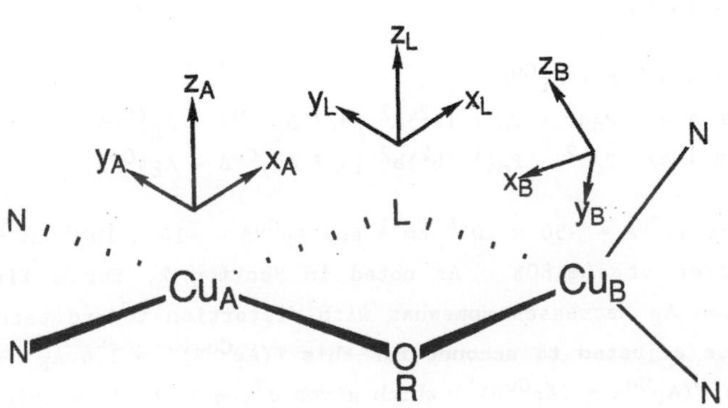

Figure 29. Spectroscopically effective active site geometry for half-met-L hemocyanin with local coordinate systems at the atoms indicated. From [68].

coordinate systems (and thus the A tensors) values of a-d in Eqn. 19 were chosen and the matrix elements of Eqn. 18 were evaluated. The g^2 tensor was diagonalized to give the principal g values, which were then compared to those obtained from the EPR simulations. Only a narrow set of a, b, c, and d values gave close agreement with physically reasonable parameters: $a^2 \approx 0.67$, $b^2 \approx 0.13$, $c^2 \approx 0.05$, $d^2 \approx 0.15$. These values give an estimate of $\alpha^2 = b^2/(a^2 + b^2) = 0.16$, indicating approximately 16% delocalization in the ground state.

A more sensitive measure of delocalization is the relative magnitude of the hyperfine coupling parameters for each copper. The simulations give A values that are projections onto the molecular g_z. A proper comparison, however, requires that the intrinsic hyperfine couplings be determined. The unitary rotation matrix that diagonalizes the g^2 tensor gives the angle between g_z and $A_z{}^{Cu_A}$, $\phi = 1.5°$, and between g_z and $A_z{}^{Cu_B}$, $\theta = 32°$, so that

$$(A_\parallel{}^{Cu_A})^2 = \cos^2(1.5°) \, (A_z{}^{Cu_A})^2 + \sin^2(1.5°) \, (A_{x,y}{}^{Cu_A})^2$$
$$(A_\parallel{}^{Cu_B})^2 = \cos^2(32°) \, (A_z{}^{Cu_B})^2 + \sin^2(32°) \, (A_{x,y}{}^{Cu_B})^2 \; .$$

The lack of resolvable features puts an upper limit on the magnitude of $A_\perp$ for each copper of $|A_{x,y}| \leq 20 \times 10^{-4}$ cm^{-1}. With this constraint, the simulated $A_\parallel$ values yield $|A_z{}^{Cu_A}| = 92 \times 10^{-4}$ cm^{-1} and 43×10^{-4} cm$^{-1} < |A_z{}^{Cu_B}| < 45 \times 10^{-4}$ cm^{-1}. Generally, A_z for copper is negative, thus we take $A_z{}^{Cu_A} = -92 \times 10^{-4}$ cm^{-1} and $A_z{}^{Cu_B} = -44 \times 10^{-4}$ cm^{-1}.

Since the hyperfine parameters of monomeric copper sites are a sensitive function of geometry (see Section 2), the most reasonable comparison between the Cu_A and Cu_B hyperfine parameters is based on A_F for each center. From Eqn. 7 the Fermi contact contributions for each of the two coppers are given in Eqn. 20 where the excited state d orbital character is taken as $1.1a^2$ or $1.1b^2$, respectively:

$$A_F{}^{Cu_A} = A_z{}^{Cu_A} - A_S{}^{Cu_A} - A_L{}^{Cu_A} \tag{20a}$$
$$= A_z{}^{Cu_A} + 4/7 \; P_d a^2 - P_d(1.1a^2)a^2 \; [3/7 \; \Delta_{g_\perp}{}^{Cu_A} + \Delta_{g_\parallel}{}^{Cu_A}]$$
$$A_F{}^{Cu_B} = A_z{}^{Cu_B} + 4/7 \; P_d b^2 - P_d(1.1b^2)b^2 \; [3/7 \; \Delta_{g_\perp}{}^{Cu_B} + \Delta_{g_\parallel}{}^{Cu_B}] \; . \tag{20b}$$

Eqns. 20 give $A_F{}^{Cu_A} \approx -50 \times 10^{-4}$ cm^{-1} and $A_F{}^{Cu_B} \approx -14 \times 10^{-4}$ cm^{-1} for a total metal character of 70%-80%. As noted in Section 2, for a fixed value of delocalization A_F decreases somewhat with distortion toward tetrahedral, and $A_F{}^{Cu_B}$ must be adjusted to account for this ($(A_F{}^{Cu_B})' \approx 1.4 \; A_F{}^{Cu_B}$). Finally, $\alpha^2 = (A_F{}^{Cu_B})'/(A_F{}^{Cu_A} + (A_F{}^{Cu_B})')$ which gives $\alpha^2{}_{EPR} \approx 0.30$, within a factor of two of the value obtained from the g^2 tensor diagonalization but 50 times greater than $\alpha^2{}_{opt}$ obtained from the IT transition intensity. This divergence of $\alpha^2{}_{opt}$ from the true delocalization is related to the use of Eqn. 21 to

903

obtain oscillator strengths for use in Eqn. 15 [72]:

$$m = <\psi_g \mid e \cdot r \mid \psi_{IT}> = e \, \alpha \, R \; . \tag{21}$$

These EPR and absorption studies have been extended over a series of half-met-L (L = NO_2^-, Cl^-, Br^-, I^-) derivatives. From Fig. 30 the EPR spectra become increasingly more complex and the IT absorption intensity increases over this series. Estimates of α^2_{opt} and α^2_{EPR} obtained from these data are summarized in Table 10. While the magnitude of α^2_{opt} and α^2_{EPR} for a given derivative are clearly very different, both correlate with the covalent nature of the tightly bound exogenous ligand. Thus the exogenous ligand must provide the pathway for electron delocalization and therefore bridge the two coppers of the coupled binuclear copper active site in hemocyanin (and tyrosinase) [39,26]. Alternatively, the parallel half-met derivative of the Type 3 site in Type 2 depleted (T2D) laccase shows only a normal four-line hyperfine pattern in the g∥ region as shown in Fig. 31 for half-met-Br^- T2D laccase

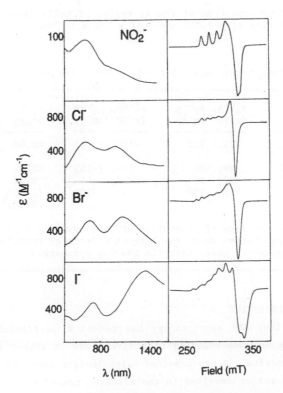

Figure 30. Low temperature optical absorption (10K) and X-band EPR (77K) spectra for the half-met-L series.

[44]. This spectral difference relative to the coupled binuclear copper active site in hemocyanin strongly suggests that exogenous ligands bind to only one copper at the Type 3 site (see Fig. 11).

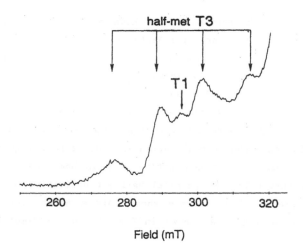

Figure 31. X-band EPR spectrum of the $g_\parallel$ region of half-met-Br⁻ T2D laccase. For the half-met T3 site $g_\parallel$ = 2.281 and $|A_\parallel|$ = 138 x 10^{-4}cm⁻¹.

TABLE 10
EPR and delocalization parameters for the half-met-L series.

L	$g_\parallel$	$g_\perp$	$A_\parallel^{Cu_A}$, $A_Z^{Cu_A}$ (x 10^{-4}cm⁻¹)	$A_\parallel^{Cu_B}$, $A_Z^{Cu_B}$ (x 10^{-4}cm⁻¹)	α^2_{EPR}	α^2_{opt}
NO₂⁻	2.302	2.106, 2.041	123, -123	<10[a], (-15)	<0.05	0
Cl⁻	2.347	2.078	96, -96	<10[a], (-15)	<0.09	5.4 x 10^{-3}
Br⁻	2.312	2.077	92, -92	32, -44	0.30	8.0 x 10^{-3}
I⁻	2.261	2.071	109, -109	93, -107	(0.30<α^2 <0.50)[b]	2.0 x 10^{-2}

a) Upper limit based on lack of resolution of this coupling in the EPR spectrum. b) This derivative shows the largest effects of noncoincidence of g and A, leading to greatest uncertainty in the Cu_B parameters.

5. SUMMARY AND PROSPECTIVE

It is clear that EPR spectroscopy has played a particularly important role in developing an understanding of active sites in copper proteins. In the blue copper proteins, it has provided major insight into the nature of the ground state wavefunction involved in the electron transfer reactivity of the active site. In the coupled binuclear copper site, EPR has elucidated key structural features and allowed structure/function correlations to be developed relating to differences in oxygen binding and activation.

Studies on the copper proteins have also led to some new insight into basic inorganic EPR spectroscopy. Two examples emphasized in this review are the contributions to the small A∥ values generally found in distorted tetrahedral cupric complexes and the experimental determination of electron delocalization in mixed-valence binuclear copper sites.

In the near future one might anticipate that EPR spectroscopy will provide new insight into the trinuclear copper cluster present in the multicopper oxidases and, combined with low temperature MCD spectroscopy of the d→d excited states, into the ligand fields of paramagnetic copper centers. The latter approach should prove particularly powerful in defining the now rather poorly understood active sites in the normal copper enzymes.

ACKNOWLEDGEMENTS

EIS gratefully acknowledges the contributions to this work from his graduate students, postdoctoral research associates, and collaborators as indicated in the references. Several figures (as referenced) are reprinted with permission of the American Chemical Society, the American Society of Biological Chemists, John Wiley & Sons, and Macmillan Magazines Ltd. This work was supported by NIH grant #DK 31450 and NSF grant #CHE 8613376.

REFERENCES

1 a) J. Peisach, P. Aisen, and W.E. Blumberg, The Biochemistry of Copper, Academic Press, New York, 1966. b) T. Vanngard in: H.M. Swartz, J.R. Bolton, and D.C. Borg (Eds), Biological Applications of Electron Spin Resonance, Wiley-Interscience, New York, 1972, pp 411-447. c) B.G. Malmstrom, B. Reinhammar, and T. Vanngard, Biochem. Biophys. Acta, 156 (1968) 67-76.

2 a) H.B Gray and E.I. Solomon, in: T.G. Spiro (Ed.), Copper Proteins, Wiley-Interscience, New York, 1981, pp. 1-39. b) E.I. Solomon, in: T.G. Spiro (Ed.), Copper Proteins, Wiley-Interscience, New York, 1981, pp. 41-108. c) E.I. Solomon, K.W. Penfield and D.E. Wilcox, Struct. Bonding (Berlin) 53 (1983) 1-57. d) E.I. Solomon, in: L. Que, Jr., (Ed.), Metal Clusters in Proteins. ACS Symposium Series No. 372, American Chemical Society, Washington, 1988, pp. 116-150. e) E.I. Solomon in: K.D. Karlin and J. Zubieta (Eds), Copper Coordination Chemistry: Biochemical and Inorganic Perspectives, Adenine Press, New York, 1983, pp 1-21.

3 a) J.S. Valentine and M.W. Pantoliano, in: T.G. Spiro (Ed.), Copper Proteins, Wiley-Interscience, New York, 1981, pp. 291-358. b) J.A. Fee, J. Peisach, and W. B. Mims, J. Biol. Chem., 256 (1981) 1910-14.

4 D.M. Dooley, Life Chem. Rep., 5 (1987) 91-154.

5 a) T. Ljones and T. Skotland, in: R. Lontie (Ed.), Copper Proteins and Copper Enzymes, Vol. II, CRC Press, Boca Raton, Florida, 1984, pp. 131-157. b) J.J. Villafranca, in: T.G. Spiro (Ed.), Copper Proteins, Wiley-Interscience, New York, 1980 pp. 263-289.

6 S. Benkovic, D. Wallick, L. Bloom, B.J. Gaffney, P. Domanico, T. Dix, and S. Pember, Biochem. Soc. Trans., 13 (1985) 436-438.

7 M.J. Ettinger and D.J. Kosman, in: T.G. Spiro (Ed.), Copper Proteins, Wiley-Interscience, New York, 1981, pp. 219-261.

8   J.P. Klinman and M. Brenner, in: T.E. King, H.S. Mason, and M. Morrison (Eds), Oxidases and Related Redox Systems. Proceedings of the 4th International Symposium on Oxidases and Related Redox Systems, Portland, Oregon, 4-8 October 1987, Alan R. Liss, Inc., New York, 1988, pp. 227-248.

9   N.J. Blackburn, M. Concannon, S.K. Shahiyan, F.E. Mabbs, and D. Collison, Biochemistry, 27 (1988) 6001-6008.

10   J.A. Tainer, E.D. Getzoff, K.M. Beem, J.S. Richardson, and D.C. Richardson, J. Mol. Biol., 160 (1982) 181-217.

11   E.I. Solomon, Comments Inorg. Chem., 3 (1984) 225-320.

12   V.T. Taniguchi, N. Sailasuta-Scott, F.C. Anson, and H.B. Gray, Pure Appl. Chem., 52 (1980) 2275-2281.

13   H.B. Gray, Chem. Soc. Rev., 15 (1986) 17-30.

14   E.I. Solomon, J.W. Hare, and H.B. Gray, Proc. Natl. Acad. Sci. (USA), 73 (1976) 1389-1393.

15   a) A. Finazzi-Agro, G. Rotilio, L. Avigliano, P. Guerrieri, V. Boffi, and B. Mondovi, Biochemistry, 9 (1970) 2009-2014. b) E.I. Solomon, P. Clendening, H.B. Gray, and F.J. Grunthaner, J. Am. Chem. Soc., 97 (1975) 3878-3879.   c) D.R. McMillin, R.A. Holwerda, and H.B. Gray, Proc. Natl. Acad. Sci. (USA), 71 (1974) 1339-1341.   d) D.R. McMillin, R.C. Rosenberg, and H.B. Gray, Proc. Natl. Acad. Sci. (USA), 71 (1974) 4760-4762.

16   a) C.A. Bates, Proc. Phys. Soc., London, 79 (1962) 69-72. b) J. Sharnoff, J. Chem. Phys., 42 (1965) 3383-3395. c) H. Yokoi, Bull. Chem. Soc. Japan, 47 (1974) 3037-3040. d) A. Bencini, D. Gatteschi, and C. Zanchini, J. Am. Chem. Soc., 102 (1980) 5234-37. e) M.A. Hitchman, Inorg. Chem, 24 (1985) 4762-65.

17   a) P.M. Colman, H.C. Freeman, J.M. Guss, M. Murata, V.A. Norris, J.A.M. Ramshaw, and M.P. Venkatappa, Nature (London), 272 (1978) 319-324. b) J.M. Guss and H.C. Freeman, J. Mol. Biol., 169 (1983) 521-563.

18   K.W. Penfield, A.A. Gewirth, and E.I. Solomon, J. Am. Chem. Soc., 107 (1985) 4519-4529.

19   a) V.T. Aikazyan and R.M. Nalbandyan, FEBS Lett. 55 (1975) 272-274. b) V.T. Aikazyan and R.M. Nalbandyan, FEBS Lett., 104 (1979) 127-130.

20   a) E.T. Adman, R.E. Stenkamp, L.C. Sieker, and L.H. Jensen, J. Mol. Biol.,123 (1978) 35-47. b) G.E. Norris, B.F. Anderson, and E.N. Baker, J. Am. Chem. Soc., 108 (1986) 2784-2785.

21   J.M. Guss, E.A. Merritt, R.P. Phizackerley, B. Hedman, M. Murata, K.O. Hodgson, and H.C. Freeman, Science, 241 (1988) 806-811.

22   a) J. Peisach, W.G. Levine, and W.E. Blumberg, J. Biol. Chem., 242 (1967) 2847-2858. b) C. Bergman, E. Gandvik, P.O. Nyman, and L. Strid, Biochem. Biophys. Res. Commun., 77 (1977) 1052-1059.

23   a) K.W. Penfield, R.R. Gay, R.S. Himmelwright, N.C. Eickman, V.A. Norris, H.C. Freeman, and E.I. Solomon, J. Am. Chem. Soc., 103 (1981) 4382-4388. b) A.A. Gewirth and E.I. Solomon, J. Am. Chem. Soc., 110 (1988) 3811-3819.

24   A.A. Gewirth, S.L. Cohen, H.J. Schugar, and E.I. Solomon, Inorg. Chem., 26 (1987) 1133-1146.

25   a) R.A. Scott, J.E. Hahn, S. Doniach, H.C. Freeman, and K.O. Hodgson, J. Am. Chem. Soc., 104 (1982) 5364-5369. b) J.E. Penner-Hahn, K.O. Hodgson, and E.I. Solomon, to be published.

26   R.S. Himmelwright, N.C. Eickman, C.D. LuBien, K. Lerch, and E.I. Solomon, J. Am. Chem. Soc., 102 (1980) 7339-7344.

27   K.E. van Holde and K.I. Miller, Quart. Rev. Biophys., 15 (1982) 1-129.

28   K. Lerch, in: H. Sigel (Ed.) Metal Ions in Biological Systems Vol. 13, Dekker, New York, 1981, pp. 143-186.

29   a) J. S. Loehr, T.B. Freedman, and T.M. Loehr, Biochem. Biophys. Res. Commun., 56 (1974) 510-515. b) T.B. Freedman, J.S. Loehr, and T.M. Loehr, J. Am. Chem. Soc., 98 (1976) 2809-2915.

30   a) J.M. Brown, L. Powers, B. Kincaid, J.A. Larrabee, and T.G. Spiro, J. Am. Chem. Soc., 102 (1980) 4210-4216. b) M.S. Co, K.O. Hodgson, T.K. Eccles, and R. Lontie, J. Am. Chem. Soc., 103 (1981) 984-986. c) G.L. Woolery, L. Powers, M. Winkler, E.I. Solomon, and T.G. Spiro, J. Am. Chem. Soc., 106 (1984) 86-92.

31 R.S. Himmelwright, N.C. Eickman, and E.I. Solomon, Biochem. Biophys. Res. Commun., 86 (1979) 628-634.

32 N.C. Eickman, R.S. Himmelwright, and E.I. Solomon, Proc. Natl. Acad. Sci. (USA), 76 (1979) 2094-2098.

33 a) E.I. Solomon, D.M. Dooley, R.H. Wang, H.B. Gray, M. Cerdonio, F. Mogno, and G.L. Romani, J. Am. Chem. Soc., 98 (1976) 1029-1031. b) D. Dooley, R.A. Scott, J. Ellinghaus, E.I. Solomon, and H.B. Gray, Proc. Natl. Acad. Sci. USA, 75 (1978) 3019-3022.

34 D.E. Wilcox and E.I. Solomon, to be published.

35 a) T.J. Thamann, J.S. Loehr, and T.M. Loehr, J. Am. Chem. Soc., 99 (1977) 4187-4189. b) J.E. Pate, R.W. Cruse, K.D. Karlin, and E.I. Solomon, J. Am. Chem. Soc., 109 (1987) 2624-2630.

36 a) R.R. Jacobson, Z. Tyeklar, A. Farooq, K.D. Karlin, S. Liu, and J. Zubieta, J. Am. Chem. Soc., 110 (1988) 3690-3692. b) K.D. Karlin, personal communication.

37 a) W.P.J. Gaykema, W.G.J. Hol, J.M. Vereijken, N.M. Soeter, H.J. Bak, and J.J. Beintema, Nature (London), 309 (1984) 23-29. b) W.P.J. Gaykema, A. Volbeda, and W.G.J. Hol, J. Mol. Biol., 187 (1985) 255-275.

38 (a) A.J.M. Schoot Uiterkamp, FEBS Lett., 20 (1972) 93-96. (b) A.J.M. Schoot Uiterkamp, H. van der Deen, H.C. Berendsen, and J.F. Boas, Biochim. Biophys. Acta, 372 (1974) 407-425.

39 R.S. Himmelwright, N.C. Eickman and E.I. Solomon, J. Am. Chem. Soc., 101 (1979) 1576-1586.

40 D.E. Wilcox, A.G. Porras, Y.T. Hwang, K. Lerch, M.E. Winkler and E.I. Solomon, J. Am. Chem. Soc., 107 (1985) 4015-4027.

41 a) R. Malkin and B.G. Malmstrom, Adv. Enzymol., 33 (1970) 177-244. b) J.A. Fee, Struct. Bonding (Berlin), 23 (1975) 1-60.

42 L. Morpurgo, I. Savini, G. Gatti, M. Bolognesi, and L. Avigliano, Biochem. Biophys. Res. Commun., 152 (1988) 623-628.

43 L. Ryden in: R. Lontie (Ed.), Copper Proteins and Copper Enzymes, Vol. III, CRC Press, Roca Raton, Florida, 1984, pp. 37-100.

44 D.J. Spira and E.I. Solomon, J. Am. Chem. Soc., 109 (1987) 6421-6432.

45 L.S. Kau, D.J. Spira, J.E. Penner-Hahn, K.O. Hodgson, and E.I. Solomon, J. Am. Chem. Soc., 109 (1987) 6433-6442.

46 a) M.D. Allendorf, D.J. Spira, and E.I. Solomon, Proc. Natl. Acad. Sci. (USA), 82 (1985) 3063-3067. b) D.J. Spira, M.D. Allendorf, and E.I. Solomon, J. Am. Chem. Soc., 108 (1986) 5318-5328.

47 A. Messerschmidt, A. Rossi, R. Ladenstein, R. Huber, M. Bolognesi, A. Marchesini, R. Petruzelli, and A. Finazzi-Agro in: T.E. King, H.S. Mason, and M. Morrison (Eds), Oxidases and Related Redox Systems. Proceedings of the 4th International Symposium on Oxidases and Related Redox Systems, Portland, Oregon, 4-8 October 1987, Alan R. Liss, Inc., New York, 1988, pp. 285-288.

48 a) M.A. Hitchman and P.J. Cassidy, Inorg. Chem., 18 (1979) 1745-1754. b) S.R. Desjardins, K.W. Penfield, S.L. Cohen, R.L. Musselman, and E.I. Solomon, J. Am. Chem. Soc., 105 (1983) 4590-4603.

49 P.J. Cassidy and M.A. Hitchman, Inorg. Chem., 16 (1977) 1568-1570.

50 K.W.H. Stevens, Proc. Roy. Soc. (London), A219 (1953) 542-555.

51 a) J.C. Slater, Phys. Rev., 81 (1951) 385-390. b) K.H. Johnson and F.C. Smith, Jr., Phys. Rev. B, 5 (1972) 831-843. c) K.H. Johnson, Adv. Quantum Chem., 7 (1973) 143-185. d) K.H. Johnson, J.G. Norman, and J.W.D. Connolly, in: F. Herman, A.D. McLean, and R.K. Nesbet (Eds), Computational Methods for Large Molecules and Localized States in Solids, Plenum, New York, 1973, pp. 161-202. e) N. Rosch, in: L. Phariseu and L. Scheire (Eds), Electrons in Finite and Infinite Structures, Plenum, New York, 1977, pp. 1-143. f) J.C. Slater, The Calculation of Molecular Orbitals, Wiley-Interscience, New York, 1979.

52 J.G. Norman, Mol. Phys., 31 (1976) 1191-1198.

53 A. Bencini and D. Gatteschi, J. Am. Chem. Soc., 105 (1983) 5535-5541.

908

54 a) C. Chow, K. Chang, and J. Willett, J. Chem. Phys, 59 (1973) 2629-2640.
    b) Note that the pure material [Nmph]$_2$[CuCl$_4$] exhibits dipolar broadened
    EPR spectra, which prevents determination of hyperfine coupling constants.
    Thus hyperfine parameters must be obtained from the disorted octahedral
    Cu-doped K$_2$PdCl$_4$ lattice.

55 A. Abragam and M.H.L. Pryce, Proc. Roy. Soc. (London), A205 (1951) 135-
    153.

56 A.J. Freeman and R.E. Watson, in: G.T. Rado and H. Suhl (Eds), Magnetism
    Vol. 2A, Academic Press, New York, 1965, pp. 167-305.

57 a) E.P. Duliba, G.C. Hurst, and R.L. Belford, Inorg. Chem., 21 (1982) 577-
    581. b) A. Schweiger and H. Gunthard, Chem. Phys., 32 (1978) 35-61. c)
    T.G. Brown and B.M. Hoffman, Mol. Phys., 39 (1980) 1073-1109.

58 a) J.E. Hahn, R.A. Scott, K.O. Hodgson, S. Doniach, S.R. Desjardins, and
    E.I. Solomon, Chem. Phys. Lett., 88 (1982) 595-598. b) T.A. Smith, J.E.
    Penner-Hahn, M.A. Berding, S.Doniach, and K.O. Hodgson, J. Am. Chem. Soc.,
    107 (1985) 5945-5955. c) R.A. Bair and W.A. Goddard, Phys. Rev., B22
    (1980) 2767-2776.

59 J.H. van Vleck, Phys. Rev., 45 (1934) 405-419.

60 A.L. Companion and M.A. Komarynsky, J. Chem. Educ., 41 (1964) 257-262.

61 B.R. McGarvey, Trans. Met. Chem., 3 (1966) 89-201.

62 E.I. Solomon, J.W. Hare, D.M. Dooley, J.H. Dawson, P.J. Stephens, and H.B.
    Gray, J. Am. Chem. Soc., 102 (1980) 168-178.

63 a) G. Felsenfeld and M.P. Printz, J. Am. Chem. Soc., 81 (1959) 6259-6264.
    b) R. Witters and R. Lontie, FEBS Lett., 60 (1975) 400-403.

64 a) J. Verplaetse, P. van Tornout, G. Defreyn, R. Witters, and R. Lontie,
    Eur. J. Biochem, 95 (1979) 327-331. b) H. Rupp, J. Verplaetse, and R.
    Lontie, Z. Naturforsch., C35 (1980) 188-192. (c) J. Verplaetse, P.
    Declercq, W. Deleersnijder, R. Witters, and R. Lontie, in: J. Lamy and
    J.Lamy (Eds) Invertebrate Oxygen Binding Proteins, Dekker, New York,
    1981, pp. 589-596. d) R. Witters, J. Verplaetse, H.R. Lijnen, and R.
    Lontie, in: J. Lamy and J. Lamy (Eds), Invertebrate Oxygen Binding
    Proteins, Dekker, New York, 1981, pp. 597-602. e) R. Lontie, C. Gielens,
    D. Groeseneken, J. Verplaetse, and R. Witters, in: T.E. King, H.S. Mason,
    and M. Morrison (Eds), Oxidases and Related Redox Systems. Proceedings of
    the 3rd International Symposium on Oxidases and Related Redox Systems,
    Albany, New York, 3-7 July 1979, Pergamon, Oxford, 1982, pp. 245-261. f)
    R. Lontie and D.R. Groesensken in: F.L. Boschke (Ed.) Topics in Current
    Chemistry, Vol. 108, Springer-Verlag, Berlin, 1983, pp. 1-33.

65 D.E. Wilcox, J.R. Long, and E.I. Solomon, J. Am. Chem. Soc., 106 (1984)
    2186-2194.

66 T.D. Smith and J.R. Pilbrow, Coord. Chem. Rev., 13 (1974) 173-278.

67 P.D.W. Boyd, A.D. Toy, T.D. Smith, and J.R. Pilbrow, J. Chem. Soc. Dalton
    Trans., (1973) 1549-1563.

68 T.D. Westmoreland, D.E. Wilcox, M.J. Baldwin, W.B. Mims, and E.I.
    Solomon, submitted for publication.

69 a) N.S. Hush, Prog. Inorg. Chem., 8 (1967) 391-444. b) N.S. Hush,
    Electrochim. Acta, 13 (1968), 1005-1023.

70 a) P.J. Stephens, Adv. in Chem. Phys., 35 (1976) 197-264. b) S.B. Piepho
    and P.N. Schatz, Group Theory in Spectroscopy, Wiley-Interscience, New
    York, 1983.

71 G. Lozos, B. Hoffman, and C. Franz, Quantum Chemistry Program Exchange
    #265.

72 K.Y. Wong, P.N. Schatz, and S.B. Piepho, J. Am. Chem. Soc., 101 (1979)
    2793-2803.

SUBJECT INDEX

Printed and bound by CPI Group (UK) Ltd, Croydon, CR0 4YY

03/10/2024

01040329-0012